MULTIVARIATE ANALYSIS — V

Multivariate Analysis — V

PROCEEDINGS OF THE FIFTH INTERNATIONAL SYMPOSIUM ON MULTIVARIATE ANALYSIS

Edited by

PARUCHURI R. KRISHNAIAH

Department of Mathematics and Statistics
University of Pittsburgh, Pittsburgh, Pa., U.S.A.

1980

NORTH-HOLLAND PUBLISHING COMPANY—AMSTERDAM · NEW YORK · OXFORD

The Fifth International Symposium on Multivariate Analysis was sponsored by the Air Force Office of Scientific Research (Grant No. AFOSR 78-3548), Army Research Office (Contract No. DAAG 2978M0053), Office of Naval Research (Grant No. N0001477G0066) and the University of Pittsburgh.

ISBN 0 444 85321 9

Published by:
NORTH-HOLLAND PUBLISHING COMPANY
AMSTERDAM · NEW YORK · OXFORD

Sole distributors for the U.S.A. and Canada:
Elsevier North-Holland, Inc.
52 Vanderbilt Avenue
New York, N.Y. 10017

Library of Congress Cataloging in Publication Data

International Symposium on Multivariate Analysis,
5th, University of Pittsburgh, 1978.
Multivariate analysis—V.

Includes index.
1. Multivariate analysis—Congresses.
I. Krishnaiah, Paruchuri R. II. Title.
QA278.I58 1978 519.5′3 79–10288
ISBN 0-444-85321-9

PRINTED IN THE NETHERLANDS

PREFACE

Many elegant and rich techniques have been developed in the last few decades in the area of multivariate analysis. With the emergence of modern computers and the development of various computer packages, more and more scientists in various disciplines are using the techniques of the multivariate analysis in the analysis of the data. But, unfortunately, there is still a very big gap between theoretical developments and applications. Also the field of multivariate analysis is vast and there is a need for interactions between workers in different areas of multivariate analysis. The Fifth International Symposium on Multivariate Analysis, like the earlier symposia, was organized to stimulate interactions between mathematical statisticians, probabilists and data analysts working on different aspects of multivariate analysis. It was felt that the symposium would give an opportunity to many workers in the world to review the past developments and to reflect upon the future directions of the field. The symposium was held at the University of Pittsburgh during the period of June 19–24, 1978.

The present volume consists of the invited papers presented at the symposium. In these papers, distinguished workers in the field from many countries discuss the current developments on a very broad spectrum of topics in the theory and applications of multivariate analysis. The topics covered include classification and pattern recognition, contingency tables, decomposition of multivariate probabilities, design and analysis of experiments, distribution theory, econometrics, estimation, limit theorems, multivariate analysis of variance, nonparametric methods, optimum properties of test procedures, psychometrics, random matrices, reduction of dimensionality, reliability, scaling methods, simultaneous test procedures, sociometry, statistical physics, stochastic control theory, time series and stochastic processes. These papers are classified into various parts in the volume but the classification is not necessarily optimum. This volume will be of great interest to mathematical statisticians, probabilists as well as scientists in other disciplines who use statistical methodology. Unfortunately, the papers presented in the contributed paper sessions and clinical sessions are not included in the volume due to lack of space but the titles of these papers are included at the end.

I wish to take this opportunity to thank many persons for their valuable help in the organization of the symposium. Dr. R. A. Smith (Provost, University of Pittsburgh) was kind enough to make the Opening Remarks. I wish to express my appreciation to Professor C. R. Rao for delivering the Inaugural Address. Thanks are due to Professors R. R. Bahadur, A. P. Basu, W. W. Cooley, S. Das Gupta, M. Dutta, S. Geisser, A. K. Gupta, S. S. Gupta, T. Jayachandran, G. Kallianpur, L. Kanal, S. Kotz, J. C. Lee, P. Masani, A. M. Mathai, R. J. Muirhead, L. D. Pitt, B. R. Rao, J. Reising, A. Rukhin, F. J. Schuurmann, P. Subbaiah, N. H. Timm, V. B. Waikar, S. Watanabe, S. J. Wolfe and S. Zacks and Drs. H. L. Harter, R. L. Launer, and R. J. Lundegard for presiding over different sessions. I am grateful to Professors A. V. Balakrishnan, R. E. Barlow, H. Bergstrom, R. N. Bhattacharya, R. D. Bock, S. Das Gupta, R. H. Farrell, T. Hida, A. J. Izenman, L. Kanal, C. G. Khatri, J. C. Lee, K. V. Mardia, G. S. Mudholkar, R. J. Muirhead, M. B. Priestley, M. Rosenberg, M. Rosenblatt, J. G. Saw, J. C. Silverstein, M. S. Srivastava, W. J. Studden, S. Watanabe and S. J. Wolfe, Dr. J. D. Carroll and Mr. C. Fang for reviewing various papers in this volume. Professors R. F. Anderson, J. C. Lee and N. H. Timm went beyond the call of duty to help me in making various arrangements for the symposium.

I am grateful to the contributors to this volume and to the North-Holland Publishing Company for their excellent cooperation. Special thanks are due to the Air Force Office of Scientific Research, Army Research Office, Office of Naval Research and the University of Pittsburgh for sponsoring this symposium. I am indebted to Professor M. M. Rao for his encouragement and suggestions in the organization of the five symposia on multivariate analysis and various other professional matters. The publication of this volume was made possible through the generosity of the Annual Giving Fund of the University of Pittsburgh.

P. R. Krishnaiah

CONTENTS

*Inaugural address.

PART I

REDUCTION OF DIMENSIONALITY AND ESTIMATION

P. R. Krishnaiah, ed., *Multivariate Analysis–V*

MATRIX APPROXIMATIONS AND REDUCTION OF DIMENSIONALITY IN MULTIVARIATE STATISTICAL ANALYSIS

C. Radhakrishna RAO
University of Pittsburgh and Indian Statistical Institute

This paper deals with problems of approximating a given matrix by a matrix in a specified subclass. In each case, solution is obtained by minimizing a suitable norm of the difference between matrices. In some cases, it is shown that the same solution minimises a wide class of norms known as unitarily invariant norms introduced by von Neumann. Applications to problems in multivariate analysis are mentioned.

Different types of generalized inverses introduced by the author (1964) have been obtained by approximating the product of a given matrix with another to the identity matrix.

1. Introduction

A number of statistical techniques specially in multivariate analysis are based on approximating a given matrix by a matrix in a specified subclass. We shall mention a few typical problems of this kind and describe methods for solving them. We denote an $m \times n$ matrix A by $A(m \times n)$.

1.1. *Problems involving matrix approximations*

(a) In the analysis of multivariate data through principal components, canonical correlations and discriminant functions, we approximate a given $A(m \times n)$ of rank r by $B(m \times n)$ of rank $k \leqslant r$. There may be a further restriction that the columns of B belong to a specified subspace.

(b) In factor analysis, an approximation to a correlation matrix R is sought in the form $FF' + \Delta$ where Δ is a diagonal matrix with non-negative diagonal elements and F is a matrix of a specified rank which is small compared to the rank of R. Again in rotation of factors, we have two given matrices A and B, not necessarily square, and the problem is to find an orthogonal transformation T (called Procrustean transformation) such that $\|A - BT\|$ is small for a suitably chosen norm.

(c) In the estimation of residuals in a linear model, one of the problems is to find a matrix B with its columns in a specified subspace such that $B'B$

is close to a given idempotent matrix. Another problem is to approximate a non-negative definite (n.n.d.) matrix A by an idempotent matrix of specified rank less than that of A.

(d) In the theory of generalized inverse of a matrix one of the problems is to find $G(n\times m)$ given $A(m\times n)$ such that $\|I_n - GA\|$ or/and $\|I_m - AG\|$ are small. In such a case G qualifies to be called a generalized inverse.

(e) In a more general context, we have a space of vector random variables (r.v.'s) and the problem is one of approximating a given r.v. by a r.v. in a specified class.

1.2. Unitarily invariant norm

In studying problems of matrix approximations of the type mentioned in Section 1.1, we have to lay down a criterion for judging how close one matrix is to another. For this purpose, we consider the class of unitarily invariant norms in the space of $m\times n$ complex matrices extending the definition given by von Neumann (1937) for square matrices.

A real valued function $\|\cdot\|$ on the space S_{mn} of $m\times n$ complex matrices is called a unitarily invariant norm if it satisfies the following conditions:

(i) $\|X\|>0 \quad (X\neq 0)$;

(ii) $\|cX\|=|c|\,\|X\|$;

(iii) $\|X+Y\|\leqslant\|X\|+\|Y\|$;

(iv) $\|VXU\|=\|X\|$ for any unitary matrices V and U of orders m and n respectively.

Consider the singular value decomposition

$$X = PDQ^* \tag{1.1}$$

where P and Q are unitary matrices and D is the diagonal matrix of singular values $\sigma_1(X)\geqslant\cdots\geqslant\sigma_r(X)$ of X where $r=\min(m,n)$. Using the condition (iv) we find

$$\|X\|=\|D\| \tag{1.2}$$

so that a unitarily invariant norm of a matrix is a symmetric function of its singular values. Symmetry follows using condition (iv).

1.3. (M, N)-invariant norm

We can define a more general norm by changing the condition (iv) of a unitarily invariant norm to

$$\text{(iv)}' \qquad \|VXU\|=\|X\| \tag{1.3}$$

for any V and U such that $V^*MV=M$ and $U^*NU=N$, where $M(m\times n)$ and $N(n\times n)$ are given positive definite (p.d.) matrices.

The following lemma provides the connection between unitarily and (M,N)-invariant norms.

Lemma 1.3.1. *Let $M^{\frac{1}{2}}$ and $N^{\frac{1}{2}}$ be Gramian square roots of M and N, and $M^{-\frac{1}{2}}$, $N^{-\frac{1}{2}}$ those of M^{-1} and N^{-1}. If $\|X\|_1$ is an (M,N)-invariant norm of X, then $\|M^{-\frac{1}{2}}XN^{-\frac{1}{2}}\|_1$ is a unitarily invariant norm of X. If $\|X\|_2$ is a unitarily invariant norm of X, then $\|M^{\frac{1}{2}}XN^{\frac{1}{2}}\|_2$ is an (M,N)-invariant norm of X.*

The lemma is proved by using the conditions (iv) and (iv)′.

1.4. Symmetric gauge function

A real valued function, defined on a space of real vectors, is called a symmetric gauge function if it satisfies the following conditions.

(i) $\phi(x)>0 \quad (x\neq 0)$
(ii) $\phi(\rho x)=|\rho|\,\phi(x)$
(iii) $\phi(x+y)\leqslant\phi(x)+\phi(y)$
(iv) $\phi(x_\pi)=\phi(x)$
(v) $\phi(Jx)=\phi(x)$.

Here x,y are real vectors, ρ any real number, π any permutation of the coordinates and J any diagonal matrix whose diagonal elements are ± 1.

It has been shown by von Neumann (1937) that any unitarily invariant norm of a matrix can be defined as

$$\|X\|=\phi(\sigma_1,\ldots,\sigma_r) \tag{1.4}$$

where ϕ is a symmetric gauge function and $\sigma_1,\ldots,\sigma_r$ are singular values of $X(m\times n)$ where $r=\min(m,n)$.

Some examples of symmetric gauge functions are

$$\begin{aligned}\phi(x_1,\ldots,x_r) &= \max_i |x_i|;\\ &=(\Sigma|x_i|^p)^{1/p} \qquad \text{for} \quad 1\leqslant p\leqslant\infty;\\ &= \max_{1<i_1<\cdots<i_k<r}\left(|x_{i_1}|+\cdots+|x_{i_k}|\right)\end{aligned} \tag{1.5}$$

The following lemma due to Ky Fan (1951) gives the conditions under which we can assert $\|X_1\|\geqslant\|X_2\|$ for *all* unitarily invariant norms.

Lemma 1.4.1. *Let* $\sigma_1 \geqslant \sigma_2 \geqslant \cdots$ *and* $\sigma_1' \geqslant \sigma_2' \geqslant \cdots$ *be the singular values of* X_1 *and* X_2 *respectively. Then* $\|X_1\| \geqslant \|X_2\|$ *for all unitarily invariant norms iff*

$$\sigma_1 + \cdots + \sigma_i \geqslant \sigma_1' + \cdots + \sigma_i', \qquad i = 1, \ldots, r. \tag{1.6}$$

Actually Ky Fan (1951) shows that (1.6) is n.s. for

$$\phi(\sigma_1, \ldots, \sigma_r) \geqslant \phi(\sigma_1', \ldots, \sigma_r') \tag{1.7}$$

to hold for any symmetric gauge function, and then the result of Lemma 1.4.1. follows from the equation (1.5). See Mirsky (1960) for a simple proof of Lemma 1.4.1.

The concepts of unitarily invariant norms and symmetric gauge functions have been found useful in studying monotonic properties of power functions and setting simultaneous confidence bounds on certain parameters in multivariate statistical analysis (see Mudholkar, (1965, 1966) and Wijsman, (1978) for further details and results on symmetric gauge functions and unitarily invariant norms).

In our study we consider matrix functions

$$L(A_1, A_2, \ldots; \Gamma_1, \Gamma_2, \ldots) \tag{1.8}$$

where $A_1, A_2, \ldots$ are given fixed matrices and $\Gamma_1, \Gamma_2, \ldots$ are matrices in specified classes $C_1, C_2, \ldots$. The problem is to find $\Gamma_1 \in C_1, \Gamma_2 \in C_2, \ldots$ which minimize

$$\|L(A_1, A_2, \ldots; \Gamma_1, \Gamma_2, \ldots)\| \tag{1.9}$$

for an appropriately chosen (M,N)-invariant norm. Generally, in statistical applications the Euclidean norm

$$\|X\|_E = (\operatorname{tr} X^*X)^{\frac{1}{2}} \tag{1.10}$$

is considered. We show that in some problems, the *same* solution minimizes (1.8) for any unitarily invariant norm.

In another paper, Rao (1979), the author considered problems of approximating a given r.v. (random variable) by a r.v. in a specified subclass. Such problems arise in the study of canonical correlations, principal component and factor analysis. In these cases, the problems are reduced to the form (1.8) and solved through methods of matrix approximations.

2. Some theorems on eigen and singular values of matrices

We represent the eigen values of a Hermitean $A(m\times m)$ by $\lambda_1(A) \geqslant \cdots \geqslant \lambda_m(A)$ and the corresponding eigen vectors by $P_1,\ldots,P_m$, which may not be unique when some of the eigen values are not distinct. The matrix $(P_1 : \cdots : P_k)$ formed by any set of the first k eigen vectors is represented by $P_{(k)}$. Similarly the matrix formed by any set of the last k eigen vectors is represented by $P_{[k]}$.

The eigen values of a Hermitean $A(m\times m)$ with respect to a p.d. matrix M (i.e., the roots of $|A-\lambda M|=0$) are represented by $\lambda_1(A|M) \geqslant \cdots \geqslant \lambda_m(A|M)$, and the corresponding eigen vectors by $P_1^{(M)},\ldots,P_m^{(M)}$. The matrices $P_{(k)}^{(M)}$ and $P_{[k]}^{(M)}$ are defined in the same way as $P_{(k)}$ and $P_{[k]}$. It is easy to see that

$$\lambda_i(A|M)=\lambda_i(M^{-\frac{1}{2}}AM^{-\frac{1}{2}}). \tag{2.1}$$

The following theorem known as Poincare Separation Theorem (PST) plays an important role in matrix approximations (see Rao, 1973, p. 64).

Theorem 2.1. (PST). *Let $A(m\times m)$ be Hermitean and $B(m\times k)$ be such that $B^*B=I_k$. Then*

(i) $\lambda_{m-k+i}(A) \leqslant \lambda_i(B^*AB) \leqslant \lambda_i(A), \quad i=1,\ldots,k.$

(ii) *The second equalities are attained iff $B=P_{(k)}T$ where T is unitary and the first equalities iff $B=P_{[k]}T$.*

The statement concerning the necessity in (ii) can be proved on the same lines as indicated in Theorem 2.2.

Note 2.1.1. If in Theorem 2.1 B satisfies the condition $B^*MB=I_k$ where M is p.d., then

$$\lambda_{m-k+i}(A|M) \leqslant \lambda_i(B^*AB) \leqslant \lambda_i(A|M). \tag{2.2}$$

Note 2.1.2. Let $G(m\times n)$ be any matrix and $P(m\times m)$ be symmetric idempotent of rank $k\leqslant m$. Then

$$\lambda_{m-k+i}(GG^*) \leqslant \lambda_i(G^*PG) \leqslant \lambda_i(GG^*) \quad i=1,\ldots,k. \tag{2.3}$$

We use Theorem 2.1. to prove two basic theorems in matrix approximations.

Theorem 2.2. *Let $A(m\times n)$ be of rank r and $B(m\times n)$ of rank k. Then*

(i) $\sigma_i(A-B)\geqslant\sigma_{i+k}(A), \qquad i+k\leqslant r,$
$\qquad\geqslant 0, \qquad\qquad\qquad i+k>r.$

(ii) *The equalities in* (i) *are attained iff $k\leqslant r$ and*

$$B=\sigma_1(A)U_1U_1^*+\cdots+\sigma_k(A)U_kU_k^*,$$

while the singular value decomposition of A is

$$A=\sigma_1(A)U_1U_1^*+\cdots+\sigma_r(A)U_rU_r^*.$$

Proof. Since $R(B)=k$, it has a rank factorization $B=CD$ where $C^*C=I_k$ and D is of order $k\times n$. Then it is seen (Rao, 1964, p. 332)

$$(A-CD)^*(A-CD)\geqslant A^*(I-CC^*)A. \tag{2.4}$$

Hence

$$\begin{aligned}\sigma_i^2(A-B)&=\lambda_i[(A-B)^*(A-B)]\\&\geqslant\lambda_i[A^*(I-CC^*)A]\geqslant\lambda_{k+i}(A^*A) \qquad \text{for} \quad k+i\leqslant r,\\&\geqslant\sigma_{k+i}^2(A) \qquad \text{by} \quad (2.3),\end{aligned} \tag{2.5}$$

noting that $(I-CC^*)$ is a projection operator (symmetric and idempotent) of rank $(m-k)$. Obviously $\sigma_i(A-B)\geqslant 0$ for $i+k>r$, which establishes part (i) of Theorem 2.2.

Sufficiency of part (ii) is trivial. To prove the necessity let all the equalities in (i) hold at $B=B_0$ (say). If $k>r$, then $\sigma_i(A-B)=0$ for all i implying $A=B_0$ which is impossible since $R(A)=r\neq k=R(B_0)$. Hence $k\leqslant r$.

Let $E=A-B_0$ and P be the projection operator onto the space generated by the columns of B_0. The inequalities in (i) obviously hold if we replace the condition $R(B)=k$ by $R(B)\leqslant k$. Then

$$\begin{aligned}\operatorname{tr}E^*(I-P)E&=\operatorname{tr}[A-B_0-P(A-B_0)]^*[A-B_0-P(A-B_0)]\\&\geqslant\operatorname{tr}(A-B_0)^*(A-B_0)\end{aligned} \tag{2.6}$$

since by choice of B_0

$$\sigma_i(A-B_0)\leqslant\sigma_i[A-B_0-P(A-B_0)].$$

But the right hand side of (2.6) is $\operatorname{tr} E^*E$. Then

$$\Rightarrow \operatorname{tr} E^*(I-P)E = \operatorname{tr} E^*E - \operatorname{tr} E^*PE \geqslant \operatorname{tr} E^*E$$

$$\Rightarrow \operatorname{tr} E^*PE = 0 \Rightarrow E^*PE = 0 \Rightarrow E^*P = 0 \quad \text{or} \quad (A-B_0)^*B_0 = 0.$$

Consider the s.v.d.

$$E = \sigma_{k+1}(A)U_{k+1}S_{k+1}^* + \cdots + \sigma_r(A)U_rS_r^*. \tag{2.7}$$

Now $E^*B_0 = 0 \Rightarrow U_i^*B_0 = 0$ in which case

$$U^*AA^*U = \Delta^2, U^*U = I_{r-k} \tag{2.8}$$

where $U = (U_{k+1} : \cdots : U_r)$ and Δ^2 is a diagonal matrix with $\sigma_{k+1}^2(A), \ldots, \sigma_r^2(A)$ in the diagonal.

Now $A = B_0 + E$ and $R(A) = R(B_0) + R(E) = R(B_0 : E)$ since $E^*B_0 = 0$. This implies that the spaces generated by the columns of A and the columns of $(B_0 : E)$ are the same. Then, there exists a matrix T such that $E = AT$, from which it follows that $U_i^*W = 0$ if W is an eigen vector of AA^* corresponding to a zero eigen value. The equation (2.8) together with $U_i^*W = 0 \Rightarrow U_{k+1}, \ldots, U_r$ are eigen vectors corresponding to the eigen values $\sigma_{k+1}^2(A), \ldots, \sigma_r^2(A)$. Then we can write the singular value decomposition of A as

$$A = \sigma_1^2(A)U_1V_1^* + \cdots + \sigma_r^2(A)U_rV_r^* \tag{2.9}$$

where U_i are orthonormal and $V_i^* = [\sigma_i(A)]^{-1}U_i^*A$ are orthonormal. From (2.7)

$$S_j^* = [\sigma_j(A)]^{-1}U_j^*E = [\sigma_j(A)]^{-1}U_j^*A = V_j^* \qquad j = k+1, \ldots, r. \tag{2.10}$$

Then B_0 is as defined in part (ii) of the theorem.

Note 2.2.1. If A, B and $A-B$ are Hermitean and n.n.d. matrices of order m and if B is at most of rank k, the following result due to Okamoto and Kanazawa (1968) becomes a special case of Theorem 2.2:

(i) $\lambda_i(A-B) \geqslant \lambda_{k+i}(A)$ for any i where $\lambda_j(A) = 0$ if $j > m$.

(ii) A necessary and sufficient condition that the equality in (i) holds for all i is that

$$B=\lambda_1(A)V_1V_1'+\cdots+\lambda_k(A)V_kV_k' \tag{2.12}$$

where $V_1,\ldots,V_k$ is a first set of eigen vectors of A.

Theorem 2.3. *Let A be an $m\times n$ matrix, and $B(m\times r)$ and $C(n\times k)$ be such that $B^*B=I_r$ and $C^*C=I_k$. Then*

$$\sigma_{t+i}\leqslant\sigma_i(B^*AC)\leqslant\sigma_i(A) \tag{2.13}$$

where $t=m+n-r-k$.

For proof of the theorem the reader is referred to Rao (1979).

Note 2.3.1. Let $A=P\Delta Q^*$ be the s.v.d. of A. Then the upper bound in (2.13) is attained when $B=P_{(r)}$ and $C=Q_{(k)}$.

Note 2.3.2. In view of the result in Note 2.3.1,

$$\|B^*AC\|\leqslant\|P^*_{(r)}AQ_{(k)}\| \tag{2.14}$$

for any unitarily invariant norm.

Note 2.3.2. If in Theorem 2.3, B and C satisfy the restrictions $B^*MB=I_r$ and $C^*NC=I_k$, where M and N are p.d. matrices, then

$$\sigma_i(B^*AC)\leqslant\sigma_i(M^{-\frac{1}{2}}AN^{-\frac{1}{2}}). \tag{2.15}$$

The upper bound in (2.15) is attained when B consists of the first r columns of $M^{-\frac{1}{2}}P$ and C consists of the first k columns of $N^{-\frac{1}{2}}Q$, where P and Q are unitary matrices in the s.v.d. of $M^{-\frac{1}{2}}AN^{-\frac{1}{2}}$.

We quote some well known theorems which are useful in solving some problems of matrix approximations.

Theorem 2.4. (Lidski, 1950 and Wielandt, 1955). *Let X, Y and Z be Hermitean matrices of order n with eigen values*

$$x_1\geqslant\cdots\geqslant x_n;\quad y_1\geqslant\cdots\geqslant y_n\quad\text{and}\quad z_1\geqslant\cdots\geqslant z_n;$$

and, if $Z = X - Y$, *then*

$$\max_{j_1 < \cdots < j_k} \sum_{i=1}^{k} (x_{j_i} - y_{j_i}) \leqslant \sum_{i=1}^{k} z_i, 1 \leqslant k \leqslant n. \tag{2.16}$$

Theorem 2.5. (von Neumann, 1937). *Let* $A(m \times n)$ *and* $B(m \times n)$ *have s.v.d.'s,* $P\Delta_1 Q^*$ *and* $R\Delta_2 S^*$. *Then*

$$-|\operatorname{tr} AUB^*V| \leqslant \operatorname{tr} \Delta_1 \Delta_2 \tag{2.17}$$

where U *and* V *are unitary matrices of orders* n *and* m, *and the equality is attained when* $U = QS^*$ *and* $V = RP^*$.

Von Neumann stated the result (2.17) for square matrices but the proof extends to rectangular matrices as stated in Theorem 2.5. See Rao and Styan (1966) and Wijsman (1978).

As a special case of (2.17) we have

$$\operatorname{tr} \Delta_1 \Delta_2 \leqslant \operatorname{tr} AB \leqslant \operatorname{tr} \Delta_1 \Delta_2. \tag{2.18}$$

Further

$$|\operatorname{tr} AU| \leqslant \operatorname{tr} \Delta_1 \tag{2.19}$$

and the equality is attained when $U = QP^*$.

When A and B are Hermitean matrices with eigen values $\alpha_1 \geqslant \cdots \geqslant \alpha_n$ and $\beta_1 \geqslant \cdots \geqslant \beta_n$ we have

$$\sum \alpha_i \beta_{n-i+1} \leqslant \operatorname{tr} AB \leqslant \sum \alpha_i \beta_i \tag{2.20}$$

due to Richter (1958), and the equality on the right hand side is attained when $B = \sum \beta_i P_i P_i'$ and the equality on the left is attained when $B = \sum \beta_{n-i+1} P_i P_i'$ where $P_1, P_2, \ldots$ are eigen vectors of A. See Theobald (1975) for a discussion of (2.20) and a n.s. condition for the attainment of the upper bound.

Theorem 2.6. (Ky Fan and Hoffman, 1955). *If* A *is a square matrix, then*

$$\lambda_i(A + A^*) \leqslant 2\sigma_i(A). \tag{2.21}$$

3. Regression type approximation

Let $A(m \times n)$ have real elements in which case A' represents the transpose of A, and $X(m \times a)$, $Y(n \times b)$ be given matrices. We shall study the problem of approximating A by a matrix of the form

$$XF_1 + F_2Y' + G \tag{3.1}$$

with the restriction that G has an assigned rank.

In the following theorems we consider the orthogonal projection operators onto the spaces generated by the columns of X and Y

$$P = X(X'MX)^- X'M$$

$$Q = Y(Y'NY)^- Y'N \tag{3.2}$$

where M and N are p.d. matrices of orders m and n, respectively where B^- denotes a generalized inverse of B, i.e. such that $BB^-B = B$.

The s.v.d. of $M^{\frac{1}{2}}AN^{\frac{1}{2}}$ is represented by

$$M^{\frac{1}{2}}AN^{\frac{1}{2}} = \sum \sigma_i R_i S_i' \tag{3.3}$$

where $\sigma_i = \sigma_i(M^{\frac{1}{2}}AN^{\frac{1}{2}})$.

Theorem 3.1. *The following results hold:*

(i) *Column regression*

$$\sigma_i\left[M^{\frac{1}{2}}(A - XF_1)N^{\frac{1}{2}}\right] \geqslant \sigma_i\left[M^{\frac{1}{2}}(A - PA)N^{\frac{1}{2}}\right] \qquad i = 1, 2, \ldots \tag{3.4}$$

and the equalities are attained when $XF_1 = PA$.

(ii) *Row regression*

$$\sigma_i\left[M^{\frac{1}{2}}(A - F_2Y')N^{\frac{1}{2}}\right] \geqslant \sigma_i\left[M^{\frac{1}{2}}(A - AQ)N^{\frac{1}{2}}\right] \qquad i = 1, 2, \ldots \tag{3.5}$$

and the equalities are attained when $F_2Y' = AQ$.

(iii) *Column-row regression*

$$\sigma_i\left[M^{\frac{1}{2}}(A - XF_1 - F_2Y')N^{\frac{1}{2}}\right] \geqslant \sigma_i\left[M^{\frac{1}{2}}(A - PA - AQ + PAQ)N^{\frac{1}{2}}\right]$$

$$i = 1, 2, \ldots \tag{3.6}$$

and the equalities are attained when

$$XF_1 = PA, F_2Y' = (A - PA)Q \quad \text{or} \quad XF_1 = P(A - AQ), F_2Y' = AQ.$$

(iv) *Lower order rank factorization*

$$\begin{aligned} \sigma_i\left[M^{\frac{1}{2}}(A-KL)N^{\frac{1}{2}}\right] &\geqslant \sigma_{k+i}\left[M^{\frac{1}{2}}AN^{\frac{1}{2}}\right], && k+i \leqslant r \\ &\geqslant 0 \qquad\qquad\qquad , && k+i > r \end{aligned} \tag{3.7}$$

where $R(A) = r \geqslant k \geqslant R(KL)$. *The equalities are attained when*

$$M^{\frac{1}{2}}KLN^{\frac{1}{2}} = \sum_{i=1}^{k} \sigma_i R_i S_i', \tag{3.8}$$

where σ_i, R_i *and* S_i *are as defined in* (3.3).

Proof. The results (3.4) of (i) and (3.5) of (ii) are easily established. It is interesting to note that the solution in (3.4) is independent of N and that in (3.5) is independent of M.

To prove (iii), observe that for given F_2

$$\begin{aligned} &N^{\frac{1}{2}}(A - XF_1 - F_2y')'M(A - XF_1 - F_2Y')N^{\frac{1}{2}} \\ &\qquad \geqslant N^{\frac{1}{2}}(A - F_2Y')'(I-P)'M(I-P)(A - F_2Y')N^{\frac{1}{2}} \end{aligned}$$

Therefore $\lambda_i(=\sigma_i^2)$ of the left hand side

$$\begin{aligned} &\geqslant \lambda_i\left[N^{\frac{1}{2}}(A - F_2Y')'(I-P)'M(I-P)(A - F_2Y')N^{\frac{1}{2}}\right] \\ &= \lambda_i\left[M^{\frac{1}{2}}(I-P)(A - F_2Y')N(A - F_2Y')'(I-P)M^{\frac{1}{2}}\right] \\ &\geqslant \lambda_i\left[M^{\frac{1}{2}}(I-P)(A - AQ)N(A - AQ)'(I-P)M^{\frac{1}{2}}\right] \\ &= \sigma_i^2\left[M^{\frac{1}{2}}(A - PA - AQ + PAQ)N^{\frac{1}{2}}\right] \end{aligned}$$

which proves (3.7). The conditions for attainment of equalities can be easily verified.

The result (iv) is a direct consequence of Theorem 2.2.

Note 3.1.1. The results (3.4)–(3.6) can be expressed in the form

$$\|\text{left hand side matrix}\| \geqslant \|\text{right hand side matrix}\| \tag{3.9}$$

for any unitarily invariant norm.

Note 3.1.2. The result (3.6) is proved for Euclidean norm by Corsten and Eijnsbergen (1972), and the result (3.7) is essentially due to Mirsky (1960). See also Corsten (1976) for a discussion of the result (3.7).

Theorem 3.2. *Let A,X,Y be as defined in* (3.1), *P,Q be the projection operators defined in* (3.2), *and*

$$\sigma_1(C)U_1V_1' + \cdots + \sigma_r(C)U_rV_r' \tag{3.10}$$

be the singular value decomposition of $C = M^{\frac{1}{2}}(I-P)A(I-Q)N^{\frac{1}{2}}$. If G is such that $R(G) \leqslant k \leqslant r = R(A)$, then

$$\sigma_i\left[M^{\frac{1}{2}}(A - XF_1 - F_2Y' - G)N^{\frac{1}{2}}\right] \geqslant \sigma_{k+i}(C), \qquad k+i=r,$$

$$\geqslant 0, \qquad k+i>r. \tag{3.11}$$

The equalities in (3.11) *are attained when*

$$G = \sigma_1(C)U_1V_1' + \cdots + \sigma_k(C)U_kV_k'$$

$$F_2Y' = (A-G)Q, \quad XF_1 = P(A-G)(I-Q). \tag{3.12}$$

Note that, as a consequence of (3.11),

$$\inf \|M^{\frac{1}{2}}(A - XF_1 - F_2Y' - G)N^{\frac{1}{2}}\| \tag{3.13}$$

where infimum is taken over all $F_1, F_2, R(G) \leqslant k \leqslant r$ is attained when F_1, F_2 and G are as chosen in (3.12) for any unitarily invariant norm.

Proof. Observe that

$$N^{\frac{1}{2}}(A - XF_1 - F_2Y' - G)'M(A - XF_1 - F_2Y' - G)N^{\frac{1}{2}}$$

$$\geqslant N^{\frac{1}{2}}(A - F_2Y' - G)'(I-P)'M(I-P)(A - F_2Y' - G)N^{\frac{1}{2}}. \tag{3.14}$$

Hence the ith eigen value of the left hand side of (3.14) is

$$\geqslant \lambda_i\left[N^{\frac{1}{2}}(A-F_2Y'-G)'(I-P)'M(I-P)(A-F_2Y'-G)N^{\frac{1}{2}}\right]$$

$$=\lambda_i\left[M^{\frac{1}{2}}(I-P)(A-F_2Y'-G)N(A-F_2Y'-G)'(I-P)'M^{\frac{1}{2}}\right]$$

$$\geqslant \lambda_i\left[M^{\frac{1}{2}}(I-P)(A-G)(I-Q)N(I-Q)'(A-G)(I-P)'M'\right]$$

$$=\sigma_i^2\left[M^{\frac{1}{2}}(I-P)(A-G)(I-Q)N^{\frac{1}{2}}\right]$$

$$\geqslant \sigma_{k+i}^2\left[M^{\frac{1}{2}}(I-P)A(I-Q)N^{\frac{1}{2}}\right], \qquad k+i\leqslant r.$$

which proves (3.11). It can be verified that the equalities are attained when G, F_1 and F_2 are as chosen in (3.12).

Note 3.2.1. The result (3.13) is proved for Euclidean norm by Corsten and Eijnsberger (1972). It is interesting to see that the result holds for any unitarily invariant norm. Corsten and Eijnsberger (1972), Corsten (1976), Krishnaiah and Schuurmann (1974) and Mandel (1969) discuss applications of the result (3.13) using Euclidean norm in testing interactions in a two-factor experiment.

4. Reduced rank regression type problem

In Theorem 3.1 [results (3.7) and (3.9)], it is shown that

$$\inf_{R(G)=k} \|M^{\frac{1}{2}}(A-G)N^{\frac{1}{2}}\| \tag{4.1}$$

is attained at a common G for all unitarily invariant norms when $R(A)=r \geqslant k$. We shall consider the minimization problem when there is a further restriction on G (in addition to its rank) that it is of the form $G=XB$ where $X(m\times s)$ of rank s is a given matrix. The optimum property stated with reference to (4.1) may not hold for any choice of M and N. We obtain some results for special choices of N and the norm to be minimized.

Let $R(XB)=k\leqslant r=R(A)$. Then, $XB=XCD$ where C is an $s\times k$ matrix such that $C'X'MXC=I$. Further let $\lambda_1\geqslant\cdots\geqslant\lambda_s$ be the eigen values and $P_1,\ldots,P_s$ the corresponding eigen vectors of $X'MANA'MX$ with respect to $X'MX$. We have the following theorem.

Theorem 4.1. *Let $\|\cdot\|_E$ denote the Euclidean norm. Then*

$$\inf_{C,D} \|M^{\frac{1}{2}}(A-XCD)N^{\frac{1}{2}}\|_E^2 = \operatorname{tr} A'MAN - (\lambda_1 + \cdots + \lambda_k) \tag{4.2}$$

and the infimum is attained at C_*, D_* defined by

$$XC_*D = XC_*C_*'X'MA, C_* = (P_1 : \cdots : P_k). \tag{4.3}$$

Proof. Observe that

$$N^{\frac{1}{2}}(A-XCD)'M(A-XCD)N^{\frac{1}{2}} \geqslant N^{\frac{1}{2}}A'(M-MXCC'X'M)AN^{\frac{1}{2}}. \tag{4.4}$$

To minimize the trace of the left hand side expression in (4.4), we need only consider the right hand expression, which is equal to

$$\operatorname{tr} A'MAN - \operatorname{tr} N^{\frac{1}{2}}A'MXCC'X'MAN^{\frac{1}{2}} \tag{4.5}$$

which leads us to maximize the second expression in (4.5) which is equal to

$$\operatorname{tr} C'X'MANA'MXC \tag{4.6}$$

where $C'(X'MX)C = I_k$. Using the result (2.2), the expression (4.6) attains the maximum value $\lambda_1 + \cdots + \lambda_k$ when C is as chosen in (4.3). With such a choice of C, equality in (4.4) is attained when D is as given in (4.3). The theorem is established for any choice of M and N.

We obtain a stronger result by making a special choice of N as shown in Theorem 4.2.

Theorem 4.2. *Let $N = (A'MA)^{-1}$. Then*

$$\inf_{C,D} \|M^{\frac{1}{2}}(A-XCD)N^{\frac{1}{2}}\| \tag{4.7}$$

is attained for any unitarily invariant norm at D and C specified in (4.3).

Proof. We start with the inequality (4.4). The ith eigen value of the left hand side expression of (4.4)

$$\geqslant \lambda_i\left[N^{\frac{1}{2}}A'(M-MXCC'X'M)AN^{\frac{1}{2}}\right] \tag{4.8}$$

$$=1-\lambda_{n-i+1}(N^{\frac{1}{2}}A'MXCC'X'MAN^{\frac{1}{2}})$$

$$\begin{cases} =1-\lambda_{n-i+1}(C'X'MANA'MXC), & i\geqslant n-k+1 \\ =1 & i<n-k+1 \end{cases}$$

$$\begin{cases} \geqslant 1-\lambda_{n-i+1}(X'MANA'MX|X'MX), & i\geqslant n-k+1 \\ =1 & i<n-k+1 \end{cases}$$

$$=\lambda_i\left[N^{\frac{1}{2}}A'(M-MXC_*C_*'X'M)AN^{\frac{1}{2}}\right] \tag{4.9}$$

where C_* is as defined in (4.3).

The equality in (4.4) is attained when $XCD=XCC'X'MA$. Then (4.4) together with (4.9) shows that

$$\lambda_i\left[N^{\frac{1}{2}}(A-XCD)'M(A-XCD)N^{\frac{1}{2}}\right]$$

$$\geqslant \lambda_i\left[N^{\frac{1}{2}}(A-XC_*D_*)'M(A-XC_*D_*)N^{\frac{1}{2}}\right]$$

which implies

$$\sigma_i\left[M^{\frac{1}{2}}(A-XCD)N^{\frac{1}{2}}\right]\geqslant \sigma_i\left[M^{\frac{1}{2}}(A-XC_*D_*)N^{\frac{1}{2}}\right]$$

when N is chosen as $(A'MA)^{-1}$. This establishes the result claimed in Theorem 4.2.

5. Generalized inverse as matrix approximation

Given a matrix A of order $m\times n$, there may not exist a matrix G such that $GA=I_n$ or $AG=I_m$. In such a case we may find G such that GA and AG are close to the identity matrices I_n and I_m. Such a G, if it exists qualifies to be called a generalized inverse (g-inverse) of A. The following theorems characterize different types of inverses introduced by Moore

(1935), Penrose (1955), Bjerhammer (1958) and Rao (1967). [See Rao and Mitra (1971) for a systematic treatment of g-inverses and their applications.]

Theorem 5.1. *Let A be a $m \times n$ matrix. Denote by P_T the orthogonal projector onto the space generated by the columns of a matrix T. The following results hold*:

(i) $$\operatorname*{Inf}_X \|I - AX\| = \|I - AG\| \tag{5.1}$$

for any unitarily invariant norm on S_{mm}, where G is any $n \times m$ matrix satisfying the equivalent conditions

$$AG = P_A \Leftrightarrow A'AG = A' \Leftrightarrow AGA = A, AG = (AG)'. \tag{5.2}$$

(ii) $$\operatorname*{Inf}_X \|I - XA\| = \|I - GA\| \tag{5.3}$$

for any unitarily invariant norm on S_{nn}, where G is any $n \times m$ matrix satisfying the equivalent conditions

$$GA = P_{A'} \Leftrightarrow AA'G' = A \Leftrightarrow AGA = A, GA = (GA)'. \tag{5.4}$$

(iii) $$\operatorname*{Inf}_X \|I - AX\| = \|I - AG\|, \operatorname*{Inf}_X \|I - XA\| = \|I - GA\|, \tag{5.5}$$

for any unitarily invariant norms on S_{mm} and S_{nn} where G is any $n \times m$ matrix satisfying the equivalent conditions

$$AG = P_A, GA = P_{A'} \Leftrightarrow AA'G' = A, A'AG = A'$$
$$\Leftrightarrow AGA = A, (AG) = (AG)', (GA) = (GA)'. \tag{5.6}$$

(iv) $$\operatorname*{Inf}_X \|I - AX\| = \|I - AG\|, \operatorname*{Inf}_Y \|I - GY\| = \|I - GA\| \tag{5.7}$$

for any unitarily invariant norms on S_{mm} and S_{nn} where G is any $n \times m$ matrix satisfying the equivalent conditions

$$AG = P_A, GA = P_G \;\Leftrightarrow\; A'AG = A', G'GA = G'$$
$$\Leftrightarrow\; AGA = A, GAG = G, (AG) = (AG)', (GA) = (GA)'. \tag{5.8}$$

Proof of (i). Let us note that

$$(I - AX)'(I - AX) \geqslant (I - P_A) \;\Rightarrow\; \sigma_i(I - AX) \geqslant \sigma_i(I - P_A) \tag{5.9}$$

which proves (5.1). The equivalence of the conditions (5.2) is established in Rao (1967) where such a G was termed as a least squares generalized inverse.

Proof of (ii). We have

$$(I-XA)(I-XA)' \geqslant (I-P_{A'}) \Rightarrow \sigma_i(I-XA) \geqslant \sigma_i(I-P_{A'}) \quad (5.10)$$

which proves (5.2). The equivalence of the conditions (5.4) is established in Rao (1967) where such a G was called a minimum norm generalized inverse.

Proof of (iii). Since (5.9) and (5.10) hold simultaneously for a choice of X such that $AX=P_A$ and $XA=P_{A'}$, the result (5.5) is established.

Proof of (iv). Since

$$(I-AX)'(I-AX) \geqslant (I-P_A)$$

$$(I-GY)'(I-GY) \geqslant (I-P_G)$$

the result (5.7) is established. The equivalence of the conditions (5.8) is established in Rao (1967).

It is interesting to note that in (5.7), A is required to be an inverse of G when G is an inverse of A in the sense of matrix approximations. The reflexive inverse so obtained is called Moore–Penrose inverse.

Note 5.1.1. G as defined by each of the equations in (5.2), (5.4), (5.6) and (5.8) exists. Let

$$A=P\begin{pmatrix}\Delta & 0\\ 0 & 0\end{pmatrix}Q' \quad (5.11)$$

be the s.v.d. of A, where Δ is nonsingular diagonal matrix with $R(\Delta)=R(A)$, and consider the matrix

$$G=Q\begin{pmatrix}\Delta^{-1} & E_2\\ E_3 & E_4\end{pmatrix}P' \quad (5.12)$$

Then:

(i) $E_2=0, E_3, E_4$ arbitrary provide *all* solutions of (5.2).
(ii) $E_3=0, E_2, E_4$ arbitrary provide *all* solutions of (5.4).

(iii) $E_2=0, E_3=0, E_4$ arbitrary provide *all* solutions of (5.6).
(iv) $E_2=E_3=E_4=0$ provide the *unique* solution of (5.8).

Note 5.1.2. In (5.1), AG is unique and in (5.3), GA is unique although in each case G may not be unique.

6. Procrustean transformations

Let A and B be two given $m\times n$ matrices. In factor analysis we seek to find a transformation matrix T of order $n\times n$ such that A is close to BT. Such a transformation is called a Procrustean transformation. (See Green, 1952, Schöneman, 1966). More generally, we consider the problem of finding transformation matrices U and T of orders m and n such that UA is close to BT.

Theorem 6.1. *Let $A=P\Delta_1 Q'$ and $B=R\Delta_2 S'$ be the s.v.d.'s of A and B respectively. Then the minimum of $\|UA-BT\|_E$ when U and T run through orthogonal matrices is attained at $U=RP'$ and $T=SQ'$.*

Proof.

$$\|UA-BT\|_E^2=\operatorname{tr}(UA-BT)'(UA-BT)$$

$$=\operatorname{tr}AA'+\operatorname{tr}B'B-2\operatorname{tr}A'U'BT$$

The result of Theorem 6.1 follows by an application of von Neumann's Theorem 2.5 [see expression (2.17)] on minimization of $\operatorname{tr}A'U'BT$.

Note 6.1.1. In Theorem 6.1, we have considered only the Euclidean norm. The result may be not true for all unitarily invariant norms.

Note 6.1.2. Using the result (2.19) we find that the minimum of $\|A-BT\|_E$ when T varies over orthogonal matrices is attained when $T=QP'$, where $P\Delta Q'$ is the s.v.d. of $A'B$.

Note 6.1.3. If in Theorem 6.1, U is orthogonal and T is arbitrary then $\min\|UA-BT\|_E$ is attained at

$$BT=P_B UA \quad \text{and} \quad U=RP' \tag{6.1}$$

where $AA' = P\Delta_1 P'$ and $P_B = R\Delta_2 R'$ are spectral decompositions of AA' and P_B, where P_B is the orthogonal projection operator onto the space of the columns of B.

Note 6.1.4. If in Theorem 6.1, A is a square matrix and B is an identity, then it is shown by Ky Fan and Hoffman (1955) that the minimum of $\|A - T\|$ is attained for any unitarily invariant norm at $T = PQ'$ where $A = P\Delta Q'$ is the s.v.d. of A.

Note 6.1.5. We may extend the result of Ky Fan and Hoffman to the case where A and T are $m \times n$ matrices and the columns of T are required to be orthogonal. In such a case the minimum of $\|A - T\|$ is attained for any unitarily invariant norm at $T = PR'$ where $A = P\Delta Q'$ is the s.v.d. of A.

Note 6.1.6. Let $\Sigma(n \times n)$ be an n.n.d. matrix of rank r with the spectral decomposition $\Sigma = \lambda_1 P_1 P_1' + \cdots + \lambda_r P_r P_r'$. Using the arguments of Ky Fan and Hoffman it can be shown that the minimum of $\|\Sigma - L\|$ when L runs through idempotent matrices of rank $k \leqslant r$ is attained at $L = P_1 P_1' + \cdots + P_k P_k'$, for any unitarily invariant norm.

Acknowledgement

I wish to thank Dr. G. P. H. Styan for drawing my attention to the wide literature on matrix approximations, which was helpful in writing this paper.

References

[1] Bjerhammer, A. (1958). A new matrix algebra. *Trans. Roy. Inst. of Technology*. Stockholm.

[2] Corsten, L. A. C. (1976). Matrix approximation: A key to multivariate methods. Presented at the Biometric Conference (Boston).

[3] Corsten, L. A. C. and van Eijnsberger, A. A. (1972). Multiplicative effects in two-way analysis of variance. *Statistica Neerlandica* **26**, 61–67.

[4] Green, B. F. (1952). The orthogonal approximation of an oblique structure in factor analysis. *Psychometrika* **17**, 429–440.

[5] Krishnaiah, P. R. and Schuurmann, F. J. (1974). On the evaluation of some distributions that arise in simultaneous tests for the equality of the latent roots of the covariance matrix. *J. Multivariate Anal.* **4**, 265–282.

[6] Ky Fan (1951). Maximum properties and inequalities for the eigen values of completely continuous operators. *Proc. Nat. Acad. Sci.* **37**, 760–766.

[7] Ky Fan and Hoffman, A. J. (1955). Some matrix inequalities in the space of matrices. *Proc. American Math. Soc.* **6**, 111–116.

[8] Lidski, V. B. (1950). On the characteristic numbers of the sum and product of symmetric matrices. *Doklady Akad. Nauk SSSR* **75**, 769–772 (in Russian).

[9] Mandel, J. (1969). The partitioning of interaction in analysis of variance. *J. Res. Nat. Bur. Stand.* **73 B**, 309–328.

[10] Mirsky, L. (1960). Symmetric gauge functions and unitarily invariant norms. *Quart. J. Math (Oxford)* **11**, 50–59.

[11] Moore, E. H. (1935). *General Analysis* Vol. 1. Amer. Philos. Soc., Philadelphia.

[12] Mudholkar, G. S. (1965). A class of tests with monotone power functions for two problems in multivariate statistical analysis. *Ann. Math. Stat.* **36**, 1794–1801.

[13] Mudholkar, G. S. (1966). On confidence bounds associated with multivariate analysis of variance and non-independence between two sets of variates. *Ann. Math. Statist.* **37**, 1736–1746.

[14] Okamoto, M. and Kanazawa, M. (1968). Minimization of eigen values of a matrix and optimality of principal components. *Ann. Math. Statist.* **39**, 859–863.

[15] Penrose, R. (1955). A generalized inverse for matrices. *Proc. Camb. Phil. Soc.* **51**, 406–413.

[16] Rao, C. R. (1964). The use and interpretation of principal component analysis in applied research. *Sankhya* A **26**, 329–358.

[17] Rao, C. R. (1967). Calculus of generalized inverse of matrices. Part I—general theory. *Sankhya* A **29**, 317–350.

[18] Rao, C. R. (1973). *Linear Statistical Inference and its Applications*. John Wiley.

[19] Rao, C. R. (1979). Separation theorems for singular values of matrices and their applications in multivariate analysis. *J. Multivariate Anal.,* (forthcoming).

[20] Rao, C. R. and Mitra, S. K. (1971). *Generalized Inverse of Matrices and its Applications*. John Wiley.

[21] Rao, C. R. and Styan, G. P. H. (1976). Notes on a matrix approximation problem and some related matrix inequalities. Discussion paper No. 137. Indian Statistical Institute, New Delhi.

[22] Richter, H. (1958). Zur Abschätzung von Matrizennormen. *Diese Nachr.* **18**, 178–187.

[23] Schöneman, P. H. (1966). The generalized solution of the orthogonal procrustes problem. *Psychometrika* **31**, 1–16.

[24] Theobald, C. R. (1975). An inequality for the trace of the product of two matrices. *Math. Proc. Camb. Phil. Soc.* **77**, 265–267.

[25] von Neumann, J. (1937). Some matrix inequalities and metrization of metric spaces. Tomsk. Univ. Rev **1**, 286–299.

[26] Wielandt, H. (1955). An extreme property of sums of eigen values. *Proc. Amer. Math. Soc.* **6**, 106–110.

[27] Wijsmann, R. A. (1978). Constructing all simultaneous confidence sets in a given class, with applications to MANOVA. To appear in *Ann. Statist.*

P. R. Krishnaiah, ed., *Multivariate Analysis–V*

RECENT RESULTS ON THE ESTIMATION OF A LINEAR FUNCTIONAL RELATIONSHIP*

T. W. ANDERSON
Department of Statistics, Stanford University, Stanford, CA 94305, USA.

1. Introduction

In the statistical literature a set of multivariate observations is said to satisfy a *linear functional relationship* if their mean values satisfy a linear equation. The case most extensively studied is that of bivariate observations with mean values satisfying a single equation and with uncorrelated errors of equal variance; sometimes the model is known as an *error-in-variables* model. In this review we shall survey some of the recent developments for this case.

The emphasis here is on the estimation of the slope of the linear relation. Two geometric representations of the problem and the maximum likelihood estimation of the slope are presented. The exact density of the maximum likelihood estimator of the slope when the population slope is zero is given, and its derivation sketched briefly. It is shown how the cumulative distribution is obtained. Asymptotic expansions of the distribution are discussed when the spread of a fixed number of points increases or alternatively when the number and spread increase together. Several tables of the distributions are given and discussed.

Since this is a survey paper details are not given; many are available in Anderson and Sawa [9]. Other references to the recent literature are given where appropriate. For further references see Anderson [5]. It may be of interest that the estimate treated here was introduced just 100 years ago (Adcock [1]).

2. The model and estimates

Let (x_g, y_g) be an observation from $N[(\mu_g, \nu_g), \sigma^2\mathbf{I}]$, $g = 1, \ldots, N$, and suppose the observations are independent. It is assumed that the means

*This work was supported by the National Science Foundation Grant SOC77-14944 at the Institute for Mathematical Studies in the Social Sciences, Stanford University.

satisfy the linear relation

$$\nu_g = \gamma + \beta\mu_g. \tag{2.1}$$

Let

$$s_{xx} = \sum_{g=1}^{N} (x_g - \bar{x})^2,\ s_{yy} = \sum_{g=1}^{N} (y_g - \bar{y})^2,\ s_{xy} = \sum_{g=1}^{N} (x_g - \bar{x})(y_g - \bar{y}) \tag{2.2}$$

where $\bar{x} = \sum_{g=1}^{N} x_g / N$ and $\bar{y} = \sum_{g=1}^{N} y_g / N$. Then the maximum likelihood estimate of the slope β is

$$\hat{\beta} = \frac{s_{yy} - s_{xx} + \sqrt{\left((s_{yy} - s_{xx})^2 + 4s_{xy}^2\right)}}{2s_{xy}}$$

$$= \frac{2s_{xy}}{s_{xx} - s_{yy} + \sqrt{\left((s_{xx} - s_{yy})^2 + 4s_{xy}^2\right)}}. \tag{2.3}$$

(Note that the least squares estimate s_{xy}/s_{xx} is the second form of (2.3) with s_{yy} and s_{xy} in the denominator replaced by 0; the least squares estimate is closer to 0 than the maximum likelihood estimate.)

Fig. 1 shows the line $\nu = \gamma + \beta\mu$ on which the means lie. The plusses denote observations, one corresponding to each mean point. The line $y = \hat{\gamma} + \hat{\beta}x$ is fitted to the observed points by minimizing the sum of squares of the distances of the observed points from the line (so-called 'orthogonal regression').

The angle θ that the line $\nu = \gamma + \beta\mu$ makes with the μ-axis can be used as a parameter instead of the slope $\beta = \tan\theta$, and the maximum likelihood estimate $\hat{\theta}$ of this parameter is the angle the fitted line $y = \hat{\gamma} + \hat{\beta}x$ makes with the x-axis, $\hat{\beta} = \tan\hat{\theta}$. The covariance matrix $\sigma^2\mathbf{I}$ is invariant with respect to rotations and so is the method of fitting the line. Hence, $\hat{\theta} - \theta$ is invariant with respect to rotations, and its distribution does not depend on θ. (Because the function $\beta = \tan\theta$ is periodic, with period π, the inverse function $\theta = \arctan\beta$ is multi-valued; we consider θ as defined modulo π. One may think of the angle θ as the unoriented direction of a diameter of a unit circle.) We shall study the distribution of $\hat{\beta}$ for $\beta = 0$. This knowledge implies the distribution of $\hat{\theta}$ for $\theta = 0$ and hence the distribution of $\hat{\theta} - \theta$ and of $\hat{\theta}$ for arbitrary θ; from this we can find the distribution of $\hat{\beta}$ for arbitrary β.

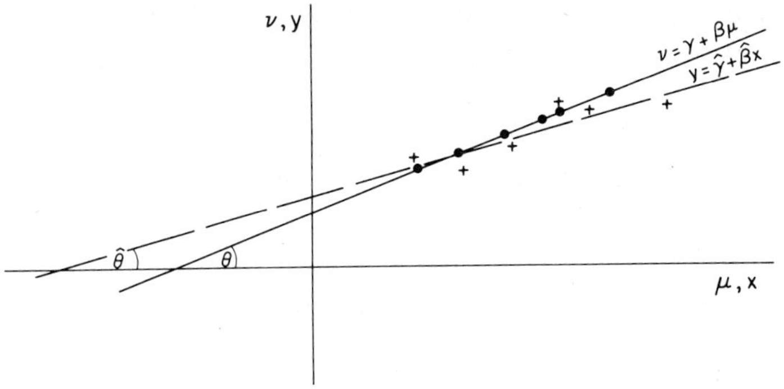

Fig. 1.

As shown by Anderson [5], the estimation of the slope of a linear functional relationship is related to the Limited Information Maximum Likelihood estimation of the coefficient of an endogenous variable in system of simultaneous equations in econometrics. Here we shall give an oversimplified explanation of the relation. Suppose v_{1t}, v_{2t} is an observation pair from a normal distribution with means $\sum_j \pi_{1j} z_{jt}$, $\sum_j \pi_{2j} z_{jt}$, where the z_{it}'s are nonstochastic independent variables and the π_{ij}'s are regression coefficients, and covariance matrix $\mathbf{\Omega}$, $t=1,\ldots,T$. The least squares estimators p_{1j}, p_{2j} of π_{1j}, π_{2j} are normally distributed with means $\mathcal{E} p_{1j} = \pi_{1j}$, $\mathcal{E} p_{2j} = \pi_{2j}$. It is assumed that $v_{1t} - \beta v_{2t}$ has mean $0, t=1,\ldots,T$; that is, that $\pi_{1j} = \beta \pi_{2j}$. The correspondence between this problem and that of the linear functional relationship is that in both models there are quantities whose means satisfy a linear equation; the number of pairs of regression coefficients corresponds to the number of degrees of freedom in the linear functional relation, $n = N-1$. (In the econometric problem the covariance matrix of the sample regression coefficients depends on $\mathbf{\Omega}$ which is usually assumed unknown and is estimated on the basis of residuals from the sample regression; the distribution of the estimator when $\mathbf{\Omega}$ is estimated differs from that when $\mathbf{\Omega}$ is known to within a proportionality factor, but the difference is negligible if T is large.)

In multivariate analysis we usually have a geometric representation in a space of dimension equal to the number of observations. The estimate $\hat{\beta}$ depends on s_{xx}, s_{xy} and s_{yy}, which in turn are computed from deviations of the observations from their means. The point $(x_1 - \bar{x}, \ldots, x_N - \bar{x})$ is in the $n = N-1$ dimensional subspace of the N-space orthogonal to the vector $(1,1,\ldots,1)$. Referred to a coordinate system in the n-space this vector is

$(X_1,\ldots,X_n)=\mathbf{X}$, say (Anderson [2], Section 3.3, for example). Thus we have the correspondence

$$(x_1-\bar{x},\ldots,x_N-\bar{x})\rightarrow(X_1,\ldots,X_n)=\mathbf{X}, \tag{2.4}$$

$$(y_1-\bar{y},\ldots,y_N-\bar{y})\rightarrow(Y_1,\ldots,Y_n)=\mathbf{Y}, \tag{2.5}$$

$$(\mu_1-\bar{\mu},\ldots,\mu_N-\bar{\mu})\rightarrow(M_1,\ldots,M_n)=\mathbf{M}, \tag{2.6}$$

$$(\nu_1-\bar{\nu},\ldots,\nu_N-\bar{\nu})\rightarrow(N_1,\ldots,N_n)=\mathbf{N}. \tag{2.7}$$

Then $\mathcal{E}\mathbf{X}=\mathbf{M}$ and $\mathcal{E}\mathbf{Y}=\mathbf{N}=\beta\mathbf{M}$ by (2.1). As a random vector, $\mathbf{X}$ has a spherical normal distribution with center $\mathbf{M}$, and $\mathbf{Y}$ has a spherical normal distribution with center $\mathbf{N}$, which is in the direction of $\mathbf{M}$.

The estimation procedure can be described in terms of this geometry; see Fig. 2. The dotted line is determined to minimize the distance squared

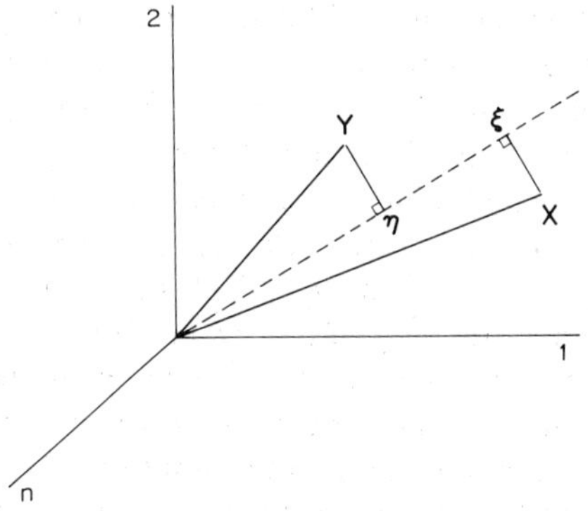

Fig. 2.

of $\mathbf{X}$ from the line plus the distance squared of $\mathbf{Y}$. Let $\boldsymbol{\xi}$ be the projection of $\mathbf{X}$ on this line and $\boldsymbol{\eta}$ be the projection of $\mathbf{Y}$ on this line. Then $\hat{\beta}=\pm\|\boldsymbol{\eta}\|/\|\boldsymbol{\xi}\|$. (In the case of least squares $\mathbf{Y}$ is projected on $\mathbf{X}$ and the estimate is the length of this projection relative to the length of $\mathbf{X}$; this number is smaller in absolute value than $|\hat{\beta}|$.)

It can be seen intuitively that the longer $\mathbf{M}$ is the more accurately the direction of $\mathbf{M}$ can be determined. Correspondingly for a given length of $\mathbf{M}$ the smaller n is the more accurately the direction of $\mathbf{M}$ can be estimated. Thus, for a given $\|\mathbf{M}\|$ we can expect that the spread of the distribution of $\hat{\beta}-\beta$ increases with the dimensionality n.

3. The exact density and distribution

The density of $\hat{\beta}$ when $\beta=0$ depends only on the standardized noncentrality parameter

$$\delta^2=\sum_{g=1}^{N}(\mu_g-\bar{\mu})^2/\sigma^2. \tag{3.1}$$

where $\bar{\mu}=\Sigma_{g=1}^{N}\mu_g/N$. (The numerator is $\|\mathbf{M}\|^2$.) Since $\hat{\beta}$ is homogeneous of degree zero, it is invariant with respect to changes of scale. Hence, we can take $\sigma^2=1$ in deriving the density and distribution of $\hat{\beta}$. The matrix

$$\mathbf{S}=\begin{pmatrix} s_{xx} & s_{xy} \\ s_{xy} & s_{yy} \end{pmatrix} \tag{3.2}$$

has the noncentral Wishart distribution with covariance matrix $\mathbf{I}$, $n=N-1$ degrees of freedom, and means covariance matrix with δ^2 as the upper left-hand element and 0 as each other element. (See Anderson and Girshick [6].) The density is

$$\frac{e^{-\frac{1}{2}\delta^2}e^{-\frac{1}{2}(s_{xx}+s_{yy})}|\mathbf{S}|^{\frac{1}{2}(n-3)}}{2^n\sqrt{\pi}\cdot\Gamma[(n-1)/2]}\sum_{j=0}^{\infty}\frac{(\delta^2/4)^j s_{xx}^j}{j!\Gamma(\frac{1}{2}n+j)}, \tag{3.3}$$

which is the product of the central Wishart density and an infinite series similar to that term in the noncentral χ^2-density. The density of the new set of variables defined by

$$s_{xx}=sy,\ s_{yy}=(1-s)y,\ s_{xy}=ry \tag{3.4}$$

is

$$\frac{e^{-\frac{1}{2}\delta^2}\left[s(1-s)-r^2\right]^{\frac{1}{2}(n-3)}}{2^n\sqrt{\pi}\,\Gamma\left[\frac{1}{2}(n-1)\right]}\sum_{j=0}^{\infty}\frac{(\delta^2/4)^j s^j}{j!\Gamma(\frac{1}{2}n+j)}e^{-\frac{1}{2}y}y^{n+j-1} \tag{3.5}$$

for $0\leqslant y<\infty$, $r^2\leqslant s(1-s)$, $0\leqslant s\leqslant 1$ and is 0 otherwise. The integral with respect to y leaves as the density of r and s

$$\frac{e^{-\frac{1}{2}\delta^2}\left[s(1-s)-r^2\right]^{\frac{1}{2}(n-3)}}{\sqrt{\pi}\cdot\Gamma[(n-1)/2]}\sum_{j=0}^{\infty}\frac{\Gamma(n+j)}{j!\Gamma(\frac{1}{2}n+j)}\left(\frac{\delta^2}{2}\right)^j s^j. \tag{3.6}$$

The density is positive in the interior of the circle with center $(\frac{1}{2},0)$ and radius $\frac{1}{2}$. This fact suggests transforming to polar coordinates

$$r=u\sin\hat{\psi},\quad s=\tfrac{1}{2}+u\cos\hat{\psi}, \tag{3.7}$$

where $-\pi\leqslant\hat{\psi}\leqslant\pi$ and $0\leqslant u\leqslant\frac{1}{2}$. Then

$$\tan\hat{\psi}=\frac{r}{s-\frac{1}{2}}=\frac{2s_{xy}}{s_{xx}-s_{yy}}=\tan 2\hat{\theta}=\frac{2\tan\hat{\theta}}{1-\tan^2\hat{\theta}}=\frac{2\hat{\beta}}{1-\hat{\beta}^2}. \tag{3.8}$$

To obtain the density of $\hat{\psi}$ (which is a function of $\hat{\beta}$) we expand $s^j=(\frac{1}{2}+u\cos\hat{\psi})^j$ according to the binomial theorem and integrate with respect to u, yielding

$$\frac{e^{-\frac{1}{2}\delta^2}}{\sqrt{\pi}\cdot 2^n}\sum_{j=0}^{\infty}\frac{\Gamma(n+j)}{\Gamma\left(\frac{1}{2}n+j\right)}\left(\frac{\delta^2}{4}\right)^j\sum_{i=0}^{j}\frac{\Gamma\left(\frac{1}{2}i+1\right)\cos^i\hat{\psi}}{\Gamma\left[\frac{1}{2}(n+1+i)\right]i!(j-i)!}. \tag{3.9}$$

The density of $\hat{\beta}=\tan\frac{1}{2}\hat{\psi}$ is found from (3.9) using (3.8). After rearrangement of terms it can be written

$$\frac{e^{-\frac{1}{2}\delta^2}}{4\sqrt{\pi}\cdot(1+\hat{\beta}^2)}\sum_{i=0}^{\infty}\frac{\Gamma[(n+i)/2]}{\Gamma\left(\frac{1}{2}n+i\right)\Gamma[(i+1)/2]}$$

$$\times{}_1F_1\left(n+i,\tfrac{1}{2}n+i;\delta^2/4\right)\left[\frac{\delta^2(1-\hat{\beta}^2)}{4(1+\hat{\beta}^2)}\right]^i, \tag{3.10}$$

where ${}_1F_1(a,b;c)$ is a confluent hypergeometric function.

4. The cumulative distribution

Since the density function when $\beta=0$ is symmetric about $\hat{\beta}=0$, we need only find the cumulative distribution function for positive arguments or its complement. The set of (r,s) for which $\hat{\beta}\geqslant z$, $0\leqslant z\leqslant 1$, is indicated in Fig. 3. The set of $\hat{\psi}$ is $\arctan(2z)/(1-z^2)\leqslant\hat{\psi}\leqslant\pi$. To integrate (3.9) with respect to $\hat{\psi}$ we use

$$\int_a^b\cos^i x\,dx=\frac{\Gamma\left[\frac{1}{2}(i+1)\right]\Gamma\left(\frac{1}{2}\right)}{\Gamma\left(\frac{1}{2}i+1\right)}$$

$$\times\left\{I_{\cos^2 a}\left[\tfrac{1}{2}(i+1),\tfrac{1}{2}\right]-I_{\cos^2 b}\left[\tfrac{1}{2}(i+1),\tfrac{1}{2}\right]\right\}, \tag{4.1}$$

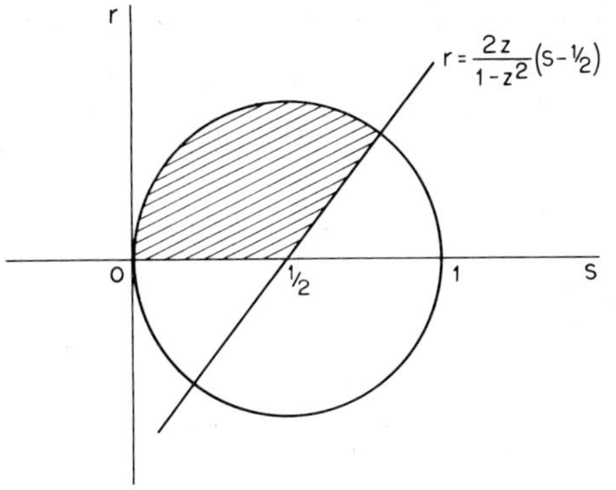

Fig. 3. $\{(r,s)|\,\hat{\beta}>z, 0\leqslant z\leqslant 1\}$

where $I_z(p,q)$ denotes the incomplete beta function.

It is also possible to go back to the density (3.6) and integrate with respect to r and s. For example, the integral over the interval $0\leqslant s\leqslant 1/(1+z^2)$, $(1/(1+z^2)$ being the ordinate of the relevant intersection of the circle and line in Fig. 3) is

$$\tfrac{1}{2}e^{-\frac{1}{2}\delta^2}\sum_{j=0}^{\infty}\left(\frac{\delta^2}{2}\right)^j\frac{1}{j!}I_{1/(1+z^2)}\left(\tfrac{1}{2}n+j,\tfrac{1}{2}n\right). \tag{4.2}$$

To carry out the integration needed for $0\leqslant s\leqslant a(s-\frac{1}{2})$, $\frac{1}{2}\leqslant s\leqslant b$ $(a>0)$, if $n-3=2q$, we can expand

$$\left[s(1-s)-r^2\right]^q s^j=\sum_{i=0}^{q}\binom{q}{i}(-1)^i r^{2i}s^{q-i+j}(1-s)^{q-i}. \tag{4.3}$$

The integral of each term gives an incomplete beta function. Since the cdf is a power series in δ^2, it is suitable for computation if δ^2 is not too large (for a given n). (See Anderson and Sawa [7] for details.)

As noted in Section 2 the distribution of the estimate of the slope for an arbitrary parameter value can be derived from the distribution for parameter value zero because the distribution of the difference between the estimator of the angle and its parameter value is independent of the parameter value. Let $\hat{\beta}_M=\tan\hat{\theta}_M$ be the maximum likelihood estimator of the slope when $\beta=\tan\theta$ is the parameter (and $\hat{\beta}=\tan\hat{\theta}$ is the estimator when the parameter value is zero). Then $\hat{\theta}_M = \hat{\theta} + \theta$ and

$$\begin{aligned}
\hat{\beta}_M - \beta &= \tan\hat{\theta}_M - \tan\theta \\
&= \tan(\hat{\theta}+\theta) - \tan\theta \\
&= \frac{\tan\hat{\theta} + \tan\theta}{1 - \tan\hat{\theta}\tan\theta} - \tan\theta \\
&= \frac{\tan\hat{\theta}(1+\tan^2\theta)}{1-\tan\hat{\theta}\tan\theta} \\
&= \frac{(1+\beta^2)\hat{\beta}}{1-\beta\hat{\beta}}. \qquad (4.4)
\end{aligned}$$

The cdf for $\hat{\beta}_M - \beta$ is found from

$$\mathbf{P}\{\hat{\beta}_M - \beta \leqslant y\} = \mathbf{P}\left\{\frac{(1+\beta^2)\hat{\beta}}{1-\beta\hat{\beta}} \leqslant y\right\}, \qquad (4.5)$$

which can be written in terms of the cdf of $\hat{\beta}$.

5. Asymptotic theory

Approximate distributions of $\hat{\beta}_M - \beta$ (or $\hat{\beta}$) can be found from asymptotic expansions of the distributions. In the econometric studies it is natural that δ^2 increases (as T increases), but N remains fixed. In Fig. 1 that means that the spread of the means (μ_g, ν_g) increases; in Fig. 2 that means that the lengths of $\mathbf{M}$ and $\mathbf{N}$ increase. In that case $\hat{\beta}_M$ has an asymptotic normal distribution with mean β and asymptotic standard deviation (ASD) of $\sqrt{(1+\beta^2)}/\delta$. More precisely, Anderson [3] gave

$$\begin{aligned}
&\mathbf{P}\left\{\frac{\delta}{\sqrt{(1+\beta^2)}}(\hat{\beta}_M - \beta) \leqslant t\right\} \\
&= \Phi(t) + \phi(t)\left[\frac{1}{\delta}P_1(t) + \frac{1}{\delta^2}P_2(t) + \frac{1}{\delta^3}P_3(t)\right] + 0(\delta^{-4}), \qquad (5.1)
\end{aligned}$$

where $\Phi(t)$ and $\phi(t)$ are the cdf and density of the standard normal distribution, respectively, and $P_j(t)$ is a polynomial in t of degree $3j-1$, $j=1,2,3$, depending on β. (See Anderson [4] for a refinement.)

It may be more appropriate to let both N and δ^2 increase. In that case $\hat{\beta}_M$ has an asymptotic normal distribution with mean β and asymptotic standard deviation of $\surd(N/\delta^2+(1+\beta^2))/\delta$. Kunitomo [10] has given

$$\mathbf{P}\left\{\frac{\surd(N/\delta^2+(1+\beta^2))}{\delta}(\hat{\beta}_M-\beta)\leqslant t\right\}$$

$$=\Phi(t)+\phi(t)\left\{\frac{1}{\sqrt{n}}P_1^*(t)+\frac{1}{n}P_2^*(t)\right\}+0(n^{-3/2}). \tag{5.2}$$

Patefield [12] has studied the distributions of the normalized difference $\hat{\theta}_M-\theta$. He has compared the approximate distributions obtained from the expansion for N fixed (a transform of (5.1)) and the normal distribution for δ and N large with empirical distributions based on simulation. Roughly speaking he finds that if N is greater than δ^2 the approximation based on N fixed and $\delta^2\to\infty$ is not very accurate, whereas the normal approximation is accurate unless δ^2 is rather small.

Table 1
Cumulative distribution of normalized estimators
$\beta=0.0$

	$\delta^2=4$			$\delta^2=10$			$\delta^2=40$		
t \ $n=$	1	7	11	1	7	11	1	7	20
0.0	0.500	0.500	0.500	0.500	0.500	0.500	0.500	0.500	0.500
0.5	0.688	0.644	0.634	0.689	0.653	0.640	0.691	0.686	0.658
1.0	0.818	0.752	0.737	0.830	0.774	0.754	0.838	0.832	0.790
2.0	0.928	0.865	0.849	0.955	0.902	0.881	0.972	0.968	0.936
3.0	0.960	0.911	0.898	0.985	0.948	0.939	0.997	0.996	0.982
4.0	0.974	0.934	0.924	0.994	0.966	0.953	0.999	0.999	0.994
MEDN[a]	0.000	0.000	0.000	0.000	0.000	0.000	0.000	0.000	0.000
U.QT	0.708	0.984	1.086	0.691	0.880	0.981	0.679	0.675	0.833
97.5%								1.961	2.727
IQR	1.416	1.968	2.771	1.381	1.769	1.961	1.358	1.463	1.667
	ASD=0.500			ASD=0.316			ASD=0.158		

[a]MEDN = Median, U.QT = Upper Quartile, 97.5% = 97.5 percentile, IQR = Interquartile range

The inequality $\tan\hat{\psi} \leqslant 2z/(1-z^2)$ for $0<z<1$ is

$$(1-z^2)s_{xy} - z(s_{xx} - s_{yy}) \leqslant 0 \tag{5.3}$$

except for a set of probability $0(e^{-\delta})$. The left-hand side of (5.3) can be written as a quadratic form in $2n$ variables which can be brought to diagonal form with an orthogonal transformation. Then the probability of (5.3) can be written

$$\mathbf{P}\{\chi_1'^2 - \chi_2'^2 \leqslant 0\}, \tag{5.4}$$

where the two variables are independent and each has a noncentral χ^2-distribution with n degrees of freedom and noncentrality parameters depending on z, β, and δ^2. In turn (5.4) can be written as the probability

Table 2
Cumulative distribution of normalized estimators
$\beta = 5.0$

	$\delta^2=4$		$\delta^2=10$	$\delta^2=40$			Normal
t \ $n=$	1	7	1	3	7	20	
−4.0	0.023	0.026	0.000	0.000	0.000	0.000	0.000
−3.0	0.023	0.026	0.000	0.000	0.000	0.000	0.001
−2.0	0.023	0.026	0.000	0.002	0.002	0.002	0.023
−1.0	0.050	0.057	0.076	0.119	0.119	0.121	0.159
−0.5	0.277	0.286	0.279	0.294	0.294	0.296	0.309
0.0	0.523	0.526	0.501	0.500	0.500	0.500	0.500
0.5	0.679	0.678	0.668	0.679	0.678	0.677	0.691
1.0	0.771	0.769	0.778	0.806	0.806	0.804	0.841
2.0	0.865	0.862	0.892	0.936	0.936	0.935	0.977
3.0	0.909	0.905	0.940	0.979	0.979	0.979	0.999
4.0	0.932	0.929	0.963	0.993	0.993	0.993	1.000
2.5%				−1.506	−1.509	−1.516	−1.960
L.QT	−.548	−.579		−.611	−.612	−.616	−.678
MEDN	−.054	−.208		0.000	0.000	0.000	0.000
U.QT	0.862	0.898		0.754	0.756	0.761	0.678
97.5%				2.827	2.835	2.861	1.960
IQR	1.410	1.477		1.366	1.368	1.377	1.356
	ASD=2.550		ASD=1.612	ASD=0.806			

that

$$(\chi_1'^2)^{1/3} - (\chi_2'^2)^{1/3} \tag{5.5}$$

is negative. The normalized form of (5.5) has a distribution with a very accurate asymptotic expansion as $\delta^2 \to \infty$ and/or $n \to \infty$; see Mudholkar, Chaubey, and Lin [11].

6. Tabulation of the distribution functions

From the distributions and asymptotic expansions described in Sections 4 and 5, Anderson and Sawa [8] have computed extensive tables of the distribution of $[\delta/\sqrt{(1+\beta^2)}](\hat{\beta}_M - \beta)$, which is the quantity with the limiting standard normal distribution. Some of that information is presented here in Tables 1 and 2. Since the density when $\beta = 0$ is symmetric the cdf is given only for nonnegative values in Table 1. It will be noted that in each table for a given n the cdf approaches the standard normal (the right-hand column of Table 2) as δ^2 increases. For small values of δ^2 the distributions are more spread out than the normal but less spread as δ^2 increases (even though the distributions are given in units of asymptotic standard deviations).

The density of $\hat{\beta}_M - \beta$ for $\beta = 5.0$ is not symmetric, but approaches symmetry as δ^2 increases. In particular the median is about 0 when δ^2 is large. The spread (for given n) decreases slightly as δ^2 increases. The interquartile range (IQR) tends to decrease as β increases.

References

[1] Adcock, R. J. (1878). A problem in least squares. *The Analyst* 5, 53–54.

[2] Anderson, T. W. (1958). *An Introduction to Multivariate Statistical Analysis*. Wiley, New York.

[3] Anderson, T. W. (1974). An asymptotic expansion of the distribution of the limited information maximum likelihood estimate of a coefficient in a simultaneous equation system. *J. Amer. Statist. Assoc.* **69**, 565–573.

[4] Anderson, T. W. (1975). An asymptotic expansion of the distribution of the maximum likelihood estimate of the slope coefficient in a linear functional relationship. Technical Report No. 176, the Economics Series, Institute for Mathematical Studies in the Social Sciences, Stanford University.

[5] Anderson, T. W. (1976). Estimation of linear functional relationships: approximate distributions and connections with simultaneous equations in econometrics. *J. Roy Statist. Soc. Ser. B* **38**, 1–36.

[6] Anderson, T. W. and Girshick, M. A. (1944). Some extensions of the Wishart distribution. *Ann. Math. Statist.* **15**, 345–375.

[7] Anderson, T. W. and Sawa, T. (1975). Distribution of a maximum likelihood estimate of a slope coefficient: The LIML estimate for known covariance matrix. Technical Report No. 174, the Economics Series, IMSSS, Stanford University.

[8] Anderson, T. W. and Sawa, T. (1977). Tables of the distribution of the maximum likelihood estimate of the slope coefficient and approximations. Technical Report No. 234, the Economics Series, IMSSS, Stanford University.

[9] Anderson, T. W. and Sawa, T. (1978). Exact and approximate distributions of the maximum likelihood estimator of a slope coefficient; the LIML estimator for a known covariance matrix. Unpublished.

[10] Kunitomo, N. (1977). Asymptotic expansions of the distributions of estimators in a linear functional relationship when the sample size is large. To be published in *J. Amer. Statist. Assoc.*

[11] Mudholkar, G. S., Chaubey, Y. P. and Lin, C. C. (1976). Approximations for the doubly noncentral-*F* distribution. *Comm. Statist. Theory Methods* **A5**, 49–63.

[12] Patefield, W. M. (1976). On the validity of approximate distributions arising in fitting a linear functional relationship. *J. Statist. Comput. Simul.* **5**, 43–60.

P. R. Krishnaiah, ed., *Multivariate Analysis–V*

ITERATIVELY REWEIGHTED LEAST SQUARES FOR LINEAR REGRESSION WHEN ERRORS ARE NORMAL/INDEPENDENT DISTRIBUTED

A. P. DEMPSTER, Nan M. LAIRD and Donald B. RUBIN*
Department of Statistics, Harvard University, Cambridge, MA, U.S.A.

This paper applies the general theory of maximum likelihood estimation from incomplete data to the problem of computing estimated regression coefficients and scale parameters. The key observation is that iteratively reweighted least squares (IRLS) algorithms generated from linear models with normal/independent (N/I) error terms are examples of what have been called EM algorithms. This correspondence in the case of N/I errors enables us to demonstrate several important facts. First, each iteration of IRLS increases the likelihood of the data. Second, under mild conditions IRLS converges to a local maximum of the likelihood. And third, simple small sample and asymptotic expressions describing the rate of convergence near a limit point can be exhibited. As intermediate steps in proving these results, special properties of weight functions associated with N/I densities are derived.

1. Introduction

The recent popularity of iteratively reweighted least squares (IRLS) for multiple linear regression analysis reflects interest in estimates which are relatively unaffected by extreme observations [1,7]. Little appears to be known about the numerical properties of IRLS algorithms used to define estimates of regression coefficients [6]. The purpose of this paper is primarily to add to the knowledge of numerical properties rather than statistical properties.

Our analysis is limited to a subclass of IRLS regression techniques, namely those associated with maximum likelihood estimation based on linear models whose error distributions come from specified nonnormal scale families. Due to our use of general theory related to maximum likelihood estimation from incomplete data [5], our discussion is mostly confined to models with normal/independent (N/I) error distributions. Within the class of symmetric error distributions, the limitation is not especially stringent in practice, because the class of N/I distributions is rich in long-tailed symmetric examples.

*This work was completed while D. B. Rubin was on leave from Educational Testing Service with a Guggenheim fellowship.

Section 2 presents background theory relating IRLS to maximum likelihood estimation from linear models with general error distributions. Section 3 extends the theory in the case of N/I error distributions by exploiting a special statistical interpretation of the weight function used to specify a particular IRLS algorithm. Section 4 makes the connection between EM algorithms and the IRLS algorithms of Section 3, thus allowing the application of EM results to prove the convergence theorems of Section 5 and to derive the rate of convergence formulas of Section 6. Section 7 applies the formulas to the t family of distributions, and illustrates with a numerical example.

2. General properties of IRLS

We begin with a precise definition and a basic property of weighted least squares, and then turn to a discussion of IRLS. We proceed from the standpoint of a statistician with data in hand who is considering a range of possible weighted least squares analyses. The data consist of an $n \times 1$ response vector $\mathbf{Y}$ and an $n \times p$ design matrix $\mathbf{X}$. Given the data, we consider a class of weighted least squares analyses, each analysis defined by an $n \times n$ diagonal matrix $\mathbf{W}$, whose nonnegative diagonal elements we call weights. Given a particular $\mathbf{W}$, the corresponding weighted least squares analysis chooses $\boldsymbol{\beta}$ to minimize $(\mathbf{Y}-\mathbf{X}\boldsymbol{\beta})^{\mathrm{T}}\mathbf{W}(\mathbf{Y}-\mathbf{X}\boldsymbol{\beta})$.

We assume throughout that $\mathbf{X}$ has rank p, so that $\mathbf{X}\boldsymbol{\beta}$ uniquely determines $\boldsymbol{\beta}$.

We restrict the class of weight matrices $\mathbf{W}$ by requiring that not too many of its diagonal elements can be zero. Specifically, the rows of $\mathbf{X}$ corresponding to the nonzero weights must have rank p. With this restriction, it is well-known that the solution of the weighted least squares problem defined above can be written

$$\mathbf{b}=(\mathbf{X}^{\mathrm{T}}\mathbf{W}\mathbf{X})^{-1}(\mathbf{X}^{\mathrm{T}}\mathbf{W}\mathbf{Y}). \tag{2.1}$$

Equation (2.1) implicitly assumes that all of the weights are finite, but it is convenient to allow a subset of the weights to be infinite, subject to the restriction that the corresponding subset of the rows of the equations $\mathbf{Y}=\mathbf{X}\boldsymbol{\beta}$ must have at least one solution. If infinite weights are allowed, we can no longer use the above definition of weighted least squares. Instead, we adopt the following convention. When some weights are infinite, $\mathbf{b}$ must satisfy $\mathbf{Y}=\mathbf{X}\mathbf{b}$ exactly for the subset of rows corresponding to infinite weights. Subject to these constraints, $\mathbf{b}$ satisfies the weighted least squares criterion defined as above using only the rows of $\mathbf{Y}$, $\mathbf{X}$, and $\mathbf{W}$ corresponding to finite weights.

Suppose we denote by $\mathcal{W}$ the class of weight matrices which satisfy the restrictions of the preceding two paragraphs for given data $(\mathbf{Y},\mathbf{X})$. That is, $\mathbf{W}\in\mathcal{W}$ is diagonal with nonnegative elements W_i such that (1) the rank of the rows of $\mathbf{X}$ corresponding to $W_i>0$ is p and (2) for rows of $\mathbf{X}$, $\mathbf{Y}$ corresponding to infinite W_i, $\mathbf{Y}=\mathbf{X}\boldsymbol{\beta}$ for at least one $\boldsymbol{\beta}$. The following lemma presents a fundamental fact about the associated class of weighted least squares estimates $\mathbf{b}$.

Lemma 1. *The class of weighted least squares* $\mathbf{b}$ *corresponding to the class of* $\mathbf{W}$ *in* $\mathcal{W}$ *is bounded by the union of a finite set of closed finite polytopes in* R^p.

A proof of Lemma 1 is given in the Appendix.

Iteratively reweighted least squares (IRLS) refers to a process of finding $\mathbf{b}^{(0)}$, $\mathbf{b}^{(1)}$, $\mathbf{b}^{(2)},\ldots,$where $\mathbf{b}^{(0)}$ is a starting value, and $\mathbf{b}^{(l+1)}$ for $l\geqslant 0$ is a weighted least squares estimate corresponding to a weight matrix $\mathbf{W}^{(l)}$, where $\mathbf{W}^{(l)}$ is determined by $\mathbf{b}^{(l)}$. Specifically, a scalar weight function $W(Z)$ is used to determine the diagonal elements $W_i^{(l)}$ of $\mathbf{W}^{(l)}$ according to the formulas

$$W_i^{(l)}=W(Z_i^{(l)}) \tag{2.2}$$

and

$$Z_i^{(l)}=(Y_i-\mathbf{X}_i\mathbf{b}^{(l)})/s^{(l)} \tag{2.3}$$

where $(Y_i,\mathbf{X}_i)$ denotes the ith row of $(\mathbf{Y},\mathbf{X})$, and $s^{(l)}$ denotes an estimate of scale at the lth iteration. Since (2.3) is undefined if $s^{(l)}=0$, we shall assume $s^{(l)}>0$ for all $l=0,\ 1,\ 2,\ldots$. In Section 4 we give conditions which ensure that $s^{(l)}>0$. For a fixed weight function $W(Z)$ and a fixed starting value $\mathbf{b}^{(0)}$, a specific version of IRLS is obtained by defining how $s^{(l)}$ is determined at each iteration. By construction, an IRLS process will continue indefinitely unless it encounters $s^{(l)}=0$ or $\mathbf{W}^{(l)}$ outside the class $\mathcal{W}$ of permissible weight matrices.

IRLS processes of the general type defined by (2.2) and (2.3) have been proposed [3,4] as an iterative technique for obtaining M-estimates [1,7]. The basis of the proposal is that IRLS can be recognized to be a process of successive substitution applied to the equations derived from differentiating an objective function. We prefer to frame our discussion in terms of M-estimates which are also maximum likelihood (ML) estimates, and hence proceed to connect IRLS and ML through the familiar likelihood equations.

Accordingly, we introduce the statistical model

$$\mathbf{Y} = \mathbf{X}\boldsymbol{\beta} + \mathbf{e}, \tag{2.4}$$

where $\boldsymbol{\beta}$ is a vector of parameters and $\mathbf{e}$ is a vector such that the components of $\sigma^{-1}\mathbf{e}$ are independently and identically distributed with known density $d(Z)$ on $-\infty < Z < \infty$. The log likelihood function associated with (2.4) is

$$L(\boldsymbol{\beta}, \sigma) = -n \log \sigma + \sum_{i=1}^{n} \log d(Z_i), \tag{2.5}$$

where

$$Z_i = (Y_i - \mathbf{X}_i\boldsymbol{\beta})/\sigma. \tag{2.6}$$

As shown below in Lemma 2, the connection between IRLS and ML depends on choosing the weight function to be

$$\begin{aligned} W(Z) &= -d'(Z)/Zd(Z), \quad \text{for } Z \neq 0 \\ &= -\lim_{Z \to 0} d'(Z)/Zd(Z), \quad \text{for } Z = 0. \end{aligned} \tag{2.7}$$

Assumptions in (2.7) are that $d(Z) > 0$ for all Z, that $d'(Z)$ exists for $Z \neq 0$, and that $W(Z)$ has a finite or infinite limit as $Z \to 0$. Also, since weights are nonnegative, we must assume that $d'(Z) \leqslant 0$ for $Z > 0$ and $d'(Z) \geqslant 0$ for $Z < 0$, whence $d(Z)$ is unimodal with a mode at $Z = 0$. To simplify the theory we assume $d'(0) = 0$.

We now relate IRLS processes defined by (2.2), (2.3) and (2.7) to first derivatives of (2.5).

Lemma 2. *For $(\boldsymbol{\beta}, \sigma)$ such that $\sigma > 0$ and $W(Z_i)$ is finite for all i, the equations derived from the* log *likelihood* (2.5) *are given by*

$$\mathbf{X}^{\mathrm{T}}\mathbf{W}\mathbf{Y} - \mathbf{X}^{\mathrm{T}}\mathbf{W}\mathbf{X}\boldsymbol{\beta} = \mathbf{0}, \tag{2.8}$$

and

$$-(\mathbf{Y} - \mathbf{X}\boldsymbol{\beta})^{\mathrm{T}}\mathbf{W}(\mathbf{Y} - \mathbf{X}\boldsymbol{\beta}) + n\sigma^2 = 0, \tag{2.9}$$

where $\mathbf{W}$ *has elements $W(Z_i)$, and $W(Z)$ is defined by* (2.7).

Proof. Differentiating (2.5) with respect to $(\boldsymbol{\beta},\sigma)$ and setting the derivatives to zero, leads to the equations

$$\sum_{i=1}^{n} [-d'(Z_i)/d(Z_i)]\mathbf{X}_i=\mathbf{0} \tag{2.10}$$

and

$$\sum_{i=1}^{n} [Z_i d'(Z_i)/d(Z_i)+1]=0. \tag{2.11}$$

Using the definition (2.7) to specify the diagonals of $\mathbf{W}$, we find that the likelihood equations (2.10) and (2.11) become the weighted least squares equations (2.8) and (2.9).

Since $\mathbf{W}$ depends on $(\boldsymbol{\beta},\sigma)$, we cannot immediately solve (2.8) and (2.9) for $(\boldsymbol{\beta},\sigma)$, but the equations clearly suggest an IRLS procedure: at each iteration substitute current values into the expression for $\mathbf{W}$; then, holding $\mathbf{W}$ fixed, solve (2.8) and (2.9) to obtain the next value of $(\boldsymbol{\beta},\sigma)$. Specifically, we take

$$\mathbf{b}^{(l+1)}=(\mathbf{X}^{\mathrm{T}}\mathbf{W}^{(l)}\mathbf{X})^{-1}\mathbf{X}^{\mathrm{T}}\mathbf{W}^{(l)}\mathbf{Y}, \tag{2.12}$$

and

$$s^{(l+1)^2}=(\mathbf{Y}-\mathbf{X}\mathbf{b}^{(l+1)})^{\mathrm{T}}\mathbf{W}^{(l)}(\mathbf{Y}-\mathbf{X}\mathbf{b}^{(l+1)})/n, \tag{2.13}$$

where $\mathbf{W}^{(l)}$ is defined by (2.2), (2.3), and (2.7).

Theorem 1. *If an instance of an* IRLS *algorithm defined by* (2.12) *and* (2.13) *converges to* $(\mathbf{b}^*,s^*)$ *where the weights are all finite and* $s^*>0$, *then* $(\mathbf{b}^*,s^*)$ *is a stationary point of* $L(\beta,\sigma)$.

Proof. Set $(l)=(l+1)=*$ in (2.12) and (2.13) and compare with (2.8) and (2.9) in Lemma 2.

Results similar to Lemma 2 and Theorem 1 with σ fixed are given in [1,4]. The robustness school generally determines $s^{(l)}$ at each iteration by some means other than maximum likelihood; in this case, only the first equations in the pairs (2.12), (2.13) and (2.8), (2.9) are relevant to the associated IRLS algorithms.

3. Normal/independent distributions

We shall see in Section 4 that IRLS has a special characterization when it is generated by a weight function $W(Z)$ with associated $d(Z)$ representing a scale mixture of normal distributions. In Section 3, we prepare the way for Section 4 by exhibiting properties of such distributions, especially as they relate to the corresponding $W(Z)$ defined by (2.7).

If u is a standard normal random variable with density $(2\pi)^{-\frac{1}{2}}e^{-\frac{1}{2}u^2}$ on $-\infty<u<\infty$, and q is a positive random variable distributed independently of u with distribution function $M(q)$, then the scaled normal random variable $Z=uq^{-\frac{1}{2}}$ is said in [1] to have an N/I distribution.

N/I distributions are a convenient family of symmetric distributions for the independent, identically distributed components of $\mathbf{e}$ in (2.4). One might suspect a close connection with weighted least squares, because knowledge of the scale factors $q_i^{-\frac{1}{2}}$ in each component $e_i=\sigma u_i q_i^{-\frac{1}{2}}$ of $\mathbf{e}$ would lead to the use of weighted least squares with weight matrix $\mathbf{W}$ whose diagonal elements are $q_1, q_2, \ldots, q_n$. As we shall see, treating these weights as missing data leads to a statistically natural derivation of IRLS.

There is a small literature on N/I distributions, most notably the paper of Andrews and Mallows [2] showing that necessary and sufficient conditions for $d(Z)$ to represent the density of a N/I distribution are

$$\left(-\frac{\mathrm{d}}{\mathrm{d}r}\right)^k \mathrm{d}(r^{\frac{1}{2}}) \geqslant 0, \tag{3.1}$$

for $r>0$ and $k \geqslant 1$.

Familiar examples of N/I distributions include the t family arising when q is distributed as χ^2 divided by its degrees of freedom. In Section 7, we use the t family for illustrative purposes because it is analytically tractable and includes distributions with very long tails as degrees of freedom go to zero. The logistic and double exponential distributions are also N/I [2]. The phrase contaminated normal is sometimes used for N/I distributions derived from a two-point distribution of q, such as a distribution which puts probability 0.9 on $q=1$ and probability 0.1 on $q=1/9$, thus indicating 10% contamination of the normal distribution with a component having 3 times the standard deviation of the main component. We will see that contaminated normals, or finite mixtures generally, are easily handled numerically.

We use only the elementary properties of N/I distributions given in Lemma 3 and Theorem 2.

Lemma 3. *A distribution is* N/I *if and only if its density is expressible in the form*

$$d(Z)=\int_0^\infty (2\pi)^{-\frac{1}{2}} q^{\frac{1}{2}} e^{-\frac{1}{2}qZ^2}\,dM(q), \tag{3.2}$$

and $d(Z)<\infty$ *for* $Z\neq 0$. *The density* $d(Z)$ *is finite at* $Z=0$ *if and only if* $E(q^{\frac{1}{2}})<\infty$.

Proof. Equation (3.2) follows immediately from the definition of an N/I density. To see that finite $d(0)$ implies finite $E(q^{\frac{1}{2}})$, assume

$$\int_0^\infty q^{\frac{1}{2}}\,dM(q)=\infty.$$

Then $d(Z)$ can be made larger than any arbitrary number ν, by taking Z sufficiently small as follows. First, choose q' so that

$$\int_0^{q'} q^{\frac{1}{2}}\,dM(q)>2\nu.$$

Then choose δ_ν so that $Z<\delta_\nu$ implies

$$e^{-\frac{1}{2}q'Z^2}>\tfrac{1}{2},$$

hence

$$e^{-\frac{1}{2}qZ^2}>\tfrac{1}{2} \quad \text{for } 0<q<q'.$$

It now follows that

$$d(Z)>\int_0^{q'} q^{\frac{1}{2}} e^{-\frac{1}{2}qZ^2}\,dM(q)>\tfrac{1}{2}\int_0^{q'} q^{\frac{1}{2}}\,dM(q)>\nu$$

for $Z<\delta_\nu$. Then $d(0)=\infty$, because as $\nu\to\infty$, $\delta_\nu\to 0$.

Theorem 2. *Suppose that* Z *is* N/I *distributed. Then for* $0<|Z|<\infty$;

(i) *The conditional distribution of* q *given* Z *exists,*

(ii) $\mathbf{E}(q^k|Z)<\infty$, *for* $k>-\frac{1}{2}$,

(iii) $W(Z)=\mathbf{E}(q|Z)$, (3.3)

(iv) $W'(Z)=-Z\,\mathrm{var}(q|Z)$, *and* (3.4)

(v) $W(Z)=W(-Z)$ *is finite, positive, and nonincreasing for* $Z>0$.

For $Z=0$:

(vi) *The conditional distribution of* q *given* Z *exists if and only if* $\mathbf{E}(q^{\frac{1}{2}})<\infty$;

(vii) $W(0)\geqslant W(Z)$ *for* $Z\neq 0$ *and* $W(0)$ *is finite if and only if* $\mathbf{E}(q^{3/2})<\infty$,

(viii) $W'(0)$ *is finite if and only if* $\mathbf{E}(q^{5/2})<\infty$.

Proof. From the joint distribution of u and q, it follows directly that the conditional distribution function of q given Z is

$$M(q|Z)=\int_0^q q^{\frac{1}{2}}e^{-\frac{1}{2}qZ^2}\,dM(q)\Big/\int_0^\infty q^{\frac{1}{2}}e^{-\frac{1}{2}qZ^2}\,dM(q), \tag{3.5}$$

for $Z\neq 0$. Differentiating the right side of (3.2) and substituting in (2.7) yields

$$W(Z)=\int_0^\infty q^{3/2}e^{-\frac{1}{2}qZ^2}\,dM(q)\Big/\int_0^\infty q^{\frac{1}{2}}e^{-\frac{1}{2}qZ^2}\,dM(q), \tag{3.6}$$

which together with (3.5) yields (3.3). Similarly, differentiating (3.6) yields (3.4). The remaining parts of Theorem 2 are immediate consequences of (3.3), (3.4), and (3.5), and limiting arguments such as that presented in the proof of Lemma 3.

Note that if the moment $\mathbf{E}(q^{\frac{1}{2}})$ is not finite, then the log likelihood (2.5) becomes infinite for $\boldsymbol{\beta}$ on any of the n hyperplanes defined by $\mathbf{Y}=\mathbf{X}\boldsymbol{\beta}$, so we henceforth assume $\mathbf{E}(q^{\frac{1}{2}})<\infty$, with the consequence that $W(Z)=E(q|Z)$ is defined, possibly infinite, for all Z.

4. EM theory

We now establish that, when IRLS is generated from an N/I error density, each iteration of IRLS can be described as an iteration of an algorithm of a more general type called EM algorithms [5]. Since we wish to apply general theory about EM in Sections 5 and 6, we briefly outline some of the general theory and notation of [5]. Since the theory of EM is easier to derive in general than in particular models, it is a useful tool for learning about IRLS.

The general theory postulates two vectors $\mathbf{y}$ and $\mathbf{x}$, which are not the same as the $\mathbf{Y}$ and $\mathbf{X}$ used throughout the paper. The general standpoint is that $\mathbf{y}$ represents the directly observed data, called *incomplete data* because there is assumed to exist further potential data which are not observed. We denote by $\mathbf{x}$ a representation of the *complete data*, including both observed and unobserved, whence it is evident that $\mathbf{y}$ is determined from $\mathbf{x}$ via a function we denote by $\mathbf{y}=\mathbf{y}(\mathbf{x})$. The statistical model is of the traditional form with a family of sampling densities $f(\mathbf{x}|\xi)$ for $\mathbf{x}$ depending on parameters ξ. The corresponding family of densities for $\mathbf{y}$ is then

$$g(\mathbf{y}|\xi)=\int f(\mathbf{x}|\xi)\,\mathrm{d}\mathbf{x}, \tag{4.1}$$

where integration is over the subspace of the sample space of $\mathbf{x}$ determined implicitly by the relation $\mathbf{y}=\mathbf{y}(\mathbf{x})$. Our objective is to maximize the likelihood of ξ given $\mathbf{y}$, i.e. (4.1) regarded as a function of ξ for fixed $\mathbf{y}$.

In our application to the model (2.4) with N/I distributed error terms, the components e_i of $\mathbf{e}$ can individually be represented as $e_i=\sigma u_i q_i^{-\frac{1}{2}}$, where the $u_i q_i^{-\frac{1}{2}}$ are independent and identically distributed with the general N/I form. If we choose to regard $\mathbf{q}=(q_1,q_2,\ldots,q_n)$ as missing information, we may represent the complete data $\mathbf{x}$ as $(\mathbf{Y},\mathbf{q})$ and regard the actual data $\mathbf{Y}$ as the incomplete data $\mathbf{y}$. The design matrix is considered fixed and known throughout. The parameter ξ is represented by $(\boldsymbol{\beta},\sigma)$, and the log of the likelihood (4.1) is given by (2.5).

Each iteration $\xi^{(l)}\to\xi^{(l+1)}$ of an EM algorithm has two steps, called an expectation-step or **E**-step, and a maximization-step or M-step. In the case of IRLS, these steps are computing the weights, and carrying out weighted least squares, respectively. Under the general version, the task of the **E**-step is to find a certain function of ξ, namely,

$$Q(\xi|\xi^{(l)})=\int \log f(\mathbf{x}|\xi)k(\mathbf{x}|\mathbf{y},\xi^{(l)})\,\mathrm{d}\mathbf{x} \tag{4.2}$$

where $k(\mathbf{x}|\mathbf{y},\xi)$ denotes the conditional density of $\mathbf{x}$ given $\mathbf{y}$ and ξ. The heuristic idea is that, although we do not have available the complete data $\log f(\mathbf{x}|\xi)$, we can find and use its conditional expectation given the data $\mathbf{y}$ and the current $\xi^{(l)}$. The M-step then finds $\xi^{(l+1)}$ which maximizes the expected log likelihood $Q(\xi|\xi^{(l)})$.

In order to apply the EM theory, we now assume q has density $m(q)$. For the linear model specified in (2.4) with N/I distributed error terms, the density of $\mathbf{x}=(\mathbf{Y},\mathbf{q})$ is expressible as the product of the conditional density of $\mathbf{Y}$ given $\mathbf{q}$, which is normal, times the marginal density of $\mathbf{q}$, so that

$$\log f(\mathbf{x}|\boldsymbol{\xi}) = -\frac{n}{2}\log 2\pi - n\log\sigma + \tfrac{1}{2}\sum_{i=1}^{n}\log q_i$$

$$-\frac{1}{2\sigma^2}\sum_{i=1} q_i(Y_i-\mathbf{X}_i\boldsymbol{\beta})^2 + \sum_{i=1}^{n}\log m(q_i). \tag{4.3}$$

The first, third, and fifth terms in (4.3) can be ignored since they do not involve $\boldsymbol{\xi}$, and hence we can take

$$Q(\boldsymbol{\xi}|\boldsymbol{\xi}^{(l)}) = \mathbf{E}\left[-n\log\sigma - \frac{1}{2\sigma^2}\sum_{i=2}^{n} q_i(Y_i-\mathbf{X}_i\boldsymbol{\beta})^2|\mathbf{Y},\boldsymbol{\xi}^{(l)}\right], \tag{4.4}$$

where $\boldsymbol{\xi}^{(l)}=(\mathbf{b}^{(l)},s^{(l)})$ for all l. The only random quantities in the conditional expectation (4.4) are the q_i and further, the expression is linear in the q_i. Moreover, conditioning q_i on $\mathbf{Y}$ and $\boldsymbol{\xi}$ is equivalent to conditioning on the e_i on the Z_i, and hence we are led back to the weight function $W(Z)$ defined in (3.3). Thus we obtain the simple form

$$Q(\boldsymbol{\xi}|\boldsymbol{\xi}^{(l)}) = -n\log\sigma - \frac{1}{2\sigma^2}\sum_{i=1} W_i^{(l)}(Y_i-\mathbf{X}_i\boldsymbol{\beta})^2, \tag{4.5}$$

where

$$W_i^{(l)} = \mathbf{E}(q_i|Z_i^{(l)}), \tag{4.6}$$

and $Z_i^{(l)}$ is given by (2.3). Clearly, if $\sigma>0$, $Q(\boldsymbol{\xi}|\boldsymbol{\xi}^{(l)})$ is finite when all $W_i^{(l)}<\infty$. If some particular $W_i^{(l)}$ is infinite, then $-Q(\boldsymbol{\xi}|\boldsymbol{\xi}^{(l)})$ is infinite unless $(Y_i-\mathbf{X}_i\boldsymbol{\beta})=0$; we will assume henceforth that in this case $W_i^{(l)}(Y_i-\mathbf{X}_i\boldsymbol{\beta})=0$. From (2.7), this is equivalent to assuming that $d'(0)=0$ since $ZW(Z)=d'(Z)/d(Z)$.

The following result essentially appears in [5] in the course of presenting examples of EM algorithms.

Theorem 3. *Starting from* $s^{(0)}>0$, $\|b^{(0)}\|<\infty$, *each instance of an* IRLS *algorithm defined by an* N/I *error distribution is an instance of the* EM *algorithm for the corresponding complete data likelihood* (4.3), *with the* q_i *regarded as missing data. Furthermore, the infinite sequence* $\boldsymbol{\xi}^{(l)}$ *is defined unless* $\mathbf{Y}=\mathbf{X}\boldsymbol{\beta}$ *for some* $\boldsymbol{\beta}$.

Proof. From (4.5) it is immediate that the E-step merely computes the $W_i^{(l)}$ from (4.6). It follows from (2.3), Lemma 1 and Theorem 2 that if $s^{(l)}>0$ each $W_i^{(l)}>0$; also, the number of infinite $W_i^{(l)}$ correspond to the zero $Z_i^{(l)}$. Hence $\mathbf{W}^{(l)}\in\mathcal{W}$ if $s^{(l)}>0$.

The M-step chooses $\xi^{(l+1)}$ to maximize $Q(\xi|\xi^{(l)})$. If all the $W_i^{(l)}$ are finite, $\xi^{(l+1)}$ satisfies (2.12) and (2.13) since $Q(\xi|\xi^{(l)})$ is simply a normal likelihood with fixed, known weights. If some $W_i^{(l)}$ are infinite, then $Q(\xi|\xi^{(l)})$ is maximized by using weighted least squares with infinite weights. That is, $Y_i=\mathbf{X}_i\mathbf{b}^{(l+1)}$ for all infinite weights, the remaining components of $\mathbf{b}^{(l+1)}$ are computed from (2.12) with the rows corresponding to infinite weights omitted, and $s^{(l+1)}$ is computed from (2.13) with $W_i^{(l)}(Y_i-\mathbf{X}_i\mathbf{b}^{(l+1)})^2=0$ for the infinite weights $W_i^{(l)}$.

Since $W^{(l)}\in\mathcal{W}$ if $s^{(l)}>0$ and $s^{(l+1)}>0$ whenever $s^{(l)}>0$ except in the trivial case with $\mathbf{Y}=\mathbf{X}\mathbf{b}^{(l+1)}$, the infinite sequence $(s^{(l)},b^{(l)})$ is defined unless $\mathbf{Y}=\mathbf{X}\boldsymbol{\beta}$ for some $\boldsymbol{\beta}$.

5. Convergence results

In this section we use the fact that the IRLS with N/I error distributions is EM in order to prove convergence.

Theorem 4. *For* N/I *distributions, each step of* IRLS *such that* $\xi^{(l+1)}\neq\xi^{(l)}$ *strictly increases the likelihood* $L(\xi)$ *defined by* (2.5).

Proof. The proof follows immediately from Theorem 1 in [5], but is reproduced here because it is brief and central to understanding EM. By definition

$$\log g(\mathbf{y}|\xi)=\log f(\mathbf{x}|\xi)-\log k(\mathbf{x}|\mathbf{y},\xi). \tag{5.1}$$

Taking expectations over the distribution $k(\mathbf{x}|\mathbf{y},\xi^{(l)})$, we have

$$L(\xi)=Q(\xi|\xi^{(l)})-H(\xi|\xi^{(l)}), \tag{5.2}$$

where

$$H(\xi|\xi^{(l)})=\mathbf{E}\left[\log k(\mathbf{x}|\mathbf{y},\xi)|\mathbf{y},\xi^{(l)}\right], \tag{5.3}$$

implying

$$L(\xi^{(l+1)})-L(\xi^{(l)})=\left[Q(\xi^{(l+1)}|\xi^{(l)})-Q(\xi^{(l)}|\xi^{(l)})\right]$$
$$-\left[H(\xi^{(l+1)}|\xi^{(l)})-H(\xi^{(l)}|\xi^{(l)})\right]. \tag{5.4}$$

The difference in H functions is non-positive by Jensen's inequality. In the case of finite weights, the difference in Q functions is strictly positive for $\xi^{(l+1)}\neq\xi^{(l)}$ because of the convexity of Q. In the case of infinite weights corresponding to a subset of rows of $\mathbf{X}$ forming a matrix of rank $r<p$, the Q functions defined on the required hyperplane of dimension $p-r$ are strictly convex, so their difference is strictly positive.

From Theorem 4 we know that $L(\xi^{(l)})$ has a finite or infinite limit point. To show convergence of the sequence $\xi^{(l)}$ to a local maximum of the likelihood, we need additional conditions which ensure the $s^{(l)}$ are bounded away from zero; Lemma 1 guarantees $\|b^{(l)}\|$ is bounded.

Lemma 4. *Suppose that the following conditions hold*:

(i) $\mathbf{E}(q^{\frac{1}{2}})<\infty$,

(ii) *At most m of the n equations $\mathbf{Y}=\mathbf{X}\boldsymbol{\beta}$ can be simultaneously satisfied for any choice of $\boldsymbol{\beta}$ where $m<n$, and*

(iii) *there exists some $a>n/(n-m)$ such that*

$$\lim_{|x|\to\infty} C_a(x)=C_a<\infty,$$

where $C_a(x)=d(x)|x|^a$,
then the sequence $s^{(l)}$ is bounded away from zero.

Proof. Suppose to the contrary that $s^{(l)}$ is not bounded away from zero, so there exists a subsequence of the $s^{(l)}$ which converges to zero. From Lemma 1, the $\mathbf{b}^{(l)}$ sequence is bounded so that we can find a subsequence $(\mathbf{b}^{(l_j)},s^{(l_j)})l_j=1,\ldots,\infty$, such that $(\mathbf{b}^{(l_j)},s^{(l_j)})\to(\mathbf{b}^*,0)$ for some $\mathbf{b}^*$. From (2.5), the likelihood function associated with the observed data $(\mathbf{Y},\mathbf{X})$ evaluated at $(\mathbf{b}^{(l_j)},s^{(l_j)})$ has the form

$$l(\mathbf{b}^{(l_j)},s^{(l_j)})=(s^{(l_j)})^{-n}\prod_{i=1}^{n} d(r_i^{(l_j)}/s^{(l_j)})$$

with $r_i^{(l_j)}=Y_i-\mathbf{X}_i\mathbf{b}^{(l_j)}$.

Since $d(x)$ is bounded above by $E(q^{\frac{1}{2}})=(2\pi)^{-\frac{1}{2}}d(0)$, we may for any l_j re-order the $r_i^{(l_j)}$ so that the last m are smallest in absolute value and write

$$l(\mathbf{b}^{(l_j)}, s^{(l_j)}) \leqslant (s^{(l_j)})^{-n} \prod_{i=1}^{n-m} d(r_i^{(l_j)}/s^{(l_j)})\left[E\left(q^{\frac{1}{2}}\right)\right]^m,$$

where equality holds if and only if the m smallest values of $|r_i^{(l_j)}|$ are zero. Since by (ii) the remaining $(n-m)r_i^{(l_j)}$ cannot be zero, we may substitute $C_a(\cdot)/|\cdot|^a$ for $d(\cdot)$ in the $(n-m)$-fold product to give

$$l(\mathbf{b}^{(l_j)}, s^{(l_j)}) \leqslant (s^{(l_j)})^{-n+a(n-m)} E\left(q^{\frac{1}{2}}\right)^m \prod_{i=1}^{n-m} C_a(r_i^{(l_j)}/s^{(l_j)})/|r_i^{(l_j)}|^a.$$

Letting $l_j \to \infty$, we have

(a) $(s^{(l_j)})^{-n+a(n-m)} \to 0$ by condition (iii),

(b) each of the $(n-m)$ $r^{(l_j)}$'s largest in magnitude must approach some $|r_i^*|>0$, and thus by condition (iii),

(c) each term in the numerator of the $(n-m)$-fold product approaches $C_a<\infty$. Hence

$$\lim_{l_j \to \infty} l(\mathbf{b}^{(l_j)}, s^{(l_j)}) = 0$$

which implies the likelihood is converging to zero and is thus decreasing, which contradicts Theorem 4.

For the class of t distributions, with r degrees of freedom, $a=(r+1)$ and $C(x)=(1/x^2+1/r)^{-(r+1)/2}$, hence condition (iii) is satisfied whenever $r> m/(n-m)$.

Using the result in Lemma 4, and Theorem 3 of [5], we are able to prove convergence.

Theorem 5. *Given the conditions of Lemma* 4, *each instance of* IRLS *defined by equations* (2.12) *and* (2.13) *converges to a local maximum of the* log *likelihood* $L(\xi)$ *defined by* (2.5).

Proof. Theorem 3 establishes that an instance of IRLS is an instance of EM. Lemma 1 proves $\mathbf{b}^{(l)}$ is bounded, and Lemma 4 proves $s^{(l)}$ is bounded above zero. By Theorem 3 in [5], an EM algorithm will converge if:

(a) the log likelihood is bounded for the sequence $\xi^{(0)} \rightarrow \xi^{(1)} \rightarrow \cdots$,
(b) $\mathbf{D}^{10}Q(\xi|\xi^{(l)})=\mathbf{0}$, $l=0,1,2,\ldots$,
(c) $\mathbf{D}^{20}Q(\boldsymbol{\theta}|\boldsymbol{\theta}^{(l)})$ has eigenvalues bounded below zero for all $\boldsymbol{\theta}$ on the line segment joining $\boldsymbol{\theta}^{(l)}$ to $\boldsymbol{\theta}^{(l+1)}$, where $\boldsymbol{\theta}$ is a 1-1 transformation of ξ.

The notation $\mathbf{D}^{20}$ denotes the matrix of second partial derivatives with respect to the first argument. Since by Lemma 4 we know that the sequence $s^{(l)}$ is bounded above zero and that $E(q^{\frac{1}{2}})<\infty$, we can conclude that the sequence $L(\xi^{(l)})$ is bounded, because

$$L(\xi)=\sum_{L=1}^{n}\log\int_{0}^{\infty}(2\pi)^{-\frac{1}{2}}\sigma^{-1}q_i^{\frac{1}{2}}e^{-q_i(Y_i-\mathbf{X}_i\beta)^2/2\sigma^2}m(q_i)\,dq_i$$

$$\leqslant \sum_{L=1}^{n}\log\int_{0}^{\infty}(2\pi)^{-\frac{1}{2}}\sigma^{-1}q_i^{\frac{1}{2}}m(q_i)\,dq_i$$

$$=-n\log 2\pi\sigma^2+n\log E\left(q^{\frac{1}{2}}\right).$$

Part (b) is an immediate consequence of the fact that $Q(\xi|\xi^{(l)})$ is a weighted normal likelihood with known, non-zero weights which may be assumed finite because $Z_i W_i$ is always finite. With $Q(\xi|\xi^{(l)})$ defined by (4.5) and (4.6) and all $W_i^{(l)}$ finite, then $Q(\xi|\xi^{(l)})$ has a unique maximum at $\xi^{(l+1)}$, defined by equations (2.12) and (2.13). Further

$$\mathbf{D}^{10}Q(\xi|\xi^{(l)})=\partial Q(\xi|\xi^{(l)})/\partial\xi=\mathbf{0} \quad \text{at} \quad \xi=\xi^{(l+1)}$$

and $Q(\xi|\xi^{(l)})$ is strictly convex with respect to the natural parameters $(\beta/\sigma^2, 1/\sigma^2)$. Choosing $\boldsymbol{\theta}$ to be the natural parameters, part (c) follows from equation (4.5) together with Lemmas 1 and 4.

In some cases it may be possible to fix σ^2 in advance. We can then apply IRLS with σ^2 fixed and $\mathbf{b}^{(l)}$ defined by equation (2.12) where $s^{(l)}=\sigma$ for all l. With this restriction, convergence is somewhat easier to prove. The following results are stated without their proofs, which are essentially the same as the proofs of the corresponding results where σ^2 is being estimated by maximum likelihood (Theorem 4 and Theorem 5).

Theorem 6. *For fixed σ^2, each step of* IRLS *as given in Theorem* 3 *strictly increases the conditional likelihood of $\boldsymbol{\beta}$ given σ.*

Theorem 7. *For fixed* σ^2, *if* $E(q^{\frac{1}{2}}) < \infty$, *each instance of* IRLS *defined by equation* (2.12) *with* $s^{(l)} = \sigma$ *converges to a local maximum of the conditional likelihood of* $\boldsymbol{\beta}$ *given* σ^2.

6. Limiting behavior

Explicit formulas are now presented which describe the behavior of IRLS close to a limit point, assuming N/I distributed errors in the linear model. The formulas are obtained by substituting the appropriate special expressions into the general expressions in [5]. Two versions of the limiting behavior are given in Theorems 8 and 9, the first holding exactly in small samples, and the second holding asymptotically in large samples.

Given that the sequence of iterates $\xi^{(l)}$ converges to ξ^*, we know from [5] that in the limit

$$\xi^{(l+1)} - \xi^* = (\xi^{(l)} - \xi^*)\mathbf{DM}(\xi^*) \tag{6.1}$$

plus terms of lower order, where

$$\mathbf{DM}(\xi^*) = \mathbf{D}^{20}H(\xi^*|\xi^*)\left[\mathbf{D}^{20}Q(\xi^*|\xi^*)\right]^{-1}. \tag{6.2}$$

Here $\mathbf{M}(\xi)$ is the vector-valued function which specifies a step of the algorithm, i.e., $\xi^{(l+1)} = \mathbf{M}(\xi^{(l)})$ for all l, while $H(\xi_1|\xi_2)$ and $Q(\xi_1|\xi_2)$ are the scalar-valued functions introduced in Sections 4 and 5.

Expression (6.2) has an interesting statistical interpretation, because $-\mathbf{D}^{20}H(\xi^*|\xi^*)$ represents the Fisher information matrix of the missing data computed at ξ^*, while $-\mathbf{D}^{20}Q(\xi^*|\xi^*) = -\mathbf{D}^2L(\xi^*) - \mathbf{D}^{20}H(\xi^*|\xi^*)$ measures the total information in the incomplete data and the missing data together. Thus, (6.2) says roughly that convergence will be rapid if the proportion of missing information is small.

From (6.2), we see that our task is to find the special forms of $\mathbf{D}^{20}Q(\xi^*|\xi^*)$ and $\mathbf{D}^{20}H(\xi^*|\xi^*)$ which apply to linear models with N/I error terms. $\mathbf{D}^{20}Q(\xi^*|\xi^*)$ is easily obtained by direct differentiation of (4.4), but $\mathbf{D}^{20}H(\xi^*|\xi^*)$ requires additional notation and theory. Specifically, we note from (4.3) that $f(\mathbf{x}|\xi)$ has the exponential family form $a(\boldsymbol{\phi})b(\mathbf{x})\exp(\boldsymbol{\phi}^T\mathbf{t})$ where the natural parameters $\boldsymbol{\phi}$ are

$$\boldsymbol{\phi}^{\mathrm{T}} = \left(-\frac{1}{2\sigma^2}, \frac{1}{\sigma^2}\boldsymbol{\beta}, -\frac{1}{2\sigma^2}\boldsymbol{\beta}^{\mathrm{T}}\boldsymbol{\beta}\right) \tag{6.3}$$

and the corresponding sufficient statistics $\mathbf{t}$ are

$$\mathbf{t}^{\mathrm{T}} = \left(\sum_{i=1}^{n} Y_i^2 q_i, \sum_{i=1}^{n} Y_i \mathbf{X}_i q_i, \sum_{i=1}^{n} \mathbf{X}_i^{\mathrm{T}} \mathbf{X}_i q_i \right). \tag{6.4}$$

Since the conditional density $k(\mathbf{x}|\mathbf{Y},\xi)$ is found from $f(\mathbf{x}|\xi)$ by fixing $\mathbf{Y}$ and renormalizing, it is evident that $k(\mathbf{x}|\mathbf{Y},\xi)$ has an exponential family form with the same $\boldsymbol{\phi}$ and $\mathbf{t}$ but with different range and normalizing factor. As noted above, $-\mathbf{D}^{20}H(\xi^*|\xi^*)$ is the Fisher information matrix of the missing data, and for exponential families it is well-known that the Fisher information matrix is easily obtained from the covariance matrix of the sufficient statistics $\mathbf{t}$.

Given the data $\mathbf{Y}$, the sufficient statistics $\mathbf{t}$ are seen from (6.4) to be linear in the q_i, and hence the covariance matrix of $\mathbf{t}$ conditional on $\mathbf{Z}$ consists of sums over $i=1,2,\ldots,n$ whose ith term is a fourth degree expression in Y_i, $\mathbf{X}_i$ weighted by the conditional variance of q_i given Y_i, or equivalently given $Z_i = (Y_i - \mathbf{X}_i\boldsymbol{\beta})/\sigma$. Thus we encounter new weights

$$V_i = V(Z_i), \tag{6.5}$$

where the weight function

$$V(Z) = \mathrm{Var}(q|Z) \tag{6.6}$$

is different from the weight function $W(Z)$ defined in (3.3), but according to (3.4) is expressible as

$$V(Z) = -(\mathrm{d}W(Z)/\mathrm{d}Z)/Z. \tag{6.7}$$

In order to provide a compact expression for weighted sums which appear in $\mathbf{D}^{20}H(\xi^*|\xi^*)$, we introduce the symbolic notation

$$[\mathbf{Z}^4]_V = \sum_{i=1}^{n} V_i Z_i^4$$

$$[\mathbf{Z}^3\mathbf{X}]_V = \sum_{i=1}^{n} V_i Z_i^3 \mathbf{X}_i$$

$$[\mathbf{Z}^2\mathbf{X}^{\mathrm{T}}\mathbf{X}]_V = \sum_{i=1}^{n} V_i Z_i^2 \mathbf{X}_i^{\mathrm{T}} \mathbf{X}_i, \tag{6.8}$$

whence we have:

Theorem 8. *If* $\xi^{(l)}$ *converges to* ξ^* *under the* EM *algorithm defined by equations* (2.12), (2.13) *and* (4.6), *then*

$$-\mathbf{D}^{20}Q(\xi^*|\xi^*)=\sigma^{*-2}\begin{bmatrix}\mathbf{X}^{\mathrm{T}}\mathbf{W}^*\mathbf{X} & \mathbf{0}\\ \mathbf{0} & 2n\end{bmatrix}, \tag{6.9}$$

$$-\mathbf{D}^{20}H(\xi^*|\xi^*)=\sigma^{*-2}\begin{bmatrix}[\mathbf{Z}^{*2}\mathbf{X}^{\mathrm{T}}\mathbf{X}]_{V^*} & [\mathbf{Z}^{*3}\mathbf{X}]_{V^*}\\ [\mathbf{Z}^{*3}\mathbf{X}^{\mathrm{T}}]_{V^*} & [\mathbf{Z}^{*4}]_{V^*}\end{bmatrix}, \tag{6.10}$$

$$\mathbf{DM}(\xi^*)=\begin{bmatrix}[\mathbf{Z}^{*2}\mathbf{X}^{\mathrm{T}}\mathbf{X}]_{V^*}(\mathbf{X}^{\mathrm{T}}\mathbf{W}^*\mathbf{X})^{-1} & \dfrac{1}{2n}[\mathbf{Z}^{*3}\mathbf{X}]_{V^*}\\ [\mathbf{Z}^{*3}\mathbf{X}^{\mathrm{T}}]_{V^*}(\mathbf{X}^{\mathrm{T}}\mathbf{W}^*\mathbf{X})^{-1} & \dfrac{1}{2n}[\mathbf{Z}^{*4}]_{V^*}\end{bmatrix}, \tag{6.11}$$

and

$$\mathbf{D}^2L(\xi^*)=-\sigma^{*-2}\begin{vmatrix}\mathbf{X}^{\mathrm{T}}\mathbf{W}^*\mathbf{X}-[\mathbf{Z}^{*2}\mathbf{X}^{\mathrm{T}}\mathbf{X}]_{V^*} & -[\mathbf{Z}^{*3}\mathbf{X}]_{V^*}\\ -[\mathbf{Z}^{*3}\mathbf{X}^{\mathrm{T}}]_{V^*} & 2n-[\mathbf{Z}^{*4}]_{V^*}\end{vmatrix} \tag{6.12}$$

where $\mathbf{Z}^*$ *is the standardized residual vector* $(\mathbf{Y}-\mathbf{X}\boldsymbol{\beta})/\sigma$ *calculated at* $\xi^*=(\boldsymbol{\beta}^*,\sigma^*)$, $\mathbf{W}^*$ *is the diagonal matrix of weights* W_i *calculated at* ξ^*, *and the symbolic expressions on the right hand side of* (6.10) *involve weights* V_i *also calculated at* ξ^*.

Proof. Formula (6.9) is an immediate consequence of differentiating (4.5). To obtain (6.10) we use a standard fact about exponential families, namely, that the Fisher information matrix is expressible as the sampling covariance matrix of the sufficient statistics **t** pre- and post-multiplied by the matrix of partial derivatives of the natural parameters $\boldsymbol{\phi}$ with respect to the actual parameters $\boldsymbol{\xi}$. Conditional on fixed **Y** and **X**, the components of **t** are simply linear expressions in random q_i, and hence

$$\mathrm{Cov}(\mathbf{t}|\mathbf{Y},\boldsymbol{\phi})=\left[\begin{array}{c:c:c}[\mathbf{Y}^4]_V & [\mathbf{Y}^3\mathbf{X}]_V & \cdots\\ \hdashline \cdots & [\mathbf{Y}^2\mathbf{X}^{\mathrm{T}}\mathbf{X}]_V & \cdots\\ \hdashline \cdots & \cdots & \cdots\end{array}\right]. \tag{6.13}$$

The partition of (6.13) corresponds to the partition of $\mathbf{t}$ in (6.4). The last row and column are omitted from the display in (6.13) because they turn out to be unnecessary, the reason being that we can shift the $\boldsymbol{\beta}$ origin to $\boldsymbol{\beta}^*$ and express $\mathbf{D}^{20}H(\xi^*|\xi^*)$ in terms of $\mathbf{Z}^*$ central at $\boldsymbol{\beta}^*$. Thus we compute the matrix of partial derivatives

$$\frac{\partial\phi}{\partial\xi}=\left[\begin{array}{c|c|c}\mathbf{0} & \sigma^{-2}\mathbf{I} & \mathbf{0}\\ \hline \sigma^{-3} & \mathbf{0} & \mathbf{0}\end{array}\right], \tag{6.14}$$

when calculated at $\boldsymbol{\beta}=\mathbf{0}$, where the row partition corresponds to $\xi=(\boldsymbol{\beta},\sigma)$ and the column partition refers to (6.3). Formula (6.10) follows immediately from (6.13) and (6.14). Formulas (6.11) and (6.12) follow directly from (5.2), (6.2), (6.9), and (6.10).

Formulas (6.9) and (6.10) together with (6.1) and (6.2) tell us how to approximate successive changes $\xi^{(l+1)}-\xi^{(l)}$ close to the limit point ξ^*. When the sample size is large, a simpler approximation, given below, is available. It is useful because it is easier to calculate, but does not hold exactly even arbitrarily close to ξ^* with actual data, as is illustrated numerically in Section 7.

To discuss large sample size, it is necessary to postulate a sequence of data sets $(\mathbf{Y},\mathbf{X})$ with increasing n and a corresponding sequence of EM algorithms each converging to its own ξ^*. The following assumptions about the sequence of data sets are made in Theorem 9:

$$\lim_{n\to\infty}\frac{1}{n}\mathbf{X}^{\mathrm{T}}\mathbf{X}=\boldsymbol{\Sigma}$$

$$\lim_{n\to\infty}\frac{1}{n}\mathbf{X}^{\mathrm{T}}\mathbf{W}^*\mathbf{X}=E(q)\boldsymbol{\Sigma}$$

$$\lim_{n\to\infty}\frac{1}{n}[\mathbf{Z}^{*2}\mathbf{X}^{\mathrm{T}}\mathbf{X}]_{V^*}=E(Z^2V(Z))\boldsymbol{\Sigma} \tag{6.15}$$

These are plausible assumptions for many real life applications, in the sense that they will hold approximately for reasonably large n.

Theorem 9. *Suppose that the assumptions of Theorem* 8 *hold for a sequence of data sets such that* (6.15) *holds. Then*

$$\lim_{n\to\infty}\mathbf{DM}(\xi^*)=\mathbf{I}-\left[\begin{array}{c|c}E(q)^{-1}\mathrm{Var}(ZW(Z))\mathbf{I} & \mathbf{0}\\ \hline \mathbf{0} & \frac{1}{2}\mathrm{Var}(Z^2W(Z))\end{array}\right] \tag{6.16}$$

and

$$\lim_{n\to\infty} -\frac{1}{n}\mathbf{D}^2L(\xi^*) = \sigma^{*-2}\left[\begin{array}{c|c} \mathrm{Var}(ZW(Z))\boldsymbol{\Sigma} & \mathbf{0} \\ \hline \mathbf{0} & \mathrm{Var}(Z^2W(Z)) \end{array}\right]. \tag{6.17}$$

Proof. From (6.9) and (6.15) it follows that

$$\lim_{n\to\infty} -\frac{1}{n}\mathbf{D}^{20}Q(\xi^*|\xi^*) = \sigma^{*-2}\left[\begin{array}{c|c} E(q)\boldsymbol{\Sigma} & 0 \\ \hline 0 & 2 \end{array}\right]$$

We now need to derive a pair of auxiliary formulas. First, since

$$E(\mathrm{Var}(Zq|Z)) = \mathrm{Var}(Zq) - \mathrm{Var}(E(Zq|Z))$$

and since $\mathrm{Var}(Zq) = E(Z^2q^2) = E([Z^2q]q) = E([Zq^{\frac{1}{2}}]^2q) = E(q)$, we have

$$E(Z^2V(Z)) = E(q) - \mathrm{Var}(ZW(Z)) \tag{6.18}$$

Similarly, since

$$E(Z^4V(Z)) = \mathrm{Var}(Z^2q) - \mathrm{Var}(E(Z^2q|Z)),$$

and since $\mathrm{Var}(Z^2q) = 2$, we have

$$E(Z^4V(Z)) = 2 - \mathrm{Var}(Z^2W(Z)) \tag{6.19}$$

From (6.15), (6.18) and (6.19), we find by substituting into (6.10) that

$$\lim_{n\to\infty} -\frac{1}{n}\mathbf{D}^{20}H(\xi^*|\xi^*)$$
$$= \sigma^{*-2}\left[\begin{array}{c|c} (E(q) - \mathrm{Var}[ZW(Z)])\boldsymbol{\Sigma} & 0 \\ \hline 0 & 2 - \mathrm{Var}(Z^2W(Z)) \end{array}\right] \tag{6.20}$$

Formulas (6.16) and (6.17) follow by substituting (6.19) and (6.20) into (6.2) and (5.2).

7. Example from t family

We now illustrate the convergence of IRLS for the t-family of distributing when $\mathbf{X}$ is an $(n \times 1)$ vector of ones. For this family, rq is distributed as χ_r^2 where r denotes the degrees of freedom. In order to use equations (6.11) and (6.16) to calculate the small and large sample rates of convergence of IRLS, we need to find W_i, V_i, $\mathrm{Var}(Y_i W_i)$, and $\mathrm{Var}(Y_i^2 W_i)$.

Lemma 5. *Let $rq \sim \chi_r^2$ and conditional on q, $Z \sim N(0, 1/q)$. Then*

$$E(q) = 1, \tag{7.1}$$

$$W(Z) = E(q|Z) = \frac{r+1}{r+Z^2}, \tag{7.2}$$

$$V(Z) = \mathrm{Var}(q|Z) = \frac{2(r+1)}{(r+Z^2)^2}, \tag{7.3}$$

$$\mathrm{Var}(ZW(Z)) = \frac{r+1}{r+3}, \tag{7.4}$$

$$\mathrm{Var}(Z^2 W(Z)) = \frac{2r}{r+3}. \tag{7.5}$$

Proof. Equation (7.1) is immediate. Equations (7.2) and (7.3) follow from the fact that conditional on Z, $q(r+Z^2) \sim \chi_{r+1}^2$. Letting $b = Z^2/(r+Z^2)$, we have

$$\mathrm{Var}(ZW(Z)) = E\left(Z^2 W(Z)^2\right) = \frac{(r+1)^2}{r} E(b(1-b))$$

and $\mathrm{Var}(Z^2 W(Z)) = (r+1)^2 \mathrm{Var}(b)$, where $b \sim \mathrm{Beta}(\frac{1}{2}, \frac{1}{2})$. Equations (7.4) and (7.5) follow.

Equations (7.4) and (7.5) together with equation (6.16) imply that in large samples the rate of convergence of IRLS for the t-distribution with r degrees of freedom will be $3/r+3$, which is the largest eigenvalue of $\mathbf{DM}$.

The data in Table 1 are a sample of 10 observations from a t-distribution with $r=3$. Also presented at the bottom of Table 1 are relevant functions of the data. The objective is to compute maximum likelihood estimates of the location (mean) β and the scale, σ. Table 2 shows the first ten cycles of IRLS starting from the sample mean and sample variance.

Table 1.
Example from t_3 with $n=10$

$N(0,1)$	χ_3^2	$y_i \sim t_3$
−0.170	4.377	−0.141
1.165	8.851	0.678
−0.029	1.863	−0.036
−0.550	7.424	−0.350
−2.134	0.546	−5.005
1.405	7.542	0.886
0.372	1.761	0.485
−1.300	0.294	−4.154
1.105	1.831	1.415
0.963	1.163	1.546

Table 2.
Successive iterations of IRLS for example of Table 1.

Iteration-l	$\beta^{(l)}$	Empirical rate for $\beta^{(l)}$	Rate for $\beta^{(l)}$ from Dm	$\sigma^{(l)2}$	Empirical rate for $\sigma^{(l)}$	Rate for $\sigma^{(l)}$ from DM
1	0.103069			1.673303		
2	0.240781	0.2414	0.2199	1.603189	−0.5172	−0.2643
3	0.277822	0.2690	0.2795	1.524210	1.1264	1.0644
4	0.292411	0.3939	0.4091	1.466860	0.7261	0.7426
5	0.300280	0.5394	0.5528	1.427958	0.6783	0.6964
6	0.305188	0.6237	0.6336	1.401828	0.6717	0.6852
7	0.308413	0.6571	0.6640	1.384252	0.6727	0.6821
8	0.310571	0.6694	0.6741	1.372393	0.6747	0.6812
9	0.312027	0.6744	0.6776	1.364371	0.6764	0.6808
10	0.313012	0.6768	0.6790	1.358934	0.6777	0.6807
11	0.313680	0.6781	0.6796	1.355244	0.6786	0.6807
12	0.314133	0.6789	0.6800	1.352738	0.6793	0.6806
13	0.314442	0.6795	0.6802	1.351035	0.6797	0.6806
14	0.314651	0.6799	0.6803	1.349876	0.6800	0.6806
15	0.314794	0.6801	0.6804	1.349088	0.6802	0.6806
16	0.314890	0.6803	0.6805	1.348552	0.6803	0.6806
17	0.314956	0.6804	0.6805	1.348188	0.6804	0.6806
18	0.315001	0.6805	0.6805	1.347939	0.6805	0.6806
19	0.315032	0.6805	0.6806	1.247771	0.6805	0.6806
20	0.315053	0.6805	0.6806	1.347656	0.6806	0.6806

$\beta^{(0)}=\bar{y}=-0.467496, \sigma^{(0)}=s^2=1.537750$

The column labelled 'Empirical rate for $\beta^{(l)}$' refers to $(\beta^{(l)}-\beta^{(l-1)})/(\beta^{(l-1)}-\beta^{(l-2)})$. The column labelled 'Rate for $\beta^{(l)}$ from DM' refers to $\Delta\beta^{(l)}/(\beta^{(l-1)}-\beta^{(l-2)})$ where

$$(\Delta\beta^{(l)},\Delta\sigma^{(l)})=(\beta^{(l-1)}-\beta^{(l-2)},\ \sigma^{(l-1)}-\sigma^{(l-2)})\,\mathbf{DM}.$$

The columns referring to the rate of convergence for $\sigma^{(l)}$ are defined similarly.

The large sample rate of convergence for this distribution is 0.5, while from formula (6.11) and the actual data, we find the small sample rate of convergence to be 0.6806. It is evident from Table 2 that our small sample theory agrees well with the empirical rate.

Appendix

Proof of Lemma 1. We begin by constructing the union of polytopes, which we then show contains $\mathbf{b}$. Each of the n rows of the equation $\mathbf{Y}=\mathbf{Xb}$ defines a hyperplane in $\mathbf{R}^p$. The n pairs of open half spaces on either side of each hyperplane partition the set of $\mathbf{b}$ not on any hyperplane into a collection $\mathcal{C}_p$ of open p-dimensional convex polytopes each characterized by a particular pattern of signs of the residuals $\mathbf{Y}-\mathbf{Xb}$. The polytopes in $\mathcal{C}_p$ are of one of two kinds, the subcollection $\mathcal{B}_p$ of polytopes which are of infinite extent, and the subcollection $\mathcal{A}_p$ which are bounded in all directions by the n hyperplanes. Next consider points $\mathbf{b}$ which lie on exactly one of the n hyperplanes. The remaining hyperplanes partition this set of $\mathbf{b}$ into a collection $\mathcal{C}_{p-1}$ of open $(p-1)$-dimensional convex polytopes which in turn can be partitioned into the subcollection $\mathcal{A}_{p-1}$ of polytopes of finite extent and the subcollection $\mathcal{B}_{p-1}$ of infinite extent. Continuing in this way we find $\mathcal{C}_{p-t}$ consisting of $(p\text{-}t)$-dimensional open polytopes inside an intersection of exactly t hyperplanes, and the partition of $\mathcal{C}_{p-t}$ into $\mathcal{A}_{p-t}$ and $\mathcal{B}_{p-t}$ of polytopes of finite and infinite extent, for $t=0,1,2,\ldots$. Define A to be the set of points $\mathbf{b}$ in $\mathbf{R}^p$ which lie in one of the polytopes in the collection $\mathcal{A}=\bigcup_t\mathcal{A}_{p-t}$. Note that A can be described as a union of closed finite convex polytopes, because all of the faces of any polytope in $\mathcal{A}_{p-t}$ are contained in $\bigcup_{t'>t}\mathcal{A}_{p-t'}$.

We now show that the set of $\mathbf{b}$ defined by (2.1) must lie in A, which we do by ruling out all points not in A. First, consider any point $\mathbf{b}$ contained in a polytope from the collection $\mathcal{B}_p$. Since the polytope is infinite, it has one-dimensional edges which extends to infinity. Consider a line through $\mathbf{b}$

parallel to such an edge. This line also must extend to infinity inside the polytope, and possesses the key property that as one moves away from infinity along the line each of the n residuals $\mathbf{Y}-\mathbf{X}\mathbf{b}$ either remains constant or shrinks in absolute value. Since $\mathbf{X}$ has rank p, not all of these residuals can be constant. Thus, as we move along the line, we eventually encounter a point $\mathbf{b}'$ where at least one of the non-zero residuals becomes zero. That is, $\mathbf{b}'$ marks the point where the line hits a face of the polytope, and therefore $\mathbf{b}'$ lies in a polytope in $\mathcal{C}_{p-t}$ for some $t>0$. The key point is that $\mathbf{b}'$ strictly dominates every $\mathbf{b}$ on the line inside the original polytope as a candidate for a weighted least square $\mathbf{b}$ under any weighting, because all of the components of $\mathbf{Y}-\mathbf{X}\mathbf{b}'$ are at least as small in absolute value as the corresponding components of $\mathbf{Y}-\mathbf{X}\mathbf{b}$, with strict inequality in at least one component.

By a similar argument, we rule out $\mathbf{b}$ in members of $\mathcal{B}_{p-1}, \mathcal{B}_{p-2}, \ldots,$ and hence we are left only with $\mathbf{b}\in A$ as potential weighted least squares solutions.

References

[1] Andrews, D. F., Bickel, P. J., Hampel, F. R., Huber P. J., Rogers, W. H. and Tukey, J. W. (1972). *Robust Estimates of Location: Survey and Advances*. Princeton University Press.

[2] Andrews, D. F. and Mallows, C. L. (1974). Scale mixtures of normal distributions. *J. Roy. Statist. Soc. Ser. B* **36**, 99–102.

[3] Andrews, D. F. (1974). Some Monte Carlo results on robust-resistant regression. In *Critical Evaluation of Physical Structural Information*, pp. 36–44. National Academy of Sciences, Washington.

[4] Beaton, A. E. and Tukey, J. W. (1974). The fitting of power series. *Technometrics* **16**, 147–185.

[5] Dempster, A. P., Laird, N. M. and Rubin, D. B. (1977). Maximum likelihood from incomplete data via the *EM* algorithm. *J. Roy. Statist. Soc. Ser. B* **39**, 1–38.

[6] Holland, P. W. and Welsch, R. E. (1977). Robust regression using iteratively reweighted least-squares. *Commun. Statist.–Theor. Meth.* **A6**, 813–827.

[7] Huber, P. J. (1964). Robust estimation of a location parameter. *Ann. Math. Statist.* **35**, 73–101.

[8] Tukey, J. W. (1960). A survey of sampling from contaminated distributions. In *Contributions to Probability and Statistics*, Olkin, I., Ed., pp. 448–485. Stanford University Press.

P. R. Krishnaiah, ed., *Multivariate Analysis–V*
© North-Holland Publishing Company (1980) 59–77.

THE STRONG UNIFORM CONSISTENCY OF KERNEL DENSITY ESTIMATES

L. P. DEVROYE* and T. J. WAGNER**
School of Computer Science, McGill University,
Montreal, P.Q., Canada and
Department of Electrical Engineering,
The University of Texas, Austin, TX 78712, U.S.A.

Let $X_1,\ldots,X_n$ be independent, identically distributed random vectors taking values in $\mathbf{R}^d$ with a common probability density f. If K is a bounded probability density on $\mathbf{R}^d$ and $\{h_n\}$ is a sequence of positive numbers then $f_n(x)=\sum_1^n K((x-X_i)/h_n)/(nh_n^d)$ is the kernel estimate of f from $X_1,\ldots,X_n$. Conditions on f, K and $\{h_n\}$ are given which insure that $\sup_x|f_n(x)-f(x)|\xrightarrow[n]{}0$ with probability one. Additionally, conditions are discussed which allow h_n to be a function of $X_1,\ldots,X_n$ and still retain the consistency properties of f_n.

1. Introduction

Let $X_1,X_2,\ldots,X_n$ be a sequence of independent, identically distributed random vectors taking values in $\mathbf{R}^d$ with a common probability density f. The kernel estimate (Rosenblatt (1957), Parzen (1962)) is given by

$$f_n(x)=\sum_{i=1}^{n} K((x-X_i)/h_n)/(nh_n^d)$$

where K, the kernel, is a bounded probability density on $\mathbf{R}^d$ and $\{h_n\}$ is a sequence of positive numbers. In this paper we are concerned mainly with conditions which insure the strong uniform consistency of f_n, that is,

$$\sup_x|f_n(x)-f(x)|\to 0 \quad \text{w.p. 1.} \tag{1.1}$$

Our results, as well as all of those that we are aware of which lead to (1.1), require that f be uniformly continuous on $\mathbf{R}^d$. The work of Schuster (1969, 1970) comes close to proving that this is a necessity for $\mathbf{R}^1$. We will be contented then just to make this assumption. If f and K are continuous on

*Research was supported in part by DOD Joint Services Electronics Program through the Air Force Office of Scientific Research (AFSC) Contract F 49620-77-C-0101.
**Research was sponsored by AFOSR Grant 77-3385.

$\mathbf{R}^d$ then $\sup_x |f_n(x)-f(x)|$ is a random variable. This remains true if f is continuous on $\mathbf{R}^d$ and if the values of K can be determined from its values on a countable dense set (e.g., for each $x \in \mathbf{R}^d$ there is a sequence $\{x_n\}$ from a countable dense set D such that $x_n \xrightarrow{n} x$ and $K(x_n) \xrightarrow{n} K(x)$). While one can easily think of kernels which do not have this property, we know of none that are interesting. Rather than explore this point further, or deal with the case that $\sup_x |f_n(x)-f(x)|$ is not a random variable, we will assume throughout this paper that it is a random variable.

Let

$$L(z) = \sup_{\|x\| > z} K(x) \qquad z \in [0, \infty)$$

and

$$L^{-1}(t) = \sup\{z : L(z) \geq t\} \qquad t \in \left[0, \sup_x K(x)\right),$$

where $\|\cdot\|$ denotes the supremum norm on $\mathbf{R}^d$. Our result may be stated as follows.

Theorem 1. *Suppose f is uniformly continuous on $\mathbf{R}^d$ and K is a bounded Riemann integrable probability density with*

$$\int_0^\infty z^{d-1} L(z)\,\mathrm{d}z < \infty. \tag{1.2}$$

If

$$h_n \xrightarrow{n} 0,$$

then (1.1) *follows from*

$$(nh_n^d)/\left(L^{-1}(\varepsilon h_n^d)\right)^d \log n \xrightarrow{n} \infty \qquad \text{for } \varepsilon > 0. \tag{1.3}$$

Remarks. For kernels with compact support the conditions (1.2) and (1.3) can be replaced by

$$nh_n^d / \log n \xrightarrow{n} \infty. \tag{1.4}$$

For kernels with

$$K(x) \leq A/\|x\|^{\alpha d} \qquad \text{for some } \alpha > 1$$

(1.2) and (1.3) can be replaced by

$$nh_n^{(1+\alpha^{-1})}/\log n \xrightarrow{n} \infty.$$

Additionally, (1.3) can always be replaced by

$$nh_n^{2d}/\log n \xrightarrow{n} \infty. \tag{1.5}$$

The condition (1.2) for K is close to the condition frequently imposed on K:

$$\|x\|^d K(x) \to 0 \quad \text{as } \|x\| \to \infty. \tag{1.6}$$

For example, (1.6) is implied by (1.2) and (1.2) follows whenever

$$\|x\|^{d+\delta} K(x) \to 0 \quad \text{as } \|x\| \to \infty$$

for some $\delta > 0$.

The above theorem does not use the restrictive assumption that K is of bounded variation on $\mathbf{R}^d$ (Nadaraya (1965), Moore and Yackel (1976) and Silverman (1978)) or that K has an integrable characteristic function (Van Ryzin (1969)). For example, the kernel which is uniform over the unit sphere satisfies neither of these assumptions and, while the kernel which is uniform over the unit cube in $\mathbf{R}^d$ has bounded variation, an orthogonal rotation of the coordinates can yield a kernel with an infinite variation while keeping $\sup_x |f_n(x) - f(x)|$ unchanged. Additionally, no moment assumptions are put on f (Deheuvels (1974), Földes and Révész (1974)) and the requirements for $\{h_n\}$, at least for kernels with compact support, are essentially the weakest possible to get (1.1) (Deheuvels (1974)).

A disadvantage of the kernel estimate is that h_n is chosen without regard to $X_1, \dots, X_n$ (Cover (1972)). One possible remedy is to replace h_n by a function of $X_1, \dots, X_n$, say $H_n = H_n(X_1, \dots, X_n)$. The resulting estimate

$$\hat{f}_n(x) = \sum_{i=1}^{n} K((x - X_i)/H_n)/(nH_n^d)$$

has been examined by Wagner (1975), primarily for $d = 1$, where several choices for H_n are also discussed. The technique used to prove Theorem 1 also yields the following result for $\hat{f}_n$. (See also the remark at the end of Section 2.)

Theorem 2. *Suppose f is uniformly continuous on $\mathbf{R}^d$ and K is a bounded Riemann integrable probability density satisfying* (1.2). *If*

$$H_n \overset{n}{\to} 0 \qquad \text{w.p. } 1, \tag{1.7}$$

and

$$nH_n^{2d}/\log n \overset{n}{\to} \infty \qquad \text{w.p. } 1 \tag{1.8}$$

then

$$\sup_x |\hat{f}_n(x) - f(x)| \overset{n}{\to} 0 \qquad \text{w.p. } 1 \tag{1.9}$$

Additionally, if the convergence in (1.7) *is in probability and if*

$$nH_n^{2d} \overset{n}{\to} \infty \text{ in probability}, \tag{1.10}$$

then

$$\sup_x |\hat{f}_n(x) - f(x)| \overset{n}{\to} 0 \text{ in probability}. \tag{1.11}$$

2. Details

Following Nadaraya (1963) it suffices to prove that

$$\sup_x |f_n(x) - Ef_n(x)| \to 0 \qquad \text{w.p. } 1 \tag{2.1}$$

since, when f is uniformly continuous on $\mathbf{R}^d$,

$$\sup_x |Ef_n(x) - f(x)| \overset{n}{\to} 0$$

whenever $h_n \overset{n}{\to} 0$.

Consider, for the moment, the kernel which is uniform over $(0,1]^d$. Then

$$\sup_x |f_n(x) - Ef_n(x)| = h_n^{-d} \sup_{A \in \mathcal{Q}} |\mu_n(A) - \mu(A)|$$

where μ_n is the empirical measure for $X_1, \ldots, X_n$, μ is the measure on the

Borel subsets of $\mathbf{R}^d$ which corresponds to f and $\mathscr{Q}$ is the class of cubes $\prod_{i=1}^{d}(x^i, x^i+h_n]$, $x=(x^1,\ldots,x^d)\in\mathbf{R}^d$. Thus

$$P\left\{\sup_x |f_n(x)-Ef_n(x)|\geqslant\varepsilon\right\}=P\left\{\sup_{A\in\mathscr{Q}} |\mu_n(A)-\mu(A)|\geqslant h_n^d\varepsilon\right\}. \tag{2.2}$$

Rather than use the inequality of Kiefer–Wolfowitz (1956) on the right-hand side of (2.2), which leads to the condition (1.5) on $\{h_n\}$, we modify a result of Vapnik–Chervonenkis (1971) allowing us to upper-bound (2.2) by $c_0 n^{2d}\exp(-cnh_n^d\varepsilon^2)$ for some c_0, $c>0$. This now leads to condition (1.4) to get (2.1) for the kernel which is uniform over $(0,1]^d$. A careful approximation of Riemann integrable kernels by linear combinations of indicators of disjoint rectangles then yields the theorem.

We begin by proving two useful lemmas concerning upper bounds for

$$P\left\{\sup_{A\in\mathscr{Q}} |\mu_n(A)-\mu(A)|\geqslant\varepsilon\right\}$$

where $\varepsilon>0$ and $\mathscr{Q}$ is a subclass of the Borel subsets of $\mathbf{R}^d$. If μ_n and μ_n' are empirical measures for two independent samples of size n then, assuming that $\sup_{\mathscr{Q}}|\mu_n(A)-\mu_n'(A)|$ and $\sup_{\mathscr{Q}}|\mu_n(A)-\mu(A)|$ are measurable, Vapnik and Chervonenkis (1971) showed that

$$P\left\{\sup_{\mathscr{Q}} |\mu_n(A)-\mu(A)|\geqslant\varepsilon\right\}\leqslant 4s(\mathscr{Q},2n)\mathrm{e}^{-n\varepsilon^2/8}$$

where

$$s(\mathscr{Q},n)=\max_{(x_1,\ldots,x_n)} N_{\mathscr{Q}}(x_1,\ldots,x_n)$$

and $N_{\mathscr{Q}}(x_1,\ldots,x_n)$ denotes the number of different sets in the class $\{\{x_1,\ldots,x_n\}\cap A : A\in\mathscr{Q}\}$. If, in addition to the measurability assumptions of Vapnik and Chervonenkis, we assume that $\sup_{\mathscr{Q}}\mu_n(A)$ is measurable we have the following lemma.

Lemma 1. *Let $\varepsilon>0$ and suppose that*

$$\sup_{\mathscr{Q}} \mu(A)\leqslant b\leqslant 1/4.$$

Then

$$P\left\{\sup_{\mathcal{Q}}|\mu_n(A)-\mu(A)|\geqslant\varepsilon\right\}$$

$$\leqslant 4s(\mathcal{Q},2n)e^{-n\varepsilon^2/(64b+4\varepsilon)}+2P\left\{\sup_{\mathcal{Q}}\mu_{2n}(A)>2b\right\} \tag{2.3}$$

for all $n\geqslant 8b/\varepsilon^2$.

Inequality (2.3) is useful for small b and for classes $\mathcal{Q}$ for which $P\{\sup_{\mathcal{Q}}\mu_{2n}(A)>2b\}$ can be upper-bounded. The following lemma provides a useful upper bound for classes $\mathcal{Q}$ whose sets have a uniform bound on their diameter. We again assume that $\sup_{\mathcal{Q}}\mu_{2n}(A)$ is a random variable and, as before, $\|\cdot\|$ will denote the sup norm on $\mathbf{R}^d$. As usual, other norms could be used instead.

Lemma 2. *Let* $\mathcal{Q}$ *be any class of Borel sets from* $\mathbf{R}^d$ *with*

$$\sup_{\mathcal{Q}}\sup_{x,y\in A}\|y-x\|\leqslant r<\infty.$$

If $S(x,r)$ *is the closed sphere centered at* x *with radius* r *and if*

$$\sup_{x\in\mathbf{R}^d}\mu(S(x,r))\leqslant b,$$

then

$$P\left\{\sup_{\mathcal{Q}}\mu_{2n}(A)\geqslant 2b\right\}\leqslant 4ne^{-nb/10} \tag{2.4}$$

for all $n\geqslant 1/b$.

As an example, which will be used in the proof of the theorem, let $\mathcal{Q}_r$ be the class of all rectangles from $\mathbf{R}^d$ with diameter not greater than r and assume that

$$\sup_{\mathcal{Q}_{2r}}\mu(A)\leqslant b\leqslant 1/4.$$

Since

$$\sup_{x\in\mathbf{R}^d}\mu(S(x,r))=\sup_{\mathcal{Q}_{2r}}\mu(A)$$

and

$$s(\mathscr{Q}_r, 2n) \leqslant (2n)^{2d} \qquad \text{for all } r$$

(see Cover (1965) for other calculations of this type) we have, from Lemmas 1 and 2, that

$$P\Big\{ \sup_{\mathscr{Q}_r} |\mu_n(A) - \mu(A)| \geqslant \varepsilon \Big\} \leqslant 4(2n)^{2d} e^{-n\varepsilon^2/(64b+4\varepsilon)} + 8ne^{-nb/10} \tag{2.5}$$

for $\varepsilon > 0$ and $n \geqslant \max(1/b, 8b/\varepsilon^2)$.

Proof of Lemma 1. The following arguments are variations of those of Vapnik and Chervonenkis (1971). Let $X_1, \ldots, X_{2n}$ be independent, identically distributed random vectors with a common probability measure μ. If μ_n' denotes the empirical measure for $X_{n+1}, \ldots, X_{2n}$ and all unlabeled supremums below are taken over $\mathscr{Q}$, then an easy modification of Lemma 1 of Vapnik and Chervonenkis yields

$$P\{\sup |\mu_n(A) - \mu(A)| \geqslant \varepsilon\} \leqslant 2P\{\sup |\mu_n(A) - \mu_n'(A)| \geqslant \varepsilon/2\} \tag{2.6}$$

where

(i) $\sup \mu(A) \leqslant b$,
(ii) $\sup |\mu_n(A) - \mu(A)|$ and $\sup |\mu_n(A) - \mu_n'(A)|$ are random variables,
(iii) $n \geqslant 8b/\varepsilon^2$.

Because

$$\begin{aligned} &P\{\sup |\mu_n(A) - \mu_n'(A)| \geqslant \varepsilon/2\} \\ &\quad \leqslant P\{\sup |\mu_n(A) - \mu_n'(A)| \geqslant \varepsilon/2;\ \sup \mu_{2n}(A) \leqslant 2b\} \\ &\qquad + P\{\sup \mu_{2n}(A) > 2b\}. \end{aligned}$$

Lemma 1 will follow from (2.6) if we can show that for any δ, $M > 0$ with $M \leqslant 1/2$

$$\begin{aligned} &P\{\sup |\mu_n(A) - \mu_n'(A)| \geqslant \delta;\ \sup \mu_{2n}(A) \leqslant M\} \\ &\quad \leqslant 2s(\mathscr{Q}, 2n) e^{-n\delta^2/(8M+2\delta)}. \end{aligned} \tag{2.7}$$

The probability on the left-hand side of (2.7) equals

$$\int_{\mathbf{R}^{2nd}} \frac{1}{(2n)!} \sum I_{[\sup|\mu_n(A)-\mu_n'(A)|\geqslant\delta]} I_{[\sup\mu_{2n}(A)\leqslant M]}\,\mathrm{d}Q$$

where I_E is the indicator of a set $E\subseteq\mathbf{R}^{2nd}$, Q is the probability measure for $X_1,\ldots,X_{2n}$ defined on the Borel subsets of $\mathbf{R}^{2nd}$ and the inner summation is taken over all $(2n)!$ permutations of $x_1,\ldots,x_{2n}$. But this integral equals

$$\int_{\mathbf{R}^{2nd}} \frac{1}{(2n)!} \sum I_{[\sup\mu_{2n}(A)\leqslant M]} \sup I_{[|\mu_n(A)-\mu_n'(A)|\geqslant\delta]}\,\mathrm{d}Q$$

$$=\int_{\mathbf{R}^{2nd}} \frac{1}{(2n)!} \sum I_{[\sup\mu_{2n}(A)\leqslant M]} \sup_{\mathscr{Q}'} I_{[|\mu_n(A)-\mu_n'(A)|\geqslant\delta]}\,\mathrm{d}Q$$

$$=\int_{\mathbf{R}^{2nd}} \frac{1}{(2n)!} \sum I_{[\sup\mu_{2n}(A)\leqslant M]} \sup I_{[|\mu_n(A)-\mu_{2n}(A)|\geqslant\delta/2]}\,\mathrm{d}Q$$

$$\leqslant\int_{\mathbf{R}^{2nd}} \sum_{A\in\mathscr{Q}'} I_{[\sup\mu_{2n}(A)\leqslant M]}\left\{\frac{1}{(2n)!}\sum I_{[|\mu_n(A)-\mu_{2n}(A)|\geqslant\delta/2]}\right\}\mathrm{d}Q \tag{2.8}$$

where $\mathscr{Q}'=\mathscr{Q}'(x_1,\ldots,x_{2n})$ is any finite subclass of $\mathscr{Q}$ which yields the same class of intersections with $\{x_1,\ldots,x_{2n}\}$ as does $\mathscr{Q}$ and where the unlabeled summations are again over all $(2n)!$ permutations of $x_1,\ldots,x_{2n}$.

If $Y_1,\ldots,Y_n$ are Bernoulli random variables with $P\{Y_1=1\}=\mathbf{p}$ then

$$P\left\{\left|\frac{1}{n}\sum_1^n Y_i-\mathbf{p}\right|\geqslant\varepsilon\right\}\leqslant 2\mathrm{e}^{-n\varepsilon((1+(b/\varepsilon))\ln(1+(b/\varepsilon))-1)}\leqslant 2\mathrm{e}^{-n\varepsilon^2/(2b+\varepsilon)} \tag{2.9}$$

provided $0\leqslant p\leqslant b\leqslant 1/2$ (Bennett (1962), Hoeffding (1963)). (The second inequality follows from $\ln(1+(a/b))\geqslant 2a/(2b+a)$ for $a,b>0$.) Hoeffding has pointed out that (2.9) remains valid if $Y_1,\ldots,Y_n$ are obtained by sampling without replacement from a sequence $y_1,\ldots,y_k$ of 0's and 1's where $k\geqslant n$ and $\sum_1^k y_i=k\mathbf{p}$. Using this last observation we have the following inequality between random variables (which holds everywhere)

$$\left\{\frac{1}{(2n)!}\sum I_{[|\mu_n(A)-\mu_{2n}(A)|\geqslant\delta/2]}\right\}\leqslant 2\mathrm{e}^{-n(\delta/2)^2/(2\mu_{2n}(A)+\delta/2)}.$$

Thus the last integral in (2.8) is upper-bounded by

$$\int \sum_{A \in \mathcal{Q}'} I_{[\sup \mu_{2n}(A) \leqslant M]} 2\mathrm{e}^{-n\delta^2/(8\mu_{2n}(A)+2\delta)} \mathrm{d}Q$$

$$\leqslant 2s(\mathcal{Q}, 2n)\mathrm{e}^{-n\delta^2/(8M+2\delta)}$$

since we can always choose $\mathcal{Q}'$ to contain no more than $s(\mathcal{Q},2n)$ sets. Inequality (2.2) and Lemma 1 now follow.

Proof of Lemma 2. If μ_{2n-1}^i is the empirical measure for $X_1,\ldots,X_{2n}$ with X_i omitted and all unlabeled supremums below are again over $\mathcal{Q}$, then

$$P\{\sup \mu_{2n}(A) > 2b\} \leqslant P\left\{\bigcup_{i=1}^{2n} \{\mu_{2n}(S(X_i,r)) > 2b\}\right\}$$

$$\leqslant P\left\{\bigcup_{i=1}^{2n} \{(2n-1)\mu_{2n-1}^i(S(X_i,r)) > 4bn-1\}\right\}$$

$$\leqslant 2n\, P\{\mu_{2n-1}^1(S(X_1,r)) > (4bn-1)/(2n-1)\}.$$

$$\leqslant 2n\, P\{\mu_{2n-1}^1(S(X_1,r)) > 3b/2\} \qquad \text{if } bn \geqslant 1$$

$$\leqslant 2n \sup_x P\{\mu_{2n-1}(S(x,r)) > 3b/2\}$$

$$\leqslant 2n \sup_x P\{\mu_{2n-1}(S(x,r)) - \mu(S(x,r)) > b/2\}$$

$$\leqslant 4n\mathrm{e}^{-(2n-1)(b/2)^2/(2b+(b/2))} \qquad \text{(from (2.9))}$$

$$\leqslant 4n\mathrm{e}^{-(2n-1)b/10} \leqslant 4n\mathrm{e}^{-nb/10},$$

which proves Lemma 2.

To prove the theorem we first approximate K by a linear combination of indicators of disjoint rectangles.

Lemma 3. *Suppose K is a nonnegative, bounded Riemann integrable function on $\mathbf{R}^d$. For each $\eta, \delta, \rho > 0$ we can find a function*

$$K^*(x) = \sum_{1}^{N} \alpha_i I_{A_i}(x)$$

where

(i) $\alpha_1,\ldots,\alpha_N$ *are nonnegative real numbers,*

(ii) $A_1,\ldots,A_N$ *are disjoint rectangles contained in* $[-\rho,\rho]^d$,

(iii) $K^*(x)\leqslant \sup_x K(x)$, $x\in\mathbf{R}^d$,

(iv) $|K^*(x)-K(x)|<\eta$ *on* $[-\rho,\rho]^d$ *except on a set* D,

(v) $D\subseteq B=\bigcup_1^M B_i$ *where* $B_1,\ldots,B_M$ *are rectangles from* $[-\rho,\rho]^d$, *whose union has Lebesgue measure less than* δ.

Proof of Lemma 3. Partition $[-\rho,\rho]^d$ into disjoint rectangles in such a way that the upper and lower sums for K over the partition differ by less than $\eta\delta$ (Spivak (1965), Chapter 3). If K_1 and K_2 are, respectively, the functions corresponding to those upper and lower sums, then

$$\begin{aligned}&\{x\in[-\rho,\rho]^d : K(x)-K_2(x)\geqslant\eta\}\\&\quad\subseteq\{x\in[-\rho,\rho]^d : K_1(x)-K_2(x)\geqslant\eta\}.\end{aligned}$$

The latter set is a union of disjoint rectangles with Lebesgue measure less than δ. Putting $K^*(x)=K_2(x)$ yields the lemma.

Proof of Theorem 1. We follow the notation of Lemma 3, assuming for the moment that η, δ and ρ are arbitrary positive numbers. The dependence of h_n on n will be suppressed where confusion is unlikely. First

$$\begin{aligned}&\sup_x|Ef_n(x)-f_n(x)|\\&\quad=\sup_x\left|h^{-d}\int K((y-x)/h)\,\mathrm{d}F(y)-h^{-d}\int K((y-x)/h)\,\mathrm{d}F_n(y)\right|\\&\quad\leqslant\sum_{i=1}^{3}\sup_x U_i(x),\end{aligned}$$

where

$$U_1(x)=h^{-d}\int|K((y-x)/h)-K^*((y-x)/h)|\,\mathrm{d}F(y),$$

$$U_2(x)=h^{-d}\left|\int K^*((y-x)/h)\,\mathrm{d}F(y)-\int K^*((y-x)/h)\,\mathrm{d}F_n(y)\right|,$$

$$U_3(x)=h^{-d}\int|K^*((y-x)/h)-K((y-x)/h)|\,\mathrm{d}F_n(y).$$

If $C \subseteq \mathbf{R}^d$, $x \in \mathbf{R}^d$ and $\alpha > 0$, let $C(x,\alpha) = \{x + \alpha z : z \in C\}$ and let

$$C_1 = S(x, \rho h)^c$$

$$C_2 = S(x, \rho h) \cap D(x,h)^c$$

$$C_3 = D(x,h)$$

where $(\ )^c$ denotes the complement of a set. Then

$$\sup_x U_1(x) \leqslant \sum_1^3 \sup_x \int_{C_i} h^{-d} |K^*((y-x)/h) - K((y-x)/h)| \, \mathrm{d}F(y).$$

Recalling that K^* is zero outside of $[-\rho, \rho]^d$ we see that the first term is upper-bounded by

$$\sup_x \int_{S(x,\rho h)^c} h^{-d} K((y-x)/h) \, \mathrm{d}F(y)$$

while the second and third terms are upper-bounded by

$$\eta M_1 h^{-d} \sup_x \lambda(S(x, \rho h) = \eta M_1 2^d \rho^d$$

and

$$2 M_1 M_2 h^{-d} \sup_x \lambda(D(x,h)) \leqslant 2 M_1 M_2 \delta$$

respectively, where $M_1 = \sup_x f(x)$, $M_2 = \sup_x K(x)$ and λ denotes the Lebesgue measure on $\mathbf{R}^d$. Thus

$$\sup_x U_1(x) \leqslant \sup_x \int_{S(x,\rho h)^c} h^{-d} K((y-x)/h) \, \mathrm{d}F(y) + \eta M_1 2^d \rho^d + 2 M_1 M_2 \delta. \tag{2.10}$$

Next,

$$\sup_x U_2(x)$$

$$\leqslant \sup_x \left| \sum_1^N \alpha_i h^{-d}(\mu_n(A_i(x,h)) - \mu(A_i(x,h))) \right|$$

$$\leqslant NM_2 h^{-d} \sup_{\mathcal{C}_n} |\mu_n(A) - \mu(A)| \tag{2.11}$$

where $\mathcal{C}_n$ is the class of rectangles whose diameter does not exceed $2\rho h$. Finally,

$$\sup_x U_3(x)$$

$$\leqslant \sum_{i=1}^3 \sup_x \int_{C_i} h^{-d} |K((y-x)/h) - K^*((y-x)/h)| \, \mathrm{d}F_n(y).$$

The first term is upper-bounded by

$$\sup_x \int_{S(x,\rho h)^c} h^{-d} K((y-x)/h) \, \mathrm{d}F_n(y)$$

while the second term is upper-bounded by

$$\eta h^{-d} \sup_x \mu_n(S(x,\rho h))$$

$$\leqslant \eta h^{-d} \sup_x |\mu_n(S(x,\rho h)) - \mu(S(x,\rho h))| + \eta h^{-d} \sup_x \mu(S(x,\rho h))$$

$$\leqslant \eta h^{-d} \sup_{\mathcal{C}_n} |\mu_n(A) - \mu(A)| + \eta M_1 (2\rho)^d,$$

Recalling that $D \subseteq B = \bigcup_{i=1}^M B_i$ the third term is bounded by

$$M_2 h^{-d} \sup_x \mu_n(B(x,h))$$

$$\leqslant M_2 h^{-d} \sup_x |\mu_n(B(x,h)) - \mu(B(x,h))| + M_2 h^{-d} \sup_x \mu(B(x,h))$$

$$\leqslant MM_2 h^{-d} \sup_{\mathcal{C}_n} |\mu_n(A) - \mu(A)| + M_1 M_2 \delta.$$

Thus

$$\sup_x U_3(x) \leqslant \sup_x \int_{S(x,\rho h)^c} h^{-d} K((y-x)/h)\,\mathrm{d}F_n(y)$$
$$+(\eta + M_2 M) h^{-d} \sup_{\mathcal{A}_n} |\mu_n(A) - \mu(A)|$$
$$+\eta M_1 2^d \rho^d + M_1 M_2 \delta. \tag{2.12}$$

From (2.10), (2.11) and (2.12) we see that

$$\sup_x |f_n(x) - Ef_n(x)|$$
$$\leqslant \sup_x \int_{S(x,\rho h)^c} h^{-d} K((y-x)/h)\,\mathrm{d}F_n(y)$$
$$+ \sup_x \int_{S(x,\rho h)^c} h^{-d} K((y-x)/h)\,\mathrm{d}F(y)$$
$$+(M_2 N + M_2 M + \eta) h^{-d} \sup_{\mathcal{A}_n} |\mu_n(A) - \mu(A)|$$
$$+3M_1 M_2 \delta + 2\eta M_1 2^d \rho^d. \tag{2.13}$$

The first two terms of (2.13) can be upper-bounded by

$$2 \sup_x \int_{S(x,\rho h)^c} h^{-d} L(\|y-x\|/h)\,\mathrm{d}F(y)$$
$$+ \sup_x \left| \int_{S(x,\rho h)^c} h^{-d} L(\|y-x\|/h)\,\mathrm{d}F_n(y) - \int_{S(x,\rho h)^c} h^{-d} L(\|y-x\|/h)\,\mathrm{d}F(y) \right|$$
$$\leqslant 2M_1 \int_\rho^\infty 2^{2d-1} t^{d-1} L(t)\,\mathrm{d}t$$
$$+ \sup_x \left| \int_{S(x,\rho h)^c} h^{-d} L(\|y-x\|/h)\,\mathrm{d}F_n(y) - \int_{S(x,\rho h)^c} h^{-d} L(\|y-x\|/h)\,\mathrm{d}F(y) \right|$$

so that

$$\begin{aligned}
&\sup_x |f_n(x) - Ef_n(x)| \\
&\leqslant (M_2 N + M_2 M + \eta) h^{-d} \sup_{\mathcal{Q}_n} |\mu_n(A) - \mu(A)| \\
&\quad + \sup_x \left| \int_{S(x,\rho h)^c} h^{-d} L(\|y-x\|/h) \, dF_n(y) \right. \\
&\qquad \left. - \int_{S(x,\rho h)^c} h^{-d} L(\|y-x\|/h) \, dF(y) \right| \\
&\quad + 2^{2d} M_1 \int_\rho^\infty t^{d-1} L(t) \, dt + 3M_1 M_2 \delta + 2\eta M_1 2^d \rho^d. \qquad (2.14)
\end{aligned}$$

By choosing ρ sufficiently large and δ and η sufficiently small the last three terms of (2.14) can be made arbitrarily small. A straightforward application of Lemmas 1 and 2 for a fixed η, δ and ρ (e.g., in (2.5) $r = 2\rho h$, $b = 2^d M_1 \rho^d h^d$ and ε is replaced by $\varepsilon h^d/(M_2 N + M_2 M + \eta)$) shows that the first term of (2.14) tends to 0 with probability one if (1.3), which implies (1.4), is satisfied. The proof will be completed then if we show that

$$\begin{aligned}
&\sup_x \left| \int_{S(x,\rho h)^c} h^{-d} L(\|y-x\|/h) \, dF_n(y) \right. \\
&\qquad \left. - \int_{S(x,\rho h)^c} h^{-d} L(\|y-x\|/h) \, dF(y) \right| \qquad (2.15)
\end{aligned}$$

tends to zero with probability one for an arbitrarily large ρ.

Let $L'(t) = L(t) I_{[t \geqslant \rho]}$ so that (2.15) becomes

$$\sup_x \left| \int h^{-d} L'(\|y-x\|/h) \, dF_n(y) - \int h^{-d} L'(\|y-x\|/h) \, dF(y) \right|. \qquad (2.16)$$

For an arbitrary integer l let

$$S_j=\left\{x:(j-1)\frac{L(\rho)}{l}<u(\|x\|)\leqslant\frac{j}{l}L(\rho)\right\}\qquad 1\leqslant j\leqslant l,$$

$$T_j=\left\{x:\frac{(j-1)}{l}L(\rho)<L(\|x\|)\leqslant L(\rho)\right\}=\bigcup_{i=j}^{l}S_i,$$

and

$$L''(x)=\sum_{j=1}^{l}(j-1)\frac{L(\rho)}{l}I_{S_j}(x)$$

so that

$$|L'(\|x\|)-L''(x)|\leqslant L(\rho)/l$$

for all x. Returning to (2.16) we see that it is bounded by

$$\sup_x\int\left|h^{-d}L''((y-x)/h)\,\mathrm{d}F_n(y)-\int h^{-d}L''((y-x)/h)\,\mathrm{d}F(y)\right|$$

$$+\sup_x\int h^{-d}|L''((y-x)/h)-L'(\|y-x\|/h)|\,\mathrm{d}F_n(y)$$

$$+\sup_x\int h^{-d}|L''((y-x)/h)-L'(\|y-x\|/h)|\,\mathrm{d}F(y)$$

$$\leqslant\frac{2L(\rho)h^{-d}}{l}+\frac{L(\rho)h^{-d}}{l}\sup_x\left|\sum_{j=1}^{l}(j-1)\big(\mu_n(S_j(x,h))-\mu(S_j(x,h))\big)\right|$$

$$\leqslant\frac{2L(\rho)h^{-d}}{l}+\frac{L(\rho)h^{-d}}{l}\sup_x\left|\sum_{j=2}^{l}\big(\mu_n(T_j(x,h))-\mu(T_j(x,h))\big)\right|$$

$$\leqslant\frac{2L(\rho)h^{-d}}{l}+L(\rho)h^{-d}\sup_x\sup_{l\geqslant j\geqslant 2}|\mu_n(T_j(x,h))-\mu(T_j(x,h))|.$$

(2.17)

Since T_j is the difference of two concentric rectangles of diameter at most $2L^{-1}(L(\rho)/l)$, (2.17) can be upper-bounded by

$$\frac{2L(\rho)h^{-d}}{l}+2L(\rho)h^{-d}\sup_{\mathcal{A}_n'}|\mu_n(A)-\mu(A)| \tag{2.18}$$

where $\mathcal{A}_n'$ is the class of rectangles with diameters at most $2hL^{-1}(L(\rho)/l)$. By taking $l=[h^{-d}]$ we can make the first term of (2.18) arbitrarily small by taking ρ large. Applying Lemmas 1 and 2 as done earlier we see that the second term of (2.18) tends to zero with probability one if (1.3) is satisfied. This completes the proof of Theorem 1.

Proof of Theorem 2. Letting

$$g_n(x)=H_n^{-d}\int K((x-y)/H_n)f(y)\,dy$$

it is straightforward to see that, with the conditions on K and f,

$$\sup_x|g_n(x)-f(x)|\xrightarrow{n}0 \quad \text{in probability} \quad \text{(w.p. 1)}$$

wherever

$$H_n\xrightarrow{n}0 \quad \text{in probability} \quad \text{(w.p. 1)}$$

(Wagner (1975)). We therefore examine the convergence of the quantity

$$\sup_x|\hat{f}_n(x)-g_n(x)|.$$

First, if $\{\mathcal{B}_n\}$ is a sequence of positive numbers for which

$$I_{[nH_n^{2d}<\mathcal{B}_n]}\xrightarrow{n}0 \quad \text{in probability} \quad \text{(w.p. 1)} \tag{2.19}$$

and

$$I_{[nH_n^{2d}\geq\mathcal{B}_n]}\sup_x|\hat{f}_n(x)-g_n(x)|\xrightarrow{n}0 \quad \text{in probability} \quad \text{(w.p. 1)} \tag{2.20}$$

then

$$\sup_x|\hat{f}_n(x)-g_n(x)|\xrightarrow{n}0 \quad \text{in probability} \quad \text{(w.p. 1)}.$$

Following the proof of Theorem 1, we see that

$$\sup_x |\hat{f}_n(x) - g_n(x)| I_{[nH_n^{2d} \geqslant \mathfrak{B}_n]} \leqslant \sum_{i=1}^{3} \sup_{x;\, nh^{2d} \geqslant \mathfrak{B}_n} U_i(x,h) \tag{2.21}$$

where the U_i are defined as before except now we make the dependence on h explicit. By following the proof of Theorem 1 we see that (2.21) can be bounded by

$$\begin{aligned}&(M_2N + M_2M + \eta)(n/\mathfrak{B}_n)^{\frac{1}{2}} \sup_{\mathcal{Q}} |\mu_n(A) - \mu(A)|\\ &+ 2^{2d}M_1\int_\rho^\infty t^{d-1}L(t)\,\mathrm{d}t + 3M_1M_2\delta + \eta M_1 2^d\rho^d\\ &+ \frac{2L(\rho)}{l}(n/\mathfrak{B}_n)^{\frac{1}{2}} + 2L(\rho)(n/\mathfrak{B}_n)^{\frac{1}{2}} \sup_{\mathcal{Q}} |\mu_n(A) - \mu(A)|\end{aligned} \tag{2.22}$$

where now $\mathcal{Q}$ is the class of *all* rectangles in $\mathbf{R}^d$. By taking

$$l = \left[(n/\mathfrak{B}_n)^{\frac{1}{2}}\right]$$

the middle four terms of (2.22) can be made arbitrarily small by choosing ρ large enough and η, δ small enough. The first and last terms of (2.22) can be combined to yield a term

$$c(n/\mathfrak{B}_n)^{\frac{1}{2}} \sup_{\mathcal{Q}} |\mu_n(A) - \mu(A)|. \tag{2.23}$$

Using the inequality of Kiefer–Wolfowitz (1956), we see that (2.23), and hence (2.20), tends to 0 in probability if $\mathfrak{B}_n \xrightarrow{n} \infty$, and tends to 0 w.p. 1 if

$$\sum_1^\infty e^{-\alpha\mathfrak{B}_n} < \infty \forall \alpha > 0.$$

Using (1.10) or (1.8), it is now easy to show the existence of sequences $\{\mathfrak{B}_n\}$ which satisfy (2.19) and (2.20). This completes the proof of Theorem 2.

Remark. If f is an arbitrary density with continuity point x, Wagner (1975) has shown

$$g_n(x) \xrightarrow{n} f(x) \qquad \text{in probability} \qquad \text{(w.p. 1)}$$

whenever

$$H_n \xrightarrow{n} 0 \qquad \text{in probability} \qquad \text{(w.p. 1).} \tag{2.24}$$

For the kernels of theorems 1 and 2, one can see, by examining the proofs of these theorems, that

$$|g_n(x) - \hat{f}_n(x)| \xrightarrow{n} 0 \qquad \text{in probability} \qquad \text{(or w.p. 1)}$$

whenever

$$nH_n^{2d} \xrightarrow{n} \infty$$

$$\left(\text{or } nH_n^{2d}/\log n \to \infty \qquad \text{w.p. 1}\right) \tag{2.25}$$

Thus, for these kernels, (2.24) and (2.25) imply

$$\hat{f}_n(x) \xrightarrow{n} f(x) \qquad \text{in probability} \qquad \text{(w.p. 1)}$$

when x is a continuity point of f.

References

[1] Bennett, G. (1962). Probability inequalities for the sum of independent random variables. *Journal of the American Statistical Association*, **57**, 33–45.

[2] Cacoullos, T. (1965). Estimation of a multivariate density. *Annals of the Institute of Statistical Mathematics*, **18**, 179–190.

[3] Cover, T. M. (1965). Geometrical and statistical properties of systems of linear inequalities with applications in pattern recognition. *IEEE Transactions on Electronic Computers*, **10**, 326–334.

[4] Cover, T. M. (1972). A hierarchy of probability density function estimates. *Frontiers of Pattern Recognition* (S. Watanabe, Editor) pp. 83–98. New York: Academic Press.

[5] Deheuvels, P. (1974). Conditions necessaires et suffisantes de convergence ponctuelle presque sure et uniforme presque sure estimateurs de la densite. *C. R. Acad. Sci. Paris Ser. A.*, **278**, 1217–1220.

[6] Földes, A. and Révész, P. (1974). A general method for density estimation. *Studia Scientiarium Mathematicarum Hungarica*, **9**, 81–92.

[7] Hoeffding, W. (1963). Probability inequalities for sums of bounded random variables. *Journal of the American Statistical Association*, **58**, 13–30.

[8] Loftsgaarden, D. O. and Quesenberry, C. P. (1965). A nonparametric estimate of a multivariate density function. *Annals of Mathematical Statistics*, **36**, 1049–1051.

[9] Moore, D. S. and Yackel, J. W. (1977). Consistency properties of nearest neighbor density function estimators. *Annals of Statistics*, **5**, 143–154.

[10] Nadaraya, E. A. (1965). On nonparametric estimates of density functions and regression curves. *Theory of Probability and Its Applications*, **10**, 186–190.
[11] Parzen, E. (1962). On the estimation of a probability density function and the mode. *Annals of Mathematical Statistics*, **33**, 1065–1076.
[12] Rosenblatt, M. (1957). Remarks on some nonparametric estimates of a density function. *Annals of Mathematical Statistics*, **27**, 832–837.
[13] Schuster, E. F. (1969). Estimation of a probability density function and its derivatives. *Annals of Mathematical Statistics*, **40**, 1187–1195.
[14] Schuster, E. F. (1970). Note on the uniform convergence of density estimates. *Annals of Mathematical Statistics*, **41**, 1347–1348.
[15] Silverman, B. W. (1978). Weak and strong uniform consistency of the kernel estimate of a density and its derivatives. *Annals of Statistics*, **6**, 177–184.
[16] Spivak, M. (1965). *Calculus on Manifolds*. W. A. Benjamin, Inc., New York.
[17] Van Ryzin, J. (1969). On strong consistency of density estimates. *Annals of Mathematical Statistics*, **40**, 1765–1772.
[18] Vapnik, V. N. and Chervonenkis, A. Ya. (1971). On the uniform convergence of the relative frequencies of events to their probabilities. *Theory of Probability and Its Applications*, **16**, 264–280.
[19] Wagner, T. J. (1975). Nonparametric estimates of probability densities. *IEEE Transactions on Information Theory*, **IT-21**, 438–440.

Note. After this paper was in proof, we learned of the result by Bertrand–Retali (Convergence uniforme d'un estimateur de la densité par la méthode du Noyau, *Rev. Roumaine Math. Pures Appl.* **23** (1978), 361–385) which implies that Theorem 1 is true if (1.2) and (1.3) are replaced by (1.4) and

$$\int_{\mathbf{R}^d} \sup\{K(u) : \|u - x\| < 1\}\, dx < \infty.$$

P. R. Krishnaiah, ed., *Multivariate Analysis–V*

DESIGNS FOR EXTRAPOLATION WHEN BIAS IS PRESENT

J. KIEFER*
Cornell University and University of California (Berkeley)

1. Introduction

We specialize and adapt the notation of Kiefer (1973) to the present setting. Let f be a k vector of real valued functions on $\mathfrak{X}^*=\mathfrak{X}\cup\mathfrak{Z}$, with $f=\binom{g}{h}$, this and the subsequent decompositions being into s and k-s components. The expectation of an observation corresponding to value x (in $\mathfrak{X}^*$) of a controllable variable is $\theta'g(x)+\beta'h(x)$. We are permitted to take an N-vector $Y=(Y_1,\ldots,Y_N)'$ of uncorrelated observations Y_i with common variance σ^2, the ith at x_i in $\mathfrak{X}$. We must estimate $\theta'g+\beta'h$ on $\mathfrak{Z}$ by a linear function $t'g$ of the components of g, where the s-vector t is a linear homogeneous estimator, $t=CY$. The problem is to choose the design $X_N=(x_1,x_2,\ldots,x_N)$ and estimation matrix C to achieve some goal of accuracy, described further, below.

When $s=k$, this is the well known problem of extrapolation of the regression (assuming the model $\theta'g$ is correct) to $\mathfrak{Z}$ based on observations on $\mathfrak{X}$, often called interpolation if $\mathfrak{Z}\subset\mathfrak{X}$; see Kiefer and Wolfowitz (1964). The estimator t is usually taken to be the best linear unbiased, or least squares (LS), estimator. In the case of polynomial regression in one variable with $\mathfrak{X}=[-1,1]$ and $\mathfrak{Z}$ a point (or possibly a half-infinite interval), an elegant solution was given by Hoel and Levine (1964) and generalized by Kiefer and Wolfowitz (1965). Similar problems for higher dimensional $\mathfrak{X}$ with $\mathfrak{Z}$ a point were considered by Studden (1971). Galil and Kiefer (1978) considered extrapolation when $\mathfrak{X}$ is a q-dimensional ball and $\mathfrak{Z}$ is a larger or smaller ball, in the case of cubic regression; for quadratic regression this is also discussed in Kiefer (1978). We write $B_q(R)$ for the q-ball of radius R centered at O and $S_q(R)$ for the surface of $B_q(R)$, i.e., the $(q-1)$-sphere of radius R.

In the case $\mathfrak{Z}=\mathfrak{X}$, the path-breaking paper of Box and Draper (1959) initiated consideration of the possibility that the "assumed model" $\theta'g$

*Research sponsored by the National Science Foundation.

might be incorrect, so that the estimated regression $t'g$ might contain bias as well as variance, due to the presence of the contaminating term $\beta'h$ in the actual, true model. Write

$$F = F(X_N) = \left[f(x_1) \vdots f(x_2) \vdots \cdots \vdots f(x_N) \right] = \begin{pmatrix} G \\ H \end{pmatrix} (k \times N)$$

and $D = D(X_N) = FC'(k \times s)$. Also write

$$\begin{aligned} Q &= N^{-1}FF' = \begin{pmatrix} M & L \\ L' & K \end{pmatrix}, \\ \Gamma &= \begin{pmatrix} \Gamma_{gg} & \Gamma_{gh} \\ \Gamma_{hg} & \Gamma_{hh} \end{pmatrix} = \int_{\mathfrak{X}} f(z) f(z)' \nu(dz) \end{aligned} \tag{1.1}$$

for a specified measure ν that reflects the "importance" at various points of $\mathfrak{X}$ of the mean-square error (MSE) at z when X_N and CY are used. This MSE, integrated over $\mathfrak{X}$ wrt ν, is $J = \sigma^2(V + B)$, where, for any $s \times (k\text{-}s)$matrix p satisfying $\Gamma_{gg} p = \Gamma_{gh}$,

$$\begin{aligned} V &= \operatorname{tr} CC'\Gamma_{gg}, \\ \sigma^2 B &= \int_{\mathfrak{X}} \{(\theta', \beta')[Dg(z) - f(z)]\}^2 \nu(dz) \\ &= (\theta', \beta') \left[D - \begin{pmatrix} I_s \\ p' \end{pmatrix} \right] \Gamma_{gg} \left[D - \begin{pmatrix} I_s \\ p' \end{pmatrix} \right]' \begin{pmatrix} \theta \\ \beta \end{pmatrix} \\ &\quad + \beta' [\Gamma_{hh} - \Gamma_{hg} \Gamma_{gg}^- \Gamma_{gh}] \beta \overset{\text{def}}{=} \sigma^2 B_1 + \sigma^2 B_2. \end{aligned} \tag{1.2}$$

Note that B_2 is unaffected by our choice of design or estimator. In the BD approach, one chooses t to be the LS estimator of θ under the assumed model $\beta = 0$; that is, $C = (GG')^- G$. Then $\sigma^2 B_1$ and V reduce to

$$\begin{aligned} \sigma^2 B_{1,\text{BD}} &= \beta'(L'M^- - p')\Gamma_{gg}(L'M^- - p')'\beta, \\ V_{\text{BD}} &= N^{-1} \operatorname{tr} \Gamma_{gg} M^-. \end{aligned} \tag{1.3}$$

The BD prescription is then that, since one cannot choose X_N to minimize J (because $\sigma^{-1}\beta$ is unknown), one chooses X_N (if possible) to minimize B, by making $L'M^- = p' = \Gamma_{hg}\Gamma_{gg}^-$, often achieved by making $L = \Gamma_{gh}$ and $M = \Gamma_{gg}$ (the design moments of certain orders thus matching those of the loss matrix Γ). The rationale for this choice, in that it is supposed to yield a value of J close to the minimum one could achieve if one knew $\sigma^{-1}\beta$, has been discussed with examples in Kiefer (1973, 1975) and Galil and Kiefer (1977a); justification for the BD approach additional to that given by Box and Draper is contained in Draper and Herzberg (1973) (hereafter DH), as described below.

The departure from the BD approach of Karson, Manson, and Hader (1969) (hereafter KMH) begins by allowing arbitrary C. Then B is minimized for each design for which $(\theta'\Gamma_{gg}+\beta'\Gamma_{hg})\Gamma_{gg}^{-}$ is estimable by choosing $C=[I_s \vdots p](FF')^{-}F$ so that $D=\binom{I_s}{p'}$ and thus $B_1=0$. Then X_N is chosen to satisfy some other criterion, usually (as hereafter) to minimize the resulting

$$V_{\mathrm{KMH}}(X_N)=N^{-1}\mathrm{tr}\begin{bmatrix}\Gamma_{gg} & \Gamma_{gh}\\ \Gamma_{hg} & \Gamma_{hg}\Gamma_{gg}^{-}\Gamma_{gh}\end{bmatrix}\begin{bmatrix}M & L\\ L' & K\end{bmatrix}^{-1}. \tag{1.4}$$

Although $J_{\mathrm{KMH}}\leqslant J_{\mathrm{BD}}$ if $LM^{-}=p'$ is satisfied, examples in the present paper illustrate that, if no design can satisfy $LM^{-}=p'$, the BD design may sometimes be better. Possible shortcomings of the KMH approach are also illustrated in Kiefer (1973). Some remarks in DH, addressed to the BD vs. KMH approach controversy, criticize the KMH approach (p. 268) on the grounds that its estimators are biased for the model fitted and that it does not permit use of certain standard least squares procedures. This does not seem completely convincing, in view of the motivation (for BD, too) of consideration of such procedures because of concern about bias: if one modifies the design to this end, why not also change the estimator? We also note the description on p. 272 of DH to the effect that the "typical" situation contemplated by BD was that in which $V/B\approx 1$; in terms of the discussion of Galil and Kiefer (1977a), this means the variable h there (a component of $\sigma^{-1}\beta$) should be of order $N^{-1/2}$ rather than of order 1 in the comparisons, strongly reinforcing the conclusion that, at least in the example considered there, other designs such as the D-optimum design may have preferred behavior in terms of J and of the maximum of the MSE function. In any event, the remainder of the present paper is devoted primarily to consideration of the BD and KMH approaches, as an early exploratory attempt to investigate procedures for extrapolation with bias. Comparisons of the MSE function, for example in terms also of its maximum, should be carried out for various designs as they were in Galil and Kiefer (1977a) for $\mathfrak{Z}=\mathfrak{X}$.

Draper and Herzberg's paper was the first to treat extrapolation with bias (i.e., considerations of the type developed in the previous two paragraphs, but no longer with $\mathcal{Z}=\mathcal{X}$). The previous formulas have been written in such a form that they apply also when $\mathcal{Z}\neq\mathcal{X}$. In DH, first degree regression is considered when $\mathcal{X}$ is a q-ball and $\mathcal{Z}$ is the line segment from a point z outside the ball to the closest point of $\mathcal{X}$. The designs considered are limited to a certain subclass of designs over which B is minimized, assuming knowledge of which of two regions β lies within. In a subsequent unpublished paper (1976, hereafter DH2), which the authors have kindly sent the present author, $\mathcal{X}$ is a q-ball, which we take to be $B_q(1)$, $\mathcal{Z}$ is the region outside $\mathcal{X}$ and inside a larger ball $B_q(R)$, ν is uniform measure on $B_q(R)-B_q(1)$, and either (a) $s=q+1$ and g is first order while h is quadratic, or else (b) g is quadratic and h is cubic. In case (a), designs are restricted to those with "equal scaling" in all coordinates, and if the BD estimator is used the best design, which minimizes both B and J independently of R, takes all observations as "far out" as possible; in the approach of the present paper, where the "scaling"restriction is not adopted, this means taking all observations on $S_q(1)$ with resulting M the same as for the uniform probability measure on $S_q(1)$. In case (b), detailed computations are carried out only for central composite rotatable designs obtained from hypercube corners, cross-polytope points, and a center-point, and it is found that a particular choice of the design parameters again minimizes both V and B independently of R; since we do not restrict the form of design we consider, our results (Section 4) will be somewhat different. Additionally, we consider also the KMH approach in both cases (a) and (b).

All our developments will be obtained in the approximate theory, in which a design measure ξ is an arbitrary probability measure on $\mathcal{X}$ and

$$Q=\begin{pmatrix} M & L \\ L' & K \end{pmatrix}=\int_{\mathcal{X}} f(x)f(x)'\xi(\mathrm{d}x). \tag{1.5}$$

Throughout the remainder of the paper, $\mathcal{X}=B_q(1)$, and in Sections 3 and 4 we let $\mathcal{Z}=B_q(R)$ with ν Lebesgue measure. This slight difference from DH2 (in which $\mathcal{Z}=B_q(R)-B_q(1)$), described in the example of Section 3.1, makes the arithmetic slightly simpler without usually affecting the character of the results materially; from a practical viewpoint, it seems likelier that we are interested in the regression on $B_q(R)$ rather than only on $B_q(R)-B_q(1)$.

The examples also illustrate such phenomena as nonuniqueness of the

BD design (Section 4.1) and possible reduction of the biased regression problem to a related problem for unbiased regression.

The restriction when $q=1$ to symmetric designs in Sections 3 and 4 is justified as in Kiefer (1974). For $q>1$ (Section 3.2), p. 30 of Galil and Kiefer (1977b) discusses the possible slight loss in efficiency due to restriction to rotationally invariant ξ: although, for uniform ν, the unrestricted minimum of V alone can be attained by a rotationally invariant ξ, there are noninvariant ξ satisfying $L'M^{-}=\Gamma_{hg}\Gamma_{gg}^{-1}$, over which V must now be minimized. Nevertheless, for the example of Section 3.2 we restrict consideration to rotationally invariant ξ for the reasons given in the cited paper.

One may well wonder about the usefulness of the model considered here for extrapolation problems. As the originators of this approach have pointed out (e.g., DH, p. 272), even when $\mathcal{X}=\mathcal{Z}$ caution is required in using this approach unless B/V is small. Obviously, if the data indicates B/V to be large, one would reconsider only fitting $t'g$, and what is really wanted is a procedure that decides whether or not the fit $t'g$ is satisfactory. When we are extrapolating to a larger region, the breakdown of the approach is then obviously all-the-more possible. Thus, the present study is only a fragmentary beginning.

2. Extrapolation or interpolation to a point

The case where $\mathcal{Z}$ is a point z seems a less important problem than that in which $\mathcal{Z}$ is larger, so our illustrations are brief.

For extrapolation to a point, the KMH development reduces the problem to a known one without bias, as we now indicate. We can suppose $\mathcal{X}$, z, f arbitrary, subject only to $g(z)$ not being the 0-vector. Since ν is supported by the single point z, the integral form of $\sigma^2 B$ in (1.2) attains its zero minimum for any D satisfying $Dg(z)=f(z)$ (for example, $D=[g(z)'g(z)]^{-1}f(z)g(z)'$). For any such D, one computes easily from (1.3) that $\sigma^{-2}J$ reduces to $V=N^{-1}f(z)'Q^{-}(\xi)f(z)$; the problem is thus that of unbiased estimation of $\binom{\theta}{\beta}'f(z)$, using the usual LS estimator of that linear parametric function. In simple polynomial settings in one variable it is the Hoel–Levine problem mentioned in Section 1.

The BD development is more complex. For simplicity, let us limit consideration here to the example of first degree regression in one variable with quadratic bias: $\mathcal{X}=[-1,1]$, $g(x)=(1,x)'$, $h(x)=x^2$. The first line of

(1.3) now becomes (with $\nu(z)=1$)

$$\beta^2\left\{\left(\mu_2-\mu_1^2\right)^{-1}\left[\mu_2^2-\mu_1\mu_3+z(\mu_3-\mu_1\mu_2)\right]-z^2\right\}^2 \tag{2.1}$$

where $\mu_i=\int_{-1}^{1}x^i\xi(dx)$; necessarily $\mu_2-\mu_1^2\neq 0$ for $\theta' g(z)$ to be estimable, if $z>1$. One can show in the case $z>1$ that there are no designs ξ for which $\theta' g(z)$ is estimable and for which (2.1) is zero. Thus, this is an example (the first of several herein) in which the BD sufficient condition $L'M^-=\Gamma_{hg}\Gamma_{gg}^-$ cannot be satisfied. The calculation of a ξ that minimizes (2.1) by maximizing the expression inside braces ignoring z^2 (it suffices to let the support of ξ contain at most three points) is intractable algebraically, but one can obtain a computer solution. We omit details. The treatment of DH assuming a restricted form of design was mentioned in Section 2. In the more general problem $\mathfrak{X}=B_q(1)$ and $z=(z_1,0,0,\ldots,0)$ with $z_1>1$, these authors restrict consideration to (exact) designs that put some number of observations at $(1,0,0,\ldots,0)$ and the remainder, equally divided, among the q vertices of a regular (q-1)-simplex perpendicular to (and centered on) the first coordinate axis. That amounts, in our present one-dimensional setting, to consideration only of designs on $\{c,1\}$ with $-1\leqslant c<1$, so that $\xi(c)+\xi(1)=1$. As one sees in the expression for B' on p. 270 of DH, the one-dimensional case under this restriction leads to the degenerate solution $\xi(1)=1$, for which $\theta' g(z)$ is not even estimable. (This is *not* the minimizer of (2.1) without such a design restriction. Also, the problem actually treated in DH, as described in Section 1, is not this one.) If $|z|<1$, a large class of designs makes (2.1) vanish, and V can be minimized among them.

3. Extrapolation to $B_q(R)$ for linear g, with quadratic bias

3.1. The case $q=1$

We suppose $\mathfrak{X}=[-1,1]$ and $\mathfrak{Z}=[-R,R]$, with $g(x)'=(1,x)$ and $h(x)=x^2$. At the outset, we let ν be any *symmetric* probability measure on $\mathfrak{Z}$. We may restrict consideration to symmetric ξ. (See Section 1.) We write $\gamma_i=\int_{\mathfrak{Z}}z^{2i}\nu(dz)$ and $\mu_j=\lambda_{j/2}=\int_{\mathfrak{X}}x^j\xi(dx)$. The regression is $\theta_0+\theta_1 x+\beta x^2$, with t_0+t_1x being the fitted regression. Thus,

$$\Gamma=\begin{bmatrix}1 & 0 & \gamma_1\\ 0 & \gamma_1 & 0\\ \gamma_1 & 0 & \gamma_2\end{bmatrix},\qquad Q=\begin{bmatrix}1 & 0 & \lambda_1\\ 0 & \lambda_1 & 0\\ \lambda_1 & 0 & \lambda_2\end{bmatrix}. \tag{3.1}$$

For the KMH development, we obtain from (1.4)

$$NV_{\mathrm{KMH}}=\frac{\lambda_2-2\lambda_1\gamma_1+\gamma_1^2}{\lambda_2-\lambda_1^2}+\frac{\gamma_1}{\lambda_1}, \tag{3.2}$$

with an obvious interpretation if Q is singular (which it cannot be when γ_1 is sufficiently large, if (3.2) is to be finite). Since (1.4) is of the form tr AQ^{-1} for nonnegative definite A where Q is the design information matrix for the problem of *quadratic* regression on $\mathfrak{X}$, it follows from symmetry of ξ and the characterization of admissible designs in Kiefer (1959) that V is minimized by a design of the form $\xi(1)=\xi(-1)=\alpha/2$, $\xi(0)=1-\alpha$. We obtain $\lambda_1=\lambda_2=\alpha$ and thus, from (3.2), $\alpha=\{1+|\gamma_1-1|[\gamma_1(\gamma_1+1)]^{-1/2}\}^{-1}$ for the minimizing value. Thus, the KMH procedure yields

$$\sigma^2 B_{\mathrm{KMH}}=\sigma^2 B_2=\beta^2(\gamma_2-\gamma_1^2), \qquad NV_{\mathrm{KMH}}=\{|\gamma_1-1|+[\gamma_1(\gamma_1+1)]^{1/2}\}^2. \tag{3.3}$$

As for the BD development, consider first an arbitrary (not necessarily symmetric) ξ. Then the BD prescription $L'M^-=p'$ (see (1.3)) becomes (for nonsingular M, with an analogue otherwise)

$$(\mu_2-\mu_1^2)^{-1}(\mu_2^2-\mu_1\mu_3,\mu_3-\mu_1\mu_2)=(\gamma_1,0) \tag{3.4}$$

which, upon substitution of the second component equation of (3.4) into the first, becomes $\mu_2=\gamma_1$. Thus, if $\gamma_1\leqslant 1$, we can make $B_{1,\mathrm{BD}}=0$ by taking $\alpha=\gamma_1$ in the design family above; if $\gamma_1>1$, we have another illustration of the unsatisfiability of $L'M^-=\Gamma_{hg}\Gamma_{gg}^-$, and it is easily shown that $\alpha=1$ is best (no asymmetric design being better in either case). Thus, we obtain from (1.3)

$$\left.\begin{aligned}\sigma^2 B_{\mathrm{BD}}&=\beta^2(\gamma_2-\gamma_1^2)\\ NV_{\mathrm{BD}}&=2\end{aligned}\right\}\quad \text{if } \gamma_1\leqslant 1$$

$$\left.\begin{aligned}\sigma^2 B_{\mathrm{BD}}&=\beta^2[\gamma_2-\gamma_1^2+(\gamma_1-1)^2]\\ NV_{\mathrm{BD}}&=1+\gamma_1.\end{aligned}\right\}\quad \text{if } \gamma_1>1 \tag{3.5}$$

We see from (3.3) and (3.5), in the case $\gamma_1\leqslant 1$ where the prescription $L'M^-=\Gamma_{hg}\Gamma_{gg}^-$ can be satisfied, confirmation of the main KMH conclusion that $B_{BD}=B_{KMH}$ and $V_{BD}\geqslant V_{KMH}$, with strict inequality unless $\gamma_1=1$. However, for $\gamma_1>1$, the fact that the original BD sufficient condition cannot be satisfied works in favor of BD! In fact, for $\gamma_1>1$ we *always* have $V_{BD}<V_{KMH}$ and $B_{BD}<B_{KMH}$, so the interesting question is, when is $J_{BD}<J_{KMH}$? We find, for $\gamma_1>1$,

$$J_{BD}<J_{KMH}\Leftrightarrow\frac{N\beta^2}{\sigma^2}<\frac{2\gamma_1+2[\gamma_1(\gamma_1+1)]^{1/2}}{\gamma_1-1}, \tag{3.6}$$

and the ratio J_{BD}/J_{KMH} may be written down from (3.3) and (3.5).

One may specialize these considerations for ν's of usual interest. For $R<1$ (interpolation), we are always in the case $\gamma_1<1$. For $R>1$, the ν of DH (uniform measure on $[-R,R]-[-1,1]$) yields $\gamma_1=(R^3-1)/3(R-1)>1$, while the uniform measure on $[-R,R]$ yields $\gamma_1=R^2/3$. Consequently, for fixed $N\beta^2/\sigma^2$, the end points of the interval of R-values for which $J_{BD}<J_{KMH}$, for the DH ν, are to the left of those for the uniform ν on $[-R,R]$. For R large (where, however, usefulness of the model is most in doubt) and either ν, we see from (3.6) that $N\beta^2/\sigma^2\approx 4$ divides the regions where each of the two methods is better, and a first approximation to J_{KMH}/J_{BD} (as $R\to\infty$ and $R^2N\beta^2/\sigma^2\to\infty$) is $(4/9)+(20\sigma^2/9N\beta^2)$. Such conclusions can also be obtained for ν's other than the uniform measure.

3.2. *The case $q>1$*

We abbreviate the previous development and, as described in Section 1, restrict consideration to rotationally invariant ξ. We also take ν to be uniform probability measure on $B_q(R)$, the ν of DH2 (uniform on $B_q(R)-B_q(1)$) requiring only minor modification. If $x^{(i)}$, $1\leqslant i\leqslant q$, are the coordinate functions, we order the components of $g(x)'$ as $(1,x^{(1)},\ldots,x^{(q)})$ and those of $h(x)'$ as $(x^{(1)2},\ldots,x^{(q)2},x^{(1)}x^{(2)},\ldots,x^{(q-1)}x^{(q)})$. Writing $\int_{\mathfrak{X}}(x^{(1)})^i\xi(dx)=\mu_i$, an effect of rotational symmetry is that $\int_{\mathfrak{X}}(x^{(1)}x^{(2)})^2\xi(dx)=\mu_4/3$. We obtain, with 1_q denoting a q-vector of 1's and J_q denoting a $q\times q$ matrix of 1's,

$$Q=\begin{bmatrix} 1 & 0 & \mu_2 1_q' & 0 \\ 0 & \mu_2 I_q & 0 & 0 \\ \mu_2 1_q & 0 & \mu_4\left(\frac{2}{3}I_q+J_q\right) & 0 \\ 0 & 0 & 0 & \frac{1}{3}\mu_4 I_{q(q-1)/2} \end{bmatrix}, \tag{3.7}$$

from which pattern Γ can also be obtained. The exact KMH computations are now somewhat messy, and we give only the asymptotic behavior for large R. We obtain designs ξ consisting of mass $1-\alpha$ at 0 and mass α spread uniformly on $S_q(1)$. The optimum BD design for R large (in fact, for $R>1$ for the ν of DH2) has $\alpha=1$; more general than the DH2 conclusion mentioned in Section 1 is the optimality of this design for a general class of convex invariant optimality criteria. For the KMH design, putting $\gamma_i=\int_{\mathfrak{X}}(x^{(1)})^{2i}\nu(\mathrm{d}x)=R^2/(q+2)$, we find the optimum $\alpha\to 1/2$ as $\gamma_1\to\infty$ (as when $q=1$, just above (3.3)). In order to compare the two designs, we consider now the *special case in which all components of* β *are zero except for the coefficient of* $(x^{(1)})^2$, which is $\bar{\beta}$ (say); other comparisons can be made similarly, for other β. Since $\gamma_2/\gamma_1^2=3(q+2)/(q+4)$, we obtain, as $R\to\infty$,

$$\sigma^2 B_{\mathrm{KMH}}\approx\frac{2(q+1)}{(q+4)}\gamma_1^2\bar{\beta}^2,\qquad N V_{\mathrm{KMH}}\approx 4q^2\gamma_1^2,$$

$$\sigma^2 B_{\mathrm{BD}}\approx\frac{3(q+2)}{(q+4)}\gamma_1^2\bar{\beta}^2,\qquad N V_{\mathrm{BD}}=1+q^2\gamma_1. \tag{3.8}$$

The analogue of the last expression of Subsection 3.1 (again, with $R\to\infty$ and $R^2N\bar{\beta}^2/\sigma^2\to\infty$) is now

$$J_{\mathrm{KMH}}/J_{\mathrm{BD}}\approx\left[2(q+1)+4q^2(q+4)\sigma^2/N\bar{\beta}^2\right]/3(q+2), \tag{3.9}$$

and the KMH design is better if (approximately, for large R) $N\bar{\beta}^2/\sigma^2>4q^2$.

4. Quadratic regression, cubic bias

To conserve space, we consider only the case $q=1$ here; the arithmetical complexity for general q is similar to that of Galil and Kiefer (1978). However, we consider an additional ν in Subsection 4.1, because of some additional features that are well illustrated by it, including that mentioned just below (4.4).

4.1. Extrapolation to a shell

For general q, ν is now uniform measure on $\mathfrak{X}=S_q(R)$; for $q=1$, we have $\nu(R)=\nu(-R)=1/2$. Now $\theta' q(x)=\theta_0+\theta_1 x+\theta_2 x^2$ and $h(x)=x^3$. We have

$$Q=\begin{pmatrix} 1 & 0 & \mu_2 & 0 \\ 0 & \mu_2 & 0 & \mu_4 \\ \mu_2 & 0 & \mu_4 & 0 \\ 0 & \mu_4 & 0 & \mu_6 \end{pmatrix} \tag{4.1}$$

and Γ of the same form with μ_{2i} replaced by R^{2i}. Thus, $L'M^-(\xi)=(0,\mu_4/\mu_2,0)$ and $p'=(0,R^2,0)$; also, $B_2=0$. Thus, from (1.3),

$$NV_{\text{BD}}=(R^2/\mu_2)+[\mu_4-2R^2\mu_2+R^4]/[\mu_4-\mu_2^2],$$

$$\sigma^2 B_{1,\text{BD}}=R^2[R^2-\mu_4/\mu_2]^2. \tag{4.2}$$

Since $m_4(\xi)/m_2(\xi)$ has the unit interval as range, we can (as in Section 3) make $B_{1,\text{BD}}=0$ iff $R\leqslant 1$. In that case, substituting $\mu_4=R^2\mu_2$ into (4.2), we obtain $NV_{\text{BD}}=2R^2/\mu_2=2\mu_4/\mu_2^2\geqslant 2$, with equality iff ξ is symmetric on two points $\pm c$; hence, $c=R$ to satisfy the previous relations, and this yields $\xi(R)=\xi(-R)=1/2$ as the unique BD design minimizing V. (Although the formula for V_{BD} in (4.2) is meaningless in this case because $\mu_4-\mu_2^2=0$, the limiting value of V_{BD} obtained here for the singular M with $\xi(\pm R)=1/2$ is correct; this is of course the only symmetric design with singular M for which $\theta' g(\pm R)$ is estimable.)

If $R>1$, $B_{1,\text{BD}}$ is minimized by taking a design of the form $\xi(-1)=\xi(1)=\alpha/2>0$, $\xi(0)=1-\alpha$, for which $m_4/m_2=1$. For such a design,

$$NV_{\text{BD}}=(R^2/\alpha)+[\alpha-2R^2\alpha+R^4]/\alpha(1-\alpha),$$

minimized by

$$\alpha=\left[1+(R^2-1)/R(1+R^2)^{1/2}\right]^{-1}.$$

Thus, we obtain

$$NJ_{\text{BD}}=\begin{cases} 2 & \text{if } R\leqslant 1, \\ \left[R(1+R^2)^{1/2}+R^2-1\right]^2+(N\beta^2/\sigma^2)R^2[R^2-1]^2 & \text{if } R>1. \end{cases} \tag{4.3}$$

The KMH approach in this case, from (1.4), since $\Gamma_{hg}\Gamma_{gg}^{-}\Gamma_{gh}=R^6=\Gamma_{hh}$ and because of the special form of Γ and of Q for ξ symmetric, yields

$$NJ_{\mathrm{KMH}}=NV_{\mathrm{KMH}}=\operatorname{tr}\Gamma Q^{-1}=(1,R,R^2,R^3)Q^{-1}(1,R,R^2,R^3)'. \tag{4.4}$$

Thus, the KMH approach in this case of quadratic regression with cubic contamination reduces to that of finding the *optimum design for unbiased extrapolation to* $\pm R$ *when the cubic model is assumed and a cubic is fitted*, a problem considered in Kiefer and Wolfowitz (1964). (Although such a reduction also occurs in the analogous linear-quadratic model of Section 3 for the present ν, it does not occur for higher degree problems such as the cubic with quartic bias.) For $R\leqslant 1$, the solution is again the two-point design of the BD approach. For $R>1$, the solution may be found numerically; for $R>1.1$ the solution here coincides with that for extrapolation to $[-R,R]$ in the cubic without bias, as tabled by Galil and Kiefer (1978). As $R\to\infty$, the design approaches the "Tchebycheff design" optimum for estimating the cubic coefficient alone, for which $\xi(\pm 1/2)=1/3$ and $\xi(\pm 1)=1/6$; moreover, $J_{\mathrm{KMH}}=V_{\mathrm{KMH}}\approx 8R^6/N$ as $R\to\infty$. We conclude from (4.3) that, as $R\to\infty$, if also $R^2N\beta^2/\sigma^2\to\infty$, the BD design is preferable to the KMH design iff (approximately) $N\beta^2/\sigma^2<8$.

4.1. *Extrapolation to $B_q(R)$*

We now suppose ν to be uniform probability measure on $B_q(R)$, once more treating in detail only the case $q=1$. We now obtain $\sigma^2\beta^{-2}B_{1,\mathrm{BD}}=(3R^2/5-\mu_4/\mu_2)^2R^2/3$, so we can have $B_{1,\mathrm{BD}}=0$ iff $R^2\leqslant 5/3$ (replacing the value 1 of Section 4.1). Substituting $\mu_4=3\mu_2R^2/5$ in this case, we obtain from (1.3)

$$NV_{\mathrm{BD}}=2R^2(R^2-\mu_2)/5\mu_2\left[3R^2/5-\mu_2\right]. \tag{4.5}$$

The treatment of (4.5) is somewhat different from that of the corresponding expression $2R^2/\mu_2$ of Section 4.1, the optimum no longer being unique. We begin by formal minimization of (4.5) wrt μ_2 (paying no attention to the restriction $\mu_4/\mu_2=3R^2/5$). The minimum, attained at $\mu_2=R^2[1-(2/5)^{1/2}]$, is $NV_{\mathrm{BD}}=[14+(160)^{1/2}]/9$. We must check that there is indeed a design ξ on $[-1,1]$ with this μ_2 and with $\mu_4/\mu_2=3R^2/5$. In fact, there are many such designs. Two of them are ξ_1 and ξ_2 defined in terms of

$\delta=3R^2/5$ by

$$\xi_1(\pm\delta^{1/2})=1/\left[2+(8/5)^{1/2}\right]=\left[1-\xi_1(0)\right]/2,$$

$$\begin{aligned}&\xi_2\left(\pm\left[\delta(1-\delta)/(1+(2/5)^{1/2}-\delta)\right]^{1/2}\right)\\&\quad=\left[(2/5)^{1/2}+1-\delta\right]^2/\left[2+(8/5)^{1/2}\right]\left[(2/5)^{1/2}+(1-\delta)^2\right]\\&\quad=\tfrac{1}{2}-\xi_2(\pm 1).\end{aligned}\tag{4.6}$$

(The design ξ_2 degenerates into ξ_1 when $\delta=1$.) It is straightforward to verify that both of these designs are indeed symmetric probability measures on $[-1,1]$ and that both satisfy $\mu_2=R^2[1-(2/5)^{1/2}]=\mu_4/\delta$ as desired. Thus, these designs yield the same performance in terms of J_{BD}. However, if one looks at higher moments one finds $\mu_6(\xi_2)>\mu_6(\xi_1)$. Although μ_6 does not appear in J_{BD}, if (for example) one contemplated the *possibility* of fitting a cubic with the resulting data, which one could do with the design $\beta\xi_1+(1-\beta)\xi_2$ for $0\leqslant\beta<1$, consideration of μ_6 becomes important. It follows from results in Kiefer (1959) that, within this family with $\delta<1$, only the design for $\beta=0$ (that is, ξ_2) is admissible for unbiased estimation problems in the cubic model, ξ_2 being "better than" other members of the family. If also quartic regression might be considered, both ξ_1 and ξ_2 are admissible (although neither can yield an unbiased estimator of the entire quartic), while $\beta\xi_1+(1-\beta)\xi_2$, with seven support points if $0<\beta<1$, is admissible only if sextic regression is considered.

For $R^2>5/3$, the optimum design is again seen to be of the form $\xi(\pm 1)=\alpha/2$, $\xi(0)=1-\alpha$, with

$$\alpha=1/\left\{1+\left[(R^4/5-2R^2/3+1)/(R^4/5+R^2/3)\right]^{1/2}\right\}$$

yielding the minimum of V_{BD}. We thus obtain (since $B_2=4R^6/175$)

$$NJ_{\mathrm{BD}}=\begin{cases}\left[14+(160)^{1/2}\right]/9+(N\beta^2/\sigma^2)4R^6/175 & \text{if } R^2\leqslant 5/3,\\ \left\{\left[R^4/5+R^2/3\right]^{1/2}+\left[R^4/5-2R^2/3+1\right]^{1/2}\right\}^2 & \\ \quad+(N\beta^2/\sigma^2)\left[4R^2/175+(1-3R^2/5)^2R^2/3\right] & \text{if } R^2>5/3.\end{cases}\tag{4.7}$$

The KMH approach is more complicated, and does not lead to a reduction such as (4.4) because Γ now has rank 4. As $R \rightarrow \infty$, though, the variance of the estimator of the cubic coefficient in the cubic model without bias becomes dominant as it did in the case of (4.4), and we obtain $NJ_{\text{KMH}} \approx 24R^6/25 + (N\beta^2/\sigma^2)4R^6/175$. If $R^2(N\beta^2/\sigma^2) \rightarrow \infty$, once more we obtain preference for the BD procedure for large R iff (approximately) $N\beta^2/\sigma^2 < 8$.

We have not here compared other designs and estimators with those of BD and KMH as was done for the $\mathcal{X} = \mathcal{F}$ (no extrapolation) on the q-simplex in Galil and Kiefer (1977a). We mention that, for sufficiently large R, the design $\xi^{(R)}$ that minimizes the maximum of the variance function on $[-R, R]$ assigns mass $[1 + (R^2 + R^4)^{1/2}/(R^2 - 1)]^{-1}$ to 0 and the remainder equally to ± 1. This design behaves very much like the BD design as $R \rightarrow \infty$. However, as we have indicated earlier, extrapolation for large R, when the model $\theta' g$ is even slightly suspect, is a problem for which any treatment may be difficult to take seriously. Even for moderate R, the KMH or BD design and estimator appear to behave reasonably well in terms of the maximum (rather than average) of the MSE over $\mathfrak{Z}$, compared with $\xi^{(R)}$. This contrasts with the result for the simplex with $\mathcal{X} = \mathfrak{Z}$ cited above, in which the D-optimum design (with LS estimator) was superior to that of BD in terms of J (and, especially, in terms of the maximum MSE) unless $N\beta^2/\sigma^2$ was fairly large. The difference stems both from a difference between the simplex and ball moment space geometries, and also from the increased role of bias in extrapolation, as exhibited in the last line of (4.7) in the powers R^4 and R^6 in which R entered into V_{BD} and B_{BD}.

5. Adaptive methods

Once the design ξ is chosen, one can look at J as a function of C and the vector $\rho = \sigma^{-1}\beta$, and formally minimize it with respect to C. If $C_{\rho,\xi}$ is the coefficient vector of the resulting linear estimator $t = C_{\rho,\xi} Y$, one could then conceivably estimate ρ from the data Y, say by an estimator $\hat{\rho}$, and then use the adaptive (nonlinear) estimator $\hat{t} = C_{\hat{\rho},\xi} Y$. Unless the observations are taken in at least two stages, the estimator $\hat{\rho}$ cannot be used to help choose the design, but one can consider the problem of finding which design ξ performs "best" in some sense (such as integrated MSE) in terms of its estimator $Y' C'_{\hat{\rho},\xi} g$. The arithmetic of computing $C_{\rho,\xi}$ is essentially that used in Section 3 of Kiefer (1973) for the problem of minimizing the average of $V + B$ in a Bayesian-like setup; but the ensuing computation of the performance of $C_{\hat{\rho},\xi}Y$, and even consideration of the choice of the estimator $\hat{\rho}$, are more difficult. This is being investigated in more meaningful contexts by J. Sacks and the present author; here we shall indicate only the first step, of calculation of $C_{\rho,\xi}$.

A natural restriction we impose is that the risk should be bounded in θ if $\beta=0$; this entails $D'=[\,I_s \,\vdots\, d\,]$ for some $s\times(k-s)$ matrix d. Thus, $GC'=I_s$ and $HC'=d'$. From (1.2), then,

$$V+B_1=\operatorname{tr}\Gamma_{gg}\{CC'+(CH'-p)\rho\rho'(HC-p')\}. \tag{5.1}$$

Upon writing

$$\begin{aligned}
U&=\Gamma_{gg}^{\frac{1}{2}}\left\{C[I_N+H'\rho\rho'H]^{\frac{1}{2}}-p\rho\rho'H[I_N+H'\rho\rho'H]^{-\frac{1}{2}}\right\},\\
R&=\Gamma_{gg}^{\frac{1}{2}}\left\{I_s-p\rho\rho'H[I_N+H'\rho\rho'H]^{-1}G'\right\},\\
L&=[I_N+H'\rho\rho'H]^{-\frac{1}{2}}G',
\end{aligned}\tag{5.2}$$

we find that the problem of minimizing (5.1) subject to $GC'=I_s$ is equivalent to the problem of minimizing tr UU' subject to $UL=R$ (assuming G of rank s so that this last can be satisfied). In fact, the *matrix minimum* of UU' subject to $UL=R$ is $U=R(L'L)^{-}L'$ (see Kiefer (1973), (1.6)), from which the minimizing $C_{\rho,\xi}$ (independent of Γ_{gg}, reflecting its yielding a matrix minimum) can be written in terms of $P_\rho=[I_N+H'\rho\rho'H]^{-1}$ as

$$C_{\rho,\xi}=p\rho\rho'P_\rho\left\{I_N-G'[GP_\rho G']^{-1}GP_\rho\right\}+[GP_\rho G']^{-1}GP_\rho. \tag{5.3}$$

This can be simplified slightly using such relations as $P_\rho=I_N-(1+\rho'HH'\rho)^{-1}H'\rho\rho'H$, but the resulting estimator is not simple: the coefficient of p in (5.3) does not generally vanish (even if $GH'=0$), and even the simpler last term in (5.3) depends in rational rather than polynomial fashion on ρ.

Another "adaptive" estimator of a different form is obtained by simply writing $\tilde{Y}=Y-H'\beta$ and finding the usual LS estimator for θ in the model $E\tilde{Y}=G'\theta$, yielding $(GG')^{-1}GY-(GG')^{-1}GH'\beta$ as estimator t in the fitted $t'g$; an adaptive estimator $\hat{\beta}$ of β can then be used. The resulting $\hat{t}$ generally differs from that of the previous paragraph, being of simpler form here. (Making GH' vanish, by changing the h of the original model through orthogonalization, of course yields the BD estimator t_{BD}, but this results in a different fitted $t'g$.) One might use for $\hat{\beta}$ the LS estimator of β in the "full" model $\theta'g + \beta'h$ (which does not generally yield $\hat{t}$ equal to the LS estimator t^* of θ in that full model), but one could also use other estimators, such as ridge regression or Stein estimators for $\hat{\beta}$.

The comparison of performance of the adaptive estimators of the types mentioned in the previous two paragraphs, of t_{BD}, and of t^*, for various actual configurations (β, σ), remains to be carried out. Ultimately this is only a beginning on the old problem (where β is not known to be small) of choosing what curve to fit.

References

[1] Box, G. E. P., and Draper, N. R. (1959). A basis for the selection of a response surface design. *J. Am. Statist. Assoc.* **54**, 622–654.

[2] Draper, N. and Herzberg, A. (1973). Some designs for extrapolation outside a sphere. *JRSS (B)* **35**, 268–276.

[3] Draper, N. and Herzberg, A. (1976). An investigation of first-order and second-order designs for extrapolation outside a hypersphere. (Unpublished).

[4] Galil, Z. and Kiefer, J. (1977a). Comparison of Box-Draper and D-optimum designs for experiments with mixtures. *Technometrics* **19**.

[5] Galil, Z. and Kiefer, J. (1977b). Comparison of rotatable designs for regression on balls, I (quadratic). *J. Statist. Planning and Inference* **1**.

[6] Galil, Z. and Kiefer, J. (1978). Extrapolation designs and Φ_p-optimum designs for cubic regression the q-ball. *J. Statist. Planning and Inference*. To appear.

[7] Hoel, P. G. and Levine, A. (1964). Optimal spacing and weighting in polynomial prediction. *Ann. Math. Statist.* **33**, 1553–1560.

[8] Karson, M. J., Manson, A. R., and Hader, R. J. (1969). Minimum bias estimation and experimental designs for response surfaces. *Technometrics* **11**, 461–476.

[9] Kiefer, J. (1959). Optimum experimental designs. *J. Roy. Statist. Soc., Ser.* B **21**, 272–319.

[10] Kiefer, J. (1973). Optimum designs for fitting biased multiresponse surfaces. In: *Multivariate Analysis III* (P. R. Krishnaiah, Ed.). Academic Press, New York, 287–297.

[11] Kiefer, J. (1974). General equivalence theory for optimum designs (approximate theory). *Ann. Statist.* **2**, 849–879.

[12] Kiefer, J. (1975). Optimal design: Variation in structure and performance under change of criterion. *Biometrika* **62**, 277–288.

[13] Kiefer, J. (1978). Asymptotic approach to families of design problems. *Commun. in Statist.* (to appear).

[14] Kiefer, J. and Wolfowitz, J. (1964). Optimum extrapolation designs I and II. *Ann. Inst. Statist. Math.* **16**, 79–108, 295–303.

[15] Kiefer, J. and Wolfowitz, J. (1965). On a theorem of Hoel and Levine on extrapolation. *Ann. Math. Statist.* **36**, 1627–1655.

[16] Studden, W. J. (1971). Optimal designs for multivariate polynomial extrapolation. *Ann. Math. Statist.* **42**, 828–832.

PART II

TIME SERIES AND STOCHASTIC PROCESSES

P. R. Krishnaiah, ed., *Multivariate Analysis–V*
© North-Holland Publishing Company (1980) 97–109.

NON-LINEAR WHITE NOISE THEORY*

A. V. BALAKRISHNAN
School of Engineering and Applied Science, University of California, Los Angeles, CA, U.S.A.

1. Introduction

In many engineering problems involving processing of observed data—aircraft flight test data, radar return data, are some typical examples—the data quality is limited ultimately by an unavoidable additive random noise component, of which about all that one can reasonably say is that it has large bandwidth compared to the information bearing signal. Traditionally it is modelled as 'white noise', a stationary (continuous time parameter) stochastic process, Gaussian distributed, with constant spectral density over all frequencies. Of course it is *not* physical in that it has infinite power; nor is it mathematically rigorous. Nevertheless, so long as only linear operations on the data are considered, the results are consistent with the asymptotic case where we allow the bandwidth to expand to infinity in any way desired. A typical example is the Kalman filter. However, serious difficulties arise of interpretation as soon as non-linear operations on the data have to be considered. For instance, a simple 'squaring' operation is difficult to define unambiguously; we get different answers that depend on how the bandwidth increases to infinity. Beginning around 1960, a purported 'rigorous' model was proposed to define non-linear operations. Let $y(t)$, $0<t$, denote the observed data, $s(t)$ the information bearing signal, and $n(t)$ the noise, so that we have:

$$y(t)=s(t)+n(t).$$

We first 'integrate' the data:

$$Y(t)=\int_0^t y(\sigma)\,\mathrm{d}\sigma=\int_0^t s(\sigma)\,\mathrm{d}\sigma+W(t),$$

where

$$W(t)=\int_0^t n(\sigma)\,\mathrm{d}\sigma.$$

*Research supported in part under AFOSR grant no. 732942, Applied Math. Divn. USAF.

The rigorists now consider the model in which $W(t)$ is a Wiener process:

$$Y(t) = \int_0^t s(\sigma)\,\mathrm{d}\sigma + W(t).$$

A whole new discipline, 'non-linear filtering', has grown based on this model, and the associated mathematical apparatus provided by the Ito integral, which allows non-linear operations to any degree of complexity. Unfortunately, from the practical point of view, the results obtained simply *cannot* be instrumented, for the fundamental reason that the actual data samples have zero probability of occurring (since the Wiener process sample paths are of unbounded variation with probability one) and the Ito integral is simply not defined thereon!

In this paper we present a theory of non-linear operations on white noise which is free from this defect. We use the term 'non-linear white noise' theory to emphasize the fact that the primary focus is on non-linear operations, and thus to distinguish the notion of white noise from the 'generalized' random process notion that most mathematicians may think of, where of course only linear operations are considered. The theory has been put to the test on actual data (aircraft flight dynamics) for which reference may be made to [1]. Here we shall only be concerned with the theory for the most part.

The basic notions, which may not be familiar to most readers because they are not standard as yet, are explained in Section 2. The white noise model itself is explained in Section 3. In Section 4 we derive the formula for the likelihood derivative for signals of finite energy in white noise. We develop a Bayes formula for the conditional expectation of the signal given the data, and show in Section 5 that the innovation process is also a white noise process. In Section 6 we derive a formula for the conditional density. We note the appearance of additional terms in these formulae, not present in the Wiener process version.

2. Basic notions

We begin by recalling relevant basic notions, some standard, and some not. For more details see [2].

Let $\mathcal{H}$ denote a real separable Hilbert space, and $\mathcal{C}$ the class of cylinder sets therein with Borel bases in finite dimensional subspaces. Let μ denote a *weak distribution*: a finitely additive probability measure on $\mathcal{C}$ which is countably additive on any class of cylinder sets with bases in the same finite dimensional subspace. We note that μ is completely specified by the characteristic function:

$$C(h)=\int_{\mathcal{H}} \exp i[h,x]\, \mathrm{d}\mu(x) \qquad h\in\mathcal{H}$$

Let P denote a finite-dimensional projection operator on $\mathcal{H}$. Any $\mathcal{C}$-measurable function of the form $f(Px)$, where $f(\cdot)$ maps $\mathcal{H}$ into $\mathcal{H}_r$ a possibly different Hilbert space (real separable), is called a *tame function*. Since probabilities of inverse images are defined, we can and shall call it a random variable. More generally, any Cauchy sequence of tame functions in probability, will be called a random variable. In this way we obtain all the random variables we would obtain by extending μ to be countably additive on a larger space, such as the algebraic dual of $\mathcal{H}$.

Let $f(\cdot)$ be any Borel measurable function on $\mathcal{H}$ into $\mathcal{H}_r$. Then of course probabilities need not be defined on inverse images, and thus we cannot define every Borel measurable function as a random variable. Suppose now that for every sequence of finite dimensional projections P_n, converging strongly to the identity, we have that $f(P_n x)$ is a Cauchy sequence in probability, and that

$$C_f(h)=\lim_n \int_{\mathcal{H}} \exp i[f(P_n x),h]\, \mathrm{d}\mu(x)$$

is independent of the particular projection sequence chosen. We shall then call $f(\cdot)$ a physical random function, or physical random variable; the probabilities of inverse images are now defined unambiguously by the limiting process, $C_f(h)$ being the corresponding characteristic function.

Let μ_G denote the weak distribution (Gauss measure) with characteristic function:

$$\exp -\tfrac{1}{2}[h,h],$$

and let L denote a linear bounded operator mapping $\mathcal{H}$ into $\mathcal{H}$. Then

$$f(x)=[Lx,x]$$

is a physical random variable if and only if $(L+L^*)$ is nuclear. Thus 'squaring' is *not*!

We need to specify our notion of the Radon–Nikodym derivative for finitely additive measures. Thus let μ_1, μ_2 denote two weak distributions.

We shall say that a function $f(\cdot)$ defined on $\mathcal{H}$ into R_1 is the Radon–Nikodym derivative of μ_2 with respect to μ_1 if $f(\cdot)$ is Borel measurable, and for every C in $\mathcal{C}$ we have:

$$\mu_2(C) = \lim_n \int_C f(P_n x)\,\mathrm{d}\mu_1(x)$$

where P_n is any sequence of finite dimensional projections converging strongly to the identity.

3. The white noise data model

Let $\mathcal{W}(T)$, $0<T$, denote the L_2 space:

$$\mathcal{W}(T) = L_2((0,T);H),$$

where H is a separable Hilbert space, in general. By *white noise* we mean the triple:

$$(\mathcal{W}(T), \mathcal{C}, \mu_G),$$

where $\mathcal{C}$ denotes the class of cylinder sets in $\mathcal{W}(T)$ as before, and μ_G is the Gauss measure defined by:

$$C(h) = \int_{\mathcal{W}(T)} \exp i[h,x]\,\mathrm{d}\mu_G = \exp -\tfrac{1}{2}[h,h].$$

Thus in our data model:

$$y(t) = s(t) + n(t) \qquad 0<t<T, \tag{3.1}$$

we assume that $n(\cdot)$ is 'white noise' in the above precise sense. The signal is of course smoother; thus we assume that it is a stochastic process: $s(t,\omega)$, ω denoting sample points in Ω, $(\Omega, \mathcal{B}, p)$ being the probability triple, in the usual notation. We assume that $s(t,\omega)$ is jointly measurable in t and ω, and furthermore that the signal has finite energy (in contrast to the noise):

$$\int_0^T E(\|s(t,\omega)\|^2)\,\mathrm{d}t < \infty.$$

We note that this is a crucial assumption for us, however reasonable! In particular, $s(\cdot,\omega)$ yields a $\mathcal{B}$-measurable map into $\mathcal{W}(T)$. Hence we can write finally for (3.1):

$$y = s(\cdot,\omega) + n. \tag{3.2}$$

Our next assumption is that the signal and noise are independent. Thus we have a finitely additive measure defined on $\mathcal{B} \times \mathcal{C}$:

$$(p \times \mu_G)(B \times C) = p(B) \cdot \mu_G(C) \qquad B \in \mathcal{B} \quad C \in \mathcal{C}.$$

In turn (3.2) induces a corresponding finitely additive measure, denoted μ_y on $\mathcal{W}(T)$, with characteristic function:

$$C_y(h) = \int_\Omega \left\{ \exp i \int_0^T [s(t,\omega), h(t)]\, \mathrm{d}t \right\} \mathrm{d}p \cdot \exp -\tfrac{1}{2}[h,h].$$

Note that $C_y(\cdot)$ is strongly continuous in h.

Our first result is that the function:

$$q(h) = \int_\Omega \left\{ \exp -\tfrac{1}{2} \left\{ \int_0^T \|s(t,\omega)\|^2 \mathrm{d}t - 2\int_0^T [s(t,\omega), h(t)]\, \mathrm{d}t \right\} \right\} \mathrm{d}p \tag{3.3}$$

is a physical random function with respect to μ_G (and also with respect to μ_y), and that it is the Radon–Nikodym derivative of μ_y with respect to μ_G. For a proof see [3].

4. Likelihood ratio

By considering T in (3.3) as a variable, and differentiating with respect to T, we can cast (3.3) in the form in which it becomes the likelihood ratio. We follow [4]. Thus we have:

$$q(h) = \exp -\tfrac{1}{2} \left\{ \int_0^T \|\hat{s}(t;h)\|^2 \mathrm{d}t - 2\int_0^T [\hat{s}(t,h), h(t)]\, \mathrm{d}t + \int_0^T P(t;h)\, \mathrm{d}t \right\} \tag{4.1}$$

where

$$\hat{s}(t;h) = \int_\Omega s(t,\omega)\, B(s(\cdot,\omega); h; t)\, \mathrm{d}p \tag{4.2}$$

$$P(t;h) = \int_\Omega \|s(t,\omega)\|^2 B(s(\cdot,\omega); h; t)\, \mathrm{d}p - \|\hat{s}(t;h)\|^2 \tag{4.3}$$

and

$$B(s(\cdot,\omega);h;t)=\frac{\exp -\frac{1}{2}\int_0^t\{\|s(\sigma,\omega)\|^2\mathrm{d}\sigma-2[s(\sigma,\omega),h(\sigma)]\,\mathrm{d}\sigma\}}{\int_\Omega\left\{\exp -\frac{1}{2}\int_0^t\{\|s(\sigma,\omega)\|^2\mathrm{d}\sigma-2[s(\sigma,\omega),h(\sigma)]\,\mathrm{d}\sigma\}\right\}\mathrm{d}p}. \tag{4.4}$$

The main point is of course that upon substituting the observed data $y(\cdot)$, in place of $h(\cdot)$,

$$\hat{s}(t;y)$$

can be interpreted as the 'conditional expectation of $s(t,\omega)$ given the observation $y(\cdot)$ up to time t'. This requires the definition of conditional expectation in the present set-up. For this we proceed as follows. Let $P(t)$, for each t, $0<t$, denote the projection on $\mathcal{W}(T)$ into itself, defined by:

$$P(t)f=g;\qquad g(s)=f(s),\quad s<t;\qquad g(s)=0,\quad s>t.$$

Suppose there is a function $Q(\cdot)$ such that for any sequence $\{P_n\}$ of finite dimensional projections converging strongly to the identity, we have that

$$\{E(s(t,\omega)\mid P_nP(t)y)\}\quad\text{and}\quad\{Q(P_nP(t)y)\}$$

are equivalent Cauchy sequences of tame functions. Then we define:

$$\hat{s}(t;h)=Q(P(t)y).$$

The claim is that the function defined in (4.2) satisfies this condition. Also, the first term on the right of equation (4.3) is the conditional expectation of

$$\|s(t,\omega)\|^2$$

and from this we can see that $P(t;h)$ is the 'conditional mean square error, given data up to time t', and is of course non-negative. Note that it is also a 'physical random variable', and is a constant in the Gaussian case, see [3].

We note also that the Cauchy sequences:

$$\{Q(P(t)P_ny)\}\quad\text{and}\quad\{Q(P_nP(t)y)\}$$

are equivalent, for all t, $0<t<T$. This is readily seen by noting that for any orthonormal system $\{\phi_i\}$ in $\mathcal{W}(T)$:

$$E\left(\|\sum_1^N \int_t^T [y(s),\phi_i(s)]\,\mathrm{d}s\,P(t)\phi_i\|^2\right)\to 0$$

5. Innovation process

We note that

$$\Phi(y)=z; \qquad z(t)=Q(P(t)y)$$

where $Q(P(t)y)$ is defined by the right-side of (4.2), maps $\mathcal{W}(T)$ into itself, and defines a physical random variable as well. Next let

$$\nu=y-\Phi(y)$$

which is then recognized as the innovation. We shall now show that ν is white noise in $\mathcal{W}(T)$ (proved for an apparent more special case in [3]). For this it is enough to show that

$$\lim_n E\left(\exp i\int_0^T [h(\sigma),\nu_n(\sigma)]\,\mathrm{d}\sigma=\exp-\tfrac{1}{2}[h,h]\right.$$

where

$$\nu_n=P_n y-\Phi(P_n y)$$

$\{P_n\}$ being again a sequence of finite dimensional projections converging strongly to the identity. Or, equivalently that:

$$\left\{\frac{\mathrm{d}}{\mathrm{d}t}\lim_n E\left(\exp i\int_0^t [h(s),\nu_n(s)]\,\mathrm{d}s\right)\right\}\Big/\lim E(\exp i\,h(s),\nu_n(s)\,\mathrm{d}s)$$

$$=-\tfrac{1}{2}[h(t),h(t)] \qquad 0<t<T$$

for all strongly continuous functions $h(\cdot)$ in $\mathcal{W}(T)$. Hence we may consider for sufficiently small Δ, $0<\Delta$:

$$\lim_n E\left(\exp i\int_0^t [h(s),\nu_n(s)]\,\mathrm{d}s\,(e^{i}\zeta n-1)/\Delta\right)$$

where

$$\zeta_n = \int_t^{t+\Delta} [h(s), \nu_n(s)]\, ds.$$

We note that these sequences may be replaced by the equivalent Cauchy sequences: replace ζ_n by

$$\eta_n = [P_n(P(t+\Delta) - P(t))h, (P(t+\Delta) - P(t))(y - z_n)]$$

where

$$z_n(t) = E(s(t,\omega)/P_nP(t)y)$$

$$= \frac{E\left(s(t,\omega) \exp -\frac{1}{2}\|P_nP(t)(y - s(\cdot,\omega))\|^2\right)}{E\left(\exp -\frac{1}{2}\|P_nP(t)(y - s(\cdot,\omega))\|^2\right)}$$

and replace

$$\int_0^t [h(s), \nu_n(s)]\, ds$$

by

$$[P_nP(t)h, P(t)(y - z_n)]$$

where the inner products are all in $\mathcal{W}(T)$.

Next we use the identity:

$$(\exp i\eta_n - 1) = i\eta_n - \eta_n^2/2 - (i/2)\int_0^{\eta_n} s^2 \exp i(\eta_n - s)\, ds.$$

Now from the definition of z_n, and the independent increment property of the noise $n(\cdot)$, it follows that:

$$E(\{\exp i[P_nP(t)h, P(t)(y - z_n)]\}\eta_n) = 0$$

$$E(\eta_n^2 \exp i[P_nP(t)h, P(t)(y - z_n)])$$

$$= E(\eta_n^2)\, E(\exp i[P_nP(t)h, P(t)(y - z_n)]).$$

It is not difficult to see that

$$E(\eta_n^2/\Delta)\to(1/\Delta)\int_t^{t+\Delta}[h(s),h(s)]\,\mathrm{d}s$$

while:

$$E(\eta_n^3/\Delta)\to\mathrm{O}(\sqrt{\Delta}).$$

6. Conditional density

We restrict ourselves to the case now of $H=R_n$, because we wish to derive a formula for the conditional density for the case where

$$S(t,\omega)=C(x(t,\omega)) \tag{6.1}$$

and $x(t,\omega)$ ($\epsilon\,R_m$, say) is a diffusion process satisfying the Ito equation:

$$\mathrm{d}x=F(x)\,\mathrm{d}t+G(x)\,\mathrm{d}W \tag{6.2}$$

where $W(t,\omega)$ is a Wiener process, with the usual conditions that guarantee existence of solution of the stochastic equation (see any standard text, e.g. Gikhman–Skorokhod). We restrict ourselves to the 'time-invariant' case (the 'coefficients' $F(x)$, $G(x)$ do not depend on the time parameter t) for simplicity. Of course it is assumed that $C(\cdot)$ is Borel measurable, R_m into R_n, and such that

$$\int_0^T E[\|C(x(t,\omega))\|^2]\,\mathrm{d}t<\infty. \tag{6.3}$$

Further we shall assume (again, for simplicity) that

$$E[\|C(x(t,\omega))\|^2]<\infty$$

for every t, $0\leqslant t\leqslant T$. Let next $f(\cdot)$ be any function mapping R_m into R_1 such that $f(\cdot)$ has compact support and C^∞ on R_m. In particular then

$$E[\|f(x(t,\omega))\|]<\infty$$

Consider now

$$\hat{f}(x(t,\cdot))=\frac{\int_{\Omega} f(x(t,\omega))\exp-\frac{1}{2}\left\{\int_0^t \|S(\sigma,\omega)\|^2\,\mathrm{d}\sigma-2\int_0^t [S(\sigma,\omega),y(\sigma)\,\mathrm{d}\sigma\right\}\mathrm{d}p}{\int_{\Omega}\exp-\frac{1}{2}\left\{\int_0^t \|S(\sigma,\omega)\|^2\,\mathrm{d}\sigma-2\int_0^t [S(\sigma,\omega),y(\sigma)]\,\mathrm{d}\sigma\right\}\mathrm{d}p}$$

Fix $y(\cdot)$ in $\mathscr{W}(T)$ in what follows. Let $r(t;x;P(t)y)$ denote for each t, the Baire function over R_m such that:

$$r(t;x(t,\omega);P_t y)$$

$$=\frac{E\left[\exp-\frac{1}{2}\left\{\int_0^t \|S(\sigma,\omega)\|^2\,\mathrm{d}\sigma-2\int_0^t [S(\sigma,\omega),y(\sigma)]\,\mathrm{d}\sigma\right\}|x(t,\omega)\right]}{\int_{\Omega}\exp-\frac{1}{2}\left\{\int_0^t \|S(\sigma,\omega)\|^2\,\mathrm{d}\sigma-2\int_0^t [S(\sigma,\omega),y(\sigma)]\,\mathrm{d}\sigma\right\}\mathrm{d}p}$$

Note that $r(t;x;P(t)y)$ is a physical random variable for each t and x. Let $\mathbf{p}(t;x)$ denote the first order probability density function of the process $x(t,\omega)$, enough assumptions being made to assure this.

Define:

$$\mathbf{p}(t;x;P_t y)=r(t;x;P_t y)\mathbf{p}(t;x).$$

It is natural to call the function on the left the 'conditional density of $x(t,\omega)$ given $y(s)$, $0\leqslant s\leqslant t$,' since

$$\hat{f}(x(t,\cdot))=\int_{R_m} f(x)p(t;x;P_t y)\,\mathrm{d}|x|.$$

We can now proceed to obtain the partial differential equation satisfied by the conditional density, in the same way as in the older literature (see, e.g. R. E. Mortensen [5]) in what was called the 'Stratanovich' sense of stochastic integral. We begin by calculating the difference quotient in direct fashion:

$$\frac{\hat{f}(x(t+\Delta;\cdot))-\hat{f}(x(t,\cdot))}{\Delta}$$

$$=\frac{\displaystyle\int_{\Omega}\frac{f(x(t+\Delta,\omega))-f(x(t,\omega))}{\Delta}\exp-\tfrac{1}{2}\left\{\int_0^{t+\Delta}\cdots\right\}\mathrm{d}p}{\displaystyle\int_{\Omega}\exp-\tfrac{1}{2}\left\{\int_0^{t+\Delta}\cdots\right\}\mathrm{d}p}$$

$$+\int_{\Omega}\frac{f(x(t,\omega))}{\Delta}\left[\frac{\exp-\tfrac{1}{2}\left\{\int_0^{t+\Delta}\cdots\right\}\mathrm{d}p}{\displaystyle\int_{\Omega}\exp-\tfrac{1}{2}\left\{\int_0^{t+\Delta}\cdots\right\}\mathrm{d}p}-\frac{\exp-\tfrac{1}{2}\left\{\int_0^{t}\cdots\right\}\mathrm{d}p}{\displaystyle\int_{\Omega}\exp-\tfrac{1}{2}\left\{\int_0^{t}\cdots\right\}\mathrm{d}p}\right].$$

Now

$$\begin{aligned}&f(x(t+\Delta,\omega))-f(x(t,\omega))\\&\quad=[\nabla f(x(t,\omega)),x(t+\Delta,\omega)-x(t,\omega)]\\&\quad\quad+\tfrac{1}{2}[x(t+\Delta,\omega)-x(t,\omega),H_f(x(t,\omega))(x(t+\Delta,\omega)-x(t,\omega))]\\&\quad\quad+\text{Higher order terms}\end{aligned}$$

(where Δ is the gradient with respect to x, and H_f is the Hessian) and upon substituting for

$$x(t+\Delta,\omega)$$

from the stochastic equation, we get in the usual way, that the first integral yields, as $\Delta\to 0$,

$$=\int_{R_m}(Lf(x))\mathbf{p}(t;x;P(t)y)\,\mathrm{d}|x|$$

$$=\int_{R_m}f(x)(L^*\mathbf{p}(t;x;P(t)y)\,\mathrm{d}|x|,$$

where L is the infinitesimal operator corresponding to the process $x(t,\omega)$

and L^* is (distributional) adjoint (we omit many of the 'fine points' involved here in terms of differential operators etc. See [6,7] for some sufficient conditions. Next, the second term yields in the limit:

$$\frac{\int_\Omega f(x(t,\omega))\left\{\left(-\tfrac{1}{2}\|S(t,\omega)\|^2+[S(t,\omega),y(t)]\right)\exp-\tfrac{1}{2}\left\{\int_0^t\cdots\right\}\mathrm{d}p\right\}}{\int_\Omega \exp-\tfrac{1}{2}\left\{\int_0^t\cdots\right\}\mathrm{d}p}$$

$$\times(-1)\frac{\int_\Omega f(x(t,\omega))\exp-\tfrac{1}{2}\left\{\int_0^t\|S(\sigma,\omega)\|^2\mathrm{d}\sigma-2\int_0^t[S(\sigma,\omega),y(\sigma)]\,\mathrm{d}\sigma\right\}\mathrm{d}p}{\int_\Omega \exp-\tfrac{1}{2}\left\{\int_0^t\cdots\right\}\mathrm{d}p}$$

$$\times\frac{\int_\Omega\left\{-\tfrac{1}{2}\|S(t,\omega)\|^2+[S(t,\omega),y(t)]\right\}\cdot\exp-\tfrac{1}{2}\left\{\int_0^t\cdots\right\}\mathrm{d}p}{\int_\Omega \exp-\tfrac{1}{2}\left\{\int_0^t\cdots\right\}\mathrm{d}p}$$

$$=\int_{R_m} f(x)\left\{+\tfrac{1}{2}\widehat{\|S(t,\omega)\|^2}-[\hat{S}(t),y(t)]\right\}p(t,x,P(t)y)\,\mathrm{d}|x|.$$

Hence

$$\begin{aligned}
&\frac{\mathrm{d}}{\mathrm{d}t}\hat{f}(x(t,\cdot))\\
&\quad=\int f(x)\left\{L^*\mathbf{p}(t;x;P(t)y)\right.\\
&\quad-\tfrac{1}{2}\left(\|C(x)\|^2-\widehat{\|S(t,\omega)\|^2}\right)\mathbf{p}(t;x;P(t)y)\\
&\quad\left.+[C(x)-\hat{S}(t),y(t)]\,\mathbf{p}(t;x;P(t)y)\right\}\mathrm{d}|x|.
\end{aligned}$$

Using

$$P(t;P(t)y)=\widehat{\|S(t,\omega)\|^2}-\|\hat{S}(t)\|^2$$

we get, finally:

$$\frac{\partial}{\partial t}\mathbf{p}(t,x,P(t)y) = L^*\mathbf{p}(t,x,P(t)y) + \left\{\left[C(x) - \hat{S}(t), y(t) - \hat{S}(t)\right]\right.$$

$$\left. - \left(\tfrac{1}{2}\|C(x) - \hat{S}(t)\|^2\right) + \tfrac{1}{2}(P(t;P(t)y))\right\}\mathbf{p}(t,x,P(t)y). \tag{6.4}$$

We may label the last two terms here as 'correction' terms to the Wiener process version (see e.g. Shiryayev [7] for the Wiener process version). Of course, as we have seen in Section 5,

$$y(t) - \hat{S}(t) \qquad 0 < t < T$$

is white noise in $L_2[0,T;H]$. The main point is that if the conditional density is to be calculated on actual observed data $y(\cdot)$, then the Wiener process version has no operational meaning, while (6.4) does. We also note that (6.4) is a stochastic partial differential equation 'bilinear' in the innovation (white noise) forcing term and can be shown to have a 'white noise' solution as in [8].

References

[1] Iliff, K. W. (1973). Dissertation. School of Engineering, University of California at Los Angeles. (UCLA-ENG-7340).

[2] Balakrishnan, A. V. (1976). *Applied Functional Analysis*. Springer–Verlag.

[3] Balakrishnan, A. V. (1977). Radon–Nikodym derivatives of a class of weak distributions on Hilbert spaces. *Journal of Applied Math. and Optimisation* **3**, 209–225.

[4] Balakrishnan, A. V. (1977). Likelihood ratios for signals in additive white Gaussian noise. *Journal of Applied Math. and Optimisation* **3**, 341–356.

[5] Mortensen, R. E. et al (1969). Sequential Processing Techniques for Trajectory Estimation. N.A.S.A. CR-1360.

[6] Rozovsky, B. L. (1975). Stochastic partial differential equations. *Mat. Sb.*. **96**(138):2, 314–341.

[7] Liptser, R. S. and Shiryayev, A. N. (1978). *Statistics of Random Processes*. Springer–Verlag.

[8] Balakrishnan, A. V. (1975). Stochastic partial differential equations. In *Variable Structure Systems* (R. Mohler and A. Ruberti, eds.). Springer–Verlag.

P. R. Krishnaiah, ed., *Multivariate Analysis–V*
© North-Holland Publishing Company (1980) 111–118.

CAUSAL ANALYSIS IN TERMS OF BROWNIAN MOTION

Takeyuki HIDA
Department of Mathematics, Nagoya University, Chikusa-ku, Nagoya, 464, Japan

Functionals of Brownian motion, call them Brownian functionals, are discussed, where development of time is involved. In order to carry out the analysis we are naturally led to introduce a concept of a normal (generalized) Brownian functional and to use partial derivatives with respect to the white noise which is the time derivative of the Brownian motion.

0. Introduction

We discuss functionals of Brownian motion $\{B(t)\}$, which we call simply *Brownian functionals*. It is, in general, easy to deal with by expressing them as functionals of white noise $\{\dot{B}(t)\}$, where $\dot{B}(t)$ is the time derivative of $B(t)$: $\dot{B}(t)=\mathrm{d}B(t)/\mathrm{d}t$. We are often requested to analyze those functionals depending on the time t and even to discuss the development of them as t goes by. For this purpose it is better to take $\{\dot{B}(t)\}$ to be the system of variables of Brownian functionals.

As is well-known a sample function of $\dot{B}(t)$ is not an ordinary function, but it is a generalized function, so that nonlinear functionals of the form $\phi(\dot{B}(t), t\in\mathbf{R})$ is hard to be defined. However, we are able to find a suitable modification so that particular functionals like polynomials in $\dot{B}(t)$'s can be defined rigorously, and we are naturally led to a class of *generalized Brownian functionals*. Our main interest is now to analyze them, where propagation of time is taken into account.

The idea behind our formulation is as follows. To approximate $\{\dot{B}(t)\}$ we take differences $\Delta_k B$, $\{\Delta_k\}$ being a partition of R. The collection $\{\Delta_k B\}$ forms a system of independent Gaussian random variables. We then form polynomials in $\Delta_k B$'s. Any function of $\Delta_k B$'s can be expressed as a linear combination of those polynomials. More concretely, the collection of all Hermite polynomials in $(\Delta_k B/\Delta_k t)$'s forms a complete orthogonal system in the Hilbert space of all functionals of $\Delta_k B$'s with finite variance. Those Hermite polynomials are of the form

$$\prod_k H_{n_k}\left(\frac{\Delta_k B}{\Delta_k t};\frac{1}{|\Delta_k t|}\right) \tag{1}$$

If we let the partition $\{\Delta_k\}$ be finer and finer, then the above system turns out to be the system of generalized random measures introduced in [4], and functionals of $\Delta_k B$'s now become functionals of the $\dot{B}(t)$, $t\in\mathbf{R}$.

By analogy with the expression $f(x_1,\ldots,x_n)$ for functions on $\mathbf{R}^n$ and with partial derivatives such as $\partial/\partial x_i f(x_1,\ldots,x_n)$, we can think of an expression

$$\phi(\dot{B}(t), t\in\mathbf{R}), \tag{2}$$

as mentioned before, and

$$\frac{\partial}{\partial\dot{B}(s)}(\dot{B}(t), t\in\mathbf{R}), \tag{3}$$

although these are formal expressions. If we note that t and s in (2) and (3) stand for the time, we understand that the passage of time is expressed explicitly. In view of this the calculus using formulas like (2) and (3) may be called a *causal analysis*.

Having been inspired by the above observation, we can start our analysis by introducing the probability distribution μ of $\{\dot{B}(t)\}$ on the space of generalized functions, on which the Hilbert space (L^2) is built up. In order to carry out our causal analysis we employ standard tools from basic analysis such as a Reproducing Kernel Hilbert Space and as Sobolev spaces. The Wiener–Itô decomposition of (L^2) plays an important role in our discussion. We shall quickly review the known results in this direction in Sections 1 and 2 (see [2] and [3] for details). Then, also in Section 2, a class of generalized Brownian functionals is introduced. A particular attention should be paid to a subclass consisting of *normal functionals*.

We then, in Section 3, come to the differential calculus with respect to the white noise, which is a development of our results in [3].

We pause finally in the last section to add a few remarks on further developments of our theory. They are concerned with an infinite dimensional Laplacian operator and with convergence of a series of generalized Brownian functionals.

1. Known results on Brownian functionals

The probability distribution μ of the white noise $\{\dot{B}(t); t\in\mathbf{R}\}$ is introduced into the space of generalized functions, say $\mathcal{S}^*$ the space of tempered distributions, by the characteristic functional

$$C(\xi)=\exp\left[-\tfrac{1}{2}\|\xi\|^2\right] \quad \left(=\mathbf{E}\{\exp\left[i\dot{B}(\xi)\right]\}\right)$$
$$\times\, \xi\in\mathcal{S}, \|\ \ \| \textit{ the } L^2(\mathbf{R})\textit{-norm}, \tag{4}$$

in such a way that

$$C(\xi)=\int_{\mathcal{S}^*}\exp\left[i\langle x,\xi\rangle\right]\mathrm{d}\mu(x), \tag{5}$$

where $\dot{B}(\xi)=\langle\dot{B},\xi\rangle$, $\langle\ ,\ \rangle$ being the canonical bilinear form that connects $\mathcal{S}$ and $\mathcal{S}^*$.

With this μ almost all x in $\mathcal{S}^*$ is now viewed as a sample function of $\dot{B}(t)$, $t\in R$. Brownian functionals with finite variance are realized as members of $(L^2)=L^2(\mathcal{S}^*,\mu)$. In particular, $\dot{B}(\xi)$ for a fixed ξ is realized by $\langle x,\xi\rangle$ in (L^2).

In order to visualize those members in (L^2) we have introduced the transformation $\mathcal{T}$ (see, e.g. [2]) defined by

$$(\mathcal{T}\phi)(\xi)=\int_{\mathcal{S}^*}\exp\left[i\langle x,\xi\rangle\right]\phi(x)\,\mathrm{d}\mu(x), \qquad \phi\in(L^2). \tag{6}$$

The collection $\mathcal{F}=\{\mathcal{T}\phi;\phi\in(L^2)\}$ can be topologized so as $\mathcal{F}$ to be isomorphic to the Hilbert space (L^2). Indeed, thus topologized, $\mathcal{F}$ turns out to be the Reproducing Kernel Hilbert Space with kernel $C(\xi-\eta)$, $(\xi,\eta)\in\mathcal{S}\times\mathcal{S}$. Let the kernel $C(\xi-\eta)$ expand into the Taylor series:

$$C(\xi-\eta)=\sum_{n=0}^{\infty}C_n(\xi,\eta), \tag{7}$$

where $C_n(\xi,\eta)=(n!)^{-1}C(\xi)(\xi,\eta)^nC(\eta)$, $(\ ,\ )$ the scalar product in $L^2(\mathbf{R})$. The kernel $C_n(\xi,\eta)$ is again positive definite, so that it defines a subspace $\mathcal{F}_n$ of $\mathcal{F}$. Thus a direct sum decomposition of $\mathcal{F}$ is obtained:

$$\mathcal{F}=\sum_{n=0}^{\infty}\oplus\mathcal{F}_n. \tag{8}$$

Apply the transformation $\mathcal{T}^{-1}$ to both sides and set $\mathcal{T}^{-1}(\mathcal{F}_n)=\mathcal{H}_n$. Then we are given the Wiener–Itô decomposition of (L^2):

$$(L^2)=\sum_{n=0}^{\infty}\oplus\mathcal{H}_n. \tag{9}$$

The subspace $\mathcal{H}_n$ is referred to as the *multiple Wiener integral* of degree n.

The following theorem plays the key role in our causal analysis.

Theorem 1. (i) *For $\phi(x)\in\mathcal{H}_n$ we have the integral representation*

$$(\mathcal{T}\phi)(\xi)=i^nC(\xi)\int\cdots\int_{R^n} F(u_1,\dots,u_n)\xi(u_1)\cdots\xi(u_n)\mathrm{d}u_1\cdots\mathrm{d}u_n, \tag{10}$$

where $F\in(L^2)\hat{}(\mathbf{R}^n)$ the class of symmetric $L^2(\mathbf{R}^n)$-functions, and the map

$$\phi\to F\in(L^2)\hat{}(\mathbf{R}^n),\quad \phi\in\mathcal{H}_n, \tag{11}$$

is one-to-one.

(ii) *Under the relationship established in (i) we have*

$$\|\phi\|_{(L^2)}=(n!)^{\frac{1}{2}}\|F\|_{L^2(\mathbf{R}^n)}. \tag{12}$$

We may therefore say that $\mathcal{F}_n$ (and hence $\mathcal{H}_n$ as well) is isomorphic to the Hilbert space $(L^2)\hat{}(\mathbf{R}^n)$ up to a multiplicative constant $(n!)^{\frac{1}{2}}$:

$$\mathcal{H}_n\cong\mathcal{F}_n\cong(n!)^{\frac{1}{2}}(L^2)\hat{}(\mathbf{R}^n) \tag{13}$$

(In [4] we dropped the multiplicative constant $(n!)^{\frac{1}{2}}$ in the isomorphism relations for simplicity, but we now think it better to put the constant explicitly.)

2. Generalized Brownian functionals and normal functionals

A naïve observation on the causal analysis has been made in the Introduction, and it tells us that the analysis would be viewed as the limit, in a suitable sense, of the analysis of functional of $\Delta_k B$'s, where the system of Hermite polynomials of the form (1) are taken to be a base. Being inspired by such an observation we are led to a generalization of a class of

Brownian functionals. They can actually be defined with the help of the integral representation of members in $\mathcal{H}_n$ established in Theorem 1.

Now, it seems to be convenient to start with a somewhat general set-up for generalization of Brownian functionals by extending the isomorphisms (13). Denote by $H^m(\mathbf{R}^n)$ the Sobolev space of order m over R^n, and set $(H^m)\hat{}(\mathbf{R}^n)=H^m(R^n)\cap(L^2)\hat{}(\mathbf{R}^n)$. Then we can give the following diagram:

$$\begin{array}{ccccccc}
\mathcal{H}_n^{(n)} & \hookrightarrow & \mathcal{H}_n & \hookrightarrow & \mathcal{H}_n^{(-n)} & \\
\updownarrow & & \updownarrow & & \updownarrow & \text{under } \mathcal{T} \\
\mathcal{F}_n^{(n)} & \hookrightarrow & \mathcal{F}_n & \hookrightarrow & \mathcal{F}_n^{(-n)} & \\
\updownarrow & & \updownarrow & & \updownarrow & \text{via (10)} \\
(n!)^{\frac{1}{2}}(H^{(n+1)/2})\hat{}(\mathbf{R}^n) & \hookrightarrow & (n!)^{\frac{1}{2}}(L^2)\hat{}(\mathbf{R}^n) & \hookrightarrow & (n!)^{\frac{1}{2}}(H^{-(n+1)/2})\hat{}(\mathbf{R}^n), &
\end{array}$$

where the vertical double-headed arrow denotes isomorphism.

A few remarks are now in order. The increment ΔB of a Brownian motion is realized in (L^2) by $\langle x,\chi_\Delta\rangle$, χ_Δ being the indicator function of Δ, or by $iC(\xi)\int\chi_\Delta(t)\xi(t)\,dt$ in $\mathcal{F}$. Hence the limit of $\Delta B/\Delta$ as $\Delta\to 0$, Δcontaining t, and that of $\langle x,\chi_\Delta\rangle/\Delta$ should approach $\dot{B}(t)$ and $x(t)$, respectively. They can be dealt with only as generalized functions, while the limit of $iC(\xi)1/\Delta\int\chi_\Delta(t)\xi(t)\,dt$ does exist in the ordinary sense; i.e., $iC(\xi)\xi(t)$ is the limit.

The diagram illustrates that $\mathcal{T}$ may be extended over $\mathcal{H}_n^{(-n)}$ without any difficulty so as to hold the relationship $\mathcal{T}(\mathcal{H}_n^{(-n)})=\mathcal{F}_n^{(-n)}$. Another remark is that, if we let the partition $\{\Delta_k\}$ be finer and finer, the limits of the Hermite polynomials appeared in (1) are expressible formally as

$$\prod_k H_{n_k}(\dot{B}(t_k);1/dt_k), \qquad t_k\text{'s are different,} \tag{14}$$

or in terms of $x\in\mathcal{S}^*$

$$\prod_k H_{n_k}(x(t_k);1/dt_k), \qquad t_k\text{'s are different.} \tag{14'}$$

As is illustrated in [4] they play a role of random measures, and are called *generalized random measures*. The integral of an $L^2(\mathbf{R}^m)$-kernel $F(u_1,\dots,u_m)$ with respect to such a generalized random measure with $\Sigma_k n_k=n$ belongs to the space $\mathcal{H}_n^{(-n)}$. The extended transformation $\mathcal{T}$ carries it to a member of $\mathcal{F}_n^{(-n)}$ expressed in the form

$$(\mathcal{T}\phi)(\xi) = i^n \Big(\prod_j n_j!\Big)^{-1} C(\xi) \int \cdots \int_{\mathbf{R}^m} F(u_1, \ldots, u_m)$$

$$\times \xi(u_1)^{n_1} \cdots \xi(u_m)^{n_m} \mathrm{d}u_1 \cdots \mathrm{d}u_m. \tag{15}$$

(See [4] for details.)

Let $\mathfrak{N}_n$ be the class of all $\mathcal{H}_n^{(-n)}$-functionals that are expressed in the form (15), and set

$$\mathfrak{N} = \sum_{n=0}^{\infty} \mathfrak{N}_n \quad \text{(algebraic sum)}. \tag{16}$$

A member of $\mathfrak{N}$ is called a *normal functional*. If in particular a normal functional belongs to $\mathcal{H}_n$, then it is called *regular*.

Remark. The name a 'normal' functional came from P. Lévy [1]. In fact, after applying $\mathcal{T}$ to a member of $\mathfrak{N}$ we are given a normal functional in Lévy's sense. The same for a regular functional.

3. Differential calculus

The differential operator $\partial/\partial x(t)$ or $\mathbf{D}_t$ can now be defined. (It has been introduced in [2] and [3], where we denote by $\mathrm{d}/\mathrm{d}\dot{B}(t)$.) For $\phi \in \mathfrak{N}_n$ write

$$(\mathcal{T}\phi)(\xi) = i^n C(\xi)\, U(\xi). \tag{17}$$

Take the functional derivative $U'_\xi(t)$ of $U(\xi)$ in the sense of Fréchet (see P. Lévy [1]) and then apply $\mathcal{T}^{-1}$ to obtain

$$\mathrm{D}_t\phi = \frac{\partial}{\partial x(t)}\phi(x) = \mathcal{T}^{-1}\left(i^{n-1} C(\xi)\, U_\xi^1(t)\right) \tag{18}$$

which is to be found in $\mathcal{H}_{n-1}^{(-n+1)}$. It should be noted that the $L^2(\mathbf{R})$-topology is used when we take the functional derivative.

We then define

$$\mathrm{D}_\alpha = \int \alpha(t)\, \mathrm{D}_t \,\mathrm{d}t, \qquad \alpha \in L^\infty(\mathbf{R}). \tag{19}$$

Theorem 2. *The vector space* $\mathfrak{N}$ *is the domain of the differential operator* D_α, $\alpha \in L^\infty(\mathbf{R})$, *and each* D_α *is a mapping of* $\mathfrak{N}_n$ *to* $\mathfrak{N}_{n-1}$. *In particular regular functionals are carried to regular ones.*

Proof. If $U(\xi)$ is expressed in the form

$$U(\xi) = \int_{\mathbf{R}^m}\cdots\int F(u_1,\ldots,u_m)\xi(u_1)^{n_1}\cdots\xi(u_m)^{n_m}\,\mathrm{d}u_1\cdots\mathrm{d}u_m,$$

then we have

$$U'_\xi(t) = \sum_j n_j \int_{\mathbf{R}^{m-1}}\cdots\int F\Big(u_1,\ldots,\overset{j}{t},\ldots,u_m\Big)$$

$$\times\,\xi(u_1)^{n_1}\cdots\xi(t)^{n_j-1}\cdots\xi(u_m)\,\mathrm{d}u_1\cdots\check{\mathrm{d}u_j}\cdots\mathrm{d}u_m,$$

from which follows our assertion.

The differential operators D_t and D_α are infinite dimensional analogue of the partial differential operator $\partial/\partial x_i$ in the analysis on $\mathbf{R}^n$, as is easily seen from above discussions. Recall now that the parameter t of D_t stands for time.

Example. Set

$$\phi(x) = \int f(t)H_2(x(t); 1/\mathrm{d}t)\,\mathrm{d}t, \qquad f \in \mathbf{C},$$

(see [3, §1]). It has interesting meaning in application, specifically in Physics. Now apply D_t to ϕ to obtain

$$\mathrm{D}_t\phi(x) = f(t)x(t).$$

If $\phi(x)$ is thought of as kinetic energy, then $\mathrm{D}_t\phi(x)$ is proportional to the momentum at the instant t.

We can further discuss many important properties of the differential operators just introduced. A systematic approach will be reported elsewhere and is omitted here.

4. Concluding remarks

What has been discussed so far may be said to be a proposed framework of the so-to-speak causal analysis, and the author hopes that many important results will be discovered in this line. At present the following two topics would be worthy of mentioning.

(i) We should like to discuss exponential functionals of polynomial in $x(t)$'s which are quite important in applications. For this purpose we must determine the topology under which the (direct) sum $\sum_{n=0}^{\infty} \mathcal{H}_n^{(-n)}$ can be defined.

(ii) The author has already introduced in [3] the infinite dimensional Laplacian operator which essentially came from the Lévy's Laplacian operator acting on functionals. It is hoped that some probabilistic as well as functional theoretic interpretations would be given to this operator.

These topics have been discussed by I. Kubo and S. Takenaka, discussions with whom the author highly appreciates.

References

[1] Lévy, P. (1951). *Problèmes Concrets d'Analyse Fonctionnelle*. Gauthier-Villars, Paris.

[2] Hida, T. (1975). Analysis of Brownian functionals. Carleton University Lecture Notes **13**, Ottawa. Second edition (1978).

[3] Hida, T. (1976). Analysis of Brownian functionals. *Math. Programming Study* **5**, 53–59.

[4] Hida, T. (1978). Generalized multiple Wiener integrals. *Proc. Japan Academy*, **Ser. A 54** (3), 55–58.

P. R. Krishnaiah, ed., *Multivariate Analysis–V*
© North-Holland Publishing Company (1980) 119–136.

AN INFINITE GEOSTOCHASTIC SYSTEM

D. A. DAWSON*
Carleton University, Ottawa, Canada

In this paper a class of infinite geostochastic models is introduced. These are models of populations subject to reproduction, spatial dispersion and a Verhulst-type crowding effect. The crowding effect is one in which the death rate increases as the local population density increases. When such a population is uniformly distributed at time zero, its limiting behavior is studied and criteria for extinction or for non-trivial steady states are obtained.

1. Introduction

In a previous paper we developed a stochastic calculus for measure-valued Markov processes which live on a compact subset of R^d. In this paper we consider the thermodynamic limit of such processes allowing us to study infinite geostochastic systems. In particular a class of such processes are considered which model spatially distributed populations which are subject to reproduction, spatial dispersion and a local Verhulst-type crowding effect. In addition to establishing the existence and uniqueness of these processes, we consider the stability question for them. In particular we investigate the extinction or the existence of non-trivial steady states when such a system is started with a uniform distribution in R^d.

2. Infinite measure-valued processes

Let $M(R^d)$ denote the family of Borel measures on R^d, $\mathbf{S}'(R^d)$ the space of tempered distributions, $C_K(R^d)$ the space of continuous functions with compact support and $\mathbf{S}(R^d)$ the space of $C^\infty(R^d)$ functions which are rapidly decreasing at infinity. For $S \,\varepsilon\, \mathbf{S}'(R^d)$ $(M(R^d))$, $\phi \varepsilon S(R^d)$ $(C_K(R^d)$, respectively), let $\langle S,\phi\rangle$ denote the action of S on ϕ.

Let $M_T(R^d) \equiv M(R^d) \cap \mathbf{S}'(R^d)$ denote the space of tempered measures on R^d with the topology induced by $\mathbf{S}'(R^d)$. Then $M_T(R^d)$ is a Lusin (or

*Supported by the National Research Council of Canada and the Killam Program of the Canada Council.

standard) space, that is, it possesses a stronger topology in which it is Polish. Let Ω be defined as the space of continuous functions from $[0,\infty)$ into $M_T(R^d)$ and let Ω_R denote the space of real-valued continuous functions on $[0,\infty)$ with the topology of uniform convergence on bounded sets. If Ω is furnished with the weakest topology making the maps $\omega\varepsilon\Omega\to\langle\omega(\cdot),\phi\rangle\ \varepsilon\Omega_R$ continuous for all $\phi\varepsilon\mathbf{S}(R^d)$, then it is also a Lusin space.

In this paper we investigate a class of $M_T(R^d)$-valued continuous Markov processes. The most convenient way to characterize such a process is as the solution to a measure-valued martingale problem on Ω (c.f. Dawson [1] for a more detailed introduction to measure-valued martingale problems). Let $\mathbf{F}$ denote the σ-algebra of Borel subsets of Ω and let

$$X(\cdot,\cdot):[0,\infty)\times\Omega\to M_T(R^d)$$

be defined by

$$X(t,\omega)\equiv\omega(t)\text{ for }t\in[0,\infty)\text{ and }\omega\in\Omega.$$

Let $\mathbf{F}_t\equiv\sigma(X(s):s\leqslant t)$ and let $\Pi(\Omega)$ denote the family of probability measures on Ω.

A martingale problem on Ω is associated with an unbounded linear operator which is defined on a linear subspace $D(L)\subset C_b(M_T(R^d))$, the space of bounded continuous functions on $M_T(R^d)$. Given $(L,D(L))$ the associated martingale problem is to find a mapping $\mu\to P_\mu$ from $M_T(R^d)$ to $\Pi(\Omega)$ such that

$$P_\mu(X(0)=\mu)=1,\text{ and} \tag{2.1.a}$$

for every $\psi\in D(L)$,

$$\psi(X(t))-\int_0^t L\psi(X(s))ds$$

is a P_μ-martingale for every $\mu\in M_T(R^d)$. (2.1.b)

We now formulate a special class of martingale problems on Ω which model distributed populations which are subject to spatial dispersion, reproduction and competition. Let Δ denote the Laplacian operator on R^d and let

$$F:M_T(R^d)\to M_T(R^d)$$

be a continuous mapping. Consider the $M_T(R^d)$-valued stochastic process, $X(t)$, such that

$$\text{for each } \phi \in \mathbf{S}(R^d) \cap D(\Delta),$$
$$\langle X_{\Delta,F}(t), \phi \rangle \text{ is a } P_\mu\text{-continuous martingale,}$$
$$\text{where } X_{\Delta,F}(t) \equiv X(t) - \int_0^t (\Delta X(s) + F(X(s)))\mathrm{d}s. \tag{2.2}$$

Furthermore, we assume that the increasing process associated with $\langle X_{\Delta,F}(t), \phi \rangle$ is given by

$$\langle\langle \phi, X_{\Delta,F}, \phi \rangle\rangle_t = \int \int_0^t \phi^2(x) X(s, \mathrm{d}x)\mathrm{d}s. \tag{2.3}$$

(2.2) together with (2.3) defines a martingale problem on Ω associated with the linear operator

$$L\psi(\mu) = \langle \psi'(\mu), \Delta\mu + F(\mu) \rangle + (\gamma/2)\langle \mu_D, \psi''(\mu) \rangle \tag{2.4}$$

where ψ', ψ'' denote the first and second Fréchet derivatives, and where μ_D denotes the measure on $R^d \times R^d$ defined by $\mu_D(A \times B) = \mu(A \cap B)$, and $D(L)$ is the algebra of functions containing all functions of the form $\psi(\langle X(t), \phi_1 \rangle, \ldots, \langle X(t), \phi_n \rangle)$ with ψ a polynomial of n variables and $\phi_1, \ldots, \phi_n \in \mathbf{S}(R^d)$.

Theorem 2.1. *Consider the martingale problem defined by* (2.1) *and* (2.4) *with* $F(\mu) \equiv \nu_0 + a\mu$ *and* $\nu_0 \in M_T(R^d)$. *Then there is a unique solution* $\{P\mu : \mu \in M_T(R^d)\}$ *which is a Markov process with transition characteristic functional*

$$L_{t,\nu}(\phi) = \exp\left(i < \nu, U_t\phi > + i \int_0^t \langle \nu_0, U_{t-u}\phi \rangle \mathrm{d}u \right) \tag{2.5}$$

where $\{U_t : t \geqslant 0\}$ *is a semigroup of nonlinear operators on* $\mathbf{S}^C(R^d)$. *If* $u(t,x) \equiv U_t\phi(x)$, *then the semigroup is obtained from the nonlinear initial value problem*

$$\partial u(t,x)/\partial t = \Delta u(t,x) + au(t,x) + i\gamma u^2(t,x),$$
$$t \geqslant 0, \quad u(0,x) = \phi(x). \tag{2.6}$$

Proof. The proof follows in essentially the same way as that of Theorem 3.1 of [1].

We next consider the above martingale problem modified by the addition of a nonlinear competition term. For this we take

$$F_g(\mu)=\nu_0-f_g(\mu)\mu+a\mu \tag{2.7.a}$$

$$f_g(\mu;x)\equiv bg(x)(\mu*\xi)(x) \tag{2.7.b}$$

where $*$ denotes the operation of convolution and where $b>0$, $0\leqslant g\leqslant 1$, $g\in C_K(R^d)$ and ξ is the indicator function of a sphere of radius R_0 centered at the origin. Then the Cameron–Martin–Girsanov formula of [1] for measure-valued martingale problems on compact subsets of R^d can be extended to obtain the following result.

Theorem 2.2. *Consider the $M_T(R^d)$-valued martingale problem given by (2.1), $a>0$ and F given by (2.7). It has a unique solution $\{P_\mu^g:\mu\in M_T(R^d)\}$. Furthermore,*

$$P_\mu^g\big|_{\mathbf{F}_t}\ll P_\mu^0\big|_{\mathbf{F}_t},$$

and the Radon–Nikodym derivative

$$R_g(t)\equiv \mathrm{d}P_\mu^g\big|_{\mathbf{F}_t}/\mathrm{d}P_\mu^0\big|_{\mathbf{F}_t}$$

is given by

$$R_g(t)=\exp\left(\int_0^t\langle f_g(X(s)),\mathrm{d}X_0(s)\rangle-\tfrac{1}{2}\int_0^t f_g^2(X(s);x)X(s,\mathrm{d}x)\mathrm{d}s\right) \tag{2.8}$$

where

$$X_0(t)=X(t)-\int_0^t(\Delta+a)X(s)\mathrm{d}s-t\nu_0.$$

Proof. Once again the proof is essentially the same as that of Theorem 5.1 and Proposition 5.3 of [1].

The function g serves as a spatial cutoff guaranteeing that the nonlinear interaction is confined to a compact subset of R^d. We next consider the limit as $g\uparrow 1$; this is analogous to the thermodynamic limit in statistical physics. Note that the limiting measures are no longer absolutely continuous with respect to $\{P_\mu^0 : \mu \in M_T(R^d)\}$ in the global sense; however they still are in a local sense to be made precise below.

Theorem 2.3. *As* $g \uparrow 1$ *(pointwise) the system of measures* $\{P_\mu^g : \mu \in M_T(R^d)\}$ *converges weakly to the system of measures* $\{P_\mu^* : \mu \in M_T(R^d)\}$ *on* Ω *which is the unique solution to the martingale problem defined by* (2.1), (2.4) *and* (2.7) *with* $g=1$.

Proof. Given $D \subset R^d$ let $\mathbf{F}_{D,t} \equiv \sigma\{X(s,B) : s \leqslant t,\ B \subset D\}$ and let $E_{D,t} \equiv E_\mu(\cdot|\mathbf{F}_{D,t})$, the conditional expectation with respect to the $\mathbf{F}_{D,t}$. Let

$$R_g^D(t) \equiv E_{D,t}(R_g(t)). \tag{2.9}$$

Note that

$$R_g^D(t) = \mathrm{d}P_\mu^g|_{\mathbf{F}_{D,t}} / \mathrm{d}P_\mu^0|_{\mathbf{F}_{D,t}}.$$

We show below that for each relatively compact open set $D \subset R^d$,

$$\begin{aligned} &R_g^D(t) \text{ converges a.s. as } g\uparrow 1 \text{ to a } \mathbf{F}_{D,t}\text{-measurable} \\ &\text{function } R_*^D(t) \text{ satisfying } E_\mu\big(R_*^D(t)\big) = 1. \end{aligned} \tag{2.10}$$

Furthermore, if $s \leqslant t$ and $D_1 \subset D_2$, then

$$E_\mu\big(R_*^{D_2}(t)|\mathbf{F}_{D_1,s}\big) = R_*^{D_1}(s), \text{ a.s.}$$

Thus by the Kolmogorov extension theorem, the $\{R_*^D(t)\}$ define unique probability measures on Ω, P_μ^*, where Ω is the space of continuous functions from $[0,\infty)$ into $M(R^d)$. To show that the measures P_μ^* are actually concentrated on $M_T(R^d)$, it suffices to note that if $\phi_n \in \mathbf{S}^+(R^d)$, $\phi_n \to 0$ in $\mathbf{S}(R^d)$ implies that

$$P_\mu^*\Big(\sup_{s\leqslant t} |\langle X(t),\phi_n\rangle| \to 0\Big) \geqslant P_\mu^0\Big(\sup_{s\leqslant t} |\langle X(t),\phi_n\rangle| \to 0\Big) = 1.$$

Now let

$$X_g(t) \equiv X(t) - \int_0^t (\Delta + a)X(s)\mathrm{d}s - tv_0 - \int_0^t F_g(X(s))X(s)\mathrm{d}s.$$

If $\mathrm{spt}(\phi) \subset K$, compact, and if $g=1$ on K, then $\langle X_g(t),\phi\rangle = \langle X_1(t),\phi\rangle$ is a

P_μ^g and a $R_g^K P_\mu^0$-martingale with increasing process given by (2.3). But since the process is concentrated on $M_T(R^d)$, this can be extended to be true for $\phi \in \mathbf{S}(R^d)$ by continuity. Thus the system $\{P_\mu^* : \mu \in M_T(R^d)\}$ solves the $g=1$ martingale problem.

It thus remains to prove (2.10). Let D denote a relatively compact subset, D^C, the complement of D, such that if $x \in D$, then the spt($\xi(\cdot - x)$) $\supset D \cup \partial D$. Then

$$R_g^D(t) = E_\mu\big(E_{D \cup \partial D, t}(R_{g \cdot 1_D}(t)) E_{D \cup \partial D, t}(R_{g \cdot 1_D C}(t)) | \mathbf{F}_{D,t}\big). \qquad (2.11)$$

Noting that $E_{D \cup \partial D, t}(R_{g \cdot 1_D}(t)) P_\mu|_{\mathbf{F}_{D,t}}$ is independent of g as long as $g=1$ on $D \cup \partial D$ and it is continuous with respect to $X(s)|_{\mathbf{F}_{\partial D}}$, then it suffices to show that

$$E_{D \cup \partial D, t}(R_{g \cdot 1_D C}(t)) P_\mu|_{\mathbf{F}_{\partial D, t}}$$

converges in law as $g \uparrow 1$.

Let $0 \leqslant t_1 \leqslant \cdots \leqslant t_n \leqslant t$ and $B_1, \ldots, B_n \subset \partial D$. We next obtain upper and lower bounds for

$$P_\mu^{g_r}(X(t_1, B_1) \geqslant x_1, \ldots, X(t_n, B_n) \geqslant x_n)$$

where it is assumed that $g_r(x) = 1$ if $x \in S^r$, a sphere of radius r centered at the origin. We accomplish this by constructing bounding measures $P_\mu^{r^+}$ and $P_\mu^{r^-}$. These are obtained from models in which the mass is of two types: regular or special. All mass immediately changes to the special type if it enters the region S_r^C and once it is special it remains special until its death. We assume that the special type of mass evolves according to the martingale problem with $g=0$. To construct the measure $P_\mu^{r^+}$ we assume that the two types of mass do not interact and $P_\mu^{r^+}$ is the law corresponding to the total mass of the two types. To obtain $P_\mu^{r^-}$ we assume that the special type acts to kill the regular type, that is, we use the total mass of the two types in the nonlinear death term $f(X_S(t) + X_R(t))$ for the regular mass. Furthermore $P_\mu^{r^-}$ is the law of the mass of the subsequent regular mass. Then it is easy to verify that

$$\begin{aligned} &P_\mu^{r^-}(X(t_1, B_1) \geqslant x_1, \ldots, X(t_n, B_n) \geqslant x_n) \\ &\leqslant P_\mu^{g_r}(X(t_1, B_1) \geqslant x_1, \ldots, X(t_n, B_n) \geqslant x_n) \\ &\leqslant P_\mu^{r^+}(X(t_1, B_1) \geqslant x_1, \ldots, X(t_n, B_n) \geqslant x_n). \end{aligned}$$

Now we can choose r_1 and $r_2 > r_1$ so that

$$P_{\mu|S^C_{r_1}}\Big(\sup_{s\leqslant t} X(s, D\cup\partial D)>\varepsilon\Big)<\varepsilon$$

and

$$P_{\mu|S^C_{r_2}}\Big(\sup_{s\leqslant t} X(s, S_{r_1})>\varepsilon\Big)<\varepsilon.$$

Then letting $X_{R,2}(t)$ denote the r_2-regular mass, X^+ the regular plus the special and X^- the r_1-regular minus that killed in interaction with the r_2-special type we obtain the following results.

If $s\leqslant t$ and $B\subset D\cup\partial D$, then $X^+(s,B)\leqslant X_{R,2}(s,B)+\varepsilon$, and $X^-(s,B)\equiv X_{R,1}(s,B)-\varepsilon b\int_0^s X_{R,1}(u,S_{r_1})\mathrm{d}u$. Thus

$$\sup_{\substack{s\leqslant t\\ B\subset D\cup\partial D}} |X^+(s,B)-X^-(s,B)|\to 0 \text{ in probability as } r_1\to\infty.$$

Hence,

$$\lim_{r\to\infty} |P^{r^+}_\mu(A)-P^{r^-}_\mu(A)|=0$$

where $A=\{X(t_1,B_1)\geqslant x_1,\ldots,X(t_n,B_n)\geqslant x_n\}$. This completes the proof of (2.10).

The same argument can be slightly modified to show that the limiting martingale problem has a unique solution.

Corollary. *Let $m_n(t;\mathrm{d}x_1,\ldots,\mathrm{d}x_n)$ denote the nth moment measures of the solution for the $g=1$ martingale problem. Then if ν_0 is absolutely continuous and if the initial measure is $\lambda\cdot$ Lebesgue, then moment densities exist and satisfy the system of equations*

$$\begin{aligned}\partial m_n(t;x_1,\ldots,x_n)/\partial t \qquad\qquad (2.12)\\ =\sum_{j=1}^n m_{n-1}(t;x_1,\ldots,x_{j-1},x_{j+1},\ldots,x_n)\nu_0(x_j)\\ +nam_n(t;x_1,\ldots,x_n)+\sum_{j=1}^n \Delta_{x_j} m_n(t;x_1,\ldots,x_n)\\ +(\gamma/2)\sum_{\substack{i,j\\ i\neq j}} m_{n-1}(t;x_1,\ldots,x_{j-1},x_{j+1},\ldots,x_n)\delta_{x_i-x_j}\\ -b\sum_{j=1}^n I^y_{x_j} m_{n+1}(t;x_1,\ldots,x_n,y),\end{aligned}$$

where Δ_{x_i} *refers to the action of the Laplacian on the variable* x_i *and*

$$I_{x_j}^y m_{n+1}(t;x_1,\ldots,x_n,y)=\int m_{n+1}(t,x_1,\ldots,x_n,y)\xi(y-x_j)\mathrm{d}y.$$

The initial condition is simply $m_n(0;x_1,\ldots,x_n)=\lambda^n$.

Proof. First note that the interaction moment measures are dominated by those of the interaction free moment measures. Thus the moment measures exist, moment densities exist and the moments of $X(t,K)$, K compact, are all finite. The system of equations is then obtained by applying the martingale equation (2.1) to functions of the form

$$\prod_{j=1}^{n}\langle X(t),\phi_j\rangle.$$

The solution of the $g=1$ martingale problem with spatially homogeneous initial measure will be called the geostochastic logistic model. It should be noted that for every $t>0$, the random measure $X(t)$ will then also be spatially homogeneous. The behavior of this system as $t\to\infty$ will be studied in the following sections.

3. The nonlinear Markov approximation

In the case $\nu_0=0$ the geostochastic logistic model always has one steady state solution, $X(t)=0$ for all t. If $\nu_0>0$, then we expect a nonzero steady state. The main problem to be investigated in the remainder of this paper is the question of the existence of a non-trivial steady state in the case $\nu_0=0$ for the geostochastic logistic model. This problem is equivalent to the extinction problem for the same model when the initial measure is a multiple of Lebesgue measure. This type of problem is closely related to a whole class of similar problems which arise in statistical physics and chemical kinetics, namely, the problem of cooperative behavior and phase transitions. One approximation method which has been employed is the mean field theory approximation which gives rise to a nonlinear 'master equation' (c.f. Nicolis and Prigogine [3]). In this approach a small distinguished volume is studied. It exchanges mass with the rest of the system and it is this exchange process which is simplified in the nonlinear master equation approach. Specifically, the flow into the distinguished volume is assumed to depend only on its expected value thus decoupling the external and internal fluctuations. For this reason this approach is known as the 'mean field theory' approach. The key assumption is that the fluctuations in the flow into the distinguished volume are statistically independent of the process occurring within this volume. In this section we study a nonlinear Markov process which arises from this type of approximate or simplified model of the geostochastic logistic model. The methods introduced are applicable to a wide range of problems.

For simplicity we consider the distinguished volume to be a sphere of radius $2R_0$, thus the same order of magnitude as the range of the interaction. In the model studied in this section all spatial structure is ignored and we consider the evolution of the total mass of the distinguished volume. Although errors are therefore introduced it is reasonable to expect this model to have the same qualitative behavior as the spatially distributed system. This latter conjecture will be discussed in the following section where the spatially distributed system is studied in more detail. Let $x(t)$, $t \geqslant 0$, denote the mass at time t. Then $x(t)$ satisfies the stochastic differential equation

$$\mathrm{d}x(t) = e(t)\mathrm{d}t + ax(t)\mathrm{d}t - bx^2(t)\mathrm{d}t + (\gamma x(t))^{\frac{1}{2}}\mathrm{d}w(t), \tag{3.1}$$
$$x(0) = x_0 \geqslant 0,$$

where $w(t)$, $t \geqslant 0$, denotes a one dimensional Wiener process and $e(t)$ denotes the mass flowing in from the outside, that is, the immigration term. Hence according to the mean field approximation,

$$e(t) = \rho m_1(t),\ \rho > 0,$$

where

$$m_n(t) \equiv E_{x_0}(x^n(t)).$$

The moment equations for the resulting nonlinear system satisfy

$$\mathrm{d}m_1(t)/\mathrm{d}t = \rho m_1(t) + am_1(t) - bm_2(t), \tag{3.2}$$
$$\mathrm{d}m_n(t)/\mathrm{d}t = \left(n\rho m_1(t) + \tfrac{1}{2}n(n-1)\gamma\right)m_{n-1}(t)$$
$$+ anm_n(t) - bnm_{n+1}(t), \qquad n \geqslant 2,$$
$$m_n(0) = x_0^n.$$

On the other hand if we consider a fixed input $e(t)=e$, then the forward Kolmogorov equation (Fokker–Planck) is

$$\partial p(t,x)/\partial t$$
$$= -\partial/\partial x\left[(e+ax-bx^2)p(t,x)\right]+(\gamma/2)\partial^2/\partial x^2(xp(t,x)), \quad (3.3)$$

$$p(0,x)=\delta(x-x_0), \text{ the Dirac delta function,}$$

where $p(t,x)$ is the probability density function of $x(t)$ at time t.

The steady state probability density function is given by

$$\mathbf{p}_e(x)=c\,x^{(2e/\gamma)-1}\exp\left(-b(x-a/b)^2/\gamma\right),\ x\geqslant 0,\ e>0, \quad (3.4)$$

and c is a normalizing constant.

Given the steady state solution, the steady state mean value is given by

$$m(e)=\int_0^\infty xp_e(x)\mathrm{d}x, \qquad e>0, \quad (3.5)$$

$$m(0)=0.$$

Note that in the case $e=0$, the steady state corresponds to $x(t)=0$ for all t. This is because 0 is an accessible boundary for the corresponding one dimensional diffusion and the system suffers extinction with probability one.

However if we consider the mean field theory model, that is, the nonlinear Markov process with $e(t)=\rho m_1(t)$, then it is possible to obtain both the degenerate steady state and a non-zero steady state. In this case the system has two 'phases', the extinct phase and the populated phase. This fact is an immediate consequence of the following theorem.

Theorem 3.1. *For γ sufficiently small (or ρ sufficiently large) the equation*

$$e=\rho m(e) \quad (3.6)$$

has a non-zero root. For γ sufficiently large (or ρ sufficiently small), the equation (3.6) *has only one root, $e=0$.*

Proof. We prove the theorem in the case $\rho=1$; the proof in the other cases is essentially the same.

The first moment equation is given by

$$dm_1(t)/dt = e + am_1(t) - bm_1^2(t) - bv(t) \tag{3.7}$$

where $v(t)$ denotes the variance at time t. Then,

$$m(e) \leqslant \left(a + (a^2 + 4be)^{\frac{1}{2}}\right)/2b \tag{3.8}$$

and hence

$$m(e) < e \qquad \text{for } e \geqslant e_0(a,b). \tag{3.9}$$

But for sufficiently small γ and $e = \gamma/2$, $p_e(x)$ is approximately Gaussian with mean a/b and variance $\gamma/2$. Hence for sufficiently small γ and $e_0 = \gamma/2$,

$$m(e_0) > e_0. \tag{3.10}$$

Noting that the function $m(e)$ is continuous on $(0, \infty)$, the first result then follows from the intermediate value theorem.

On the other hand for sufficiently large γ, $p_e(x)$ can be stochastically dominated by

$$\begin{aligned} p(x) &= cc_1 x^{(2e/\gamma)-1}, \qquad 0 < x \leqslant \gamma^{\frac{1}{2}} a/b, \quad c_1 < 1 \\ &= c\left(\gamma^{\frac{1}{2}} a/b\right)^{(2e/\gamma)-1} \exp(-b(x - a/b)^2/\gamma), \qquad x \geqslant \gamma^{\frac{1}{2}} a/b, \end{aligned}$$

where c is a normalizing constant. But then it is easy to verify that

$$m_A(e) = \int xp(x)dx = 0\left(\gamma^{-\frac{1}{2}}\right)$$

and

$$\sup_{e \leqslant e_0(a,b)} |dm_A(e)/de| \to 0 \quad \text{as } \gamma \to \infty.$$

Hence we can choose γ such that $m_A(e) < e$ for $0 < e \leqslant e_0(a,b)$ and hence

$$m(e) < e \qquad \text{for all } e > 0.$$

and the proof of the second assertion is complete.

We can interpret this result in another way. The total positive feedback term is given by $e(t)+ax(t)$. Then a nondegenerate steady state is possible only if the positive feedback signal is sufficiently weakly correlated with $x(t)$.

4. The method of cumulant inequalities

In many applications (c.f. Nicolis and Prigogine [3]) the system of moment equations forms an infinite hierarchy of the form (2.12). In order to solve such a system various truncation methods have been introduced. For example, one of these is the cumulant neglect method in which all cumulants whose order is equal to or larger than a given order are set identically equal to zero and the resulting finite system of equations is solved. For many problems this method produces reasonable results; however there is no rigorous justification for it. In this section we exploit certain cumulant inequalities in order to obtain bounds on the moments. This method will be applied to the extinction problem for the spatially distributed geostochastic logistic model. In particular using it we establish the non-extinction for the geostochastic model for sufficiently small γ. The method can potentially be used for a larger class of problems; however we will illustrate the idea by outlining the main steps in this simple case.

We begin by applying the method to the simplified model introduced in Section 3 where the qualitative behavior has already been obtained from the exact solution. Consider the nonlinear Markov process described by (3.1) where

$$e(t)=(1-\rho)a_0m_1(t) \quad \text{and} \quad a=\rho a_0, \qquad 0<\rho<1.$$

If we let $v(t)\equiv\mathrm{Var}(x(t))$ and $K_3(t)$ denote the third cumulant of $x(t)$, then from (3.2) it follows that

$$\mathrm{d}m_1(t)/\mathrm{d}t=a_0m_1(t)-bm_1(t)^2-bv(t), \tag{4.1.a}$$

$$\mathrm{d}v(t)/\mathrm{d}t=2\rho a_0v(t)-4bm_1(t)v(t)-2bK_3(t)+\gamma m_1(t). \tag{4.1.b}$$

Recall that

$$K_3(t)=E\big((x(t)-m_1(t))^3\big).$$

But since $x(t)\geqslant 0$, $x(t)-m_1(t)\geqslant -m_1(t)$. Hence

$$K_3(t)\geqslant -m_1(t)E\big((x(t)-m_1(t))^2\big)=-m_1(t)v(t). \tag{4.2}$$

Thus since $b>0$,

$$dv(t)/dt \leqslant 2\rho a_0 v(t) - 2bm_1(t)v(t) + \gamma m_1(t). \tag{4.3}$$

Now consider the solution of the pair of equations

$$d\hat{m}(t)/dt = a_0\hat{m}(t) - b\hat{m}(t)^2 - b\hat{v}(t), \tag{4.4.a}$$

$$d\hat{v}(t)/dt = 2\rho a_0\hat{v}(t) - 2b\hat{m}(t)\hat{v}(t) + \gamma a_0/b, \tag{4.4.b}$$

with initial conditions

$$\hat{m}(0) = \hat{m}_1(0) \leqslant a_0/b,\ \hat{v}(0) = v(0) = 0. \tag{4.4.c}$$

Proposition 4.1.
(i) *For sufficiently small γ the system (4.4) has a fixed point $(\hat{m}_0, \hat{v}_0)$ which is asymptotically stable.*
(ii) *If $m_1(0) \leqslant \hat{m}_0$, then $m(t) \geqslant \hat{m}(t)$ and $v(t) \leqslant \hat{v}(t)$ for all $t \geqslant 0$.*

Proof. The steady state is given by

$$\hat{v}_0 = \gamma a_0/((2b\hat{m}_0 - 2\rho a_0)b) \tag{4.5}$$

$$\hat{m}_0 = \left(a_0 + \left(a_0^2 - 4b^2\hat{v}_0\right)^{\frac{1}{2}}\right)/2b.$$

It is easy to verify that these equations have a solution for sufficiently small γ which is an asymptotically stable fixed point of the system (4.4). This can best be seen from the phase portrait (Fig. 1) for small γ (and or sufficiently small ρ).

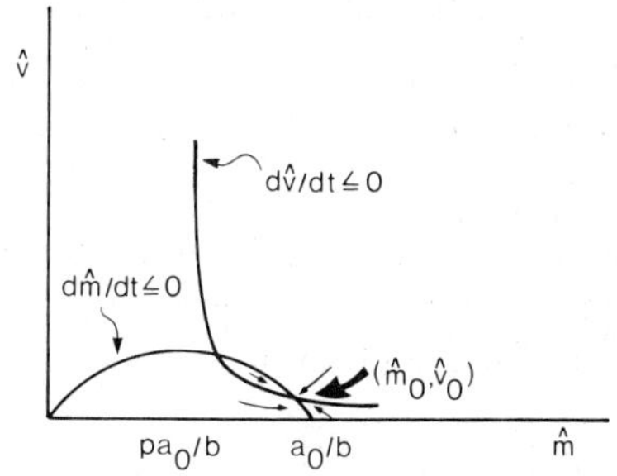

Fig. 1.

To prove (ii) we proceed as follows. Clearly, $\mathrm{d}\hat{m}(0)/\mathrm{d}t \leqslant \mathrm{d}m_1(0)/\mathrm{d}t$ and $\mathrm{d}\hat{v}(0)/\mathrm{d}t \geqslant \mathrm{d}v_1(0)/\mathrm{d}t$. Hence there exists $\varepsilon > 0$ such that

$$\hat{m}(t) \leqslant m_1(t) \quad \text{and} \quad \hat{v}(t) \geqslant v(t) \qquad \text{for } 0 < t < \varepsilon.$$

Now let $t_0 \equiv \{\inf t : \hat{m}(t) > m_1(t) \text{ or } \hat{v}(t) < v(t)\}$ or ∞ if the corresponding set is empty. But if $\hat{m}(t_0) = m_1(t_0)$ and $\hat{v}(t_0) > v(t_0)$, then $\mathrm{d}\hat{m}(t_0)/\mathrm{d}t < \mathrm{d}m_1(t_0)/\mathrm{d}t$; if $\hat{v}(t_0) = v(t_0)$ and $\hat{m}(t_0) < m_1(t_0)$, then $\mathrm{d}\hat{v}(t_0)/\mathrm{d}t > \mathrm{d}v(t_0)/\mathrm{d}t$; and if $\hat{m}(t_0) = m_1(t_0)$ and $\hat{v}(t_0) = v(t_0)$, then $\mathrm{d}\hat{v}(t_0)/\mathrm{d}t > \mathrm{d}v(t_0)/\mathrm{d}t$, $\mathrm{d}\hat{m}(t_0)/\mathrm{d}t = \mathrm{d}m_1(t_0)/\mathrm{d}t$ and $\mathrm{d}^2\hat{m}(t_0)/\mathrm{d}t^2 < \mathrm{d}^2 m_1(t_0)/\mathrm{d}t^2$. Thus if $t_0 < \infty$, a contradiction is obtained. Thus the proof of (ii) is complete.

Corollary. *For sufficiently small γ, the nonlinear stochastic system does not experience ultimate extinction.*

Remark 4.1. If the cumulant neglect method ($K_3 = 0$) is applied to the $\rho = 1$ case, a 'false' asymptotically stable fixed point is obtained. This shows that qualitatively incorrect conclusions can be drawn from the cumulant neglect method.

We now return to the spatially homogeneous geostochastic logistic model described in Section 2.

Proposition 4.2. *Consider the spatially homogeneous geostochastic logistic model constructed in Theorem 2.3. Then for sufficiently small γ, the system does not go to extinction, that is, for every compact set K of positive Lebesgue measure, there exists $\varepsilon > 0$ such that*

$$\lim_{t \to \infty} P(X(t,K) > \varepsilon) > 0.$$

Outline of proof. From the moment equations (2.12) we obtain

$$\mathrm{d}m_1(t)/\mathrm{d}t = a m_1(t) - b I_x^0 v(t,x) - b V m_1(t)^2 \tag{4.6.a}$$

where $m_1(t)$ is the density of the spatially homogeneous process $X(t)$ and $v(t,x)$ is the covariance density ($v(t,x) \equiv m_2(t;y,x+y) - m_1(t)^2$). $V \equiv c_d R_0^d$, the volume of a sphere of radius R_0, and

$$I_y^x v(t,y) \equiv \int_{\|y-x\| \leqslant R_0} v(t,y)\,\mathrm{d}y.$$

Similarly,

$$\begin{aligned}\partial v(t,x)/\partial t = {} & 2av(t,x) + 2\Delta v(t,x) \\ & -2bVm_1(t)v(t,x) - 2m_1(t)bI_y^x v(t,y) \\ & -2bI_z^x K_3(t;0,x,z) + \gamma m_1(t)\delta(x), \end{aligned} \tag{4.6.b}$$

where $\delta(x)$ is the Dirac δ-function and $K_3(t;0,x,z)$ is the third cumulant density function at time t.

A argument similar to the proof of (4.2) shows that

$$\begin{aligned}K_3(t;0,x,z) & \geqslant -m_1(t)\min(v(t,x),v(t,z)) \\ & \geqslant -m_1(t)v(t;x\vee z)\end{aligned} \tag{4.7}$$

where

$$\begin{aligned}x\vee z & = z \text{ if } \|z\| > \|x\| \\ & = x \text{ if } \|x\| \geqslant \|z\|.\end{aligned}$$

Hence (4.6.b) becomes the differential inequality

$$\partial v(t,x)/\partial t \leqslant A(m_1(t))v(t,x) + \gamma m_1(t)\delta(x) \tag{4.8}$$

where

$$A(m)u(x) \equiv 2(a - bVm)u + 2\Delta u + 2mbD(u;x) \tag{4.9}$$

and

$$D(u;x) \equiv \int_{\substack{\|z-x\| \leqslant R_0 \\ \|z\| < \|x\|}} (u(x) - u(z))\,\mathrm{d}z \tag{4.10}$$

Noting that in the case being considered, $u(x)$ is spherically symmetric, and letting $r = \|x\|$, we obtain

$$\Delta u(r) = \partial^2 u/\partial r^2 + ((\mathrm{d}-1)/r)\partial u/\partial r,$$

and

$$D(u;r) = \int_{r-R_0}^{r} (\partial u(s)/\partial s) g_d(r;s)\,\mathrm{d}s.$$

For example, if $d=1$, then

$$g_1(r;s)=s-(r-R_0)+(r-R_0)\wedge 0 \qquad \text{for } r-R_0\leqslant s\leqslant r,$$

$$=0, \qquad \text{otherwise.}$$

For simplicity we restrict our attention to the case $d=1$ for the remainder of the proof; the general case is obtained by an appropriate modification of this argument.

In order to obtain the required result it clearly suffices to show that the system given by (4.6.a) and (4.8) has an asymptotically stable pair $(m_0, v_0(\cdot))$ such that

$$m_1(t)\geqslant m_0,$$

and

$$v(t,x)\leqslant v_0(x) \qquad \text{for all } x \text{ and } t.$$

In order to emphasize the main idea behind this method, we first look at the same problem for a slightly simpler system of differential inequalities.

Consider the system

$$dm_1(t)/dt = am_1(t) - bI_x^0 v(t,x) - bVm_1(t)^2, \tag{4.11.a}$$

$$\partial v(t,x)/\partial t \leqslant 2(a-bVm_1(t))v(t,x)+2\partial^2 v(t,x)/\partial x^2$$
$$+\tfrac{1}{2}m_1(t)bVR_0 g(x)(\partial v(t,x)/\partial x)+\gamma m_1(t)\delta(x) \tag{4.11.b}$$

where here

$$g(x)=+1 \qquad \text{if } x\geqslant 0$$
$$=-1 \qquad \text{if } x\leqslant 0.$$

The system (4.11) has a steady state $(m_0, v_0(\cdot))$ if the inequality (4.11.b) is replaced by an equality and it is given by:

$$v_0(x)=A\exp(-k_0|x|) \tag{4.12.a}$$

where k_0 is the largest root of the equation

$$2(a-bVm_0)+2k^2-\tfrac{1}{2}m_0 bVR_0 k=0, \tag{4.12.b}$$

and

$$A=\tfrac{1}{2}\gamma m_0/k_0, \tag{4.12.c}$$

and

$$m_0=(a-b\gamma(1-\exp(-k_0R_0))k_0^{-2})/bV. \tag{4.12.d}$$

Note that for fixed m_0, (4.12.b) has 2 real roots if and only if

$$m_0^2b^2V^2R_0^2-64(a-bVm_0)\geqslant 0. \tag{4.13}$$

It is easy to verify that a pair (m_0,k_0) satisfying the above equations exists for sufficiently small γ. Furthermore this steady state is stable with respect to perturbations which are dominated by an exponential function $B.\exp(-h|x|)$ for h larger than the smallest root of (4.12.b). One can also note that such a solution does not exist if γ is sufficiently large.

Noting that a maximum principle holds for (4.11.b), it follows that $m_1(t)\geqslant m_0$, $v(t,x)\leqslant v_0(x)$ for all t and x.

Let us now return to the system given by (4.6.a) and (4.8). If the inequality (4.8) is replaced by an equality, then there is a steady state solution $(m_0,v_0(\cdot))$ for sufficiently small γ which is stable with respect to perturbations of the form considered above. (Note that $v_0(\cdot)$ has the same exponential form except in a neighborhood of the origin.) The only new difficulty is that the classical maximum principle does not hold for this system. However letting

$$w(t,x)\equiv v_0(x)-v(t,x),$$

we have

$$\partial w(t,x)/\partial t=A(m_0)w(t,x)+f(t,x)$$

where $w(0,x)\geqslant 0$, and $f(t,x)\geqslant 0$. Then it can be shown that

$$\partial/\partial t\left(\int w(t,x)\,\mathrm{d}x\right)\geqslant 0$$

and hence

$$\int w(t,w)\,\mathrm{d}x\geqslant 0 \qquad \text{for all } t\geqslant 0.$$

This completes the outline of the proof.

Remark 4.2. Note that the result of Theorem 3.1 to the effect that for large γ, the nonlinear stochastic logistic model suffers extinction with probability one suggests the same result for the spatially distributed system. Thus it is reasonable to conjecture that for sufficiently large γ,

$$X(t, K) \to 0 \text{ in probability as } t \to \infty, \tag{4.14}$$

for every compact set $K \subset R^d$. This is to be expected since the mean field assumption is more favourable to survival in the spatially distributed system than if the mass flowing into a compact region is in fact positively correlated with it.

Remark 4.3. The method outlined above is potentially applicable to a variety of problems. However to facilitate rigorous proofs more work needs to be done on systems of differential–integral inequalities of the type (4.6)–(4.8).

References

[1] Dawson, D. A. (1978) Geostochastic calculus. *Can. J. Statist.* **6**, 143–168.

[2] Goel, N. S. and Richter-Dyn, N. (1974). *Stochastic Models in Biology*. Academic Press, New York.

[3] Nicolis, G. and Prigogine, I. (1977). *Self-organization in Non-equilibrium Systems*. Wiley-Interscience.

P. R. Krishnaiah, ed., *Multivariate Analysis–V*

A STOCHASTIC EQUATION FOR THE CONDITIONAL DENSITY IN A FILTERING PROBLEM

G. KALLIANPUR
University of Minnesota, Minneapolis, MN, U.S.A.

1. Introduction

In recent years much effort has been devoted to the study of the stochastic differential equations of non-linear filtering, in particular, the equations satisfied by the conditional density whose existence is normally ensured by assumptions natural to the problem. The usual procedure is to derive the equation for the conditional density from that of the optimal filter. This approach, however, makes it necessary to deal with questions of differentiability of the density. The difficulties are non-trivial as can be seen from a discussion of one of the simplest such problems in the book by Liptser and Shiryaev [9, Theorem 8.7]. In the important special case when the signal Markov process is independent of the noise in the observation process model, Kunita has obtained an alternative equation for the optimal non-linear filter [8]. The derivation of the equation and the proof of the uniqueness of the solution are both very simple. Moreover, the equation does not involve the differential operator of the signal process. The purpose of the present paper is to use Kunita's equation (with appropriate modifications to be noted below) to obtain a stochastic equation for the conditional density which avoids the difficulties referred to above.

In the rapidly growing literature on the subject more general filtering models have been considered. (See Rozovskii [11]). In Rozovskii's paper which is devoted to stochastic partial differential equations, the signal and observation processes are governed by a stochastic differential equation. The signal is not assumed to be independent of observation noise but the coefficients in the defining equations are uniformly bounded and satisfy differentiability conditions. (The boundedness condition persists in our model (1)). The paper [11] also considers an equation similar to ours and obtains smoothness properties for the density. Our problem, which is more special than in [11] enables us to make full use of the Bayes Formula and to establish the result under different and somewhat weaker conditions. The main result (Theorem 1 of Section 3) is proved without the assumption that the signal process is given by a stochastic differential equation. The conditions under which Theorem 1 holds are, however, satisfied by $\hat{C}$-diffusion processes (see Section 4).

To keep the exposition as brief as possible, all details such as those involving measurability questions are omitted. The reader is referred to Liptser and Shiryaev [9], Kunita [8], Kallianpur and Striebel [7] and Meyer [10] for detailed information on the structure of the filtering model considered in this paper. We are indebted to Rozovskii [11] for reference [1].

2. Existence of a solution

The signal process X_t is assumed to a right-continuous, temporally homogeneous Markov process taking values in $\mathbf{R}^d$ and defined on a complete probability space $(\Omega, \mathsf{A}, \mathbf{P})$. The transition probability function $\mathbf{P}_t(x,B)$, $(B \in \mathsf{B}(\mathbf{R}^d))$ is a $\hat{C}$-function in the terminology of Dynkin and the associated semigroup $(\mathbf{P}_t), t \geqslant 0$ is a Feller semigroup. [3, Vol. 1, pp. 76–77]. We denote by $\hat{C}$, the space of all real continuous functions on $\mathbf{R}^d$ vanishing at infinity. The infinitesimal operator A of the semigroup is a $\hat{C}$-infinitesimal operator in the sense of [3]. The observation process Z_t is defined by

$$Z_t = \int_0^t h(X_s)\,\mathrm{d}s + W_t \tag{2.1}$$

where $h = (h^1, \ldots, h^N)$, $h^i \in C(\mathbf{R}^d)$ and (W_t) is an N-dimensional Wiener process which is stochastically independent of (X_t). (We write $\perp\!\!\!\perp$ to denote stochastic independence). Let $\mathsf{F}_t^Z = \sigma(X_s, 0 \leqslant s \leqslant t) v$ {all $\mathbf{P}$-null sets in $\mathfrak{F}$} and, for $f \in C(\mathbf{R}^d)$ let $\mathbf{E}^t(f)$ stand for the conditional expectation $\mathbf{E}[f(X_t)|\mathsf{F}_t^Z]$. It is convenient at this point to introduce the N-dimensional innovation Wiener process defined by

$$\nu_t = Z_t - \int_0^t \mathbf{E}^s(h)\,\mathrm{d}s. \tag{2.2}$$

In the integral in (2.2), a progressively measurable version of the conditional expectation is taken so that the definition is meaningful. Then (ν_t, F_t^Z) is an N-dimensional, Wiener martingale (Fujisaki, Kallianpur and Kunita [4]). It is shown in [4] that $\mathbf{E}^t(f)$ satisfies the following stochastic differential equation:

$$\mathbf{E}^t(f)=\mathbf{E}^0(f)+\int_0^t \mathbf{E}^s(Af)\,ds+\int_0^t(\mathbf{E}^s(fh)-\mathbf{E}^s(f)\mathbf{E}^s(h),d\nu_s) \tag{2.3}$$

for all $f\in D(A)$.

H. Kunita has obtained another equation satisfied by $\mathbf{E}^t(f)$,

$$\mathbf{E}^t(f)=\mathbf{E}^0(\mathbf{P}_t f)+\int_0^t(\mathbf{E}^s((\mathbf{P}_{t-s}f)h)-\mathbf{E}^s(\mathbf{P}_{t-s}f)\mathbf{E}^s(h),d\nu_s) \tag{2.4}$$

for all $f\in\hat{C}$.

Let M be the space of probability measures on $\mathbf{R}^d$ endowed with the topology of weak convergence. Suppose that (β_t), $t\geq 0$ is an N-dimensional Wiener process defined on a probability space $(\Omega,\mathsf{A},\mathbf{P})$. Let π_0 be the fixed element of M given by $\pi_0(f)=\mathbf{E}f(X_0)$. Following Kunita we shall say that an M-valued stochastic process (π_t) is a solution of the equation

$$\pi_t(f)=\pi_0(\mathbf{P}_t f)+\int_0^t(\pi_s((\mathbf{P}_{t-s}f)h)-\pi_s(\mathbf{P}_{t-s}f)\pi_s(h),d\beta_s) \tag{2.5}$$

if it satisfies the following conditions.

(a) $\pi_t(f)(\omega)$ is (t,ω)-measurable for each $f\in\hat{C}$;

(b) $\pi_s \perp\!\!\!\perp \sigma[\beta_v-\beta_u, s\leq u\leq v\leq s']$ for every $s,s'(0\leq s\leq s')$;

(c) $\pi_t(f)(\omega)$ satisfies (5) a.s. for each t.

It is known that there is a version of the conditional expectation $\mathbf{E}^t(f)$ which satisfies (a), (b) and (c), so that equation (2.5) has a solution $\pi_t(f)=\mathbf{E}^t(f)$. In proving this result it is convenient to use the Bayes formula of Kallianpur and Striebel [6] and the result of J. M. C. Clark [2] that $\mathsf{F}_t^\nu=\mathsf{F}_t^Z$ for every t, a consequence of the independence of (X_t) and (W_t) and the boundedness of h. The uniqueness of the solution of (2.5) is proved in the same manner as in [8]. It is to be noted here that Kunita takes the state space of (X_t) to be compact, metric but the changes in the assumptions on the signal process do not cause any difficulties.

As mentioned in the Introduction the equation for the conditional density derived from (2.5) involves fewer restrictions than the equation which corresponds to (2.4) because of the obvious differentiability questions that arise in connection with the latter equation.

We begin by stating the assumptions on the transition probability function $\mathbf{P}_t(x,B)$ of (X_t):

(A.1) For $t>0$, the transition probability density $\mathbf{p}_t(x,y)$ exists and is continuous in (t,x,y).

(A.2) A fixed initial distribution μ is assumed which has a probability density $\mathbf{p}_0$ with respect to Lebesgue measure.

(A.3) $\mathbf{p}_0(x)$ is bounded and continuous on $\mathbf{R}^d$; $\mathbf{p}_0(x) \leqslant C_0$.

(A.4) There exists a postive constant C_1 such that

$$\int_{\mathbf{R}^d} \mathbf{p}_t(x,y)\,\mathrm{d}x \leqslant C_1 \qquad \text{for all } t>0 \quad \text{and} \quad y \in \mathbf{R}^d.$$

(A.5) For $x \in \mathbf{R}^d$,

$$\lim_{t \downarrow 0} \int_{\mathbf{R}^d} \mathbf{p}_0(y)\mathbf{p}_t(x,y)\,\mathrm{d}y = \mathbf{p}_0(x)$$

uniformly over bounded sets in $\mathbf{R}^d$. It follows from assumptions (A.1) and (A.2) that X_t has a probability density $\mathbf{p}_t$ given by

$$\mathbf{p}_t(y) = \int \mathbf{p}_t(x,y)\mathbf{p}_0(x)\,\mathrm{d}x, \qquad (t>0).$$

Furthermore, using (A.1), (A.2), (A.4) and (A.5) one can easily show that $\mathbf{p}_t(x)$ is a bounded, continuous function of (t,x) for $t \geqslant 0$ and $x \in \mathbf{R}^d$.

For ease of writing, the outline of the proofs will be given for $d=N=1$. The statements of the theorems as well as the assumptions are given for general d and N.

For the particular model of filtering considered in this paper it is not hard to show, from the Bayes formula of [6], that the probability density $\mathrm{d}/\mathrm{d}x\mathbf{P}[X_t \leqslant x|\mathsf{F}_t^Z]$ exists a.s. and is given by

$$\tilde{\varphi}_t(x,\omega) = \mathbf{p}_t(x)\mathbf{E}\left[\rho_t(u,\omega)|\mathsf{F}^{X_t}\right](x) \tag{2.6}$$

where ρ_t is the density in the expression for the Bayes formula

$$\rho_t(u,\omega) = \exp\left[\int_0^t (h_s(u) - \hat{h}_s(\omega))\,\mathrm{d}\nu_s(\omega) - \frac{1}{2}\int_0^t (h_s(u) - \hat{h}_s(\omega))^2\,\mathrm{d}s\right] \tag{2.7}$$

where $\hat{h}_s(\omega)=\mathbf{E}[h(X_s)|\mathsf{F}_s^Z](\omega)$. The formula (2.6) is given in Liptser and Shiryaev [9, p. 296] and the particular form (2.7) is found in Meyer's article in [10]. Here $u\in\Omega_X$, the probability space on which the signal process is given. It can be shown that there is a (t,x,ω)-measurable version of φ which we shall work with from now on and denote by φ. The following result is used in the proof of Theorem 1.

Lemma. *For $m>1$ and n an arbitrary positive integer there exists a positive constant $L_{m,n}$ such that*

$$\mathbf{E}\prod_{i=1}^{n}(\varphi_t(x_i,\omega))^m \leqslant L_{m,n}\prod_{i=1}^{n}(p_t(x_i))^m \tag{2.8}$$

where $x_i\in\mathbf{R}^d$ and $t\in[0,T]$.

Proof. From (2.6), $\varphi_t^m(x,\omega)=\mathbf{p}_t^m(x)g_t^m(x)$ where $g_t(x)=\mathbf{E}[\rho_t(u,\omega)|X_t(u)=x]$.

$$\mathbf{E}\prod_{i=1}^{n}\varphi_t^m(x_i,\omega)=\prod_{i=1}^{n}\mathbf{p}_t^m(x_i)\mathbf{E}\left[\prod_{i=1}^{n}g_t^m(x_i)\right].$$

Now if $a_i\geqslant 0(i=1,\dots,n)$, using Hölder's inequality and the inequality relating the geometric mean and the arithmetic mean we have

$$(a_1\cdots a_n)^{1/n}\leqslant\frac{a_1+\cdots+a_n}{n}\leqslant(a_1^n+\cdots+a_n^n)^{1/n}n^{-1/n}$$

so that

$$a_1\cdots a_n\leqslant\frac{a_1^n+\cdots+a_n^n}{n}.$$

Hence setting $a_i=g_t^m(x_i,\omega)$ we have

$$\mathbf{E}\prod_{i=1}^{n}g_t^m(x_i,\omega)\leqslant\frac{1}{n}\sum_{i=1}^{n}\mathbf{E}[\,g_t^{mn}(x_i,\omega)\,].$$

$$\begin{aligned}\mathbf{E}[\,g_t^{mn}(x,\omega)\,]&=\mathbf{E}\{(\mathbf{E}[\rho_t(u,\omega)|X_t(u)=x\,])^{mn}\}\\&\leqslant\mathbf{E}[\rho_t^{mn}(u,\omega)\,].\end{aligned}$$

Writing $\bar{h}_s = h_s - \hat{h}_s$ and recalling that $|\bar{h}_s| \leqslant 2\|h\|$ where $\|h\| = \sup|h_s(u)|$, for $s \in [0,T]$, $u \in \Omega_x$ and $\omega \in \Omega$ we obtain

$$\rho_t^{mn}(u,\omega) = \exp\left[mn \int_0^t \bar{h}_s \, \mathrm{d}\nu_s - \frac{mn}{2} \int_0^t \bar{h}_s^2 \, \mathrm{d}s \right]$$

$$\leqslant \exp 2(m^2n^2 + mn) T \|h\|^2 \cdot \exp\left[\int_0^t (mn\bar{h}_s) \, \mathrm{d}\nu_s - \frac{1}{2} \int_0^t (mn\bar{h}_s)^2 \, \mathrm{d}s \right].$$

It follows that $\mathbf{E}(\rho_t^{mn}) \leqslant \exp 2(m^2n^2 + mn) T \|h\|^2$, $\mathbf{E} \prod_{i=1}^n g_t^{mn}(x_i,\omega) \leqslant L_{m,n} \equiv \exp 2(m^2n^2 + mn) T \|h\|^2$ and we have the estimate (8). The above proof has been given for $N=1$. For $N>1$ we only have to use the N-dimensional version of (2.7) which can be deduced from the vector-valued form of the Bayes formula given in Kallianpur–Striebel [7]. The constant L_m is the same except that now $\|h\|^2 = \sum_{i=1}^n \|h^i\|^2$ where $\|h^i\| = \sup|h_s^i(u)|$. Taking $n=1$ and $m=2$ in the Lemma we obtain

$$\sup_{\substack{0 \leqslant t \leqslant T \\ x \in \mathbf{R}^d}} \mathbf{E}[\varphi_t(x,\omega)]^2 < \infty. \tag{2.9}$$

To find the equation satisfied by the density we start from the fact that $\pi_t(f) = \int f(x) \varphi_t(x,\omega) \, \mathrm{d}x, (f \in \hat{C})$ is a solution of (2.5) with the initial condition $\pi_0(f) = \int f(x) \mathbf{p}_0(x) \, \mathrm{d}x$. Introduce the notation

$$R_{t-s}(\varphi; x,\omega) = \int \mathbf{p}_{t-s}(y,x) \varphi_s(y,\omega) \left[h(y) - \int h(y') \varphi_s(y',\omega) \, \mathrm{d}y' \right] \mathrm{d}y \tag{2.10}$$

and

$$Q_t(\varphi; x,\omega) = \int \mathbf{p}_0(y) \mathbf{p}_t(y,x) \, \mathrm{d}x + \int_0^t \mathbf{R}_{t-s}(\varphi; x,\omega) \, \mathrm{d}\nu_s(\omega). \tag{2.11}$$

The stochastic integral in (2.11) can be taken to be (t,x,ω) measurable. The desired equation for φ is obtained in the same manner as in Liptser and Shiryaev [9]. We indicate an outline of the argument. The integrand in the stochastic integral on the right hand side of (2.5) may be written as

$$\int\left[\int f(y)\mathbf{p}_{t-s}(x,y)\,\mathrm{d}y\right]h(x)\varphi_s(x,\omega)\,\mathrm{d}x$$

$$-\int\left[\int f(y)\mathbf{p}_{t-s}(x,y)\,\mathrm{d}y\right]\varphi_s(x,\omega)\,\mathrm{d}x\int h(y)\varphi_s(y,\omega)\,\mathrm{d}y. \quad (2.12)$$

The first term in (2.12) equals

$$\int f(x)\left[\int \mathbf{p}_{t-s}(y,x)h(y)\varphi_s(y,\omega)\,\mathrm{d}y\right]\mathrm{d}x \qquad \text{(a.s.)},$$

the interchange of integration being permissible by Fubini's theorem since

$$\mathbf{E}\int\int |h(y)\cdot f(x)|\mathbf{p}_{t-s}(y,x)\varphi_s(y,\omega)\,\mathrm{d}y\,\mathrm{d}x$$

$$\leqslant \|h\|\,\|f\|\mathbf{E}\int\int \mathbf{p}_{t-s}(y,x)\varphi_s(y,\omega)\,\mathrm{d}y\,\mathrm{d}x = \|h\|\,\|f\|.$$

The other term in (2.12) is similarly treated and thus the right hand side stochastic integral in (2.5) equals $\int f(x)\mathbf{R}_{t-s}(x,\omega)\,\mathrm{d}x$. It remains to verify that

$$\int_0^t\left[\int f(x)\mathbf{R}_{t-s}(x,\omega)\,\mathrm{d}x\right]\mathrm{d}\nu_s(\omega)$$

$$= \int f(x)\int_0^t \mathbf{R}_{t-s}(x,\omega)\,\mathrm{d}\nu_s(\omega)\,\mathrm{d}x \qquad \text{a.s.} \quad (2.13)$$

for all $f \in C_0(\mathbf{R})$, the set of all continuous functions in $\mathbf{R}$ with compact support. This is carried out as in [9] and we omit the details. It must be noted, of course, that the stochastic integral $\int_0^t \mathbf{R}_{t-s}(x,\omega)\,\mathrm{d}\nu_s(\omega)$ exists because (the measurability criterion being fulfilled), $\mathbf{E}[\mathbf{R}_{t-s}(x,\omega)]^2 \leqslant 4\|h\|^2\mathbf{E}[\int \mathbf{p}_{t-s}(y,x)\varphi_s(y,\omega)\,\mathrm{d}y]^2 \leqslant C_2$, a constant independent of x and $t-s$ by (A.4) and the Lemma. The interchange of integration in $\pi_0(\mathbf{P}_t f)$ is obvious and we obtain the equation

$$\int f(x)\left[\varphi_t(x,\omega)-Q_t(\varphi;x,\omega)\right]\mathrm{d}x=0 \qquad \text{a.s.} \tag{2.14}$$

for all f in $C_0(\mathbf{R})$ and $t>0$. It follows that for every $t>0$

$$P\left[\omega:\varphi_t(x,\omega)=Q_t(\varphi;x,\omega)\right]=1 \qquad \text{for a.e.x.} \tag{2.15}$$

Definition. A process $u_t(x,\omega)$ is a solution of the stochastic equation

$$u_t(x,\omega)=\int \mathbf{p}_0(y)\mathbf{p}_t(y,\omega)\,\mathrm{d}y$$
$$+\int_0^t\left[\int \mathbf{p}_{t-s}(y,x)u_s(y,\omega)\left\{h(y)-\int h(y')u_s(y',\omega)\,\mathrm{d}y'\right\}\mathrm{d}y\right]\mathrm{d}\beta_s(\omega) \tag{2.16}$$

if it satisfies the following conditions:

(a) For almost all ω, $u_t(x,\omega)$ is a continuous function of (t,x) on $[0,T]\times\mathbf{R}^d$;

(b) $u_0(x,\omega)=p_0(x)$ a.s. for every x;

(c) For each (t,x), $u_t(x,\omega)$ satisfies (2.16) a.s.

The conditional density $\varphi_t(x,\omega)$ which satisfies (2.14) is not quite a solution of (2.16) in the sense of the definition just given.

It can be shown that there exists a (t,x)-continuous version of φ which is, indeed, a solution of (2.16). We give here an outline of the argument.

Let us denote by $J_t(\varphi,x,\omega)$ a measurable and separable version of the stochastic integral $\int_0^t \mathbf{R}_{t-s}(\varphi,x,\omega)\,\mathrm{d}\nu_s(\omega)$. (For a discussion of separability in the multiparameter case, see Gikhman and Skorokhod [5].) The method is to obtain inequalities which enable us to invoke an extension to multi-parameter processes of a criterion of Kolmogorov's for the existence of sample continuous versions. (See Blagoveschenskii and Freidlin [1].) We need the following additional assumptions:

Let α,γ be positive numbers such that $0<\gamma<1/4$ and $1<\alpha<d/(d+2\gamma-1/2)$.

(B.1) For $0\leqslant s<t\leqslant T$, and $x,x'\in\mathbf{R}^d$,

$$\int |\mathbf{p}_{t-s}(y,x)-\mathbf{p}_{t-s}(y,x')|^\alpha\,\mathrm{d}y\leqslant|x-x'|^{\alpha\gamma}H_1(t,s).$$

(B.2) For $0 \leqslant s < t < t' \leqslant T$ and $x \in \mathbf{R}^d$,

$$\int |\mathbf{p}_{t'-s}(y,x) - \mathbf{p}_{t-s}(y,x)|^{\alpha} \, dy \leqslant |t-t'|^{\alpha\gamma} H_2(t,s).$$

The functions H_1 and H_2 do not involve x or x' and satisfy the condition

$$\int_0^t [H_i(t,s)]^{4/\alpha} \, ds \leqslant A_i < \infty \qquad \text{for } t \in [0,T] \quad (i=1,2).$$

There exists a number δ($\delta \in (0,1)$ and depending on α and d) such that for all t in $[0,T]$,

$$\text{(B.3)} \qquad \int_0^t \left[\int \mathbf{p}_s(y,x)^{\alpha} \, dy \right]^{4/\alpha} ds \leqslant C_3 t^{\delta},$$

C_3 being a constant independent of x and $t \in [0,T]$. With the help of the Lemma and assumptions (B.1)–(B.3) it is not hard to derive the following inequalities.

$$\mathbf{E}[J_{t'}(\varphi,x,\omega) - J_t(\varphi,x,\omega)]^{2k} \leqslant C_4\left[|t'-t|^{2k\gamma} + |t'-t|^{\frac{1}{2}(1+\delta)k}\right], \tag{2.17}$$

$$\mathbf{E}[J_t(\varphi,x',\omega) - J_t(\varphi,x,\omega)]^{2k} \leqslant C_5 |x'-x|^{2k\gamma}, \tag{2.18}$$

where $k > 1$.

From (2.17) and (2.18) we have

$$\mathbf{E}[J_{t'}(\varphi,x',\omega) - J_t(\varphi,x,\omega)]^{2k} \leqslant C_6\left[|t'-t|^{2k\gamma} + |t'-t|^{\frac{1}{2}(1+\delta)k} + |x'-x|^{2k\gamma}\right]. \tag{2.19}$$

Let $n = [(\alpha-1)(d+1)/(\alpha l)] + 1$ where $l = \min(\gamma, (1+\delta)/4)$ and the constants α, γ and δ are the ones mentioned in (B.1)–(B.3). Now choose the number k such that $2k = n\alpha/(\alpha-1)$. Then, writing $\theta = (t,x)$ and $\theta' = (t',x')$, ($\theta, \theta' \in [0,T] \times \mathbf{R}^d$) we rewrite (2.19) in the form

$$\mathbf{E}[J_{t'}(\varphi,x',\omega) - J_t(\varphi,x,\omega)]^{2k} \leqslant C_7 |\theta - \theta'|^{d+1+r}$$

where $r > 0$, and $|\theta' - \theta|^2 = (t'-t)^2 + \sum_{i=1}^{d} (x'^i - x^i)^2$. Thus the extended

Kolmogorov criterion is verified and we conclude from [1] the existence of a continuous version of $\varphi_t(x,\omega)$, i.e., there exists a process $\bar{J}_t(x,\omega)$ which is continuous in (t,x) **P** a.s. such that

$$\mathbf{P}\left[\omega : \bar{J}_t(x,\omega) = J_t(\varphi,x,\omega)\right] = 1 \qquad \text{for all } (t,x), \quad t>0, \tag{2.20}$$

and such that for each $t>0$ and $x \in \mathbf{R}^d$, $\bar{J}_t(x,.)$ is F_t-measurable. For each x and ω define

$$\bar{\varphi}_t(x,\omega) = \int \mathbf{p}_0(y)\mathbf{p}_t(y,x)\,\mathrm{d}y + \bar{J}_t(x,\omega) \tag{2.21}$$

for $t>0$ and $\bar{\varphi}_0(x,\omega) = \mathbf{p}_0(x)$.

From (2.11) and (2.20) it is seen that

$$\mathbf{P}\left[\omega : \varphi_t(x,\omega) = Q_t(\varphi;x,\omega)\right] = 1 \tag{2.22}$$

for $t>0$ and $x \in \mathbf{R}^d$. It follows from (2.22) and (2.15) that for $t>0$ and a.e.x, we have

$$\mathbf{P}\left[\omega : \bar{\varphi}_t(x,\omega) = \varphi_t(x,\omega)\right] = 1. \tag{2.23}$$

For $t>0$, $\bar{\varphi}_t(x,\omega) \geqslant 0$, **P** a.s. for every x and

$$\int \bar{\varphi}_t(x,\omega)\,\mathrm{d}x = 1, \mathbf{P} \text{ a.s.}$$

From the above relations and using the measurability of both $\bar{\varphi}$ and φ it can be shown that there exists $M \subset [0,T] \times \Omega$ such that $m \times P(M^c) = 0$ (where m is the Lebesgue measure) and

$$\bar{\varphi}_t(x,\omega) = \varphi_t(x,\omega) \qquad \text{for a.e.x} \tag{2.24}$$

if $(t,\omega) \in M$. It then follows that

$$\int_0^t R_{t-s}(\bar{\varphi};x,\omega)\,\mathrm{d}\nu_s(\omega) = \int_0^t \mathbf{R}_{t-s}(\varphi;x,\omega)\,\mathrm{d}\nu_s(\omega), P \text{ a.s.}$$

for every x and $t>0$. Hence

$$\mathbf{P}\left[\omega : Q_t(\bar{\varphi};x,\omega) = Q_t(\varphi;x,\omega)\right] = 1 \tag{2.25}$$

for $t>0$ and all x. From (2.22) and (2.25) it immediately follows that, for each $t>0$

$$\mathbf{P}\left[\omega:\bar{\varphi}_t(x,\omega)=Q_t(\bar{\varphi};x,\omega)\right]=1 \tag{2.26}$$

for $x\in\mathbf{R}^d$. In other words, $\bar{\varphi}$ is a continuous version of the conditional density which is a solution of the stochastic equation (2.16) in which β has been taken to be the innovation ν. Now if β is an arbitrary Wiener process given on a probability space $(\tilde{\Omega},\tilde{\mathrm{A}},\tilde{P})$ the existence of a solution of (2.16) is deduced from the case just considered by proceeding exactly as in Kunita (Theorem 2.1, [8]).

3. Uniqueness

We shall show that the stochastic equation (2.16) has a unique solution in the class of continuous processes $\varphi_t'(x,\omega)$ subject to the following restrictions:

(C.1) $\int|\varphi_t'(x,\omega)|\mathrm{d}x\leqslant M<\infty$ a.s., for every $t>0$.

(C.2) For each $t>0$: $E\int_0^t[\int\mathbf{p}_{t-s}(t,x)\varphi_s'(y,\omega)\,\mathrm{d}y]^2\mathrm{d}s$, is finite and locally integrable in x.

Let $\varphi_t'(x,\omega)$ satisfying (C.1) and (C.2) be a solution of (2.16) with the initial condition $\varphi_0'(x,\omega)=\mathbf{p}_0(x)$ a.s. for all x. For $f\in C$ define $\pi_t'(f)(\omega)=\int f(x)\varphi_t'(x,\omega)\,\mathrm{d}x$. Consider each term on the right hand side of (2.16). It is easy to see that $\int f(x)(\int p_0(y)p_t(y,x)\,\mathrm{d}y)\,\mathrm{d}x=\pi_0'(P_tf)$. We observe that the measurability and the F_t-adaptability (for each (t,x)) of $R_{t-s}(\varphi';x,\omega)$ can be verified and that using (C.2) along with (A.1)–(A.5) it can be shown that

$$\int f(x)\left[\int_0^t\mathbf{R}_{t-s}(\varphi';x,\omega)\,\mathrm{d}\beta_s(\omega)\right]$$
$$=\int_0^t\left[\int f(x)\mathbf{R}_{t-s}(\varphi';x,\omega)\,\mathrm{d}x\right]\mathrm{d}\beta_s(\omega) \tag{3.1}$$

a.s. for $f\in C_0(R)$. To see this we essentially repeat arguments already used in the previous Section. Because of (C.1) and (C.2) the Fubini theorem can again be applied to yield the result

$$\int f(x)\mathbf{R}_{t-s}(\varphi';x,\omega)\,dx$$
$$=\pi_s'[(\mathbf{P}_{t-s}f)h](\omega)-\pi_s'(\mathbf{P}_{t-s}f)(\omega)\pi_s'(h)(\omega)\quad \text{a.s.}$$

Hence equation (2.16) for $\varphi_t'(x,\omega)$ is recast in the form

$$\pi_t'(f)=\pi_0(\mathbf{P}_tf)+\int_0^t\{\pi_s'((\mathbf{P}_{t-s}f)h)-\pi_s'(\mathbf{P}_{t-s}f)\pi_s'(h)\}\,d\beta_s \tag{3.2}$$

and this equation remains valid for all $f\in C_0$. Next observe that $\varphi_t(x,\omega)$ the conditional density of the optimal filter satisfies conditions (C.1) and (C.2) and the above argument shows that (3.2) holds with π_t' replaced by $\pi_t(f)(\omega)=\int f(x)\varphi_t(x,\omega)\,dx$. Thus π_t and π_t' are solutions of equation (2.5). By the uniqueness of the solution of (2.5), it follows that

$$P[\omega:\pi_t(f)(\omega)=\pi_t'(f)(\omega)\,\forall t\in[0,T]]=1 \tag{3.3}$$

for all $f\in\hat{C}$. Since $\hat{C}$ contains a countable dense subset we conclude from (3.3) that there exists Ω_0 with $P(\Omega_0)=1$ such that $\omega\in\Omega_0$ implies

$$\pi_t(f)(\omega)=\pi_t'(f)(\omega)\qquad \text{for all } f\in\hat{C}\quad\text{and}\quad t\in[0,T]. \tag{3.4}$$

If Ω_1 denotes the set of ω's for which (C.1) and (C.2) hold and Ω_2 is the set on which $\varphi_t(x,\omega)$ and $\varphi_t'(x,\omega)$ are continuous functions of (t,x) it follows that for $\omega\in\Omega_0\cap\Omega_1\cap\Omega_2$ we have

$$\varphi_t(x,\omega)=\varphi_t'(x,\omega)\qquad\text{for all } (t,x).$$

This concludes the proof of uniqueness.

Remark. The above conditions imply uniqueness in the class of solutions satisfying (C.1) and

(C.3) $\sup_{(t,x)}\mathbf{E}|\varphi_t'(x,\omega)|^2<\infty.$

Note that condition (C.3) is satisfied by φ. (See (2.9).) We have thus established the following result.

Theorem 1. *Let the assumptions* (A.1)–(A.5) *and* (B.1)–(B.3) *be satisfied. Then for the stochastic filtering model* (1) *there exists a conditional density* $\varphi(t,x,\omega)$ *which is continuous in* (t,x) $\mathbf{P}$ a.s. *and which is a solution of equation* (2.16).

Furthermore, this solution is unique in the class of processes satisfying conditions (C.1) *and* (C.2) [*or, equivalently, conditions* (C.1) *and* (C.3)].

4. Example

We give an example of a class of signal processes for which the main result of this paper is applicable.

Theorem 2. *Let* (X_t) *be a diffusion* $\hat{C}$*-process* (see Dynkin [3], Vol. 1 for the definition) *whose generator is given by a differential operator*

$$(1) \quad Lf(x)=\sum_{i,j=1}^{d} a_{ij}(x)\frac{\partial^2 f(x)}{\partial x^i \partial x^j}+\sum_{i=1}^{d} b_i(x)\frac{\partial f(x)}{\partial x^i}-c(x)f(x)$$

whose coefficients satisfy the following conditions:

(2) *The functions* a_{ij}, b_i *and* c *are bounded and satisfy a Hölder condition on* $\mathbf{R}^d$.

(3) *There exists a positive constant* λ *such that for all* x *in* $\mathbf{R}^d$ *and all* $(\xi^1,\ldots,\xi^d)$,

$$\sum_{i,j=1}^{d} a_{ij}(x)\xi^i\xi^j \geqslant \lambda \sum_{i=1}^{d} (\xi^i)^2.$$

(4) $c(x)\geqslant 0$ *for all* x *in* $\mathbf{R}^d$.

It will be assumed further that X_0 has a density $\mathbf{p}_0(x)$ which is bounded and continuous.

Then, for the stochastic filtering model (1) *with* (X_t) *as the signal process, equation* (2.16) *has a solution which is unique in the class specified in Section 3.*

Proof. It is sufficient to show that the conditions imposed on the signal process in Section 2 are satisfied. Then from (0.24) and Theorem 5.11 of Dynkin [3] it follows that the transition semigroup $(\mathbf{P}_t)$ of (X_t) is a Feller semigroup and that the transition probability density $\mathbf{p}_t(x,y)$, $(t>0)$ exists, is continuous in (t,x,y) (which is assumption (A.1)) and has the properties $(0.24)C_1$ and $(0.24)C_2$ of [3, Vol. 2, p. 229]. The assumptions (A.4) and (A.5) are easily seen to be satisfied. The verification of (B.1) and (B.2) follows after straightforward analysis which uses the estimates for $\mathbf{p}_t(x,y)$ given in $(0.24)C_2$ of [3].

References

[1] Blagoveschenskii, Yu N. and Freidlin, M. I. (1961). Certain properties of diffusion processes depending on a parameter. *Soviet Math. Dokl.*, **2**, 633–636.

[2] Clark, J. M. C. (1969). Conditions for one-to-one correspondence between an observation process and its innovation. Technical Report, Centre for Computing and Automation, Imperial College.

[3] Dynkin, E. B. (1965). *Markov Processes—I and II.* Springer-Verlag, New York, Heidelberg and Berlin.

[4] Fujisaki, M., Kallianpur, G. and Kunita, H. (1972). Stochastic differential equations for the non-linear filtering problem. *Osaka Journal of Mathematics*, **9**, 19–40.

[5] Gikhman, I. I. and Skorokhod, A. V. (1969). *Introduction to the Theory of Random Processes* (English Translation). W. B. Saunders Company.

[6] Kallianpur, G. and Striebel, C. (1968). Estimation of stochastic systems: arbitrary system process with additive white noise observation errors. *Ann. Math. Statist.*, **39**, 785–801.

[7] Kallianpur, G. and Striebel, C. (1969). Stochastic differential equations in statistical estimation problems. In *Multivariate Analysis-II* (P. R. Krishnaiah, ed.), pp. 367–388. Academic Press, New York.

[8] Kunita, H. (1971). Asymptotic behavior of the nonlinear filtering errors of Markov processes. *J. Multivariate Anal.*, **1**, 365–393.

[9] Liptser, R. S. and Shiryaev, A. N. (1977). Statistics of Random Processes I: General Theory, (English Translation). Springer-Verlag, New York, Heidelberg and Berlin.

[10] Meyer, P. A. (1973). Sur un problème de filtration. *Seminaire de Probabilités VII.* Springer-Verlag, New York, Heidelberg and Berlin.

[11] Rozovskii, B. L. (1975). On stochastic partial differential equations. *Math. U.S.S.R. Sbornik*, **25**, 295–322.

P. R. Krishnaiah, ed., *Multivariate Analysis–V*

SIMPLE MODELS FOR POSITIVE-VALUED AND DISCRETE-VALUED TIME SERIES WITH ARMA CORRELATION STRUCTURE

P. A. W. LEWIS*
Department of Operations Research, Naval Postgraduate School, Monterey, CA 93940, U.S.A.

Three models for positive-valued and discrete-valued stationary time series are discussed. All have the property that for a range of specified marginal distributions the time series have the same correlation structure as the usual linear, autoregressive-moving average (ARMA) model. The models differ in the range of marginal distributions which can be accomodated and in the simplicity and flexibility of each model. Specifically the EARMA-type processes can be extended from the exponential distribution to a rather narrow range of continuous distributions; the DARMA-type processes can be defined usefully for any discrete marginal distribution and are simple and flexible. Finally the marginally controlled semi-Markov generated process can be defined for any continuous or discrete positive-valued distribution and is therefore very flexible. However the model suffers from some complexity and parametric obscurity.

1. Introduction

In much of the current work on the analysis of stationary time series there is an implicit assumption that the marginal distribution of the time series is normal. The assumption is implicit in that the marginal distribution is not considered to be of interest per se in the analysis, and also in that the statistical procedures which are used are very definitely based on normality assumptions. The stationary model on which much of this time series analysis is based is the mixed autoregressive moving average process,

$$a_0X_i + a_iX_{i-1} + \cdots + a_pX_{i-p} = b_0\varepsilon_i + b_1\varepsilon_i + \cdots + b_q\varepsilon_{i-q}$$

$$i = 0, \pm 1, \pm 2, \ldots, \qquad (1.1)$$

sometimes called the ARMA(p,q) or Box–Jenkins process. The process (1.1) is specified quite generally as a linear combination of i.i.d. random

*Research supported by National Science Foundation Grant AF476 and Office of Naval Research Grant NR-42-284 at the Naval Postgraduate School.

variables $\{\varepsilon_i\}$ of unspecified distribution, the linear, additive structure determining the correlation structure of the stationary sequence $\{X_i\}$ under well-known restrictions on the parameters. If one wants $\{X_i\}$ to be a time series with normally distributed marginal distribution, this can be accomplished by taking the ε_i's to be normally distributed. The model is then completely specified.

There are, however, many situations in which observations occur serially and in which the marginal distribution is patently non-normal. For example, data on the number of occurrences of all known diseases in each week is kept by the National Center for Health Statistics. The data is not only discrete count data, but for many diseases it is mostly on the order of $0, 1, 2, 3$, and very seldom above this.

It has been suggested that such non-normal data be handled by data transformations and this is probably appropriate if the data is only slightly non-normal. In other cases it seems reasonable to start afresh and develop models from scratch. In this paper we summarize attempts to do this for stationary time series which are known to be non-normal because of either positivity or discreteness or both. The essence of the models is that the marginal distribution is specified, as well as the correlation structure. More generally the models are required to be simple and flexible in the following senses:

(a) The models should be specified in terms of easily observed and measured quantifiers. When the models are stationary, these quantifiers would typically be

(i) the marginal distribution, and
(ii) second-order moments (correlations).

(b) The models should be parametrically parsimonious and hopefully parametrically meaningful.

(c) The models should be easy to generate on computers, i.e., they should be structurally simple; in fact it might be preferable for the models to have linear structure.

(d) The models should be easy to fit to data, both informally and formally.

The model (1.1) certainly has most of the above features, but it is not known in general how to specify the distribution of ε_i so as to produce a given, continuous marginal distribution for the X_i's. Moreover, it is clearly not possible to do this at all if the X_i's are discrete random variables.

The work described in this paper on non-normal time series is joint work with D. P. Gaver, P. A. Jacobs and A. J. Lawrance. Although the work has much broader connotation, it will be described in the context in which it arose, that of the description of stochastic point processes, or series of events occurring in time. One way in which these point processes can be described is as a sequence of intervals between events $\{X_i\}$, which are of course positive-valued random variables. In the common case of a Poisson point process the X_i's have an exponential distribution. However, as in the case of epidemics, point processes are generally observed as counts of events in successive fixed intervals and these are non-negative discrete valued random variables. For the Poisson process these counts are independent and Poisson distributed and this serves as the null model in the analysis of count data from point processes.

Three distinct models are discussed in the context of the analysis and description of point processes. All of them satisfy the requirements discussed above to some degree. The EARMA-type process described first has recently been extended to have a complete ARMA-type correlation structure, but the process cannot be extended to all continuous marginal distributions. Marginally controlled semi-Markov generated processes, on the other hand, give a complete analog to (1.1), but they do not have linear structure. They can also be extended to give processes with discrete marginal distributions. A simpler, random linear structure has been derived, however, which gives discrete processes with ARMA structure. These are DARMA-type processes and come closer than the other processes to fulfilling the requirements of simplicity and flexibility.

Further details on the processes are to be found in the references.

2. Interval models: sequences of continuous positive-valued random variables

Univariate point processes in continuous time can be described equally well through the structure of the intervals between events $\{X_i\}$, where the X_i's are continuous and positive-valued random variables, or the counting process $\{N(t)\}$, where $N(t)$ gives the number of events in $(0, t]$ and is discrete and non-negative. We discuss the modelling of the intervals $\{X_i\}$ first. Of course the applications of the models are much broader; the X_i's might for instance be the magnitudes of successive shocks in a sequence of earthquakes or the successive response times of a computer to messages sent via a terminal.

2.1. *The first-order autoregressive exponential model* (EAR(1))

In a Poisson process the intervals $\{X_i\}$ are independent and identically distributed (i.i.d.) random variables with exponential distribution

$$F_X(x)=1-e^{-\lambda x}, \qquad \lambda>0; \quad x\geqslant 0. \tag{2.1}$$

Several attempts have been made to generalize the Poisson process by making the X_i dependent, but with exponential or conditionally exponential marginal distributions (Cox, 1955). The simplest and most successful attempt in the sense of broad applicability (Gaver and Lewis, 1978) gives a process called the EAR(1) model, derived from the following consideration.

A first-order autoregressive stochastic sequence is defined by the stochastic difference equation (a special case of (1.1))

$$X_i=\rho X_{i-1}+\varepsilon_i, \qquad i=0,\pm1,\pm2,\ldots; \quad |\rho|<1, \tag{2.2}$$

where the ε_i are assumed to be an i.i.d. stationary random sequence. If the ε_i are normally distributed, so are the X_i. What must the distribution of the ε_i be in order for the X_i sequence to be stationary with an exponential(λ) distribution? The answer is surprisingly easy (Gaver and Lewis, 1978).

Let $0\leqslant\rho<1$, and let $\{E_i\}$ be an i.i.d. exponential(λ) sequence. Now let ε_i be equal to zero with probability ρ and equal to E_i with probability $1-\rho$. Then we have

$$X_i=\begin{cases}\rho X_{i-1} & \text{probability } \rho,\\ \rho X_{i-1}+E_i & \text{probability } (1-\rho),\end{cases} \tag{2.3}$$

$$=\rho X_{i-1}+V_iE_i, \tag{2.4}$$

where $\{V_i\}$ is an i.i.d. binary sequence and $P\{V_i=0\}=1-P\{V_i=1\}=\rho$. Moreover if we let $X_0=E_0$, and define X_i as in (2.3), the resulting sequence is stationary for $i=0,1,\ldots$.

The point process with the interval structure (4.3) is called the EAR(1) point process. It is a tractable model, and most of its important properties are given in Gaver and Lewis (1978). In particular we have that $\rho(k)=\rho^k$. This model is in a sense degenerate because it contains runs of X_i in which values are exactly ρ times the previous value; it could, however, be a reasonable model for point processes observed in computer systems (e.g., inter-arrival times of requests to a storage subsystem) in which the intervals have exponential marginal distributions but are dependent. Note that as defined the model can only provide sequences $\{X_i\}$ with positive serial correlations. We can, however, define the process to include negative correlations (Gaver and Lewis, 1978); there is also a way to obviate the degeneracy (Lawrance, 1978).

Simple generalizations of this first-order, autoregressive, Markovian exponential process are the following.

2.2. *The moving average exponential model* (EMA(q))

We define another stationary sequence $\{X_i\}$, using the $\{E_i\}$ sequence above, according to

$$X_0 = E_0, \tag{2.5}$$

$$X_i = \beta E_i + U_i E_{i-1}, \qquad i=1,\dots; \quad 0 \leqslant \beta \leqslant 1, \tag{2.6}$$

where $\{U_i\}$ is an i.i.d. binary sequence in which $U_i = 1$ with probability $(1-\beta)$. This is a first order exponential moving average process (EMA(1)) (Lawrance and Lewis, 1977) which is one-dependent; in particular

$$\rho(1) = \beta(1-\beta) \tag{2.7}$$

$$\rho(k) = 0, \qquad k = 2, 3, \dots . \tag{2.8}$$

Properties of the EMA(1) process are given by Lawrance and Lewis (1977).

It is easy to see that we can make E_{i-1} in (2.6) a random linear combination of E_{i-1} and E_{i-2} to get an EMA(2) process, and can continue the process back q steps to obtain an EMA(q) process. The general EMA(q) model takes the form

$$X_i = \begin{cases} \beta_q E_i & \text{w.p. } b_{q+1}, \\ \beta_q E_i + \beta_{q-1} E_{i-1} & \text{w.p. } b_q, \\ \cdots & \cdots \\ \beta_q E_i + \beta_{q-1} E_{i-1} + \cdots + \beta_1 E_{i-q+1} & \text{w.p. } b_2, \\ \beta_q E_i + \beta_{q-1} E_{i-1} + \cdots + \beta_1 E_{i-q+1} + E_{i-q} & \text{w.p. } b_1, \end{cases} \tag{2.9}$$

for $0 \leqslant \beta_1, \beta_2, \ldots, \beta_q \leqslant 1$; $i = 0, \pm 1, \pm 2, \ldots$ and

$$b_i = \begin{cases} \beta_q & i = q+1, \\ (1-\beta_q)\cdots(1-\beta_i)\beta_{i-1} & q \geqslant i \geqslant 2, \\ (1-\beta_q)\cdots(1-\beta_i) & i = 1. \end{cases} \tag{2.10}$$

Note that the β_i's can be obtained uniquely from the b_i's; there are $q+1$ b_i's but only q β's, since the sum of the b_i's is equal to one.

This model is clearly only q dependent; in particular the correlations for the EMA(q) process are

$$\rho^{(q)}(k) = \operatorname{corr}(X_i, X_{i-k}) = \begin{cases} \sum_{v=1}^{q-k+1} b_v b_{v+k} & 1 \leqslant k \leqslant q, \\ 0 & q+1 \leqslant k < \infty. \end{cases} \tag{2.11}$$

Thus the serial correlations are just lagged products of the b_i sequence and the formula (2.11) is completely analogous to the formula for the serial correlations of the standard MA(q) process; see Box and Jenkins (1970, p. 68). It can be seen from (2.11) that all the correlations are nonnegative and it may be further shown that they are bounded above by $\frac{1}{4}$.

2.3. *The* EARMA(1, 1) *model*

By making E_{i-q} in (2.9) autoregressive over the previous E_i's, we obtain a mixed qth order moving-average, first order autoregressive process which we denote by EARMA($1, q$). Consider explicitly the case $q = 1$. The first order moving-average and first order autoregressive process EARMA(1, 1) is given by

$$X_i = \beta E_i + U_i A_{i-1}, \tag{2.12}$$

with

$$A_{i-1} = \rho A_{i-2} + V_i E_{i-1}, \tag{2.13}$$

for $i = 1, 2, 3, \ldots$ and $A_{-1} = E_{-1}$ with U_i and V_i as defined above. This sequence of random variables is not Markovian.

The second-order correlation structure of the process is given by

$$\rho(k) = \rho^{k-1} c(\beta, \rho), \tag{2.14}$$

where

$$c(\beta,\rho)=\beta(1-\beta)+\rho(1-\beta)(1-2\beta). \tag{2.15}$$

The point process whose intervals have the EARMA(1, 1) structure is discussed in detail in Jacobs and Lewis (1977). In particular, for $\beta=1$ it is a Poisson process. The process is very simple to generate on a computer and is very useful for modelling dependent sequences in queueing systems (Jacobs, 1978; Lewis and Shedler, 1978).

2.4. The pth-order autoregressive model EAR(p)

Quite recently ways have been found to obtain exponential sequences $\{X_i\}$ which have autoregressive structure of order p, and to combine these with the moving average process to get a mixed autoregressive-moving average process EARMA(p,q); see Lewis and Lawrance (1978). Another method of defining pth-order autoregressive exponential sequences, which is closely related to the DARMA(p,q) process discussed later, and which we have only just begun to explore, is described here.

This pth-order exponential autoregressive model can be written as

$$X_i=\alpha_{S_i}X_{i-S_i}+\varepsilon_{i,S_i}, \tag{2.16}$$

where the S_i's are i.i.d. discrete random variables taking values $1,2,\ldots,p$, and ε_{i,S_i} is defined to be 0 w.p. α_j, and E_i w.p. α_j if $S_i=j$. If one assumes stationarity and that X_{i-1}, X_{i-2}, are marginally exponential(λ), then X_i is a random mixture of E_i and $X_{i-1},\ldots,X_{i-p}$ and is exponential(λ). The correlation equations from this process are variants of the familiar Yule–Walker equations. The model is more tractable than the pth-order autoregressive process given in Lewis and Lawrance (1978) and is probably simpler to extend to other distributions than the exponential.

A drawback of these EARMA-type processes is that the serial correlations are all positive, although the scheme given in Gaver and Lewis (1978) for a negatively correlated EARMA1 process can probably be extended to the complete EARMA(p,q) process.

2.5. The semi-Markov generated point process with fixed marginal distribution

The question arises as to whether there are interval processes $\{X_i\}$ with exponential marginal distributions and, for example, ARMA(1, 1) second-order correlation structure and which cover a broader range of correlation than the EARMA(1, 1) process (though perhaps at a cost of more complicated structure).

We discuss briefly one such process. It is a special case of the semi-Markov generated point process introduced by Cox (1962) and extended by Haskell and Lewis (1978). We first describe the two-state semi-Markov generated model. In this model there are two types of intervals with distributions $F_1(x)$ and $F_2(x)$, sampled in accordance with a two-state Markov chain for which the one-step transition matrix is

$$\mathbf{P}=\begin{pmatrix} \alpha_1 & 1-\alpha_1 \\ 1-\alpha_2 & \alpha_2 \end{pmatrix} \tag{2.18}$$

and the stationary vector is

$$\boldsymbol{\Pi}=\boldsymbol{\Pi}\mathbf{P}=\left(\frac{1-\alpha_2}{2-\alpha_1-\alpha_2}, \frac{1-\alpha_1}{2-\alpha_1-\alpha_1}\right). \tag{2.19}$$

When we form the point process we assume that no information is available about the type of interval, i.e., that in the actual bivariate point process of transitions we suppress knowledge of the type of transition. Then the distribution of an interval between transitions (events) X_i in the stationary point process is

$$F_X(x)=\pi_1 F_1(x)+\pi_2 F_2(x) \tag{2.20}$$

and the correlation between X_i and X_{i+k} is

$$\rho(k)=M^k, \qquad k=1,2,\ldots, \tag{2.21}$$

where M is a positive constant and $\beta=\alpha_1+\alpha_2-1=\alpha_1(1-\alpha_2)$. Thus the correlation structure is that of an ARMA(1, 1) process. For a derivation of this result see Cox and Lewis (1966, Ch. 7, 194–196). Lewis and Shedler (1973) used this process to model the page exception process in a multiprogrammed computer system. The problem is to deal with the mixture distribution (2.20) for the marginal distribution of intervals; this seems to limit the utility of the model. However, there is a way around it which produces a marginally controlled semi-Markov generated process.

To obtain an exponential marginal distribution, consider the following device (Jacobs and Lewis, 1977). Fix x_0, where $0<x_0<\infty$, and let

$$F_1(x)=\begin{cases} \dfrac{\int_0^x \mathrm{e}^{-\lambda u}\,\mathrm{d}u}{1-\mathrm{e}^{-\lambda x_0}} & 0\leqslant x\leqslant x_0, \\ 1 & x>x_0; \end{cases}$$

$$F_2(x)=\begin{cases} 0 & x\leqslant x_0, \\ \dfrac{\int_0^x \mathrm{e}^{-\lambda u}\,\mathrm{d}u}{\mathrm{e}^{-\lambda x_0}} & x>x_0; \end{cases} \tag{2.22}$$

then $F(x)$, the marginal distribution of an interval, is exponential(λ) if we set $\pi_1=1-\exp(-\lambda x_0)$. There is one degree of freedom left in the matrix P; in addition to λ, we have free parameters π_1 (or x_0) and α_1 although the range of α_1 is restricted. What then is the range of β, and can it be negative?

Straightforward manipulation shows that

$$\beta=\frac{\pi_1-\alpha_1}{\pi_1-1}, \tag{2.23}$$

which lies in absolute value between zero and one but can be negative; therefore the serial correlations can be negative. Thus the model appears to be broader than the EARMA(1, 1) model. The question of comparing the two models when β is positive has not yet been explored; it requires higher order interval correlations, as discussed by Brillinger (1972).

2.6. *Generalizations*

The marginally controlled semi-Markov generated sequence $\{X_i\}$ discussed above can be extended in such a way that X_i will have any distribution, say $F(x)$. Thus we let

$$F_1(x)=\begin{cases} \dfrac{F(x)}{F(x_0)} & 0\leqslant x\leqslant x_0, \\ 1 & x>x_0; \end{cases}$$

$$F_2(x)=\begin{cases} 0 & x\leqslant x_0, \\ \dfrac{F(x)-F(x_0)}{1-F(x_0)} & x>x_0; \end{cases} \tag{2.24}$$

then the marginal distribution of an interval is equal to $F(x)$, from (2.30), if we set $\pi_1 = F(x_0)$. Note that the model is very non-linear and the correlation structure is a complicated function of the functional form of $F(x)$.

The much simpler EARMA structure can be extended to some extent. Random variables for which the equation (2.2) has a proper solution are called self-decomposable random variables or random variables of type L. This class includes random variables with Gamma, Cauchy, Pareto, double exponential and perhaps many other distributions. For these random variables, a pth-order-autoregressive process can be defined as at (2.16). The unique feature of the exponential process is that the ε_i which makes X_i exponential(λ) in (2.2) is again an exponential(λ) random variable, albeit mixed with an atom at zero. This property, shared with the double exponential and normal random variables, is what makes it simple to define a moving-average type process, as at (2.9).

3. Count models: sequences of discrete-valued random variables

As remarked earlier, most data on point processes is recorded as numbers of events in successive fixed-length intervals. Despite this fact, most point process models assume that exact times of events are known and it is not simple to derive from these models the statistics of the counts in fixed intervals. Thus in this area in particular flexible models for discrete-valued random variables are needed.

Another application might be to modelling of air pollution data in which concentrations of various chemicals in the air is indicated on a scale of zero to ten. In general this situation requires multivariate time series, but space prohibits discussion of multivariate versions of the DARMA-type processes discussed in this section.

3.1. *The first-order autoregressive discrete model* (DAR(1))

Again we denote the sequence of discrete-valued random variables by $\{X_i\}$. If the X_i are counts in a Poisson process then the X_i's are i.i.d. Poisson-distributed random variables. Once dependence is observed in data it is useful to assume, as a first cut, that the dependence is Markovian and use a Markov chain model in which the distribution of X_{i+1} depends only on the value of X_i and is specified by the transition matrix $\mathbf{P}$ with elements

$$P(k,j) = P\{X_{i+1}=j|X_i=k\}, \tag{3.1}$$

with j and k taking values in the space E, a discrete subset of the real line. Under suitable conditions there is a stationary distribution $\boldsymbol{\pi}$ for $\{X_i\}$ given by the equation

$$\boldsymbol{\pi}=\boldsymbol{\pi}\mathbf{P}. \tag{3.2}$$

The Markov chain model (3.2) is by virtue of its place in the statistician's toolbox the discrete counterpart of the AR(1) process. However the AR(1) process has one dependency parameter ρ, plus any parameters which specify the distribution of the ε_i's. The Markov chain on the other hand can have an infinite number of parameters and in many cases $\boldsymbol{\pi}$ cannot be obtained explicitly from (3.2). This is awkward for statistical analysis. A solution is given by constructing the DAR(1) model (discrete autoregressive model of under one) which is an analog of the EAR(1) model, as follows.

Let Y_i be an i.i.d. sequence of random variables taking values in the space E, and let V_i be an i.i.d. binary sequence for which $P\{V_i=1\}=\rho$. Then

$$X_i=V_iX_{i-1}+(1-V_i)Y_i \qquad i=0,\pm1,\pm2,\ldots; \quad 0\leqslant\rho<1. \tag{3.3}$$

$$=\begin{cases} X_{i-1} & \text{w.p. } \rho, \\ Y_i & \text{w.p. } (1-\rho). \end{cases} \tag{3.4}$$

If X_0 has distribution π, then so does X_1 since it is a mixture of two random variables, X_0 and Y_1, with distribution π. Consequently all the $X_i, i=1,2,\ldots$ have marginal distribution π.

Note that $\{X_i\}$ is a Markov chain with transition probabilities

$$P(k,j)=P\{X_{i+1}=j|X_i=k\}=\begin{cases}(1-\rho)\pi(j) & k\neq j,\\ \rho+(1-\rho)\pi(j) & k=j;\end{cases} \tag{3.5}$$

in fact it is a Markov chain in which the correlation structure is specified by one parameter ρ, and with specified marginal (stationary) distribution π. Thus π may be a Poisson distribution and then the DAR(1) model is a 2-parameter (λ,ρ) Markov chain. The analogy with the AR(1) model is clear.

As with the EAR(1) model the serial correlations are $\rho(k)=\rho^k\geqslant0$. Extensions to negatively correlated sequences are given in Jacobs and Lewis (1978c).

3.2. *The* pth-*order autoregressive discrete model* (DAR(p))

First order Markov dependence is a special kind of dependence which is attractive because of analytical tractability considerations, but it is not necessarily met with in practice. One immediate consequence of the Markovian property is that runs of distinct values, say $X_i = j$, have a length which is geometrically distributed (Jacobs and Lewis, 1978a) and this is easily checked in data. If the data fails to have this property, what other types of dependency can be utilized?

A first direction might be to go to higher order (say pth-order) autoregression, which is an explicit pth-order Markov structure, and the DAR(1) model can be extended in this direction. Thus in addition to the assumptions at (3.3) let A_i be an i.i.d. sequence of random variables taking values in $\{1,2,\ldots,p\}$, with $P\{A_i = j\} = \alpha_j$. Then the DAR(p) process is defined as

$$X_i = V_i X_{i-A_i} + (1 - V_i) Y_i, \qquad i = 0, \pm 1, \pm 2, \ldots \tag{3.6}$$

so that X_i is (exclusively) either one of the previous p values $X_{i-1},\ldots,X_{i-p}$, or the error term Y_i. Properties of this model are developed extensively in Jacobs and Lewis (1978c). When $\alpha_1 = 1$, and all other α_j's are zero it is the DAR(1) model.

Yule-Walker equations for the correlations in the stationary DAR(p) process are given in Jacobs and Lewis (1973c) as well as stationarity conditions. In particular for $p = 2$ we have the limiting result

$$\nu(k,j) = \lim_{i\to\infty} P\{X_{i+1} = k, X_{i+2} = j\}$$

$$= \begin{cases} \{1-\rho(1)\}\pi(k)\pi(j) & k \neq j, \\ \rho(1)\pi(j) + \{1-\rho(1)\}\pi(j)^2 & k = j, \end{cases} \tag{3.7}$$

where $\rho(1) = \text{corr}(X_i, X_{i+1})$ in the stationary process. Thus, if we let X_0 and X_{-1} have the joint distribution $\nu(k,j)$, a stationary, second-order autoregressive process with any marginal distribution can be generated. A scheme for obtaining sequences which are possibly negatively correlated is given in Jacobs and Lewis (1978c).

3.3. *The* qth *order moving average discrete model* (DMA(q))

The other alternative to Markovian dependence (of any order) which is usually considered in time series analysis is the finite-length dependence produced by the moving-average part of the ARMA(p,q) process (1.1). This type of behavior is easily produced for discrete random variables by a random index model of the type

$$X_i = Y_{i-S_i}, \tag{3.8}$$

where S_i are i.i.d. random variables with $P\{S_i \leqslant k\} = b_k$. Thus we may write

$$X_i = Y_{i-k} \quad \text{w.p. } b_k - b_{k-1}, \qquad k=0,\dots,q; \quad b_{-1}=0. \tag{3.9}$$

The autoregressive process DAR(p) is also a random index model, but the random indices are not independent. The correlation structure of this DMA(q) process is easily found to be

$$\begin{aligned} \rho^{(q)}(k) = \operatorname{corr}(X_i X_{i-k}) &= \sum_{v=0}^{q-k} b_v b_{v+k} \qquad 1 \leqslant k \leqslant q, \\ &= 0 \qquad\qquad\qquad k > q. \end{aligned} \tag{3.10}$$

This is the exact analog of (2.11) for the EMA(q) process and the corresponding formula for the MA(q) process. Note that the DMA(q) process is not Markovian. Runs properties of the process are given in Jacobs and Lewis (1978a); the runs are not geometrically distributed.

3.4. *Mixed autoregressive-moving average discrete models*

As in the case of the ARMA(p,q) model (1.1), it is useful to have both autoregressive, Markovian dependence and moving average dependence combined into one model. In Jacobs and Lewis (1978a) this was done by replacing the Y_{i-q} term in (3.8) by a discrete autoregression (3.3) over $Y_{i-q}, Y_{i-q-1}, \dots$. Clearly this can be extended by replacing Y_{i-q} by a p-th order autoregression (3.6) over $Y_{i-q}, Y_{i-q-1}, \dots$ to obtain a DARMA(p,q) model which is the analog of the EARMA(p,q) model of Lawrance and Lewis (1978). This is not a complete analog of the ARMA(p,q) model in that there is no cross-over of the autoregression and the moving average, but it is in fact possible to do this to obtain a model called NDARMA(p,q) as follows:

Let

$$X_i = V_i X_{i-A_i} + (1-V_i) Y_{i-S_i} \qquad i=0, \pm 1, \pm 2, \dots, \tag{3.11}$$

where the A_i are i.i.d. random variables taking values in $\{1,2,\dots,p\}$ with

$P\{A_i=j\}=\alpha_j$; the S_i are i.i.d. random variables taking values in $\{0,\ldots,q\}$ with $P\{S_i\leqslant k\}=F(k)$ and the V_i's are i.i.d. Bernoulli random variables with $P\{V_i=1\}=\rho$.

The model works because a mixture of *dependent* random variables, all with marginal distribution π, has distribution π; thus if $X_{i-1},\ldots,X_{i-p}$ have marginal distribution π, then so will X_i since it is a mixture of the dependent random variables $X_{i-1},\ldots,X_{i-p}$ and $Y_i,\ldots,Y_{i-q}$. Note that when $\rho=0$ we have the DMA(q) process; if in addition $F(0)=1$ the sequence is i.i.d. since $X_i=Y_i$. When $1>\rho\neq 0$, $F(0)=1$ we have the DAR(p) process. Thus the parameters are such that interesting special cases fall out easily. Moreover the ρ parameter measures the degree of mixture of Markovian and moving average dependence, and the distributions of the A_i's and S_i's give a picture of where the dependence is lagged over previous X_i or Y_i values.

The model (3.11) has not yet been fully explored. At first sight it seems preferable to the DARMA(p,q) model, possibly because of the compactness of (3.11) and its close analogy to ARMA(p,q) models. The DARMA(p,q) and NDARMA(p,q) models are, however, distinct and in fact preliminary investigation of the (1.1) case shows that the DARMA(1,1) model (Jacobs and Lewis, 1978a) has a broader correlation structure than does the NDARMA(1,1). On the other hand the autoregression is not explicit in the DARMA(1,1) model. Both models, therefore, will probably be useful in modelling discrete data such as occur in sampled point processes.

3.5. *The marginally controlled semi-Markov generated process*

In the structure of the 2-state marginally controlled semi-Markov generated process detailed at (2.24) no assumption was made about continuity of $F(x)$. Thus $F(x)$ could be discrete, giving a sequence $\{X_i\}$ with known discrete marginal distribution $F(x)$ and ARMA(1,1) correlation structure. By going to an n-state semi-Markov model, a process with ARMA(p,q) correlation structure can be generated (Haskell and Lewis, 1978) with n a function of p and q, and the procedure to obtain a given marginal distribution is just an extension of (2.24). Thus we have, in terms of the quantification of the process by marginal distribution and correlation structure, a direct competitor to the DARMA-type processes.

Comparison of the two types of discrete processes is interesting and points up the simplicity of the DARMA-type processes. In particular the correlation structure of the DARMA(p,q) process is explicit in form if not in detail and the process is a simple, random linear combination of random variables generated from an i.i.d. sequence Y_i. This is clearly not true for the marginally controlled semi-Markov generated process; the recognition that its correlation structure is ARMA-type is accidental and not intuitive. Deeper comparison of these processes in terms, say, of the range of correlation the model will encompass will be instructive. Here again the DARMA-type processes have an advantage; their correlation structure is independent of the marginal distribution π.

4. Summary and conclusions

We have outlined in this paper three models for discrete-valued and positive-valued time series, all of which to some degree satisfy the criteria of flexibility or simplicity or both set forth in the introduction. Perhaps the main point about the models is that they are designed to accomodate situations in which the marginal distributions in the stationary processes are given and are non-normal.

Properties of these models such as mixing and asymptotic results, higher-order moments, distributions of runs for the discrete models and sums of random variables and point spectra are considered in the references.

There are many other properties of the processes which are still to be explored. Statistical estimation, except in an ad hoc manner and for the Markovian cases, is difficult and has yet to be examined. Extensions to multivariate cases is of great interest for real applications and has been done to some degree in the context of queues with correlated service and arrival times (Jacobs, 1978, and Lewis and Shedler, 1978). The DARMA-type processes, in particular, can be easily extended to coupled equations in the same way as linear processes are extended in econometric models. They might therefore find use in modelling multivariate situations such as the number of cars passing different points in a road evaluated in successive fixed time intervals.

Finally an important problem is to extend the models so as to include inhomogeneity, particularly of the seasonal type, and the effects of concomittant or auxilliary variables. Several schemes are under consideration for these extensions of the models.

References

[1] Box, G. E. P. and Jenkins, G. (1970). *Time Series Analysis, Forecasting and Control*. Holden Day, San Francisco.

[2] Brillinger, D. R. (1972). The spectral analysis of stationary interval functions. *Proc. 6th Berkeley Symp. Math. Stat. and Prob.* University of California Press, Berkeley, 483–514.

[3] Cox, D. R. (1955). Some statistical methods connected with series of events. *J. R. Statist. Soc. B* **17**, 129–164.

[4] Cox, D. R. (1962). *Renewal Theory*. Methuen, London.

[5] Cox, D. R. and Lewis, P. A. W. (1966). *The Statistical Analysis of Series of Events*. Methuen, London and Wiley, New York.

[6] Gaver, D. P. and Lewis, P. A. W. (1978). First order autoregressive Gamma sequences. Naval Postgraduate School Report NPS5-78-016.

[7] Haskell, R. and Lewis, P. A. W. (1978). Interval spectra of semi-Markov generated point processes. To appear.

[8] Jacobs, P. A. (1978). A closed cyclic queuing network with dependent exponential service times. *J. Appl. Prob.* **15**, 573–589.

[9] Jacobs, P. A. and Lewis, P. A. W. (1977). A mixed autoregressive-moving average exponential sequence and point process (EARMA(1, 1)). *Adv. Appl. Prob.*, **9**, 87–104.

[10] Jacobs, P. A. and Lewis, P. A. W. (1978a). Discrete time series generated by mixtures. I: Correlational and runs properties. *J. R. Statist. Soc. B*, **40**(1), 94–105.

[11] Jacobs, P. A. and Lewis, P. A. W. (1978b). Discrete time series generated by mixtures. II: Asymptotic properties. *J. R. Statist. Soc. B* **40**(2), 222–228.

[12] Jacobs, P. A. and Lewis, P. A. W. (1978c). Discrete time series generated by mixtures. III: Autoregressive processes (DAR(p)). To appear.

[13] Lawrance, A. J. (1978). Some autoregressive models for point processes. *Proc. Bolyai Janos Math. Soc. Colloquium on Point Processes and Queuing Theory*, to appear.

[14] Lawrance, A. J. and Lewis, P. A. W. (1977). A moving average exponential point process (EMA1). *J. Appl. Prob.*, **14**, 98–113.

[15] Lawrance, A. J. and Lewis, P. A. W. (1978). An exponential autoregressive-moving average process EARMA(p, q): Definition and correlational properties. Naval Postgraduate School Report NPS55-78-1. To appear in *J. R. Statist. Soc. B*.

[16] Lewis, P. A. W. and Shedler, G. S. (1973). Empirically derived micro-models for sequences of page exceptions. *IBM J. Res. Dev.*, **17**, 86–100.

[17] Lewis, P. A. W. and Shedler, G. S. (1978). Analysis and modelling of point processes in computer systems. *Bull. Int. Statist. Inst.*, **47**(2), 193–210.

P. R. Krishnaiah, ed., *Multivariate Analysis–V*

ON THE PREDICTION OF PERIODICALLY CORRELATED STOCHASTIC PROCESSES

A. G. MIAMEE
Arya-Mehr University of Technology, Isfahan, Iran

and

H. SALEHI*
Michigan State University, East Lansing, MI 48824

Periodically correlated stochastic processes are studied. A Wold–Cramer concordance for such processes is established. Algorithms for determining the linear predictor and the prediction error vector for these processes are obtained.

1. Introduction

Periodically correlated stochastic processes which are a natural generalization of stationary random processes have been studied by several authors including L. I. Gudzinko [7], E. G. Gladyšev [6], and H. Ogura [15]. Most often random processes encountered in engineering, physics, and some other fields are assumed to be stationary. However, in many important situations such as the study of linear and nonlinear systems in communication engineering and the study of time series in economics, periodically correlated nonstationary processes are more suitable tools in describing some important phenomena [3, 7, 15]. Periodically correlated random processes constitute a subclass of the so called harmonizable processes which were introduced by M. Loève [9] and studied by several authors such as H. Cramér [4], Yu A. Rozanov [18], J. L. Abréu [1], H. Niemi [14] and A. G. Miamee and H. Salehi [12]. In these papers concepts of spectral representation and spectral distribution for harmonizable processes are introduced, and some basic facts concerning these processes are derived. Nevertheless, since the class of harmonizable processes is so broad, unlike the case of stationary processes, these studies are not as conclusive. But for periodically correlated processes some questions such as spectral distribution have been satisfactorily settled, in fact it had been

*Supported in part by NSF MCS 77-02661

shown [6], that the spectral distribution of a periodically correlated process of period q is concentrated on the $2q-1$ equidistant straight line segments contained in the square $[0,2\pi]\times[0,2\pi]$ which are parallel to its main diagonal.

In this paper we pursue the study of these periodically correlated processes and establish necessary and sufficient conditions for the Wold and Cramér decompositions to be concordant. We also obtain an algorithm for determining the predictor and prediction error vector of such processes. Such an algorithm has been proved to be significant and useful in the prediction theory, and for the stationary processes has been made available by E. J. Akutowicz [2] for univariate and by N. Wiener and P. Masani [19] for the multivariate case.

Section 2 contains the notations and terminologies. In Section 3 some results concerning the Wold-Cramér concordance of periodically correlated processes are established. Section 4 is devoted to an algorithm for determining the predictor and prediction error vector.

2. Preliminaries

Let (Ω,β,P) be a probability space. $\mathcal{H}=L_2(\Omega,\beta,P)$ denotes the Hilbert space of all complex-valued random variables on Ω with zero expectation and finite variance. The inner product and norm of $\mathcal{H}$ are given by

$$(x,y)=\int_\Omega x(\omega)\,\overline{y(\omega)}\,dP(\omega) \qquad x,y\in\mathcal{H}$$

and

$$\|x\|=\sqrt{\int_\Omega |x(\omega)|^2\,dP(\omega)}.$$

Following [11] for $q\geqslant 1$, $\mathcal{H}^q$ denotes the cartesian product of $\mathcal{H}$ with itself q times, i.e. The set of all column vectors $X=(x^0,x^1,\ldots,x^{q-1})^T$ with $x^i\in\mathcal{H}$, for all $i=0,1,\ldots,q-1$. As usual we endow the space $\mathcal{H}^q$ with a Gramian structure: For X and Y in $\mathcal{H}^q$ their Gramian (X,Y) is defined to be the $q\times q$ matrix

$$(X,Y)=\left[(x^i,y^j)\right]_{i,j=0}^{q-1}.$$

One can easily verify that

$$(X,X)\geqslant 0; \qquad (X,X)=0 \Leftrightarrow X=0;$$

$$\left(\sum_{k=1}^{m} A_k X_k, \sum_{l=1}^{n} B_l Y_l\right)=\sum_{k=1}^{m}\sum_{l=1}^{n} A_k(X_k,Y_l)B_l^*,$$

for any $X,X_k,Y_l\in\mathcal{H}^q$ and any $q\times q$ matrices A_k,B_l. We say that X is orthogonal to Y in $\mathcal{H}^q$ if $(X,Y)=0$. It is well known that $\mathcal{H}^q$ is a Hilbert space under the following inner product

$$((X,Y))=\operatorname{trace}(X,Y)=\sum_{j=0}^{q-1}(x^j,y^j).$$

A closed subset $\overline{\mathfrak{M}}$ of $\mathcal{H}^q$ is called a subspace of $\mathcal{H}^q$ if it is a manifold, i.e., $Ax+By\in\overline{\mathfrak{M}}$ whenever $X,Y\in\overline{\mathfrak{M}}$ and A,B are $q\times q$ matrices. It is easy to see that $\overline{\mathfrak{M}}$ is a subspace of $\mathcal{H}^q$ if and only if there exists a subspace $\mathfrak{M}$ of $\mathcal{H}$ such that $\overline{\mathfrak{M}}=\mathfrak{M}^q$. For any $x\in\mathcal{H}$, its orthogonal projection on a subspace $\mathfrak{M}$ of $\mathcal{H}$ is denoted by $(x|\mathfrak{M})$. Given a vector $X=(x^0,x^1,\ldots,x^{q-1})^T\in\mathcal{H}^q$, its projection on a subspace $\overline{\mathfrak{M}}=\mathfrak{M}^q$ is the vector $(X|\overline{\mathfrak{M}})$ whose ith component is $(x^i|\mathfrak{M})$ for each $i=0,1,\ldots,q-1$. For random variables $\{x_j\}_{j\in J}$ in $\mathcal{H}$ we denote by $\mathfrak{S}\{x_j, j\in J\}$ the subspace spanned by x_j, for all j in the indexed set J. Similarly for random vectors $\{X_j\}_{j\in J}$ in $\mathcal{H}^q$, $\mathfrak{S}\{X_j, j\in J\}$ is the subspace of $\mathcal{H}^q$ spanned by all $X_j, j\in J$.

A sequence $x_n, -\infty<n<\infty (X_n, -\infty<n<\infty)$ of elements of $\mathcal{H}(\mathcal{H}^q)$ is called a univariate (q-variate) stochastic process. For convenience we may abbreviate $x_n, -\infty<n<\infty$ by x_n or simply x. A stochastic process x_n is called periodically correlated of period q if the correlation function

$$R(n,\tau)=(x_{n+\tau},x_n)$$

is periodic in n of period q. (We should mention that when $q=1$ the process is called stationary). Since $R(n,\tau)$ is periodic in n with period q, one can write

$$R(n,\tau)=\sum_{k=0}^{q-1} R_k(\tau)\exp\left(\frac{2\pi ikn}{q}\right).$$

For convenience we extend the definition of these functions $R_k(\tau), k=0,1,2,\ldots,q-1$, to all integers by $R_k(\tau)=R_{k+q}(\tau)$. It is shown in [6] that each $R_k(\tau)$ has the representation

$$R_k(\tau)=\frac{1}{2\pi}\int_0^{2\pi}e^{-i\tau\lambda}\,dF_k(\lambda),$$

where each $F_k(\cdot)$ is a complex-valued measure on $[0,2\pi]$. Let $\mathcal{F}(\cdot)$ be the $q\times q$ matrix valued measure, given on intervals by

$$\mathcal{F}(\lambda_1,\lambda_2]=\left[F_{k-j}((\lambda_1+2\pi j)/q,(\lambda_2+2\pi j)/q]\right]_{j,k=0}^{q-1},\lambda_1\leqslant\lambda_2. \quad (1)$$

It is proved in [6] that $\mathcal{F}(\cdot)$ is a nonnegative definite hermitian matrix valued measure. It is also shown in [6] that

$$R(n,\tau)=\frac{1}{4\pi^2}\int_0^{2\pi}\int_0^{2\pi}e^{-i(n+\tau)\lambda_1+in\lambda_2}\,dF(\lambda_1,\lambda_2), \quad (2)$$

where the spectral measure $F(\cdot,\cdot)$ is given by

$$F(A,B)=\sum_{k=-q+1}^{q-1}\int_{A\cap(B-2\pi k/q)}dF_k(\lambda) \quad (3)$$

(B-a is the set of all b-a with $b\in B$). In other words the spectral measure $F(\cdot,\cdot)$ is concentrated on $2q-1$ straight line segments $\lambda_1-\lambda_2=2\pi k/q, k=-q+1,\ldots,q-1$, contained inside the square $0\leqslant\lambda_1,\lambda_2\leqslant 2\pi$, and the measures $F_k(\cdot)$ give the mass of $F(\cdot,\cdot)$ on these lines according to (3). Representation (2) and formula (3) in particular show that any periodically correlated process is harmonizable. With any $\mathcal{H}$-valued process x_n we associate the $\mathcal{H}^q$-valued process X_n whose ith coordinate x_n^i is given by $x_{nq+i}, i=0,1,2,\ldots,q-1$. This correspondence establishes a one-to-one linear transformation S_q from the $\mathcal{H}$-valued processes onto the $\mathcal{H}^q$-valued processes. In other words we have

$$(S_qx)_n=\begin{bmatrix} x_{nq}\\ x_{nq+1}\\ \vdots\\ x_{nq+q-1}\end{bmatrix}.$$

From now on we will simply write S instead of S_q. One can easily verify that x is periodically correlated with period q if and only if Sx is a q-variate stationary stochastic process. (For the theory of multivariate stationary stochastic processes, see [11]). This associated q-variate stationary stochastic process has a spectral measure F which is a $q \times q$ nonnegative definite matrix valued measure such that

$$((Sx)_n, (Sx)_0) = \frac{1}{2\pi} \int_0^{2\pi} \bar{e}^{in\lambda} \, dF(\lambda).$$

The following theorem which slightly improves Theorem 2 of [6] gives the relation between this measure F and the earlier measure $\mathcal{F}$ given in (1). Proof of the theorem is simple and hence is omitted.

Theorem 2.1. *With the above notations we have*

$$dF(\lambda) = qU^*(\lambda) \, d\mathcal{F}(\lambda) \, U(\lambda); \tag{4}$$

in the sense that given any set A

$$F(A) = \int_A qU^*(\lambda) \, d\mathcal{F}(\lambda) \, U(\lambda).$$

(For the definition and properties of this type of integral one may refer to [17]), here U is a unitary valued function whose (j,k)th entry is

$$q^{-\frac{1}{2}} \exp\left[\frac{2\pi i jk + ik\lambda}{q} \right].$$

3. Wold decomposition

In this section a Wold decomposition for a periodically correlated process is established and some necessary and sufficient conditions for its concordance with its Cramér decomposition is obtained.

Let $x_n(X_n)$ be a univariate (q-variate) process. We define its past and present at n by $\mathfrak{M}(x,n) = \mathfrak{S}\{x_m, m \leqslant n\}(\mathfrak{M}(X,n) = \mathfrak{S}\{x_m^i, m \leqslant n, 0 \leqslant i \leqslant q-1\})$. The process $x_n(X_n)$ is called regular if $\mathfrak{M}(x, -\infty) = \bigcap_n \mathfrak{M}(x,n) = 0$ $(\mathfrak{M}(X, -\infty) = \bigcap_n \mathfrak{M}(X,n) = 0)$ and singular if for all n, $\mathfrak{M}(x,n) = \mathfrak{M}(x,n+1), (\mathfrak{M}(X,n) = \mathfrak{M}(X,n+1))$.

A process $x_n(X_n)$ is called subordinated to $y_n(Y_n)$ if for each n, $\mathfrak{M}(x,n) \subset \mathfrak{M}(y,n)(\mathfrak{M}(X,n) \subseteq \mathfrak{M}(Y,n))$. The Wold decomposition is well-known for any $\mathcal{H}$-valued stochastic process [4]. If x_n is assumed to be periodically correlated with period q then it is easy to verify that the resulting Wold decomposition components are also periodically correlated with the same period. Hence:

Theorem 3.1. (Wold decomposition Theorem). *Let x_n be a periodically correlated process of period q. Then there exists a unique decomposition*

$$x_n = y_n + z_n, \qquad -\infty < n < \infty$$

such that

(a) *y_n and z_n are periodically correlated with period q,*

(b) *y_n and z_n are orthogonal to each order,*

(c) *y_n is regular and z_n is singular,*

(d) *y_n and z_n are subordinated to x_n.*

We will later need to know the relation between the Wold components y_n and z_n of x_n as described in Theorem 3.1 and the Wold components Y_n and Z_n of its associated stationary process $X = Sx$. Since $\mathfrak{M}(x, -\infty) = \mathfrak{M}(X, -\infty)$ the following theorem revealing this relation immediately follows.

Theorem 3.2. *With the above notations we have $Y = Sy$ and $Z = Sz$.*

Let x_n be a periodically correlated process of period q. We define the rank of x_n, denoted by rank(x), to be the number of integers i in $0 \leqslant i \leqslant q-1$ (or equivalently in $n \leqslant i \leqslant n+q-1$ for any integer n) such that $x_i \notin \mathfrak{M}(x, i-1)$. This is what H. Cramér [4] calls the cardinality of innovations of x_n in the set $\{0, 1, 2, \ldots, q-1\}$.

The following theorem exhibits the relation between the rank of a periodically correlated process x and the rank of its associated q-variate stationary stochastic process Sx.

Theorem 3.3. *Let x_n, $-\infty < n < \infty$ be a periodically correlated process with period q. Then*

$$\text{rank}(x) = \text{rank}(Sx).$$

Proof. Let r be rank(x). First we assume that $0 < r < q$. Let n_1 be the smallest nonnegative integer less than $q-1$ such that $x_{n_1} \in \mathfrak{M}(x, n_1 - 1)$. Then $x_{n_1} = \Sigma_{0 \leqslant i \leqslant n_1 - 1} \alpha_i x_i + h$, for some complex numbers $\alpha_0, \alpha_1, \ldots, \alpha_{n_1 - 1}$

and some h in $\mathfrak{M}(x,-1)$. Hence, denoting by P the orthogonal projection on $\mathfrak{M}(x,-1)$, we have $Px_{n_1}=\Sigma_{0\leqslant i\leqslant n_1-1}\alpha_i Px_i+h$. Thus $\hat{x}_{n_1}=\Sigma_{0\leqslant i\leqslant n_1-1}\alpha_i\hat{x}_i$ where for each integer $i, \hat{x}_i=x_i-Px_i$. This shows that $\hat{x}_0,\hat{x}_1,\ldots,\hat{x}_{n_1}$ are linearly dependent. It is not hard to show that $s=n_1$ is the smallest nonnegative integer for which $\hat{x}_0,\hat{x}_1,\ldots,\hat{x}_s$ are linearly dependent. In fact suppose s is the smallest nonnegative integer less than n_1 such that $\hat{x}_0,\hat{x}_1,\ldots,\hat{x}_s$ are linearly dependent. Either $s=0$ which means $\hat{x}_0=0$ and $0=s<n_1$. Hence $0<n_1$ and $x_0\in\mathfrak{M}(x,-1)$ which contradicts the choice of n_1, or $0<s<n_1$ in which case we have $\hat{x}_s=\Sigma_{i=0}^{s-1}\alpha_i\hat{x}_i$. Hence $x_s-Px_s=\Sigma_{i=0}^{s-1}\alpha_i(x_i-px_i)$,

$$x_s=\sum_{i=0}^{s-1}\alpha_i x_i+P\left(x_s-\sum_{i=0}^{s-1}\alpha_i x_i\right)=\sum_{i=0}^{s-1}\alpha_i x_i+h,$$

with h in $\mathfrak{M}(x,-1)$. Thus $x_s\in\mathfrak{M}(x,s-1)$ which contradicts the choice of n_1. Now let n_2 be the smallest integer larger than n_1 such that $x_{n_2}\in\mathfrak{M}(x,n_2-1)$. By an argument similar to the one given above one can show that $\hat{x}_0,\hat{x}_1,\ldots,\hat{x}_{n_1-1},\hat{x}_{n_1+1},\ldots,\hat{x}_{n_2}$ are linearly dependent and in fact $s=n_2$ is the smallest integer larger than n_1 for which $\hat{x}_0,\hat{x}_1,\ldots,\hat{x}_{n_1-1},\hat{x}_{n_1+1},\ldots,\hat{x}_s$ are linearly dependent. Continuing this argument we obtain a sequence $n_1<n_2<\cdots<n_{q-r}$, with r being the rank of x_n, such that $x_n\in\mathfrak{M}(x,n-1)$ for all $n\in\{n_1,n_2,\ldots,n_{q-r}\}$ and $x_n\notin\mathfrak{M}(x,n-1)$ for all other integers n in $\{0,1,2,\ldots,q-1\}$. The above argument also shows that the number of linearly independent vectors in $\{\hat{x}_0,\hat{x}_1,\ldots,\hat{x}_{q-1}\}$ is at most r. We claim this number is exactly r. Because otherwise there should exist an integer s less than q and greater than n_{q-r} such that $\hat{x}_s=\Sigma_{k=0}^{s-1}\alpha_k\hat{x}_k$ which implies

$$x_s=\sum_{0<k<s-1}\alpha_k x_k+h,$$

with $h\in\mathfrak{M}(x,-1)$. But this means that $x_s\in\mathfrak{M}(x,s-1)$. Since $s>n_{q-r}$ this is a contradiction to the properties of $n_1,n_2,\ldots,n_{q-r}$ mentioned above. Thus the number of linearly independent vectors in $\{\hat{x}_0,\hat{x}_1,\ldots,\hat{x}_{q-1}\}$ is exactly equal to $r=\text{rank}(x)$. But this number is exactly the rank of the q-variate stationary stochastic process Sx [16]. This completes the proof in this case.

If $r=0$ then obviously $\text{rank}(Sx)=0$. If $r=q$ then rank $(Sx)=q$, because otherwise there exists a nonnegative integer less than q such that

$$\hat{x}=\sum_{i=0}^{s-1}\alpha_i\hat{x}_i.$$

This implies that $x_s \in \mathfrak{M}(x, s-1)$ which contradicts $r=q$. This completes the proof of the lemma.

The problem of Wold–Cramér concordance plays an important role in the prediction theory of stochastic processes. For a univariate stationary stochastic process a complete solution to this problem may be found in [5]. For a full rank q-variate stationary stochastic process, the problem was solved by N. Wiener and P. Masani [19] and for the non full rank case an answer was given by J. B. Robertson [16]. We will need the following result of J. B. Robertson [16] on concordance which for completeness we state it here.

Theorem 3.4. *Let X_n be a q-variate stationary stochastic process. Let $X_n = Y_n + Z_n$ be its Wold decomposition and F_X, F_Y and F_Z be the spectral measures of the process X_n, Y_n and Z_n respectively. Let $F_X = F^a + F^s$ be the Cramér decomposition of the spectral measure F_X. Then $F_Y = F^a$ and $F_Z = F^s$ if and only if $rank(X) = rank(F'_X)$, a.e.*

As we mentioned before the spectral measure F_x of a periodically correlated process x_n of period q is concentrated on $2q-1$ straight line segments $\lambda_1 - \lambda_2 = 2k\pi/q$, $-q+1 \leqslant k \leqslant q-1$, inside the square $0 \leqslant \lambda_1, \lambda_2 \leqslant 2\pi$. Let

$$F_x = F_x^a + F_x^s \tag{5}$$

be its Cramér decomposition with respect to the one dimensional Lebesgue measure on these $2q-1$ straight line segments and let $x_n = y_n + z_n$ be the Wold decomposition of x_n. Then $F_x = F_y + F_z$, where F_y and F_z are the spectral measure of the periodically correlated components y_n and z_n of x_n. The following theorem gives necessary and sufficient conditions for the concordance of these decompositions.

Theorem 3.5. (Wold–Cramér concordance). *With the above notations the Wold and Cramér decompositions are concordant, i.e., $F_y = F_x^a$ and $F_z = F_x^s$, if and only if* rank(x) = rank$(\mathfrak{F}'_x(\lambda))$, *a.e. Here $\mathfrak{F}'_x(\lambda)$ is the derivative of the measure $\mathfrak{F}_x$ given in relation* (1).

Proof. Let X_n be the associated q-variate stationary process of x_n, i.e. let $X = Sx$, and let $F_x = F_x^a + F_x^s$ and $F_X = F_X^a + F_X^s$ be the Cramér decomposition of F_x and F_X respectively. Suppose $x_n = y_n + z_n$ and $X_n = Y_n + Z_n$ are the Wold decompositions of x_n and X_n respectively. The proof of the theorem consists of the following four steps.

(i) By (4), rank $(F_x'(\lambda)) = \text{rank}(\mathfrak{F}_x'(\lambda))$, a.e.

(ii) By Lemma 3.3, $\text{rank}(x) = \text{rank}(Sx)$.

(iii) By Theorem 3.4, $F_Y = F_X^a$ and $F_Z = F_X^s$ if and only if $\text{rank}(X) = \text{rank}(F_X')$ a.e.

(iv) $F_x^a = F_y$ and $F_x^s = F_z \Leftrightarrow F_X^a = F_Y$ and $F_X^s = F_Z$, or equivalently

$$F_x^a = F_y \Leftrightarrow F_X^a = F_Y.$$

To show (iv) we first note that by Lemma 3.2, $Y = Sy$ and $Z = Sz$. Now suppose that $F_y = F_x^a$. Hence F_y and F_x^a have the same spectral intensity on each of the $2q-1$ straight line segments mentioned above. But the spectral intensities of F_x and F_y are given by formula (3) in terms of the measures F_k of x_n and those of y_n respectively, and an inspection of (3) shows that each F_k of y_n is equal to the absolutely continuous component of the corresponding F_k of x_n. Therefore by (1) we have

$$(\mathfrak{F}_y)_{jk} = (\mathfrak{F}_x^a)_{jk}, \qquad 0 \leqslant j,k \leqslant q-1.$$

Now by (4), for $0 \leqslant j,k \leqslant q-1$, we have

$$\begin{aligned}(dF_y)_{jk} &= q \sum_{l,m=0}^{q-1} U_{jl}^*(d\mathfrak{F}_y)_{lm} U_{mk} \\ &= q \sum_{l,m=0}^{q-1} U_{jl}^*(d\mathfrak{F}_x^a)_{ml} U_{mk} \\ &= q \sum_{l,m=0}^{q-1} U_{jl}^*(d\mathfrak{F}_x^a)_{lm} U_{mk} = (dF_X^a)_{jk},\end{aligned}$$

which means $F_Y = F_X^a$. The proof of the converse is similar and hence is omitted.

Remark 3.6. The spectral measure of a stationary process is defined on $[0, 2\pi]$ and can be realized as a measure on the main diagonal $\lambda_1 = \lambda_2$ of $0 \leqslant \lambda_1, \lambda_2 \leqslant 2\pi$, and for the Wold-Cramér concordance one considers its Cramér decomposition with respect to the one dimensional Lebesgue measure on this diagonal. For the purpose of concordance of a periodically correlated process x_n of period q we were lead to the Cramér decomposition $F_x = F_x^a + F_x^s$ of F_x with respect to the one dimensional Lebesgue measure of the $2q-1$ straight line segments mentioned before. The question which remains open is to formulate a Wold-Cramér concordance theorem for a general harmonizable process and to see what kind of Cramér decomposition would be appropriate here.

4. An algorithm for determining the predictor and the predictor error vector.

The problem of finding an algorithm for determining the best linear predictor is important in most applications of the theory of stochastic processes. In this section we will give such an algorithm for periodically correlated processes.

Let x_n be a periodically correlated stochastic process of period q. For any given time n the best linear predictor of x_n, in the least square sense, with respect to the past of x_n process up to time $n-1$ is simply $(x_n|\mathfrak{M}(x, n-1))$. So our problem is to get an algorithm for computing $(x_n|\mathfrak{M}(x, n-1)$ for all integers n. But because of periodicity it suffices to compute these predictors just for integers $n=0,1,2,\ldots,q-1$. Let X_n be the associated q-variate stationary stochastic process of x_n. Let F_x and F_X as usual denote the spectral measures of x_n and X_n respectively. Suppose $\mathfrak{F}_x$ is as given in (1). For the purpose of this section we assume that the spectral measure F_x of x_n and hence its Fourier coefficients, i.e. the correlation function of x_n are known and we give our algorithms in terms of these known quantities. According to relations (1), (3) and the definition of $X = Sx$ we can compute $\mathfrak{F}_x$ and F_X and hence we may assume these to be known as well.

For the rest of this section we assume that $\mathfrak{F}_x$ satisfies the following assumption and we obtain our algorithms under this assumption.

Assumption. $\mathfrak{F}_x$ is absolutely continuous with respect to the Lebesgue measure and its derivative $\mathfrak{F}'(\lambda)$ is invertible, a.e. Furthermore $\mathfrak{F}_x' \in L_\infty$ and $(\mathfrak{F}_x')^{-1} \in L_1$.

Now from relation (4) of Theorem 2.1 one concludes that F_X is absolutely continuous, $(F_X')^{-1}$ exists, a.e., $F_X' \in L_\infty$, and $(F_X')^{-1} \in L_1$. Under these assumptions the density F_X' admits a factorization of the form

$$F_X'(\lambda) = \Phi(\lambda)\Phi^*(\lambda),$$

where the optimal factor Φ and its reciprocal Φ^{-1} have square integrable entries and have only nonnegative Fourier coefficients:

$$\Phi(\lambda)=\sum_{k=0}^{\infty} C_k e^{ik\lambda}, \qquad \Phi^{-1}(\lambda)=\sum_{k=0}^{\infty} D_k e^{ik\lambda}.$$

The matrices C_k and D_k are computable as can be seen in [10]. Let $E_k=\sum_{n=1}^{k} C_n D_{k-n}$. It is shown in [10] that

$$\left(X_O|\overline{\mathfrak{M}}(X,-1)\right)=\sum_{k=1}^{\infty} E_k X_{-k}, \tag{6}$$

where for each integer n, $\overline{\mathfrak{M}}(X,n)=\mathfrak{S}\{X_i, i\leqslant n\}$.

Since $\overline{\mathfrak{M}}(X,-1)=(\mathfrak{M}(x,-1))^q$ we can rewrite relation (6) in the form

$$\left(x_0^j|\mathfrak{M}(X,-1)\right)=\sum_{k=1}^{\infty}\sum_{i=0}^{q-1}(E_k)_{ji}X_{-k}^i, \qquad j=0,1,\ldots,q-1.$$

which means

$$\left(x_j|\mathfrak{M}(x,-1)\right)=\sum_{k=1}^{\infty}\sum_{i=0}^{q-1}(E_k)_{ji}x_{-qk+i}, \qquad j=0,1,\ldots,q-1. \tag{7}$$

As mentioned above, to get an algorithm for the predictors, we have to compute the projections $(x_j|\mathfrak{M}(x,j-1))$ for each j, $0\leqslant j\leqslant q-1$. For $j=0$, this is simply given by $\sum_{k=1}^{\infty}\sum_{i=0}^{q-1}(E_k)_{0i}X_{-qk+i}$ according to (7). So it remains to determine $(x_j|\mathfrak{M}(x,j-1)$ for each j, $1\leqslant j\leqslant q-1$. Take an arbitrary fixed j with $1\leqslant j\leqslant q-1$. Obviously there exists scalars $\alpha_0,\alpha_1,\ldots,\alpha_{j-1}$ such that

$$\left(x_j|\mathfrak{M}(x,j-1)\right)=\left(x_j|\mathfrak{M}(x,-1)\right)+\sum_{k=0}^{j-1}\alpha_k x_k.$$

It remains to determine these scalar α_k's. Clearly the vector

$$x_j-\left(x_j|\mathfrak{M}(x,j-1)\right)=x_j-\left(x_j|\mathfrak{M}(x,-1)\right)-\sum_{k=0}^{j-1}\alpha_k x_k$$

is orthogonal to all x_l, $0\leqslant l\leqslant j-1$. Hence we get the following system of j

equations in j unknowns $\alpha_0, \alpha_1, \ldots, \alpha_{j-1}$:

$$0=(x_l,x_j)-\left(x_l,(x_j|\mathfrak{M}(x,-1))-\sum_{k=0}^{j-1}\alpha_k(x_l,x_k)\right. \qquad 0\leqslant l\leqslant j-1,$$

or

$$\sum_{k=0}^{j-1}\alpha_k R(k,l-k) = \sum_{k=1}^{\infty}\sum_{i=0}^{q-1}(E_k)_{ji}R(i,l+qk-i)-R(j,l-j),$$

$$0\geqslant l\geqslant j-1,$$

and by solving this system of linear equation we can compute α_k's and hence the predictors.

We can also obtain an algorithm for determining the prediction error vector $G_x=[g_j]_{j=0}^{q-1}$, where for each j, $0\leqslant j\leqslant q-1$, $g_j=\|x_j-(x_j|\mathfrak{M}(x,j-1)\|^2$.

From $x_j-(x_j|\mathfrak{M}(x,j-1))=x_j-(x_j|\mathfrak{M}(x,-1)+(x_j|\mathfrak{M}(x,-1))-(x_j|\mathfrak{M}(x,j-1)$ it is clear that $g_j=\|x_j-(x_j|\mathfrak{M}(x,-1)\|^2-\|(x_j|\mathfrak{M}(x,j-1)-(x_j|\mathfrak{M}(x,-1)\|^2$.

The first term is the (j,j)th entry of prediction error matrix G_X of the stationary process $X=Sx$, and hence is computable [10]. The second term can be calculated from our algorithm obtained for the linear predictor.

References

[1] Abréu, J. L. (1970). A note on harmonizable and stationary sequences. *Bol. Soc. Mat. Mexicana* **15**, 48–51.

[2] Akutowicz, E. J. (1957). On an explicit formula for linear least square prediction. *Math. Scand.* **5**, 261–266.

[3] Box, G. E. P. and Jenkins, G. M. (1976). *Time Series Analysis: Forecasting and Control*. Holden-Day, San Francisco. Second Edition.

[4] Cramér, H. (1962). On some classes of nonstationary stochastic processes, In *Proc. IV Berkeley Symp. Math. Stat. Prob.*, Vol II 57–78. Univ. Calif. Press.

[5] Doob, J. L. (1953). *Stochastic Processes*. Wiley, New York.

[6] Gladyšev, E. G. (1961). On periodically correlated random sequences. *Soviet Math. Dokl.* **2**, 385–388.

[7] Gudzinko, L. I. (1959). On periodically nonstationary processes. *Radiotekn. i Elektron* **6**, 1062–1064. (Russian)

[8] Kolmogorov, A. N. (1941). Stationary sequences in Hilbert space. *Bull. Math. Univ. Moscow* **2**, 1–40. (Russian)

[9] Loève, M. (1963). *Probability Theory*. Van Nostrand, New York. Third edition.

[10] Masani, P. (1960). The prediction theory of multivariate stochastic processes, III. *Acta Math.* **104**, 142–162.

[11] Masani, P. (1966). Recent trends in prediction theory. *Multivariate Analysis* (P. R. Krishnaiah, Ed.), pp. 351–382. Academic Press, New York.

[12] Miamee, A. G. and Salehi, H. (1978). Harmonizability, V-boundedness and stationary dilation of stochastic processes. *Indiana Univ. Math. J.* **27**, 37–50.

[13] Nadkarni, M. G. (1965). Vector valued weakly stationary stochastic processes and factorability of matrix valued functions, and strong mixing and uniformly ergodic Gaussian processes. Thesis, Brown University, Providence, RI.

[14] Niemi, H. (1975). On stationary dilations and the linear prediction of certain stochastic processes. *Soc. Sci. Denn. Comment, Phys.—Math.* **45**, 111–130.

[15] Ogura, H. (1971). Spectral representation of a periodic nonstationary random process. *IEEE Trans. information* theory **IT–17**, 143–149.

[16] Robertson, J. B. (1968). Orthogonally decompositon of multivariate weakly stationary stochastic processes. *Canad. J. Math.* **20**, 368–383.

[17] Robertson, J. B. and Rosenberg, M. (1968). The decomposition of matrix-valued measures. *Michigan Math. J.* **15**, 353–368.

[18] Rozanov, Yu. A. (1959). Spectral analysis of abstract functions. *Theor. Probability Appl.* **4**, 271–287.

[19] Wiener, N. and Masani, P. (1958). The prediction theory of multivariate stationary processes, II. *Acta Math.* **99**, 93–137.

P. R. Krishnaiah, ed., *Multivariate Analysis–V*

MULTIPLE TIME SERIES MODELING II

Emanuel PARZEN and H. J. NEWTON*
Institute of Statistics,
Texas A & M University, College Station, TX, U.S.A.

This paper defines the problem of time series modeling as model identification (determining the predictor variables) and parameter identification (estimating the prediction filter and the prediction error covariance matrix). Various auto-regression and cross-regression representations are defined for a stationary multiple time series. The role of basic regression and latent value algorithms is discussed. It is suggested that principal component analysis of spectral density matrices may not be useful in practice, whereas autoregressive methods are. The problem of defining an index time series is discussed; an approach is described in terms of the notion of predictable components.

1. Introduction

The problem of multiple time series modeling (or equivalently, the problem of system identification) has an extensive theory; see, for example, Akaike (1976), Brillinger (1975), Mehra and Lainiotis (1976), Priestley (1976), (1978), Priestley, Subba Rao and Tong (1973), and Subba Rao (1975). However, its practical implementation still seems to be at its beginning stages. This paper, a sequel to Parzen (1967), (1969), (1977), aims to describe some practical approaches to empirical modeling of multiple stationary time series which have been implemented in our extensive library of computer programs called TIMESBOARD.

The theory and practice of statistical data analysis involves answers to two kinds of questions:

(1) *Parameter identification:* what are the most likely parameter values (assuming a probability model for the data which is specified up to a finite number of parameters), and

(2) *Model identification:* what are models that adequately fit the observed sample probabilities, moments, order statistics (quantile functions), and correlations.

Empirical multiple time series analysis is concerned with finding relations among (normal) time series $Y_1(t),\ldots,Y_d(t)$, $X_1(t),\ldots,X_r(t)$, where Y_j

*Research supported by the Office of Naval Research.

is used to denote a 'dependent' or 'output' variable and X_k is used to denote an 'independent' or 'input' variable. In classical multivariate analysis observations at different times t are assumed *independent* and one can confine the search for relationship to relationship between contemporaneous variables. In time series analysis observations at different times t are correlated and a major goal is to identify the time lags at which significant relationships exist.

When a multiple time series $\{Y(t), t=0, \pm 1, \pm 2, \ldots\}$ is assumed to be normal, stationary, and have zero means, its probability law is specified by its covariance matrix function

$$R(v)=\mathbf{E}[Y(t)Y^*(t+v)], v=0, \pm 1, \ldots;$$

we use the following notation on matrices: * to denote complex conjugate transpose, ′ transpose, and $-$ complex conjugate.

The assumption of zero means is to be interpreted that $Y(\cdot)$ has been pre-processed (by a detrending procedure, say) to approximately obey this assumption. One then says that $Y(\cdot)$ has been detrended.

2. Basic regression and latent value algorithms

Our approach to multiple time series modeling is based on using in suitable ways certain algorithms of conventional regression and multivariate analysis. Let $\mathbf{X}$ and $\mathbf{Y}$ be jointly normal *complex*-valued random *vectors* with zero means and partitioned covariance matrix

$$\Gamma=\begin{bmatrix}\Gamma_{\mathbf{XX}} & \Gamma_{\mathbf{XY}}\\ \Gamma_{\mathbf{YX}} & \Gamma_{\mathbf{YY}}\end{bmatrix}=\mathbf{E}\begin{bmatrix}\mathbf{X}\\ \mathbf{Y}\end{bmatrix}[\mathbf{X}^* \quad \mathbf{Y}^*]$$

It should be noted that in time series analysis $\mathbf{Y}$ could denote the values of a time series at the present time t, and $\mathbf{X}$ could denote the values of the time series at past times.

2.1 Regression of Y on X

One can express $\mathbf{Y}$ as a part $\mathbf{Y}^\mu$ (linearly) explained by $\mathbf{X}$ and a residual $\mathbf{Y}^\nu=\mathbf{Y}-\mathbf{Y}^\mu$. One seeks the *regression coefficient* matrix B such that

$$\mathbf{Y}^\mu=\mathbf{E}[\mathbf{Y}|\mathbf{X}]=B\mathbf{X}$$

and one seeks the residual covariance matrix Σ such that

$$\Sigma = \mathbf{E}[\mathbf{Y}^{\nu}\{\mathbf{Y}^{\nu}\}^{*}].$$

Since $\mathbf{E}[\mathbf{Y}^{\mu}\mathbf{X}^{*}] = \mathbf{E}[\mathbf{Y}\mathbf{X}^{*}]$, by definition of conditional expectation, it follows that

$$B = \Gamma_{\mathbf{YX}}\Gamma_{\mathbf{XX}}^{-1}$$

$$\Sigma = \Gamma_{\mathbf{YY}} - \Gamma_{\mathbf{YX}}\Gamma_{\mathbf{XX}}^{-1}\Gamma_{\mathbf{XY}}$$

Using matrix pivoting procedures one can transform Γ to a partitioned matrix

$$\begin{bmatrix} \Gamma_{\mathbf{XX}}^{-1} & B' \\ -B & \Sigma \end{bmatrix}$$

which may be the computationally most convenient way to compute B and Σ. If Γ is not known theoretically but is estimated from data one estimates B and Σ using the foregoing formulas with an estimator of Γ.

2.2. *Subset regression*

A basic technique of model identification is reduction of the dimension of the input vector **X**, in order to find a 'parsimonious' set of explanatory variables represented by the components of the vector **X** of as low dimension as possible. The techniques for doing this (called Subset Regression or regression model identification procedures) choose a 'parsimonious' set of variables from among all possible subsets of components of **X** by computing, and comparing, the Σ matrix for each possible definition of **X**. These procedures are easiest to interpret when **Y** is scalar; the techniques of Predictable Components enables one to transform an output vector **y** to independent scalar variables $W_1, \ldots, W_d$.

2.3. *Principal components*

The problem of modeling a d-dimensional random vector **Y** can perhaps be reduced to the problem of modeling a scalar random variable by transforming to d uncorrelated random variables $W_1, \ldots, W_d$ which are linear combinations of **Y**. Let $V_1, \ldots, V_d$ and $\lambda_1, \ldots, \lambda_d$ be the latent vectors and latent values of $\Gamma_{\mathbf{YY}}$ satisfying: $\Gamma_{\mathbf{YY}}\mathbf{V}_j = \lambda_j \mathbf{V}_j$, $j = 1, \ldots, d$; $\lambda_1 > \lambda_2 > \cdots > \lambda_d$; and $\mathbf{V}_j'\mathbf{V}_j = 1$. Then $W_j = \mathbf{V}_j'\mathbf{Y}$ is called the jth principal component of **Y**; it can be characterized as the linear combination of **Y** (whose coefficients have sum of squares equal to 1) which has maximum variance and is uncorrelated with W_k, $k<j$. Note that the variance of W_j is λ_j. A more useful technique for modeling seems to be canonical correlations and the components to which they lead, which we call predictable components.

2.4. Predictable components and canonical correlations

To gain insight into the regression representation

$$\mathbf{Y}=\mathbf{Y}^{\mu}+\mathbf{Y}^{\nu}$$

where $\mathbf{Y}^{\mu}=B\mathbf{X}$ and $\mathbf{Y}^{\nu}$ has covariance matrix Σ, one should transform **Y** to d uncorrelated random variables $\mathbf{W}_1,\ldots,\mathbf{W}_d$ defined as follows: $\mathbf{W}_1$ is the most predictable linear combination of **Y** from **X**; $\mathbf{W}_2$ is the next most predictable, and so on. Writing $\mathbf{W}=m'\mathbf{Y}$, the unpredictability measure of W is

$$\lambda=\frac{\operatorname{Var}[m'\mathbf{Y}^{\nu}]}{\operatorname{Var}[m'\mathbf{Y}]}=\frac{m'\Sigma m}{m'\Gamma_{\mathbf{YY}}m}$$

It is shown in Rao (1973), p. 74 that the vector m which minimizes λ is the latent vector corresponding to the latent value λ of

$$\Gamma_{\mathbf{YY}}^{-1}\Sigma=I-\Gamma_{\mathbf{YY}}^{-1}\Gamma_{\mathbf{YX}}\Gamma_{\mathbf{XX}}^{-1}\Gamma_{\mathbf{XY}}.$$

Equivalently m and λ satisfy

$$\Sigma m=\lambda\Gamma_{\mathbf{YY}}m.$$

Let $\mathbf{V}_j$ be the latent vector satsifying

$$\Gamma_{\mathbf{YY}}^{-1}\Gamma_{\mathbf{YX}}\Gamma_{\mathbf{XX}}^{-1}\Gamma_{\mathbf{XY}}\mathbf{V}_j=(1-\lambda_j)V_j,\qquad j=1,\ldots,d$$

where $1-\lambda_1<1-\lambda_2<\cdots<1-\lambda_d$. Then $\mathbf{V}_j'\mathbf{Y}$ has predictor $\mathbf{V}_j'\mathbf{Y}^{\mu}=\mathbf{V}_j'B\mathbf{X}$ $=\mathbf{U}_j'\mathbf{X}$ where $\mathbf{U}_j=B'\mathbf{V}_j$ satisfies

$$\Gamma_{\mathbf{XX}}^{-1}\Gamma_{\mathbf{XY}}\Gamma_{\mathbf{YY}}^{-1}\Gamma_{\mathbf{YX}}U_j=(1-\lambda_j)\mathbf{U}_j.$$

The prediction error $\mathbf{V}_j'\mathbf{Y}^{\nu}$ has variance

$$\mathbf{V}_j'\Sigma\mathbf{V}_j=\lambda_j\mathbf{V}_j'\Gamma\mathbf{V}_j.$$

Thus $1-\lambda_j$ can be interpreted as the square of a correlation coefficient, called a canonical correlation (see Brillinger (1975)).

We call $\mathbf{W}_j=\mathbf{V}_j'\mathbf{Y}$ the jth predictable component of $\mathbf{Y}$.

2.5. *Subset regression of predictable components*

Instead of predicting $\mathbf{W}_j=\mathbf{V}_j'\mathbf{Y}$ by the linear combination $\mathbf{U}_j'\mathbf{X}$, where $\mathbf{U}_j=\mathbf{B}'\mathbf{V}_j$, one seeks for each $j=1,\ldots,d$ a parsimonious predictor of $\mathbf{W}_j$ given $\mathbf{X}$, using subset regression techniques. One might also interpret some of the coefficients of $\mathbf{V}_j$ as being zero if they are small enough. In this way one finds relations

$$c_{11}\mathbf{Y}_1+\cdots+c_{1d}\mathbf{Y}_d=b_{11}\mathbf{X}_1+\cdots+b_{1r}\mathbf{X}_r+\varepsilon_1$$

$$c_{21}\mathbf{Y}_1+\cdots+c_{2d}\mathbf{Y}_d=b_{21}\mathbf{X}_1+\cdots+b_{2r}\mathbf{X}_r+\varepsilon_2$$

and so on, where each equation has as few non-zero coefficients as possible, the variances of ε_1, $\varepsilon_2,\ldots$ are successively larger, and $\varepsilon_1,\ldots,\varepsilon_d$ are approximately uncorrelated.

3. Autoregression and cross regression representations

Modeling a multiple time series $\mathbf{Y}(t)$ begins with representations as a sum

$$\mathbf{Y}(t)=\mathbf{Y}^{\mu}(t)+\mathbf{Y}^{\nu}(t)$$

where $\mathbf{Y}^{\mu}(t)$ is the predictor of $\mathbf{Y}(t)$ given a certain information base, and $\mathbf{Y}^{\nu}(t)$ is always defined by $\mathbf{Y}^{\nu}(t)=\mathbf{Y}(t)-\mathbf{Y}^{\mu}(t)$.

We define $\mathbf{Y}^{\mu}(t)$ to be the conditional expectation of $Y(t)$ given a set of explanatory or predictor variables which we denote PREDVAR:

$$\mathbf{Y}^{\mu}(t)=\mathbf{E}\left[\mathbf{Y}(t)|\text{PREDVAR}\right].$$

Instead of a regression matrix B such that $\mathbf{Y}^{\mu}=B\cdot\text{PREDVAR}$, one must specify a linear operator or filter, called the PREDFIL or prediction filter, which transforms the PREFVAR series to $\mathbf{Y}^{\mu}(\cdot)$. In symbols

$$\mathbf{Y}^{\mu}=\text{PREDFIL}\cdot\text{PREDVAR}$$

or

PREDVAR —— | PREDFIL | —— $\mathbf{Y}^{\mu}(t)$.

Instead of an error covariance matrix Σ, one specifies a sequence, called PREDSIGMA, of prediction error covariance matrices

$$\Sigma(v)=\mathbf{E}[\mathbf{Y}^{\nu}(t)\{\mathbf{Y}^{\nu}(t+v)\}^{*}]$$

which describes the probability law of $\mathbf{Y}^{\nu}(\cdot)$ when it is normal. We denote $\Sigma(0)$ by Σ. Often $\Sigma(v)=0$ for $v\neq 0$; we then call $Y^{\nu}(\cdot)$ multiple white noise.

Parameter identification is concerned with the estimation of PREDFIL *and* PREDSIGMA *for a given choice of* PREDVAR. *Model identification is concerned with choosing* PREDVAR: as in subset regression, this is often *done by suitable comparisons of* PREDSIGMA corresponding to different choices of PREDVAR.

Some typical choices for PREDVAR (to predict $Y(t)$ at a fixed time t) are:

Case auto (infinite). The predictor variables consist of the infinite past

$$\mathbf{Y}(t-1),\ \mathbf{Y}(t-2),\ldots\ :\ \mathbf{Y}^{\mu}(t)=-\sum_{j=1}^{\infty} A_{\mathbf{YY}}(j)\mathbf{Y}(t-j).$$

Case auto (m). The predictor variables consist of the finite past

$$\mathbf{Y}(t-1),\ldots,\mathbf{Y}(t-m):\quad \mathbf{Y}^{\mu}(t)=-\sum_{j=1}^{m} A_{\mathbf{YY}}(j)\mathbf{Y}(t-j).$$

Case cross (infinite). The predictor variables consist of the infinite past and future values of 'input' variables $\mathbf{X}(s)$, $s=0,\ \pm 1,\ldots,$ which are also a normal, zero mean, stationary multiple time series:

$$\mathbf{Y}^{\mu}(t)=\sum_{k=-\infty}^{\infty} A_{\mathbf{YX}}(k)\mathbf{X}(t-k).$$

Case cross (finite). The predictor variables consist of a finite set of past and future values of $X(s)$, $t-m_1\leqslant s\leqslant t+m_2$ where m_1 and m_2 are fixed positive integers:

$$\mathbf{Y}^{\mu}(t)=\sum_{k=-m_1}^{m_2} A_{\mathbf{YX}}(k)\mathbf{X}(t-k).$$

Case auto-cross (infinite past). The predictor variables consist of the infinite past $\mathbf{Y}(t-1)$, $\mathbf{Y}(t-2)\cdots$ of the 'output' variables, and the infinite past $\mathbf{X}(t-1)$, $\mathbf{X}(t-2),\ldots$ of the 'input' variables:

$$\mathbf{Y}^{\mu}(t)=-\sum_{j=1}^{\infty} A_{\mathbf{YY}}(j)\mathbf{Y}(t-j)+\sum_{k=1}^{\infty} A_{\mathbf{YX}}(k)\mathbf{X}(t-k).$$

Case auto-cross (m). The predictor variables consist of $Y(t-1),\ldots,$ $Y\ (t-m)$, $X(t-1),\ldots,X(t-m)$:

$$Y^{\mu}(t)=-\sum_{j=1}^{m} A_{\mathbf{YY}}(j)\mathbf{Y}(t-j)+\sum_{k=1}^{m} A_{\mathbf{YX}}(k)\mathbf{X}(t-k).$$

Case auto-cross-cross (m). The predictor variables add $\mathbf{X}(t)$ to $\mathbf{Y}(t-1),\ldots,\mathbf{Y}(t-m)$, $\mathbf{X}(t-1),\ldots,\mathbf{X}(t-m)$:

$$\mathbf{Y}^{\mu}(t)=-\sum_{j=1}^{m} A_{\mathbf{YY}}(y)\mathbf{Y}(t-j)+\sum_{k=0}^{m} A_{\mathbf{YX}}(k)\mathbf{X}(t-k).$$

Formulas for PREDFIL and PREDSIGMA are discussed in the sequel. To identify the foregoing models, one basic approach is to model a multiple time series $\mathbf{Z}(t)$ defined as follows: when $\mathbf{Y}(\cdot)$ is modeled using its own past, $\mathbf{Z}(t)=\mathbf{Y}(t)$; when modeling $\mathbf{Y}(\cdot)$ also using the past of $\mathbf{X}(\cdot)$,

$$\mathbf{Z}(t)=\begin{bmatrix}\mathbf{X}(t)\\ \mathbf{Y}(t)\end{bmatrix}.$$

4. The spectral method

The spectral density matrix $f(\omega)$, $-\pi\leqslant\omega\leqslant\pi$, of a stationary multiple time series $Y(\cdot)$ with covariance matrix function $R(v)$ is defined by

$$f(\omega)=\frac{1}{2\pi}\sum_{v=-\infty}^{\infty} \mathrm{e}^{-iv\omega}R(v)$$

which leads to the representation

$$R(v)=\int_{-\pi}^{\pi} \mathrm{e}^{iv\omega}f(\omega)\,\mathrm{d}\omega.$$

An estimator of $f(\omega)$ is denoted $\hat{f}(\omega)$.

For a fixed frequency ω, $f(\omega)$ is a covariance matrix and $\hat{f}(\omega)$ is a sample covariance, whose distribution can usually be assumed to be approximately a complex Wishart matrix (a review of complex multivariate distributions and their applications in some problems of inference on multiple time series is given by Krishnaiah (1976)).

It is appealing to carry out the techniques of conventional multivariate analysis on $\hat{f}(\omega)$ at each frequency ω; the theory for this is given by Brillinger (1975). *From a practical point of view one would like to interpret in the time domain the analysis carried out in the spectral domain.*

Spectral methods are appropriate for estimating PREDVAR and PREDSIGMA in the *cross* case, in which $\mathbf{Y}^{\mu}(\cdot)$ is only a function of a series $\mathbf{X}(\cdot)$. Then one estimates the spectral density matrix

$$f_{\mathbf{Z}}(\omega)=\begin{bmatrix} f_{\mathbf{XX}}(\omega) & f_{\mathbf{XY}}(\omega) \\ f_{\mathbf{YX}}(\omega) & f_{\mathbf{YY}}(\omega) \end{bmatrix}.$$

Define the filter transfer function

$$B(\omega)=\sum_{k=-\infty}^{\infty} \mathrm{e}^{-i\omega k} A_{\mathbf{YX}}(k).$$

One can show that

$$B(\omega)=f_{\mathbf{YX}}(\omega) f_{\mathbf{XX}}^{-1}(\omega).$$

By matrix pivoting techniques one can transform $f_{\mathbf{Z}}(\omega)$ to

$$\begin{bmatrix} f_{\mathbf{XX}}^{-1}(\omega) & B^{*}(\omega) \\ -B(\omega) & f_{\varepsilon\varepsilon}(\omega) \end{bmatrix},$$

where

$$f_{\varepsilon\varepsilon}(\omega)=f_{\mathbf{YY}}(\omega)-f_{\mathbf{YX}}(\omega) f_{\mathbf{XX}}^{-1}(\omega) f_{\mathbf{XY}}(\omega).$$

Finally $A_{\mathbf{YX}}(j)$ and Σ are found by

$$A_{\mathbf{YX}}(j)=\frac{1}{2\pi}\int_{-\pi}^{\pi} B(\omega) \mathrm{e}^{ij\omega}\,\mathrm{d}\omega,$$

$$\Sigma=\int_{-\pi}^{\pi} f_{\varepsilon\varepsilon}(\omega)\,\mathrm{d}\omega.$$

When estimating $A_{YX}(j)$ from an estimator of $B(\omega)$ one can use a regression approach (see Parzen (1967a) and equation (2) below).

To estimate PREDFIL and PREDSIGMA in the *cross* case, one needs to first estimate $f_Z(\omega)$ which can be done by two methods: windowed covariances and autoregressive. We prefer the latter and describe it in the next section. The remainder of this section discusses some difficulties in implementing in the frequency domain the 'principal components' method of conventional multivariate analysis.

Index time series. The multiple time series analyst aims to display the information in a d-dimensional time series by summarizing it in a series of reduced dimension. In particular one seeks to find a univariate time series which in some sense best summarizes the information in a d-dimensional series. One approach to this problem is through principal components in the frequency domain.

One method of defining an index series is to form a scalar series

$$W(t)=\sum_{j=-\infty}^{\infty} b^{\mathrm{T}}(j)\mathbf{Y}(t-j)$$

and an approximating series to $\mathbf{Y}(t)$ denoted

$$\mathbf{Y}^{\alpha}(t)=\sum_{j=-\infty}^{\infty} c(j)\mathbf{W}(t-j),$$

where the coefficient matrices $b(\cdot)$ and $c(\cdot)$ are determined to minimize the mean square approximation error $\mathbf{E}(\mathbf{Y}(t)-\mathbf{Y}^{\alpha}(t))^{*}(\mathbf{Y}(t)-\mathbf{Y}^{\alpha}(t))$. Brillinger (1975) shows that the $b(\cdot)$ and $c(\cdot)$ are given by

$$b(j)=\frac{1}{2\pi}\int_{-\pi}^{\pi} V_1(\omega)e^{ij\omega}\,d\omega=c(j), \tag{1}$$

where $\mathbf{V}_1(\omega)$ is the eigenvector corresponding to the largest eigenvalue of the spectral density matrix $f_{\mathbf{YY}}(\omega)$ of $\mathbf{Y}$. Further, the minimum mean square approximation error is equal to

$$\int_{-\pi}^{\pi}\sum_{j=2}^{d}\lambda_j(\omega)\,d\omega,$$

where $\lambda_1(\omega),\ldots,\lambda_d(\omega)$ are the eigenvalues of $f_{\mathbf{YY}}(\omega)$.

To estimate the $b(\cdot)$, Brillinger suggests using a nonparametric estimator of $f_{\mathbf{YY}}(\cdot)$, finding its largest eigenvector, and approximating the integral in (1) by rectangular sums (which can be done via the fast Fourier transform algorithm). We show how to use a regression approach to estimate $b(j)$ from an estimator of $\mathbf{V}_1(\omega)$. However it involves a complex valued regression problem with heteroskedastic errors having singular covariance matrices. Further $\mathbf{V}_1(\omega)$ is only determined up to a complex-valued factor of modulus 1.

Theorem (Brillinger (1975), p. 351). *Let $f_{\mathbf{YY}}^{(T)}(\omega)$ be a nonparametric estimator of $f_{\mathbf{YY}}(\omega)$ with weighting function $K(\cdot)$, and let B_T be a nonnegative bandwidth parameter. Let $\lambda_1^{(T)}(\omega),\ldots,\lambda_d^{(T)}(\omega)$ and $\mathbf{V}_1^{(T)}(\omega),\ldots,\mathbf{V}_d^{(T)}(\omega)$ be the eigenvalues and eigenvectors of $f_{\mathbf{YY}}^{(T)}(\omega)$. Then*

$$\begin{aligned}\lim_{T\to\infty} B_T T \operatorname{Cov}\left(V_1^{(T)}(\omega_j), V_1^{(T)}(\omega_k)\right)\\ &= \delta_{jk} 2\pi \int K^2(\omega)\,d\omega\, \lambda_1(\omega) \sum_{l=2}^{d} \frac{\lambda_l(\omega)}{\left[\lambda_1(\omega)-\lambda_l(\omega)\right]^2} V_1(\omega) V_1^*(\omega)\\ &= C(\omega),\end{aligned}$$

where δ_{jk} is the Kronecker delta and $\omega=\omega_j=\omega_k$. Thus this limiting $(d\times d)$ covariance matrix is of rank at most $d-1$ and is thus singular. Further, it is complex valued.

Since $V_1(\cdot)$ and $b(\cdot)$ are Fourier pairs, we can write

$$V_1^{(T)}(\omega) = V_1(\omega) + \varepsilon(\omega),$$

where

$$V_1(\omega) = \sum_{j=-\infty}^{\infty} b(j) e^{-ij\omega},$$

and

$$\operatorname{Cov}\left(\varepsilon(\omega_j), \varepsilon(\omega_k)\right) \to \delta_{jk} C(\omega).$$

Thus one could determine estimators of the $b(\cdot)$ by a regression approach, i.e. minimize the errors in

$$V_1^{(T)}(\omega_k)=\sum_{j=-M}^{M} b_{T,M}(j)e^{-ij\omega_k}+\varepsilon_{T,M}(\omega_k), \qquad k=1,\ldots,Q, \quad (2)$$

giving rise to the equation

$$v=X\beta+\varepsilon$$

where $v=(V_1^{(T)}(\omega_1),\ldots,V_1^{(T)}(\omega_Q))^T$, the (j,k)th block element of X is $X_{jk}=e^{i(M-k+1)\omega_j}1_d$, $j=1,\ldots,Q$, $k=1,\ldots,2M+1$, $\beta=(b_{T,M}^T(-M),\ldots,b_{T,M}^T(M))^T$, and $\mathrm{Cov}\,\varepsilon$ is asymptotically the block diagonal matrix C having $C(\omega_k)$ as the kth block diagonal element. Thus one could find the complex valued weighted least squares estimator of β using a generalized inverse of C. But this would lead to complex valued estimators of the $b(\cdot)$. Note that if one assumes $C(\omega_k)=I_d$, then one obtains the estimators given in Brillinger.

In an attempt to obtain real valued estimators of the $b(\cdot)$, one could perform a regression on the real and imaginary parts of $V_1^{(T)}(\omega)$. This would require knowing the covariance matrix of the vector of real and imaginary parts of the complex errors. That is we have a complex random variable $Z=X+iY$ where $E(ZZ^*)=\Sigma$ and we want $\mathrm{Cov}\begin{bmatrix} X \\ Y \end{bmatrix}$. If one adds the assumption that $E(ZZ')=0$, then

$$\mathrm{Cov}\begin{pmatrix} X \\ Y \end{pmatrix}=\frac{1}{2}\begin{bmatrix} \mathrm{Re}\,\Sigma & -\mathrm{Im}\,\Sigma \\ \mathrm{Im}\,\Sigma & \mathrm{Re}\,\Sigma \end{bmatrix}$$

With this condition, the normal equations can be solved using a generalized inverse of the covariance matrices.

However, all of the above methods lead to nonunique estimators since they are of the form (where $w(\cdot)$ are weighting matrices)

$$\hat{b}(j)=\sum_k w(k)V_1^{(T)}(\omega_k)$$

and if $V_1^{(T)}(\omega_k)$ is a unit length eigenvector of $f_{YY}^{(T)}(\omega_k)$, then so is $e^{i\theta}V_1^{(T)}(\omega_k)$ for any θ.

Bloomfield (1978) proposes a procedure for making a unique identification of $V_1(\omega)$. However it is still difficult to interpret the index series $W(t)$ formed from the filter whose frequency transfer function is $V_1(\omega)$.

In the worked example in Brillinger (1975), p. 355, the multiple time series consists of components which have been pre-whitened by the removal of seasonal effects. The spectral density matrix $f_{YY}(\omega)$ is then real, and $b(j)=0$ for $j\neq 0$. The index series $W(t)$ is a linear function only of the values of $Y(t)$ at the same time t, and indeed turns out to be their average.

5. The autoregressive method

Given a sample realization $Z(1),\ldots,Z(T)$ from a d-dimensional time series $Z(\cdot)$, the pth order autoregressive approximating spectral estimator is defined (see Parzen (1969)):

$$\hat{f}_p(\omega)=\frac{1}{2\pi}G_p^{-1}(e^{i\omega})\Sigma_p G_p^{-*}(e^{i\omega}),\qquad \omega\in[-\pi,\pi],$$

where $G_p(z)$ is the complex matrix polynomial

$$G_p(z)=\sum_{j=0}^{p}A_p(j)z^j;$$

$A_p(0)=I$; and $A_p(1),\ldots,A_p(p)$ and Σ_p are found from the sample Yule–Walker equations

$$\sum_{j=0}^{p}A_p(j)R_T(j-v)=\delta_{v,0}\Sigma_p,\qquad v=1,\ldots,p.$$

$R_T(v)$ is the sample covariance function defined by

$$R_T(v)=\frac{1}{T}\sum_{t=1}^{T-v}Z(t)Z^*(t+v)=R_T^*(-v).$$

An optimal order $\hat{p}$ is determined as the value of p which minimizes an order determining criterion such as Akaike's AIC criterion (see Akaike (1976)) or

$$\mathrm{CAT}(p)=\mathrm{tr}\left[\frac{d}{T}\sum_{j=1}^{p}\frac{T-jd}{T}\Sigma_j^{-1}-\frac{T-pd}{T}\Sigma_p^{-1}\right],$$

where $\mathrm{tr}(A)$ is the trace of the matrix A, introduced in Parzen (1977).

To make inferences about PREDFIL and PREDSIGMA, one uses mainly the time domain parameters $\hat{p}$, $A_{\hat{p}}(\cdot)$, $\Sigma_{\hat{p}}$, $R_T(\cdot)$; the use of the frequency domain parameters $f_{\hat{p}}(\omega)$ and spectral parameters derived from $f_{\hat{p}}(\omega)$ was illustrated in the previous section. We now summarize the autoregressive method of modeling Y using the representation $Y = Y^{\mu} + Y^{\nu}$.

The *auto* case assumes $Y^{\mu}(t)$ is only a function of past $Y(\cdot)$; then its parameters can be obtained by applying the autoregressive method to $Y(\cdot)$.

The *cross* case assumes $Y^{\mu}(t)$ is only a function of a series $X(\cdot)$; then estimators are obtained by applying the autoregressive method to form an estimator $\hat{f}_Z(\omega)$ from which one forms $\hat{B}(\omega)$ and $\hat{f}_{\varepsilon\varepsilon}(\omega)$ which are used to estimate PREDFIL and PREDSIGMA (the formulas are given in the preceding section).

The *autocross* case assumes $Y^{\mu}(t)$ is a function of past $Y(\cdot)$ and past $X(\cdot)$; then estimators $\hat{p}$, $A_{YY,\hat{p}}(j)$, $A_{YX,\hat{p}}(j)$ and $\Sigma_{YY,\hat{p}}$ can be found by fitting an autoregressive approximant to

$$Z(t) = \begin{bmatrix} X(t) \\ Y(t) \end{bmatrix}: \qquad \sum_{j=0}^{\hat{p}} A_{\hat{p}}(j) Z(t-j) = \varepsilon(t)$$

where $\varepsilon(t) = (\varepsilon^*_X(t), \varepsilon^*_Y(t))^*$, and

$$A_{\hat{p}}(j) = \begin{bmatrix} A_{XX,\hat{p}}(j) & A_{XY,\hat{p}}(j) \\ A_{YX,\hat{p}}(j) & A_{YY,\hat{p}}(j) \end{bmatrix},$$

$$\Sigma_{\varepsilon\varepsilon,\hat{p}} = \begin{bmatrix} \Sigma_{XX,\hat{p}} & \Sigma_{XY,\hat{p}} \\ \Sigma_{YX,\hat{p}} & \Sigma_{YY,\hat{p}} \end{bmatrix}.$$

Thus $Y(t) = Y^{\mu}(t) + Y^{\nu}(t)$, where

$$Y^{\mu}(t) = -\sum_{j=1}^{\hat{p}} A_{YY,\hat{p}}(j) Y(t-j) + \sum_{j=1}^{\hat{p}} A_{YX,\hat{p}}(j) X(t-j), \tag{1}$$

and $\operatorname{Var} Y^{\nu}(t) = \Sigma_{YY,\hat{p}}$. Similarly one can form $X^{\mu}(t)$ and $X^{\nu}(t) = X(t) - X^{\mu}(t)$.

The *auto-cross-cross* case adds $X(t)$ to the predictor set for $Y(t)$. Parzen (1969) shows that one can do this by forming a predictor of $Y^{\nu}(t)$, denoted by

$$Y^{\nu+}(t): \quad Y^{\nu+}(t) = \Sigma_{YX,\hat{p}} \Sigma_{XX,\hat{p}}^{-1} X^{\nu}(t)$$

where $X^{\nu}(t) = X(t)$-predictor of $X(t)$ given past X and Y. The predictor of $Y(t)$ given the past of Y, and the past and present of X, is

$$Y^{\mu+}(t) = Y^{\mu}(t) + Y^{\nu+}(t)$$

where $Y^{\mu}(t)$ is given by (1). The estimated covariance matrix of $Y(t) - Y^{\mu+}(t)$ is given by

$$\Sigma_{YY,\hat{p}} - \Sigma_{YX,\hat{p}} \Sigma_{XX,\hat{p}}^{-1} \Sigma_{XY,\hat{p}}.$$

It is important to note that one can vary which components of Z belong to Y and which to X quite simply by using the autoregressive method.

As a simple illustration of how joint autoregressive modeling provides an approach to fitting and estimating relations between time series, consider two scalar series $Y(t)$ and $X(t)$ satisfying

$$\begin{aligned} Y(t) &= \gamma_0 X(t) + \gamma_1 X(t-1) + \eta(t) \\ X(t) - \rho X(t-1) &= \delta(t) \end{aligned} \tag{2}$$

where $\delta(\cdot)$ and $\eta(\cdot)$ are independent white noise processes. Then

$$Z(t) = \begin{bmatrix} X(t) \\ Y(t) \end{bmatrix} \quad \text{obeys} \quad A(0)Z(t) + A(1)Z(t-1) = \begin{bmatrix} \delta(t) \\ \eta(t) \end{bmatrix},$$

where

$$A(0) = \begin{bmatrix} 1 & 0 \\ -\gamma_0 & 1 \end{bmatrix}, \quad A(1) = \begin{bmatrix} -\rho & 0 \\ -\gamma_1 & 0 \end{bmatrix}.$$

One may write

$$Z(t) + AZ(t-1) = \varepsilon(t), \quad \operatorname{Var} \varepsilon = \Sigma,$$

where

$$A = -\begin{bmatrix} \rho & 0 \\ \rho\gamma_0 + \gamma_1 & 0 \end{bmatrix}$$

$$\Sigma = \begin{bmatrix} \sigma_\delta^2 & \sigma_\delta^2 \gamma_0 \\ \gamma_0 \sigma_\delta^2 & \gamma_0^2 \sigma_\delta^2 + \sigma_\eta^2 \end{bmatrix}.$$

If $Z(t)$ obeys an autoregressive scheme of order 1, and its coefficient matrix A has zeroes in its second column, then one can represent the time series by the regression model in equation (2).

6. Predictable components

To model and analyze a univariate time series Y, one approach is the method of *decomposition*, which attempts to decompose the time series as a sum of terms (representing trend, seasonal, and irregular as an example). To model and analyze a d-dimensional multiple time series $Y(t)$, one can adopt an analogous approach which writes Y as a linear transformation of a d-dimensional vector

$$W(t)=\begin{bmatrix} W_1(t) \\ \cdots \\ W_d(t) \end{bmatrix}$$

whose components are uncorrelated in the sense that its zero lag covariance matrix is a diagonal matrix D:

$$R_W(0)=E[W(t)W^*(t)]=D.$$

One chooses the components $W_1(t),\ldots,W_d(t)$ on the basis of *predictability* using a set PREDVAR of prediction variables. One chooses: $W_1(t)$ to be the most predictable linear combination of $Y(t)$; $W_2(t)$ to be the most predictable linear combination of $Y(t)$ which is uncorrelated with $W_1(t)$; $W_3(t)$ to be the most predictable linear combination of $Y(t)$ which is uncorrelated with $W_1(t)$ and $W_2(t)$; and so on.

For any linear combination $c'Y(t)$, its *unpredictability* criterion is

$$\lambda=\frac{\operatorname{Var}[c'Y^{\nu}(t)]}{\operatorname{Var}[c'Y(t)]}=\frac{c'\Sigma c}{c'R_Y(0)c}$$

The coefficient vector c with smallest value of λ is the latent vector corresponding to the *smallest* latent value λ_1 of $R_Y^{-1}(0)\Sigma$. Further λ_1 equals the unpredictability criterion λ of the most predictable linear combination of $Y(t)$.

For a specified set PREDVAR of prediction variables, one needs only Σ and $R_Y(0)$ to form the coefficient vectors needed to form the predictable components

$$W_j(t) = c_j' Y(t), \qquad j = 1, \ldots, d,$$

where $c_1, \ldots, c_d$ are the latent vectors corresponding to the ordered latent values $\lambda_1 < \lambda_2 < \cdots < \lambda_d$ of $R_Y^{-1}(0)\Sigma$:

$$R_Y^{-1}(0)\Sigma c_j = \lambda_j c_j.$$

When λ_1 is close to 0 (say, smaller than $8/T$) then we consider $W_1(t)$ to be a *highly smooth* time series and therefore an *index* time series. When λ_d is close to 1 (say, $1-\lambda_d$ smaller than $8/T$) we consider $W_d(t)$ to be a *white noise* series. For further details on the interpretations to be placed on the predictable components, see Box and Tiao (1977). The ideas of this section are an extension of their ideas.

Comparing the graphs of the time series $Y_1(t), \ldots, Y_d(t)$ with $W_1(t), \ldots, W_d(t)$ is very helpful in clarifying how much $W_1(t)$ summarizes the common smooth behavior of the original time series.

In summary, given a set PREDVAR of prediction variables for a multiple time series $Y(t)$, one can form the prediction filter PREDFIL, the prediction mean square error matrix PREDSIGMA, and the predictable components $W_1(t), \ldots, W_d(t)$. They may be interpretable. Further by comparing these estimated parameters for different choices of PREDVAR, one may be able to determine which sets of 'input' (prediction) variables are most informative about the 'output' variables $Y(t)$.

The present paper does not examine in detail the prediction filters, even though these should be estimated for a complete model identification; rather it emphasizes the insight to be obtained by determining the index time series $W_1(t)$. An illustrative example has been worked out; it involves monthly ozone level time series discussed in Parzen and Pagano (1978).

References

[1] Akaike, H. (1976). Canonical correlation analysis of time series and the use of an information criterion. In *System Identification* (Eds. R. K. Mehra and D. G. Mainiotis), pp. 27–96. Academic Press, New York.

[2] Bloomfield, P. (1978). Principal components of time series. Presented at Ames (Iowa) Special Meeting on Time Series Analysis, organized by the Institute of Mathematical Statistics.

[3] Box, G. E. P. and Tiao, G. C. (1977). A canonical analysis of multiple time series. *Biometrika*, **64**, 355–365.

[4] Brillinger, D. R. (1975). *Time Series Data Analysis and Theory*. Holt, Rinehart and Winston, New York.

[5] Krishnaiah, P. R. (1976). Some recent developments on complex multivariate distributions. *J. Multivariate Anal.*, **6**, 1–30.

[7] Mehra, R. K. and Lainiotis, D. G. (1976). *System Identification: Advances and Case Studies*. Academic Press, New York.

[8] Parzen, E. (1967). Empirical multiple time series analysis. *Proc. Fifth Berkeley Sympos. on Math. Statist. and Probability*, (Eds. L. LeCam and J. Neyman), Vol. I, pp. 305–340.

[9] Parzen, E. (1967a). Time series analysis for models of signal plus white noise. In *Spectral Analysis of Time Series* (Ed. B. Harris), 233–257, Wiley, New York.

[10] Parzen, E. (1969). Multiple time series modeling. *Multivariate Analysis–II* (Ed. P. Krishnaiah), pp. 389–409. Academic Press, New York.

[11] Parzen, E. (1977). Multiple time series: determining the order of approximating autoregressive schemes. *Multivariate Analysis–IV* (Ed. P. R. Krishnaiah), pp. 283–295. North-Holland, Amsterdam.

[12] Parzen, E. and Pagano, M. (1978). Statistical time series analysis of worldwide total ozone for trends. Unpublished manuscript.

[13] Priestley, M. B. (1968). System identification, Kalman filtering, and stochastic control. Presented at Ames (Iowa), Special Meeting on Time Series Analysis, organized by the Institute of Mathematical Statistics.

[14] Priestley, M. B. (1976). Applications of multivariate techniques in the study of multivariate stochastic systems. Paper presented at the Sixth International Conference on Stochastic Processes at Tel Aviv, Israel.

[15] Priestley, M. B., Subba Rao, T. and Tong, H. (1973). Identification of the structure of multivariate stochastic systems. *Multivariate Analysis–III* (Ed. P. R. Krishnaiah), pp. 351–368. Academic Press, New York.

[16] Rao, C. R. (1973). *Linear Statistical Inference and Its Applications*. Wiley, New York.

[17] Subba Rao, T. (1975). An innovation approach to the reduction of the dimensions in a multivariate stochastic system. *Internat. J. Control*, 673–680.

PART III

LIMIT THEOREMS AND NONPARAMETIC METHODS

P. R. Krishnaiah, ed., *Multivariate Analysis–V*

ON ARRANGEMENTS WHICH DETERMINE LIMIT THEOREMS FOR EIGENVALUES OF A SAMPLE COVARIANCE MATRIX

Harald BERGSTRÖM
Department of Mathematics, Chalmers University of Technology and University of Göteborg, Göteborg, Sweden

1. Introduction

Consider independent normal random variables $X_{i,j}, i=1,2,\ldots,p, j=1,2,\ldots,n$ with mean value 0 and unit variance. We form sums

$$Y_{r,s}=\sum_{j=1}^{n} X_{r,j}X_{s,j} \tag{1.1}$$

and $p\times p$ matrices $(Y_{r,s})$ having $Y_{r,s}$ as coordinate in the rth row and sth column. Clearly these matrices are symmetrical. The distribution of $S_p^{(n)}=(Y_{r,s})$ is the well-known Wishart distribution with n degrees of freedom. Instead of $S_p^{(n)}$ we shall deal with the normalized matrix $(1/n)S_p^{(n)}$. Denote its eigen-values $\lambda_{p,1}^{(n)},\ldots,\lambda_{p,p}^{(n)}$. They are random variables. Let $F_p^{(n)}$ be the empirical distribution function of the $\lambda_{i,p}^{(n)}$

$$F_p^{(n)}(x)=\frac{1}{p}\cdot\text{number of } \lambda_{p,i}, \text{ which are } \leqslant x$$

and let $\lambda_p^{(n)}$ be an eigenvalue taken at random with probability $1/p$. Then

$$E[\lambda_p^{(n)}]^k=\frac{1}{p}\sum_{j=1}^{p} E[\lambda_{p,j}^{(n)}]^k=E\int x^k\,\mathrm{d}F_p^{(n)}(x). \tag{1.2}$$

Arharov [1] has given the first five moments of $\lambda_p^{(n)}$ and Marcenko and Pasteur [3] have shown that $F_p^{(n)}$ converges in probability to a certain distribution function if p and n depend on the same integervalued parameter N in such a way that p/n tends to a limit as $N\to+\infty$. They used Stieltjes transforms. Independently Dag Jonsson [2] has proved this convergence by the help of the moment method and a combinatorial method.

I shall here give a short survey of Jonsson's method. Further I shall present a new proof of the combinatorial part of the proof. It turns out that one essential number which appears in Jonsson's proof is the number of arrangements of elements in a sequence of k elements, where some or all belong to s different classes, and the arrangements should have certain very elementary general properties. This number then has an interesting general significance.

The following main theorem was proved by Marcenko-Pasteur and Jonsson. I give it in Jonsson's version.

Theorem 1 (Jonsson). *Let $F_p^{(n)}$ be defined as above and let $p=p(N)$ and $n=n(N)$ depend on the integervalued parameter N in such a way that $p(N)/n(N)\to y(N\to+\infty)$. Then for $0<y\leqslant 1$ the empirical distribution $F_p^{(n)}$ converges in probability to the distribution function F_y as $N\to+\infty$, where F_y has the density function*

$$f_y(x)=\begin{cases}[4-(1+y-x)^2]^{1/2}/2\pi yx & \textit{for } (1-\sqrt{y})^2<x<(1+\sqrt{y})^2,\\ 0 & \textit{otherwise}.\end{cases} \tag{1.3}$$

Jonsson carries through his proof in three steps. In the first one he proves that the kth moment of the distribution function F_y is the polynomial

$$b_k(y)=\sum_{s=1}^{k} b_{k,s}y^{k-s},\quad b_{k,s}=\frac{1}{s}\binom{k}{s-1}\binom{k-1}{s-1}. \tag{1.4}$$

The number of arrangements, which I mentioned above and which is given in Lemma 2.2, is the number

$$\binom{k}{s-1}\binom{k-1}{s-1}(s-1)!.$$

Jonsson deals with the moments $\lambda_p^{(n)}$ in the second step. The kth moment $\lambda_p^{(n)}$ is equal to

$$\begin{aligned}E[\lambda_p^{(n)}]^k&=\frac{1}{p}E\sum_{j=1}^{p}\lambda_{p,j}^{(n)}=\frac{1}{pn^k}E\,\mathrm{trace}[S_p^{(n)}]^k\\ &=\frac{1}{pn^k}\sum_{r_1=1}^{p}\cdots\sum_{r_k=1}^{p}EY_{r_1r_2}Y_{r_2r_3}\cdots Y_{r_kr_1}.\end{aligned} \tag{1.5}$$

The terms in this multiple sum are polynomials in n and different chains $(r_1, r_2, \ldots, r_1)$ of indices correspond to polynomials of different degree. Hence the multiple sum is a polynomial in n and p. By an ingenious but very complicated method Jonsson succeeds to determine the coefficients of the terms $p^{s'}n^{s''}$ of highest degree, which highest degree is $k+1$. Only these terms matter when n and p pass to $+\infty$. So he finds that the terms of highest degree is the polynomial

$$\sum_{s=1}^{k} b_{k,s} p^{k-s+1} n^s. \tag{1.6}$$

If we divide this polynomial by pn^k and let $n \to +\infty$ we get the polynomial (1.4). Hence the expectation of the 'stochastic' moments of $F_p^{(n)}(x)$ tend to the corresponding moments of the distribution function $F_y(x)$.

In the third step Jonsson proves that $F_p^{(n)}(x)$ converges in probability to $F_y(x)$. Then he first proves that $\lambda_k^{(n)}$ converges in probability to $b_k(y)$, i.e. that

$$M_{k,p}^{(n)} = \frac{1}{pn^k} \operatorname{trace}\left[S_p^{(n)} \right]^k$$

converges in probability to $b_k(y)$. He does this by proving that var $M_{k,p}^{(n)}$ tends to 0 as $N \to +\infty$, and this variance can be estimated in a rather simple way. He also proves that the moments (1.4) satisfy Carleman's uniqueness criterion. By standard methods he can then prove the convergence in probability of $F_p^{(n)}(x)$ to $F_y(x)$. Under additional conditions he can also prove the almost sure convergence of $F_p^{(n)}(x)$ to $F_y(x)$ and also that this convergence is uniform with respect to x.

In the next section I shall give a new proof of the combinatorial part of the proof of the main theorem and I will do this in the form of three lemmas.[1]

2. Computations of moment

It follows from (1.5) that an essential part of the proof in the main theorem is the computation of

$$E \operatorname{trace}\left[S_p^{(n)} \right]^k = \operatorname{trace} E\left[S_p^{(n)} \right]^k. \tag{2.1}$$

[1]It may be of interest to compare Jonsson's proof of the combinatorial part with an analogue proof of some statements in Ulf Grenander and Jack Silverstein [2].

Now

$$S_p^{(n)} = \sum_{j=1}^{n} A_j, \tag{2.2}$$

where A_j is a $p \times p$ matrix whose coordinate in the rth row and the sth column is $X_{r,j}X_{s,j}$. Hence using the laws for summation and multiplication of matrices we get

$$E[S_p^{(n)}]^k = E\left(\sum_{j=1}^{n} A_j\right)^k = \sum E \prod_{m=1}^{k} A_{j_m}, \tag{2.3}$$

where the sum should be taken over all possible formally different products (n^k in number) and where the j_m takes values $1,2,\ldots,n$. We divide the product on the right hand side into classes in the following way. Consider the set of numbers $(1,2,\ldots,s), s \leqslant k$. We say that the product $\prod_{m=1}^{k} A_{j_m}$ belongs to the class C_s if any j_m belongs to $(1,2,\ldots,s)$ and if for any $i \in (1,2,\ldots,s)$ there is some $j_m = i$. To the class C_s there corresponds a certain contribution to the right hand side of (2.3). Denote this contribution by

$$\sum_{C_s} E \prod_{m=1}^{k} A_{j_m}. \tag{2.4}$$

From the set $(1,2,\ldots,n)$ we can form $\binom{n}{s}$ different subsets of s numbers. Since the random variables X_{r_i} are independent and have the same distribution, we conclude that to any such subset there corresponds the same contribution (2.4). Hence

$$E[S_p^{(n)}]^k = \sum_{s=1}^{n} \binom{n}{s} \sum_{C_s} E \prod_{m=1}^{k} A_{j_m}. \tag{2.5}$$

We now turn to the computation of (2.4). Since we only care for the trace of the matrix (2.4) we shall study the mapping

$$\prod_{m=1}^{k} A_{j_m} \to \operatorname{trace} E \prod_{m=1}^{k} A_{j_m} = E \operatorname{trace} \prod_{m=1}^{k} A_{j_m}. \tag{2.6}$$

This mapping depends on the arrangement of the factors A_{j_m}. If for instance

$$\prod_{m=1}^{k} A_{j_m} = \prod_{i=1}^{s} A_i^{\nu_i}, \qquad \sum_{i=1}^{s} \nu_i = k,$$

we get by the independence of the $A_i^{\nu_i}$

$$E \prod_{m=1}^{k} A_{j_m} = \prod_{i=1}^{s} (E A_i^{\nu_i}). \tag{2.7}$$

Let $a_{r,s}^{(i)}$ be the coordinate in the rth row and the sth column of A_i^t. Then

$$a_{r,s}^{(i)} = \sum_{r_2=1}^{p} \cdots \sum_{r_t=1}^{p} {}_{,}X_{r,i}X_{r_2,i}X_{r_2,i}X_{r_3,i}X_{r_3,i} \cdots X_{r_t,i}\, X_{r_t,i}X_{s,i} \tag{2.8}$$

$$= \left(\sum_{j=1}^{p} X_{j,i}^2 \right)^{t-1} X_{r,i}X_{s,i}$$

and

$$E a_{r,s}^{(i)} = 0 \quad \text{for } r \neq s,$$

$$= E\left(\sum_{j=1}^{p} X_{j,i}^2 \right)^{t-1} X_{s,i}^2 = \frac{1}{p} E\left(\sum_{i=1}^{p} X_{r,i}^2 \right)^2 \quad \text{for } r = s.$$

The sum $\sum_{j=1}^{p} X_{j,i}^2$ has the χ^2-distribution with p degrees of freedom and its tth moment is

$$p(p+2)(p+4) \cdots [p+2(t-1)].$$

Thus

$$E a_{r,s}^{(i)} = M_p^{(t)} = (p+2)(p+4) \cdots [p+2(t-1)],$$

and

$$E A_i^{\nu_i} = M_p^{(\nu_i)} \mathfrak{J},$$

where $\mathfrak{I}$ denotes the unit matrix. By (2.7) we then get

$$\prod_{i=1}^{s} EA_i^{\nu_i} = \prod_{i=1}^{s} M_p^{(\nu_i)} \mathfrak{I}. \tag{2.9}$$

Thus in this case the mapping (2.6) reduces to

$$\prod_{m=1}^{k} A_{j_m} \to p \prod_{i=1}^{s} M_p^{(\nu_i)}. \tag{2.10}$$

We shall show that we have this mapping not only in the special case (2.7) but for a certain well-defined class T_s of products, determined by the arrangements of the factors. Note that the order of the polynomial on the right hand side of (2.10) is p^{k-s+1}. We shall also prove that the order of $\prod_{m=1}^{k} A_{j_m}$ of a product in C_s is of lower order than p^{k-s+1} when it does not belong to T_s.

We are now going to define T_s and shall then describe the arrangements of the factors for a product in T_s.

If $j_m \neq i$ for $m = m_1$ and $j_m = i$ for $m = m_1+1, m_1+2, \ldots, m_1+\nu_i, j_{m_1+\nu_i+1} \neq i, m_1+\nu_i+1 \leqslant k$, we call $A_i, i = m_1+1, \ldots m_1+\nu_i$ a stationary A_i-sequence of length ν_i and we denote this sequence by $[A_i]$. However, we consider these stationary sequences in their cyclical connection. Hence if $j_{m_1} \neq i, j_{m_1+1} = i, j_{m_1+2} = i, \ldots, j_k = i, j_1 = i, \ldots, j_{m_2} = i, j_{m_2+1} \neq i, m_2 < m_1$ we say that we have a stationary A_i-sequence of length $k - m_1 + m_2$. If all $A_{j_m} = A_i$, the stationary A_i-sequence is of course of the length k. We say that the factors in the product have a *particular* arrangement if it has the following properties.

For some $i_1 \in (1, 2, \ldots, s)$ the product contains only one stationary A_{i_1}-sequence. If this is cancelled but the remaining factors have unchanged order, they form a product which again contains only one stationary A_{i_2}-sequence and then the factors in this new product has again a particular arrangement. This means that all factors will be cancelled if we in this way successively cancel any A_i-sequence appearing only once. We now define T_s as that subclass of C_s which only contains products with particular arrangement of the factors.

Lemma 2.1. *For a product of k factors in the class T_s the mapping* (2.6) *is the mapping* (2.10) *and hence the expectation of the trace of the product is a polynomial in p of the degree $k-s+1$.*

Proof. Let for a certain i the product contain only one A_i-sequence. Then the product must have one of the following forms

$$CA_i^{\nu_i}D \quad \text{or} \quad A_i^{\nu_i'}CA_i^{\nu_i''}$$

where C and D does not contain A_i and hence $CD \in T_{s-1}$ in the first case and $C \in T_{s-1}$ in the second case. Denote the coordinate in the rth row and sth column of C, $A_i^{\nu_i}$, D, $A_i^{\nu_i'}$ and $A_i^{\nu_i''}$ by respectively $c_{r,s}$, $a_{r,s}^{(\nu_i)}$, $d_{r,s}a_{r,s}^{(\nu_i')}$ and $a_{r,s}^{(\nu_i'')}$. Then the coordinate in the rth row and sth column of $CA_i^{\nu_i}D$ has the expectation

$$E\left\{\sum_{r_1=1}^{p}\sum_{r_2=1}^{p} c_{r,r_1}a_{r_1,r_2}^{(\nu_i)}d_{r_2,s}\right\} = \sum_{r_1=1}^{p} Ea_{r_1,r_1}^{(\nu_i)}Ec_{r,r_1}d_{r_1,s} \tag{2.11}$$

since $Ea_{r_1,r_2}^{(\nu_i)}=0$ for $r_1 \neq r_2$ according to (2.8). Further by (2.9) we get

$$Ea_{r_1,r_1}^{(\nu_i)} = M_p^{(\nu_i)}$$

and thus by (2.11)

$$E(CA_1^{\nu_i}D) = M_p^{(\nu_i)}E(CD) \tag{2.12}$$

where $CD \in T_{s-1}$. The matrices $A_i^{\nu_i'}$ are symmetrical, i.e. $a_{r,s}^{(\nu_i)} = a_{s,r}^{(\nu_i)}$. Hence

$$\text{trace } A_i^{\nu_i'}CA_i^{\nu_i''} \sum_{r_1=1}^{p}\sum_{r_2=1}^{p}\sum_{r_3=1}^{p} a_{r_1,r_3}^{(\nu_i')}c_{r_2,r_3}a_{r_3,r_1}^{(\nu_i'')}$$

$$= \sum_{r_2=1}^{p}\sum_{r_3=1}^{p}\sum_{r_1=1}^{p} a_{r_2,r_1}^{(\nu_i')}a_{r_1,r_3}^{(\nu_i'')}c_{r_2 r_3}$$

$$= \text{trace } A_i^{\nu_i}C$$

with $\nu_i = \nu_i' + \nu_i''$. Then by (2.9)

$$E\,\text{trace } A^{\nu_i'}CA^{\nu_i''} = E\,\text{trace} A_i^{\nu_i}C = M_p^{\nu_i}E\,\text{trace } C \tag{2.13}$$

where $C \in T_{s-1}$. Using (2.12) and (2.13) we get Lemma 2.2. by induction.

Lemma 2.2. *A product of k factors in the class C_s but not in the class T_s is a polynomial in p of degree less than $k-s+1$.*

Proof. By the reduction principle used above we can reduce Etrace of a product if it contains only one stationary A_i-sequence for given i and then

$$E\,\text{trace product} = M_p^{(\nu_i)}E\,\text{trace new product} \tag{2.14}$$

where $M_p^{(\nu_i)}$ is a polynomial in p of degree $\nu_i - 1$ and the new product has $k-\nu_i$ factors and belongs to the class C_{s-1}. In this way we can reduce a product until it cannot be further reduced. Now a matrix product which cannot be further reduced must contain at least two stationary A_i-sequences for every class. According to (2.14) it is sufficient to prove the lemma for a matrix product which cannot be further reduced. We may suppose that (2.6) is a product which cannot be reduced. Consider then.

$$E\,\text{trace} \prod_{m=1}^{k} A_{j_m} = \sum_{r_1=1}^{p} \cdots \sum_{r_k=1}^{p} E\{X_{r_1,j_1}X_{r_2,j_1}X_{r_2,j_2}X_{r_3,j_2}\cdots X_{r_k,j_k}X_{r_1,j_k}\}. \tag{2.15}$$

Here we look upon $r_1,\ldots,r_k$ as variables taking values $1,2,\ldots,p$. A term in the multiple sum is different from 0 only if for every i, the $X_{r_j,i}$ appears in even powers which means that there must be relations between the variables r_j. Clearly the points (r_m,j_m) make a 'walk' in the plane as m takes the values $m=1,2,\ldots,k$, and then the values j_m are indices for the matrices in (2.6) and hence given for this product. Since there are at least two stationary A_i-sequences for any given i, the point (r_m,j_m), as $m=1,2,\ldots,$ must pass into (r_{m_1},i) for the first time for some m_1, then take values (r_{m_1},i) for $m=m_1+1,\ldots,m_2$ then take the value (r_{m_2+1},i), $i_1\neq i$ and return to a point with second coordinate i for the last time, say at (r_{m_3},i) and then take values $(r_{m_3+1},i)\cdots(r_{m_4},i)$. We look upon $r_m, m=1,2,\ldots,k$ as different variables which however may take the same value. Since the $X_{m,i}$ must appear in even powers in a product in (2.15) in order to give a contribution to the sum there must be relations between r_{m_1}, r_{m_2}, r_{m_3} and r_{m_4} if the product should give a contribution to the sum, $r_{m_1}=r_{m_2}, r_{m_3}=r_{m_4}$ for instance, anyhow at least two such relations. Certainly every variable r_m appears twice in a product in (2.15). However, it is easily seen that at most one relation between two r_m's for a given i can be relation for $i_1\neq i$ and at most for one $i_1\neq i$. Thus in the summation in (2.15) at most $k-s$ variables r_m are free, though we may have different combinations of relations between them. The number of such combinations only depends on k and s but not on p. Hence the sum in (2.15) is a polynomial in p at most of the degree $k-s$.

Lemma 2.3. *The number of products belonging to T_s and having k factors is*

$$\binom{k}{s-1}\binom{k-1}{s-1}[(s-1)!].$$

Remark. Note that this is the number of particular arrangements of k elements from s classes, the elements in a class being indistinguishable, every class being represented in the arrangement.

Proof. Consider a product $\prod_{m=1}^{k} A_{j_m}$. We call the values $1, 2, \ldots k$ of m the places of the factors in the product. As above we deal with stationary A_i-sequences. We shall look upon such a sequence as starting at some place $\beta_i, 1 \leqslant \beta_i \leqslant k$. Hence if $j_k = 1$ the corresponding A_1-sequence may start at β and the A_1 may fill the places $\beta, \beta+1, \ldots k, 1, 2, \ldots, \beta$. We call β_i the starting place of the sequence A_i. However there may be several A_i-sequences. But there is a first A_i-sequence with the starting place $\alpha_i, 1 \leqslant \alpha_i \leqslant k$. Consider now the situation when the first stationary sequence is an A_1-sequence with the starting place α_1, the second stationary sequence being not an A_1-sequence and with starting place at α_2 is an A_2-sequence etc. Thus the 'first' stationary sequences have the order:

1:	A_1-sequence with starting place α_1
2:	A_2-sequence with starting place α_2
3:	A_3-sequence with starting place α_3
$s-1$:	A_{s-1}-sequence with starting place α_{s-1}

Then we have a configuration as below

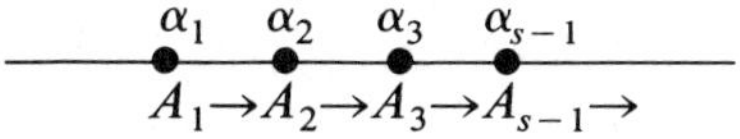

We require that all A_1 follow after $m = \alpha$ all A_2 after $m = \alpha_2$ etc. all A_{s-1} and A_s after $m = \alpha_{s-1}$ and eventually before $m = \alpha_1$. The α_i must satisfy the inequality

$$1 \leqslant \alpha_1 < \alpha_2 < \cdots < \alpha_{s-1} \leqslant k \tag{2.16}$$

There are exactly $\binom{k}{s-1}$ different vector solutions $(\alpha_1, \ldots \alpha_{s-1})$ of this equation. Let the number of factors A_i be ν_i. Then

$$\nu_1 + \nu_2 + \cdots + \nu_s = k \tag{2.17}$$

This equation has exactly $\binom{k-1}{s-1}$ different vector solutions $(\nu_1, \nu_2, \ldots, \nu_s)$. (See [4, p. 49]). The $(s-1)$ first stationary sequences can be ordered in

$(s-1)!$ different ways. Hence our lemma follows if we show that there is exactly one product in T_s for given α_i and ν_i satisfying (2.16) and (2.17) and given permutation of the $(s-1)$ first stationary sequences, or it is then sufficient to consider the case when the first stationary sequences have the order as in the figure. We divide the products in T_s and with the order of first stationary sequences as in the figure into two classes $T_s^{(1)}$ and $T_s^{(2)}$ and say that the product belongs to the class $T_s^{(1)}$ if it contains one stationary A_s-sequence, to $T_s^{(2)}$ if it contains at least two stationary A_s-sequences. If the 'first' stationary sequences for the different A_i's are ordered as in the figure, then the particular arrangement must satisfy the following conditions if it belongs to $T_s^{(2)}$:

(a) The interval $[\alpha_i, \alpha_{i+1}), 1 \leqslant i \leqslant s-1$ contains A_i as first element and then on the place α_i and besides A_i it can contain only A_s and A_j with $1 \leqslant j < i$. The interval $[\alpha_{i-1}, k)$ contains A_{i-1} on the place α_{i-1} and otherwise it can contain any A_j. The interval $[1, \alpha_i)$ (if not empty) can contain an $A_j, j \neq s$ only if $m_k = j, m_1 = j$

(b) According to the definition of the particular arrangement, a product in T_s cannot have factors A_i and $A_j, i \neq j$ repeated in the order

$$\cdots A_i \cdots A_j \cdots A_i \cdots A_j.$$

Using this property we conclude

(c) If an element A_i has a place between two stationary A_j sequences,

$$[A_j] \cdots A_i \cdots [A_j],$$

then all A_i must lie between these sequences. Note that there are at least two stationary A_s-sequences for a product in $T_s^{(2)}$.

Using (a), (b) and (c) we find: If $\nu_i < \alpha_{i+1} - \alpha_i, 1 \leqslant i < i+1 \leqslant s-1$ then A_i-elements must occupy the places $\alpha_i, \alpha_i + 1, \ldots, \alpha_i + \nu_i - 1$ leaving the hole $[\alpha_i + \nu_i, \alpha_{i+1})$

If $\nu_i \geqslant \alpha_{i+1} - \alpha_i$, then A_i elements must occupy all places in $[\alpha_i, \alpha_{i+1}]$ eventually leaving the surplus $\nu_i - (\alpha_{i+1} - \alpha_i)$ of A_i-elements which can fill holes in intervals after $[\alpha_i, \alpha_{i+1})$.

These two statements are easy to verify. Consider the first one. According to (a) there can only be elements, A_j, $i \leqslant j$ and A_s in $[\alpha_i, \alpha_{i+1}]$. We cannot have any such element $\neq A_i$ between A_i on the place α_i and another A_i since then we should have an arrangement $\cdots [A_j] \cdots A_i \cdots [A_j] \cdots A_i$ if

$j<i$ and correspondingly, if $j \in s$, since there are at least two stationary A_s-sequences. The other statement is shown in the same way.

Having thus shown how and where we get holes and surplus of elements we find that (a), (b) and (c) give us a procedure for filling holes with a surplus of elements. Clearly a hole in $[\alpha_1, \alpha_2)$ can only be filled with A_s-elements. In the same way, if there are holes in $[\alpha_i, \alpha_{i+1})$ for $i=1,2,\ldots,$ $j-1$, all these holes must be filled by A_s-elements. Then let $[\alpha_j, \alpha_{j+1})$ be the first interval which has not a hole. If $\nu_j = \alpha_{j+1} - \alpha_j$, then clearly all elements A_i will be used for the places in the interval but no hole is left. Next let $\nu_j > \alpha_{j+1} - \alpha_j$, j being the first index for which this is true. Further let $\nu_i \geqslant \alpha_{i+1} - \alpha_i$ for $i=j,\ldots,t-1, \nu_t < \alpha_{t+1} - \alpha_t$, $(t+1 \leqslant s-1)$. Then we have A_t-elements on the places $\alpha_t, \alpha_{t+1}, \ldots, \alpha_t + \nu_t - 1$. If $\nu_{t-1} > \alpha_t - \alpha_{t-1}$, then we must have elements A_{t-1} on the places $\alpha_t + \nu_t, \alpha_t + \nu_t + 1$ and so on until the hole is filled or until we have consumed the surplus of A_{t-1}-elements. In the last case we have to continue to fill the hole with A_{t-2}-elements if there is a surplus of such elements, otherwise with A_{t-3}-elements if there are such elements left. In this way we continue until the hole is covered, or if there are not elements $A_i, i<t$ enough to cover the hole, to fill the remaining part with A_s-elements. If there is a surplus of elements $A_i, i<t$, when the hole $[\alpha_t \alpha_{t-1})$ is covered we pass to the next hole and fill it with elements $A_i, i<t$ in the same order as before. At last we fill the parts $[\alpha_{i-1}, k)$ and $[1, \alpha_1)$. Clearly this procedure is possible only if

$$\begin{cases} \nu_s + \nu_1 \geqslant \alpha_2 - \alpha_1 \\ \nu_s + \nu_1 + \nu_2 \geqslant \alpha_3 - \alpha_1 \\ \cdots \\ \nu_s + \nu_1 + \nu_2 + \cdots + \nu_{s-2} \geqslant \alpha_{s-1} - \alpha_1 \end{cases}$$

Then $\nu_{s-1} + \nu_s = k - (\alpha_{s-1} - \alpha_1)$ is the number of elements A_s and $A_i, i<s$ left to fill the intervals $[\alpha_{i-1}, k]$ and $[1, \alpha_1)$ (The last one may be empty.). Thus a product in T_s and with the arrangement as in the figure belongs to $T_s^{(2)}$ if and only if the relations (2.18) are satisfied. Otherwise it belongs to $T_s^{(1)}$. In the latter case we cancel the only A_s-sequence noting the places of its elements. Then we are in the same situation as before but facing particular arrangements of $k - \nu_s$ elements from $s-1$ classes. Thus our proof follows by induction. We have to consider the case $s=1$ particularly but then there obviously exists exactly one product.

The proof presented here is constructive i.e. it shows how particular arrangements can be formed for given α_i and ν_i satisfying (2.16) and (2.17).

References

[1] Arharov, L. V. (1971). Limit theorems for the characteristic roots of a sample covariance matrix. *Soviet Math. Dokl.* **12**, 1206–1209.

[2] Ulf Grenander and Jack Silverstein, Spectral analysis of networks with random topologies, SIAM J. Appl. Math. **32** (2) (1977).

[3] Jonsson, Dag (1976). Some limit theorems for the eigenvalues of a sample covariance matrix. Depart. of Math. Report No. 1976:6. Uppsala University.

[4] Marcěnko, V. A. and Pasteur, L. A. (1967). Distribution of eigenvalues of some sets of random matrices. *Math. USSR-Sbornik*, **1**, 507–536.

[5] Perron, O. (1931). *Algebra I*.

P. R. Krishnaiah, ed., *Multivariate Analysis–V*

MAXIMAL DEVIATION THEORY OF DENSITY AND FAILURE RATE FUNCTION ESTIMATES BASED ON CENSORED DATA*

J. R. BLUM and V. SUSARLA**
University of Arizona, University of Wisconsin-Milwaukee, Tucson AZ, U.S.A.

Limit theorems are obtained for the maximum (over a finite interval) of a normalized deviation of the density estimate when the data is censored on the right. The results presented here generalize the available maximal deviation results on density estimates by the kernel method introduced by Rosenblatt. As in Bickel and Rosenblatt [2], the results presented here can be used in tests for goodness-of-fit, and tests of hypotheses concerning the unknown density, and for finding confidence curves for the density when the observations are censored on the right. We also provide similar limit theorems for an estimate of the failure rate function, again based on right censored observations.

1. Motivation and description of the problem.

In certain survival analysis or medical follow-up studies (say a cancer treatment study over a fixed time period or an open ended time period), one does not observe the true random survival times $X_1,\ldots,X_n$ of the n patients who entered the study, but rather observes only the right censored version of $X_1,\ldots,X_n$ due to dropping out of the patients from the study, or due to live withdrawal of patients from the study. That is, one observes only the random vectors $(\delta_1,Z_1),\ldots,(\delta_n,Z_n)$ where

$$\delta_i = I_{[X_i \leqslant Y_i]} \quad \text{and} \quad Z_i = \min\{X_i, Y_i\} \tag{1}$$

for $i=1,\ldots,n$ where $Y_1,\ldots,Y_n$ are n random variables with supports in $(0,\infty)$. There has been extensive literature on the estimation of the common distribution F of the i.i.d. random variables $X_1,\ldots,X_n$ whenever $Y_1,\ldots,Y_n$ are independent random variables. (See, for example, Breslow and Crowley [3].)

*Research supported in part by the National Science Foundation under Grant No. MCS76-05952; revised under Grant No. MCS77-26809.

**Research supported by National Institute of General Medical Sciences of NIH of HEW under Grant No. 5R01GM23129-02.

Gehan [5] points out that the problem of estimation of f, the density of F with respect to Lebesgue measure on the Borel σ-field B in $\mathbf{R}^+ = (0, \infty)$, is an important problem in the analysis of survival data in its own right and also in the estimation of the failure rate function or the hazard function corresponding to F. (In this context, see also the books by Barlow and Proschan [1] and Gross and Clark [6].) In this paper, among other results, we obtain the asymptotic distribution of the maximal deviation of an estimator of f whenever $Y_1, \ldots, Y_n$ are i.i.d. according to a common distribution G with support in $(0, \infty)$. The motivation for our problem has been stated in the context of medical situations, but the problem has immediate applications to any other life testing situation such as time to failure of a component, time to replacement, etc.

The estimator $\hat{f}_n$ to be discussed in this paper is given by

$$\hat{f}_n(x) = (fG^*)_n(x) / G_n^*(x) \tag{2}$$

where

$$nb(fG^*)_n(x) = \sum_{j=1}^{n} [\delta_j = 1] k((x - Z_j)/b) \tag{3}$$

and

$$G_n^*(x) = \prod_{j=1}^{n} \{(1 + N^+(Z_j))/(2 + N^+(Z_j))\}^{[\delta_j = 0, Z_j \leq x]} \tag{4}$$

where k is a real valued function satisfying certain conditions, $b = b(n) \downarrow 0$ as $n \uparrow \infty$, [A] denotes the indicator function of the measurable set A, and

$$n\mathbf{N}^+(x) = \text{number of } Z_j (j = 1, \ldots, n) > x. \tag{5}$$

The motivation for (2) comes from the observations that

(a) it can be shown that $(fG^*)_n(x)$ is a good estimator of $f(x)G^*(x)$ (with $G^* = 1 - G$) by following the standard arguments used in the density estimation methods, and

(b) $G_n^*(x)$ can be seen to be a good estimator of $G^*(x)$ by modifying the Kaplan and Meier [7] product limit estimator of a distribution function based on right censored data. The asymptotic distribution of

$$M_n = \sup_{0 \leq x \leq 1} \frac{\surd(nb) |\hat{f}_n(x) - (bG^*(x))^{-1} E[[\delta_1 = 1] k((x - Z_1)/b)|}{\surd(f(x)/G^*(x))} \tag{6}$$

will be determined as $n \to \infty$ under appropriate conditions on f, $G^*(=1-G)$ and K. We shall obtain our results by replacing $\hat{f}_n$ in (6) by $\tilde{f}_n$ defined by

$$\tilde{f}_n(x) = (fG^*)_n(x)/G^*(x) \tag{7}$$

and then use a result of Rosenblatt [10].

We assume throughout the paper that f, $G^*(=1-G)$, and k satisfy some or all of the following conditions ((A3), $(A3)^+$, (A4), (B1) and (B2). These are similar to those in Bickel and Rosenblatt [2] and Rosenblatt [10]) for the results to be proved in this section.

(A1) $G(1) < 1$, and G is continuous on $[0,1]$

(A2) $f > 0$ and continuous on $[0,1]$

(A3) $fG^* > 0$ and continuously differentiable on $[0,1]$

$(A3)^+$ $fG^* > 0$ and continuously differentiable up to (and including) order l on $[0,1]$.

(A4) $\sqrt{(fG^*)}$ is absolutely continuous on $[-\varepsilon, 1+\varepsilon]$ for some $\varepsilon > 0$, and $(fG^*)'/\sqrt{fG^*}$ is bounded in absolute value. Moreover,

$$\int_{[u \geqslant 3]} |u|^{3/2} (\log \log |u|)^{1/2} [|k'(u)| + |k(u)|] du < \infty$$

(B1) k assigns mass one to the real line, is bounded and vanishes outside $[-A, A]$. Moreover k is absolutely continuous on $[-A, A]$ with derivative k' such that $\int |k'(t)|^l dt < \infty$ for $l = 1, 2$.

(B2) $\int_{-A}^{A} k(u) du = 1$, and $\int_{-A}^{A} u^i k(u) du = 0$ for $i = 1, \ldots, l-1$, $l \geqslant 2$.

The plan of the rest of the paper is as follows: In the next section, we obtain the asymptotic distribution of M_n of (6). Section 3 treats the maximal deviation problem in the context of the problem of estimation of the failure rate function $f/(1-F)$. In Section 4, we propose a new method of estimation of f (and hence also of $f/(1-F)$). We conclude the paper with a few remarks.

To facilitate ease of notation, we assume without further mentioning that all limits are as $n \to \infty$, $b(n)$ is abbreviated by b, [A] stands for the indicator function of A, x is always in $[0,1]$, $G^* = 1-G$, and that suppressed arguments are clear from the context. $O_p(\varepsilon_n)$ and $o_p(\varepsilon_n)$ have the usual meaning. That is, $A_n = O_p(\varepsilon_n)$ means that A_n/ε_n are bounded random variables in probability for large n. $o_p(\varepsilon_n)$ is similarly defined.

2. Maximal deviation theory for M_n of (2).

As pointed out in the previous section, we first obtain the following lemma which says that the asymptotic distribution of M_n of (6) is the same as that of $\tilde{M}_n$ where

$$\tilde{M}_n = \sup_{0\leqslant x\leqslant 1} \frac{\sqrt{(nb)}|(fG^*)_n(x) - E[(fG^*)_n(x)]|}{\sqrt{(f(x)G^*(x))}}. \tag{8}$$

Lemma 1. *Let* (A1), (A2) *and* (B1) *hold. If* $b=n^{-\delta}$ *with* $0<\delta<1$, *then*

$$\sqrt{(nb)} \sup_{0\leqslant x\leqslant 1} |\hat{f}_n(x) - \frac{(fG^*)_n(x)}{G^*(x)}| = o_p(1).$$

Proof. By the definitions of $\hat{f}_n$, $(fG^*)_n$, and G_n^* given by (2), (3) and (4) respectively, it follows that

$$\hat{f}_n(x) - \frac{(fG^*)_n(x)}{G^*(x)} = (fG^*)_n(x)\left(\frac{1}{G_n^*(x)} - \frac{1}{G^*(x)}\right) \tag{9}$$

Therefore

$$|\text{lhs of (9)}| \leqslant |(fG^*)_n(x)||(1/G_n^*(x)) - (1/G^*(x))|.$$

By Lemma 4.1 of Susarla and Van Ryzin [11], the condition (A1) along with the fact that F has density f which is continuous and positive at 1 implies that

$$\sup\{|(1/G_n^*(x)) - (1/G^*(x))| : 0\leqslant x\leqslant 1\}$$
$$= O(n^{-\beta}) \text{ almost surely (a.s.)}$$

for every β with $0<2\beta<1$. Since $b=n^{-\delta}$ with $\beta>0$, it follows that

$$\sqrt{(nb)} \sup\{|(1/G_n^*(x)) - (1/G^*(x))| : 0\leqslant x\leqslant 1\} = o_p(1). \tag{10}$$

In view of (9) and (10), we shall complete the proof by showing that $(fG^*)_n(x)$ is bounded in probability for large n.

Let F_n be the empirical distribution of $X_1, \ldots, X_n$ whose common distribution is F. We obtain by integration by parts and the conditions on k that

$$|b^{-1}\int |k((x-t)/b)| \mathrm{d}(F_n - F)|$$
$$\leqslant b^{-1}(\text{variation of } k) \sup\{|F_n(t) - F(t)| : -\infty<t<\infty\}.$$

Since (B1) implies that the variation of $k<\infty$, and since $\sup\{|F_n(t)-F(t)|: -\infty<t<\infty\}=O(\ln n/\sqrt{n})$ a.s., the above inequality shows that

$$\sup\left\{(nb)^{-1}\sum_{j=1}^{n}|k((x-x_j)/b)|: 0\leqslant x\leqslant 1\right\}$$

$$\leqslant O(\ln n/\sqrt{n})+\sup\left\{b^{-1}\int|k((x-t)/b)|\mathrm{d}F(t): 0\leqslant x\leqslant 1\right\}$$

$$\leqslant O(\ln n/\sqrt{n})+c \quad \text{a.s.}$$

for some constant c. The last inequality follows since f is uniformly bounded on [0, 1], and k satisfies (B1). This completes the proof of the lemma.

In view of the above lemma, the asymptotic distribution of M_n of (6) (in which we are interested) can be obtained from that of (8) provided that the latter one has an asymptotic distribution. Consequently we shall obtain the asymptotic distribution of $\tilde{M}_n$.

Now we can apply the theorems of Rosenblatt [10] to $\tilde{M}_n$ with $(fG^*)_n$ defined by (3). We note that $(fG^*)_n$ is the kernel estimate of the density fG^* which corresponds to the sub-distribution function $\tilde{F}(t)=P[\delta_1=1, Z_1\leqslant t]$. (This is the reason why our conditions (A3), (A3)$^+$ and (A4) correspond exactly to the ones given by Bickel and Rosenblatt [2] and Rosenblatt [10].) The fact that $\tilde{F}$ is only a sub-distribution function is irrelevant here in obtaining the asymptotic distribution of $\tilde{M}_n$ of (8) because the kernel function k has compact support by condition (B1). Therefore, the following result can be obtained from Theorem 1 and the corollary to Theorem 2 of Rosenblatt [10].

Theorem 1. *Let* (A3), (A4), *and* (B1) *hold. Let* $b(n)=n^{-\delta}$ *with* $0<\delta<1$. *Then*

$$\left[p\left[(2\delta\log n)^{\frac{1}{2}}\left(\frac{\tilde{M}_n}{(\lambda(k))^{\frac{1}{2}}}-d_n\right)<t\right]\to e^{-2e^{-t}}\right] \tag{11}$$

with

$$\lambda(k)=\int_{-\infty}^{\infty}k^2(t)\,\mathrm{d}t \tag{12}$$

and

$$d_n = (2\delta \log n)^{\frac{1}{2}} + \frac{1}{(2\delta \log n)^{\frac{1}{2}}} \left\{ \log \frac{K_1(k)}{\sqrt{\pi}} + \tfrac{1}{2}(\log \delta + \log \log n) \right\} \tag{13}$$

with

$$K_1(k) = \frac{k^2(A) + k^2(-A)}{2\lambda(k)} \qquad \text{if } K_1(k) > 0. \tag{14}$$

Otherwise

$$d_n = (2\delta \log n)^{\frac{1}{2}} + \frac{1}{(2\delta \log n)^{\frac{1}{2}}} \log \left\{ \frac{1}{\pi} (K_2(k)/2)^{\frac{1}{2}} \right\} \tag{15}$$

with

$$2\lambda(k) K_2(k) = \int_{-\infty}^{\infty} (k'(t))^2 \, dt. \tag{16}$$

If, in addition, (A3)$^+$, (B2) *and* $1 < \delta(2l+1) < 2l+1$ *also hold, then* $E[(fG^*)_n(x)]$ *which appears in* (11) *via* $\tilde{M}_n$ *can be replaced by* $f(x)G^*(x)$.

As a corollary to the above theorem and Lemma 1, we obtain the following result concerning the asymptotic distribution of M_n of (2).

Theorem 2. *Let* (A1), (A2), (A3), (A4) *and* (B1) *hold and let* $b(n) = n^{-\delta}$ *with* $0 < \delta < 1$. *Then*

$$M_n = \sqrt{(nb)} \sup_{0 \leq x \leq 1} \frac{|\hat{f}_n(x) - (bG^*(x))^{-1} E[[\delta_1 = 1] k((x - Z)/b)|}{\sqrt{(f(x)/G^*(x))}} \tag{17}$$

satisfies (11) *with* $\tilde{M}_n$ *replaced by* M_n. *If, in addition* (A3)$^+$ *and* (B2) *hold and* $1 < \delta(2l+1) < 2l+1$, *then* $(bG^*(x))^{-1} E[[\delta_1 = 1] k((x - Z_j)/b)]$ *appearing in* (17) *may be replaced by* $f(x)$ *and* (11) *still holds.*

Next we provide a function k satisfying the conditions of Theorem 2.

Example 1. Let the kernel function k be given by

$$k(u)=\begin{cases}\frac{1}{2} & -1\leqslant 2u\leqslant 1\\ 0 & \text{otherwise}\end{cases} \tag{18}$$

Notice that $\int_{-1}^{1}k(u)\,du=1=1-\int_{-1}^{1}uk(u)du$. Consequently, we have $l=2$ in the above results. Hence for any δ with $1<5\delta<5$, the conclusions of the theorem and the corollary hold provided that f, and G satisfy the remaining conditions imposed on them.

Remark 1. The above theorem and its corollary specialize to Theorem 2 of Rosenblatt [10] in the one-dimensional case. This specialization can be obtained by taking $G(x)\equiv 0$ for $x\leqslant 1$.

Remark 2. The estimator $\hat{f}_n$ presented here satisfies not only the conclusion of the above corollary, but can be shown to be mean square consistent, and a.s. consistent with and without rates depending upon the conditions on k, f, G, and δ. Such results can be obtained as special cases of the general delta sequence approach considered by Coplien [4] for the estimation of a density with right censored data.

3. Estimation of a failure rate function with censored data

In this section, we consider estimation of

$$r(x)=\frac{f(x)}{1-F(x)}=\frac{f(x)}{F^*(x)} \qquad 0\leqslant x\leqslant 1. \tag{19}$$

The function r is usually known as the failure rate function or the hazard function corresponding to F. The estimator $\hat{r}_n$ of r is defined by

$$\hat{r}_n(x)=\frac{(fG^*)_n(x)}{H_n^*(x)} \tag{20}$$

where $(fG^*)_n$ is defined by (3) and

$$nH_n^*(x)=\text{number of } Z_j(j=1,\ldots,n)>x. \tag{21}$$

The object of this section is to obtain the following theorem concerning the asymptotic distribution of

$$R_n = \sup_{0 \leqslant x \leqslant 1} \sqrt{(nb)}(H^*(x)r^{-1}(x))^{\frac{1}{2}}$$
$$\left[r_n(x) - (bH^*(x))^{-1}E\left[\left[\delta_1 = 1\right]k((x - Z_1)/b)\right]\right] \tag{22}$$

where $H^* = F^* G^*$.

Theorem 3. *Let* (A1), (A2), (A3), (A4) *and* (B1) *hold and* $b = n^{-\delta}$ *with* $0 < \delta < 1$. *Then*

$$P\left[(2\delta \log n)^{\frac{1}{2}}\left(\frac{R_n}{(\lambda(k))^{\frac{1}{2}}} - d_n\right) < t\right] \to e^{-2e^{-t}} \tag{23}$$

where $\lambda(k)$ *and* d_n *are given by* (12) *through* (16). *If, in addition,* $(A3)^+$, (B2) *and* $1 < \delta(2l+1) < 2l+1$ *also hold, then* $(bH^*(x))^{-1}E[[\delta_1 = 1]k((x - Z_1)/b]$ *can be replaced by* $r(x)$.

Proof. The proof of Theorem 3 follows directly from Theorem 2 and the following observation. Consider the difference

$$r_n(x) - \frac{(fG^*)_n(x)}{H^*(x)} = (fG^*)_n(x)((1/H_n^*(x)) - (1/H^*(x))). \tag{24}$$

Since sup $\{|H_n^*(x) - H^*(x)| : 0 \leqslant x \leqslant 1\} = O(\ln n/\sqrt{n})$ a.s., and that sup $\{|(fG^*)_n(x)| : 0 \leqslant x \leqslant 1\}$ was shown to be a bounded random variable a.s. in Lemma 1 under the assumed conditions, it follows that the asymptotic distribution of R_n is the same as that of $\tilde{M}_n$ of (8). Hence the desired conclusions follow from Theorem 2.

Remark 3. The estimator $r_n(x)$ can be shown to be a.s. consistent, mean-square consistent and asymptotically normal. Such results may be obtained from those of Coplien [4].

Remark 4. Again, by taking $G(x) = 0$ for $x \leqslant 1$, our results specialize to those of Rice and Rosenblatt [8].

Example 2. The kernel k in Example 1 given at the end of Theorem 2 also satisfies the conditions of Theorem 3.

4. Another estimator for f (and hence for r).

An alternative estimator $\hat{\hat{f}}_n$ is given by

$$b\hat{\hat{f}}_n(x) = \int_{-\infty}^{\infty} k((x-t)/b)\, \mathrm{d}\hat{\hat{F}}_n(t) \tag{25}$$

where $\hat{\hat{F}}_n$ is an estimator of F, and k is a kernel function as defined in the earlier sections. For example, one can take $\hat{\hat{F}}_n$ defined by

$$1-\hat{\hat{F}}_n(t) = \prod_{j=1}^{n} \{N^+(Z_j)/(N^+(Z_j)+1)\}^{[Z_j \leqslant t, \delta_j = 1]}. \tag{26}$$

This is the Kaplan–Meier [7] maximum likelihood estimator of $1-F$. The ratio of $\hat{\hat{f}}_n$ to $1-\hat{\hat{F}}_n$ provides a new estimator for r. We have verified that Theorems 2 and 3 hold when k is the indicator function of $[-2^{-1}, 2^{-1}]$ if we take $1-\hat{\hat{F}}_n$ to be equal to zero whenever the rhs of (26) is not defined properly.

In terms of final asymptotic results, we conjecture that $\hat{f}_n$ of (2) and $\hat{\hat{f}}_n$ of (25) should behave the same way. That is, both $\hat{\hat{f}}_n$ and $\hat{f}_n$ should have the same type of maximal deviation theory results, and both should be uniformly (in x in $[0,1]$) a.s. and mean square consistent, with and without rates depending on the conditions satisfied by f, G and k. It is probably easier to compute (2) than (25).

5. Concluding remarks.

The results presented here can be used to find confidence curves, tests of hypothesis, and goodness-of-fit tests for f and r, as in Bickel and Rosenblatt [2], and Rice and Rosenblatt [8], when the data is censored on the right. The idea here is to replace G^* and f in the second parts of Theorems 2 and 3 by their estimators G_n^* and $\hat{f}_n$ respectively and then noting that such a replacement has an error which goes to zero in probability uniformly for x in $[0,1]$. Of course, this replacement might require some further conditions analogous to those in Bickel and Rosenblatt [2]. Just as in Section 3 of Rice and Rosenblatt [8], we can replace the supremum on $[0,1]$ in M_n of (6) in R_n of (22) by the supremum on $[0,\alpha_n]$ with $\alpha_n \uparrow \infty$ as $n \uparrow \infty$ under certain additional conditions of fG^*, G^*, and α_n. We hope to provide such details along with simulation models and numerical examples involving cancer survival data as they become available. Also, the estimators presented here can be used to treat certain empirical Bayes problems of Robbins [9] when the observations are censored on the right. The approximation obtained for $\hat{f}_n$ of (2) can be used to obtain the asymptotic distribution of $\int_0^1 (\hat{f}_n(x)-c_n(x))^2/f(x)\,dx$ for an appropriate choice of c_n. Such a result would generalize Theorem 4.1 of Bickel and Rosenblatt [2]. We hope to consider these problems as well as questions on efficiencies of tests based on these asymptotic distributions at a later date.

References

[1] Barlow, Richard E. and Proschan, Frank (1974). *Statistical Theory of Reliability and Life Testing Probability Models*. Holt, Rinehart and Winston, New York.

[2] Bickel, P. J. and Rosenblatt, M. (1973). On some global measures of the deviations of density function estimates. *Ann. Statist.* **1**, 1071–1095.

[3] Breslow, N. and Crowley, J. (1974). A large sample study of the lifetable and product limit estimates under random censorship. *Ann. Statist.* **2**, 437–453.

[4] Coplien, K. (1978). Thesis at the University of Wisconsin-Milwaukee.

[5] Gehan, E. (1969). Estimating survival functions from the life table. *J. Chron. Dis.* **21**, 629–644.

[6] Gross, Alan J. and Clark, Virginia A. (1975). *Survival Distributions: Reliability Applications in Biomedical Sciences*. Wiley, New York.

[7] Kaplan, E. L. and Meier, P. (1958). Nonparametric estimation from incomplete observations. *J. Amer. Stat. Assoc.* **53**, 457–481.

[8] Rice, John, and Rosenblatt, Murray (1976). Estimation of the log survivor function and hazard function. *Sankhya, Series A* **38**, Part 1, 60–78.

[9] Robbins, H. (1964). The empirical Bayes approach to statistical decision problems. *Ann. Math. Statist.* **35**, 1–20.

[10] Rosenblatt, M. (1976). On the maximal deviation of k-dimensional density estimates. *Ann. Prob.* **4**, 1009–1015.

[11] Susarla, V. and Van Ryzin (1978). Large sample theory for a Bayesian nonparametric survival curve estimator based on censored samples. *Ann. Statist.* **6**.

P. R. Krishnaiah, ed., *Multivariate Analysis–V*

SOME ASYMPTOTIC RESULTS ON THE MULTIVARIATE RANK STATISTICS

Marie HUŠKOVÁ
Department of Statistics, Charles University, Prague 8, Sokolovská 83, Czechoslovakia

Consider a p-dimensional vector of simple linear rank statistics $\mathbf{S}_c$ with the joint distribution function G_c. Under some conditions the upper estimation of the integral $\int f \mathrm{d}(G-\Phi)$, where Φ is the normal distribution and f is an arbitrary bounded real-valued function on $\mathbf{R}^p$, is obtained. The rate of convergence of the distribution G_c and of some quadratic forms of $\mathbf{S}_c$ to the normal and χ^2-distribution, resp., is received as a special case. The proving method is related to that developed by Bhattacharya and R. R. Rao [5], the results of the author for the univariate case are also utilized.

1. Introduction

Let $\mathbf{X}_j=(X_{1j},\ldots,X_{pj})$, $j=1,\ldots,N$, be independent p-dimensional random variables with continuous distribution functions and R_{ij} be the rank of X_{ij} in the sequence of $X_{i1},\ldots,X_{iN}$. Put

$$\mathbf{S}_c=(S_{1c},\ldots,S_{pc})', \tag{1.1}$$

$$S_{ic}=\sum_{j=1}^{N} c_{ij}a_{Ni}(R_{ij}), \qquad 1\leqslant i\leqslant p, \tag{1.2}$$

with c_{ij} being regression constants and $a_{Ni}(j)$ scores. Let G_c be the distribution function of $\mathbf{S}_c$ and $\Phi(\cdot\,;\boldsymbol{\mu},\boldsymbol{\Sigma})$ be the normal distribution with mean $\boldsymbol{\mu}$ and variance matrix $\boldsymbol{\Sigma}$.

The purpose of this paper is to estimate the error of

$$\int f(x)\,\mathrm{d}(G_c(\mathbf{x})-\Phi(\mathbf{x};E\mathbf{S}_c,\operatorname{var}\mathbf{S}_c)) \tag{1.3}$$

for an arbitrary function f being Borel measurable defined on R^p. The motivation of this paper comes from the book by R. N. Bhattacharya and R. R. Rao [5] where the problem of estimation of (1.3) for Q_c being the distribution function of a sum of independent random vectors is studied with some refinements.

The estimation of (1.3) obtained here is the same (i.e., depends on f and c_{ij} in the same way—compare Lemma 3 and Theorem) as for Q_c being the distribution function of

$$\sum_{j=1}^{N} (c_{1j}Y_{1j},\ldots,c_{pj}Y_{pj})',$$

where $(Y_{1j},\ldots,Y_{pj})', j=1,\ldots,N$, are independent random vectors.

By a particular choice of f we obtain the rate of convergence of the distribution function of $\mathbf{S}_c$ and $\mathbf{S}_c'(\operatorname{var}\mathbf{S}_c)^{-1}\mathbf{S}_c$ to the normal one and χ^2-distribution, resp.

As for the related papers the rate of convergence for $p=1$ (simple linear rank statistics) was studied both under hypothesis and alternatives in several papers (e.g. [1, 2, 6, 7, 9, 11]). The multivariate case was treated for instance by Jensen (Friedman's χ^2-statistic [10]). The present author considered the problem of the convergence rate for quadratic rank statistics using another method than here (comparison of the results is given in Section 3).

The proving method used in the present paper is related to that developed by R. N. Bhattacharya and R. R. Rao [5] for the sum of independent random vectors and to that suggested for $p=1$ in [7]. We derive the estimation of (1.3) under the hypothesis of randomness and make several remarks on its estimation both under contiguous and general alternatives.

2. Assumptions and main theorem

We adopt the following assumptions:

(i) $\mathbf{X}_1,\ldots,\mathbf{X}_N$ are independent p-dimensional identically distributed random vectors with continuous distribution function;

(ii) the scores $a_{Ni}(j)$ are in either of the following forms:

$$a_{Ni}(j)=\phi_i(j/(N+1)), \qquad 1\leqslant i\leqslant p, \quad 1\leqslant j\leqslant N, \tag{2.1}$$

$$a_{Ni}(j)=E\phi_i(U_{Ni}^{(j)}), \qquad 1\leqslant i\leqslant p, \quad 1\leqslant j\leqslant N, \tag{2.2}$$

where $U_{Ni}^{(j)}$ denotes the jth order statistic in the sequence $F_i(X_{i1}),\ldots,F_i(X_{iN})$;

(iii) the score-generating functions ϕ_i are nonconstant defined on $\langle 0,1\rangle$, their first derivatives ϕ_i are absolutely continuous and the second ones ϕ_i

square integrable over (0, 1);

(iv) there exists constants d_1, d_2 (not depending on c_{ij} and N) such that

$$0 < d_1 \leqslant \lambda_{1c} \leqslant \lambda_{pc} \leqslant d_2 \tag{2.3}$$

where λ_{1c} and λ_{pc} denotes the minimal and maximal eigenvalue of the variance matrix $\operatorname{var}\mathbf{S}_c$.

For a given measure μ on the Borel σ-field on $\mathbf{R}^p$ define (according to Bhattacharya [4])

$$\tilde{\omega}_f(\varepsilon, \mu) = \sup\left(\int_{\mathbf{R}^p} \omega_{f_y}(\mathbf{x}, \varepsilon)\mu(d\mathbf{x}); \mathbf{y} \in \mathbf{R}^p \right), \tag{2.4}$$

where $f_y(\mathbf{x}) = f(\mathbf{x}+\mathbf{y})$ and

$$\omega_f(\mathbf{x}, \varepsilon) = \sup(|f(\mathbf{x}) - f(\mathbf{y})|; \mathbf{y} \in \mathbf{R}^p,\ \|\mathbf{x}-\mathbf{y}\| < \varepsilon), \quad \mathbf{x} \in \mathbf{R}^p, \quad \varepsilon > 0, \tag{2.5}$$

with $\|\cdot\|$ denoting the Euclidean norm.

Our main assertion is the following:

Theorem. *Let f be a bounded Borel measurable function defined on $\mathbf{R}^p$. Then under assumptions* (i)–(iv) *there exist constants D_1, D_2, D_3 (not depending on N, c_{ij}, and f) such that*

$$\left| \int_{\mathbf{R}^p} f(\mathbf{x})\, d(G_c(\mathbf{x}) - \Phi(\mathbf{x}; E\mathbf{S}_c, \operatorname{var}\mathbf{S}_c)) \right|$$

$$\leqslant D_1 \sup_{\mathbf{x}\in\mathbf{R}^p} |f(\mathbf{x})| \sum_{i=1}^{p} \sum_{j=1}^{N} |c_{ij}^*|^3$$

$$+ D_2 \tilde{\omega}_f\left(D_3 \sum_{i=1}^{p} \sum_{j=1}^{N} |c_{ij}^*|^3, \Phi(\cdot\,; E\mathbf{S}_c, \operatorname{var}\mathbf{S}_c) \right), \tag{2.6}$$

where

$$c_{ij}^* = (c_{ij} - \bar{c}_i) \Big/ \left(\sum_{j=1}^{N} (c_{ij} - \bar{c}_i)^2 \right)^{\frac{1}{2}}, \bar{c}_i = N^{-1} \sum_{j=1}^{N} c_{ij}. \tag{2.7}$$

3. Remarks and applications

Through this section (unless otherwise stated) it is supposed assumptions (i)–(iv) are satisfied and f_v are constants not depending on N, c_{ij} and the function f.

(1) The assertion of the Theorem remains true if we replace $\operatorname{var}\mathbf{S}_c$ by $\operatorname{var}\mathbf{T}_c$ defined by (4.6) below.

(2) Let $A \subset \mathbf{R}^p$ be a Borel measurable set. Then the inequality (2.6) reduces to

$$|G_c(A) - \Phi(A; E\mathbf{S}_c, \operatorname{var}\mathbf{S}_c)|$$

$$\leqslant f_1 \sum_{i=1}^{p} \sum_{j=1}^{N} |c_{ij}^*|^3 + f_2 \sup_{\mathbf{y} \in \mathbf{R}^p} \Phi((\delta A)^\gamma + \mathbf{y}; E\mathbf{S}_c \operatorname{var}\mathbf{S}_c),$$

where (δA) is the topological boundary of A and $(\delta A)^\gamma$ is the set of all points having distances from A less than

$$\gamma = f_3 \sum_{i=1}^{p} \sum_{j=1}^{N} |c_{ij}^*|^3,$$

(c_{ij}^* are defined by (2.7)).

(3) Denote by $\mathcal{C}$ the class of all Borel measurable convex subsets of $\mathbf{R}^p$. Then

$$\sup_{A \in \mathcal{C}} |G_c(A) - \Phi(A; E\mathbf{S}_c, \operatorname{var}\mathbf{S}_c) \leqslant f_4 \sum_{i=1}^{p} \sum_{j=1}^{N} |c_{ij}^*|^3.$$

Particularly,

$$\sup_{\mathbf{x} \in \mathbf{R}^p} |G_c(\mathbf{x}) - \Phi(\mathbf{x}; E\mathbf{S}_c, \operatorname{var}\mathbf{S}_c)| \leqslant f_4 \sum_{i=1}^{p} \sum_{j=1}^{N} |c_{ij}^*|^3.$$

(4) The class of sets $\{\mathbf{S}_c(\operatorname{var}\mathbf{S}_c)^{-1}\mathbf{S}_c < \mathbf{x}\}$ for all $x \in \mathbf{R}^1$ is a subclass of $\mathcal{C}$. Denoting H_c and $F(x;p)$ the distribution of the random variable $\mathbf{S}_c'(\operatorname{var}\mathbf{S}_c)^{-1}\mathbf{S}_c$ and the central χ^2-distribution with p degrees of freedom, respectively, one can assert

$$\sup_{x \in \mathbf{R}^1} |H_c(x) - F(x;p)| \leqslant f_5 \sum_{i=1}^{p} \sum_{j=1}^{N} |c_{ij}^*|^3.$$

(5) Quadratic rank statistics for testing the multivariate hypothesis of randomness. Consider $\mathbf{S}_c'\mathbf{\Sigma}_c^{-1}\mathbf{S}_c$, where $\mathbf{\Sigma}_c$ is a matrix with the elements

$$\sigma_{iv}=(N-1)^{-1}\sum_{j=1}^{N} c_{ij}c_{vj}\sum_{k=1}^{N}(a_{Ni}(R_{ik})-\bar{a}_{Ni})(a_{Nv}(R_{vk})-\bar{a}_{Nv}),$$

$\bar{a}_{Ni}=N^{-1}\sum_{j=1}^{N}a_{Ni}(j)$. If the matrix $\mathbf{\Sigma}_c$ is not random, then $\mathbf{\Sigma}_c=\operatorname{var}\mathbf{S}_c$ and the convergence rate is done in the previous remark. If $\mathbf{\Sigma}_c$ is random the result coincides with that in [8], i.e., the rate is

$$\sum_{i=1}^{p}\sum_{j=1}^{N}|c_{ij}^{*}|^{3+\delta}N^{\delta},\qquad \delta>0.$$

(6) Rank statistics for k-sample problem. Let $\{X_{i1},\ldots,X_{in_i}\}$, $i=1,\ldots,p$, be p-independent random samples with the same continuous distribution function. Define

$$\mathcal{K}_c=(N-1)^{-1}\sum_{i=1}^{p} n_i^{-1}\left(\sum_{j=1}^{n_i}a_N(R_{ij})-\bar{a}_N\right)^2\left(\sum_{j=1}^{N}(a_N(j)-\bar{a}_N)^2\right)^{-1},$$

where $N=\sum_{i=1}^{p}n_i$, $\bar{a}_N=N^{-1}\sum_{j=1}^{N}a_N(j)$. Assume that $a_N(j)$ are given either (2.1) or (2.2) and assumptions (iii), (iv) are satisfied, then there exists a constant f_6 such that

$$\sup_{x\in\mathbf{R}^1}|P(\mathcal{K}_c<x)-F(x;p-1)|\leqslant f_6\max_{1\leqslant i\leqslant p} n_i^{-\frac{1}{2}}.$$

(7) Contiguous and general alternatives. Combining the proving methods of our Theorem together with that of Theorem 3.2 in [7] (for contiguous alternatives in the univariate case) we obtain under assumptions (i)–(v), (vii) in [8] the estimation coinciding with (2.6), where $\sum_{i=1}^{p}\sum_{j=1}^{N}|c_{ij}^{*}|^3$ is replaced by $\sum_{i=1}^{p}\sum_{j=1}^{N}(|c_{ij}^{*}|^3+|\theta_{ij}^{*}|^3)$ (θ_{ij}^{*} denotes the parameter the distribution of X_{ij} depends on). Similarly, the estimation of (1.3) can be obtained combining the proof of our Theorem and Theorem A[9] (for general alternatives in the univariate case).

(8) Signed rank statistics. For the case of the vector of signed rank statistics (utilized for construction of test statistics for symmetry problem) an analogous assertion both under hypothesis and alternatives can be derived.

4. Proof of theorem

First we give some additional notations and prove some lemmas. Throughout this section d_v's are constants not depending on N, c_{ij} and the function f.

Without loss of generality we may assume

$$c_{ij}^* = c_{ij}, \qquad 1 \leqslant i \leqslant p, \quad 1 \leqslant j \leqslant N, \tag{4.1}$$

$$\int_0^1 \phi_i(u)\,du = 0, \qquad \int_0^1 \phi_i^2(u)\,du = 1 \qquad 1 \leqslant i \leqslant p. \tag{4.2}$$

Define

$$T_{ic} = \sum_{j=1}^{N} c_{ij}\phi_i\big(F_i(X_{ij})\big), \qquad 1 \leqslant i \leqslant p, \tag{4.3}$$

$$T_{ic}^* = (N+1)^{-1} \sum_{j=1}^{N} \sum_{\substack{h=1 \\ h \neq j}}^{N} c_{ij} g_i(X_{ij}, X_{ih}), \qquad 1 \leqslant i \leqslant p, \tag{4.4}$$

$$g_i(X_{ij}, X_{ih}) = \big(u(X_{ij} - X_{ih}) - F_i(X_{ij})\big)\phi_i'\big(F_i(X_{ij})\big)$$
$$- E\big\{\big[\big(u(X_{ij} - X_{ih}) - F_i(X_{ij})\big)\phi_i'\big(F_i(X_{ij})\big)\big]/X_{ih}\big\},$$
$$1 \leqslant j,\ h \leqslant N, \quad j \neq h, \qquad 1 \leqslant i \leqslant p \tag{4.5}$$

$$\mathbf{T}_c = (T_{1c}, \ldots, T_{pc})', \tag{4.6}$$

$$\mathbf{T}_c^* = (T_{1c}^*, \ldots, T_{pc}^*)', \tag{4.7}$$

$$\mathbf{V}_c = \operatorname{var} \mathbf{T}_c, \tag{4.8}$$

where $u(x) = 1$ if $x \geqslant 0$ and $u(x) = 0$ if $x < 0$.

Notice that

$$ET_{ic} = ET_{ic}^* = 0, \qquad 1 \leqslant i \leqslant p, \tag{4.9}$$

$$E\big(g_i(X_{ij}, X_{ih})/X_{ij}\big) = E\big(g_i(X_{ij}, X_{ih})/X_{ih}\big) = 0,$$
$$1 \leqslant i \leqslant p, \quad 1 \leqslant j,h \leqslant N \tag{4.10}$$

In the rest of this section we shall write shortly $\sum_{i,j}$ instead of $\sum_{i=1}^{p} \sum_{j=1}^{N}$.

Lemma 1. *Under assumption* (i), (iii), (iv), (4.1), (4.2) *and if scores are given by* (2.1) *there exist constants* d_1, d_2 *such that*

$$P\left(\|\mathbf{S}_c - \mathbf{T}_c - \mathbf{T}_c^*\| \geq N^{-\frac{1}{2}}\right) \leq d_1 N^{-\frac{1}{2}}, \tag{4.11}$$

$$P\left(\sum_{i=1}^{p}\left(\sum_{j=1}^{N} c_{ij}\left(a_{Ni}(R_{ij})\right) - \phi_i\left(U_{iN}^{(j)}\right)\right)^2 \geq N^{-1}\right) \leq d_2 N^{-\frac{1}{2}},$$

where $U_{iN}^{(j)}$ *are defined below* (2.2).

Proof. The assertion follows from Lemma 2.4 [7] and the following inequalities

$$P\left(\|\mathbf{S}_c - \mathbf{T}_c - \mathbf{T}_c\| \geq N^{\frac{1}{2}}\right) \leq \sum_{1=1}^{p} P\left(|S_{ic} - T_{ic} - T_{ic}| \geq (Np)^{-\frac{1}{2}}\right),$$

$$P\left(\sum_{i=1}^{p}\left(\sum_{j=1}^{N} c_{ij}\left(a_{Ni}(R_{ij}) - \phi_i\left(U_{iN}^{(j)}\right)\right)\right)^2 \geq N^{-1}\right)$$

$$\leq \sum_{i=1}^{p} P\left(\left|\sum_{j=1}^{N} c_{ij}\left(a_{Ni}(R_{ij}) - \phi_i\left(U_{iN}^{(j)}\right)\right)\right| \geq (Np)^{-\frac{1}{2}}\right).$$

Lemma 2. *Under the assumptions of Lemma* 1 *there exist constants* d_3, d_4, d_5 *such that*

$$\lambda_{1c} - \sum_{i,j} |c_{ij}|^3 d_3 \leq \lambda_{1c}^* \leq \lambda_{pc}^* \leq \sum_{i,j} |c_{ij}|^3 d_4 + \lambda_{pc}, \tag{4.12}$$

$$|E\phi_i(R_{ij}/(N+1))| \leq d_5 N^{-1}, \tag{4.13}$$

where λ_{1c} *and* λ_{1c}^* *denote the minimal eigenvalue of* $\operatorname{var}\mathbf{S}_c$ *and* $\operatorname{var}\mathbf{T}_c$, *resp., and* λ_{pc} *and* λ_{pc}^* *the maximal one of* $\operatorname{var}\mathbf{S}_c$ *and* $\operatorname{var}\mathbf{T}_c$, *resp.*

Proof. Using standard tools and noticing

$$E\phi_i(R_{ij}/(N+1)) = N^{-1}\sum_{j=1}^{N} \phi_i(j/(N+1)), \qquad 1 \leq i \leq p,$$

$$\int_0^1 \phi_i(u)\,\mathrm{d}u = 0, \qquad 1 \leq i \leq p,$$

we get easily (4.13).

Obviously,

$$|\operatorname{cov}(S_{ic}, S_{vc}) - \operatorname{cov}(T_{ic}, T_{vc})|$$

$$\leqslant (\operatorname{var} S_{ic} \operatorname{var}(S_{vc} - T_{vc}))^{\frac{1}{2}} (\operatorname{var} T_{vc} \operatorname{var}(S_{1c} - T_{ic}))^{\frac{1}{2}}$$

and

$$(\operatorname{var}(S_{vc} - T_{vc}))^{\frac{1}{2}} \leqslant (\operatorname{var}(S_{vc} - T_{vc} - T^*_{vc}))^{\frac{1}{2}} + (\operatorname{var} T^*_{vc})^{\frac{1}{2}}.$$

Thus by the proof of Lemma 2.4 [7] and Lemma 2.6 [7] we have that there exists a constant b_1 such that

$$|\operatorname{cov}(S_{ic}, S_{vc}) - \operatorname{cov}(T_{ic}, T_{vc})| \leqslant b_1 N^{-\frac{1}{2}}, \qquad 1 \leqslant i, v \leqslant p,$$

which implies

$$\sup_{\|\mathbf{x}\| \leqslant 1} \|(\operatorname{var} \mathbf{T}_c - \operatorname{var} \mathbf{S}_c)\mathbf{x}\| \leqslant p^{\frac{1}{2}} b_1 N^{-\frac{1}{2}},$$

where $\sup_{\|\mathbf{x}\| \leqslant 1}$ denotes the supremum over all p-dimensional vector with the Euclidean norm smaller or equal 1.

Hence the inequalities

$$\lambda^*_{1c} \geqslant \lambda_{1c} - \sup_{\|\mathbf{x}\| \leqslant 1} \|(\operatorname{var} \mathbf{T}_c - \operatorname{var} \mathbf{S}_c)\mathbf{x}\|,$$

$$\lambda^*_{pc} \leqslant \lambda_{pc} + \sup_{\|\mathbf{x}\| \leqslant 1} \|(\operatorname{var} \mathbf{T}_c - \operatorname{var} \mathbf{S}_c)\mathbf{x}\|$$

give (4.12).

Applying Theorem 16.2 by Bhattacharya [5] to the sums of independent random vectors $\mathbf{T}_c$ we obtain:

Lemma 3. *Under assumptions of Lemma* 1 *for all bounded Borel-measurable functions on* $\mathbf{R}^k$ *there exist constants* d_6, d_7 *such that*

$$\left|\int f(\mathbf{x})\, d(Q_c(\mathbf{x}) - \Phi(\mathbf{x}; \mathbf{0}, \mathbf{V}_c))\right|$$

$$\leqslant \sup_{\mathbf{x} \in \mathbf{R}^p} |f(\mathbf{x})| d_6 \sum_{i,j} |c_{ij}|^3 \lambda^{*-\frac{3}{2}}_{1c} + 2\tilde{\omega}_f \left(\lambda^{*\frac{1}{2}}_{1c} \lambda^{*-\frac{3}{2}}_{pc} \sum_{i,j} |c_{ij}|^3, \quad \Phi(\cdot\,; \mathbf{0}, \mathbf{V}_c) \right),$$

where Q_c *denotes the distribution function of* $\mathbf{T}_c$, $\tilde{\omega}_f$ *and* $\mathbf{V}_c$ *is given by* (2.5) *and* (4.8), *resp. The assertion remains true replacing* $\tilde{\omega}_f(\cdot,\cdot)$ *by* $\tilde{\omega}_g(d_7 \sum_{i,j}|c_{ij}|^3, \Phi(\cdot\,;\mathbf{0},\mathbf{V}_c))$, *where* $g(\mathbf{x})=f(\mathbf{V}_c^{-\frac{1}{2}}\mathbf{x})$.

Define

$$\mathbf{R}_j=\left(c_{1j}\phi_1(F_1(X_{1j})),\ldots,c_{pj}\phi_p(F_p(X_{pj}))\right)', \qquad 1\leqslant j\leqslant N. \quad (4.14)$$

To obtain the estimation of derivatives of the characteristic function of $\mathbf{T}_c$ we need the following three lemmas the proofs of which are only sketched for they are very similar to the proofs of Lemmas 2.7, 2.8[7].

Lemma 4. *Under assumptions of Lemma* 1 *there exist constants* d_8, d_9 *such that if*

$$\|\mathbf{t}\|<2p^{\frac{1}{2}}\log\Big(\sum_{i,j}|c_{ij}|^3\Big)^{-1}$$

then

$$|Ee^{\mathrm{i}\mathbf{t}'\mathbf{T}_c}\mathbf{t}'\mathbf{T}_c^*|\leqslant d_8\|\mathbf{t}\|^4N^{-\frac{1}{2}}\sum_{v,j}|c_{vj}|^3,$$

$$|Ee^{\mathrm{i}\mathbf{t}'\mathbf{T}_c}\prod_{v=1}^{p}T_{vc}^{\alpha_v}T_{vc}^{*\beta_v}|\leqslant d_9\|\mathbf{t}\|^{p+3}\sum_{i,j}|c_{ij}|^3N^{-\frac{1}{2}},$$

where $\alpha_1,\ldots,\alpha_p,\beta_1,\ldots,\beta_p$ *are nonnegative integers such that* $\sum_{v=1}^{p}(\alpha_v+\beta_v)\leqslant p+2$ *and at least one of* β_v *is nonzero.*

Proof. Obviously,

$$\exp(\mathbf{t}'\mathbf{R}_j)=1+\mathrm{i}\mathbf{t}'\mathbf{R}_j-\eta_j\|\mathbf{t}'\mathbf{R}_j\|^2/2, \qquad 1\leqslant j\leqslant N, \quad (4.15)$$

where $\mathbf{R}_j$ is defined by (4.14) and $|\eta_j|\leqslant 1$. Hence for $\|\mathbf{t}\|\leqslant 2p^{\frac{1}{2}}\cdot(\log\sum_{i,j}|c_{ij}|^3)^{-1}$ the Taylor expansion for $\log E\exp(\mathrm{i}\mathbf{t}'\mathbf{R}_j)$, $1\leqslant j\leqslant N$, can be established. Slightly modifying Theorem 8.9 [5] we have

$$\left|E\exp\left\{i\sum_{j\in B}\mathbf{t}'\mathbf{R}_j\right\}\right|\leqslant\exp\left\{-\frac{5}{24}\|\mathbf{t}\|^2\left(\lambda_{1c}^*-p\max_{\substack{1\leqslant i\leqslant p\\1\leqslant j\leqslant N}}c_{ij}^2(N-\#B)\right)\right\}, \quad (4.16)$$

where B denotes a subset of $\{1,\ldots,N\}$ and $\#$ denotes its cardinal number. Using these facts we obtain the assertion by direct computation.

Lemma 5. *Under assumptions of Lemma 1 there exist constants d_{10}, d_{11} such that if*

$$t \leqslant d_{10}\Big(\sum_{i,j}|c_{ij}|^3\Big)^{-1},$$

then

$$\Big|\mathrm{E}\,e^{it'T_c}\prod_{v=1}^{p}T_{vc}^{\alpha_v}T_{vc}^{*\beta_v}\Big| \leqslant \|\mathbf{t}\|^{\sum_{v=1}^{p}(\alpha_v+\beta_v)}d_{11}\sum_{v=1}^{p}(\alpha_v+2\beta_v)^{\sum_{v=1}^{p}(\alpha_v+2\beta_v)}$$
$$\exp\Big\{-\frac{5}{24}\|\mathbf{t}\|^2\Big[\lambda_{1c}^{*}-2p\sum_{v=1}^{p}(\alpha_v+2\beta_v)\max_{\substack{1\leqslant i\leqslant p\\ 1\leqslant j\leqslant N}}c_{ij}^2\Big]\Big\}$$

holds, where $\alpha_1,\ldots,\alpha_p,\beta_1,\ldots,\beta_p$ are nonnegative integers. If, moreover,

$$\sum_{v=1}^{p}(\alpha_v+2\beta_v)\leqslant\frac{1}{2}\lambda_{1c}^{*}\log\Big(\sum_{i,j}|c_{ij}|^3\Big)^{-1},$$

then

$$\Big|\mathrm{E}\,e^{it'T_c}\prod_{v=1}^{p}T_{vc}^{\alpha_v}T_{vc}^{*\beta_v}\Big|\leqslant(\|t\|d_{11}')^{\sum_{v=1}^{p}(\alpha_v+\beta_v)}$$
$$\times\Big(\sum_{v=1}^{p}(\alpha_v+2\beta_v)^{\sum_{v=1}^{p}(\alpha_v+2\beta_v)}\Big)\exp\Big\{-\frac{5}{48}\|\mathbf{t}\|^2\lambda_{1c}^{*}\Big\}.$$

Proof. Denote

$$\alpha_q^{*}=\sum_{v=1}^{q-1}\alpha_v,\quad \beta_q^{*}=\sum_{v=1}^{q-1}\beta_v,\qquad 1\leqslant q\leqslant p,\quad \alpha_0^{*}=0,\quad \beta_0^{*}=0,$$

$$B=(j_1,\ldots,j_{\alpha_p+1},v_1,\ldots,v_{\beta_p+1},m_1,\ldots,m_{\beta_p+1}).$$

Then one can write

$$\mathrm{E}e^{it'\mathbf{T}_c}\prod_{v=1}^{p}T_{vc}^{\alpha_v}T_{vc}^{*\beta_v}$$

$$=\sum_B \mathrm{E}\exp\Big\{it'\sum_{j\notin B}\mathbf{R}_j\Big\}\mathrm{E}\exp\Big\{it'\sum_{j\in B}\mathbf{R}_j\Big\}h(B)(N+1)^{-\beta^*_{p+1}},$$

where $\sum_B$ denotes the sum over all B such that $1\leqslant j_w, v_z, m_z\leqslant N$, $v_z\neq m_z$, $w=1,\ldots,\alpha^*_{p+1}$, $z=1,\ldots,\beta^*_{p+1}$, $h(B)$ is a product of some c_{ij} $g_i(X_{ij},X_{ih}),\phi_i(F_i(X_{ij}))$. Thus by (4.2) and (4.10) we get that

$$\mathrm{E}h(B)=0$$

if among members of B exists at least one member differing from the others. The proof can now be finished, similarly as that of Lemma 6 [9], using the last fact, (4.14) and (4.15).

Lemma 6. *Under assumptions of Lemma* 1 *there exists a constant* d_{12} *such that*

$$\left|\mathrm{E}(\mathbf{t}'\mathbf{T}_c)^{2k}\prod_{v=1}^{p}T_{vc}^{\alpha_v}T_{vc}^{*\beta_v}\right|\leqslant(\|\mathbf{t}\|d_{12})^{2k}(2k)^{2k+p+1}N^{-(k+\sum_{v=1}^{p}\beta_v)},$$

where $k,\alpha_1,\ldots,\alpha_p,\beta_1,\ldots,\beta_p$ *are nonnegative integers satisfying* $\sum_{v=1}^{p}(\alpha_v+\beta_v)\leqslant p+1$ *and at least one* β_v *is nonzero.*

Proof. By Jensen inequality we have

$$\left|\mathrm{E}(\mathbf{t}'\mathbf{T}_c^*)^{2k}\prod_{v=1}^{p}T_{vc}^{\alpha_v}T_{vc}^{*\beta_v}\right|$$

$$\leqslant\prod_{v=1}^{p}(\mathrm{E}T_{vc}^{*2A})^{\beta_v/2A}\prod_{v=1}^{p}(\mathrm{E}T_{vc}^{2A})^{\alpha_v/2A}\prod_{v=1}^{p}(\mathrm{E}\|\mathbf{T}_c^*\|^{2A})^{k/A}\|\mathbf{t}\|^{2k},\tag{4.17}$$

where $A=\sum_{v=1}^{p}(\alpha_v+\beta_v)+2k$. By definition (see (4.6)), $\mathbf{T}_c$ is a sum of independent random vectors and thus by Lemma 1 [9]

$$\mathrm{E}T_{vc}^{2A}\leqslant b_2^A A^A,\tag{4.18}$$

where b_2 is a constant (not depending on c_{ij} and k), and by Lemma 2.6 [7]

$$\mathrm{E}T_{vc}^{*2A} \leqslant b_3^A N^{-A}(2A)^{2A}, \tag{4.19}$$

where b_3 is a constant. The assertion can now be concluded easily from (4.17–19).

Denote by Q_c and Q_c^* the distribution function of $\mathbf{T}_c$ and $\mathbf{T}_c^*$, resp., and by $\hat{}$ characteristic functions.

Lemma 7. *Under assumptions of Lemma* 1 *there exist constants* d_{13}, d_{14} *such that*

$$\int_{\|\mathbf{t}\| < d_{13}(\Sigma_{i,j}|c_{ij}|^3)^{-1}} |D^\alpha(\hat{Q}_c - \hat{Q}_c^*)(\mathbf{t})\,\mathrm{d}\mathbf{t} \leqslant d_{14} \sum_{i,j} |c_{ij}|^3,$$

where $D^\alpha = (\partial/\partial t_1)^{\alpha_1} \cdots (\partial/\partial t_p)^{\alpha_p}$, $\sum_{v=1}^p \alpha_v \leqslant p+1$.

Proof. Using the Taylor expansion for $\exp \mathrm{i}\mathbf{t}'\mathbf{T}_c$ one can write

$$D^\alpha(\hat{Q}_c - \hat{Q}_c^*)(\mathbf{t}) = (i)^{\Sigma_{v=1}^p \alpha_v} \mathrm{E}\exp\{\mathrm{i}\mathbf{t}'\mathbf{T}_c\} \left\{ \prod_{v=1}^p T_{vc}^{\alpha_v} - \prod_{v=1}^p (T_{vc} + T_{vc}^*)^{\alpha_v} \right\}$$

$$\left\{ \sum_{k=0}^{2k^*-1} \frac{(\mathrm{i}\mathbf{t}'\mathbf{T}_c)^k}{k!} + \eta \frac{\|\mathbf{t}'\mathbf{T}_c\|^{2k^*}}{(2k^*)!} \right\}, \qquad |\eta| \leqslant 1,$$

for an arbitrary natural k^*. The general terms that are sufficient to estimate are either of the following forms:

$$\mathrm{E}\exp\{\mathrm{i}\mathbf{t}'\mathbf{T}_c\} \sum_{v=1}^p T_{vc}^{\alpha_v} T_{vc}^{*\beta_v} (\mathbf{t}'\mathbf{T}_c^*)^k, \qquad \mathrm{E}\left| \prod_{v=1}^p T_{vc}^{\alpha_v} T_{vc}^{*\beta_v} \|\mathbf{t}'\mathbf{T}_c^*\|^{2k} \right|.$$

Their estimations are given by Lemmas 4–6. The proof can be concluded in the same way as that of Theorem 2.1 [7].

Proof of the main theorem. Denoting by Q_c and Q_c^* the distribution functions of $\mathbf{T}_c$ and $\mathbf{T}_c + \mathbf{T}_c^*$, resp. Denoting the probability measure induced by random variables $\mathbf{X}_1,\ldots,\mathbf{X}_N$ by P_N one can write

$$\left|\int_{\mathbf{R}^p} f(\mathbf{x})\,\mathrm{d}(G_c(\mathbf{x})-Q_c^*(\mathbf{x}))\right|$$

$$\leqslant \left|\int_{\|\mathbf{S}_c-\mathbf{T}_c-\mathbf{T}_c^*\|<\varepsilon} (f(\mathbf{S}_c)-f(\mathbf{T}_c+\mathbf{T}_c^*))\,\mathrm{d}P_N\right|$$

$$+\left|\int_{\|\mathbf{S}_c-\mathbf{T}_c-\mathbf{T}_c^*\|\geqslant\varepsilon} (f(\mathbf{S}_c)-f(\mathbf{T}_c+\mathbf{T}_c^*))\,\mathrm{d}P_N\right| \tag{4.20}$$

$$\leqslant \int \omega_f(\mathbf{x},\varepsilon)\,\mathrm{d}Q_c^*(\mathbf{x})+2\sup_{\mathbf{x}\in\mathbf{R}^p}|f(\mathbf{x})|P(\|\mathbf{S}_c-\mathbf{T}_c-\mathbf{T}_c^*\|\geqslant\varepsilon)$$

$$\leqslant \tilde{\omega}_f(\varepsilon,Q_c)+2\sup_{\mathbf{x}\in\mathbf{R}^p}|f(\mathbf{x})|P(\|\mathbf{S}_c-\mathbf{T}_c-\mathbf{T}_c^*\|\geqslant\varepsilon),$$

$\varepsilon>0$ arbitrary. Obviously,

$$\int_{\mathbf{R}^p} f(\mathbf{x})\,\mathrm{d}(Q_c^*(\mathbf{x})-\Phi(\mathbf{x};\mathbf{0},\mathbf{V}_c))$$

$$=\int_{\mathbf{R}^p} f(\mathbf{x})d(Q_c^*-Q_c)(\mathbf{x})+\int_{\mathbf{R}^p} f(\mathbf{x})d(Q_c(\mathbf{x})-\Phi(\mathbf{x};\mathbf{0},\mathbf{V}_c)). \tag{4.21}$$

The estimation of the second member on the right we get applying Lemma 3 (Q_c denotes the distribution of the sum of independent vectors $\mathbf{T}_c$):

$$\left|\int_{\mathbf{R}^p} f(\mathbf{x})d(Q_c(\mathbf{x})-\Phi(\mathbf{x};\mathbf{0},\mathbf{V}_c))\right|$$

$$\leqslant \sup_{\mathbf{x}\in\mathbf{R}^p}|f(\mathbf{x})|b_1\sum_{i,j}|c_{ij}|^3\lambda_{1c}^{*-\frac{3}{2}}$$

$$+b_2\tilde{\omega}_f\left(b_3\lambda_{2c}^{*\frac{1}{2}}\lambda_{1c}^{*-\frac{3}{2}}\sum_{i,j}|c_{ij}|^3\Phi(\cdot;\mathbf{0},\mathbf{V}_c)\right), \tag{4.22}$$

where b_v's are constants not depending on N, c_{ij} and $\mathbf{V}_c$; λ_{1c}^* and λ_{2c}^* are the minimal and maximal eigenvalues of $\mathbf{V}_c$, resp. By Lemma 7 in [4] and

Lemma 3 in [3],

$$\left\|\int f(\mathbf{x})\,\mathrm{d}(Q_c^*(\mathbf{x})-Q_c(\mathbf{x}))\right\|\leqslant b_4 \sup_{\mathbf{x}\in\mathbf{R}^p}|f(\mathbf{x})|(\|(Q_c^*-Q_c)*K_\eta\|+\tilde{\omega}_f(2\eta;Q_c)) \tag{4.23}$$

where $K_\eta, \eta>0$, is a probability measure with the properties

$$K_1\{\mathbf{x};\|\mathbf{x}\|<1\}\leqslant 3/4, \qquad \int_{\mathbf{R}^p}\|\mathbf{x}\|^{p+1}K_1(\mathrm{d}\mathbf{x})<+\infty,$$

$$\hat{K}_\eta(\mathbf{t})=0 \quad \text{if } \|\mathbf{t}\|>b_5/\eta, \qquad K_\eta(A)=K_1(\eta^{-1}A)$$

with b_4, b_5 being constants not depending on N, c_{ij} and η. Replacing in Lemma 3 $f(\mathbf{x})$ by $\omega_f(\mathbf{x};2\eta)$ we obtain

$$\tilde{\omega}_{f*}(2\eta,Q_c)\leqslant\tilde{\omega}_{f*}(2\eta,\Phi(\cdot\,;\mathbf{0},\mathbf{V}_c))+b_6\sup_{\mathbf{x}\in\mathbf{R}^p}|f(\mathbf{x})|\sum_{i,j}|c_{ij}|^3$$

$$+3\tilde{\omega}_{f*}\Big(2\eta+b_7\sum_{i,j}|c_{ij}|^3,\Phi(\cdot\,;\mathbf{0},\mathbf{V}_v)\Big). \tag{4.24}$$

Put

$$\eta=b_8\sum_{i,j}|c_{ij}|^3, \tag{4.25}$$

where b_8 is a suitable constant. From Lemma 8 in [4] and Lemma [7], similarly as in the proof of Theorem in [4] one can derive

$$\|(Q_c-Q_c^*)*K_\eta\|\leqslant b_9\max\Big\{\int D^{\beta_1}(\hat{Q}_c-\hat{Q}_c^*)(\mathbf{t})D^{\beta_2}K\eta(\mathbf{t})\,\mathrm{d}\mathbf{t},$$

$$0\leqslant\beta_1+\beta_2\leqslant p+1\}$$

$$\leqslant b_{10}\max\Big\{\int_{\|\mathbf{t}\|<(b_{11}/\eta)}D^{\beta}(\hat{Q}_c-\hat{Q}_c^*)(\mathbf{t})\,\mathrm{d}\mathbf{t};0\leqslant\beta\leqslant p+1\Big\}.$$

Thus by Lemma 7 we have

$$\|(Q_c-Q_c^*)*K_\eta\|\leqslant b_{12}\sum_{i,j}|c_{ij}|^3.$$

which together with (4.21–25) implies

$$\left|\int f(\mathbf{x})\, d(Q_d^*(\mathbf{x}) - \Phi(\mathbf{x}; \mathbf{0}, \mathbf{V}_c))\right|$$

$$\leqslant b_{13} \sup_{\mathbf{x}\in\mathbf{R}^p} |f(\mathbf{x})| \sum_{i,j} |c_{ij}|^3 + b_{14}\tilde{\omega}_f\left(b_{15}\sum_{i,j}|c_{ij}|^3, \phi(\cdot\,; \mathbf{0}, \mathbf{V}_c)\right). \quad (4.26)$$

Now replacing in (4.26) $f(\mathbf{x})$ by $\omega_f(\mathbf{x}; \varepsilon)$ we have

$$\tilde{\omega}_f(\varepsilon, Q_c) \leqslant \tilde{\omega}_f(\varepsilon, \Phi(\cdot, \mathbf{0}, \mathbf{V}_c)) + b_x \sup_{\mathbf{x}\in\mathbf{R}^p} |f(\mathbf{x})| \sum_{i,j} |c_{ij}|^3$$

$$+ 3\tilde{\omega}_f\left(\varepsilon + b_{17}\sum_{i,j}|c_{ij}|^3, \phi(\cdot\,; \mathbf{0}, \mathbf{V}_c)\right).$$

The assertion of our Theorem follows from the last inequality (4.20) and (4.26), where we put $\varepsilon = N^{-\frac{1}{2}}$.

References

[1] Von Bahr, B. (1976), Remainder term estimate in a combinatorial limit theorem. *Z. Wahrscheinlichkeitstheorie verw. Gebiete* **35**, 131–139.

[2] Bergström, H. and Puri, M. L. (1977). Convergence and remainder terms in linear rank statistics. *Ann. Statist.* **5**, 671–680.

[3] Bhattacharya, R. N. (1972). Recent results on refinements of the central limit theorems. *Proc. Sixth Berkeley Symp. Math. Statist. Prob.* **2**, 453–484.

[4] Bhattacharya, R. N. (1975). On errors of normal approximation. *Ann. Prob.* **3**, 815–828.

[5] Bhattacharya, R. N. and Rao, R. Ranga (1976). *Normal Approximation and Asymptotic Expansions*. Wiley and Sons.

[6] Erickson, R. V. and Koul, H. L. (1976). L_1 rates of convergence for linear rank statistics. *Ann. Statist.* **4**, 772–774.

[7] Hušková, M. (1977). Rate of convergence of linear rank statistics under hypothesis and alternatives. *Ann. Statist.* **5**, 658–670.

[8] Hušková, M. (1977). Rates of convergence of quadratic rank statistics. *J. Multivariate Analysis* **7**, 63–73.

[9] Hušková, M. (1979). The rate of convergence of simple linear rank statistics. *Hájek's Memorial* (to appear).

[10] Jensen, D. R. (1974). The joint distribution of Friedman's χ^2-statistics. *Ann. Statist.* **2**, 311–323.

[11] Jurečková, J. and Puri, M. L. (1975). Order of normal approximation for rank test statistic distribution. *Ann. Prob.* **3**, 526–533.

P. R. Krishnaiah, ed., *Multivariate Analysis–V*

SOME LIMIT THEOREMS FOR PARTIAL SUMS OF STATIONARY SEQUENCES

M. ROSENBLATT*
University of California, San Diego, La Jolla, CA 92093, U.S.A.

The central limit theorem (asymptotic normality) for partial sums of independent, identically distributed random variables with positive variance suitably centered and normalized is a classical result of probability theory. There are by now a variety of conditions on stationary dependent sequences that are sufficient for analogous results to hold. A number of these are discussed. Some small consideration is also given to contexts in which the dependence is strong enough so that one has a non-normal and even nonstable limiting distribution.

1. Introduction

If $X_1, X_2, \ldots, X_n, \ldots$ are independent, identically distributed random variables with mean $\mathbf{E}X_i \equiv 0$ and variance $\sigma^2 > 0$, by the classical central limit theorem

$$\sum_{j=1}^{[nt]} X_j/(\sqrt{(n)}\sigma) = S_n/(\sqrt{(n)}\sigma) \tag{1}$$

is asymptotically normal with mean zero and variance t as $n \to \infty$. Here $[u]$ is the greatest integer less than or equal to u.

The comparable dependent situation is that in which one has a stationary sequence $\{X_j, j = \ldots, -1, 0, 1, \ldots\}$. Various measures of dependence have been introduced. We shall mention two of these. Let $\mathfrak{B}_n = \mathfrak{B}(X_j, j \leqslant n)$ be the Borel field of events generated by X_j, $j \leqslant n$ and $\mathfrak{F}_m = \mathfrak{B}(X_j, j \geqslant m)$ the Borel field generated by X_j, $j \geqslant m$. The stationary sequence $X = (X_j)$ is *strongly mixing* if

$$\sup_{\substack{A \in \mathfrak{B}_0 \\ B \in \mathfrak{F}_s}} |P(AB) - P(A)P(B)| = \alpha(s) \downarrow 0 \tag{2}$$

as $s \to \infty$. Let ${}_nL$ and L_m be the spaces of random variables with finite

*Research partially supported by the Office of Naval Research.

second moments measurable with respect to $\mathcal{B}_n$ and $\mathcal{F}_m$ respectively. Further corr(X, Y) denotes the correlation of the random variables X, Y. We shall say that *maximal correlation* of past and future is zero if

$$\sup_{\substack{X \in {}_0L \\ Y \in L_s}} |\mathrm{corr}(X,Y)| = \beta(s) \downarrow 0. \tag{3}$$

The maximal correlation condition (3) is obviously more stringent than the strong mixing condition. The following theorem of Ibragimov makes use of the maximal correlation condition [6].

Theorem 1. *Let $X=(X_j)$ be a stationary sequence satisfying the maximal correlation condition* (3) *with* $\mathbf{E}X_j \equiv 0$, $\mathbf{E}|X_j|^{2+\delta} < \infty$ *for some* $\delta > 0$, *and* $\sigma_n^2 = E|\sum_{j=1}^n X_j|^2 \to \infty$ *as* $n \to \infty$. *Then*

$$\sum_{j=1}^{n} X_j / \sigma_n$$

is asymptotically $N(0,1)$ *as* $n \to \infty$. *If* $\sigma_n^2 \to \infty$ *as* $n \to \infty$, *then* $\sigma_n^2 = nh(n)$ *with* $h(n)$ *slowly varying.*

An example recently constructed by Bradley [1] shows that one cannot relax $E|X_j|^{2+\delta} < \infty$, $\delta > 0$, to $E|X_j|^2 < \infty$. In his example all other conditions are satisfied together with $E|X_j|^2 < \infty$ and a subsequence of normalized partial sums are shown to converge weakly to a nontrivial compound Poisson distribution.

There are various formulations of a central limit theorem for processes with the strong mixing property [5, 7]. However, there may be limiting distributions other than the normal. But a limiting distribution must be stable as is indicated by the following theorem (see [5] for a proof).

Theorem 2. *Let $X=(X_j)$ be a stationary strongly mixing sequence. If the normalized partial sums*

$$B_n^{-1} \sum_{j=1}^{n} X_j - A_n, \tag{4}$$

with $B_n \to \infty$ *as* $n \to \infty$, *have a nondegenerate limiting distribution it must be stable. If the stable distribution has exponent* α, *then* $B_n = n^{1/\alpha} h(n)$ *where* h *is slowly varying as* $n \to \infty$.

Notice that if $\mathbf{E}|X_j|^{2+\delta}<\infty$ for some $\delta>0$, in the case of the maximal correlation condition, $\sigma_n^2=nh(n)$ with slowly varying h. One can produce examples of strongly mixing processes with bounded random variables (for instantaneous functions of countable state Markov chains see [2]) such that $\sigma_n^2\cong n^{1+\delta}$, $\delta>0$, and the normalized partial sums have a non-normal stable limiting distribution.

It is of some interest to see what the strong mixing condition means for a stationary Markov sequence. Let

$$P(x,A)=P[X_{n+1}\in A\,|\,X_n=x] \tag{5}$$

be the transition function of a stationary Markov sequence with invariant instantaneous probability measure μ

$$\mu(A)=P(X_n\in A)$$

$$\int \mu(\mathrm{d}x)P(x,A)=\mu(A). \tag{6}$$

Given an integrable function f (with respect to μ) let

$$(Tf)(x)=\int P(x,\mathrm{d}y)f(y)=\mathbf{E}[f(X_{n+1})\,|\,X_n=x]. \tag{7}$$

Also let

$$\|f\|_\infty=\operatorname*{ess\,sup}_x |f(x)|, \tag{8}$$

and

$$\|f\|_1=\int \mu(\mathrm{d}x)|f(x)|=\mathbf{E}|f(X_n)|. \tag{9}$$

The ess sup is understood to be taken with respect to the measure μ. One can then obtain the following result [7].

Theorem 3. *A stationary Markov sequence with transition function $P(\cdot,\cdot)$ and invariant probability measure μ is strongly mixing if and only if*

$$\lim_{n\to\infty}\sup_{f\perp 1}\frac{\|T^nf\|_1}{\|f\|_\infty}=0. \tag{10}$$

It should be noted that the maximal correlation condition for a stationary Markov sequence is

$$\|T^n\| = \sup_{f \perp 1} \frac{\|T^n f\|_2}{\|f\|_2} \to 0 \qquad \text{as } n \to \infty, \tag{11}$$

where $\|f\|_2 = \{E|f(X_n)|^2\}^{\frac{1}{2}}$. This is just the condition that the L^2 norm of the operator T^n, $\|T^n\|$, acting on the space of square integrable functions with mean zero tend to zero. Actually if

$$\|T^{n_0}\| = \delta < 1$$

for an integer $n_0 \geqslant 1$, then $\|T^{kn_0}\| \leqslant \delta^k$ so that the norm tends to zero exponentially. The maximal correlation condition does not imply such exponential decay of the mixing coefficient if the process is nonMarkovian (see [1]). One should also note that one can have

$$\sup_{f \perp 1} \frac{\|Tf\|_1}{\|f\|_\infty} = 1,$$

and yet have strong mixing for a stationary Markov sequence. It is also possible to have

$$\sup_{f \perp 1} \frac{\|Tf\|_1}{\|f\|_\infty} = \rho < 1,$$

and yet not have strong mixing for a Markov sequence. Also examples of strongly mixing Markov sequences can be given such that (2) approaches zero as slowly as may be desired as $n \to \infty$.

2. Random rotations on the circle

A simple class of Markov sequences are given by the random rotations. Not all of the usual questions concerning the central limit theorem and allied problems for these processes have been answered. But a number of these questions can be answered.

Let η be a regular probability measure on the Borel subsets of $[0, 1)$. The transition probability function is given by

$$P(x, A) = \eta((A - x) \text{ modulo one}), \tag{12}$$

where B modulo one $=\{x, 0 \leqslant x < 1$: for some integer $k, x+k \in B\}$. The invariant probability measure is taken to be the uniform distribution on $[0,1)$. Notice that the process $X=(X_n)$ with this transition function satisfies

$$X_{n+1}=X_n+Y_n \text{ modulo one}, \tag{13}$$

where Y_n is independent of X_n and has distribution η. Let the Fourier–Stieltjes coefficients of η be

$$\hat{\eta}_k=\int_0^1 \exp(2\pi ikx)\eta(\mathrm{d}x). \tag{14}$$

It can then be shown that for such processes strong mixing and the maximal correlation condition are equivalent to each other and to

$$\sup_{k\neq 0}|\hat{\eta}_k|<1. \tag{15}$$

As noted earlier, for the case of countable state Markov sequences the maximal correlation condition is properly stronger than strong mixing. Because of the exponential decay of the mixing coefficient when one has strong mixing for random rotation models, the following theorem is obtained [7].

Theorem 4. *Let $X=(X_j)$ be a strongly mixing process of random rotations. Then given any function $f\in L^2$, $f\not\equiv 0$, with mean zero it follows that*

$$\sigma_n^2=\mathbf{E}\left|\sum_{j=1}^{n} f(X_j)\right|^2 \cong n\sum_{\nu}|f_\nu|^2\frac{1-|\hat{\eta}_\nu|^2}{|1-\hat{\eta}_\nu|^2}, \tag{16}$$

as $n\to\infty$, where the f_ν are the Fourier coefficients of f. Further

$$\sum_{j=1}^{n} f(X_j)/\sigma_n$$

is asymptotically $N(0,1)$ as $n\to\infty$.

If X is not a strongly mixing process of random rotations, there may still be functions $f\in L^2$ such that $\sum_{j=1}^{n} f(X_j)/\sigma_n$ is asymptotically $N(0,1)$. Of course this is the case in which

$$\limsup_{k\to\infty}|\hat{\eta}_k|=1.$$

The following result of Heyde [4] is helpful in getting some insights into this question.

Theorem 5. *Let $X=(X_j)$ be a stationary ergodic process with mean zero and finite second moment. Assume that*

$$Y=\sum_{r=0}^{\infty}\{\mathbf{E}(X_r|\mathcal{B}_0)-\mathbf{E}(X_r|\mathcal{B}_{-1})\} \tag{17}$$

converges in L^2 on the probability space and

$$\mathbf{E}Y^2=\sigma^2=\lim_{n\to\infty}\frac{1}{n}\mathbf{E}\left|\sum_{j=1}^{n}X_j\right|^2>0.$$

Then $S_n/(\sqrt{(n)}\sigma)$ is asymptotically $N(0,1)$ as $n\to\infty$.

Let us consider $X=(X_j)$ a process of random rotations with transition probability function (12) and invariant measure the uniform distribution on $[0,1)$. Let f be in L^2 with $\mathbf{E}f(X)\equiv 0$. Then

$$\mathbf{E}\left|\sum_{k=0}^{n-1}f(X_k)\right|^2=\sum_{\nu}|f_\nu|^2\left\{\sum_{k=-n}^{n}\hat{\eta}_\nu^{(k)}(n-|k|)\right\}, \tag{18}$$

where

$$\eta^{(k)}=\begin{cases}\eta^k & \text{if } k\geqslant 0\\ (\bar{\eta})^j & \text{if } k=-j\leqslant 0.\end{cases}$$

But

$$\sum_{k=-n}^{n}\eta^{(k)}\left(1-\frac{|k|}{n}\right)=\frac{1}{2\pi n}\int_{-\pi}^{\pi}\frac{\sin^2\frac{n}{2}\lambda}{\sin^2\frac{\lambda}{2}}\,\frac{1-|\eta|^2}{|1-\eta e^{i\lambda}|^2}\,d\lambda, \tag{19}$$

if $|\eta|<1$. For $\eta=e^{-i\mu}$ we understand by (19) the limit of the expression with $\eta=\eta(r)=re^{-i\mu}$ as $r\uparrow 1-$, that is

$$\frac{1}{n}\frac{\sin^2\frac{n}{2}\mu}{\sin^2\frac{\mu}{2}}.$$

Since $\mathbf{E}f(X)\equiv 0$ we have $f_0=0$. Relations (18) and (19) imply that the variance $\sigma_n^2=\mathbf{E}|\sum_{k=0}^{n-1}f(X_j)|^2$ will grow at a rate at least as fast as cn (with $c>0$, a constant) unless $|\hat{\eta}_\nu|\equiv 1$ for all ν such that $f_\nu\neq 0$. Since expression (19) is less than or equal to $(1+|\eta|)/(1-|\eta|)$, it follows that

$$\sum_\nu |f_\nu|^2 \frac{1}{1-|\hat{\eta}_\nu|}<\infty \tag{20}$$

is sufficient condition for σ_n^2 to grow at a rate no faster than cn $(c>0)$ as $n\to\infty$. If (20) is satisfied then

$$\frac{1}{n}\sigma_n^2\to\sum_\nu |f_\nu|^2|1-\hat{\eta}_\nu|^{-2}(1-|\hat{\eta}_\nu|^2)=\sigma^2,$$

as $n\to\infty$. Also, the necessary and sufficient condition for

$$\sum_{k=0}^{\infty}\{\mathbf{E}(f(X_k)|X_0)-\mathbf{E}(f(X_k)|X_{-1})\} \tag{21}$$

to converge is that $|\hat{\eta}_\nu|<1$ for all ν for which $f_\nu\neq 0$ and that the variance of (21)

$$\sum_\nu |f_\nu|^2|1-\hat{\eta}_\nu|^{-2}(1-|\hat{\eta}_\nu|^2)<\infty. \tag{22}$$

Using Theorem 5, we have the following result.

Theorem 6. *Let $X=(X_j)$ be a process of random rotations with invariant uniform distribution on $[0,1)$. Take $f\in L^2$, $f\not\equiv 0$, with mean zero. Then (22) is a sufficient condition that $\sigma_n^2=\mathbf{E}|\sum_{k=0}^{n-1}f(X_k)|^2$ grow no faster than cn as $n\to\infty$. If one also has*

$$0<\sigma^2=\sum|f_\nu|^2|1-\eta_\nu|^{-2}(1-|\hat{\eta}_\nu|^2)<\infty,$$

then $\sigma_n^2\cong\sigma^2 n$ as $n\to\infty$ and $\sum_{k=0}^{n-1}f(X_k)/(\sigma\sqrt{n})$ is asymptotically $N(0,1)$ as $n\to\infty$. If η is symmetric about zero ($\hat{\eta}_\nu$ is then real for all ν), condition (22) is necessary and sufficient for $\sigma_n^2\cong\sigma^2 n$ as $n\to\infty$ if $\sigma^2>0$.

The fact that condition (22) is necessary and sufficient for $\sigma_n^2\cong\sigma^2 n$ when η is symmetric follows by a simple application of Fatou's lemma.

Cogburn [7] has shown that the only possible partial limit laws of stationary asymptotically negligible instantaneous real-valued functions

$$\sum_{j=1}^{n} g_n(X_j) \tag{23}$$

of a Markov sequence $X=(X_j)$ with no cyclically moving sets are infinitely divisible if and only if the sequence is strongly mixing. It is perhaps of some interest to look at a simple example of a noninfinitely divisible limiting distribution that one can get for special partial sums (23) when dealing with, for example, some ergodic but not strongly mixing processes of random rotations. Consider $X=(X_j)$ a process of random rotations with uniform invariant measure and measure η symmetric about zero that is not strongly mixing. It then follows that

$$\overline{\lim_{\nu\to\infty}}\, \hat{\eta}_\nu = 1.$$

Choose a sequence $\nu_n \to \infty$ such that $n(1-\hat{\eta}_{\nu_n}) \to 0$. Then if $g_n(x) = \exp(2\pi i \nu_n x)$, it follows

$$\mathbf{E}\left|\sum_{k=0}^{n-1} g_n(X_k)\right|^2 \cong n^2,$$

and

$$\mathbf{E}\left|\sum_{k=0}^{n-1} \mathbf{E}[\, g_n(X_k)|X_0]\right|^2 \cong n^2.$$

But one can then determine the asymptotic distribution of

$$\sum_{k=0}^{n-1} \cos(2\pi \nu_n X_k), \tag{24}$$

as $n\to\infty$ suitably normalized. Under the assumptions made above the asymptotic distribution of (24) is the same as that of

$$n \cos 2\pi \nu_n X_0,$$

and so, asymptotically, after normalization by n, one would obtain the same distribution as

$$\cos 2\pi X,$$

where X is uniformly distributed on $[0,1]$.

3. Final remarks

Notice that Theorem 2 implies that if normalized and centered partial sums (1) of a stationary strongly mixing sequence are asymptotically standard normal ($N(0,t)$), then the normalization $B_n = n^{\frac{1}{2}} h(n)$ with $h(n)$ slowly varying as $n \to \infty$. In the paper [8], sufficient conditions are given for asymptotic normality for normalized partial sums of stationary sequences that are not strongly mixing. In such a case the normalization B_n may be $n^{1/\alpha}$ with $1 < \alpha < \infty$. These processes have longer range dependence than the strongly mixing processes. Recently there has been interest in the limit laws for normalized partial sums for such stationary sequences with long range dependence. A large class of limit laws that are not stable, have all moments and naturally have scale renormalization properties have been determined (see [3], [9] and [10, 11]). Most of those determined thus far require a normalization $n^{1/\alpha}$ with $1 < \alpha \leqslant 2$. One suspects that all such limit laws have not been determined. If the normalized sums

$$B_n^{-1} \sum_{j=1}^{[nt]} X_j - A_n$$

converge weakly to a process $Z(t)$, we say that we have a scale renormalization property if $Z(t)$ has the same distribution as $C^{-H} Z(ct)$, $c > 0$, for some number H.

References

[1] Bradley, Jr., R. C. Doctoral thesis at University of California, San Diego.

[2] Davydov, Yu. A. (1973). Mixing conditions for Markov chains. *Theor. Probability Applic.* **18**, 312–328.

[3] Dobrushin, R. L. Gaussian and their subordinated self-similar random generalized fields, to appear in Ann. Prob. (1979).

[4] Heyde, C. C. (1974). On the central limit theorem for stationary processes. *Z. Wahrscheinlichkeitstheorie verw. Gebiete* **30**, 315–320.

[5] Ibragimov, I. A. and Linnik, Yu. V. (1971). *Independent and Stationary Sequences of Random Varibles*. Wolters-Noordhof, Groningen.

[6] Ibragimov, I. A. (1975). A note on the central limit theorem for dependent random variables. *Theor. Probability Applic*. **22**, 135–141.

[7] Rosenblatt, M. (1971). *Markov Processes. Structure and Asymptotic Behavior*. Springer, Berlin.

[8] Rosenblatt, M. (1976). Fractional integrals of stationary processes and the central limit theorem. *J. Appl. Prob*. **13**, 723–732.

[9] Rosenblatt, M. (1960). Independence and dependence. *Proc. 4th Berkeley Symposium on Mathematical Statistics and Probability*, 431–443.

[10] Taqqu, M. S. (1977). Law of the iterated logarithm for sums of nonlinear functions of Gaussian variables that exhibit a long range dependence. *Z. Wahrscheinlichkeitstheorie verw. Gebiete* **40**, 203–238.

[11] Taqqu, M. S. A representation for self-similar processes. Unpublished manuscript.

P. R. Krishnaiah, ed., *Multivariate Analysis–V*
© North-Holland Publishing Company (1980) 249–261.

A LIMIT THEOREM FOR SUMS OF I.I.D. RANDOM VARIABLES WITH SLOWLY VARYING TAIL PROBABILITY

S. WATANABE
Kyoto University, Japan

1. Introduction

Let $B(t)$ be a one-dimensional Brownian motion and $M(t)$ and $1(t)$ be defined by

$$M(t) = \operatorname*{Max}_{0 \leqslant s \leqslant t} B(s) \tag{1.1}$$

$$1(t) = \lim_{\varepsilon \downarrow 0} \frac{1}{2\varepsilon} \int_0^t I_{[0,\varepsilon)}(B(s))\, \mathrm{d}s : \text{the local time at } x=0. \tag{1.2}$$

Set

$$X(t) = M(1^{-1}(t)), \tag{1.3}$$

where $1^{-1}(t)$ is the right-continuous inverse function of $t \rightsquigarrow 1(t)$. The process $X(t)$ or its inverse $X^{-1}(t) = 1(M^{-1}(t))$ play important roles in some limit theorems for stochastic processes. For instance, the process $X(t)$ is obtained as a limit process of maxima of i.i.d. random variables whose tail probability is regularly varying with index $\alpha = -1$ and this may be regarded as a special case of well-known Gnedenko's theorem ([2]). Also, Kasahara and Kotani [5] found recently that the process $X^{-1}(t)$ appears as a limit process of occupation times of a two-dimensional Brownian motion and a class of one-dimensional diffusion processes. As another example, we will show in this paper that $X(t)$ appears as a limit process for a certain class of sums of i.i.d. random variables. This may be regarded as an extreme case of the usual limit theorems involving the stable process of index α as α tends to 0.

2. The process $M(1^{-1}(t))$

We will show that the process $X(t)=M(1^{-1}(t))$ is a Markov process on $[0,\infty)$ and find its transition semigroup and the infinitesimal generator.

Let $\hat{C}[0,\infty)$ be the Banach space of all real continuous functions $f(x)$ on $[0,\infty)$ such that $\lim_{x\to\infty}f(x)=0$ with the norm $\|f\|=\operatorname{Max}_{x\in[0,\infty)}|f(x)|$. Let $\{T_t\}_{t\geqslant 0}$ be a one-parameter semigroup of operators on $\hat{C}[0,\infty)$ defined by

$$
\begin{aligned}
(T_t f)(x) &= f(x)\mathrm{e}^{-t/x}+\int_x^\infty \frac{t}{y^2}\mathrm{e}^{-t/y}f(y)\,\mathrm{d}y, && x>0,\\
&= \int_0^\infty \frac{t}{y^2}\mathrm{e}^{-t/y}f(y)\,\mathrm{d}y, && x=0 \text{ and } t>0,\\
&= f(0), && x=0 \text{ and } t=0.
\end{aligned}
\tag{2.1}
$$

Clearly, it is a Feller semigroup (i.e. strongly continuous non-negative contraction semigroup) and hence, a nice Markov process (Hunt process) corresponds to it. Now, the process $X(t)$ is identified with this Markov process.

Theorem 2.1. *The process $X(t)$ defined by* (1.3) *is a Markov process corresponding to the semigroup defined by* (2.1); *namely we have*

$$
\mathbf{E}[f(X(t))\,|\,X(u),u\leqslant s]=(T_{t-s}f)(X(s)) \text{ a.s.,} \tag{2.2}
$$

for every $t>s\geqslant 0$ and $f\in\hat{C}[0,\infty)$.

Proof. This theorem follows at once if we decompose the Brownian motion into excursions at $x=0$. We adopt the formulation given in [3] and [8].

Let

$$
\begin{aligned}
W^+(W^-)=\{w;\,&[0,\infty)\ni t\rightsquigarrow w(t)\in[0,\infty),(\text{resp.}\,(-\infty,0]),\\
&\text{continuous}, w(0)=0,\\
&\exists\sigma(w)>0\,|\,w(t)>0 \quad (\text{resp. }<0) \quad \text{if } 0<t<\sigma(w)\\
&\qquad\text{and}\\
&\qquad w(t)=0 \qquad\qquad \text{if} \quad t\geqslant\sigma(w)\}.
\end{aligned}
$$

Let $\mathcal{B}(W^+)(\mathcal{B}(W^-))$ be the σ-field generated by Borel cylinder sets. Let

$$p^0(t,x,y)=\frac{1}{\sqrt{(2\pi t)}}\left(e^{-(x-y)^2/2t}-e^{-(x+y)^2/2t}\right),$$

$$t>0,\ x,y\in[0,\infty) \text{ or } x,y\in(-\infty,0],$$

$$K^+(t,x)=\sqrt{\left(\frac{2}{\pi t^3}\right)}\,x\,e^{-x^2/2t},\qquad t>0,\quad x\in[0,\infty)$$

$$K^-(t,x)=\sqrt{\left(\frac{2}{\pi t^3}\right)}(-x)\,e^{-x^2/2t},\qquad t>0,\quad x\in(-\infty,0].$$

Then, we can determine σ-finite measures n^+ and n^- on $(W^+,\mathcal{B}(W^+))$ and $(W^-,\mathcal{B}(W^-))$ respectively by

$$n\pm\{w;w(t_1)\in dx_1,w(t_2)\in dx_2,\ldots,w(t_n)\in dx_n,\sigma(w)>t_n\}$$
$$=K\pm(t_1,x_1)p^0(t_2-t_1,x_1,x_2)\cdots p^0(t_n-t_{n-1},x_{n-1},x_n)\,dx_1\,dx_2\cdots dx_n,$$
$$0<t_1<t_2<\cdots<t_n.$$

Let n be the measure on the sum $W=W^+\cup W^-$ such that $n|_{W^\pm}=n^\pm$. A stationary Poisson point process p on W with the characteristic measure n is called a *Poisson point process of Brownian excursions*; that is, each sample of p is a mapping $D_p(\subset(0,\infty))\ni t\rightsquigarrow p(t)\in W$ defined on a countable subset D_p of $(0,\infty)$ and the corresponding counting measure $N_p(E)=\#\{s\in D_p;\ (s,p(s))\in E\}$, $E\subset(0,\infty)\times W$, is a Poisson random measure with the mean measure $\bar{N}_p(E)=\int_E dt\,n(dw)$. A Brownian motion $B(t)$ is constructed from a given Poisson point process of Brownian excursions by the following steps; first, set

$$A(t)=\sum_{s\leqslant t,s\in D_p}\sigma[p(s)].$$

Then $t\rightsquigarrow A(t)$ is strictly increasing and $\lim_{t\uparrow\infty}A(t)=\infty$ a.s. and its inverse $\phi(t)=A^{-1}(t)$ is a continuous increasing function. For each $t\geqslant 0$, set $s=\phi(t)$. If $A(s-)<A(s)$, then $s\in D_p$ and we set $B(t)=p(s)(t-A(s-))$. If $A(s-)=A(s)$, we set $B(t)=0$. We can identify $B(t)$ with a one-dimensional Brownian motion starting at 0 and $\phi(t)$ with the local time of $B(t)$ given as (1.2), cf. [8].

From this construction, it is clear that, if we define a Poisson point process $\tilde{p}$ on $(0,\infty)$ by

$$D_{\tilde{p}}=\{s\in D_p; p(s)\in W^+\}$$

and

$$\tilde{p}(s)=\operatorname*{Max}_{0\leqslant u\leqslant\sigma[p(s)]} p(s)[u], \qquad s\in D_{\tilde{p}},$$

then we have

$$M(1^{-1}(t))=\operatorname*{Max}_{s\leqslant t, s\in D_{\tilde{p}}} \tilde{p}(s). \tag{2.3}$$

The characteristic measure $\tilde{n}(\mathrm{d}y)$ of $\tilde{p}$ is computed as

$$\tilde{n}([y,\infty))=n^+\left\{w; \operatorname*{Max}_{0\leqslant s\leqslant\sigma(w)} w(s)\geqslant y\right\}$$

$$=\lim_{t\downarrow 0} n^+\left\{\sigma(w)>t, \operatorname*{Max}_{t\leqslant s\leqslant\sigma(w)} w(s)\geqslant y\right\}$$

$$=\lim_{t\downarrow 0}\int_{[0,\infty)} K^+(t,x)P_x(\sigma_y<\sigma_0)\,\mathrm{d}x$$

$$=\lim_{t\downarrow 0}\int_0^\infty \surd\left(\frac{2}{\pi t^3}\right)x\,\mathrm{e}^{-x^2/2t}\frac{x\wedge y}{y}\,\mathrm{d}x=\frac{1}{y}$$

where P_x is the Wiener measure starting at x and σ_a is the first hitting time to a. Hence $\tilde{n}(\mathrm{d}y)=\mathrm{d}y/y^2$. Now it is easy to verify (2.2) from (2.3) by noticing the well known structure of Poisson point processes.

For the later use, we will discuss on the infinitesimal generator of the semigroup $\{T_t\}$. Let

$$\mathfrak{D}=\{f\in\hat{C}[0,\infty); f\in C^1 \mid \text{for some constants } a_1,a_2 (0<a_1<a_2),$$

$$f(x)=f(0) \text{ for } 0\leqslant x\leqslant a_1 \text{ and } f(x)=0 \text{ for } x\geqslant a_2\} \tag{2.4}$$

and for $f \in \mathfrak{D}$, set

$$(Af)(x) = \int_x^\infty [f(y) - f(x)] \frac{dy}{y^2}$$

$$= \int_x^\infty f'(y) \frac{dy}{y}. \tag{2.5}$$

Theorem 2.2. *The infinitesimal generator, in Hille–Yosida's sense, of T_t is the smallest closed extension of the operator $(A, \mathfrak{D})$ given by* (2.4) *and* (2.5).

Proof. That the infinitesimal generator is an extension of $(A, \mathfrak{D})$ is easy to see. In order to show that $\mathfrak{D}$ is a core of it, take $g \in \mathfrak{D}$ and set $f(x) = -\int_x^\infty g'(y)y/(\lambda y + 1)\,dy$. Then $f \in \mathfrak{D}$ and $g(x) = \lambda f(x) - \int_x^\infty f'(y)/y\,dy = \lambda f(x) - Af(x)$. This proves that $(\lambda - A)(\mathfrak{D}) \supset \mathfrak{D}$ and since $\mathfrak{D}$ is dense in $\hat{C}[0,\infty)$ this proves that $\mathfrak{D}$ is a core of the generator.

3. A limit theorem related to the process $M(1^{-1}(t))$

Let $\xi_1, \xi_2, \ldots, \xi_n, \ldots$ be a sequence of non-negative i.i.d. random variables and $S_n = \xi_1 + \xi_2 + \cdots + \xi_n$. Suppose $P[\xi > x] \sim 1/L(x)$ as $x \uparrow \infty$ where $L(x)$ is a strictly increasing continuous function on $[0,\infty)$ such that $L(0) = 0$, $\lim_{x\uparrow\infty} L(x) = \infty$ and is slowly varying i.e., $\lim_{x\uparrow\infty} L(\lambda x)/L(x) = 1$ for every $\lambda > 0$. Examples of such functions are $L(x) = \log(x+1)$, $= \log\{\log(x+1) + 1\}$, $= \exp(\{\log(x+1)\}^\beta) - 1$, $0 < \beta < 1$, etc. We will show that the sequence of processes $(1/n)L(S_{[nt]})$ converges, as $n \to \infty$, to the process $M(1^{-1}(t))$ in a certain sense implying the convergence of every finite dimensional distribution. For each $n = 1, 2, \ldots$, the discrete time process $m \leadsto (1/n)L(S_m)$ is a Markov chain on $[0,\infty)$ with one-step transition operator $T^{(n)}$ on $\hat{C}[0,\infty)$ given by

$$(T^{(n)}f)(x) = \mathbf{E}\left[f\left(\frac{1}{n} L\{L^{-1}(nx) + \xi\} \right) \right], \tag{3.1}$$

and we will show the convergence of the semigroups, namely,

Theorem 3.1. *For every* $t_0 > 0$ *and* $f \in \hat{C}[0,\infty)$,

$$\sup_{0 \leqslant t \leqslant t_0} \|(T^{(n)})^{[nt]} f - T_t f\| \to 0 \quad \text{as } n \to \infty, \tag{3.2}$$

where $\{T_t\}$ *is given in* (2.1).

Clearly, (3.2) implies that every finite dimensional distribution of $(1/n)L(S_{[nt]})$ converges to that of $M(1^{-1}(t))$.

Proof. By Trotter's theorem ([6]), it is sufficient to prove that, for $f \in \mathfrak{D}$,

$$\|n(T^{(n)}f - f) - Af\| \to 0 \quad \text{as} \quad n \to \infty. \tag{3.3}$$

Without loss of generality, we may assume that $L(1) = 1$ and by a well known representation theorem of slowly varying functions [7], we may assume that $L(x)$ is of the form

$$L(x) = \exp\left(\int_1^x \frac{\varepsilon(t)}{t}\,\mathrm{d}t\right), \qquad x \geqslant 1, \tag{3.4}$$

where $\varepsilon(t)$ is continuous, $\varepsilon(t) > 0$ and $\lim_{t\uparrow\infty} \varepsilon(t) = 0$. Then, the inverse function $L^{-1}(x)$ of $x \rightsquigarrow L(x)$ is expressed in the form

$$L^{-1}(x) = \exp\left(\int_1^x \frac{\tilde{\varepsilon}(t)}{t}\,\mathrm{d}t\right), \qquad x \geqslant 1, \tag{3.5}$$

where $\tilde{\varepsilon}(t)$ is continuous, $\tilde{\varepsilon}(t) > 0$ and $\lim_{t\uparrow\infty} \tilde{\varepsilon}(t) = \infty$. Indeed,

$$\tilde{\varepsilon}(t) = \frac{1}{\varepsilon(L^{-1}(t))}$$

or

$$\varepsilon(t) = \frac{1}{\tilde{\varepsilon}(L(t))}\,.$$

Now, we will show (3.3). Let $f \in \mathfrak{D}$ and suppose that $f(0) = f(x)$, $x \in [0, a_1]$

and $f(x)=0$, $x \geqslant a_2$, $(0<a_1<a_2)$. By setting $G(y)=P[\xi \leqslant y]$,

$$\begin{aligned}
n(T^{(n)}f-f)(x) &= n\int_0^\infty \left[f\left(\frac{1}{n}L\{L^{-1}(nx)+y\}\right)-f(x)\right] \mathrm{d}G(y) \\
&= n\int_0^{L^{-1}(nx+1)-L^{-1}(nx)} \\
&\quad \times \left[f\left(\frac{1}{n}L\{L^{-1}(nx)+y\}\right)-f(x)\right] \mathrm{d}G(y) \\
&\quad + n\int_{L^{-1}(nx+1)-L^{-1}(nx)}^{\infty} \\
&\quad \times \left[f\left(\frac{1}{n}L\{L^{-1}(nx)+y\}\right)-f(x)\right] \mathrm{d}G(y) \\
&:= I_1+I_2.
\end{aligned}$$

Clearly $I_1=0$ if $x \leqslant \frac{1}{2}a_1$ and $n \geqslant 2/a_1$. Noting

$$|f(x)-f(y)| \leqslant M|x-y| \qquad (M=\|f'\|),$$

$$\begin{aligned}
|I_1| &\leqslant M\int_0^{L^{-1}(nx+1)-L^{-1}(nx)} [L(L^{-1}(nx)+y)-nx]\,\mathrm{d}G(y) \\
&\leqslant M\int_{(x_0,\infty)} \mathrm{d}G(y) + \int_0^{x_0} [L(L^{-1}(nx)+y)-nx]\,\mathrm{d}G(y).
\end{aligned}$$

Using the next lemma and letting $x_0 \uparrow \infty$, we see that $I_1 \to 0$ uniformly in $x \geqslant \frac{1}{2}a_1$ (and hence uniformly in $x \in [0,\infty)$) as $n \to \infty$.

Lemma 1. *$L(L^{-1}(nx)+y)-nx \to 0$ as $n\to\infty$ uniformly in $x \geqslant \frac{1}{2}a_1$ and $0 \leqslant y \leqslant x_0$.*

Proof. Using (3.4),

$$\begin{aligned}
L(L^{-1}(nx)+y)-nx &= nx\left[\exp\left(\int_{L^{-1}(nx)}^{L^{-1}(nx)+y} \frac{\varepsilon(t)}{t}\,\mathrm{d}t\right)-1\right] \\
&\leqslant nx\left[\exp\left(\frac{y\varepsilon_n(x)}{L^{-1}(nx)}\right)-1\right] \leqslant \text{const.}\left(\frac{nx}{L^{-1}(nx)}\right)x_0\varepsilon_n(x)
\end{aligned}$$

where

$$\varepsilon_n(x) = \operatorname*{Max}_{L^{-1}(nx) \leq t \leq L^{-1}(nx) + x_0} \varepsilon(t)$$

and the assertion is immediate.

Therefore, it is sufficient to show that $I_2(=I_2(x))$ converges to $\int_x^\infty f'(y)/y \, dy$ uniformly in x as $n \to \infty$. The uniform convergence on $[0, a_1/2] \cup [a_2, \infty]$ is easy. Hence it is sufficient to show that the convergence holds uniformly in $x \in [\frac{1}{2}a_1, a_2]$. Now, by setting $1 - G(y) = H(y)$ and $H_n(y) = nH(L^{-1}(ny) - L^{-1}(nx))$, we have

$$I_2 = -\int_{x+1/n}^{\infty} [f(y) - f(x)] \, dH_n(y)$$

$$= -\left(f\left(x + \frac{1}{n}\right) - f(x)\right) H_n\left(x + \frac{1}{n}\right) + \int_{x+1/n}^{a_2} f'(y) H_n(y) \, dy.$$

Lemma 2. *$L^{-1}(nx+1) - L^{-1}(nx) \to \infty$ uniformly in $x \geq \frac{1}{2}a_1$ as $n \to \infty$.*

Proof. By (3.5),

$$L^{-1}(nx+1) - L^{-1}(nx) = L^{-1}(nx)\left[\exp\left(\int_{nx}^{nx+1} \frac{\tilde{\varepsilon}(t)}{t} \, dt\right) - 1\right]$$

$$\geq L^{-1}(nx) \frac{\tilde{\varepsilon}_n(x)}{nx+1}$$

where $\tilde{\varepsilon}_n(x) = \mathrm{Min}_{nx \leq t \leq nx+1} \tilde{\varepsilon}(t)$ and the assertion is obvious.

By the assumption that $H(y) \sim 1/L(y)$ as $y \to \infty$ and Lemma 2, we have that, for every $\varepsilon > 0$ there exists n_0 such that if $n \geq n_0$

$$\frac{(1-\varepsilon)n}{L(L^{-1}(ny) - L^{-1}(nx))} \leq H_n(y) \leq \frac{(1+\varepsilon)n}{L(L^{-1}(ny) - L^{-1}(nx))}$$

$$\text{for all} \quad x \geq \tfrac{1}{2}a_1 \quad \text{and} \quad y \geq x + \frac{1}{n}. \tag{3.6}$$

Now, by (3.6),

$$\left|\left(f\left(x+\frac{1}{n}\right)-f(x)\right)H_n\left(x+\frac{1}{n}\right)\right|\leqslant M\frac{1}{n}H_n\left(x+\frac{1}{n}\right)\to 0 \tag{3.7}$$

uniformly in $x\geqslant \frac{1}{2}a_1$ as $n\to\infty$. Therefore, it is sufficient to prove that

$$\int_{x+1/n}^{a_2} f'(y)H_n(y)\,\mathrm{d}y\to\int_x^{a_2} f'(y)\frac{\mathrm{d}y}{y} \tag{3.8}$$

uniformly in $x\in\left[\frac{1}{2}a_1,a_2\right]$ as $n\to\infty$.

Lemma 3. *Let $\delta>0$ be given. Then*

$$\frac{n}{L(L^{-1}(ny)-L^{-1}(nx))}\to\frac{1}{y}$$

uniformly in $x\in[\frac{1}{2}a,a_2]$ and $y\in[x+\delta,a_2]$ as $n\to\infty$.

Proof. By setting $\tilde{\varepsilon}_n(x)=\mathrm{Min}_{nx\leqslant t\leqslant n(x+\delta)}\tilde{\varepsilon}(t)$,

$$\frac{ny}{L(L^{-1}(ny)-L^{-1}(nx))}=\frac{ny}{L\left(L^{-1}(ny)\left(1-\dfrac{L^{-1}(nx)}{L^{-1}(ny)}\right)\right)}$$

and

$$\begin{aligned}1\geqslant 1-\frac{L^{-1}(nx)}{L^{-1}(ny)}&=1-\exp\left(-\int_{nx}^{ny}\frac{\tilde{\varepsilon}(t)}{t}\,\mathrm{d}t\right)\\&\geqslant 1-\exp\left(-\int_{nx}^{n(x+\delta)}\frac{\tilde{\varepsilon}(t)}{r}\,\mathrm{d}t\right)\\&\geqslant 1-\exp\left(-\tilde{\varepsilon}_n(x)\log\frac{x+\delta}{x}\right)\\&\geqslant \tfrac{1}{2}\end{aligned}$$

for $n\geqslant n_1$. Hence the assertion follows (cf. [7, Theorem 1.1]).

Set

$$\int_{x+1/n}^{a_2} f'(y)H_n(y)\,dy = \int_{x+1/n}^{x+\delta} f'(y)H_n(y)\,dy + \int_{x+\delta}^{a_2} f'(y)H_n(y)\,dy := I_3 + I_4.$$

Then, by (3.6) and Lemma 3,

$$I_4 \to \int_{x+\delta}^{a_2} f'(y)\frac{dy}{y} = \int_{x}^{a_2} f'(y)\frac{dy}{y} + 0(\delta)$$

uniformly in $x \in \left[\frac{1}{2}a_1, a_2\right]$ as $n \to \infty$. Therefore, our proof will be complete if we can show the following

Lemma 4. *$|I_3| \leqslant K\delta + \alpha_n$ where K is a constant and $\alpha_n \to 0$ as $n \to \infty$ uniformly in $x \in [\frac{1}{2}a, a_2]$.*

Proof. In the following, $K_1, K_2, \ldots$ are positive constants.

$$L^{-1}(ny) - L^{-1}(nx) = L^{-1}(nx)\left(\exp\left(\int_{nx}^{ny} \frac{\tilde{\varepsilon}(t)}{t}\,dt\right) - 1\right) \geqslant L^{-1}(nx)\int_{nx}^{ny} \frac{\tilde{\varepsilon}(t)}{t}\,dt.$$

Let y_n be defined by $\int_{nx}^{ny_n} \frac{\tilde{\varepsilon}(t)}{t}\,dt = 1$. Since $\tilde{\varepsilon}(t) \to \infty$, $y_n = x + \delta_n(x)$ where $\delta_n(x) \to 0$ uniformly in $x \in [\frac{1}{2}a_1, a_2]$. If $y \geqslant y_n$, then $L^{-1}(ny) - L^{-1}(nx) \geqslant L^{-1}(nx)$ and hence $L(L^{-1}(ny) - L^{-1}(nx)) \geqslant nx$. Hence

$$H_n(y) \leqslant \frac{n(1+\varepsilon)}{L(L^{-1}(ny) - L^{-1}(nx))} \leqslant \frac{1+\varepsilon}{x} \leqslant K_1$$

and

$$\left|\int_{y_n}^{x+\delta} f'(y)H_n(y)\,dy\right| \leqslant K_2\delta.$$

Next, if $x+\frac{1}{n}\leqslant y\leqslant y_n$, then $z_n:=\int_{nx}^{ny}\frac{\tilde{\varepsilon}(t)}{t}\,\mathrm{d}t\leqslant 1$ and therefore,

$$\begin{aligned}
L(L^{-1}(ny)-L^{-1}(nx)) &= \exp\left(\int_1^{L^{-1}(ny)-L^{-1}(nx)}\frac{\varepsilon(t)}{t}\,\mathrm{d}t\right)\\
&\geqslant \exp\left(\int_1^{L^{-1}(nx)z_n}\frac{\varepsilon(t)}{t}\,\mathrm{d}t\right)\\
&= nx\exp\left(-\int_{L^{-1}(nx)z_n}^{L^{-1}(nx)}\frac{\varepsilon(t)}{t}\,\mathrm{d}t\right)\\
&\geqslant nx\exp\left(-\bar{\varepsilon}_n(x)\int_{L^{-1}(nx)z_n}^{L^{-1}(nx)}\frac{\mathrm{d}t}{t}\right)\\
&= nx\,z_n^{\bar{\varepsilon}_n(x)}
\end{aligned}$$

where

$$\bar{\varepsilon}_n(x)=\operatorname*{Max}_{L^{-1}(nx)z_n\leqslant t\leqslant L^{-1}(nx)}\varepsilon(t)$$

which tends to 0 as $n\to\infty$ because of $L^{-1}(nx)z_n\geqslant K_3L^{-1}(nx)(1/n)\to\infty$. Therefore, noting that we may assume $\tilde{\varepsilon}(t)\geqslant 1$,

$$\begin{aligned}
\frac{nx}{L(L^{-1}(ny)-L^{-1}(nx))} &\leqslant \left(\int_{nx}^{ny}\frac{\tilde{\varepsilon}(t)}{t}\,\mathrm{d}t\right)^{-\bar{\varepsilon}_n(x)}\\
&\leqslant \left(\int_{nx}^{ny}\frac{\mathrm{d}t}{t}\,\mathrm{d}t\right)^{-\bar{\varepsilon}_n(x)}\\
&= \left(\log\frac{y}{x}\right)^{-\bar{\varepsilon}_n(x)}\\
&\leqslant \frac{K_4}{(y-x)^{\bar{\varepsilon}_n(x)}}.
\end{aligned}$$

We may assume that $\bar{\varepsilon}_n(x)<\frac{1}{2}$ and hence by (3.6),

$$\left|\int_{x+1/n}^{y_n}f'(y)H_n(y)\,\mathrm{d}y\right|\leqslant K_5\int_{x+1/n}^{y_n}\frac{\mathrm{d}y}{(y-x)^{\frac{1}{2}}}\to 0$$

uniformly in $x\in[\frac{1}{2}a_1,a_2]$.

Remark 3.1. If we define, instead of (3.1):

$$(\tilde{T}^{(n)}f)(x)=\mathbf{E}\left[f\left(x\vee\frac{1}{n}L(\xi)\right)\right]$$

then we can prove (3.3) for $\tilde{T}^{(n)}$ almost immediately; $n(\tilde{T}^{(n)}f-f)(x) = \int_x^\infty f'(y)\tilde{H}_n(y)\,\mathrm{d}y$ where $\tilde{H}_n(y)=n(1-G(L^{-1}(ny)))$ and clearly $\tilde{H}_n(y)\to 1/y$ uniformly in $y\in[a_1,a_2]$ as $n\to\infty$. This implies that

$$\frac{1}{n}L\Big(\operatorname*{Max}_{1\leqslant k\leqslant[nt]}\xi_k\Big)\to M(1^{-1}(t))\quad \text{as } n\to\infty$$

and thus we are led to a special case of Gnedenko's well known theorem on the maxima of i.i.d. random variables: our theorem does not seem to be directly derived from it, however.

Remark 3.2. Our theorem may be regarded as an extreme case of limit theorems for non-negative i.i.d. random variables with regularly varying tail probability with exponent $-\alpha$ as α tends to 0. The following remark due to S. Kotani explains this situation; let $X_t^{(\alpha)}$ be a one-sided stable process with exponent α. Then the process $(X_t^{(\alpha)})^\alpha$ converges to the process $M(1^{-1}(t))$ as $\alpha\downarrow 0$. This is seen most simply as follows. Let $1(t,x)$ be the sojourn time density (the local time) of a one-dimensional Brownian motion $B(t)$: $\int_0^t I_{\mathbf{E}}(B(s))\,\mathrm{d}s=2\int_{\mathbf{E}}1(t,x)\,\mathrm{d}x$ and let $T(t,x)=1(1^{-1}(t,0),x)$. Then,

$$X^{(\alpha)}(t)=\int_0^\infty T(t,x)x^{(1-2\alpha)/\alpha}\,\mathrm{d}x$$

is a one-sided stable process of exponent α [4, p. 226] and hence

$$\begin{aligned}\lim_{\alpha\downarrow 0}X^{(\alpha)}(t)^\alpha &= \lim_{\alpha\downarrow 0}\left[\int_0^\infty T(t,x)x^{(1-2\alpha)/\alpha}\,\mathrm{d}x\right]^{\alpha/(1-2\alpha)}\\ &=\operatorname{Max}\big(\operatorname{range}\big[B(s);s\leqslant 1^{-1}(t,0)\big]\big)\\ &=M(1^{-1}(t)).\end{aligned}$$

Cf. Cressie [1] for the corresponding fact for stable laws.

References

[1] Cressie, N. (1975). A note on the behavior of the stable distributions for small index α. *Z. Wahrscheinhlichkeitstheorie verw. Gebiete*. **33**, 61–64.

[2] Gnedenko, B. V. (1943). Sur la distribution limite du terme maximum d'une série aléatoire. *Ann. Math*. **44**, 423–453.

[3] Itô, K. (1972). Poisson point processes attached to Markov processes. In *Proc. Sixth Berkeley Symp.*, pp. 225–239, University of California Press.

[4] Itô, K. and McKean, H. P. (1965). *Diffusion Processes and Their Sample Paths*. Springer, 1965.

[5] Kasahara, Y. and Kotani, S. On limit processes for a class of additive functionals of recurrent diffusion processes, to appear in *Z. Wahrscheinlichkeitstheorie verw. Gebiete*.

[6] Kato, T. (1966). *Perturbation Theory for Linear Operators*, Springer.

[7] Seneta, E. (1976). *Regularly Varying Functions*. Lecture Notes in Math. 508, Springer, 1976.

[8] Watanabe, S. (1978) Martingales and point processes. *Stochastic Analysis* (A. Friedman and M. Pinsky, eds.). Academic Press.

PART IV

CHARACTERISTIC FUNCTIONS, DISTRIBUTION THEORY AND RANDOM MATRICES

P. R. Krishnaiah, ed., *Multivariate Analysis–V*

COMPANION MATRICES ASSOCIATED WITH RANDOM ALGEBRAIC POLYNOMIALS

Mark J. CHRISTENSEN
Georgia Institute of Technology, School of Mathematics, Atlanta, GA 30332, U.S.A.

and

A. T. BHARUCHA-REID*
Wayne State University, Department of Mathematics, Detroit, MI 48202, U.S.A.

1. Introduction

A *random matrix* is a matrix whose elements are random variables or random functions. Random matrices arise in mathematical statistics (indeed; they play a central role in multi-variate analysis), and in mathematical physics (quantum mechanics, Ising models, etc.). We refer to Girko [6] and Mehta [9] for authoritative accounts of random matrices and their applications. Random matrices also play an important role in the theory of random equations, where they arise in connection with systems of algebraic, difference, and differential equations; and also in connection with integral equations with random degenerate kernels (cf. Bharucha-Reid [2]).

The characteristic polynomial of a random matrix is a *random algebraic polynomial*; that is, an algebraic polynomial whose coefficients are random variables. The study of random algebraic polynomials, which leads to interesting probabilistic generalizations of classical results on algebraic polynomials, while of independent interest is buttressed in its import by the fact that many problems in the applied mathematical sciences lead to random algebraic polynomials.

In Section 2 we show that associated with every random algebraic polynomial is a particular kind of random matrix, the so-called companion matrix. For deterministic polynomials it is known that the solution set of the polynomial and the set of eigenvalues of the associated companion matrix are the same. In this note we report the results of a numerical investigation of the relationship between the solution set of a random algebraic polynomial and the set of eigenvalues of the associated random companion matrix.

*Research supported by Army Research Office Grant No. DAAG29-77-G-0164.

2. Companion matrices

For any given algebraic equation

$$F_n(z)=z^n+a_{n-1}z^{n-1}+\cdots+a_0=0, \tag{2.1}$$

where $z\in Z$ (the complex plane), we have the associated *companion matrix*

$$C=\begin{bmatrix} -a_{n-1} & -a_{n-2} & -a_{n-3} & \cdots & -a_1 & -a_0 \\ 1 & 0 & 0 & \cdots & 0 & 0 \\ 0 & 1 & 0 & \cdots & 0 & 0 \\ \cdot & \cdot & \cdot & \cdots & \cdot & \cdot \\ 0 & 0 & 0 & \cdots & 1 & 0 \end{bmatrix}, \tag{2.2}$$

(cf. Marden [8, Ch. VII]; Young and Gregory [11, pp. 219–220]; and Wilkinson [10, pp. 12–13]). It is well-known that the spectrum $\sigma(c)$ of C coincides with the solution set $S(F_n(z))$ of (2.1); that is

$$\sigma(c)=S(F_n(z))=\{z:F_n(z)=0\}\subset Z. \tag{2.3}$$

It follows from (2.3) that

$$|\zeta|\leqslant r(C)\leqslant\|C\|, \tag{2.4}$$

for any solution (root) ζ of (2.1), where $r(C)$ is the spectral radius of C (i.e., $r(C)=\sup_{\zeta\in\sigma(C)}|\zeta|$), and $\|C\|$ is the operator norm of C (i.e., $\|C\|=\sup_{|x|=1}\|Cx\|$), considering C to be an operator on an n-dimensional Banach space X. Fujii and Kubo [5] have shown that if X is assumed to be the n-dimensional unitary space with orthonormal basis $\{b_1,b_2,\ldots,b_n\}$ and inner product $(\cdot,\cdot)$, then the companion matrix C admits the representation

$$C=E-b_1\otimes h, \tag{2.5}$$

where

$$E=\begin{bmatrix} 0 & 0 & 0 & \cdots & 0 & 0 \\ 1 & 0 & 0 & \cdots & 0 & 0 \\ 0 & 1 & 0 & \cdots & 0 & 0 \\ \cdot & \cdot & \cdot & \cdots & \cdot & \cdot \\ 0 & 0 & 0 & \cdots & 1 & 0 \end{bmatrix}$$

and $h=\bar{a}_{n-1}b_1+\cdots+\bar{a}_0b_n$ (here $\bar{a}_k$, $k=0,\ldots,n-1$, denotes the conjugate transpose of a_k). In (2.5) we have used the tensor product operation $(x_1 \otimes x_2)y = (y, x_2)x_1$ for $y \in X$. Fujii and Kubo have used (2.4) and (2.5) to obtain bounds for roots of algebraic equations.

We now consider the *random algebraic polynomial*

$$F_n(z,\omega) = z^n + a_{n-1}(\omega)z^{n-1} + \cdots + a_0(\omega), \tag{2.6}$$

where $z \in Z$ and the coefficients $a_i(\omega)$, $i = 0, \ldots, n-1$, are z-valued, mean zero random variables on a probability space $(\Omega, \mathcal{Q}, \mu)$, and $a_n(\omega) = 1$ a.s. Associated with $F_n(z,\omega)$ is the *random companion matrix*

$$C(\omega) = \begin{bmatrix} -a_{n-1}(\omega) & -a_{n-2}(\omega) & -a_{n-3}(\omega) & \cdots & -a_1(\omega) & -a_0(\omega) \\ 1 & 0 & 0 & \cdots & 0 & 0 \\ 0 & 1 & 0 & \cdots & 0 & 0 \\ \cdot & \cdot & \cdot & \cdots & \cdot & \cdot \\ 0 & 0 & 0 & \cdots & 1 & 0 \end{bmatrix}. \tag{2.7}$$

We also have

$$\sigma(((\omega)) = S(F_n(z,\omega)) = \{z : F_n(z,\omega) = 0\} \quad \text{a.s.}, \tag{2.8}$$

and

$$|\zeta(\omega)| \leqslant r(C(\omega)) \leqslant \|C(\omega)\| \quad \text{a.s.} \tag{2.9}$$

It is known that

(1) any root of a random algebraic polynomial is measurable, hence a random variable (cf. [1, 3]);

(2) the spectral radius of a random matrix is a well-defined random variable (cf., [2, p. 85], and Grenander [7, p. 161]); and

(3) $\|C(\omega)\|$ is a nonnegative real-valued random variable (cf. [2, p. 77]). Hence there is no problem establishing (2.9). Similarly, we have

$$C(\omega) = E - b_1 \otimes h(\omega), \tag{2.10}$$

where E is as defined before, and $h(\omega) = \bar{a}_{n-1}(\omega)b_1 + \cdots + \bar{a}_0(\omega)b_n$. Bounds for the roots of random algebraic polynomials can be obtained using (2.9) and (2.10).

Although we have the theoretical relation (2.8), it is of interest to investigate the extent to which (2.8) holds when numerical methods are used to determine the roots of $F_n(z,\omega)$ and the eigenvalues of $C(\omega)$. In Section 3 we present an algorithm and code which, using commonly available software, enables us to generate a random algebraic polynomial, calculate its roots, calculate the eigenvalues of the associated random companion matrix, and then determine the total and average deflections between the roots and eigenvalues. For simplicity we define the *deflection* by the Minkowski 1-norm:

$$\mathrm{d}(x,y)=|x|+|y|$$

as computed between nearest-neighbor pairs. All subroutines used are contained in the IMSL scientific library [12]. Subroutine GGNDR generates pseudo-random normal deviates, subroutine ZRPOLY computes the zeros of an algebraic polynomial with real coefficients; and subroutine EIGRF computes the eigenvalues and (optionally) eigenvectors of a real matrix.

3. Algorithm and code

The following program, written for a polynomial of degree 20, can easily be modified to handle polynomials of any degree.

```
      PROGRAM SRWUP(INPUT,OUTPUT,TAPE6=OUTPUT)
      REAL A(20,20),R(20),B(21),DELTA,S
      COMPLEX X(20),W(20),Z(20,20)
      INTEGER N,IER,IJOB,IZ,IA,ISEED,P,L
      DIMENSION WK(444)
      N=20
      IA=N
      P=19
      IJOB=0
      ISEED=23456
      CALL GGNOR(ISEED,N,R)
      DO 77 I=1,N
      B(I+1)=10.0*R(I)
77    CONTINUE
      B(1)=1
      CALL ZRPOLY(B,N,X,IER)
```

```
      DO 99 I=1,N
      PRINT *,X(I)
99    CONTINUE
      DO 88 I=1,P
      DO 66 J=1,N
      L=J-1
      IF (I.EQ.L) GOTO 55
      A(I,J)=0.0
      GOTO 66
55    A(I,J)=1
66    CONTINUE
88    CONTINUE
      DO 11 I=1,N
      A(N,I)= -10.0*R(N-(I-1))
11    CONTINUE
      CALL EIGRF(A,N,IA,IJOB,W,Z,IZ,WK,IER)
      DO 22 I=1,20
      PRINT *,W(I)
22    CONTINUE
      S=0.0
      DO 33 I=1,N
      DO 44 J=1,N
      DELTA=ABS(REAL(X(I))-REAL(W(J)))+ABS(AIMAG(X(I))-
      AIMAG(W(J)))
      IF (DELTA.GE.0.00001) GOTO 44
      S=S+DELTA
      GOTO 33
44    CONTINUE
33    CONTINUE
      PRINT *,S
      PRINT *, IER
      STOP
      END
```

4. A numerical example

Using the above program, we considered a random polynomial of degree 20. The output is given below.

Roots of the polynomial

(9,446105618511, 0.)
(− 1.078068112626, .1178910069071)
(− 1.078068112626, −.1178910069071)
(−.9329475446899, .430962367617)
(−.9329475446899, −.430962367617)
(−.7794190754537, .685050.565282)
(−.7794190754537, −.6850501575282)
(.3188872802251, 1.233525372171)
(.3188872802251, − 1.233525372171)
(−.1946092923604, .9823801020167)
(−.1946092923604, −.9823801020167)
(.541264353815, .9530125508606)
(.541264353815, −.9530125508606)
(−.1797276227758, .7853839416825)
(−.1797276227758, −.7853839416825)
(.970886613797, .4265379327365)
(.970886613797, −.4265379327375)
(.105069539117, 0.)
(.594450392011, .4438465834284)
(.594450392011, −.4438465834284)

Eigenvalues of the companion matrix

(9,446105618509, 0.)
(− 1.078068112625, .1178910069077)
(− 1.078068112625, −.1178910069077)
(−.9329475446882, .430962367272)
(−.9329475446882, −.430962367672)
(−.7794190754518, .6850501575275)

(.7794190754518, −.6850501575275)
(.3188872802247, 1.233525372169)
(.3188872802247, − 1.233525372169)
(−.1946092923599, .9823801020154)
(−.1946092923599, −.9823801020154)
(.5412643538145, .9530125508595)
(.5412643538145, −.9530125508595)
(−.1797276227757, .7853839416824)

(−.1797276227757, −.7853839416824)
(.9708866137961, .4265379327371)
(.9708866137961, −.4265379327371)
(1.050695391169, 0.)
(.594450392011, .4438465834283)
(.594450392011, −.4483465834283)

The *total deflection* is equal to 2.954525513132 E−11; and the *average deflection* is hence equal to 1.5 E−12.

For a polynomial of degree 50, the total deflection was 3.147686999938 E−9; and the average deflection was 6.0 E−11. It is also of interest to note that using another program which used the IMSLIB random number generator and polynomial solver, together with the MFSLIB (Boeing Aircraft) eigenvalue solver, LATRYN, the total deflection was 3.270361759178 E−11, and the average deflection was 1.5 E−12 for a polynomial of degree 20.

5. Discussion

As stated in Section 2, we initiated this investigation in order to ascertain if the theoretical relation (2.8) holds when computational methods are used to determine the roots of a random algebraic polynomial and the eigenvalues (point spectrum) of the associated companion matrix operator. Because companion matrices are sparse for large values of n, we conjectured there would be nontrivial differences between the roots of a random algebraic polynomial and the eigenvalues of the associated random companion matrix. As it turns out, this conjecture was not valid.

In this paper we presented results based on only one realization of a random algebraic polynomial (equivalently, the associated random companion matrix); however, in view of earlier results obtained by the authors [4] on the stability of the roots of random algebraic polynomials, if many realizations (or samples) are considered there is (at least in the case of tight coefficients) a remarkable stability exhibited by the roots of the different realizations of a random algebraic polynomial.

Hence, we conclude that in computational studies of the roots of random algebraic polynomials, it really doesn't make any difference whether a polynomial subroutine or a matrix eigenvalue subroutine is used. Clearly, for polynomials of high degree, the companion matrix approach is much more expensive in terms of computer time.

References

[1] Bharucha-Reid, A. T. (1970). Random algebraic equations. In *Probabilistic Methods in Applied Mathematics*, Vol. 2 (Bharucha-Reid, A. T., Ed.), pp. 1–52. Academic Press, New York.

[2] Bharucha-Reid, A. T. (1972). *Random Integral Equations*. Academic Press, New York.

[3] Bharucha-Reid, A. T. *Random Algebraic Polynomials*. Academic Press, New York. To appear.

[4] Christensen, M. J. and Bharucha-Reid, A. T. Stability of the roots of random algebraic polynomials. To appear.

[5] Fujii, M. and Kubo, F. (1973). Operator norms as bounds for of algebraic equations. *Proc. Japan Acad.* **49**, 805–808.

[6] Girko, V. L. (1975). *Random Matrices* (Russian). Izdat. Obĕd. "Višča Škola" pri Kiev. Gosudarstv. Univ., Kiev.

[7] Grenander, U. (1963). *Probabilities on Algebraic Structures*. Wiley, New York.

[8] Marden, M. (1960). *The Geometry of the Zeros of a Polynomial in a Complex Variable*. 2nd ed. American Mathematical Society, Providence RI.

[9] Mehta, M. L. (1968). *Random Matrices and the Statistical Theory of Energy Levels*. Academic Press, New York.

[10] Wilkinson, J. H. (1965). *The Algebraic Eigenvalue Problem*. Clarendon Press, Oxford.

[11] Young, D. M. and Gregory, R. T. (1972). *A Survey of Numerical Mathematics*, Vol. I. Addison-Wesley, Reading MA.

[12] *International Mathematical and Statistical Libraries*, Library 1 (Edition 5), Reference Manual (1975).

P. R. Krishnaiah, ed., *Multivariate Analysis–V*

RECENT RESULTS ON THE DECOMPOSITION OF MULTIVARIATE PROBABILITIES

Roger CUPPENS
Université Paul Sabatier, Toulouse, France

We give here an exposure of the works which appeared on the theory of decomposition of multivariate probabilities since the edition of the author's 'Decomposition of Multivariate Probabilities' [3].

1. Introduction

Since the works by Linnik and Lukacs in the late fifties, the studies on the theory of decompositions of probabilities have never decreased. For example, there has appeared on this subject three books [3, 21, 24] since 1970 and two expository notes [23, 30] since 1975. Nevertheless, these notes are concerned with probability laws on general structures and give only an incomplete account on multivariate probabilities. We give here a report on the works on decompositions of multivariate laws which appeared since the publication of [3].

We recall now some definitions and notations.[1] Let $\mathcal{P}_n$ be the set of all the probabilites defined on the Borel sets of R^n. Then $\mathcal{P}_n$ is a semigroup for the convolution $*$ defined by

$$(\mathbf{p}_1 * \mathbf{p}_2)(B) = \int \mathbf{p}_1(B-u)\mathbf{p}_2(\mathrm{d}u).$$

This semigroup is unitary, the identity being δ_0 where for any $m \in R^n, \delta_m$ is the probability

$$\delta_m(B) = \begin{cases} 0 & \text{if } m \notin B, \\ 1 & \text{if } m \in B. \end{cases}$$

(The probabilities δ_m are the invertible elements of $\mathcal{P}_n$).

For a semigroup $(A, \circ)$ with identity, we can introduce indecomposable and infinitely divisible elements: $a \in A$ is indecomposable if $a = b \circ c$ implies that one of the elements b or c is invertible; $a \in A$ is infinitely divisible if for any $\mathbf{p} \in N^*$ there exists some $b_{\mathbf{p}}$ such that $a = b_{\mathbf{p}} \circ \cdots \circ b_{\mathbf{p}}$ (p times).

[1] For the notations which we do not recall here, see [3].

Defining the characteristic function $\hat{\mathbf{p}}$ of a probability $\mathbf{p}$ by

$$\hat{\mathbf{p}}(t)=\int e^{i(t,x)}\mathbf{p}(\mathrm{d}x)$$

for any $t\in R^n[(t,x)$ is the inner product of t and $x]$, we know that $\mathbf{p}\to\hat{\mathbf{p}}$ is an isomorphism of the semigroup $\mathcal{P}_n$ on the semigroup of n-variate characteristic functions (for the ordinary multiplication of functions). We have then the following characterization of infinitely divisible elements of $\mathcal{P}_n$ [3, p. 91]:

$\mathbf{p}\in\mathcal{P}_n$ is infinitely divisible if and only if its characteristic function $\hat{\mathbf{p}}$ admits the representation

$$\hat{\mathbf{p}}(t)=\exp\left[i(\alpha,t)-P(t)+\int K(t,u)\mu(\mathrm{d}u)\right] \tag{1}$$

for any $t\in R^n$ where $\alpha\in R^n$, P is a nonnegative quadratic form, μ is a finite measure and K is the kernel

$$K(t,u)=\begin{cases}\left[e^{i(t,u)}-1-\dfrac{i(t,u)}{1+\|u\|^2}\right]\dfrac{1+\|u\|^2}{\|u\|^2} & \text{if } u\neq 0,\\ 0 & \text{if } u=0.\end{cases}$$

This representation is unique. We call this representation the (α,P,μ)–Lévy–Hinčin representation of $\hat{\mathbf{p}}$ and μ the Poisson measure of $\mathbf{p}$.

On the other hand, a characterization of indecomposable elements of $\mathcal{P}_n$ is not known. The fundamental results on the decomposition theory of multivariate probabilities are the two Hinčin theorems [3, Sections 4.5 and 4.6]:

(1). *Any probability* $\mathbf{p}$ *of* $\mathcal{P}_n$ *can be written as* $\mathbf{p}=\mathbf{q}*\mathbf{r}$ *where* q *has no indecomposable factor and for* $\mathbf{r}$ *there exists some sequence* $\{\mathbf{r}_j\}$ *of indecomposable probabilities such that* $\lim_{j\to\infty}(\mathbf{r}_1*\cdots*\mathbf{r}_j)=r$ *(in the sense of weak convergence).*

(2). *An n-variate probability which has no indecomposable factor is infinitely divisible.*

At this time, most of the studies in this field are concerned with the problem of characterization of the class I_0^n of probabilities which have no indecomposable factor: if $\hat{\mathbf{p}}$ admits an (α,P,μ)–Lévy–Hinčin representation, what condition on P and μ implies that $\mathbf{p}$ belongs to I_0^n?

Other interesting problems are those of the study of the class of indecomposable probabilities and the study, for a given probability, of the set of its indecomposable factors.

2. Probabilities with an absolutely continuous Poisson measure

In the univariate case, Cuppens has proved the following result ([3], theorem 7.2.2): let $\hat{\mathbf{p}}$ be a characteristic function admitting an $(\alpha, 0, \mu)$–Lévy–Hinčin representation. If μ is absolutely continuous (with respect to the Lebesgue measure) with a continuous Radon–Nikodym derivative, then $\mathbf{p}$ belongs to I_0^1 if and only if there exists some $m > 0$ such that μ is concentrated on $[m, 2m]$ or $[-2m, -m]$. Recently, Mase [25] has proved that this theorem still holds without any hypothesis on the derivative of μ.

In the multivariate case, Livsič has given the following result ([22]; cf. also [3, theorem 7.2.1]): let $\hat{\mathbf{p}}$ be an n-variate characteristic function admitting an $(\alpha, 0, \mu)$–Lévy–Hinčin representation. If μ is absolutely continuous (with respect to the Lebesgue measure) with a continuous Radon–Nikodym derivative, then $\mathbf{p}$ belongs to I_0^n if and only if μ is concentrated on a set A satisfying

$$A^* \cap \left[\bigcup_{k=2}^{\infty} (k)A \right] = \emptyset$$

where A^* is the convex hull of A.

In this case, Mase [26] has shown that the hypothesis of the continuity of the derivative of μ can be replaced by the weaker one: the set $A = \{x\colon \mathrm{d}\mu/\mathrm{d}\lambda(x) > 0\}$ is open. We do not know if this condition can be removed. Nevertheless, we can easily prove [5] the following theorem containing Mase's result:

Let $\hat{\mathbf{p}}$ be an n-variate characteristic function admitting an $(\alpha, 0, \mu)$–Lévy–Hinčin representation with an absolutely continuous measure μ. If

$$(A^*)^\circ \cap \left[\bigcup_{k=2}^{\infty} (k)A \right] = \emptyset$$

where $A = S(\mu)$, then $\mathbf{p}$ belongs to I_0^n. Conversely, if $\mathbf{p}$ belongs to I_0^n, then

$$B^* \cap \left[\bigcup_{k=2}^{\infty} (k)B \right] = \emptyset$$

where $B = (S(\mu))^\circ$.

Since there exist closed sets with positive Lebesgue measure and empty interior, these two conditions are very far from each other. I think that the sufficient condition is also necessary, but I am unable to prove it.

In this direction, we quote another work by Mase [27] where he proves in the univariate case some conditions which are either necessary or sufficient for the belonging to I_0^1. We state some of them: Let $\hat{\mathbf{p}}$ be a univariate characteristic function admitting an $(\alpha, 0, \mu)$–Lévy–Hinčin representation and let us denote by μ_{ac} and μ_c respectively the absolutely continuous and the continuous part of μ.

(1) If $\mu_{ac}(R_+) > 0$ and $\mu_c(R_-) > 0$, then $\mathbf{p}$ does not belong to I_0^1;

(2) if α and β are two positive numbers belonging to $S(\mu_{ac})$ and if $\mu_c(]0, \beta - \alpha[) > 0$, then p does not belong to I_0^1.

It would be interesting to study the extension of these results to the multivariate case.

3. Probabilities with a discrete Poisson measure

The main results in this direction are concerned with the univariate case.

Let us recall ([3], p. 122) that a set Λ of real numbers is a Linnik set if it satisfies the following condition: for any $(a, b) \in \Lambda^2$ such that $\operatorname{sgn} a = \operatorname{sgn} b$, then either ab^{-1} or ba^{-1} is an integer greater than 1.

Linnik ([21], Ch. IV) has proved the following result: if a univariate probability has a non degenerate normal factor and no indecomposable factor, then its Poisson measure is concentrated on a Linnik set. In the same direction, Fryntov [8] has stated the following result: let $\hat{\mathbf{p}}$ be a univariate characteristic function admitting an (α, P, μ)–Lévy–Hinčin representation. If $\mathbf{p}$ belongs to I_0^1 and if the measure μ satisfies the conditions

$$\liminf_{r \to \infty} \frac{\log \mu([\alpha/r, 1/r])}{\log r} > 1,$$

$$\liminf_{r \to \infty} \frac{\log \mu([-1/r, -\alpha/r])}{\log r} > 1$$

for some $\alpha \in [0, 1]$, then μ is concentrated on a Linnik set. Unfortunately, the proof of this result has not yet been published.

As an application of this result, we recall the following Cuppens's result ([3, Theorem 7.4.2]): if the characteristic function $\hat{\mathbf{p}}$ admits an $(\alpha, 0, \mu)$–Lévy–Hinčin representation where the measure μ is concentrated on an independent set and satisfies

$$\int \frac{\|x\|^2}{1+\|x\|^2}\mu(\mathrm{d}x) < +\infty, \tag{2}$$

then $\mathbf{p}$ belongs to I_0^n. Fryntov's results shows that in this result the hypothesis (2) is necessary.

Another research direction is the extension of Paul Lévy's results on finite products of Poisson laws ([3, Section 8.4 and Appendix B]). For stating the results, we introduce the following definition:

If A is a set of real numbers and $m \in R^*$, we put

$$A^m = \{k \in A : k/m > 1\},$$

$$A_m^+ = \{k \in A : 0 < k/m < 1\},$$

$$A_m^- = \{k \in A : k/m < 0\}.$$

Then we say that a set A satisfies condition (L) if for any $m \neq 0$

$$m \notin (Z)A^m \cap (N)A_m^+$$

if $A_m^- = \emptyset$ and

$$m \notin (Z)A^m \cap (N)\big(A_m^+ \cup \big[(Z)A_m^- + ((N)A_m^+)_m\big]\big)$$

if $A_m^- \neq \emptyset$ (if $A \subset Z$, this last condition is equivalent to

$$m \notin (Z)A^m \cap (N)(A_m^+ \cup \{\operatorname{sgn} m \cdot d_m\})$$

where d_m is the g.c.d. of A_m^-). The independent sets and the positive Linnik sets satisfy condition (L). We have then the following necessary and sufficient conditions:

(1) (Fryntov [8,9]). Let $\hat{\mathbf{p}}$ be some univariate characteristic function admitting an $(\alpha, 0, \mu)$–Lévy–Hinčin representation where μ is concentrated on the set of positive rationals and satisfies condition (2) and for some $K > 0$ the estimation

$$\mu(\{x \colon |x| > y\}) = 0(\exp(-Ky^2)) \quad y \to +\infty.$$

Then p belongs to I_0^1 if and only if $D(\mu) = \{x : \mu(\{x\}) \neq 0\}$ satisfies condition (L).

(2) (Fryntov and Čistjakov [10]). Let $\hat{\mathbf{p}}$ be some univariate characteristic function admitting an $(\alpha,0,\mu)$–Lévy–Hinčin representation where μ is concentrated on Z and satisfies condition (2) and the estimation

$$\limsup_{k\to+\infty}\frac{|k|\log|k|}{\log(1/\mu(\{k\}))}<+\infty.$$

Then $\mathbf{p}$ belongs to I_0^1 if and only if $D(\mu)$ satisfies condition (L).

A result on measures concentrated on Q is not known as well as the possible extensions to multivariate case.

4. Indecomposable laws

Important advances on this subject have been given by Kudina. In [16], she studies the univariate case and gives the following result: if $\mathbf{p}$ is a decomposable univariate probability with $\operatorname{lext}\mathbf{p}=m, \operatorname{rext}\mathbf{p}=M$[2], there exist $k,l\in[m,M]$ such that for any $\varepsilon>0$

$$\mathbf{p}([k-2\varepsilon,k+2\varepsilon])\mathbf{p}([l-2\varepsilon,l+2\varepsilon])\geqslant\mathbf{p}([M-\varepsilon,M])\mathbf{p}([m,m+\varepsilon]).$$

This result implies the indecomposability of the beta law.

In [17, 18] she studies the general case and proves the following result: let $\mathbf{p}$ be an n-variate probability with a compact support $S(\mathbf{p})$ and k and l be the points common to $S(\mathbf{p})$ and to the hyperplane orthogonal to the diameter of $S(\mathbf{p})$.

(a) If $\mathbf{p}$ is decomposable, there exist c and d belonging to $S(p)\backslash\{k,l\}$ such that $\mathbf{p}(\{c\})\mathbf{p}(\{d\})\geqslant\mathbf{p}(\{k\})\mathbf{p}(\{l\})$.

(b) If k and l belong to $D(\mathbf{p})$ and if $D(\mathbf{p})$ is indecomposable, then $\mathbf{p}$ is indecomposable.

This result implies the following theorems:

Theorem 1. *Let A be an arbitrary closed set of $R^n (n\geqslant 1)$. There exists some indecomposable probability with support A.*

Theorem 2. *Let A be a non compact closed set of R and ρ and σ be two given numbers $(1<\rho\leqslant+\infty, 0\leqslant\sigma\leqslant+\infty)$. There exists some indecomposable probability $\mathbf{p}$ such that $S(p)=A$ and $\hat{\mathbf{p}}$ is an entire function with order ρ and type σ.*

[2]We follow here the notations of [24]: if p is a probability defined on the Borel sets of R, then $\operatorname{lext}p=\inf S(p)$, $\operatorname{rext}p=\sup S(p)$.

Theorem 3. *Let A be a closed set of R^n which is not both compact and enumerable. The set of indecomposable probabilities with support A is dense (in the sense of weak convergence) in the set of all probabilities with support A.*

Theorem 4. *If A is an infinite closed set, the set of indecomposable probabilities with support contained in A is dense in the set of all probabilities with support contained in A.*

These two last results make precise the result by Parthasarathy, Ranga Rao and Varadhan ([3], p. 76) that the set of indecomposable probabilities is dense in the set of all the probabilities. The example of the probabilities $\mathbf{p}=C_0\delta_0+C_1\delta_m+C_2\delta_{2m}(m\in R^n)$ which are indecomposable if and only if $C_1^2<4C_0C_2$ proves that a statement analogous to (4) is false for finite sets.

On the other hand, we must notice in the univariate case two studies on the factors of infinitely divisible laws which do not belong to I_0^1. It is known ([24], p. 179) that the non-normal stable laws on R have indecomposable factors [The same result for stable laws on R^n is not known]. Shimizu [34] gives an extension to dissymetric functions of the well known Polya's theorem on the convex characteristic functions and obtains some classes of noninfinitely divisible factors for stable laws. Ilinskiĭ [13] states some interesting results on the set of indecomposable factors for some infinitely divisible laws including geometric laws, exponential laws, Laplace law and symmetric stable laws with exponent not greater than 1.

We give now an outline of Shanbhag's works [31–33] where he studies various extensions of Paul Lévy's result on indecomposability of Wishart's law ([3], p. 92). We can state Lévy's result as following: if X and Y are independent normal random variables, then $(X^2,2XY,Y^2)$ is an indecomposable random vector. Shanbhag [31] proves that this result still holds if X and Y are arbitrary non degenerate independent random variables. He even shows [32] that if U and W are independent random variables, $V^2=4UW$, then (U,V,W) is decomposable if and only if U and W are special binomial laws. On the other hand, he shows that if X is some non degenerate normal random variable, then (X^2,X) is an indecomposable vector with all its projections infinitely divisible.

5. Spherically symmetric probabilities

We suppose in this section that $n\geqslant 2$.

An n-variate measure is called spherically symmetric around the point $a\in R^n$ if it is invariant for any rotation around the point a. It is easily shown that a bounded measure μ is spherically symmetric around the origin if and only if its Fourier-Stieltjes transform $\hat{\mu}$ depends only on $\|t\|$ and that a probability $\mathbf{p}$ with an (α, P, μ)–Lévy–Hinčin representation is spherically symmetric around the point α if and only if $P(t)=k\|t\|^2$ and μ is spherically symmetric around the origin.

If $\mathbf{p}$ is a probability spherically symmetric around the origin, we can define the radial projection of $\mathbf{p}$ as the probability $\sigma_n(\mathbf{p})$ defined on the Borel sets of R_+ by

$$\sigma_n(\mathbf{p})(B)=p(\{x: \|x\|\in B\}).$$

Kingman [15] has shown that σ_n is an isomorphism of the semigroup (for usual convolution) of all spherically symmetric probabilities on the semigroup $\mathfrak{S}_n$ of probabilities defined on R_+ for the operation of radial convolution $\underset{n}{*}$ defined for any continuous function ϕ by

$$\int_0^\infty \phi(x)(\mu_1 \underset{n}{*} \mu_2)(\mathrm{d}x)$$

$$=\int_0^\infty \int_0^\infty \left\{ \int_{-1}^{+1} \phi\left[(x^2+y^2+2\lambda xy)^{1/2} \right] f_n(\lambda)\,\mathrm{d}\lambda \right\} \mu_1(\mathrm{d}x)\mu_2(\mathrm{d}y)$$

where

$$f_n(\lambda)=\frac{\Gamma(n/2)}{\sqrt{(\pi)}\,\Gamma((n-1)/2)}(1-\lambda^2)^{(n-3)/2}.$$

Bingham [1] proved for semigroups $\mathfrak{S}_n$ the analogues of Hinčin's theorems:

(1). *Any probability of* $\mathfrak{S}_n$ *is the radial convolution of two (perhaps degenerate) probabilities* $\mathbf{q}$ *and* $\mathbf{r}$. $\mathbf{q}$ *has no indecomposable factor (for the radial convolution) while, for* $\mathbf{r}$, *there exists some sequence* $\{\mathbf{r}_k\}$ *of indecomposable probabilities (for the radial convolution) such that* $\mathbf{r}_1 \underset{n}{*} \cdots \underset{n}{*} \mathbf{r}_k$ *converges weakly to* $\mathbf{r}$ *when* $k \to +\infty$;

(2). *Any probability of* $\mathfrak{S}_n$ *with no indecomposable factor (for the radial convolution) is infinitely divisible for the radial convolution.*

Ostrovskiĭ [28, 29] has shown that the probabilities of $\mathfrak{S}_n$ which have no indecomposable factors for the radial convolution are the Rayleigh probabilities (which are the image by σ_n of spherically symmetric normal probabilities). From this result it follows that all the spherically symmetric probabilities of I_0^n are normal.

Kudina [19] has shown the following result: if $\mathbf{p} \in \mathbb{S}_n$ is decomposable for the radial convolution and has a compact support, then

(a) either $\mathbf{p}_d + \mathbf{p}_s = 0$ or $\operatorname{rext}(\mathbf{p}_d + \mathbf{p}_s) < \operatorname{rext}(\mathbf{p}) = l$ (here $\mathbf{p}_d$ and $\mathbf{p}_s$ are respectively the discrete and the singular part of $\mathbf{p}$);

(b) $\mathbf{p}_{ac}$ satisfies

$$\mathbf{p}_{ac}(\{x : \mathbf{r} < x < l\}) = \mathrm{O}((l-r)^{(n-1)/2}) \quad r \to l - \mathrm{O}.$$

Moreover if $n \geqslant 4$ the derivative of $\mathbf{p}_{ac}$ is continuous and satisfies

$$\mathbf{p}'_{ac}(\mathbf{r}) = \mathrm{O}((l-\mathbf{r})^{(n-3)/2}) \quad (r \to l - \mathrm{O}).$$

On the other hand, she has shown [20] that if $\mathbf{p}$ is a spherically symmetric probability satisfying

$$\mathbf{p}(\{x : \|x\| > \mathbf{r}\}) = \mathrm{O}(\exp(-\mathbf{r}^{2+\varepsilon})) \tag{3}$$

for some $\varepsilon > 0$ when $\mathbf{r} \to +\infty$, then all the components of $\mathbf{p}$ are spherically symmetric. Moreover, in [21], she has shown that this result still holds if (3) is replaced by

$$\mathbf{p}(\{x : \|x\| > \mathbf{r}\}) = \mathrm{O}(\exp(-K\mathbf{r}^2))$$

for any $K > 0$ (the example of normal laws proves that this last condition must be satisfied for any $K > 0$ for the validity of this result).

From these two results, it follows that the uniform probability on a ball of R^n is indecomposable if $n \geqslant 3$ (the result is not known for $n = 2$). This result reinforces the following Letac's conjecture: if the uniform probability on an open convex set A of R^n is decomposable, then A is a cube (that is a product of intervals). For uniform probabilities on cubes, all the factors are described by Lepetit's theorem [3, Theorem 5.3.2].

6. α-decompositions

We recall [3, Ch. 9] that if the characteristic function $\hat{\mathbf{p}}$ of an n-variate probability $\mathbf{p}$ has no real zeros, the probabilities $\mathbf{p}_1, \ldots, \mathbf{p}_m$ are α-factors of $\mathbf{p}$ if there exist some constants $\alpha_1, \ldots, \alpha_m$ such that

$$\hat{\mathbf{p}}(t) = \prod_{j=1}^{m} (\hat{\mathbf{p}}_j(t))^{\alpha_j}$$

for any $t \in R^n$. We denote by I_α^n the set of n-variate probabilities which have only infinitely divisible α-factors. Clearly, $I_\alpha^n \subset I_0^n$, but we do not know (even in the case $n=1$) if $I_\alpha^n \neq I_0^n$.

Čistjakov [2], solving a problem stated by Dugué twenty years ago, proved the following result: let $\mathbf{q}$ be an α-factor of a univariate probability $\mathbf{p}$. If $\hat{\mathbf{p}}$ is the limit of an analytic function in $\{t \in C: |\mathrm{Re}\, t| < \rho, 0 < \mathrm{Im}\, t < \beta\}$, then $\hat{\mathbf{q}}$ is also the limit of an analytic function in the same domain. This result can be extended easily to the multivariate case [4, Theorem 1].

Using this result and a method due to Fryntov [6, 7] in the case $n=1$, Cuppens [4], proved some results on I_α^n. We state some of them:

(1) Let A be an open set satisfying

$$A^* \cap \left[\bigcup_{k=2}^{\infty} (k) A \right] = \emptyset$$

where A^* is the convex hull of A and μ be a bounded measure concentrated on A. If $p = \omega \exp \mu [\omega = \exp(-\mu(R^n))]$, then $\mathbf{p}$ belongs to I_α^n (From this result, it follows that Mase's result stated in section 1 still holds for I_α^n).

(2) Let $A \in F_\sigma^n$ an independent set contained in a half-space $\{x \in R^n : (x, \theta) > 0\}$ for some $\theta \in R^n$ and μ be a bounded measure concentrated on A. If $\mathbf{p} = \omega \exp \mu [\omega = \exp(-\mu(R^n))]$, then $\mathbf{p}$ belongs to I_α^n.

(3) Let A be an enumerable independent set of R^n and μ be a bounded measure concentrated on A. If $\mathbf{p} = \omega \exp \mu [\omega = \exp(-\mu(R^n))]$, then $\mathbf{p}$ belongs to I_α^n

7. Multivariate characteristic functions

Finally we quote some recent results on multivariate characteristic functions.

Ilinskiĭ [14] proved the following result: Let A be some closed set of R^n, $\mathfrak{U}$ be a covering of A^c by simply connex open sets and $\omega(.; U)$ $(U \in \mathfrak{U})$ be a family of real-valued continuous functions satisfying the following conditions

$$t \in U_1, -t \in U_2 \Rightarrow \omega(t; U_1) = -\omega(t; U_2) \pmod{2\pi},$$

$$t \in U_1 \cap U_2 \Rightarrow \omega(t; U_1) = \omega(t; U_2) \pmod{2\pi}.$$

There exists some n-variate characteristic function f satisfying

$$\{x: f(x)=0\}=A,$$

$$f(t)=|f(t)|\exp(i\omega(t;U)) \qquad \forall t\in U$$

if and only if the following conditions are satisfied:

(1) the set A is such that $0\notin A, A=-A$;

(2) for any point $\tau\in U$, there exists some neighborhood $V_\tau\subset U$ of τ where the representation

$$\omega(t;U)=\sum_{k\in Z^n} c_k(\tau)\exp[i\lambda(\tau)(t,k)] \qquad \forall t\in V_\tau$$

is valid for some $\lambda(\tau)\in R$, $c_k(\tau)\in C$ and $\sum_{k\in Z^n}|c_k(\tau)|<+\infty$.

In particular, a closed set A of R^n is the set of zeros of some n-variate characteristic function if and only if $0\notin A$ and $A=-A$.

Wolfe [35] studied the derivatives of n-variate characteristic functions and obtained the following result:

Let $\mathbf{p}$ be a probability defined on the Borel sets of R^n and $m\in N$. The following conditions are equivalent:

(i) the partial derivatives of order m of $\hat{\mathbf{p}}$ at the origin exist;

(ii) there exists some polynomial P_m with degree m such that $\hat{\mathbf{p}}(t)=P_m(t)+o(\|t\|^m)$ $(t\to 0)$;

(iii) $\int\|x\|^m\mathbf{p}(\mathrm{d}x)<+\infty$ if m is even and $\mathbf{p}(\{x:\|x\|>T\})=o(T^{-m})$ $(T\to+\infty)$ and $\lim_{T\to\infty}\int_{\|x\|\leqslant T}x^k\mathbf{p}(\mathrm{d}x)$ exists for any $k=(k_1,\dots,k_n)\in N^n$ such that $\sum k_j=m$ if m is odd.

In this case, if $k=(k_1,\dots,k_n)\in N^n$ with $\sum k_j=m$, $P_m(t)=\sum_k a_k t^k$, then

$$a_k=\frac{1}{k!}D^k\hat{\mathbf{p}}(0)=\frac{i^m}{k!}\lim_{T\to\infty}\int_{\|x\|\leqslant T}x^k\mathbf{p}(\mathrm{d}x)$$

(if m is even, we can replace in this last relation $\lim_{T\to\infty}\int_{\|x\|\leqslant T}$ by $\int$).

Finally, we must notice Ginzburg's work [11] on multivariate entire characteristic functions. First, he shows a result on the order of these functions:

Let us suppose that the characteristic function $\hat{\mathbf{p}}$ of an n-variate probability $\mathbf{p}$ is an entire function with finite order ρ and for $u\in R^n$, $\|u\|=1$, let $\rho(u)$ be the order of the characteristic function of $\mathrm{pr}_u(\mathbf{p})$. Then

(1) if u is a linear combination of $u_1,\ldots,u_k$, we have

$$\rho(u) \leqslant \sup_{1 \leqslant j \leqslant k} \rho(u_j);$$

(2) $\rho(u)$ takes at most n distinct values and $\rho(u) < \sup_{v \in R^n;\, \|v\|=1} \rho(v)$ holds on a linear space with dimension at most $n-1$;

(3) $\rho = \sup_{v \in R^n;\, \|v\|=1} \rho(v)$.

Then he studied the hypersurfaces of conjugate orders of these functions. These hypersurfaces are defined as following: If ϕ is an n-variate entire function, let

$$B(\phi) = \{(a_1,\ldots,a_n) \in R_+^n : \ln M(r_1,\ldots,r_n;\phi) > r_1^{a_1} + \cdots + r_n^{a_n} \quad \text{if } \|(r_1,\ldots,r_n)\| \text{ is great enough}\}$$

where

$$M(r_1,\ldots,r_n;\phi) = \sup_{t_j = r_j (j=1,\ldots,n)} |\phi(t_1,\ldots,t_n)|.$$

The boundary $S(\phi)$ of $B(\phi)$ is the hypersurface of conjugate orders of ϕ. Ginzburg proved the following result: The hypersurface of conjugate orders of an entire characteristic function is the boundary of an hyperoctant of the kind $\{(a_1,\ldots,a_n) \in R^n:\ a_j > \rho_j, j=1,\ldots,n\}$ where the ρ_j are constants greater than 1.

References

[1] Bingham, N. H. (1971). Factorization theory and domains of attraction for generalized convolution algebras. *Proc. London Math. Soc.* (3) **23**, 16-30.

[2] Čistjakov, G. P. (1975). A Dugué's problem (in Russian). *Dokl. Akad. Nauk SSSR* **224**, 775-778.

[3] Cuppens, R. (1975). *Decomposition of Multivariate Probabilities*. Academic Press, New York.

[4] Cuppens, R. (1976). On α-factors of multivariate infinitely divisible probabilities. *J. Multivariate Anal.* **6**, 455-471.

[5] Cuppens, R. Decompositions of multivariate infinitely divisible probabilities with absolutely continuous Poisson measure (to appear).

[6] Fryntov, A. E. (1974). The expansion of ridge and characteristic functions (in Ukrainian). *Dopovidi Akad. Nauk Ukrain. RSR Ser. A* n° **8**, 721-724.

[7] Fryntov, A. E. (1974). The α-factors of infinitely divisible laws (in Russian). *Mat. Fiz. Funk. Analysis* **5**, 51-63.

[8] Fryntov, A. E. (1975). The factorization of infinitely divisible laws (in Russian). *Teor. Verojatnost. i Primenen.* **20**, 661-664.

[9] Fryntov, A. E. (1976). Decompositions of the composition of an enumerable set of Poisson laws (in Russian). *Mat. Sb.* **99**, 176-191.

[10] Fryntov, A. E. and Čistjakov, G. P. (1977). The belonging to the class I_0 of lattice probability laws (in Russian). *Izv. Akad. Nauk SSSR Ser. Mat.* **41**, 462-475.

[11] Ginzburg, B. N. (1974). The growth of entire characteristic functions of multivariate probability laws (in Russian). *Teor. Funkcii Funkcional. Anal. i Priložen.* **20**, 38-40.

[12] Hayman, W. (1956). A generalisation of Stirling' formula. *J. Reine Angew. Math.* **196**, 67-95.

[13] Ilinskiĭ, A. I. (1974). The indecomposable components of certain infinitely divisible laws (in Russian). *Dokl. Akad. Nauk SSSR* **215**, 529-531.

[14] Ilinskiĭ, A. I. (1975). The zeros and the argument of a characteristic function (in Russian). *Teor. Verojatnost. i Primenen.* **20**, 421-427.

[15] Kingman, J. F. C. (1963). Random walks with spherical symmetry. *Acta Math.* **109**, 11-53.

[16] Kudina, L. S. (1972). Indecomposability of the arcsine law (in Russian). *Dokl. Akad. Nauk SSSR* **204**, 25-26.

[17] Kudina, L. S. (1972). Indecomposable laws with a preassigned spectrum. *Teor. Funkcii Funkcional. Anal. i Priložen.* **16**, 206-212.

[18] Kudina, L. S. (1973). The closure of the set of indecomposable distributions with a fixed spectrum (in Russian). *Teor. Funkcii Funkcional. Anal. i Priložen.* **17**, 51-56.

[19] Kudina, L. S. Decompositions of radially symmetric distributions (in Russian). *Teor. Verojatnost. i Primenen.* **20**, 656-660.

[20] Kudina, L. S. The components of radially symmetric laws (in Russian). *Teor. Funkcii Funkcional. Anal. i Priložen.* **25**, 77-81.

[21] Linnik, Ju. V. and Ostrovskii, I. V. (1972). *Decomposition of Random Variables and Vectors* (in Russian). Izdat. Nauka, Moscow.

[22] Livsič, L. Z. (1974) Certain conditions for the absence of indecomposable components of infinitely divisible laws (in Russian). *Teor. Funkcii Funkcional. Anal. i Priložen.* **19**, 53-66; ibid. **20**, 86-101.

[23] Livsič, L. Z., Ostrovskiĭ, I. V. and Čistjakov, G. P. (1975). The arithmetic of probability laws. Probability theory, Mathematical statistics, Theoretical cybernetics. Vol. **12**, pp. 5-42. Akad. Nauk SSSR Vsesojuz. Inst. Naucn. i Teh.-Informacii, Moscow.

[24] Lukacs, E. (1970). *Characteristic Functions*. 2nd ed. Griffin, London.

[25] Mase, S. (1975). Decomposition of infinitely divisible characteristic functions with absolutely continuous Poisson spectral measures. *Ann. Inst. Statist. Math.* **27**, 289-298.

[26] Mase, S. (1975). Decomposition of multivariate infinitely divisible characteristic functions with absolutely continuous Poisson spectral measures. *J. Multivariate Anal.* **5**, 415-424.

[27] Mase, S. (1977). Decomposition of infinitely divisible characteristic functions without gaussian component. *Ann. Inst. Statist. Math.* **29**, 275-286.

[28] Ostrovskiĭ, I. V. (1973). A description of the class I_0 in a special semigroup of probability measures (in Russian). *Dokl. Akad. Nauk SSSR* **209**, 788-791.

[29] Ostrovskiĭ, I. V. (1973). A description of the class I_0 in a special semigroup of probability measures (in Russian). *Mat. Fiz. Funk. Analysis* **4**, 3-12.

[30] Ostrovskiĭ, I. V. (1977). The arithmetic of probability distributions. *J. Multivariate Anal.* **7**, 475-490.

[31] Shanbhag, D. N. (1974). An extension of Lévy's result concerning indecomposability of Wishart's distribution. *Proc. Cambridge Phil. Soc.* **75**, 109-113.

[32] Shanbhag, D. N. (1975). Some results on the decomposability of the distribution of quadratic expressions. *Math. Proc. Cambridge Philos. Soc.* **77**, 553-558.

[33] Shanbhag, D. N. (1976). On the structure of the Wishart distribution. *J. Multivariate Anal.* **6**, 347-355.

[34] Shimizu, R. (1972). On the decompositions of stable characteristic functions. *Ann. Inst. Statist. Math.* **24**, 347-353.

[35] Wolfe, S. J. (1973). On the finite series expansion of multivariate characteristic functions. *J. Multivariate Anal.* **3**, 328-335.

P. R. Krishnaiah, ed., *Multivariate Analysis–V*

INVARIANT POLYNOMIALS WITH TWO MATRIX ARGUMENTS, EXTENDING THE ZONAL POLYNOMIALS

A.W. DAVIS
C.S.I.R.O. Division of Mathematics and Statistics, Adelaide, South Australia.

A class of invarant polynomials in two matrices is defined, generalizing the zonal polynomials. Basic properties of the polynomials are derived, and their applications to distribution theory indicated.

1. Introduction

The theory of group representations and zonal spherical functions was applied by James [7] to the evaluation of certain integrals arising in multivariate distribution theory. The integrals were expanded as power series in the zonal polynomials of the real positive definite symmetric matrices, which had been previously studied by Hua [6]. Subsequently, Constantine [2] showed that Herz's [5] hypergeometric functions of matrix argument could be expanded in terms of zonal polynomials, and expressed a number of noncentral multivariate distributions in terms of these functions. A survey of the area was given by James [8]. Since then, many authors have utilized and developed these techniques, so that an extensive literature now exists.

However, there remain related distributional problems which cannot be solved in terms of zonal polynomials, including

(a) the cumulative distribution functions of the basic noncentral distributions (Constantine [2]),

(b) the latent roots of the noncentral Wishart distribution (Pillai [12, p.30]), and

(c) the doubly noncentral multivariate F distribution.

The latter arises in particular from a formal approach to the MANOVA distributions in multivariate Edgeworth populations (Davis [3]), and was the initial stimulus for this investigation. In the above and certain other situations, integrals arise having the form (see Section 6)

$$\int_{O(m)} C_\kappa(AH'XH)C_\lambda(BH'YH)\mathrm{d}H \tag{1.1}$$

where A,B,X,Y are $m\times m$ symmetric matrices, $\mathrm{d}H$ is the invariant Haar measure over the group $O(m)$ of $m\times m$ orthogonal matrices, and C_κ, C_λ are the zonal polynomials indexed by the ordered partitions κ, λ of the integrals k, l respectively into $\leqslant m$ parts.

In seeking to evaluate (1.1), we observe that since $\mathrm{d}H$ is invariant under left translation $H\to\mathcal{H}H(\mathcal{H}\in O(m))$, (1.1) is a homogeneous polynomial of degree k, l in the elements of X, Y respectively, invariant under the *simultaneous* transformations

$$X\to\mathcal{H}'X\mathcal{H},\quad Y\to\mathcal{H}'Y\mathcal{H},\quad \mathcal{H}\in O(m). \tag{1.2}$$

Invariance of $\mathrm{d}H$ under right translation implies that the same holds for A,B. We thus require a basis for polynomials having the property (1.2) which will facilitate the evaluation of (1) the integrals (1.1) and (2) expectations with respect to the Wishart distribution. Similar requirements in the one matrix case are fulfilled by the zonal polynomials.

Section 2 defines a class of invariant polynomials with two matrix arguments as a direct extension of the zonal polynomials. Elementary properties of the polynomials are derived in Section 3, and in Section 4 an expansion of (1.1) is given which constitutes the cornerstone of the subsequent theory (equation (4.13)). Derivation of this result leads to the definition of a class of 'orthogonal' invariant polynomials whose properties are discussed in Section 5. Owing to restrictions on space, their use in multivariate distribution theory can only be briefly indicated (Section 6) but it is hoped that these applications, together with details about the construction of the polynomials, will be presented in subsequent papers. In particular, the application to MANOVA distributions in nonnormal cases requires the low-degree invariant polynomials. This probably constitutes the main practical justification for studying these polynomials at the present stage, since the problems associated with construction, and the convergence of infinite series in the invariant polynomials, are considerably more serious than in the case of the zonal polynomials.

We finally note that the invariant polynomials of this paper are not zonal polynomials, and lack certain properties of the latter. In particular, there is no guarantee that they will possess an analogue of the Laplace-Beltrami operator for the zonal polynomials (James [9]). Hence, although one could define hypergeometric functions of two matrix arguments generalizing e.g. those of Appell in the scalar case, they may not satisfy useful systems of differential equations as do the hypergeometric functions of Herz and Constantine (Muirhead [11]).

2. Invariant polynomials in two matrix arguments

Let $X=(x_{ij})$, $Y=(y_{ij})$ be $m\times m$ complex symmetric matrices; $P_k[X]$ the class of homogeneous polynomials of degree k in the x_{ij}; $P_{k,l}[X,Y]$ the class of homogeneous polynomials of degree k, l in the x_{ij}, y_{ij} respectively; $\boldsymbol{\xi}_k$ the vector of all monomials $\Pi_{i\leqslant j}x_{ij}^{k_{ij}}$ $(\Sigma_{i\leqslant j}k_{ij}=k,\ k_{ij}\geqslant 0)$; $Gl(m,R)$ the group of $m\times m$ real nonsingular matrices L. Then $\boldsymbol{\xi}_k$ is a basis for $P_k[X]$, and defining a similar basis $\boldsymbol{\eta}_l$ for $P_l[Y]$,

$$P_{k,l}[X,Y]=P_k[X]\otimes P_l[Y] \tag{2.1}$$

in the sense that $\boldsymbol{\xi}_k\otimes\boldsymbol{\eta}_l$ is a basis for $P_{k,l}[X,Y]$. The simultaneous congruence transformations

$$X\to L'XL,\quad Y\to L'YL,\quad L\in Gl(m,R), \tag{2.2}$$

produce linear transformations in $P_k[X]$, $P_l[Y]$ and $P_{k,l}[X,Y]$ with matrices $T_{2k}(L)$, $T_{2l}(L)$ and $T_{2k,2l}(L)$, respectively, such that

$$\begin{aligned}&\boldsymbol{\xi}_k\to T_{2k}(L)'\boldsymbol{\xi}_k,\quad \boldsymbol{\eta}_l\to T_{2l}(L)'\boldsymbol{\eta}_l,\\&\boldsymbol{\xi}_k\otimes\boldsymbol{\eta}_l\to T_{2k,2l}(L)'(\boldsymbol{\xi}_k\otimes\boldsymbol{\eta}_l).\end{aligned} \tag{2.3}$$

These constitute representations of $Gl(m,R)$ in the respective vector spaces,

$$T_{2k}(L_1L_2)=T_{2k}(L_1)T_{2k}(L_2),\quad \text{etc.}, \tag{2.4}$$

of polynomial degrees $2k, 2l, 2f (f=k+l)$ respectively in the elements of L.

It is a classical result (e.g. Boerner [1, Ch. 5] that a vector space in which a polynomial representation of $Gl(m,R)$ is defined may be decomposed into a direct sum of irreducible invariant subspaces, each carrying an irreducible representation of $Gl(m,R)$. The inequivalent representations of degree $2k$, say, may be indexed by the ordered partitions $\mathcal{K}$ of $2k$ into $\leqslant m$ parts; we shall denote these by $\tilde{T}_{\mathcal{K}}(L)$. Then A.T. James (unpublished lecture notes) has shown that

(a) $\tilde{T}_{\mathcal{K}}(L)$ occurs in the decomposition of $\tilde{T}_{2k}(L)$ if it contains the identity representation when L is restricted to the subgroup $O(m)$;

(b) The only such $\tilde{T}_{\mathcal{K}}(L)$ are those for which $\mathcal{K}=2\kappa$, where κ is any ordered partition of k into $\leqslant m$ parts;

(c) $\tilde{T}_{2\kappa}(L)$ contains the identity representation exactly once when restricted to $O(m)$, and

(d) $\tilde{T}_{2\kappa}(L)$ occurs with multiplicity one in the decomposition of $T_{2k}(L)$, so that

$$T_{2k}(L)=\bigoplus_{\kappa}\tilde{T}_{2\kappa}(L),\quad T_{2l}(L)=\bigoplus_{\lambda}\tilde{T}_{2\lambda}(L) \tag{2.5}$$

where κ, λ run through the ordered partitions of k, l respectively into $\leqslant m$ parts. The decompositions (2.5) were originally obtained by Littlewood and Ffoulkes (see references in [7]). Correspondingly, we have the decompositions into irreducible invariant subspaces

$$P_k[X]=\bigoplus_{\kappa}\mathcal{V}_{\kappa}[X],\quad P_l[Y]=\bigoplus_{\lambda}\mathcal{V}_{\lambda}[Y], \tag{2.6}$$

where in virtue of (c) $\mathcal{V}_{\kappa}[X]$ for example contains exactly one one-dimensional subspace generated by a polynomial in X invariant under $X\to H'XH$ ($H\in O(m)$), which when suitably normalized is the zonal polynomial $C_{\kappa}(X)$. From (2.3) and (2.5)

$$\begin{aligned}T_{2k,2l}(L)&=T_{2k}(L)\otimes T_{2l}(L)\\&=\bigoplus_{\kappa}\bigoplus_{\lambda}\{\tilde{T}_{2\kappa}(L)\otimes\tilde{T}_{2\lambda}(L)\},\end{aligned} \tag{2.7}$$

and correspondingly

$$P_{k,l}[X,Y]=\bigoplus_{\kappa}\bigoplus_{\lambda}\{\mathcal{V}_{\kappa}[X]\otimes\mathcal{V}_{\lambda}[Y]\}. \tag{2.8}$$

The Kronecker products $\tilde{T}_{2\kappa}(L)\otimes\tilde{T}_{2\lambda}(L)$ are also representations of $Gl(m,R)$ having polynomial degree $2f$, but are not in general irreducible; they may therefore be decomposed into direct sums of representations

Table 1
Decomposition of Kronecker products of irreducible representations

$2f$	2κ	2λ	Φ					
4	2	2	4	31	2^2			
6	4	2	6	51	42			
	2^2	2	42	321	2^3			
8	6	2	8	71	62			
	42	2	62	53	521	4^2	431	42^2
	2^3	2	42^2	32^21	2^4			

$\tilde{T}_\Phi(L)$, Φ being an ordered partition of $2f$ into $\leqslant m$ parts. Rules for determining the particular $\tilde{T}_\Phi(L)$ occurring in such a decomposition, together with their multiplicities, are given for example by Robinson [13, Section 3.3]. Table 1 presents the decompositions for some low order κ, λ.

We thus obtain the decomposition

$$P_{k,l}[X,Y]=\bigoplus_{\kappa}\bigoplus_{\lambda}\bigoplus_{\Phi}\mathcal{V}_{\Phi}^{\kappa,\lambda}[X,Y] \tag{2.9}$$

into irreducible subspaces, where Φ runs through the irreducible representations in the decomposition of $\tilde{T}_{2\kappa}\otimes\tilde{T}_{2\lambda}$. If $\Phi=2\phi$, where ϕ is a partition of $f=k+l$, then by (c) $\mathcal{V}_{2\phi}^{\kappa,\lambda}[X,Y]$ contains a one-dimensional subspace generated by a polynomial $\Gamma_{\phi}^{\kappa,\lambda}(X,Y)$, say, which is invariant under (1.2). The question of normalization of this polynomial will be discussed in Section 4. We note that

(i) Subspaces $\mathcal{V}_{\Phi}^{\kappa,\lambda}$ occur with Φ not of the form 2ϕ; these contain no invariant polynomials;

(ii) A representation 2ϕ may occur in (2.9) with multiplicity greater than 1 for a given κ, λ, so that strictly an additional subscript should be used; however, for notational simplicity we shall omit this, and write for example $\phi'\equiv\phi$ to indicate equivalent representations. A more important consequence is that the $\mathcal{V}_{2\phi}^{\kappa,\lambda}$, and hence the corresponding polynomials $\Gamma_{\phi}^{\kappa,\lambda}$, are not uniquely defined when 2ϕ occurs with multiplicity >1 for a given κ, λ. However, the direct sum of equivalent subspaces

$$\mathcal{U}_{2\phi}^{\kappa,\lambda}=\bigoplus_{\phi'\equiv\phi}\mathcal{V}_{2\phi'}^{\kappa,\lambda} \tag{2.10}$$

is uniquely defined. A sufficient resolution of the non-uniqueness problem will be described in Section 4.

Multiplicity first occurs in the case of the polynomials of degree 6, with $k=l=3$, $\kappa=\lambda=[2,1]$, when $2\phi=2[3,2,1]$ occurs with multiplicity 3.

3. Elementary properties of the $\Gamma_{\phi}^{\kappa,\lambda}$.

(a) $$\Gamma_{\phi}^{\kappa,\lambda}(X,X)=\{\Gamma_{\phi}^{\kappa,\lambda}(I,I)/C_\phi(I)\}C_\phi(X), \tag{3.1}$$

(b) $$\Gamma_{\phi}^{\kappa,\lambda}(X,I)=\{\Gamma_{\phi}^{\kappa,\lambda}(I,I)/C_\kappa(I)\}C_\kappa(X), \tag{3.2}$$

with a corresponding result for $\Gamma_{\phi}^{\kappa,\lambda}(I,Y)$;

(c) $$\Gamma_{\kappa}^{\kappa,0}(X,Y) \overset{\text{def}}{=} C_{\kappa}(X), \Gamma_{\lambda}^{0,\lambda}(X,Y) \overset{\text{def}}{=} C_{\lambda}(Y), \tag{3.3}$$

(d) $$\int_{O(m)} \Gamma_{\phi}^{\kappa,\lambda}(A'H'XHA, A'H'YHA)\mathrm{d}H$$

$$= \Gamma_{\phi}^{\kappa,\lambda}(X,Y)C_{\phi}(AA')/C_{\phi}(I) \tag{3.4}$$

(e) If $VV' \sim W_m(n,\Sigma,O)$ (the central Wishart distribution with n degrees of freedom and population covariance Σ), and $RR' = \Sigma$, then

$$E_V \Gamma_{\phi}^{\kappa,\lambda}(V'XV, V'YV) = 2^f\left(\tfrac{1}{2}n\right)_{\phi} \Gamma_{\phi}^{\kappa,\lambda}(R'XR, R'YR) \tag{3.5}$$

where $(a)_\phi$ is the generalized hypergeometric coefficient (Constantine [2]).

For brevity we shall merely indicate the salient points in the proofs.

(a) Setting $Y = X$ defines a group homomorphism of $P_{k,l}[X,Y]$ onto $P_f[X]$; since $\mathcal{V}_{\Phi}^{\kappa,\lambda}[X,Y]$ may be mapped into the null space, $\Gamma_{\phi}^{\kappa,\lambda}(X,X)$ may be identically zero.

(b) Setting $Y = I$ maps $\mathcal{V}_{\kappa}[X] \otimes \mathcal{V}_{\lambda}[Y]$ onto $\mathcal{V}_{\kappa}[X]$.

(c) $\tilde{T}_0(L)$ is the identity representation.

(d) Choose $\Gamma_{\phi}^{\kappa,\lambda}(X,Y)$ as the first basis vector in $\mathcal{V}_{2\phi}^{\kappa,\lambda}[X,Y]$, and the remainder in subspaces invariant under the non-identity representations of $O(m)$; then James has shown that the (1,1) element of $\tilde{T}_{2\phi}(A)$ is $C_\phi(AA')/C_\phi(I)$ and that

$$\int_{O(m)} \tilde{T}_{2\phi}(H)\mathrm{d}H = \begin{bmatrix} 1 & 0 & \cdots & 0 \\ 0 & 0 & \cdots & 0 \\ \vdots & \vdots & & \vdots \\ 0 & 0 & \cdots & 0 \end{bmatrix}.$$

No result corresponding to (3.4) holds if A is replaced by B in one argument of $\Gamma_{\phi}^{\kappa,\lambda}$.

(e) $W = R^{-1}VV'R'^{-1} \sim W_m(n,I,O)$, and $V = RW^{\frac{1}{2}}H_V$, where $H_V \in O(m)$; the result follows using (d).

4. Evaluation of the integral (1.1)

The evaluation of (1.1) constitutes the most important single result for the invariant polynomials introduced in this paper, and the derivation also yields a sufficient resolution of the non-uniqueness mentioned in Section 4.

This is effected by the construction of a new set of invariant polynomials $C_\phi^{\kappa,\lambda}(X,Y)$ which are linear combinations of $\Gamma_{\phi'}^{\kappa,\lambda}$ for $\phi'\equiv\phi$, and are orthogonal in a sense to be defined below (extending the orthogonality of the zonal polynomials, James [7]).

Since $C_\kappa(AX)C_\lambda(BY)$ is invariant under (2.2) together with the contragredient transformation

$$A\to L^{-1}AL^{-1'}, B\to L^{-1}BL^{-1'}, \quad L\in Gl(m,R), \tag{4.1}$$

an argument of James [7] Section 4 (see also Hannan [4] Section 2.2) may be adapted to show that

$$\int_{O(m)} C_\kappa(AH'XH)C_\lambda(BH'YH)\mathrm{d}H = \sum_{\phi\in\kappa\cdot\lambda}\ \sum_{\phi'\equiv\phi} q_{\phi,\phi'}^{\kappa,\lambda}\Gamma_\phi^{\kappa,\lambda}(A,B)\Gamma_{\phi'}^{\kappa,\lambda}(X,Y) \tag{4.2}$$

where $\phi\in\kappa\cdot\lambda$ is an abbreviation for $2\phi\in2\kappa\otimes2\lambda$. Since $C_\kappa(AX)C_\lambda(BY)$ may be shown to span $\mathcal{V}_\kappa[X]\otimes\mathcal{V}[Y]$ as A, B vary over the symmetric matrices, the integral (1.1) must span the polynomials in this space invariant under (1.2), and hence the bilinear form in (4.2) must have a non-singular matrix $Q_{\kappa,\lambda}=(q_{\phi,\phi'}^{\kappa,\lambda})$ ($q_{\phi,\phi'}^{\kappa,\lambda}=0$ unless $\phi\equiv\phi'$). At this stage, however, $Q_{\kappa,\lambda}$ is unknown.

We now follow an approach based on Saw [14]. Let $U=(u_{ij})$ be an $m\times m$ matrix of independent unit normal variables, and consider

$$\begin{aligned} g &= E_U \operatorname{etr}\{\alpha AU'XU+\beta BU'YU\}, \qquad \alpha,\beta \text{ real},\\ &= \sum_{k,l=0}^{\infty} \alpha^k\beta^l E_{k,l}/k!l!, \end{aligned} \tag{4.3}$$

where, since $U=W^{\frac{1}{2}}H$, $W\sim W_m(m,I,O)$,

$$\begin{aligned} E_{k,l} &= \sum_{\kappa,\lambda} E_W\int_{O(m)} C_\kappa(AH'W^{\frac{1}{2}}XW^{\frac{1}{2}}H)C_\lambda(BH'W^{\frac{1}{2}}YW^{\frac{1}{2}}H)\mathrm{d}H\\ &= \boldsymbol{\Gamma}_{k,l}(A,B)'\tilde{Q}_{k,l}\Gamma_{k,l}(X,Y). \end{aligned} \tag{4.4}$$

Here $\tilde{\boldsymbol{\Gamma}}_{k,l}(X,Y)$ denotes the vector of all $\Gamma_\phi^{\kappa,\lambda}(X,Y)$ for fixed k, l and

$$\tilde{Q}_{k,l}=\operatorname{diag}\{\tilde{Q}_{\kappa,\lambda}\}, \quad \tilde{Q}_{\kappa,\lambda}=\left(2^f\left(\tfrac{1}{2}m\right)_\phi q_{\phi,\phi'}^{\kappa,\lambda}\right). \tag{4.5}$$

On the other hand, writing $G=\alpha A\otimes X+\alpha B\otimes Y$,

$$g=|I_{m^2}-G|^{-\frac{1}{2}}=\sum_{f=0}^{\infty}1.3\cdots(2f-1)C_f(G)/f! \tag{4.6}$$

whence

$$E_{k,l}=\boldsymbol{\pi}_{k,l}(A,B)'\Delta_{k,l}\boldsymbol{\pi}_{k,l}(X,Y) \tag{4.7}$$

where $\Delta_{k,l}$ is a diagonal matrix with positive diagonal elements, and $\boldsymbol{\pi}_{k,l}(X,Y)$ denotes the vector of all *distinct* products of traces

$$(\operatorname{tr}X^{a_1}Y^{b_1}X^{c_1}\cdots)^{r_1}(\operatorname{tr}X^{a_2}Y^{b_2}X^{c_2}\cdots)^{r_2}\cdots \tag{4.8}$$

of total degree k,l in the elements of the symmetric matrices X, Y respectively (i.e. account must be taken of the symmetry of X and Y, and the properties $\operatorname{tr}XY=\operatorname{tr}YX$, $\operatorname{tr}Z'=\operatorname{tr}Z$).

Example. When $k=2$, $l=1$ it is found from $C_3(G)$ or otherwise that

$$\boldsymbol{\pi}_{2,1}(X,Y)'=((\operatorname{tr}X)^2\operatorname{tr}Y,\operatorname{tr}XY\operatorname{tr}X,\operatorname{tr}X^2\operatorname{tr}Y,\operatorname{tr}X^2Y),$$

$$\Delta_{2,1}=5\operatorname{diag}(1,4,2,8). \tag{4.9}$$

It seems clear that the distinct products (4.8) for a given k, and l (e.g. the components in (4.9) for $k=2$, $l=1$) are functionally independent, although this may be difficult to prove rigorously. If we assume independence then it follows from (4.4) and (4.7) that

(a) the products (4.8) constitute an elementary basis for the invariant polynomials, generalizing the familiar basis of the zonal polynomials constructed from the $s_i=\operatorname{tr}(X^i)$. The number of such terms should thus equal the sum of the multiplicities of the irreducible representations $\tilde{T}_{2\phi}$ occurring in $T_{2k}\otimes T_{2l}$.

(b) $E_{k,l}=\boldsymbol{\sigma}_{k,l}(A,B)'\boldsymbol{\sigma}_{k,l}(X,Y)$ iff the components of $\boldsymbol{\sigma}_{k,l}(X,Y)$ are '$\Delta_{k,l}$-orthonormal' linear combinations of the (4.8) in the sense that

$$\boldsymbol{\sigma}_{k,l}(X,Y)=S\boldsymbol{\pi}_{k,l}(X,Y),\quad S\Delta_{k,l}^{-1}S'=I. \tag{4.10}$$

(c) from (4.4) and (4.5) the $\Gamma_\phi^{\kappa,\lambda}$ are $\Delta_{k,l}$-orthogonal for inequivalent ϕ.

(d) for ϕ' equivalent to ϕ, choose *any set* of $\Delta_{k,l}$-orthonormal polynomials $\tilde{\Gamma}_{\phi'}^{\kappa,\lambda}(X,Y)$ say in the unique invariant subspace $\mathfrak{U}_{2\phi}^{\kappa,\lambda}$ equation (2.10)). Then from (4.2) and (4.5) the integral is given by

$$\sum_{\phi\in\kappa.\lambda} \tilde{\Gamma}_{\phi'}^{\kappa,\lambda}(A,B)\tilde{\Gamma}_{\phi}^{\kappa,\lambda}(X,Y)/2^f\left(\tfrac{1}{2}m\right)_{\phi'}, \tag{4.11}$$

noting that the denominators are constant for equivalent ϕ. Renormalizing to

$$C_{\phi}^{\kappa,\lambda}(X,Y)=\sqrt{z_\phi}\,\tilde{\Gamma}_{\phi}^{\kappa,\lambda}(X,Y) \tag{4.12}$$

where $z_\phi = C_\phi(I_m)/2^f(\tfrac{1}{2}m)_\phi$ is independent of m (Constantine [2]), and the sign of the square root is chosen to make the coefficient of $(\operatorname{tr}X)^k(\operatorname{tr}Y)^l$ in $C_\phi^{\kappa,\lambda}$ positive if it is non-zero, we finally obtain

$$\int_{O(m)} C_\kappa(AH'XH)C_\lambda(BH'YH)\,\mathrm{d}H = \sum_{\phi\in\kappa.\lambda} C_{\phi}^{\kappa,\lambda}(A,B)C_{\phi}^{\kappa,\lambda}(X,Y)/C_\phi(I_m). \tag{4.13}$$

The $\Delta_{k,l}$-orthogonality property provides the required resolution of the non-uniqueness, and considerably facilitates the construction of the polynomials. The $C_\phi^{\kappa,\lambda}$ are of course not unique if 2ϕ has multiplicity >1 in $2\kappa\otimes 2\lambda$, corresponding to the fact that (4.13) is invariant under 'orthogonal' transformations within the $\mathfrak{U}_{2\phi}^{\kappa,\lambda}$. Details of construction must be deferred to a subsequent paper. Table 2 presents some low degree $C_\phi^{\kappa,\lambda}$; the

Table 2
Polynomials $C_\phi^{\kappa,\lambda}(X,Y)$

f	k	l	$C_\phi^{\kappa,\lambda}$	multiplier	$\operatorname{tr}X\operatorname{tr}Y$	$\operatorname{tr}XY$		
2	1	1	$C_2^{\kappa,\lambda}$	1/3	1	2		
			$C_{1^2}^{1,1}$	2/3	1	−1		
					$(\operatorname{tr}X)^2\operatorname{tr}Y$	$\operatorname{tr}XY\operatorname{tr}X$	$\operatorname{tr}X^2\operatorname{tr}Y$	$\operatorname{tr}X^2Y$
3	2	1	$C_3^{2,1}$	1/15	1	4	2	8
			$C_{21}^{2,1}$	2/5	1	−1	2	−2
			$C_{21}^{1^2,1}$	$1/\sqrt{5}$	1	2	−1	−2
			$C_{1^3}^{1^2,1}$	1/3	1	−2	−1	2

complete set to $f=k+l=6$ has been tabulated and is available from the author. It is convenient for tabulation purposes to define $C_\phi^{\kappa,\lambda}(X,Y)=C_\phi^{\lambda,\kappa}(Y,X)$ when $\kappa\neq\lambda$, but this symmetry may neither be convenient nor possible to maintain in cases of multiplicity >1 when $\kappa=\lambda$. We do not therefore assume symmetry to hold in general.

5. Properties of the $C_\phi^{\kappa,\lambda}$

Since the $C_\phi^{\kappa,\lambda}$ are linear combinations of the $\Gamma_{\phi'}^{\kappa,\lambda}$ for $\phi'\equiv\phi$, we have from Section 3:

$$C_\phi^{\kappa,\lambda}(X,X)=\theta_\phi^{\kappa,\lambda}C_\phi(X), \text{ where } \theta_\phi^{\kappa,\lambda}=C_\phi^{\kappa,\lambda}(I,I)/C_\phi(I) \tag{5.1}$$

may be zero.

$$C_\phi^{\kappa,\lambda}(X,I)=\{\theta_\phi^{\kappa,\lambda}C_\phi(I)/C_\kappa(I)\}C_\kappa(X) \tag{5.2}$$

and similarly for $C_\phi^{\kappa,\lambda}(I,Y)$.

$$C_\kappa^{\kappa,0}(X,Y)\overset{\text{def}}{=}C_\kappa(X), C_\lambda^{0,\lambda}(X,Y)\overset{\text{def}}{=}C_\lambda(Y). \tag{5.3}$$

The basis (4.8) allows us to write without ambiguity

$$\int_{O(m)}C_\phi^{\kappa,\lambda}(AH'XH,AH'YH)\mathrm{d}H=C_\phi^{\kappa,\lambda}(X,Y)C_\phi(A)/C_\phi(I), \tag{5.4}$$

$$E_W C_\phi^{\kappa,\lambda}(XW,YW)=2^f\left(\tfrac{1}{2}n\right)_\phi C_\phi^{\kappa,\lambda}(X\Sigma,Y\Sigma) \tag{5.5}$$

where $W\sim W_m(n,\Sigma,\text{O})$. Hence using (4.13) and (5.1):

$$E_W\{C_\kappa(XW)C_\lambda(YW)\}=\sum_{\phi\in\kappa\cdot\lambda}2^f\left(\tfrac{1}{2}n\right)_\phi\theta_\phi^{\kappa,\lambda}C_\phi^{\kappa,\lambda}(X\Sigma,Y\Sigma). \tag{5.6}$$

$$C_\phi^{\kappa,\lambda}(\alpha X,\beta Y)=\alpha^k\beta^l C_\phi^{\kappa,\lambda}(X,Y). \tag{5.7}$$

From (4.13)

$$C_\kappa(X)C_\lambda(Y)=\sum_{\phi\in\kappa\cdot\lambda}\theta_\phi^{\kappa,\lambda}C_\phi^{\kappa,\lambda}(X,Y),\tag{5.8}$$

$$(\operatorname{tr}X)^k(\operatorname{tr}Y)^l=\sum_{\kappa,\lambda;\phi\in\kappa\cdot\lambda}\theta_\phi^{\kappa,\lambda}C_\phi^{\kappa,\lambda}(X,Y),\tag{5.9}$$

$$C_\kappa(X)C_\lambda(X)=\sum_{\phi\in\kappa\cdot\lambda}\left(\theta_\phi^{\kappa,\lambda}\right)^2C_\phi(X)\tag{5.10}$$

whence in the usual notation (e.g. Khatri and Pillai [10])

$$g_{\kappa,\lambda}^{\phi}=\sum_{\phi'\equiv\phi}\left(\theta_{\phi'}^{\kappa,\lambda}\right)^2.\tag{5.11}$$

Thus $g_{\kappa,\lambda}^{\phi}\geqslant 0$. If $g_{\kappa,\lambda}^{\phi}>0$ so that not all $\theta_{\phi'}^{\kappa,\lambda}=0(\phi'\equiv\phi)$, we may choose the first $C_\phi^{\kappa,\lambda}$ in $\mathfrak{U}_{2\phi}^{\kappa,\lambda}$ to be proportional to the component of $(\operatorname{tr}X)^k(\operatorname{tr}Y)^l$ in this space with $\theta_\phi^{\kappa,\lambda}=+\sqrt{g_{\kappa,\lambda}^{\phi}}$, the remaining $C_{\phi'}^{\kappa,\lambda}$ having $\theta_{\phi'}^{\kappa,\lambda}=0$ $(\phi'\equiv\phi)$.

From (4.13) in an obvious notation

$$\int_{O(m)}\operatorname{etr}(AH'XH+BH'YH)\mathrm{d}H=\sum_{\kappa,\lambda;\phi}^{\infty}C_\phi^{\kappa,\lambda}(A,B)C_\phi^{\kappa,\lambda}(X,Y)/k!l!C_\phi(I).\tag{5.12}$$

This expansion may be used to derive a number of useful results, including the following.

$$\int_{O(m)}C_\phi^{\kappa,\lambda}(A'H'XHA,B)\mathrm{d}H=C_\phi^{\kappa,\lambda}(A'A,B)C_\kappa(X)/C_\kappa(I),\tag{5.13}$$

with a corresponding result for $C_\phi^{\kappa,\lambda}(A,B'H'YHB)$.

Laplace transform

$$\int_{R>0}\operatorname{etr}(-RW)|R|^{t-p}C_\phi^{\kappa,\lambda}(ARA',B)\mathrm{d}R=\Gamma_m(t,\kappa)|W|^{-t}C_\phi^{\kappa,\lambda}(AW^{-1}A',B),\tag{5.14}$$

where $p=\frac{1}{2}(m+1)$, and $\Gamma_m\,(t,\kappa)$ is defined in [2]. A similar result holds for $C_\phi^{\kappa,\lambda}(A,BRB')$. From (5.5), $C_\phi^{\kappa,\lambda}(XR,YR)$ has Laplace transform

$\Gamma_m(t,\phi)|W|^{-t}C_\phi^{\kappa,\lambda}(XW^{-1},YW^{-1}).$

Binomial expansion

$$C_\phi(X+Y)=\sum_{\kappa,\lambda(\phi\in\kappa\cdot\lambda)}\binom{f}{k}\theta_\phi^{\kappa,\lambda}C_\phi^{\kappa,\lambda}(X,Y), \tag{5.15}$$

and in particular

$$C_f(X+Y)=\sum_{k+l=f}\binom{f}{k}C_f^{k,l}(X,Y) \tag{5.16}$$

so that $\binom{f}{k}$ $C_f^{k,l}(X,Y)$ is given by the terms of degree k, l in X, Y respectively in $C_f(X+Y)$. It follows from (4.6) that the diagonal elements of $\Delta_{k,l}$ in Section 4 are proportional to the coefficients of the terms (4.8) in $C_f^{k,l}(X,Y)$ so that the orthogonality of the $C_\phi^{\kappa,\lambda}$ generalizes that of the C_ϕ (James [7])

$C_{1^f}^{1^k,1^l}$ may similarly be obtained from $C_{1^f}(X+Y)$.

6. Latent roots of the noncentral Wishart distribution

By way of illustration, we shall apply the invariant polynomials to this distributional problem, which was mentioned in the Introduction. Let $S\sim W_m(n,\Sigma,\Omega)$; then for real q we may write

$$f(S)=f_q(S)|q\Sigma|^{-\frac{1}{2}n}\operatorname{etr}\left\{-\tfrac{1}{2}(\Sigma^{-1}-qI)S\right\}{}_0F_1\left(\tfrac{1}{2}n;\tfrac{1}{2}\Sigma^{-\frac{1}{2}}\Omega\Sigma^{-\frac{1}{2}}S\right), \tag{6.1}$$

where $f_q(S)$ is the $W_m(n,q^{-1}I,0)$ density function, and ${}_0F_1$ is the Bessel function of matrix argument [2]. Hence the joint density of the latent roots $\Lambda=\operatorname{diag}(l_1,\ldots,l_m)$ of S is given by

$$f(\Lambda)=\left\{\pi^{\frac{1}{2}m^2}/\Gamma_m\left(\tfrac{1}{2}m\right)|q\Sigma|^{\frac{1}{2}n}\right\}f_q(\Lambda)\Pi_{i<j}(l_i-l_j)$$

$$\int_{O(m)}\operatorname{etr}\left\{-\tfrac{1}{2}(\Sigma^{-1}-qI)H'\Lambda H\right\}{}_0F_1\left(\tfrac{1}{2}n;\tfrac{1}{2}\Sigma^{-\frac{1}{2}}\Omega\Sigma^{-\frac{1}{2}}H'\Lambda H\right)\mathrm{d}H$$

$$=\left\{\pi^{\frac{1}{2}m^2}/\Gamma_m\left(\tfrac{1}{2}m\right)|q\Sigma|^{\frac{1}{2}n}\right\}f_q(\Lambda)\Pi_{i<j}(l_i-l_j)$$

$$\sum_{\kappa,\lambda;\phi}^{\infty}(-1)^k\theta_\phi^{\kappa,\lambda}C_\phi\left(\tfrac{1}{2}\Lambda\right)C_\phi^{\kappa,\lambda}(\Sigma^{-1}-qI,\Omega\Sigma^{-1})/k!l!\left(\tfrac{1}{2}n\right)_\lambda C_\phi(I). \tag{6.2}$$

The factoring out of $f_q(\Lambda)$ was suggested by Prof. K. C. S. Pillai. The expansion may be numerically useful if qI can be chosen close to Σ^{-1}, and this point is currently being investigated.

References

[1] Boerner, H. (1963). *Representations of Groups.* North-Holland, Amsterdam.

[2] Constantine, A. G. (1963). Some non-central distribution problems in multivariate analysis. *Ann. Math. Statist.* **34**, 1270–1285.

[3] Davis, A. W. (1976). Statistical distributions in univariate and multivariate Edgeworth populations. *Biometrika* **63**, 661–670.

[4] Hannan, E. J. (1965). Group representations and applied probability. *J. Appl. Prob.* **2**, 1–68.

[5] Herz, C. S. (1955). Bessel functions of matrix argument. *Ann. Math.* **61**, 474–523.

[6] Hua, L. K. (1955). On the theory of functions of several complex variables. III. A complete orthonormal system in the hyperbolic space of symmetric and skew symmetric matrices. *Acta Math. Sinica.* **5**, 205–242 (In Chinese). *Amer. Math. Soc. Transl.* (2), **32**, 205–242.

[7] James, A. T. (1960). The distribution of the latent roots of the covariance matrix. *Ann. Math. Statist.* **31**, 151–158.

[8] James, A. T. (1964). Distributions of matrix variates and latent roots derived from normal samples. *Ann. Math. Statist.* **35**, 475–501.

[9] James, A. T. (1968). Calculations of zonal polynomial coefficients by use of the Laplace-Beltrami operator. *Ann. Math. Statist.* **39**, 1711–1718.

[10] Khatri, C. G. and Pillai, K. C. S. (1968). On the noncentral distribution of two test criteria in multivariate analysis of variance. *Ann. Math. Statist.* **39**, 215–226.

[11] Muirhead, R. J. (1970). Systems of partial differential equations for hypergeometric functions of matrix argument. *Ann. Math. Statist.* **41**, 991–1001.

[12] Pillai, K. C. S. (1976). Distributions of characteristic roots. Mimeo Series No. 462, Department of Statistics, Purdue University.

[13] Robinson, G. De B. (1961). *Representation Theory of the Symmetric Group*. Edinburgh University Press, Edinburgh.

[14] Saw, J. G. (1977). Zonal polynomials: an alternative approach. *J. Multivariate Analysis* **7**, 461–467.

P. R. Krishnaiah, ed., *Multivariate Analysis–V*

CALCULATION OF COMPLEX ZONAL POLYNOMIALS*

R. H. FARRELL
Cornell University, Ithaca, NY, U.S.A.

1. Introduction

James [3, p. 487] defined zonal polynomials $\tilde{C}_{(m)}(X)$ of $n\times n$ matrices with complex number entries by the formula

$$\tilde{C}_{(m)}(X)=\chi_{(m)}(1)\chi_{(m)}(X). \tag{1.1}$$

The formula is computable for all $n\times n$ matrices X although James [3] defined $\tilde{C}_{(m)}$ only for Hermitian matrices. In (1.1) the number $\chi_{(m)}(1)$ is the dimension of the representation of the symmetric group on m letters in the ideal of its group algebra determined by the Young's diagram for the partition (m) of m, and $\chi_{(m)}(X)$ is the polynomial character of GL(n) determined by the irreducible representation of signature (m). The basic properties, stated by James, without proof,

$$(\operatorname{tr} X)^m=\Sigma_{(m)}\tilde{C}_{(m)}(X) \tag{1.2}$$

and

$$\tilde{C}_{(m)}(I)\int \tilde{C}_{(m)}(UXU^*Y)\,\mathrm{d}U=\tilde{C}_{(m)}(X)\tilde{C}_{(m)}(Y), \tag{1.3}$$

are contained in formulas (89) and (92) of [3]. In (1.3) U^* is the conjugate transpose of U and integration is with respect to Haar measure of unit mass on the unitary group. One purpose of this paper is to continue the development in Farrell [2, Chapter 12] to more completely establish (1.2) and (1.3) for James' definition (1.1). This discussion is carried out in Section 2 of this paper.

This paper was partly inspired by Saw [8]. For zonal polynomials of real symmetric matrices Saw introduced quadratic forms and was able to establish the identity of a quadratic form in zonal polynomial variables

*Research supported in part by the National Science Foundation under Contract No. MCS 75–22481.

with a quadratic form in s-functions as variables. With additional information about the coefficient matrices this leads to a direct way of computing zonal polynomials as functions of the monomial symmetric functions. The generalization of Saw's idea to the calculation of the characters of GL(n) is discussed in Section 3. It should be noted that whereas the use of quadratic forms in the study of zonal polynomials of (real) symmetric matrices was new, the idea of doing this in the study of group characters is old. We shall have occasion to refer back to the quadratic form in Weyl [9, p. 211, (7.12)].

James lists without reference, apparently meaning Littlewood [6] as his source, four formulas for the primative polynomial characters of GL(n). Two formulas (36) and (37), are readily computable in the sense that the number of terms obtained upon expanding the determinant depends only on the number of non zero parts of the partition (m) of m. A fifth formula, not mentioned by James (Weyl [9, p. 203, (6.5)]), turns out to be the same as James' formula (36) but Weyl expressed his result in a form more readily computable. See Section 4 for further discussion.

In Section 5 tables are given. The discussion of Sections 2–4 introduces four sets of symmetric functions, the primative characters, the s-functions, the monomial symmetric functions, and the variables $h_n = p_n, n \geqslant 1$, in terms of which James and Weyl (and Littlewood) wrote determinants. We have computed for $m=2$, 3, 4, and 5 and the 2, 3, 5, and 7 partitions respectively which correspond, coefficient matrices which interrelate the four sets of symmetric functions.

Expression of the primative group characters of GL(n) in terms of s-functions is almost equivalent to the computation of the values of characters of the symmetric groups. See the end of Section 3. Extensive tables of the characters for the symmetric groups are available in, for example, Littlewood [6].

2. Some theory

Polynomials e homogeneous of degree m in the entries of an $n \times n$ matrix X are uniquely representable in the form

$$e(X) = \operatorname{tr} E_0 \otimes_1^m X \tag{2.1}$$

with E_0 a bi-symmetric matrix and $\otimes_1^m X$ the Kronecker product of X with itself m times.

The following are equivalent. If U is unitary then $e(UXU^*)=e(X)$, for all $X\in\mathrm{GL}(n)$; $e(XY)=e(YX)$ for all X, $Y\in\mathrm{GL}(n)$; the matrix E_0 is in the center of $\mathfrak{A}$, the algebra of bi-symmetric matrices. Note that if $e(UXU^*)=e(X)$ then $e(XU)=e(UX)$ for all X in $\mathrm{GL}(n)$ and U unitary. Since e is a polynomial in the matrix entries, by Weyl [9, p. 177, Lemma (7.1.A)], $e(XY)=e(YX)$ holds for all Y in $\mathrm{GL}(n)$. If X and Y in $\mathrm{GL}(n)$ implies $e(XY)=e(YX)$ then, recalling (2.1), $\mathrm{tr}\, E_0\otimes X\otimes Y=\mathrm{tr}(\otimes Y E_0)\otimes X=\mathrm{tr}(E_0\otimes Y)\otimes X$. Since the matrices in parentheses are bi-symmetric and since this holds for all X in $\mathrm{GL}(n)$ it follows that $E_0\otimes Y==\otimes Y E_0$ for all Y in $\mathrm{GL}(n)$. But the linear span of the matrices $\otimes Y$ is the algebra $\mathfrak{A}$ of bi-symmetric matrices so E_0 is in the center of $\mathfrak{A}$. That E_0 in the center implies unitary invariance is easy.

Functions e with the property $e(XY)=e(YX)$ are called 'central' by Loomis [7, p. 157], in a discussion of the group algebras of compact topological groups. In reading Theorem 2.3 of this section and its proof the reader should compare the discussion with Loomis [7, p. 156–160].

We will first state three Theorems, and finish this Section with the proofs. We have tried to keep notations consistent with those in Farrell [2].

Theorem 2.1. *If $P_{\varepsilon_{(m)}}$ is an idempotent irreducible in the center of $\mathfrak{A}$, the algebra of bi-symmetric matrices, and if $\chi_{(m)}$ is the corresponding primative character of* $\mathrm{GL}(n)$, *then*

$$\mathrm{tr}\, P_{\varepsilon_{(m)}}\otimes_1^m X=\chi_{(m)}(1)\,\chi_{(m)}(X). \tag{2.2}$$

$\chi_{(m)}(1)$ is the dimension of the representation of the symmetric group on m letters in the ideal determined by the Young's diagram for the partition (m) of m. Consequently

$$(\mathrm{tr}\, X)^m=\Sigma_{(m)}\chi_{(m)}(1)\,\chi_{(m)}(X). \tag{2.3}$$

Theorem 2.2. *If E_0 is in the center of $\mathfrak{A}$ then there exists a polynomial f of m variables such that*

$$\mathrm{tr}\, E_0\otimes_1^m X=f(\mathrm{tr}\, X,\mathrm{tr}\, X^2,\ldots,\mathrm{tr}\, X^m). \tag{2.4}$$

Theorem 2.3. *If (m) is a partition of m into not more than n parts, define $D_{(m)}$ by, if $X\in\mathrm{GL}(n)$, then*

$$D_{(m)}(X)=\mathrm{tr}\, P_{\varepsilon_{(m)}}\otimes_1^m X. \tag{2.5}$$

Then, with integration by Haar measure of unit mass on the unitary group,

(i) if $(m)_1 \neq (m)_2$, $\int D_{(m)_1}(U) D_{(m)_2}(XU^{-1}) \mathrm{d}U = 0$;

(ii) if e is a polynomial, homogeneous of degree m, in the variables of X, and if $e(X) = e(UXU^{-1})$ for all X in $\mathrm{GL}(n)$ and $U\, n \times n$ unitary, then there exists a partition (m) of m such that $\int e(U) D_{(m)}(U^{-1}) \mathrm{d}U \neq 0$;

(iii) if $e \neq 0$ is invariant as in (ii) then for all X and Y in $\mathrm{GL}(n)$, $e(XY) = e(YX)$;

(iv) If $\gamma_1 \neq 0$ and if e has the splitting property that $\int e(XUYU^{-1}) = e(X)e(Y)/\gamma_1$ then e is invariant in the sense of (ii) and there exists a number $\gamma_2 \neq 0$ and a partition (m) of m such that $e = \gamma_2 D_{(m)}$;

(v) if e and f are central in the sense of Loomis then the convolution $e * f$ is central (See Loomis [7, p. 157]);

(vi) the convolution of $D_{(m)}$ with itself satisfies $D_{(m)} * D_{(m)} = \omega D_{(m)}$ where $\omega = \| D_{(m)} \|_2^2 / D_{(m)}(I)$.

Discussion of Theorem 2.1. The proof of this theorem takes a considerable proportion of the book by Weyl [9]. We give a brief outline of steps of the argument.

The group representations being considered are maps into the endomorphisms of $M(E^m, C)$, the space of multilinear m-forms with coefficients in the complex numbers. E is an n-dimensional vector space over C. If T^* is a linear transformation of E with adjoint T acting on the dual space of E then $\pi(T)$ is defined by $(\pi(T)f)(x_1, \ldots, x_n) = f(T^*x_1, \ldots, T^*x_m)$. As is shown in [9], the enveloping algebra, the least algebra containing all the endomorphisms $\pi(T)$, $T \in \mathrm{GL}(n)$, is the algebra $\mathfrak{A}$ of bi-symmetric transformations (which can be defined in a coordinate free way or as matrices relative to the cannonical basis of the dual of E). Relative to the cannonical basis of $M(E^m, C)$, we can speak of bi-symmetric matrices, as does Weyl.

As is shown in Weyl [9, Ch. III], an irreducible representation is obtained as follows. Let J be an idempotent that commutes with every transformation in $\mathfrak{A}$. Then the mapping $T \to \pi(T)J$ is an irreducible representation of $\mathrm{GL}(n)$ in the endomorphisms of $J\,M(E^m, C)$ if and only if J is irreducible in the commutator algebra of $\mathfrak{A}$.

The symmetric group on m letters acts on $M(E^m, C)$ by $(\sigma f)(x_1, \ldots, x_m) = f(x_{\sigma(1)}, \ldots, x_{\sigma(m)})$ and the algebra of transformations which envelops the transformations σ, which we will call $\mathfrak{S}$, is a homomorphic image of the group algebra of the symmetric group (if the dimension n of E is less than m then the kernal contains more than 0). Then $\mathfrak{A}$ is the commutator algebra of $\mathfrak{S}$ and $\mathfrak{S}$ is the commutator algebra of $\mathfrak{A}$. See Farrell [2, Section 12.6]. Further, $\mathfrak{A}$ and $\mathfrak{S}$, as H^* algebras (see Loomis [7] for a definition), are completely reducible, so each irreducible idempotent of $\mathfrak{S}$ is contained in an ideal of $\mathfrak{S}$ generated by an idempotent of the center of $\mathfrak{S}$. Since each ideal of $\mathfrak{S}$ is a full matrix algebra, the irreducible idempotents in an ideal are all conjugate and are conjugate to the Young's symmetrizers. See Weyl [9] and Farrell [2, Section 12.4].

The quantity $(\chi_{(m)}(1))^2$ is the dimension of the ideal (which is a full matrix algebra) in $\mathfrak{S}$ determined by the partition (m), so the idempotent of the center $P_{\varepsilon_{(m)}}$ is the sum of $\chi_{(m)}(1)$ irreducible idempotents. Thus the character $\operatorname{tr}\pi(T)\, P_{\varepsilon_{(m)}}$ is the sum of $\chi_{(m)}(1)$ copies of the primative character $\pi(T)J$ with J an irreducible idempotent in the ideal determined by $P_{\varepsilon_{(m)}}$.

Since $\mathfrak{A}$ is completely reducible, the unit of $\mathfrak{A}$, which is the identity matrix, is the sum of the idempotents $P_{\varepsilon_{(m)}}$ taken over the partitions (m) of m into not more than n parts. $\pi(T)$ is an endomorphism of $M(E^m, C)$ which relative to the canonical basis has matrix $\otimes_1^m T$ so that the character $\operatorname{tr}\pi(T)P_{\varepsilon_{(m)}} = \operatorname{tr} P_{\varepsilon_{(m)}} \otimes_1^m T$. Summing these over the partitions (m) gives the last statement of Theorem 2.1.

Proof of Theorem 2.2. The polynomial $e(X) = \operatorname{tr} E_0 \otimes_1^m X$, with E_0 in the center of $\mathfrak{A}$, is unitarily invariant in the sense that if $X \in \mathrm{GL}(n)$ and if U is $n \times n$ unitary then $e(X) = e(UXU^{-1})$. If X is also unitary then by Weyl [9, 179, Theorem (7.1.C)], there exists a unitary matrix U such that UXU^{-1} is diagonal. This implies that if X is unitary $e(X)$ is a symmetric function of the eigenvalues of X and hence there exists a polynomial f of m variables such that $e(X) = f(\operatorname{tr} X, \operatorname{tr} X^2, \ldots, \operatorname{tr} X^m)$. Both sides of this identity are polynomials, which are equal for all unitary X. By Weyl [9, p. 177, Lemma (7.1.A)], equality holds for all X in $\mathrm{GL}(n)$.

Proof of Theorem 2.3. *Part (i).* The functions $D_1 = D_{(m)_1}$ and $D_2 = D_{(m)_2}$ have the splitting property described in (iv). See Farrell [2, Section 12.7]. Then if Y is $n \times n$ unitary and if $X \in \mathrm{GL}(n)$,

$$(D_1(Y)/D_1(I))\int D_1(U)D_2(XU^{-1})\,\mathrm{d}U$$

$$=\int\int D_1(YVUV^{-1})D_2(XU^{-1})\,\mathrm{d}V\,\mathrm{d}U$$

$$=\int\int D_1(U)D_2(XU^{-1}V^{-1}YV)\,\mathrm{d}U\,\mathrm{d}V$$

$$=(D_2(Y)/D_2(I))\int D_1(U)D_2(XU^{-1})\,\mathrm{d}U. \tag{2.6}$$

Since this identity holds for all Y which are $n\times n$ unitary and since D_1 and D_2 are polynomials of matrix entries, by Weyl [9, Lemma 7.1.A], the identity holds for all Y in $\mathrm{GL}(n)$. Since D_1 and D_2 are distinct polynomials, the integral must be zero.

Part (ii). $P_{\varepsilon_{(m)}}$ is a real symmetric matrix. If Y is unitary then

$$D_{(m)}(Y^{-1})=\overline{\operatorname{tr} P_{\varepsilon_{(m)}}\otimes_1^m Y^t}=\overline{\operatorname{tr} P_{\varepsilon_{(m)}}\otimes_1^m Y}=\overline{D_{(m)}(Y)}\,. \tag{2.7}$$

Hence the convolution (integration with respect to unitary matrices)

$$D_{(m)}*D_{(m)}(I)=\|D_{(m)}\|_2^2>0. \tag{2.8}$$

Since the polynomial e is invariant, $e=\Sigma_{(m)}a_{(m)}D_{(m)}$. See Farrell [2, Section 12.7]. By the orthogonality established in Part (i), if $e\neq 0$ and (m) is a partition for which $a_{(m)}\neq 0$, then it follows that $\int e(U)D_{(m)}(U^{-1})\neq 0$.

Part (iii). This was proven in the discussion preceding the statement of Theorem 2.1.

Part (iv). From

$$\gamma_1^{-1}e(X)e(VYV^{-1})=\int e(XU(VYV^{-1})U^{-1})\,\mathrm{d}U=e(X)e(Y)/\gamma_1, \tag{2.9}$$

follows the invariance of the function e. Hence there is a partition (m) of m such that $e*D_{(m)}\neq 0$. Then for unitary matrices X,

$$\gamma_1^{-1}e(X)\int e(Y)D_{(m)}(Y^{-1})\,\mathrm{d}Y$$

$$=\int\int e(XUYU^{-1})D_{(m)}(Y^{-1})\,\mathrm{d}U\,\mathrm{d}Y$$

$$=\int\int e(U^{-1}XUY)D_{(m)}(Y^{-1})\,\mathrm{d}U\,\mathrm{d}Y$$

$$=\int\int e(Y)D_{(m)}(Y^{-1}U^{-1}XU)\,\mathrm{d}U\,\mathrm{d}Y$$

$$=(D_{(m)}(X)/D_{(m)}(I))\int e(Y)D_{(m)}(Y^{-1})\,\mathrm{d}Y. \tag{2.10}$$

Since the integral is not zero we obtain

$$e(X)=\gamma_1(D_{(m)}(I))^{-1}D_{(m)}(X). \tag{2.11}$$

Both sides of (2.11) are polynomials of matrix entries and the identity holds for all $n\times n$ X which are unitary. By Weyl [9, Lemma 7.1.A], equality holds for all X in GL(n).

Part (v). Let e and f be central and X, Y be $n\times n$ unitary matrices. Then

$$(e*f)(XY)=\int e(XYU^{-1})f(U)\,\mathrm{d}U$$

$$=\int e(XU^{-1})f(UY)\,\mathrm{d}U=\int e(XU^{-1})f(YU)\,\mathrm{d}U$$

$$=\int e(XU^{-1}Y)f(U)\,\mathrm{d}U=(e*f)(YX). \tag{2.12}$$

(2.12) holds for all X and Y unitary and since $e*f$ is a polynomial of the variables, by Weyl [9, Lemma 7.1.A], the result follows.

Part (vi). In the proof of part (ii), see (2.8) we have shown $D_{(m)}*D_{(m)}(I)>0$. By part (v) the convolution $D_{(m)}*D_{(m)}$ is an invariant function. Also it is clear that the convolution evaluated $D_{(m)}*D_{(m)}(X)$ is a polynomial homogeneous of degree m in the entries of X. Hence there exist numbers $a_{(m)}$ such that

$$D_{(m)}*D_{(m)}=\Sigma_j a_{(m)_j}D_{(m)_j}. \tag{2.13}$$

By the orthogonality established in Part (i) and the associative law for convolutions, if $(m)\neq(m)_i$ then from (2.13)

$$0=(D_{(m)_i}*D_{(m)})*D_{(m)}=\Sigma_j a_{(m)_j}D_{(m)_i}*D_{(m)_j}=a_{(m)_i}D_{(m)_i}*D_{(m)_i}. \tag{2.14}$$

Hence the coefficient $a_{(m)_i}=0$. But the convolution in (2.13) is not the zero function so there exists a constant $\delta\neq 0$ with

$$D_{(m)}*D_{(m)}(X)=\delta D_{(m)}(X), X\in \mathrm{GL}(n). \tag{2.15}$$

3. Quadratic forms

We will consider in this section complex valued random variables Z for which the real and imaginary parts of Z are independent normal $(0,1)$ random variables. The generating function to be discussed uses $n\times n$ matrices X consisting of n^2 independently and identically distributed components each distributed as Z. We can then write $X=X_1+iX_2$ in which the $n\times n$ matrices X_1 and X_2 are independent random matrices each distributed as a matrix of n^2 mutually independent normal $(0,1)$ random variables. We will refer to X as an $n\times n$ matrix of independent complex normal $(0,1)$ random variables. A direct calculation suffices to verify the following lemma.

Lemma 3.1. *If X is an $n\times n$ matrix of independent complex normal $(0,1)$ random variables and if U is an $n\times n$ unitary matrix, then X and UX have the same distribution, and, X and XU have the same distribution.*

Lemma 3.2. *If $Z=Z_1+iZ_2$, and Z_1 and Z_2 are independent (real) normal $(0,1)$ random variables, then the moment generating function is*

$$Ee^{t|Z|^2}=1/(1-2t). \tag{3.1}$$

The following generating function should be compared with the one obtained by Saw [8].

Theorem 3.3. *Let X be an $n\times n$ matrix of independent complex normal $(0,1)$ random variables and X^* be the conjugate transpose of X. Let A, B be $n\times n$ Hermitian matrices. Suppose the eigenvalues of A are $a_1\geqslant\cdots\geqslant a_n$ and those of B are $b_1\geqslant\cdots\geqslant b_n$.*

If $|\delta|<(a_1b_1)^{-1}$ then

$$E\exp\operatorname{tr}\delta AXBX^*=\sum_{f=0}^{\infty}(2^f\delta^f/f!)\sum_{(f)}n(\pi_1,\ldots,\pi_f)J_{(m)}(A)J_{(m)}(B), \tag{3.2}$$

where π_i is the number of times i occurs in the partition (f) and $n(\pi_1,\ldots,\pi_f) = f!/(1^{\pi_1}2^{\pi_2}\cdots f^{\pi_f}(\pi_1!)\cdots(\pi_f!))$.

Proof. By the unitary invariance of the normal distribution, Lemma 3.1, $E\exp\operatorname{tr}\delta AXBX^* = E\exp\operatorname{tr}\delta(UAU^{-1})X(VBV^{-1})X^*$. Choose U and V to diagonalize A and B with the diagonal elements in descending order. Then, if the ij element of X is X_{ij},

$$\operatorname{tr}(UAU^{-1})X(VBV^{-1})X^* = \sum_{i=1}^{n}\sum_{j=1}^{n} a_i b_j X_{ij}\bar{X}_{ij}. \tag{3.3}$$

Hence for the expectation, using (3.1), we obtain

$$(3.3) = \prod_{i=1}\prod_{j=1}(1-2\delta a_i b_j)^{-1}. \tag{3.4}$$

One should compare this and the following calculations with [9, p. 211]. For $|\delta|<(a_1b_1)^{-1}$ we may take natural logarithms of the right side of (3.4) and expand in series about 1 to obtain

$$\begin{aligned}(3.4) &= \sum_{i=1}^{n}\sum_{j=1}^{n}\sum_{h=1}^{\infty}(2\delta a_i b_j)^h/h \\ &= \sum_{h=1}^{\infty}(2\delta)^h h^{-1}\left(\sum_{i=1}^{n} a_i^h\right)\left(\sum_{i=1}^{n} b_i^h\right) \\ &= \sum_{h=1}^{\infty}(2\delta)^h h^{-1} s_h(A)s_h(B),\end{aligned} \tag{3.5}$$

where we write

$$s_h(A) = \operatorname{tr} A^h, h \geqslant 1. \tag{3.6}$$

Taking the exponential of (3.5) and expanding in a series gives

$$E\exp\operatorname{tr}\delta AXBX^* = \sum_{m=0}^{\infty}(1/m!)\left(\sum_{h=1}^{\infty} 2^h\delta^h h^{-1} s_h(A)s_h(B)\right)^m. \tag{3.7}$$

If (f) is a partition of f into m parts and h occurs in this partition π_h times, by the multinomial theorem, the term

$$\prod_{h=1}^{m} \left(2^h\delta^h h^{-1} s_h(A) s_h(B)\right)^{\pi_h} \tag{3.8}$$

occurs with multinomial coefficient $m!/(\pi_1!\cdots\pi_m!)$. Therefore if we write

$$J_{(f)}(A) = (\operatorname{tr} A)^{\pi_1}(\operatorname{tr} A^2)^{\pi_2}\cdots(\operatorname{tr} A^f)^{\pi_f}, \tag{3.9}$$

and use (3.2) we obtain the conclusion of (3.2).

Using (1.1), (1.2) and (1.3) together with the unitary invariance of the complex normal matrix X, we obtain

$$\begin{aligned} &E \exp \operatorname{tr} \delta AXBX^* \\ &= E \sum_{m=0}^{\infty} (\delta^m/m!)(\operatorname{tr} AXBX^*)^m \\ &= \sum_{m=0}^{\infty} (\delta^m/m!) \sum_{(m)} E\tilde{C}_{(m)}(AXBX^*) \\ &= \sum_{m=0}^{\infty} (\delta^m/m!) \sum_{(m)} \tilde{C}_{(m)}(A)\tilde{C}_{(m)}(B)E\tilde{C}_{(m)}(XX^*)/(\tilde{C}_{(m)}(I))^2. \end{aligned} \tag{3.10}$$

As observed in (1.1) the value of the primative character is

$$\chi_{(m)}(A) = \tilde{C}_{(m)}(A)/\chi_{(m)}(1). \tag{3.11}$$

so that

$$\begin{aligned} &E \exp \operatorname{tr} \delta AXBX^* \\ &= \sum_{m=0}^{\infty} (\delta^m/m!) \sum_{(m)} \chi_{(m)}(A)\chi_{(m)}(B)E\tilde{C}_{(m)}(XX^*)/(\chi_{(m)}(I))^2. \end{aligned} \tag{3.12}$$

Equate (3.2) and (3.12) as series in δ convergent in an interval about $\delta=0$ to obtain

$$\sum_{(f)} 2^f n(\pi_1,\ldots,\pi_f) J_{(f)}(A) J_{(f)}(B)$$

$$= \sum_{(f)} \chi_{(f)}(A)\chi_{(f)}(B) E\tilde{C}_{(f)}(XX^*)/(\chi_{(f)}(I))^2. \qquad (3.13)$$

Compare this with [9, p. 211, (7.12)], and obtain as a result the following lemma.

Lemma 3.4.

$$E\tilde{C}_{(m)}(XX^*) = f!2^f(\chi_{(m)}(I))^2.$$

We restate Weyl's identity in our notation as

$$(1/f!)\sum_{(f)} n(\pi_1,\ldots,\pi_f) J_{(f)}(A) J_{(f)}(B) = \sum_{(f)} \chi_{(f)}(A)\chi_{(f)}(B). \quad (3.14)$$

The number $n(\pi_1,\ldots)$ introduced in (3.2) is the number of permutations of f symbols in the conjugacy class $\pi_1,\ldots$. See [9, p. 209].

That the quadratic forms (3.14) can be used to write the characters $\chi_{(f)}(A)$ in terms of the monomial symmetric functions was observed by Saw [9]. We follow the notation introduced in Farrell [2, Section 12.12]. If two partitions $(f)_1$ and $(f)_2$ of f are given by $f_{11} \geqslant f_{12} \geqslant \cdots \geqslant f_{1n}$ and $f_{21} \geqslant f_{22} \geqslant \cdots \geqslant f_{2n}$ then we say $(f)_1 \geqslant (f)_2$ if and only if for the least integer i such that $f_{1i} \neq f_{2i}$ it holds that $f_{1i} \geqslant f_{2i}$. The monomial symmetric function $M_{(f)}(A)$ of the partition (f) given by $f_1 \geqslant \cdots \geqslant f_n$ is the sum of the distinct terms $a_1^{f_1}\cdots a_n^{f_n}$ taken over all permutations of the subscript indices $a_1,\ldots,a_n$, the eigenvalues of A. Over the complex numbers A has n eigenvalues although the normal form of A may not be diagonal. The implication of Theorem 2.2 is that the values of the characters ignore details of the normal forms other than the eigenvalues. The functions $J_{(f)}$ were defined in (3.9). If the ordered partitions of f into not more than n parts are $(f)_1 \geqslant (f)_2 \geqslant \cdots (f)_r$, where r is the number of partitions, we write

$$\chi^t=(\chi_{(f)_1},\dots,\chi_{(f)_r});$$

$$M^t=(M_{(f)_1},\dots,M_{(f)_r}); \text{ and}$$

$$J^t=(J_{(f)_1},\dots,J_{(f)_r}). \tag{3.15}$$

Lemma 3.5. *There exist $r\times r$ lower triangular matrices* A_0 *and* B_0 *such that*

$$\chi=A_0^tM \text{ and } J=B_0M. \tag{3.16}$$

Remark. That B_0 is lower triangular is a standard result about symmetric functions. That A_0^t is upper triangular follows from the evaluation of $C_{(m)}(X)=\operatorname{tr}P_{\varepsilon(m)}\otimes_1^m X$ for diagonal matrices. See Farrell [2, Section 12.8].

From (3.14), following the statement of Lemma 3.4, if Λ_1 is the diagonal $r\times r$ matrix with $(f),(f)$ entry $(f!)^{-1}n(\pi_1,\dots,\pi_f)$, then

$$J^t(A)\Lambda_1J(B)=\chi^t(A)\chi(B);$$

$$M^t(A)B_0^t\Lambda_1B_0M(B)=M^t(A)A_0A_0^tM(B);$$

$$B_0^t\Lambda_1B_0=A_0A_0^t. \tag{3.17}$$

The last equality follows since the vectors $M(B), B\in\mathrm{GL}(n)$, are a system of rank r.

It was the idea of Saw [8], that if B_0 is known, then from (3.17) A_0, a lower triangular matrix, is uniquely determined and computable easily. By (3.16) this would express the primative characters as linear combinations of the monomial symmetric functions.

Write

$$\chi=C_0J. \tag{3.18}$$

The $r\times r$ matrices C_0 computed for degrees of homogeneity 2, 3, 4, and 5 are given in Table 3, p. 318. From (3.17) it follows that

$$\chi^t\chi=J^tC_0^tC_0J, \quad \text{hence} \quad \Lambda_1=C_0^tC_0. \tag{3.19}$$

Since Λ_1 is diagonal the columns of C_0 are orthogonal and

$$C_0^\tau\chi=C_0^\tau C_0J=\Lambda_1J \quad \text{and} \quad (C_0\Lambda_1^{-1})^t\chi=J. \tag{3.20}$$

From the above, $C_0\Lambda_1^{-\frac{1}{2}}$ is an orthogonal matrix. From Weyl [9, p. 213, (7.16)], we obtain

$$J_{(f)'}(X)=\sum_{(f)}\chi((f),k')\chi_{(f)}(X), \tag{3.21}$$

to be read as follows. If in the partition $(f)'$ i occurs k_i' times then $s_1^{k_1'}\cdots s_f^{k_f'}$ is given by the right side of (3.21) where $\chi((f),k')$ is the value of the character at $(f)'$ of the representation with signature (f) of the symmetric group in its group algebra, and as before $\chi_{(f)}(X)$ is the value at X of the character for GL(n). Thus from (3.20) is obtained

Lemma 3.6. *If $\chi=C_0J$ then $C_0\Lambda_1^{-\frac{1}{2}}$ is an orthogonal matrix (see* (3.17) *for Λ_1) and $(C_0\Lambda_1^{-1})^t$ is the matrix of character values $\chi((f),k)$ with (f) the column index and the class signature k the row index in $(C_0\Lambda_1^{-1})^t$.*

In Table 3, giving some examples of C_0, the row index is the partition (f) and the column index is the class signature $k=(k_1,\ldots,k_f)$ so that the number $c((f),k)(n(k_1,\ldots,k_f)/f!)^{-1}$ is the value of the character. For the partition $f,0,\ldots,0$ the character is identically 1 which shows the first row of C_0 has (f), k entry $n(k_1,\ldots,k_f)/f!$, which are the coefficients of p_f defined in Section 4. The last column of C_0 results from permutations of class $1,1,\ldots,1$ for which $n(1,\ldots,1)=f!$. Hence the last column of C_0 contains the values $\chi((f),1^f)/f!$ and the values $\chi((f),1^f)$ are the dimensions of the representations of the symmetric group on f letters.

We have defined Λ_1 to be the diagonal matrix with entries $n(k_1,\ldots,k_f)/f!$. The matrix introduced by Saw [8] is the matrix Λ with entries

$$2^{f-(k_1+\cdots+k_f)}n(k_1,\ldots,k_f). \tag{3.22}$$

We then have the following summary of the known results.

Summary. The new idea, due to Saw [8], is to index coordinates of vectors with ordered partitions and to introduce matrices to transform these vectors. The vector of the characters χ is given by

$$\chi=C_0J=A_0^tM,\ \text{where}\ J=B_0M. \tag{3.23}$$

Quadratic forms which result are

$$\chi^t\chi = J^t\Lambda_1 J = J^t C_0^t C_0 J = M^t A_0 A_0^t M. \tag{3.24}$$

Hence $C_0\Lambda_1^{-1/2}$ is orthogonal and from (3.23)

$$\Lambda_1^{-1} C_0 \chi = J \tag{3.25}$$

so that $(C_0\Lambda_1^{-1})^t$ is the matrix of values of the characters of the symmetric group. Since

$$M^t A_0 A_0^t M = M^t B_0^t \Lambda_1 B_0 M, \tag{3.26}$$

the lower triangular matrix A_0 is uniquely determined by the lower triangular matrix B_0. Since by (3.24) $C_0^t C_0 = \Lambda_1$, it follows from (3.23) that

$$J = \Lambda_1^{-1} C_0^t A_0^t M = B_0 M; B_0 = \left(A_0 C_0 \Lambda_1^{-1}\right)^t. \tag{3.27}$$

Hence as B_0 determines A_0 it also uniquely determines C_0.

The vector $C^t = (C_{(f)_1}, \ldots, C_{(f)_r})$ of zonal polynomials of real symmetric matrices (see (3.22)) satisfy equations

$$C^t D C = J^t \Lambda J = M^t A_0 C_0 \Lambda_1^{-1} \Lambda \Lambda_1^{-1} C_0^t A_0^t M. \tag{3.28}$$

Since $C = A_1^t M$, where A_1 is a lower triangular $r \times r$ matrix, the identity

$$(3.28) = M^t A_1 D A_1^t M \tag{3.29}$$

uniquely determines A_1 except for proportionalities. The diagonal matrix D has as diagonal entries the numbers $EC_{(f)}(XX^t)/(C_{(f)}(I))^2$.

If $C_0 = EF$, E upper, F lower triangular, then E and F are determined uniquely up to a diagonal matrix. That is, C_0 determines A_0 and B_0 uniquely except for proportionalities given by a diagonal matrix.

The variables $p_n, n \geqslant 1$, are defined in (4.3). Write $P^t = (P_{(f)_1}, \ldots, P_{(f)_r})$ for the symmetric functions of degree f, with $P_{(f)} = \prod_{h=1}^n (P_h)^{\pi_h}$. Then, if $\chi = E_0 P$, as is remarked below, E_0 is a lower triangular matrix. Write $P = F_0^t J$. It can be seen that F_0 is lower triangular. Then $\chi = E_0 F_0^t J = C_0 J$ and $E_0 F_0^t = C_0$ gives the lower, upper triangular decomposition of C_0.

If X in (4.3) is a rank one symmetric matrix with trace equal one then the values of the $p_n, n \geqslant 0$, are $p_n = 1$. Hence in this case P is a vector of all ones and J is a vector of all ones. The matrix F_0^t of the preceding paragraph then has row sums equal one, uniquely determining the normalization of F_0, hence of E_0. In this case the monomial symmetric functions M is the vector $M^t=(1,0,\dots,0)$ so that in $J=B_0M$ the first column of B_0 has as entries one. This uniquely determines the normalization of B_0, hence of A_0.

4. Computational formulas

James [3, p. 481], defines $h_i=h_i(X)$ as the symmetric function which is the sum of all monomial terms of degree i in the eigenvalues of $X, i\geqslant 1$, with $h_0=1$ and if $i<0$ then $h_i=0$. Given a partition of f which is $f_1\geqslant f_2\geqslant\cdots\geqslant f_n$ into not more than n parts, the ij element of a $n\times n$ matrix can be defined to be h_{f_i-i+j} with i the row index and j the column index. The matrix so defined is the transpose of the matrix in (36) of James [3], whose formula for the primative character $\chi_{(f)}$ is

$$\chi_{(f)}(X)=|h_{f_i-i+j}|, \tag{4.1}$$

the determinant of the indicated matrix.

The corresponding formula in Weyl [9, p. 203, (6.5)] is

$$\chi_{(f)}(X)=|p_{f_i-i+j}| \tag{4.2}$$

where the terms $p_i, i\geqslant 0$ are the coefficients of the powers of z in

$$|I-zX|^{-1}=\sum_{h=0}^{\infty} p_h z^h. \tag{4.3}$$

If the eigenvalues of X are $x_1,\dots,x_n$ then the left side of (4.3) is $\prod_{h=1}^{n}(1-zx_h)^{-1}$. An argument similar to that of Section 3 gives

$$|I-zX|^{-1}=\sum_{f=0}^{\infty}(1/f!)z^f\sum_{(f)} n(\pi_1,\dots,\pi_f)J_{(f)}(X), \tag{4.4}$$

with $J_{(f)}(X)$ defined just prior to Lemma 3.5. Hence if $f\geqslant 1$,

$$p_f(X)=(1/f!)\textstyle\sum_{(f)} n(\pi_1,\dots,\pi_f)J_{(f)}(X), \tag{4.5}$$

the sum taken over all partitions of f. Here the coefficients are non-integral.

An alternative is to write, for z near zero,

$$(1-zx_i)^{-1}=\sum_{h=0}^{\infty}(zx_1)^h; \tag{4.6}$$

$$\prod_{i=1}^{n}(1-zx_i)^{-1}=\prod_{i=1}^{n}\left(\sum_{h=0}^{\infty}z^hx_i^m\right)=\sum_{f=0}^{\infty}z^fh_f \tag{4.7}$$

where h_f is the sum of all monomial terms of degree f. In particular it follows that $p_f=h_f, f>0$ and we define $p_0=1$ and $p_i=0$ for $i<0$. Then determinants (4.1) and (4.2) are in fact the same and if evaluated for the partition of f with a single nonzero summand f, one obtains for this partition (f) that $\chi_{(f)}=h_f=p_f$.

If the partition (f) has two or more nonzero summands then evaluation of (4.1) or (4.2) requires the formation of products of the terms p_f. Since the linear forms in the functions $J_{(f)}$ are a semigroup under multiplication, computations are much easier with the functions p_f than with the h_f expressed as sums of monomials.

Table 1 gives expressions for p_1, p_2, p_3, p_4, and p_5 as linear forms in the functions $J_{(f)}$. These linear forms were used for the computations.

Table 2 gives the explicit evaluation of the determinant $|p_{f_i-i+j}|$ for the 17 partitions of 2, 3, 4, and 5. By substitution of the values p_f into the formulas of Table 2 one obtains the primative characters for the 17 indicated partitions in terms of the functions $J_{(f)}$. See Table 3. Note that in each of the four examples, the last column, the coefficients of s_1^2, s_1^3, s_1^4, and s_1^5, where $s_i=(\operatorname{tr} X^i)$, are the values of $\chi_{(m)}(1)$. From Lemma 3.5 the matrix C_0 in Table 3 is equal to $A\mathrm{t}_0\mathrm{B}_0^{-1}$ and if C_0 is multiplied by B_0 one obtains the upper triangular matrix A_0^{τ} which gives the primative characters in terms of the monomial symmetric functions. The matrices B_0 for $f\leqslant 12$ are given in David, Kendall and Barton [1], but note that they use *augmented* monomial symmetric functions.

The result of multiplication of $A_0^{\tau}\mathrm{B}_0^{-1}=C_0$ computed from the determi-

Table 1

$$\begin{aligned}
p_1&=s_1\\
2p_2&=s_2+s_1^2\\
6p_3&=2s_3+3s_2s_1+s_1^3\\
24p_4&=6s_4+8s_3s_1+6s_2s_1^2+3s_2^2+s_1^4.\\
120p_5&=24s_5+30s_4s_1+20s_3s_2+20s_3s_1^2+15s_2^2s_1+10s_2s_1^3+s_1^5
\end{aligned}$$

Table 2
Primative characters in terms of the variables p_n.

	p_2	p_1^2
2	1	
1^2	-1	1

	p_3	p_2p_1	p_1^3
3	1		
21	-1	1	
1^3	1	-2	1

	p_4	p_3p_1	p_2^2	$p_2p_1^2$	p_1^4
4	1				
31	-1	1			
2^2	0	-1	1		
21^2	1	-1	-1	1	
1^4	1	2	1	-3	-1

	p_5	p_4p_1	p_3p_2	$p_3p_1^2$	$p_2^2p_1$	$p_2p_1^3$	p_1^5
5	1						
41	-1	1					
32	0	-1	1				
31^2	1	-1	-1	1			
2^21	0	1	-1	-1	1		
21^3	-1	1	2	-1	-2	1	
1^5	1	-2	-2	3	3	-4	1

nants (4.2) by the matrices B_0 is shown in Table 4. Note, as explained earlier, that the values of $\chi_{(f)}(1)/(f!)$ form the last column of each matrix. The matrices B_0 (with unaugmented monomials) are given in Table 5.

James [4] showed that for zonal polynomials of real symmetric matrices the coefficients giving the polynomials as sums of monomials are nonnegative. The corresponding result for the polynomial characters of GL(n) follows from the discussion in Farrell [2, Section 12.8]. That these coefficients should be integral when unaugmented monomial symmetric functions are used is a result unknown to the author.

Following the idea of Saw [8], the author computed $B_0'\Lambda_1 B_0$ for $m=5$ and took the lower triangular part A_0. The answer indeed was the same as that given in Table 4. Work wise this computation seemed more difficult

Table 3

$f!\chi = FJ$ $f!x$ primative characters in terms of s-functions

		s_2	s_1^2					
$f=2$	2	1	1					
	1^2	−1	1					
		s_3	s_2s_1	s_1^3				
$f=3$	3	2	3	1				
	21	−2	0	2				
	1^3	2	−3	1				
		s_4	s_3s_1	s_2^2	$s_2s_1^3$	s_1^4		
$f=4$	4	6	8	3	6	1		
	31	−6	0	−3	6	3		
	2^2	0	−8	6	0	2		
	21^3	6	0	−3	−6	3		
	1^4	−6	8	3	−6	1		
		s_5	s_4s_1	s_3s_2	$s_3s_1^2$	$s_2^2s_1$	$s_2s_1^3$	s_1^5
$f=5$	5	24	30	20	20	15	10	1
	41	−24	0	−20	20	0	20	4
	32	0	−30	20	−20	15	10	5
	31^2	24	0	0	0	−30	0	6
	2^21	0	30	−20	−20	15	−10	5
	21^3	−24	0	20	20	0	−20	4
	1^5	24	−30	−20	20	15	−10	1

Note: The matrix $C_0 = (1/f!)F$.

than the computation from the determinants when using pencil and paper techniques.

It will be seen on Table 2 that the coefficient matrices have ones on the diagonal and are lower triangular. This was known to Weyl, who named this situation 'a linear substitution of arithmetic recursive type' and who formally stated [9, Theorem 7.6.D p. 205] the fact that the matrices would be lower triangular.

Table 4
$\chi = A_0^t M$

		2	1^2
$f=2$	2	1	1
	1		1

		3	21	1^3
$f=3$	3	1	1	1
	21		1	2
	1^3			1

		4	31	2^2	21^2	1^4
$f=4$	4	1	1	1	1	1
	31		1	1	2	3
	2^2			1	1	2
	21^2				1	3
	1^4					1

		5	41	32	31^2	2^21	21^3	1^5
$f=5$	5	1	1	1	2	2	1	1
	41		1	1	4	4	3	4
	32			1	2	4	3	5
	31^2				2	2	3	6
	2^21					2	2	5
	21^3						1	4
	1^5							1

Table 5
$J = B_0 M$ Matrices B_0

s_2	1						
s_1^2	1	2					
s_3	1						
s_2s_1	1	1					
s_1^3	1	3	6				
s_4	1						
s_3s_1	1	1					
s_2^2	1	0	2				
$s_2s_1^2$	1	2	2	2			
s_1^4	1	4	6	12	24		
s_5	1						
s_4s_1	1	1					
s_3s_2	1	0	1				
$s_3s_1^2$	1	2	1	4			
$s_2^2s_1$	1	1	2	0	4		
$s_2s_1^3$	1	3	4	12	12	6	
s_1^5	1	5	10	40	60	60	120

References

[1] David, F. N. Kendall M. G. and Barton, D. E., (1966). *Symmetric Function and Allied Tables*. Cambridge University Press.

[2] Farrell, R. H. (1976). *Techniques of Multivariate Calculation.* Vol. 520, Lecture Notes in Mathematics, Springer-Verlag, Berlin.

[3] James. A. T. (1964). Distributions of matrix variates and latent roots derived from normal samples. *Ann. Math. Statist*. **35**, 475–501.

[4] James, A. T. (1968). Calculation of zonal polynomial coefficients by use of the Laplace–Beltrami operator. *Ann. Math. Statist.* **39**, 1711–1718.

[5] Helgason, S. (1962). *Differential Geometry and Symmetric Spaces*. Academic Press, New York.

[6] Littlewood, D. E. (1950). *The Theory of Group Characters and Matrix Representations of Groups*. The Clarendon Press, Oxford.

[7] Loomis, L. H. (1953). *An Introduction to Abstract Harmonic Analysis*. D. van Nostrand Company, Inc.

[8] Saw, J. G. (1977). Zonal polynomials: An alternative approach. *J. Multivariate Analysis* **7** 461–467.

[9] Weyl, H. (1946) *Classical Groups*. Princeton University Press.

P. R. Krishnaiah, ed., *Multivariate Analysis–V*

ON A CHARACTERIZATION OF THE GENERALIZED MULTINOMIAL DISTRIBUTIONS

B. GYIRES
Kossuth L. University, Debrecen, Hungary

The author gave the definition and dealt with the properties of the generalized multinomial distributions in his papers [1] and [2] respectively. In this paper the following theorem will be proved:

Finite numbers of independent v-dimensional random vector-variables are generalized multinomial random vector-variables if and only if the sum of these is a generalized multinomial random vector-variable.

This theorem is a generalization of the one of Shanbhag and Basawa concerning the well-known multinomial random vector variables ([5]).

The proof of the theorem is based on a lemma regarding the incomposable property of a characteristic function of a v-dimensional random vector-variable.

1. The theorem

Let $v \geqq 2$ and n be arbitrary positive integers. Let

$$A = \begin{bmatrix} p_{11} & \cdots & p_{1v} \\ \vdots & \cdots & \vdots \\ p_{n1} & \cdots & p_{nv} \end{bmatrix} \tag{1}$$

be a stochastic matrix, i.e. let

$$p_{jk} \geqq 0, \qquad \sum_{k=1}^{v} p_{jk} = 1$$

$$(j = 1,\ldots,n; \quad k = 1,\ldots,v).$$

In the following the matrices obtained by the permutation of the rows of A are regarded as the same matrix, and we denote these also by A. Let S_n be the set of $n \times v$ stochastic matrices.

If $A \in S_n$, $B \in S_m$ then the sum of A and B, denoted by $A+B$, is defined as follows:

$$A+B=\begin{pmatrix} A \\ B \end{pmatrix} \in S_{n+m}.$$

Obviously this operation is commutative, and associative.

If $\mathfrak{M}=(a_{jk})$ is a $n \times n$ matrix with complex numbers as its elements, then the permanent of $\mathfrak{M}$, denoted by Per $\mathfrak{M}$, is defined as follows:

$$Per\, \mathfrak{M} = \sum_{(i_1,\ldots,i_n)} a_{1i_1} \cdots a_{ni_n},$$

where $(i_1,\ldots,i_n)$ runs over the full symmetric group.

Let Γ be the set of vectors $(\beta_1,\ldots,\beta_v)$, where all components are non-negative integers satisfying the condition $\beta_1+\cdots+\beta_v=n$. Let $C_{\beta_1\cdots\beta_v}(A)$ denote the $n \times n$ matrix which consists of certain columns of $A \in S_n$. Namely the k-th column of A appears β_k-times $(k=1,\ldots,v)$ in $C_{\beta_1\cdots\beta_v}(A)$, where $(\beta_1,\ldots,\beta_v)$ runs over Γ.

Let R_v be the v-dimensional real vector space with row vectors as its elements.

The generalized multinomial distributed random vector-variables were defined by the author in the paper [2] as follows:

Definition. The random vector-variable $\eta(A)=(\eta_k(A)) \in R_v$ defined on the probability space $(\Omega, \mathcal{Q}, P)$ is called a *generalized multinomial distributed random vector-variable* generated by the matrix $A \in S_n$, if

$$\mathbf{P}(\eta_k(A)=\beta_k\ (k=1,\ldots,v))$$

$$=\frac{1}{\beta_1!\ldots\beta_v!}\, Per\, C_{\beta_1\cdots\beta_v}(A), \quad (\beta_1,\ldots,\beta_v) \in \Gamma.$$

If all rows of A are equal, then $\eta(A)$ is the well-known multinomial distributed random vector-variable ([4], 31-32).

Let $\varphi_A(t)$, $t=(t_j) \in R_v$ denote the characteristic function of $\eta(A)$, where the matrix $A \in S_n$ is given by (1). The author proved [1, Corollary 1] that

$$\varphi_A(t)=\prod_{j=1}^{n}\left(p_{j1}e^{it_1}+\cdots+p_{jv}e^{it_v}\right). \tag{2}$$

The aim of this paper is to prove the following characteristic property of the generalized multinomial distributed random vector-variables:

Theorem. *Let $\eta_j \in R_v$ $(j=1,\dots,m)$ be independent non-degenerate random vector-variables. Then the relations*

$$\eta_j = \eta(A_j), \qquad A_j \in S_{n_j} \quad (j=1,\dots,m)$$

hold if and only if

$$\eta_1 + \cdots + \eta_m = \eta(A), \quad A \in S_n,$$

where $A_1 + \cdots + A_m = A$, and $n_1 + \cdots + n_m = n$.

In the special case of the multinomial distributed random vector-variables, and if $m=2$, we get the result of Shanbhag and Basawa ([5]) from our theorem.

2. The proof of the theorem

(a) Let t^* be the transpose of $t \in R_v$. Let $\alpha = (\alpha_j) \in R_v$. Then $e^{i\alpha t^*}$, $t \in R_v$ is the characteristic function of a degenerate distribution.

A characteristic function $f(t)$, $t \in R_v$ is said to be decomposable if can be written in the form $f(t) = f_1(t) f_2(t)$, where $f_1(t)$ and $f_2(t)$, $t \in R_v$ are both characteristic function of non-degenerate distribution.

A characteristic function which admits only the product representation in which one of the factors is a characteristic function of a degenerate distribution is called indecomposable.

For the proof of the theorem we need a lemma which we discuss in this section.

Lemma. *Let $p_k \geqq 0$ $(k=1,\dots,v)$, $p_1 + \cdots + p_v = 1$. Then the characteristic function*

$$\varphi(t) = p_1 e^{it_1} + \cdots + p_v e^{it_v}, \quad t = (t_j) \in R_v \tag{3}$$

is indecomposable.

Proof. Namely let

$$\varphi(t) = \varphi_1(t) \varphi_2(t), \quad t \in R_v, \tag{4}$$

where $\varphi_1(t)$ and $\varphi_2(t)$ are characteristic functions. Denote $\varphi(t_j)$, $\varphi_1(t_j)$, $\varphi_2(t_j)$ the characteristic functions are obtained by the substitution of

$$t_1 = \cdots = t_{j-1} = t_{j+1} = \cdots = t_v = 0$$

for $\varphi(t)$, $\varphi_1(t)$ and $\varphi_2(t)$ respectively. On the basis of the expressions (3) and (4) we get

$$\varphi(t_j) = p_j e^{it_j} + 1 - p_j = \varphi_1(t_j)\varphi_2(t_j), \tag{5}$$
$$t_j \in R_1 \quad (j=1,\ldots,v).$$

Since function (5) is a characteristic function of a discrete distribution with two discontinuity points, it is indecomposable [3, 124]. Thus without loss of generality let

$$\varphi_1(t_j) = \begin{cases} (p_j e^{it_j} + 1 - p_j)e^{-i\alpha_j t_j} & (j=1,\ldots,m), \\ e^{i\alpha_j t} & (j=m+1,\ldots,v), \end{cases} \tag{6}$$

$$\varphi_2(t_j) = \begin{cases} e^{i\alpha_j t_j} & (j=1,\ldots,m), \\ (p_j e^{it_j} + 1 - p_j)e^{-i\alpha_j t_j} & (j=m+1,\ldots,v). \end{cases} \tag{7}$$

Denote $\Psi_1(t_1,\ldots,t_m)$ as the characteristic function of the distribution for which the projections of the first m numbers of the axis are determined by the first formula of (6). The second formula of (6) gives the projections on the other axis. These are degenerate characteristic functions and the random variables determined by them are independent from other ones. Therefore

$$\varphi_1(t) = \Psi_1(t_1,\ldots,t_m) \prod_{j=m+1}^{v} e^{i\alpha_j t_j}, \tag{8}$$

and similarly

$$\varphi_2(t) = \Psi_2(t_{m+1},\ldots,t_v) \prod_{j=1}^{m} e^{i\alpha_j t_j}. \tag{9}$$

Here $\Psi_2(t_{m+1},\ldots,t_v)$ is the characteristic function of the distribution for which the projections on the last $v-m$ numbers of the axis are determined by the characteristic function of the second row of (7).

Taking the relation (4) into consideration we have that

$$\varphi(t)=\Psi_1(t_1,\ldots,t_m)\Psi_2(t_{m+1},\ldots,t_v)\prod_{j=1}^{v} e^{i\alpha_j t_j}.$$

If we substitute now first $t_{m+1}=\cdots=t_v=0$, then $t_1=\cdots=t_m=0$, we get

$$\Psi_1(t_1,\ldots,t_m)\prod_{j=1}^{m} e^{i\alpha_j t_j}=1+\sum_{j=1}^{m} p_j(e^{it_j}-1),$$

$$\Psi_2(t_{m+1},\ldots,t_v)\prod_{j=m+1}^{v} e^{i\alpha_j t_j}=1+\sum_{j=m+1}^{v} p_j(e^{it_j}-1).$$

We see on the basis of the expressions (8) and (9) that

$$\varphi_1(t)=\prod_{j=m+1}^{v} e^{i\alpha_j t_j}\prod_{j=1}^{m} e^{-i\alpha_j t_j}\left[1+\sum_{j=1}^{m} p_j(e^{it_j}-1)\right], \tag{10}$$

$$\varphi_2(t)=\prod_{j=1}^{m} e^{i\alpha_j t_j}\prod_{j=m+1}^{v} e^{-i\alpha_j t_j}\left[1+\sum_{j=m+1}^{v} p_j(e^{it_j}-1)\right]. \tag{11}$$

We use now the relation (4) also. Then we get the identity

$$\left[\sum_{j=1}^{m} p_j(e^{it_j}-1)\right]\left[\sum_{j=m+1}^{v} p_j(e^{it_j}-1)\right]\equiv 0, \quad t=(t_j)\in R_v.$$

Here both factors can not be simultaneously identically equal to zero. Otherwise $p_j=0$ $(j=1,\ldots,v)$ is in contradiction to our condition for p_j $(j=1,\ldots,v)$. Thus if the first factor is not identically equal to zero, then the identity

$$\sum_{j=m+1}^{v} p_j(e^{it_j}-1)\equiv 0, \qquad t_j\in R_1 \quad (j=m+1,\ldots,v)$$

holds. From here $p_{m+1}=\cdots=p_v=0$, i.e. $m=v$. Using this fact in the expressions (10) and (11) we conclude that

$$\varphi_1(t)=\varphi(t)e^{-i\alpha t^*}, \qquad \varphi_2(t)=e^{i\alpha t^*}$$

$$\alpha=(\alpha_j)\in R_v, \quad t\in R_v$$

and this is the statement of our lemma.

(b) In this section we give the proof of the theorem.

If the random vector-variables

$$\eta_j = \eta_j(A), \qquad A_j \in S_{n_j} \quad (j=1,\dots,m)$$

are independent, then it follows from (2) that

$$\varphi_\eta(t) = \prod_{j=1}^{m} \varphi_{A_j}(t) = \varphi_A(t),$$

where

$$\eta_1 + \cdots + \eta_m = \eta, \qquad A_1 + \cdots + A_m = A,$$

i.e., $\eta = \eta(A)$, $A \in S_n$.

Let now the matrix $A \in S_n$ be given by (1) and let $n \geqq m$. If η_j $(j=1,\dots,m)$ are independently distributed non-degenerate v-dimensional random vector-variables and if $\eta = \eta(A)$, then

$$\prod_{j=1}^{m} \varphi_{\eta_j}(t) = \varphi_A(t), \quad t \in R_v.$$

Since on the basis of the lemma the factors of $\varphi_A(t)$ in (2) are indecomposable characteristic functions, there exist positive integers n_j $(j=1,\dots,m)$ so that $\varphi_{\eta_j}(t)$ is equal to the product of n_j numbers of the factors of $\varphi_A(t)$. Taking into consideration that the random vector-variables η_j $(j=1,\dots,m)$ are independent, the characteristic functions $\varphi_{\eta_j}(t)$ $(j=1,\dots,m)$ have different factors of $\varphi_A(t)$, and $\sum_{j=1}^{m} n_j = n$. In other words there exist matrices $A_j \in S_{n_j}$ $(j=1,\dots,m)$ so that $\sum_{j=1}^{m} A_j = A$ and $\eta_j = \eta(A_j)(j=1,\dots,m)$. Thus the proof of the theorem is completed.

References

[1] Gyires, B. (1973). Discrete distributions and permanents. *Publ. Math. Debrecen* **20**, 93–106.

[2] Gyires, B. (1977). On the asymptotic behaviour of the generalized multinomial distributions. *Publ. Math. Debrecen* **24**, 163–171.

[3] Lukács, E. (1960). *Characteristic Functions*. Charles Griffin., London.

[4] Lukács, E. and Laha, R. G. (1964). *Applications of Characteristic Functions*. Hafner, New York.

[5] Shanbhag, D. N. and Basawa, J. U. (1971). On a characterization property of the multinomial distribution. *Ann. Math. Stat*. **42**, 2200.

P. R. Krishnaiah, ed., *Multivariate Analysis–V*
© North-Holland Publishing Company (1980) 327–335.

ON THE NON-NULL DISTRIBUTION OF A TEST CRITERION FOR TESTING EQUALITY OF POPULATIONS

A. M. MATHAI and P. N. RATHIE
Department of Mathematics, McGill University, Montreal, Canada
Departamento de Estatistica, Universidade Estadual de Campinas, Sao Paulo, Brasil

Testing equality of normal populations is one of the most important problems in a variety of practical situations. In this paper we consider the non-null distribution of the likelihood ratio test for testing equality of k univariate normal populations. The exact non-null moments are derived by using a simpler technique and the various particular cases are discussed in detail. In one of the particular cases the exact non-null moments are available in terms of Lauricella's function F_D and the exact density is available in a series involving H-functions. The most general case is also given.

1. Introduction

Consider k independent univariate normal populations $N(\mu_i,\sigma_i^2)$, $i=1,\dots,k$. Consider the problem of testing the hypothesis H_0: $\mu_1=\mu_2=\cdots=\mu_k$, $\sigma_1^2=\cdots=\sigma_k^2$, against the negation of H_0. Let there be simple random samples of sizes N_i, $i=1,\dots,k$ from these k populations respectively. The likelihood ratio test criterion for testing H_0 can be easily seen to be the following:

$$\lambda=\left\{N^{N/2}\Big/\prod_{i=1}^{k}N_i^{N_i/2}\right\}\left[\prod_{i=1}^{k}(s_i^2)^{N_i/2}\Big/\left\{\sum_{i=1}^{k}s_i^2+\sum_{i=1}^{k}N_i(\bar{x}_i-\bar{x})^2\right\}^{N/2}\right]$$

where $s_i^2=\Sigma_{j=1}^{N_i}(x_{ij}-\bar{x}_i)^2$, $\bar{x}_i=\Sigma_{j=1}^{N_i}x_{ij}/N_i$, $N=N_1+\cdots+N_k$ and $\bar{x}=\Sigma_{i=1}^{k}N_i\bar{x}_i/N$.

The hth moment of λ, that is $\mathbf{E}(\lambda^h)$, when the null hypothesis is assumed to be true is called the hth null moment and if H_0 is not assumed to be true then it is the hth non-null moment. Here we will consider the exact non-null moments in the most general case and then derive the null moments as particular cases. Since s_i^2/σ_i^2, $i=1,\dots,k$ are independent chi-square variables and $\bar{x}_i$, $i=1,\ldots,k$ are independently normally distributed and since the s_i^2 and $\bar{x}_i$ are mutually independent we get the hth non-null moment by integrating out over the joint densities of these chi-square and normal variables.

The problem of testing equality of mean vectors and equality of covariance matrices was considered by Wilks (1932) and approximations to the distributions of the likelihood ratio criterion were considered by Chang, Krishnaiah and Lee (1977).

2. Moments

The hth non-null moment of λ is given by

$$\mathrm{E}(\lambda^h)=\left\{\frac{N^{N/2}}{\prod_{i=1}^{k} N_i^{N_i/2}}\right\}^h \int_{s_1^2>0}\cdots\int_{s_k^2>0}\int_{\bar{x}_1}\cdots\int_{\bar{x}_k}$$

$$\times\left\{\prod_{i=1}^{k}\frac{(s_i^2)^{N_i/2}}{\sum_{i=1}^{k} s_i^2+\sum_{i=1}^{k} N_i(\bar{x}_i-\bar{x})^2}\right\}^h$$

$$\times\prod_{i=1}^{k}\frac{(s_i^2)^{n_i/2-1}\mathrm{e}^{-s_i^2/(2\sigma_i^2)}}{(2\sigma_i^2)^{n_i/2}\Gamma(n_i/2)}\prod_{i=1}^{k}\frac{N_i^{\frac{1}{2}}\mathrm{e}^{-N_i(\bar{x}_i-\mu_i)^2/(2\sigma_i^2)}}{(2\pi\sigma_i^2)^{\frac{1}{2}}}$$

$$\times\,\mathrm{d}s_1^2\cdots\mathrm{d}s_k^2\,\mathrm{d}\bar{x}_1\cdots\mathrm{d}\bar{x}_k. \tag{2.1}$$

Using,

$$\left\{\sum_{i=1}^{k} s_i^2+\sum_{i=1}^{k} N_i(\bar{x}_i-\bar{x})^2\right\}^{-Nh/2}$$

$$=(\Gamma(Nh/2))^{-1}\int_0^\infty y^{Nh/2-1}\mathrm{e}^{-\left\{\sum_{i=1}^{k} s_i^2+\sum_{i=1}^{k} N_i(\bar{x}_i-\bar{x})2\right\}y}\,\mathrm{d}y \tag{2.2}$$

for $\mathrm{R}(Nh/2)>0$ where $\mathrm{R}(\cdot)$ denotes the real part of $(\cdot)$ and integrating out the terms containing $s_1^2,\ldots,s_k^2$, we have

$$\mathbf{E}(\lambda^h)=(\Gamma(Nh/2))^{-1}\left\{\frac{N^{N/2}}{\prod_{i=1}^{k} N_i^{N_i/2}}\right\}^h$$

$$\times\prod_{i=1}^{k}\frac{\Gamma(N_ih/2+n_i/2)}{\Gamma(n_i/2)(2\sigma_i^2)^{n_i/2}}$$

$$\times\int_0^\infty y^{Nh/2-1}\prod_{i=1}^{k}\left(y+1/2\sigma_i^2\right)^{-(N_ih/2+n_i/2)}$$

$$\times\int_{\bar{x}_1}\cdots\int_{\bar{x}_k}\mathrm{e}^{-\sum_{i=1}^{k}N_i(\bar{x}_i-\bar{x})^2y}\prod_{i=1}^{k}\frac{N_i^{\frac{1}{2}}\,\mathrm{e}^{-N_i(\bar{x}_i-\mu_i)^2/(2\sigma_i^2)}}{(2\pi\sigma_i^2)^{\frac{1}{2}}}\,\mathrm{d}y\,\mathrm{d}\bar{x}_1\cdots\mathrm{d}\bar{x}_k \tag{2.3}$$

where $n_i=N_i-1$ for $i=1,\ldots,k$.

In order to evaluate the integrals involving $\bar{x}_i$, $i=1,\ldots,k$ we use the following simplifications. It is easy to see that

$$\sum_{i=1}^{k}N_i(\bar{x}_i-\bar{x})^2=\bar{X}'A\bar{X} \tag{2.4}$$

where $\bar{X}'=(\bar{x}_1,\ldots,\bar{x}_k)$ and $A=(a_{ij})$ with $a_{ii}=N_i-N^{-1}N_i^2$ and $a_{ij}=-N_iN_j/N$, $i\neq j$.

Let $w_i=(\bar{x}_i-\mu_i)(N_i/2\sigma^2)^{\frac{1}{2}}$. Then $\bar{X}=\mu+DW$, where $\mu=(\mu_1,\ldots,\mu_k)'$, $W=(w_1,\ldots,w_k)'$ and $D=\operatorname{diag}(\beta_1,\ldots,\beta_k)$ with $\beta_i=\sigma_i(2/N_i)^{\frac{1}{2}}$. Clearly $D=D'$. Hence

$$\bar{X}'A\bar{X}=\mu'A\mu+2\mu'ADW+W'DADW \tag{2.5}$$

Also

$$\prod_{i=1}^{k}\mathrm{e}^{-N_i(\bar{x}_i-\mu_i)^2/(2\sigma_i^2)}\left(N_i/2\pi\sigma_i^2\right)^{\frac{1}{2}}\mathrm{d}\bar{x}_1\cdots\mathrm{d}\bar{x}_k=\mathrm{e}^{-W'W}\mathrm{d}W/(\pi)^{k/2} \tag{2.6}$$

The W-integral of (2.3) is

$$\int_W e^{-W'(I+yD'AD)W-2y\mu'ADW}\,dW$$

$$= e^{-y\mu'A\mu+y^2\mu'AD(I+yDAD)^{-1}DA\mu}|I+yDAD|^{-\frac{1}{2}}(\pi)^{k/2} \tag{2.7}$$

Hence (2.3), (2.5) and (2.7) yield

$$E(\lambda^h)=\left\{\frac{\{\Gamma(Nh/2)\}^{-1}N^{N/2}}{\prod_{i=1}^{k} N_i^{N_i/2}}\right\}^h \prod_{i=1}^{k}\frac{\Gamma(N_ih/2+n_i/2)}{(2\sigma_i^2)^{n_i/2}\Gamma(n_i/2)}g(y) \tag{2.8}$$

where

$$g(y)=\int_0^\infty y^{Nh/2-1}\prod_{i=1}^{k}\left(y+(2\sigma_i^2)^{-1}\right)^{-(N_ih/2+n_i/2)}$$

$$\times|I+yDAD|^{-\frac{1}{2}}\exp\{-y\mu'A\mu+y^2\mu'AD(I+yDAD)^{-1}DA\mu\}\,dy$$

(2.9)

3. The non-null case

In this case first we consider a particular case of (2.8) when $\mu_1=\cdots=\mu_k=\mu$ (say) and σ_i's are unspecified. That is, we are looking for the alternatives in a class of independent univariate normal populations where the mean values are assumed to be the same. Towards the end of this section, we will deal with the most general case.

Particular case ($\mu_1=\cdots=\mu_k=\mu$). In this case $\mu'A=0$ and we have

$$|I+yDAD|=(1+\lambda_1y)(1+\lambda_2y)\cdots(1+\lambda_{k-1}y) \tag{3.1}$$

where $\lambda_1,\ldots,\lambda_{k-1}$ are the eigenvalues of DAD. It is easy to see that A is of rank $k-1$. Hence (2.9) reduces to

$$g(y)=\int_0^\infty y^{Nh/2-1}\prod_{i=1}^{k}\left(y+(2\sigma_i^2)^{-1}\right)^{-N_ih/2-n_i/2}\prod_{i=1}^{k-1}(1+\lambda_i y)^{-\frac{1}{2}}\,\mathrm{d}y \tag{3.2}$$

Making the transformation $u=1/(1+y)$, we have

$$g(y)=\frac{\Gamma(\frac{1}{2}Nh)\Gamma(n/2+(k-1)/2)}{\prod_{i=1}^{k-1}\lambda_i^2\Gamma(Nh/2+n/2+(k-1)/2)}$$

$$F_D\Big(\tfrac{1}{2}n+\tfrac{1}{2}(k-1);\tfrac{1}{2}N_1h+\tfrac{1}{2}n_1,\ldots,\tfrac{1}{2}N_kh+\tfrac{1}{2}n_k,\tfrac{1}{2},\ldots,\tfrac{1}{2};$$

$$\tfrac{1}{2}N_1h+\tfrac{1}{2}n+\tfrac{1}{2}(k-1);$$

$$1-\frac{1}{2\sigma_1^2},\ldots,1-\frac{1}{2\sigma_k^2},1-\lambda_1^{-1},\ldots,1-\lambda_{k-1}^{-1}\Big),$$

$$\sigma_i^2>\tfrac{1}{4},\quad i=1,\ldots,k;\quad \lambda_i^2>\tfrac{1}{4},\quad i=1,\ldots,k-1 \tag{3.3}$$

where $F_D(\,)$ is the Lauricella's hypergeometric function defined by (see Appell and Kampé de Fériet (1926) or Mathai and Saxena (1978))

$$F_D(a;b_1,\ldots,b_p;c;x_1,\ldots,x_p)=\{\Gamma(c)/\Gamma(a)\Gamma(c-a)\}\int_0^1 u^{a-1}(1-u)^{c-a-1}$$

$$\times(1-ux_1)^{-b_1}\cdots(1-ux_p)^{-b_p}\,\mathrm{d}u,$$

$$R(a,c-a)>0,\quad |x_i|<1,b_i>0,\quad i=1,\ldots,p \tag{3.4}$$

Thus

$$\mathrm{E}(\lambda^h)=\left\{\frac{N^{N/2}}{\prod_{i=1}^{k} N_i^{N_i/2}}\right\}^h\left\{\prod_{i=1}^{k}\frac{\Gamma\left(\frac{1}{2}N_ih+\frac{1}{2}n_i\right)}{\Gamma\left(\frac{1}{2}n_i\right)(2\sigma_i^2)^{n_i/2}}\right\}$$

$$\times\left\{\Gamma\left(\tfrac{1}{2}n+\tfrac{1}{2}(k-1)\right)\prod_{i=1}^{k-1}\frac{\lambda_i^{-\frac{1}{2}}}{\Gamma\left(\frac{1}{2}Nh+\frac{1}{2}n+\frac{1}{2}(k-1)\right)}\right\}$$

$$\times F_D\Bigg(\tfrac{1}{2}n+\tfrac{1}{2}(k-1);\tfrac{1}{2}N_1h+\tfrac{1}{2}n_1,\ldots,\tfrac{1}{2}N_kh+\tfrac{1}{2}n_k,\tfrac{1}{2},\ldots,\tfrac{1}{2};$$

$$\tfrac{1}{2}Nh+\tfrac{1}{2}n+\tfrac{1}{2}(k-1);1-\frac{1}{2\sigma_1^2},\ldots,1-\frac{1}{2\sigma_k^2},1-\frac{1}{\lambda_1}\ldots,$$

$$1-\frac{1}{\lambda_{k-1}}\Bigg),\qquad \sigma_i^2>\tfrac{1}{4},\quad i=1,\ldots,k;$$

$$\lambda_i^2>\tfrac{1}{4},\quad i=1,\ldots,k-1. \tag{3.5}$$

Note 1. If $\sigma_i^2>\frac{1}{4}, i=1,\ldots,k$ and $\lambda_i^2>\frac{1}{4}, i=1,\ldots,k-1$ are not satisfied, then put $y=\alpha z$ in (3.2) to get

$$g(y)=\alpha^{-n/2}\int_0^{\infty} z^{Nh/2-1}\prod_{i=1}^{k}\left(z+(2\sigma_i^2\alpha)^{-1}\right)^{-N_ih/2-n_i/2}$$

$$\times\prod_{i=1}^{k-1}(1+(\lambda_i\alpha)z)^{-\frac{1}{2}}\,\mathrm{d}z \tag{3.6}$$

Now α can be chosen such that $\alpha>1/(4\sigma_i^2), i=1,\ldots,k$ and $\alpha^2>1/(4\lambda_i^2), i=1,\ldots,k-1$ so that the definition of F_D makes sense.

Verification of the null case. For $\mu_1 = \cdots = \mu_k$,

$$|I+yDAD| = \prod_{i=1}^{k} (2\sigma_i^2)(N_1 \ldots N_k)^{-1}|D^{-2}+yA|$$

$$= \left(\prod_{i=1}^{k} 2\sigma_i^2\right)\Big[(2\sigma_1^2)^{-1}\{y+(2\sigma_2^2)^{-1}\}\cdots\{y+(2\sigma_k^2)^{-1}\}$$

$$+\frac{N_2}{N}y\{-(2\sigma_1^2)^{-1}+(2\sigma_2^2)^{-1}\}\{y+(2\sigma_3^2)^{-1}\}$$

$$\cdots\{y+(2\sigma_k^2)^{-1}\}+\cdots+\frac{N_k}{N}y\{-(2\sigma_1^2)^{-1}$$

$$+(2\sigma_k^2)^{-1}\}\{y+(2\sigma_2^2)^{-1}\}\cdots\{y+(2\sigma_{k-1}^2)^{-1}\}\Big]$$

Hence, when $\sigma_1^2 = \cdots = \sigma_k^2 = \sigma^2$, (3.7) yields $|I+yDAD| = (1+2\sigma^2 y)^{k-1}$.

Thus $|I+yDAD|=0$ implies that the eigenvalues are given by $\lambda_i = 2\sigma^2$, $i=1,\ldots,k-1$. Hence using Mathai and Saxena (1978), we have

$$F_D\left(\tfrac{1}{2}n+\tfrac{1}{2}(k-1);\tfrac{1}{2}N_1h+\tfrac{1}{2}n_1,\ldots,\tfrac{1}{2}N_kh+\tfrac{1}{2}n_k,\tfrac{1}{2},\ldots,\tfrac{1}{2};\right.$$

$$\left.\tfrac{1}{2}Nh+\tfrac{1}{2}n+\tfrac{1}{2}(k-1);1-(2\sigma^2)^{-1},\ldots,1-(2\sigma^2)^{-1}\right)$$

$$= {}_2F_1\left(\tfrac{1}{2}n+\tfrac{1}{2}(k-1),\tfrac{1}{2}Nh+\tfrac{1}{2}n+\tfrac{1}{2}(k-1);\tfrac{1}{2}Nh+\tfrac{1}{2}n\right.$$

$$\left.+\tfrac{1}{2}(k-1);1-(2\sigma^2)^{-1}\right)$$

$$= {}_1F_0\tfrac{1}{2}n+\tfrac{1}{2}(k-1);\ -;1-(2\sigma^2)^{-1} = (2\sigma^2)^{\frac{1}{2}n+\frac{1}{2}(k-1)}$$

Now (3.5) reduces to

$$E(\lambda^h) = \left\{\frac{N^{N/2}}{\prod_{i=1}^{k} N_i^{N_i/2}}\right\}^h \left\{\prod_{i=1}^{k}\frac{\Gamma\left(\tfrac{1}{2}N_ih+\tfrac{1}{2}n_i\right)}{\Gamma\left(\tfrac{1}{2}n_i\right)}\right\}$$

$$\times\left\{\Gamma\left(\tfrac{1}{2}n+\tfrac{1}{2}(k-1)\right)/\Gamma\left(\tfrac{1}{2}Nh+\tfrac{1}{2}n+\tfrac{1}{2}(k-1)\right)\right\} \qquad (3.8)$$

This is the null case for the unequal sample sizes. When the sample sizes are equal this agrees with the result obtained by Neyman and Pearson (1931). The density and distribution function of λ in this case have been given recently by Jain, Rathie and Shah (1975) in terms of G-function as well as in forms suitable for computation.

Density function. The density function of λ for the case $\mu_1 = \cdots = \mu_k = \mu$, denoted by $f(\lambda)$, is given by

$$f(\lambda) = (2\pi i\lambda)^{-1}\int_L \mathbf{E}(\lambda^s)\lambda^{-s}\,ds \tag{3.9}$$

where L is a suitably chosen contour.

Using (3.9), (3.5) and the following definition of $F_D(\ )$ (see Mathai and Saxena (1978))

$$F_D(a; b_1,\ldots,b_n; c; x_1,\ldots,x_n)$$

$$= \sum_{m_1=0}^{\infty}\cdots\sum_{m_n=0}^{\infty}\left\{\frac{(a)_{m_1+\cdots+m_n}(b_1)_{m_1}\cdots(b_n)_{m_n}x_1^{m_1}\cdots x_n^{m_n}}{(c)_{m_1+\cdots+m_n}m_1!\cdots m_n!}\right\},$$

$$|x_i|<1, \quad i=1,\ldots,n,$$

we can represent the density in series involving H-functions. In the most general case we simplify (2.9) as follows: Since DAD is symmetric we can always find an orthogonal matrix Q such that $Q'Q=I$ and $QDADQ' = \Lambda = \operatorname{diag}(\lambda_1,\ldots,\lambda_{k-1})$. Then $y\mu'A = \operatorname{tr} y\Lambda C = y(\lambda_1 C_{11} + \lambda_2 C_{22} + \cdots + \lambda_{k-1}C_{k-1\,k-1})$ where $C = QD^{-1}\mu\mu' D^{-1}Q'$ with the leading diagonal elements $C_{11},\ldots,C_{k-1\,k-1}$. In this case the hth moment reduces to the following form.

$$\mathbf{E}(\lambda^h) = \left\{\frac{N^{N/2}}{\prod_{i=1}^{k} N_i^{N_i/2}}\right\}^h \frac{\prod_{i=1}^{k}\Gamma\left(\frac{1}{2}N_i h + \frac{1}{2}n_i\right)}{(2\sigma_i^2)^{\frac{1}{2}n_i}\Gamma\left(\frac{1}{2}n_i\right)}$$

$$\times\left\{\frac{\Gamma\left(\frac{1}{2}Nh + r\right)\Gamma\left(\frac{1}{2}n + \frac{1}{2}(k-1)\right)}{\Gamma\left(\frac{1}{2}Nh + r + \frac{1}{2}n + \frac{1}{2}(k-1)\right)}\right\}\left\{\frac{\prod_{i=1}^{k-1}\lambda_i^{-\frac{1}{2}}}{\Gamma\left(\frac{1}{2}Nh\right)}\right\}$$

$$\times \sum_{r=0}^{\infty} \sum_{R} \Big\{ (-1)^r C_{11}^{r_1} \cdots C_{k-1k-1}^{r_{k-1}} F_D\Big(\tfrac{1}{2}n + \tfrac{1}{2}(k-1);$$

$$\tfrac{1}{2}N_1 h + \tfrac{1}{2}n_1, \ldots, \tfrac{1}{2}N_k h + \tfrac{1}{2}n_k, r_1 + \tfrac{1}{2}, \ldots, r_{k-1} + \tfrac{1}{2};$$

$$\tfrac{1}{2}Nh + r + \tfrac{1}{2}n + \tfrac{1}{2}(k-1); 1 - (2\sigma_1^2)^{-1}, \ldots, 1 - (2\sigma_k^2)^{-1},$$

$$1 - \lambda_1^{-1}, \ldots, 1 - \lambda_{k-1}^{-1}\Big) / r_1! \cdots r_{k-1}! \Big\}$$

for $\sigma_i^2 > \frac{1}{4}, i = 1, \ldots, k; \lambda_i^2 > \frac{1}{4}, i = 1, \ldots, k-1,$

where $R = (r_1, \ldots, r_{k-1}), r_1 + \cdots + r_{k-1} = r.$

The density function can be obtained by following the same technique of the particular case mentioned earlier in this section.

References

[1] Appell, P. and Kampé de Fériet, M. J. (1926). *Fonctions Hypergéométriques et Hypersphériques, Polynomes d'Hermite*. Gauthier-Villars.

[2] Chang, T. C., Krishnaiah, P. R. and Lee, J. C. (1977). Approximations to the distributions of the likelihood ratio statistics for testing the hypotheses on covariance matrices and mean vectors simultaneously. In *Applications of Statistics* (P. R. Krishnaiah, editor). North-Holland Publishing Company.

[3] Jain, S. K., Rathie, P. N. and Shah, M. C. (1975). The exact distributions of certain likelihood ratio criteria. *Sankhya, Ser. A* **37**, 150–163.

[4] Mathai, A. M. (1970). Exact distribution of a criterion for testing the hypothesis that several multivariate populations are identical. *J. Indian Statist. Assoc.* **8**, 1–17.

[5] Mathai, A. M. and Saxena, R. K. (1978). *The H-function With Applications in Statistics and Other Disciplines*. Wiley Halsted, New York and Wiley Eastern, New Delhi.

[6] Neyman, J. and Pearson, E. S. (1931). On the problem of k-samples. *Bull. Acad. Pol. Sci.* **3**, 460–481.

[7] Wilks, S. S. (1932). Certain generalizations in the analysis of variance *Biometrika* **24**, 471–494.

P. R. Krishnaiah, ed., *Multivariate Analysis–V*
© North-Holland Publishing Company (1980) 337–347.

ASYMPTOTIC EXPANSIONS FOR HYPERGEOMETRIC FUNCTIONS*

M. S. SRIVASTAVA
Indian Statistical Institute and University of Toronto

and

E. M. CARTER
University of Guelph

Asymptotic expansions for many hypergeometric functions are derived with remainder terms of order $O(n^{-2})$. These expansions are useful in deriving the asymptotic distribution of test statistics used in MANOVA and canonical correlation analysis.

1. Introduction

In many testing problems in multivariate analysis the exact distribution of test statistics under the alternate hypothesis involve hypergeometric functions of matrix arguments. (See e.g., James (1964) and Constantine (1963).) For example if F is a non-central F matrix with degrees of freedom n_1 and n_2, and noncentrality matrix n_2A, then the distribution of $f_1 > \cdots > f_p$, the characteristic roots of F, is given by

$$\pi^{\frac{1}{2}p^2}\Gamma_p^{-1}\left(\tfrac{1}{2}n_1\right)\Gamma_p^{-1}\left(\tfrac{1}{2}n_2\right)\Gamma_p\left(\tfrac{1}{2}(n_1+n_2)\right)\Gamma_p^{-1}\left(\tfrac{1}{2}p\right)$$

$$\prod_{i<j}(f_i-f_j)\prod_{i=1}^{p}\left[f_i^{\frac{1}{2}(n_1-p-1)}(1+f_i)^{-\frac{1}{2}(n_1+n_2)}\right]$$

$$ {}_1F_1\left(\tfrac{1}{2}(n_1+n_2)\right);\ \tfrac{1}{2}n_1:\ \tfrac{1}{2}n_2A,\ D_f(I+D_f)^{-1} \tag{1}$$

where $\Gamma_p(b)=\pi^{\frac{1}{4}p(p-1)}\prod_{i=1}^{p}\Gamma(b-\frac{1}{2}(i-1))$ and $D_f=\operatorname{diag}(f_1,\ldots,f_p)$.

The choice of the non-centrality matrix as n_2A arises from assuming a fixed alternative in testing problems. This and many other similar problems lead us to investigate the asymptotic behaviour of the hypergeometric

*Supported by the National Research Council of Canada.

functions. In this paper we give the asymptotic expansion of

$$ {}_0F_0\left(\tfrac{1}{2}nA,B\right), \qquad {}_0F_1\left(\tfrac{1}{2}l;\tfrac{1}{2}nA,\tfrac{1}{2}nB\right), $$

$$ {}_1F_0\left(\tfrac{1}{2}n;-A,B\right), \qquad {}_1F_1\left(\tfrac{1}{2}n;\tfrac{1}{2}l:\tfrac{1}{2}nA,B\right), $$

$$ {}_2F_1\left(\tfrac{1}{2}n,\tfrac{1}{2}n;\tfrac{1}{2}l:A,B\right). $$

The procedure given here can be applied in many other situations such as in ${}_1F_1(\frac{1}{2}(n_1+n_2);\frac{1}{2}n_1;\frac{1}{2}n_2A,B)$; here we can let n_1 and n_2 both go to infinity such that $n_1/n_2\to\lambda$. Also the case of ${}_0F_1(\frac{1}{2}n:\frac{1}{2}nA,\frac{1}{2}nB)$ can be obtained (See Carter and Srivastava (1977).)

2. Expansion of hypergeometric functions

In this section we give the asymptotic expansion of certain hypergeometric functions. For convenience we adopt the following notation: Let $\Gamma=(\Gamma_1,\Gamma_2)$ be a $p\times p$ orthogonal matrix with Γ_1 a $p\times q$ semi-orthogonal matrix, and let $\mathrm{d}\Gamma$ be the non-normalized invariant measure with $\int \mathrm{d}\Gamma=\pi^{\frac{1}{2}p^2}\Gamma_p^{-1}(\frac{1}{2}p)$. Then

$$ \mathrm{d}\Gamma_1=\int_D \mathrm{d}\Gamma, \quad \text{where} \quad D=\{\Gamma_2:\Gamma_2\Gamma_2'=I-\Gamma_1\Gamma_1'\}. $$

$\delta\Gamma$ will be used to denote the normalized invariant measure. We also define $\alpha_p(D_b)=\Pi_{1\leqslant i<j\leqslant p}(b_i-b_j)$ and $D_b=\mathrm{diag}(b_1,\ldots,b_p)$.

Although some of the following results were derived by various authors, these results may be obtained by the procedures given in the proof of Theorem 3.

2.1. Expansion of ${}_0F_0$, where

$$ {}_0F_0\left(\tfrac{1}{2}nA,B\right)=\int_{HH'=I}\left(\mathrm{etr}\,\tfrac{1}{2}nAHBH'\right)\delta H $$

Theorem 1. *Let* $B=\mathrm{diag}(b_1,\ldots,b_p)$ $b_1>\cdots>b_p$ *and let* $A=\mathrm{diag}(a_1,\ldots,a_p)$, $a_1>\cdots>a_q>a_{q+1}=\cdots a_p=0$.

Then

$$
{}_0F_0\left(\tfrac{1}{2}nA,B\right)=(2\pi/n)^{\frac{1}{4}q(q-1)+\frac{1}{2}q(p-q)}\pi^{-\frac{1}{2}pq}\Gamma_q\left(\tfrac{1}{2}p\right)\left[\alpha_q(D_b)\alpha_q(D_a)\right]^{-\frac{1}{2}}
$$

$$
\times\prod_{i=1}^{q}\prod_{j=q+1}^{p}\left[(b_i-b_j)a_i\right]^{-\frac{1}{2}}\exp\left[\tfrac{1}{2}n\sum_{i=1}^{q}a_ib_i\right]
$$

$$
\times\left\{1+n^{-1}\left[\sum_{1\leqslant i<j\leqslant q}c_{ij}+\sum_{i=1}^{q}\sum_{j=q+1}^{p}d_{ij}\right]+\mathrm{O}(n^{-2})\right\},
$$

where $c_{ij}=(a_i-a_j)^{-1}(b_i-b_j)^{-1}$ *and* $d_{ij}=(b_i-b_j)^{-1}a_i^{-1}$.

Usually the case of $a_{q+1}=\cdots=a_p=0$ is of little interest. However, the result is used in later theorems.

Proof. See James (1969), Khatri and Srivastava (1978) or Chikuse (1976), Chattopadhyay and Pillai (1973).

2.2. Asympototic expansion of ${}_1F_0$, *where*

$$
{}_1F_0\left(\tfrac{1}{2}n;-A,B\right)=\int_{HH'=I}|I+AHBH'|^{-\frac{1}{2}n}\delta H
$$

Theorem 2. *Let* $A=\mathrm{diag}(a_1,\ldots,a_p)$ *and* $B=\mathrm{diag}(b_1,\ldots,b_p)$ *with* $a_1>\cdots>a_q>a_{q+1}=\cdots=a_p=a\geqslant 0$ *and* $0<b_1<\cdots<b_p$. *Then for large n*

$$
{}_1F_0\left(\tfrac{1}{2}n;-A,B\right)=(2\pi/n)^{\frac{1}{4}q(q-1)+\frac{1}{2}q(p-q)}\prod_{1\leqslant i<j\leqslant p}(b_j-b_i)^{-\frac{1}{2}}
$$

$$
\times\prod_{i=1}^{q}\left[(1+a_ib_i)^{-\frac{1}{2}(n-p+1)}(a_i-a)^{\frac{1}{2}(p-q)}\right]
$$

$$
\times\prod_{i=1}^{q-1}\prod_{j=i+1}^{q}\left[(a_i-a_j)^{-\frac{1}{2}}\right]\prod_{j=q+1}^{p}\left[(1+ab_j)^{\frac{1}{2}(p-q-1)}\right]
$$

$$
\times\left\{1+n^{-1}\left[\frac{1}{2}\sum_{1\leqslant i<j\leqslant q}c_{ij}+\frac{1}{2}\sum_{j=q+1}^{p}\sum_{i=1}^{q}c_{ij}\right.\right.
$$

$$
\left.\left.+\frac{1}{24}q(q-1)(4q+1)+3(p^2-q^2)\right]+\mathrm{O}(n^{-2})\right\},
$$

where $c_{ij}=(1+a_ib_i)(1+a_jb_j)(a_i-a_j)(b_j-b_i)$ *for* $1\leqslant i<j\leqslant q$ *and* $c_{ij}=(1+a_ib_i)(1+ab_j)(a_i-a)(b_j-b_i)$ *for* $i=1,\ldots,q;\ j=q+1,\ldots,p$.

Proof. The proof is given in Khatri and Srivastava (1978), Chang (1970), Chattopadhyay and Pillai (1973), Chattopadhyay, Pillai and Li (1976), Sugiura (1976) and Constantine and Muirhead (1976), Chang (1973), Li, Pillai and Chang (1970), Muirhead (1978).

2.3. Asymptotic expansion of ${}_0F_1$, *where*

$$ {}_0F_1\left(\tfrac{1}{2}l;\tfrac{1}{2}nA,\tfrac{1}{2}nB\right)=\int \operatorname{etr}\left(\tfrac{1}{2}nA\Gamma_1[B:0]\Gamma_2'\right)\delta\Gamma_1\delta\Gamma_2. $$

We shall consider the case when the last $p-q$ roots of A are zero, that is A is of the form

$$ A=\begin{pmatrix} A_1 & 0 \\ 0 & 0 \end{pmatrix}, $$

where $A_1=\operatorname{diag}(a_1,\ldots,a_q)$. Then we have the following Theorem.

Theorem 3. *Let* $A_1=\operatorname{diag}(a_1,\ldots,a_q)$, $A=\begin{pmatrix} A_1 & 0 \\ 0 & 0 \end{pmatrix}$ *and* $B=\operatorname{diag}(b_1,\ldots,b_p)$ *with* $a_1>\cdots>a_q>0$ *and* $b_1>\cdots>b_p>0$. *Then for* $l\geqslant p>q$ *and* l *integral*

$$ {}_0F_1\left(\tfrac{1}{2}l:\tfrac{1}{2}nA^2,\tfrac{1}{2}nB^2\right)= $$

$$ \pi^{-\frac{1}{2}q(p+l)}\Gamma_q\left(\tfrac{1}{2}p\right)\Gamma_q\left(\tfrac{1}{2}l\right)(2\pi/n)^{\frac{1}{2}q(q-1)+\frac{1}{2}q(l+p-2q)} $$

$$ \times\left(\exp n\sum_{i=1}^{q}a_ib_i\right)\prod_{i=1}^{q}a_i^{-\frac{1}{2}(p-q)}\left[\alpha_q(D_a^2)\alpha_q(D_b^2)\right]^{-\frac{1}{2}} $$

$$ \times\prod_{i=1}^{q}\left[a_i^{-\frac{1}{2}(l-q)}b_i^{-\frac{1}{2}(l-p)}\right]\prod_{i=1}^{q}\prod_{j=q+1}\left(b_i^2-b_j^2\right)^{-\frac{1}{2}} $$

$$ \times\left\{1+n^{-1}\left[\frac{1}{2}\sum_{1\leqslant i<j\leqslant q}c_{ij}+\frac{1}{2}\sum_{j=q+1}^{p}d_{ij}\right.\right. $$

$$ \left.\left.-\frac{1}{8}(l-p)(l-p)(l-p-2)\sum_{i=1}^{q}(a_ib_i)^{-1}\right]+O(n^{-2})\right\}, $$

where $c_{ij} = (a_i b_i + a_j b_j)(a_i - a_j)^{-1}(b_i - b_j)^{-1}$ *for* $i \neq j$; $i,j = 1,\ldots,q$ *and* $d_{ij} = b_i a_i^{-1}(b_i^2 - b_j^2)^{-1}$ *for* $i = 1,\ldots,q$; $j = q+1,\ldots,p$.

Proof. The proof follows the procedure given in Carter and Srivastava (1977) and Khatri and Srivastava (1978). That is, we express ${}_0F_1(\frac{1}{2}l; \frac{1}{2}nA^2, \frac{1}{2}nB^2)$ as

$$\pi^{-\frac{1}{2}q(p+1)}\Gamma_q\left(\tfrac{1}{2}p\right)\Gamma_q\left(\tfrac{1}{2}l\right)\int \operatorname{etr} nA\Gamma_1\begin{pmatrix} B \\ 0 \end{pmatrix}\Gamma_2' \, d\Gamma_1 \, d\Gamma_2 \tag{5}$$

where Γ_1 and Γ_2 are $q \times l$ and $q \times p$ semi-orthogonal matrices. We then make the transformation

$$\begin{aligned}\Gamma_1 = \Big[& -I + 2\left(I - Q_{11}(4n)^{-\frac{1}{2}} + Q_{12}Q_{12}'(4n)^{-1} + Q_{13}Q_{13}'(4n)^{-1}\right)^{-1}, \\ & \times\left(I - Q_{11}(4n)^{-\frac{1}{2}} + Q_{12}Q_{12}'(4n)^{-1} + Q_{13}Q_{13}'(4n)^{-1}\right)^{-1} Q_{12} n^{-\frac{1}{2}}, \\ & \times\left(I - Q_{11}(4n)^{-\frac{1}{2}} + Q_{12}Q_{12}'(4n)^{-1} + Q_{13}Q_{13}'(4n)^{-1}\right)^{-1} Q_{13} n^{-\frac{1}{2}} \Big]\end{aligned}$$

and

$$\begin{aligned}\Gamma_2 = \Big[& -I + 2\left(I - P_{11}(4n)^{-\frac{1}{2}} + P_{12}P_{12}'(4n)^{-1}\right)^{-1}, \\ & \times\left(I - P_{11}(4n)^{-\frac{1}{2}} + P_{12}P_{12}'(4n)^{-1}\right)^{-1} P_{12} n^{-\frac{1}{2}} \Big]\end{aligned}$$

where Q and P are $q \times q$ skew symmetric matrices, Q_{12} and P_{12} are $q \times (p-q)$ matrices, and Q_{13} is a $p \times (l-p)$ matrix. The Jacobian of the transformation is given by

$$\begin{aligned} & J(\Gamma_1, \Gamma_2 \to Q, Q_{12}, Q_{13}, P_{11}, P_{12}) = \\ & |I - P_{11}(4n)^{-\frac{1}{2}} + P_{12}P_{12}'(4n)^{-1}|^{-(p-1)} \\ & \times |I - Q_{11}(4n)^{-\frac{1}{2}} + Q_{12}Q_{12}'(4n)^{-1} + Q_{13}Q_{13}'(4n)^{-1}|^{-(l-1)}.\end{aligned}$$

Asymptotically as $n\to\infty$ we have $(q_{ij},p_{ij})'\sim N_2(\mathbf{0},\Sigma_{ij})$ where

$$\Sigma_{ij}=(a_i^2-a_j^2)^{-1}(b_i^2-b_j^2)^{-1}\begin{pmatrix} a_ib_i+a_jb_j & a_ib_j+a_jb_i \\ a_ib_j+a_jb_i & a_ib_i+a_jb_j \end{pmatrix}$$

$$1\leqslant i<j\leqslant q$$

and

$$\Sigma_{ij}=a_i^{-1}(b_i^2-b_j^2)^{-1}\begin{pmatrix} b_i & b_j \\ b_j & b_i \end{pmatrix}$$

$$i=1,\ldots,q,\quad j=q+1,\ldots,p.$$

Also $q_{ij}\sim N(\mathbf{0},(a_ib_i)^{-1})$ $i=1,\ldots,q, j=p+1,\ldots,l$. Using the above densities we expand (5) in powers of n^{-1} and take the expectation of terms involving q_{ij} and p_{ij}.

2.4. Expansion of ${}_1F_1(\frac{1}{2}m;\frac{1}{2}l;\frac{1}{2}nA,B)$

Theorem 4. *Let*

$$A=\begin{pmatrix} A_1 & 0 \\ 0 & 0 \end{pmatrix},$$

where $A_1=\operatorname{diag}(a_1,\ldots,a_q)$, $a_1>\cdots>a_q>0$, *and* $B=\operatorname{diag}(b_1,\ldots,b_p)$, $b_1>\cdots>b_p>0$. *Then as* $m\to\infty$, $n\to\infty$ *such that* $m/n\to\lambda$,

$$\begin{aligned}
{}_1F_1\left(\tfrac{1}{2}m;\tfrac{1}{2}l:\tfrac{1}{2}nA,B\right)=&\Gamma_q^{-1}\left(\tfrac{1}{2}m\right)\left(\tfrac{1}{2}n\right)^{\frac{1}{2}q\left[m-p-l+\frac{1}{2}(q+1)\right]}\\
&\times\Gamma_q\left(\tfrac{1}{2}l\right)\Gamma_q\left(\tfrac{1}{2}p\right)\pi^{-\frac{1}{4}q(q-2l+3)}\left[\alpha_q(D_a)\alpha_q(D_b)\right]^{-\frac{1}{2}}\\
&\times\prod_{i=1}^{q}a_i^{-\frac{1}{2}(m-q+1)}\gamma_i^{\frac{1}{2}m-\frac{1}{4}(l+p-2)}\\
&\times\prod_{i=1}^{p}b_i^{-\frac{1}{4}(l-p)}\prod_{i=1}^{q}\prod_{i=q+1}^{p}(b_i-b_j)^{-\frac{1}{2}}\\
&\times\prod_{i=1}^{q}(8\eta_i)^{\frac{1}{2}}\left[\exp-\tfrac{1}{2}n\sum_{i=1}^{q}\gamma_ia_i^{-1}+n\sum_{i=1}^{q}(b_i\gamma_i)^{\frac{1}{2}}\right]\\
&\left\{1+n^{-1}\left[\sum_{i=1}^{q}d_i+\tfrac{1}{2}\sum_{1\leqslant i<j\leqslant q}d_{ij}+\tfrac{1}{2}\sum_{1\leqslant i<j\leqslant q}c_{ij}\right.\right.\\
&\left.\left.+\sum_{i=1}^{q}\sum_{j=q+1}^{p}c_{ij}-\frac{1}{8}(l-p)(l-p-2)\sum_{i=1}^{q}(b_i\gamma_i)^{-\frac{1}{2}}+\mathrm{O}(n^{-2})\right]\right\}
\end{aligned}$$

where

$$\delta_i^2 = 1 + 4\lambda(a_i b_i)^{-1}, \gamma_i^{\frac{1}{2}} = \tfrac{1}{2}[1+\delta_i]\left(a_i b_i^{\frac{1}{2}}\right),$$

$$\eta_i^{-1} = a_i b_i \delta_i (1+\delta_i),$$

$$d_i = \tfrac{1}{4}(l+p)^2 \eta_i - 2(l+p)\eta_i^2 + \tfrac{1}{2}(l+p)\eta_i + \frac{20}{3}\eta_i^3 - 3\eta_i^2$$

$$d_{ij} = a_i a_j \left(\gamma_i^{\frac{1}{2}} - \gamma_j^{\frac{1}{2}}\right)^{-1} \left(b_i^{\frac{1}{2}} - b_j^{\frac{1}{2}}\right)^{-1}$$

$$c_{ij} = \left[(b_i\gamma_i)^{\frac{1}{2}} + (b_j\gamma_j)^{\frac{1}{2}}\right]\left(b_i^{\frac{1}{2}} - b_j^{\frac{1}{2}}\right)^{-1}\left(\gamma_i^{\frac{1}{2}} - \gamma_j^{\frac{1}{2}}\right)^{-1} \qquad 1 \leqslant i < j \leqslant q$$

and

$$c_{ij} = b_i^{\frac{1}{2}} \gamma_i^{-\frac{1}{2}} (b_i - b_j)^{-1}, \qquad i = 1, \ldots, q, \quad j = q+1, \ldots, p.$$

Proof. First we use the integral representation

$$ {}_1F_1\left(\tfrac{1}{2}m; \tfrac{1}{2}l : \tfrac{1}{2}nA, B\right) = \int_{S_1>0} \prod_{i=1}^{q} a_i^{-\frac{1}{2}m} \Gamma_q^{-1}\left(\tfrac{1}{2}m\right)\left(\tfrac{1}{2}n\right)^{\frac{1}{2}mq} |S_1|^{\frac{1}{2}(m-q-1)}$$
$$\times \operatorname{etr}\left(-\tfrac{1}{2}nS_1A_1^{-1}\right) {}_0F_1\left(\tfrac{1}{2}l : \tfrac{1}{2}nS, \tfrac{1}{2}nB\right) \mathrm{d}S_1,$$

where

$$S = \begin{pmatrix} S_1 & 0 \\ 0 & 0 \end{pmatrix}.$$

Then we make the transformation $S_1 \to \Gamma D_{S_1} \Gamma'$ where Γ is a $q \times q$ orthogonal matrix and $D_{S_1} = \operatorname{diag}(s_1, \ldots, s_q)$ $s_1 > \cdots > s_q > 0$. We now have

$$ {}_1F_1\left(\tfrac{1}{2}m; \tfrac{1}{2}l : \tfrac{1}{2}nA, B\right) = \int_{s_i>0} \prod_{i=1}^{q} a_i^{-\frac{1}{2}m} \Gamma_q^{-1}\left(\tfrac{1}{2}m\right)\left(\tfrac{1}{2}n\right)^{\frac{1}{2}mq} \Gamma_q^{-1}\left(\tfrac{1}{2}q\right)\pi^{\frac{1}{2}q^2}$$
$$\times \prod_{1 \leqslant i < j \leqslant q} (s_i - s_j) \prod_{i=1}^{q} s_i^{\frac{1}{2}(m-q-1)}$$
$$\times {}_0F_0\left(-\tfrac{1}{2}nA_1^{-1}, D_{S_1}\right) {}_0F_1\left(\tfrac{1}{2}l, \tfrac{1}{2}nS, \tfrac{1}{2}nB\right).$$

We now apply Theorem 1 and Theorem 3 to the above hypergeometric functions to obtain

$$
\begin{aligned}
&{}_1F_1\left(\tfrac{1}{2}m;\tfrac{1}{2}l:\tfrac{1}{2}nA,B\right)\\
&=\Gamma_q^{-1}\left(\tfrac{1}{2}m\right)\left(\tfrac{1}{2}n\right)^{\frac{1}{2}mq}\Gamma_q\left(\tfrac{1}{2}l\right)\Gamma_q\left(\tfrac{1}{2}p\right)\\
&\quad\pi^{-\frac{1}{2}q(l+p)}(2\pi/n)^{\frac{3}{4}q(q-1)+\frac{1}{2}q(l+p-2q)}\left[\alpha_q(D_a)\alpha_q(D_b)\right]^{-\frac{1}{2}}\\
&\quad\prod_{i=1}^{q}\prod_{j=q+1}^{p}(b_i-b_j)^{-\frac{1}{2}}\prod_{i=1}^{q}a_i^{-\frac{1}{2}(m-q+1)}\prod_{i=1}^{p}b_i^{-\frac{1}{4}(l-p)}\\
&\quad\int_D\prod_{i=1}^{q}\left[s_i^{\frac{1}{2}(m-q-1)-\frac{1}{4}(l-q)}\exp\left(-\tfrac{1}{2}ns_ia_i^{-1}+ns_i^{\frac{1}{2}}b_i^{\frac{1}{2}}\right)\right]\\
&\quad\left\{1+n^{-1}\left[\tfrac{1}{2}\sum_{1\leq i<j\leq q}a_ia_j(s_i-s_j)^{-1}(a_i-a_j)^{-1}\right.\right.\\
&\qquad+\tfrac{1}{2}\sum_{1\leq i<j\leq q}\left((b_is_i)^{\frac{1}{2}}+(b_js_j)^{\frac{1}{2}}\right)(s_i-s_j)^{-1}(b_i-b_j)^{-1}\\
&\qquad+\tfrac{1}{2}\sum_{i=1}^{q}\sum_{j=q+1}^{p}(b_is_i^{-1})^{\frac{1}{2}}(b_i-b_j)^{-1}\\
&\qquad\left.\left.-\tfrac{1}{8}(l-p)(l-p-2)\sum_{i=1}^{q}(b_is_i)^{-\frac{1}{2}}\right]+O(n^{-2})\right\}
\end{aligned}
$$

where $D=\{(s_1,\ldots,s_q):s_1>\cdots>s_q>0\}$.

We now let $s_i=\gamma_i(1+x_in^{-\frac{1}{2}})$ and then expand in terms of n^{-1}. Asymptotically we have the x_i's independently distributed as $N(0,8\eta_i)$. Integrating out x_i, $i=1,\ldots,q$ we obtain the desired result.

2.5. *Expansion of* ${}_2F_1(\frac{1}{2}n,\frac{1}{2}n,\frac{1}{2}l:A,B)$

Theorem 5. *Let*

$$A=\begin{pmatrix}A_1 & 0\\ 0 & 0\end{pmatrix},$$

$A_1=\operatorname{diag}(a_1,\ldots,a_q)$ $a_1>\cdots>a_q>0$ *and* $B=\operatorname{diag}(b_1,\ldots,b_p)$ $b_1>\cdots>b_p>$

0 with $l \geqslant p \geqslant q, l, p, q$ integral. Then as $n \to \infty$

$$
\begin{aligned}
&{}_2F_1\left(\tfrac{1}{2}n, \tfrac{1}{2}n; \tfrac{1}{2}l : A, B\right) = \\
&\quad 2^{\frac{1}{2}q(p+l+1)} n^{nq - \frac{1}{2}q(3p+3l-2)} \Gamma_q^{-2} \tfrac{1}{2}(n) \\
&\quad \times \prod_{i=1}^{q} a_i^{-\frac{1}{4}(l+p-2q)} \left(1-(a_ib_i)^{\frac{1}{2}}\right)^{-n+\frac{1}{2}(l+p-1)} \Bigg] \prod_{i=1}^{p} b_i^{-\frac{1}{4}(l-p)} \\
&\quad \times \left[\alpha_q(D_a)\alpha_q(D_b)\right]^{-\frac{1}{2}} \prod_{i=1}^{q} \prod_{j=q+1}^{p} (b_i - b_j)^{-\frac{1}{2}} \\
&\quad \times \Bigg\{ 1 + n^{-1} \Bigg[\tfrac{1}{2} \sum_{1 \leqslant i < j \leqslant q} \left((a_ib_i)^{\frac{1}{2}} + (a_jb_j)^{\frac{1}{2}}\right)\left(1-(a_ib_i)^{\frac{1}{2}}\right)\left(1-(a_jb_j)^{\frac{1}{2}}\right) \\
&\quad (a_i - a_j)^{-1}(b_i - b_j)^{-1} + \tfrac{1}{2} \sum_{i=1}^{q} \sum_{j=q+1}^{p} (b_ia_i^{-1})^{\frac{1}{2}}\left(1-(a_ib_i)^{\frac{1}{2}}\right)(b_i - b_j)^{-1} \\
&\qquad + \tfrac{1}{8} \sum_{i=1}^{q} (a_ib_i)^{\frac{1}{2}} - \tfrac{1}{8}(l-p)(l-p-2) \sum_{i=1}^{q} (a_ib_i)^{-\frac{1}{2}} \\
&\qquad + \tfrac{1}{8}q(l-p)(l-p-2) + \tfrac{1}{3}q + \tfrac{1}{8}q(p+l-2)(p+l+2) \Bigg] O(n^{-2}) \Bigg\}.
\end{aligned}
$$

Proof.

$$
\begin{aligned}
&{}_2F_1\left(\tfrac{1}{2}n, \tfrac{1}{2}n; \tfrac{1}{2}l : A, B\right) \\
&\quad = \int_{U>0, V>0} (n/2)^{nq} \Gamma_q^{-1}\left(\tfrac{1}{2}n\right) \Gamma_q^{-1}\left(\tfrac{1}{2}n\right) |U|^{\frac{1}{2}(n-q-1)} |V|^{-\frac{1}{2}(q+1)} \\
&\qquad \times \prod_{i=1}^{q} a_i^{-\frac{1}{2}n} \operatorname{etr}\left[-\tfrac{1}{2}nUV^{-1} - nVA^{-1}\right] {}_0F_1\left(\tfrac{1}{2}l; \tfrac{1}{2}nU, \tfrac{1}{2}nB\right) \mathrm{d}U \mathrm{d}V \\
&\quad = \int_D \left(\frac{n}{2}\right)^{nq} \pi^{q^2} \Gamma_q^{-2}\left(\tfrac{1}{2}q\right) \Gamma_q^{-2}\left(\tfrac{1}{2}n\right) \alpha_q(D_U)\alpha_q(D_V) \\
&\qquad \times \prod_{i=1}^{q} \left[u_i^{\frac{1}{2}(n-q-1)} v_i^{-\frac{1}{2}(q+1)} a_i^{-\frac{1}{2}n}\right] {}_0F_0\left(-\tfrac{1}{2}nV^{-1}, U\right) \\
&\qquad \times {}_0F_0\left(-\tfrac{1}{2}nA^{-1}, V\right) {}_0F_1\left(\tfrac{1}{2}l; \tfrac{1}{2}nU, \tfrac{1}{2}nB\right) \prod^{q} \mathrm{d}u_i \, \mathrm{d}v_i,
\end{aligned}
$$

where $D = \{(u_i, v_i, i = 1, \ldots, q) : u_1 > \cdots > u_q > 0;\ v_1 > \cdots > v_q > 0\}$.

By applying Theorems 1 and 3 to the above hypergeometric functions we obtain

$$
\begin{aligned}
&{}_2F_1\left(\tfrac{1}{2}n,\tfrac{1}{2}n;\tfrac{1}{2}l:A,B\right)\\
&=(n/2)^{nq}\Gamma_q^{-2}\left(\tfrac{1}{2}n\right)\pi^{-\frac{1}{2}(p+l)q}\Gamma_q\left(\tfrac{1}{2}l\right)\Gamma_q\left(\tfrac{1}{2}p\right)\\
&\quad\times(2\pi/n)^{q(q-1)+\frac{1}{2}(p+l-2q)}\left[\alpha_q(D_a)\alpha_q(D_b)\right]^{-\frac{1}{2}}\\
&\quad\times\prod_{i=1}^{q}\prod_{j=q+1}^{p}(b_i-b_j)^{-\frac{1}{2}}\prod_{i=1}^{q}\left[u_i^{\frac{1}{2}(n-q-1)-\frac{1}{4}(l+p-2q)}v_i^{-1}a_i^{-\frac{1}{2}(n-q+1)}\right]\\
&\quad\times\prod_{i=1}^{p}b_i^{-\frac{1}{4}(l-p)}\exp-\tfrac{1}{2}n\sum_{i=1}^{q}\left[u_iv_i^{-1}+v_ia_i^{-1}-2u_i^{\frac{1}{2}}b_i^{\frac{1}{2}}\right]\\
&\quad\left\{1+n^{-1}\left[\tfrac{1}{2}\sum_{1\leqslant i<j\leqslant q}(v_iv_j)(u_i-u_j)^{-1}(v_i-v_j)^{-1}\right.\right.\\
&\qquad+\tfrac{1}{2}\sum_{1\leqslant i<j\leqslant q}a_ia_j(v_i-v_j)^{-1}(a_i-a_j)^{-1}\\
&\qquad+\tfrac{1}{2}\sum_{1\leqslant i<j\leqslant q}\left((u_ib_i)^{\frac{1}{2}}+(u_jb_j)^{\frac{1}{2}}\right)(u_i-u_j)^{-1}(b_i-b_j)^{-1}\\
&\qquad+\tfrac{1}{2}\sum_{i=1}^{q}\sum_{j=q+1}^{p}\left(u_i^{-\frac{1}{2}}b_i^{\frac{1}{2}}\right)(b_i-b_j)^{-1}-\tfrac{1}{8}(l-p)(l-p-2)\\
&\qquad\left.\left.\sum_{i=1}^{q}(u_ib_i)^{-\frac{1}{2}}\right]+O(n^{-2})\right\}.
\end{aligned}
$$

We now let $v_i=u_i^{\frac{1}{2}}a_i^{\frac{1}{2}}(1+n^{-\frac{1}{2}}y_i)$ and $u_i=a_i(1-a_i^{\frac{1}{2}}b_i^{\frac{1}{2}})^{-2}(1+n^{-\frac{1}{2}}x_i)$. By integrating over x_i and y_i we obtain the desired result.

Remark. The editor and the referee of this paper sent to us papers [8], [9] and [13] which are still under print. The papers [8] and [9] contain the first term for the ${}_1F_1$ and ${}_2F_1$ distribution and are obtained by using Hsu's theorem [10]. Paper [13] derives distributions of the functions of characteristic roots by using perturbation technique. This paper contains results upto $O(n^{-2})$.

References

[1] Carter, E. M. and Srivastava, M. S. (1977). Asymptotic expansion of the roots of the non-central Wishart distribution. Submitted for publication.

[2] Chang, T. C. (1970). On an asymptotic representation of the distribution of the characteristic roots of $S_1S_2^{-1}$ when roots are not all distinct. *Ann. Math. Statist.* **41**, 440–445.

[3] Chang, T. C. (1973). On an asymptotic distribution of the characteristic roots of $S_1S_2^{-1}$ when roots are not all distinct. *Ann. Inst. Statist. Math.* **25**, 447–452.

[4] Chattopadhyay, A. K. and Pillai, K. C. S. (1973). Asymptotic expansions for the distributions of the characteristic roots when the parameter matrix has several multiple roots. In: *Multivariate Analysis III* (P. R. Krishnaiah, Ed.) pp. 117–127. Academic Press, New York, 117–127.

[5] Chattopadhyay, A. K., Pillai, K. C. S. and Li, H. C. (1976). Maximization of an integral of a matrix function and asymptotic expansions of distributions of latent roots of two matrices. *Ann. Statist.* **4**, 796–806.

[6] Chikuse, Y. (1976). Asymptotic distribution of the latent roots of the covariance matrix with multiple population roots. *J. Multiv. Anal.* **6**, 237–49.

[7] Constantine, A. G. and Muirhead, R. J. (1976). Asymptotic expansions for distributions of latent roots in multivariate analysis. *J. Multiv. Anal.* **6**, 369–391.

[8] Glynn, W. J. (1978). An asymptotic representation of the density of latent roots in MANOVA with an application to testing. To appear.

[9] Glynn, W. J. and Muirhead, R. J. (1978). Inference in canonical correlation analysis. To appear in *J. Multiv. Anal.*

[10] Hsu, L. C. (1948). A theorem on the asymptotic behavior of a multiple integral. *Duke Math. J.* **15**, 623–632.

[11] James, A. T. (1969). Tests of equality of latent roots of the matrix. In *Multivariate Analysis III* (P. R. Krishnaiah, Ed.) pp. 205–218. Academic Press, New York, 205–218.

[12] Khatrai, C. G. and Srivastava, M. S. (1978). Asymptotic expansions for distributions of characteristic roots of covariance matrices. *South African Statist. J.* **12**, 161–186.

[13] Krishnaiah, P. R. and Lee, J. C. (1977). On the asymptotic joint distributions of certain functions of the eigenvalues of four random matrices. Presented at the Internation Statistical conference at New Delhi; to appear in *J. Multiv. Anal.* **9**.

[14] Li, H. C., Pillai, K. C. S. and Chang, T. C. (1970). Asymptotic expansions for distributions of roots of two matrices from classical and complex Gaussian populations. *Ann. Math. Statist.* **41**, 1541–1556.

[15] Muirhead, R. J. (1978). Latent roots and matrix variates: a review of some asymptotic results. *Ann. Statist.* **6**, 5–33.

[16] Sugiura, N. (1976). Asymptotic expansions of the distributions of the latent roots and the latent vector of the Wishart and multivariate *F* matrices. *J. Multiv. Anal.* **6**, 500–525.

P. R. Krishnaiah, ed., *Multivariate Analysis–V*

BIVARIATE EXTREMES: FOUNDATIONS AND STATISTICS

J. TIAGO de OLIVEIRA
Center of Statistics and Applications (I.N.I.C.), Lisbon Faculty of Sciences, Lisbon, Portugal

1. Introduction

Bivariate and multivariate extremes are useful in many concrete problems as the largest ages of death for men and women each year for some country (whose bivariate distribution splits naturally in the product of the margins, corresponding to independence), the floods (or droughts) in two points of the same river each year, bivariate extreme meteorological data (temperature, pressure, e.g.) each week, etc.

The purpose of the bivariate theory of extremes is to study the asymptotic probabilistic behaviour and also to find statistical models to describe concrete situations. As done for the univariate extremes, we will use asymptotic theory which, in general, gives a reasonable fit to the data.

Statistical decision theory for bivariate extremes (and, evidently, for multivariate extremes) is, yet, in the very beginning. In general, we do not dispose of the best tests, best estimators, etc. Also, for the different models we have not defined the domains of applications. It should, also, be remarked that procedures to separate between statistical models are not yet available. The few papers known until now choose, from the beginning, a model, or compare two of them by the use of Kolmogoroff (or other) test —see Gumbel and Goldstein [5] for instance.

2. Asymptotic behaviour of the sample pairs of maxima

Let $(X_1, Y_1), \ldots, (X_n, Y_n)$ be a sample of n random independent pairs whose distribution function is $F(x,y)$. The distribution function of $(\max X_i, \max Y_i)$ is $\mathbf{P}(\max X_i \leqslant x, \max Y_i \leqslant y) = \mathbf{P}[(X_1 \leqslant x, Y_1 \leqslant y), \ldots, (X_n \leqslant x, Y_n \leqslant y)] = F^n(x,y)$; remark that, in general, $(\max X_i, \max Y_i)$ is not a sample point.

The use of $F^n(x,y)$ is not easy and for relatively large samples it is convenient to resort to asymptotic theory, i.e., to use a distribution function $L(x,y)$, such that for convenient sequences λ_n, $\delta_n>0$, λ'_n, $\delta'_n>0$ we have

$$\mathbf{P}\left(\frac{\max Y_i-\lambda_n}{\delta_n}\leqslant x, \frac{\max Y_i-\lambda'_n}{\delta'_n}\leqslant y\right)=F^n(\lambda_n+\delta_n x,\lambda'_n+\delta'_n y)\overset{w}{\to}L(x,y).$$

As the margins have asymptotic distribution functions of the Gumbel, Weibull or Fréchet models (i.e., $\Lambda(x)=\exp(-e^{-x})$; $\Psi_\alpha(x)=\exp-(-x)^\alpha$, if $x\leqslant 0$, $\Psi_\alpha(x)=1$, if $x\geqslant 0$ $(\alpha>0)$ and $\Phi_\alpha(x)=0$ if $x\leqslant 0$, $\Phi_\alpha(x)=\exp-x^{-\alpha}$ if $x\geqslant 0$ $(\alpha>0)$, respectively, in reduced form) which are easily convertible one to another, we will always suppose that the margins are of the reduced form of Gumbel (asymptotic) model. Let us recall that the mean value and variance for Λ are γ and $\pi^2/6$. In the case $F^n(\lambda_n+\delta_n x,\lambda'_n+\delta'_n y)$ converges, the asymptotic distribution function is

$$\Lambda(x,y)=[\Lambda(x)\Lambda(y)]^{k(y-x)}=\exp[-(e^{-x}+e^{-y})k(y-x)]$$

where $k(\cdot)$, the *dependence function*, expresses the asymptotic connection between $\max X_i$ and $\max Y_i$. For details see Finkelshteyn [3], Geffroy [4] and Tiago de Oliveira [10, 11], results coordinated in Tiago de Oliveira [16].

An important property of $\Lambda(x,y)$ is the stability relation

$$\Lambda^c(x,y)=\Lambda(x-\log c,y-\log c),\quad c>0.$$

As a consequence of the Boole double-inequality for the probability of an intersection we have, using stability,

$$\Lambda(x)\Lambda(y)\leqslant\Lambda(x,y)\leqslant\min[\Lambda(x),\Lambda(y)]$$

so that the dependence function $k(w)$ must then satisfy

$$\left(\frac{1}{2}\leqslant\right)\frac{\max(1,e^w)}{1+e^w}\leqslant k(w)\leqslant 1,$$

which implies $k(\pm\infty)=1$.

The upper bound, $k(w)=1$, corresponds to independence and the lower bound, $k(w)=\max(1,e^w)/(1+e^w)$, corresponds to the diagonal case where, for reduced margins, we have $X=Y$ with probability one.

As $\Lambda(x,y)$ is a continuous function, the convergence of $F^n(\lambda_n+\delta_n x, \lambda_n'+\delta_n' y)$ is uniform so that $F^n(x,y)$ is approximated by

$$\Lambda\left(\frac{x-\lambda}{\delta},\frac{y-\lambda'}{\delta'}\right)$$

for convenient parameters of location λ, λ' and dispersion δ, $\delta'(>0)$. It is, exactly, the approximation

$$F^n(x,y)\approx\Lambda\left(\frac{x-\lambda}{\delta},\frac{y-\lambda'}{\delta'}\right)$$

which justifies the practical use of $\Lambda(\cdot,\cdot)$ as a distribution of the observed sample pairs of maxima.

As $\Lambda(x,y)=\exp[-(e^{-x}+e^{-y})k(y-x)]$ is a distribution function, $k(w)$ must satisfy some relations; in the case of existence of planar density

$$p_\Lambda(x,y)=\partial^2\Lambda/\partial x\partial y,$$

$k(w)$ must satisfy the conditions:

$$k(-\infty)=k(+\infty)=1$$

$$[(1+e^{w})k(w)]'\geqslant 0$$

$$[(1+e^{-w})k(w)]'\leqslant 0$$

$$(1+e^{-w})k''(w)+(1-e^{-w})k'(w)\geqslant 0.$$

Note that k'' must exist almost everywhere. The models with planar density will be called differentiable models.

But it should be noted that for three, of the five, bivariate models known until now, a planar density does not exist; those models are called nondifferentiable.

Note that if we exchange the variables X, Y we have, for bivariate extremes,

$$\tilde{\Lambda}(x,y)=\mathbf{P}(Y\leqslant x, X\leqslant y)=\Lambda(y,x),$$

so that $\tilde{k}(w)=k(-w)$; in the case where the variables are exchangeable we must have

$$\Lambda(x,y)=\Lambda(y,x)\quad\text{or}\quad k(w)=k(-w).$$

Another important property which may be very useful for the generation of bivariate extreme models is convexity: if $k_1(w)$ and $k_2(w)$ are dependence functions, then the mixtures $\alpha k_1(w)+(1-\alpha)k_2(w)$ $(0\leqslant\alpha\leqslant 1)$ are also dependence functions; this corresponds to the fact that if (X_1, Y_1) and (X_2, Y_2) are two independent bivariate extreme pairs with dependence functions $k_1(w)$ and $k_2(w)$ then the pair

$$X=\max[X_1+\log\alpha, X_2+\log(1-\alpha)]$$

$$Y=\max[Y_1+\log\alpha, Y_2+\log(1-\alpha)]$$

has the dependence function $\alpha k_1(w)+(1-\alpha)k_2(w)$ or the distribution function $\Lambda_1^\alpha(x,y)\Lambda_2^\alpha(x,y)$.

As Sibuya [9] has shown, bivariate extremes have positive association, i.e., large (small) values of one component are more associated with large (small) values of the other component than in independence; consequently, the correlation coefficients are all non-negative as it will be shown.

In some cases the (reduced) difference $Y-X$ can be important. Its distribution function is

$$D(w)=(1+e^{-w})^{-1}+k'(w)/k(w)$$

(k' exists almost everywhere, even in the non-differentiable case). There is a biunique correspondence between $k(w)$ and $D(w)$. In the case of independence we have $D(w)=(1+e^{-w})^{-1}$ and in the diagonal case we have $D(w)=H(w)$, the Heavside jump function. In the differentiable case $D(w)$ satisfies relations easily deduced from the ones for k.

Let us consider now the correlation coefficients. As the (classical) correlation coefficient exists always, by Schwartz inequality, the computation of the variance of $Y-X$ shows that its expression is

$$\rho=1-\frac{3}{\pi^2}\int_{-\infty}^{+\infty} w^2\,dD(w)=-\frac{6}{\pi^2}\int_{-\infty}^{+\infty}\log k(w)\,dw.$$

ρ is always non-negative because $k(w)\leqslant 1$; note that $k(-w)$ leads to the same correlation coefficient. The nullity of ρ implies $k(w)=1$, that is, independence.

The non-parametric correlation coefficients, see Fraser [2], have the expressions:

grade correlation coefficient: $x=12\int_{-\infty}^{+\infty}\frac{e^{w}}{(1+e^{w})^{2}[1+k(w)]^{2}}dw-3$

difference-sign correlation coefficient: $\tau=1-\int_{-\infty}^{+\infty}D(w)[1-D(w)]\,dw$

Spearman correlation coefficient: $\rho_s=\frac{(n-2)x+3\tau}{n+1}$

It can, also, be shown that all those correlation coefficients are non-negative and that the nullity of each of them implies independence.

In what regards regression, as the mean value and variance of the margins are γ and $\pi^2/6$, the regression lines are

$$Ly(x)=\gamma+\rho(x-\gamma) \quad \text{and} \quad Lx(y)=\gamma+\rho(y-\gamma).$$

As the conditional distribution function of Y if $X=x$ is

$$G(y/x)=\Lambda(x,y)\exp(e^{-x})\cdot\{k(y-x)+(1+e^{-y+x})k'(y-x)\},$$

we get for the general regression curve

$$\bar{y}(x)=\int_{-\infty}^{+\infty}y\,dG(y/x), \quad \text{as} \quad \gamma=\int_{-\infty}^{+\infty}y\,d\Lambda(y),$$

$$\bar{y}(x)=\gamma-\int_{-\infty}^{+\infty}\{G(y/x)-\Lambda(y)\}\,dy$$

which leads finally to

$$\bar{y}(x)=\gamma-\int_{-\infty}^{+\infty}(\exp\{-[(1+e^{w})k(w)-1]e^{-x}\}k(w)$$

$$-\exp\{-e^{-w}e^{-x}\})(1+e^{-w})\,dw.$$

The general regression curve $\bar{x}(y)$ is obtained substituting

$$\tilde{k}(w)=k(-w) \quad \text{for } k(w)$$

in the previous formula.

Let us recall that Geffroy [4] proved that if the initial distribution function $F(x,y)$ is such that

$$\lim_{x\to w_x, y\to w_y} \frac{1-F(+\infty,y)-F(x,+\infty)+F(x,y)}{1-F(x,y)}=0$$

(where w_x and w_y denote the right ends of x and y) then the asymptotic distribution function of maxima splits in the product of the margins (independence). As this condition is very general, tests for independence should play an important role in the applications of the theory of bivariate extremes. Also as

$$\Lambda(x)\Lambda(y) \leqslant \Lambda(x,y) \leqslant \min[\Lambda(x),\Lambda(y)]$$

we have for the index of dependency

$$\sup_{x,y}[\Lambda(x,y)-\Lambda(x)\Lambda(y)] \leqslant \sup_{x,y}[\min(\Lambda(x),\Lambda(y))-\Lambda(x)\Lambda(y)]=1/4.$$

Consequently, for small samples, it will be difficult to separate between the models. We will see this for some empirical data.

Evidently, in applications, we can not use standard margins but margins with location and dispersion parameters.

There are some results concerning independence. The most important is the following: if (X, Y) is a bivariate extreme pair then X/Y is extremal if and only if (X, Y) is independent (see Tiago de Oliveira [11 or 16] for the proof of this theorem and others).

$\Lambda(x,y)$ can be characterized as the only bivariate distribution function such that $\max(X-a, Y-b)-\lambda(a,b)$, for convenient $\lambda(a,b)$, has a Gumbel (reduced) distribution whatever may be a and b.

Finally it should be recalled that the corresponding distribution function for minima, also with Gumbel reduced margins, is

$$\Omega(x,y)=1-\Lambda(-x)-\Lambda(-y)+\Lambda(-x,-y).$$

3. Differentiable models

The differentiable models known until now are the mixed and the logistic ones. Others can be obtained through the use of the mixture property, as for the mixed model; the class of twice differentiable dependence functions is closed for convexity. Remark that as $k(w)=k(-w)$ in the two models, the random variables X and Y are exchangeable.

For simplicity of notation in this section, the density $p_{\Lambda}(x,y)=\partial^2\Lambda/\partial x\partial y$ of $\Lambda(x,y/\theta)$ will be denoted by $p_\theta(x,y)$.

The *mixed model*, with standard Gumbel margins, has the distribution function

$$\Lambda(x,y/\theta)=\exp\left[-(e^{-x}+e^{-y})+\frac{\theta}{e^{x}+e^{y}}\right]\quad(0\leqslant\theta\leqslant 1).$$

Its dependence function

$$k(w/\theta)=1-\theta e^{w}/(1+e^{w})^2$$

is the mixture of

$$k_1(w)=1\quad\text{and}\quad k_2(w)=1-e^{w}/(1+e^{w})^2.$$

The distribution function of the (reduced) difference $Y-X$ is

$$D(w/\theta)=\mathbf{P}(Y-X\leqslant w)=\frac{e^{w}}{1+e^{w}}\,\frac{(1+e^{w})^2-\theta}{(1+e^{w})^2-\theta\cdot e^{w}}.$$

For $\theta=0$ we have independence and for $\theta=1$ we have dependence but not the diagonal case. It is easy to see that the index of dependency has the expression

$$\sup[\Lambda(x,y/\theta)-\Lambda(x)\Lambda(y)]=\theta(1-\theta/4)^{4/\theta}(4-\theta)$$

whose maximum, attained at $\theta=1$, is 0.105.

For small samples, the separation of this model from the independence case is, naturally, a difficult one.

Let $(x_1,y_1),\ldots,(x_n,y_n)$ be a sample of n independent reduced pairs from a population with a mixed model distribution with standard margins (no parameters). The likelihood is then

$$L(x_i,y_i/\theta)=\prod_1^n p_\theta(x_i,y_i)$$

where the density has the expression

$$p_\theta(x,y)=\mathrm{e}^{-x-y}\Lambda(x,y/\theta)\left\{1-\theta\frac{\mathrm{e}^{2x}+\mathrm{e}^{2y}}{(\mathrm{e}^x+\mathrm{e}^y)^2}+2\theta\frac{\mathrm{e}^{2x}+\mathrm{e}^{2y}}{(\mathrm{e}^x+\mathrm{e}^y)^3}\right.$$

$$\left.+\theta\frac{2\mathrm{e}^{2x+2y}}{(\mathrm{e}^x+\mathrm{e}^y)^4}\right\}$$

Evidently $p_0(x,y)=\Lambda'(x)\Lambda'(y)$, where $\Lambda'(x)=\mathrm{e}^{-x}\exp(-\mathrm{e}^{-x})$ and $\Lambda'(y)$ are the densities of the margins.

The locally most powerful test for independence ($\theta=0$ vs. $\theta>0$) is

$$\left.\frac{L'(x_i,y_i/\theta)}{L(x_i,y_i/\theta)}\right|_{\theta=0}=\Sigma v(x_i,y_i)>c$$

where the prime in L denotes differentiation with respect to θ and

$$v(x,y)=p_\theta'(x,y/\theta)|_{\theta=0}/\Lambda'(x)\Lambda'(y)=p_\theta'(x,y)|_{\theta=0}/p_0(x,y)$$

$$=2\frac{\mathrm{e}^{2x}+\mathrm{e}^{2y}}{(\mathrm{e}^x+\mathrm{e}^y)^3}-\frac{\mathrm{e}^{2x}+\mathrm{e}^{2y}}{(\mathrm{e}^x+\mathrm{e}^y)^2}+\frac{1}{\mathrm{e}^x+\mathrm{e}^y}$$

The mean value of $v(X,Y)$ is zero but the variance is infinite; the distribution of $\Sigma v(x_i,y_i)$ is not known nor, even, the asymptotic distribution.

The maximum likelihood estimator equation is easily obtained from the expression of $p_\theta(x,y)$ and is numerically solvable. Note that not only the Cramér–Rao bound for $\theta=0$ is zero (being the inverse of the variance of $v(x,y)$ times n) but also $n\,\mathrm{Var}(\hat{\theta})\to 0$ as $\theta\to 0$ which shows that $\hat{\theta}$ has a special behaviour at $\theta=0$ unknown yet.

We have then, to use resort to the usual methods of the data analysis, although inefficient.

Let us begin by considering independence testing.

Of all methods considered until now (correlation coefficients, strip, quadrants), see Tiago de Oliveira [17] for details, the best to test independence is to use of the (classical) correlation coefficient. Its population value is

$$\rho = \frac{6}{\pi^2}\left[\arccos(1-\theta/2)\right]^2$$

which variates from $\rho(1)=0$ to $\rho(1)=2/3$. Denoting by

$$\rho^* = \Sigma(x_i-\bar{x})(y_i-\bar{y})/\sqrt{\left(\Sigma(x_i-\bar{x})^2 \cdot \Sigma(y_i-\bar{y})^2\right)}$$

the sample correlation coefficient, we shall accept independence if

$$\sqrt{n}\cdot\rho^* \leqslant \lambda_\alpha\left(\int_{-\infty}^{\lambda_\alpha} \frac{1}{\sqrt{2\pi}} e^{-t^2/2} dt = 1-\alpha\right)$$

and reject otherwise.

An estimator of θ is given by

$$\rho^* = \frac{6}{\pi^2}\left[\arccos(1-\theta^*/2)\right]^2$$

or

$$\theta^* = 2\left(1-\cos\sqrt{\left(\frac{\pi^2}{6}\rho^*\right)}\right) \quad \text{if } \rho^* \leqslant 2/3$$

The difficult expression of $V(\rho^*)$ leads us to the use of some other method for the obtention of confidence intervals, the most useful and simple being the quadrants method, see Gumbel and Mustafi [6]. In this model the probability that both components are smaller or larger than the margin medians is

$$p(\theta) = \Lambda(\tilde{\mu}, \tilde{\mu}/\theta) = 4^{-(1-\theta/4)}$$

where $\tilde{\mu} = -\log\log 2$ is the median of the reduced margins. If we denote by N^* the number of pairs whose components are both smaller or larger than the sample medians, we know that N^*/n is asymptotically normally distributed with mean value $p(\theta)$ and variance $p(\theta)(1-p(\theta))/n$. Note that all those statistics are independent of the margin parameters, avoiding thus the need for statistical decision for the margins if we are interested only in their interconnections.

The logistic model

$$\Lambda(x,y/\theta)=\exp\{-[e^{-x/(1-\theta)}+e^{-y/(1-\theta)}]^{(1-\theta)}\} \quad \text{for } 0\leqslant\theta<1$$

has a planar density, with independence for $\theta=0$; for $\theta=1$ we have the diagonal case.

As the dependence function is

$$k(w/\theta)=[1+e^{-w/(1-\theta)}]^{1-\theta}/(1+e^{-w}),$$

we have

$$D(w/\theta)=[1+e^{-w/(1-\theta)}]^{-1}.$$

From

$$\sup_{x,y}[\Lambda(x,y/\theta)-\Lambda(x,y)]=(1-2^{-\theta})\cdot 2^{-\theta(2^{\theta}-1)}$$

whose maximum value for θ is $1/4$ we see that this model admits alternatives more far away from independence, as will happen in the other models. The density has the expression

$$\Lambda(x,y/\theta)\Big\{(e^{-x/1-\theta}+e^{-y/1-\theta})^{-2\theta}$$
$$+\frac{\theta}{1-\theta}(e^{-x/1-\theta}+e^{-y/1-\theta})^{-1-\theta}\Big\}e^{-(x+y)(1-\theta)}.$$

The locally most powerful test for independence leads us to the use of the test statistic

$$v(x,y)=-x-y+xe^{-x}+ye^{-y}+(e^{-x}+e^{-y}-\log 2)+\frac{1}{e^{-x}+e^{-y}}$$

whose mean value is zero and the variance is infinite. The situation is strongly similar to the one of the mixed model, the only difference being the expression of the correlation coefficient $p(\theta)=\theta(2-\theta)$, increasing from $\rho(\theta)=0$ to $\rho(1)=1$ and of the probability for the quadrants method

$$p(\theta)=\exp\{\log 2\cdot(1-2^{1-\theta})\}.$$

The general regression in those models and its comparison to linear regression has to be considered as was done for the non-differentiable models (Tiago de Oliveira [15]). For those models the best use of the (univariate) sample of reduced differences $y_i - x_i$ or its estimation in the case the margin parameters is not known.

4. Nondifferentiable models

The *biextremal* and *Gumbel* models are the non-differentiable ones considered (in papers) until now; they were dealt with in Tiago de Oliveira [13] and [14] and, concerning regression, both in Tiago de Oliveira [15]. A forthcoming paper, Tiago de Oliveira [18], will introduce the *natural model*, a generalization of the biextremal model.

The biextremal model, which appears naturally in extremal processes—see Tiago de Oliveira [12] and references therein—has the distribution function

$$\Lambda(x,y/\theta)=\exp\left[-\max(e^{-x}+(1-\theta)e^{-y},e^{1-y})\right] \quad (0\leqslant\theta\leqslant1)$$

with the dependence function

$$k(w/\theta)=\frac{1-\theta+\max(\theta,e^{w})}{1+e^{w}}=1-\frac{\min(\theta,e^{w})}{1+e^{w}}$$

and the distribution function of the (reduced) difference is $D(w/\theta)=0$

$$\text{if } w<\log\theta, =\left[1+(1-\theta)e^{w})^{-1}\right] \quad \text{if} \quad w\geqslant\log\theta.$$

From the expression of $D(w/\theta)$ we see that $Y\geqslant X+\log\theta$, with $P(Y=X+\log\theta)=D(\log\theta/\theta)=\theta$. This model appears naturally if we consider (X,Z) as an independent pair and take the new pair (X,Y) with

$$Y=\max(X+\log\theta, Z+\log(1-\theta)).$$

As $k(w)\neq k(-w)$ the random variables X and Y are not exchangeable.

It is obvious that it exists a singular part concentrated on the line $y=x+\log\theta$.

As for $\theta=0$ and $\theta=1$ we obtain the independence and diagonal cases we have, as for the logistic model,

$$\sup_{\theta,x,y}\left[\Lambda(x,y/\theta)-\Lambda(x,y)\right]=1/4.$$

The best method to estimate θ is to take $\theta^* = \min(e^{y_i - x_i}, 1)$ its variance being

$$\sim 2(1-\theta)^n/n^2 \quad \text{if} \quad \theta > 0$$

and

$$\sim 1/n^2 \quad \text{if} \quad \theta = 0$$

as follows from

$$\mathbf{P}(\theta^* \leqslant z) = 0 \quad \text{if } z < \theta,$$

$$= 1 - \left(\frac{1-\theta}{z+1-\theta}\right)^n \quad \text{if } \theta \leqslant z < 1,$$

$$= 1 \quad \text{if } z > 1.$$

The best test for independence, using θ^*, at the significance level α, is to accept $\theta = 0$ vs $\theta > 0$ if $\theta^* \leqslant \alpha^{-1/n} - 1$. When we are not dealing with reduced margins, the reduced values should be estimated using the estimators of the margin parameters; the distribution of θ^* with estimated

$$x_i^* = (x_i - \lambda_x^*)/\delta_x^*, y_i^* = (y_i - \lambda_y^*)/\delta_y^*$$

is not known.

The correlation coefficient of the biextremal model is given by

$$\rho(\theta) = -\frac{6}{\pi^2} \int_0^\theta \frac{\log t}{1-t} \, dt$$

and increases from $\rho(0) = 0$ to $\rho(1) = 1$. It was shown in Tiago de Oliveira [15] that the linear regression is a very good approximation to the general regression.

The Gumbel model has the distribution function

$$\Lambda(x,y/\theta) = \exp\{-[e^{-x} + e^{-y} - \theta \min(e^{-x}, e^{-y})]\}$$

with the dependence function

$$k(w/\theta) = 1 - \theta \frac{\min(1, e^w)}{1 + e^w}$$

and the distribution function of (reduced) difference

$$D(w/\theta)=\frac{1-\theta}{1-\theta+e^{w}} \quad \text{if } w<0,$$

$$=\frac{e^{w}}{1-\theta+e^{w}} \quad \text{if } w\geqslant 0$$

with a jump of $\theta/(2-\theta)$ at $w=0$, given thus the probability $\mathbf{P}(Y=X)$. The dependence function is a mixture of the independence and the diagonal cases. Also for this model we have

$$\sup_{\theta,x,y}\left[\Lambda(x,y/\theta)-\Lambda(x)\Lambda(y)\right]=1/4.$$

This model is the conversion for Gumbel margins of the bivariate exponential model given in Marshall and Olkin [7]. If we denote by nf_n the number of points (x_i,y_i) with $x_i=y_i$ and by $T_n=\frac{1}{n}\sum\max(e^{-x_i},e^{-y_i})$ the maximum likelihood estimator—using Radon-Nikodym densities with respect to $\frac{1}{2}[\Lambda(x,y/0)+\Lambda(x,y/1)]$—is given by

$$\hat{\theta}=\left(T_n-1+\sqrt{\left((T_n-1)^2+4f_nT_n\right)}\right)/2T_n$$

taking $\hat{\theta}=0$ if the expression is negative. $\hat{\theta}^*$, not truncated, is asymptotically normal with mean value θ and variance

$$\frac{\theta(1-\theta)(2-\theta)}{n(1+\theta)}.$$

For the test of independence ($\theta=0$) as $f_n=0$ and the variance of $\hat{\theta}$ is null, we can use

$$T_n,\ \sqrt{n}(2\cdot T_n-1)$$

being asymptotically standard normal.

The correlation coefficient is

$$\rho(\theta)=\frac{12}{\pi^2}\int_0^{\theta}\frac{\log(2-t)}{1-t}\,dt$$

and, as for the biextremal model, the linear regression is a very good approximation to the general one.

Finally, let us consider the natural model. If we take independent random (reduced) Gumbel variables Z and T and the new random pair (X, Y) with

$$X = \max(Z-a, T-b), \quad Y = \max(Z-c, Y-d)$$

is natural. If we impose to the margins to be reduced we get

$$e^{-a} + e^{-b} = e^{-c} + e^{-d} = 1$$

so that $a, b, c, d \geqslant 0$. Taking $a, d \geqslant 0$ as the parameters we get

$$\Lambda(x, y / a, d) = \mathbf{P}(X \leqslant x, Y \leqslant y) = \exp\{-(e^{-x} + e^{-y}) k(y-x)\}$$

where

$$\begin{aligned} k(w) &= 1/(1+e^{w}) \quad \text{if } w \leqslant a + \log(1-e^{-d}), \\ &= (e^{-a} + e^{-d-w})/(1+e^{-w}) \\ &\qquad \text{if } a + \log(1-e^{-d}) \leqslant w \leqslant -d - \log(1-e^{-a}) \\ &= 1/(1+e^{-w}) \quad \text{if } -d - \log(1-e^{-a}) \leqslant w. \end{aligned}$$

Note that:

(a) $a + \log(1-e^{-d}) \leqslant w \leqslant -d - \log(1-e^{-a})$;

(b) the random pair (X, Y) is contained in the strip $a + \log(1+e^{-d}) \leqslant Y - X \leqslant -d - \log(1-e^{-a})$ which contains the origin and $e^{-a} + e^{-d} \geqslant 1$;

(c) the exchange of a and d corresponds to the exchange of X and Y;

(d) the independence and diagonal cases are contained in the model for $a = d = 0$ and $a = d = \log 2$ respectively;

(e) for $a = 0$, $d = -\log(1-\theta)$ we obtain the biextremal model and for $a = -\log(1-\theta)$, $d = 0$ we get the 'dual' of the biextremal model, corresponding to the exchange between X and Y in the biextremal model;

(f) the left and right tails of $k(w / a, d)$ coincide with the ones of the diagonal case;

(g) the natural model is not closed for convexity.

The correlation coefficient has the expression

$$\rho(a, d) = 1 - \frac{6}{\pi^2}\left[\frac{1}{2}(a + \log(1-e^{-d}))^2 + a\left[\log(e^a - 1)(e^d - 1)\right] + \int_{e^a}^{e^d/e^d - 1} \frac{\log \xi}{\xi - 1}\, d\xi\right]$$

from which the regression lines $Ly(x)$ and $Lx(y)$ can be obtained; the general regression curve of Y in X is

$$\bar{y}(x/a,d)=x-d-\log(1-e^{-a})$$
$$-e^{-a}\left[\exp(1-e^{-a})e^{-x}\right]\int_{(1-e^{-a})e^{-x}}^{\exp(-a-x-\log(e^{d}-1))}\frac{e^{-t}}{t}\,dt$$

the regression curve $\bar{x}(y/a,d)$ is obtained by the exchange of a and d.

No statistical decision was dealt with until now; but it is evident, from previous reasoning that the qualities

$$a^*+\log(1-e^{-d^*})=\min(y_i-x_i)$$
$$d^*+\log(1-e^{-a^*})=-\max(y_i-x_i)$$

obtained equating the extremes of y_i-x_i to the previous bounds, can give the relations

$$e^{-a^*}=\frac{e^{\max(y_i-x_i)}-1}{e^{\max(y_i-x_i)}-e^{\min(y_i-x_i)}}$$
$$e^{-d^*}=\frac{e^{\max(y_i-x_i)}(1-e^{\min(y_i-x_i)})}{e^{\max(y_i-x_i)}-e^{\min(y_i-x_i)}}$$

which have solutions $0\leqslant e^{-a^*}$, $e^{-d^*}\leqslant 1$ if $\min(y_i-x_i)\leqslant 0\leqslant \max(y_i-x_i)$, with probability

$$1-\{D^n(0)+(1-D(0))^n\}\to 1 \quad \text{if } 0<D(0)=\frac{e^{-a}+e^{-d}}{2}<1;$$

if $\max(y_i-x_i)<0$, with probability $D^n(0)$, as $e^{-a^*}<0$ and $e^{-d^*}>1$ the convenient truncation leads to the estimators $a^*=+\infty$, $d^*=0$; if $\min(y_i-x_i)>0$, with probability $(1-D(0))^n$, as $e^{-a^*}>1$ and $e^{-d^*}<0$, the truncated estimators are $a^*=0$ and $d^*=+\infty$.

5. Some applications

Applications of bivariate extremes parallel the ones for univariate extremes. Although very important, few studies have been made until now.

A paper by Gumbel and Goldstein [5] has shown that oldest ages of death in Sweden for men and women for 54 years (1905/1958) are independent (as could be expected) and that the floods in two points of Ocmulgee River are dependent. Gumbel and Mustafi [6] have shown that the logistic fits well to data of Fox River. Amaral and Gomes [1] have shown that the logistic model gives the best fit.

For data relating to telecommunications, Posner, Rodemich, Aslock and Lurie [8] fitted the mixed model but advanced the idea that the choice of the dependence function—the choice of the bivariate extreme models—is not very important.

The unpublished paper by Amaral and Gomes [1] analyses all those data. For the telecommunication data the best fit is obtained for logistic and biextremal models, although the fit by the mixed and Gumbel were also good, which confirms the idea advanced by Posner and all. As a consequence of this study the biextremal and the logistic models seem to be the most important ones; this fact for the biextremal could, in some way, be expected because it is associated to the extremal processes as the binormal to Gaussian processes. Note, although, that the natural model, in cause of development now, was not considered.

6. Multivariate extensions

Multivariate extremes theory is still in a more elementary stage than the theory of bivariate extremes. We dispose only of the general expression of the distribution functions, few inequalities and an independence result.

It was shown that the asymptotic distribution function $\Lambda(x_1,\ldots,x_m)$ of multivariate extremes with Gumbel reduced margins, is of the form

$$\Lambda(x_1,\ldots,x_m)=\exp\{-(\mathrm{e}^{-x_1}+\cdots+\mathrm{e}^{-x_m})k(x_2-x_1,\ldots,x_m-x_1)\}$$

where the dependence function k must satisfy the conditions for Λ to be a distribution function; $k=1$ corresponds to independence. As the margins are $\Lambda(x_i)$, using a Fréchet inequality and the stability, we get the inequality

$$\Lambda(x_1)\cdots\Lambda(x_m)\leqslant\Lambda(x_1,\ldots,x_m)\leqslant\min(\Lambda(x_1),\ldots,\Lambda(x_m)).$$

When bivariate margins

$$\Lambda_{ij}(x_j,x_j)=\exp[-(\mathrm{e}^{-x_i}+\mathrm{e}^{-x_j})k_{ij}(x_j-x_i)]$$

are known and compatible we get the double inequality

$$\left\{\prod_{i\neq j}\Lambda_{ij}(x_i,x_j)\right\}^{1/(2(m-1))}\leqslant\Lambda(x_1,\ldots,x_m)$$

$$\leqslant\frac{\left[\prod_{i\neq j}\Lambda_{ij}(x_i,x_j)\right]^{1/2}}{\left[\prod_{i}\Lambda(x_i)\right]^{m-2}}.$$

If the bivariate margins are independent, i.e., if

$$\Lambda_{ij}(x_i,x_j)=\Lambda(x_i)\Lambda(x_j),$$

then

$$\Lambda(x_1,\ldots,x_m)=\Lambda(x_1)\cdots\Lambda(x_m).$$

This very important result reduces general independence to the verification of independence for bivariate margins.

Some other inequalities are also known.

The generalization of the biextremal model is the multiextremal distribution with standard margins.

$$\Lambda(x_1,\ldots,x_m)$$

$$=\exp\left[-\sum_{1}^{m}a_i\max\left(\frac{e^{-x_i}}{a_1+\cdots+a_i},\ldots,\frac{e^{-x_m}}{a_1+\cdots+a_m}\right)\right],\qquad a>0$$

and the generalization of Gumbel model is

$$\Lambda(x_1,\ldots,x_m)=\exp|-\Big(a\sum_{i}e^{-x_i}+a'\sum_{i<j}\max(e^{-x_i},e^{-x_j})$$

$$+\cdots+a^{(m-1)}\max(e^{-x_1},\ldots,e^{-x_m})$$

with the a conveniently related to have bivariate Gumbel margins. But statistical decision for those multivariate models was not developed yet; only reduction to bivariate margins can, in some cases, give techniques to obtain estimators, tests, etc. The generalization of the natural model, although, is under study.

References

[1] Amaral, M. Antónia and Gomes, M. Ivette (1975). The fitting of bivariate extreme models (to be published).

[2] Fraser, D. A. S. (1957). *Non-parametric methods in statistics*. John Wiley and Sons, New York.

[3] Finkelshteyn, B. V. (1953). Limiting distribution of extremes of a variational series of a two-dimensional random variable. *Dokl. Ak. Nauk. SSSR*, **91**, 209–214 (in Russian).

[4] Geffroy, J. (1958/59). Contributions à l'étude des valeurs extrèmes. *Publ. Int. Stat. Paris*, **7**, 37–121 and **8**, 124–184.

[5] Gumbel, E. J. and Goldstein, Neil (1964). Analysis of empirical bivariate extremal distributions. *J. Amer. Stat. Assoc.*, **59**, 794–816.

[6] Gumbel, E. J. and Mustafi, C. K. (1967). Some analytical properties of bivariate extremal distributions. *J. Amer. Stat. Assoc.*, **62**, 569–588.

[7] Marshall, A. W. and Olkin, I. (1967). A multivariate exponential distribution. *J. Amer. Stat. Assoc.*, **62**, 30–44.

[8] Posner, E. C., Rodemich, E. R., Ashlock, J. C. and Lurie, S. (1969). Applications an estimator of high efficiency in bivariate extreme theory. *J. Amer. Stat. Assoc.*, **64**, 1403–1414.

[9] Sibuya, M. (1960). Bivariate extremal statistics. *Ann. Inst. Stat. Math.*, **11**, 195–210.

[10] Tiago de Oliveira, J. (1958). Extremal distributions. *Rev. Fac. Ciências* Lisboa, 2 ser., A, Mat., **7**, 215–227.

[11] Tiago de Oliveira, J. (1962/63). Structure theory of bivariate extremes, extensions. *Est. Mat., Estat. e Econ.*, **7**, 165–195.

[12] Tiago de Oliveira, J. (1968). Extremal processes; definition and properties. *Publ. Inst. Univ. Paris.*, **27**, 25–36.

[13] Tiago de Oliveira, J. (1970). Biextremal distributions; statistical decision. *Trab. Estad. y Inv. Oper.*, **21**, 107–117.

[14] Tiago de Oliveira, J. (1971). A new model of bivariate extremes; statistical decision. *Studi de Probabilità, Statistica e Ricerca Operativa in onore di Guiseppe Pompili*, pp. 437–449. Tip. Oderisi, Gubbio.

[15] Tiago de Oliveira, J. (1974). Regression in the non-differentiable bivariate extreme models. *J. Amer. Stat. Assoc.*, **69**, 816–818.

[16] Tiago de Oliveira, J. (1975). Bivariate and multivariate extremes distribution. *Statistical distributions in Scientific Work*, Vol. **1**, (G. Patil et al., eds.) pp. 355–361. Dr. Reidel Publ. Co.

[17] Tiago de Oliveira, J. (1975). Statistical decision for extremes. *Trab. Estad. y Inv. Oper.*, **26**, 433–471.

[18] Tiago de Oliveira, J. (1978). Familles naturelles de distributions bivariées des extrêmes (to be published).

PART V

TESTS OF HYPOTHESES AND CLASSIFICATION

P. R. Krishnaiah, ed., *Multivariate Analysis–V*

MULTIVARIATE REGRESSION ANALYSIS WITH SPHERICAL ERROR

D. A. S. FRASER and Kai W. NG
University of Toronto, Toronto, Canada

In a recent publication Arnold Zellner [8] has examined the regression model using the multivariate student distribution for the error: tests and confidence regions are derived for the regression coefficients and error variance and are found to correspond closely to those under normal theory; flat and conjugate priors are also used for a Bayesian analysis. This paper examines tests and confidence methods for the multivariate regression model with error distribution given by the multivariate student or more generally by spherical error. The model is based on recent results concerning the identification of distribution form and the analysis uses the parameter and distribution factorization methods developed in D. A. S. Fraser and Jock MacKay [2] and organized for multivariate regression in D. A. S. Fraser and Kai W. Ng [4].

0. Introduction

The formation and validity of statistical models has been discussed in Fraser [5]. As part of this, the distribution form can in certain contexts be identified objectively and correspondingly the context requires the statistical model to use explicitly the distribution describing that distribution form. A model with an explicit (and objective) distribution for error or variation is a structural-type model (Fraser, [5]). This paper examines multivariate regression with a multivariate student or more generally spherical distribution for the explicit error distribution.

The determination of tests and confidence regions for structural models has been developed in D. A. S. Fraser and Jock MacKay [2], and organized for the multivariate regression model in D. A. S. Fraser and Kai W. Ng [4]. The use of the more extensive structural model leads to essentially unique tests and confidence regions for the primary parameters of the model: the order in which parameters are examined is the one available option; and the uniqueness is present without the introduction of the usual reduction principle such as sufficiency and conditionality. The factorization methods developed in the cited papers produced the unique tests and confidence regions. These methods are used in this paper to produce in a direct, straightforward manner the appropriate tests and confidence regions for multivariate regression with student or spherical error.

1. Preliminaries

Arnold Zellner [8] has examined the common regression model using a multivariate student distribution for the error. The multivariate-t distribution in standardized form can be written as

$$f_\lambda(\mathbf{z}) = \frac{\Gamma((\lambda+n)/2)}{\Gamma(\lambda/2)\pi^{n/2}}(1+\lambda^{-1}\mathbf{z}'\mathbf{z})^{-(\lambda+n)/2}\lambda^{-n/2}$$

$$= \frac{A_\lambda}{A_{\lambda+n}}(1+\lambda^{-1}\mathbf{z}'\mathbf{z})^{-(\lambda+n)/2}\lambda^{-n/2} \tag{1}$$

where the use of $A_\lambda = 2\pi^{\lambda/2}/\Gamma(\lambda/2)$ for the surface volume of a unit sphere in $\mathbf{R}^\lambda$ leads to simpler and tidier formulas. Some arguments for the use of students distribution in special applications are given in Zellner [8]. When we proceed to the matrix generalization for the multivariate models we will avoid the nuisance of the scaling factor $\sqrt{\lambda}$ in the density expression and use an alternative standardization given by

$$f_\lambda(\mathbf{z}) = \frac{A_\lambda}{A_{\lambda+n}}(1+\mathbf{z}'\mathbf{z})^{-(\lambda+n)/2}; \tag{2}$$

this corresponds to a reparametrization of the scale parameters in the full statistical model.

Some of the results for the multivariate student distributions extend easily to a general spherical distribution for the error. For this we take

$$f(\mathbf{z}) = g(\mathbf{z}'\mathbf{z}), \tag{3}$$

where $g(\cdot)$ is a non-negative function over the non-negative real numbers. This is the general form of a distribution that is invariant under rotation in the sample space $\mathbf{R}^n$.

Consider the regression model

$$\mathbf{y} = X\boldsymbol{\beta} + \sigma\mathbf{z}, \tag{4}$$

where $\mathbf{y}$ is an $n\times 1$ vector of observable responses, X is an $n\times r$ design matrix with full rank $r<n$, $\boldsymbol{\beta}$ is an $r\times 1$ vector of regression parameters,

$\sigma>0$ is a scale parameter, and $\mathbf{z}$ is an $n\times 1$ vector of unobservable random variables representing the (standardized) error with distribution (1). Note that the shape parameter λ in (1) is generally unknown. For properties of the multivariate-t distribution, see Johnson and Kotz [6], chapter 37 and the references cited within.

Let $\mathbf{y}^0$ be the observed response. Then we have an *inference base*

$$\mathcal{I}=(\mathfrak{M},\mathbf{y}^0), \tag{5}$$

where the *model* $\mathfrak{M}$ consists of (4) and (1), delineating the generation of the *data* $\mathbf{y}^0$ from the random error $\mathbf{z}$. For detailed discussion of an inference base, see Fraser [5]. We start the analysis with the basic question: what do we know about the unobservable $\mathbf{z}$ through the observable $\mathbf{y}$? Or, in other words, what portion of the unobservable $\mathbf{z}$ can be observed and what is the appropriate probabilistic description for the unobservable portion? To answer this, we use some *transparent* coordinates system for the sample space $\mathbf{R}^n$ such that a part of the coordinates represents the observable portion of $\mathbf{z}$ and the remaining part represents the unobservable portion of $\mathbf{z}$. It should be emphasized that there are many choices of the transparent coordinates and that they are all one-to-one equivalent, giving the same probabilistic description.

For each vector $\mathbf{z}$ in the sample space $\mathbf{R}^n$, let $\mathbf{b}(\mathbf{z})$ be the $r\times 1$ vector of regression coefficients of $\mathbf{z}$ on the space spanned by the columns of X, $\mathbf{d}(\mathbf{z})$ be the unit residual vector, and $s(\mathbf{z})$ the residual length:

$$\begin{aligned}\mathbf{b}(\mathbf{z})&=(X'X)^{-1}X'\mathbf{z}\\ \mathbf{s}^2(\mathbf{z})&=(\mathbf{z}-X\mathbf{b}(\mathbf{z}))'(\mathbf{z}-X\mathbf{b}(\mathbf{z}))\\ \mathbf{d}(\mathbf{z})&=s^{-1}(\mathbf{z})(\mathbf{z}-X\mathbf{b}(\mathbf{z})).\end{aligned} \tag{6}$$

This amounts to writing the $\mathbf{R}^n$ as the algebraic direct sum of the subspace $\mathcal{L}(X)$ and its orthogonal complement $\mathcal{L}^{\perp}(X)$, where in $\mathcal{L}(X)$ we use the coordinates relative to the basis $(\mathbf{x}_1,\mathbf{x}_2,\ldots,\mathbf{x}_r)$ while in $\mathcal{L}^{\perp}(X)$ we use $(n-r)$ dimensional spherical coordinates with $s(\cdot)$ as radius and $\mathbf{d}(\cdot)$ as directional unit vector. Thus the volume element in terms of the new coordinates is given by

$$d\mathbf{z}=|X'X|^{1/2}d\mathbf{b}\cdot s^{(n-r)-1}ds\,da, \tag{7}$$

where da is the area element on the unit sphere in $\mathcal{L}^{\perp}(X)$.

In terms of these coordinates for $\mathbf{R}^n$, the model $\mathfrak{M}$ consisting of (4) and (1) is given by

$$\begin{aligned}
\mathbf{d}(\mathbf{y}) &= \mathbf{d}(\mathbf{z}) \\
\mathbf{b}(\mathbf{y}) &= \boldsymbol{\beta} + \sigma \mathbf{b}(\mathbf{z}) \\
s(\mathbf{y}) &= \sigma s(\mathbf{z}) \qquad (8) \\
f_\lambda(\mathbf{z})\, d\mathbf{z} &= \frac{A_\lambda}{A_{\lambda+n}} \frac{|X'X|^{1/2} s^{n-r-1}}{(1+\lambda^{-1}(s^2+\mathbf{b}'X'X\mathbf{b}))^{(\lambda+n)/2}}\, d\mathbf{b}\, d s\, d a,
\end{aligned}$$

where we have used (7) and the relation

$$\mathbf{z} = X\mathbf{b}(\mathbf{z}) + s(\mathbf{z})\mathbf{d}(\mathbf{z}) \tag{9}$$

It is clear in this presentation of the model that the observable portion of the random $\mathbf{z}$ is given by $\mathbf{d}(\mathbf{z}) = \mathbf{d}(\mathbf{y})$ and that the appropriate probabilistic description for the unobservable portion of $\mathbf{z}$ is the conditional distribution of $\mathbf{b}(\mathbf{z})$, $s(\mathbf{z})$ given $\mathbf{d}(\mathbf{z})$. This leads to the quantities

$$\left\{ \begin{aligned} \frac{\mathbf{b}(\mathbf{y}) - \boldsymbol{\beta}}{\sigma} &= \mathbf{b} \\ \frac{s(\mathbf{y})}{\sigma} &= s \end{aligned} \right. \quad \text{or} \quad \left\{ \begin{aligned} \frac{\mathbf{b}(\mathbf{y}) - \boldsymbol{\beta}}{s(\mathbf{y})} &= \mathbf{t} \\ \frac{s(\mathbf{y})}{\sigma} &= s \end{aligned} \right. \tag{10}$$

to be examined conditionally given the observed $\mathbf{d}(\mathbf{z}) = \mathbf{d}(\mathbf{y}^0)$. Thus we have answered the basic question posed earlier concerning the random error $\mathbf{z}$.

The shape parameter λ is the only parameter involved in the error distribution (1) and the observed portion of the error $\mathbf{z}$ is $\mathbf{d}(\mathbf{z})$, so the probability of the observed $\mathbf{d}(\mathbf{z}) = \mathbf{d}(\mathbf{y}^0)$, as a function of λ, is *the likelihood function of* λ. With a given value of λ, the inference of $\boldsymbol{\beta}$ and σ are based on the conditional distribution of the quantities (10).

2. Inference of component parameters

Consider first the inference of the parameters in the order λ, σ, $\boldsymbol{\beta}$. The factorization procedure from Fraser and MacKay [2] separates the error variable $\mathbf{z}$ in the corresponding order $\mathbf{b}$, s, $\mathbf{d}$ with the marginal distribution for **d**, conditional distribution for s given **d** and the conditional distribution for **b** given **d**, s. In this case, the error distribution (1) becomes

$$f_\lambda(\mathbf{z})\,d\mathbf{z} = \frac{A_\lambda}{A_{\lambda+n}} \frac{|X'X|^{1/2} s^{n-r-1} \lambda^{-n/2}}{(1+\lambda^{-1}(s^2+\mathbf{b}'X'X\mathbf{b}))^{(\lambda+n)/2}}\, d\mathbf{b}\, ds\, d\mathbf{a}$$

$$= \frac{A_{\lambda+n-r}}{A_{\lambda+n-r+r}} \frac{(\lambda+s^2)^{-r/2}|X'X|^{1/2}}{\left(1+(\lambda+s^2)^{-1}\mathbf{b}'X'X\mathbf{b}\right)^{(\lambda+(n-r)+r)/2}}\, d\mathbf{b} \tag{11}$$

$$\times \frac{A_\lambda A_{n-r}}{A_{\lambda+n-r}} \frac{s^{n-r-1}\lambda^{-(n-r)/2}}{(1+s^2/\lambda)^{(\lambda+n-r)/2}}\, ds \tag{12}$$

$$\times \frac{1}{A_{n-r}}\, d\mathbf{a}; \tag{13}$$

and the spherical error distribution (3) becomes

$$g(\mathbf{z}'\mathbf{z})d\mathbf{z} = h^{-1}(s^2) g(s^2+\mathbf{b}'X'X\mathbf{b})|X'X|^{1/2} d\mathbf{b} \tag{11a}$$

$$\times A_{n-r} h(s^2) s^{n-r-1}\, ds \tag{12a}$$

$$\times \frac{1}{A_{n-r}}\, d\mathbf{a}, \tag{13a}$$

where the non-negative function $h(\cdot)$ over the non-negative real values is obtained through the normalization of the conditional distribution (11a) given **d**, s.

In both cases the marginal probability distribution for the observable **d** is a uniform distribution on the unit sphere in $\mathfrak{L}^{\perp}(X)$, not depending on λ. Thus there is no discriminatory information available concerning the values of the shape parameter λ alone. With several runs of the regression model we can test the goodness of fit for the spherical error (against non-spherical error) using the tests of uniform distributions on the surface of unit hyperspheres. See Prentice [7] for such tests.

For the parameter σ given λ we use the relation

$$\frac{s(\mathbf{y})}{\sigma} = s(\mathbf{z}),$$

where the random variable $s = s(\mathbf{z})$ has the distribution (12) in the case of

Student's error and (12a) in the case of spherical error. Note that if we write the above relation as

$$\frac{s^2(\mathbf{y})}{(n-r)\sigma^2} = \frac{s^2}{n-r}$$

then $s^2/(n-r)$ has an $F(n-r,\lambda)$ distribution for the case of Student's error, and it becomes a Chi-square with $(n-r)$ degrees of freedom as $\lambda \to \infty$, which corresponds to the case of normal error. So the tests or confidence intervals for σ^2 are based on F-distribution. It agrees with Zellner's results for σ^2 in [8].

For the tests and confidence regions of $\boldsymbol{\beta}$ we use the relation

$$\frac{\mathbf{b}(\mathbf{y}) - \boldsymbol{\beta}}{\sigma} = \mathbf{b}$$

and the distribution (13) in the case of Student's error and (13a) in the case of spherical error. Note that in the case of Student's error, the distribution (13) is a multivariate Student $(\lambda + n - r; \mathbf{0}, (\lambda + s^2)^{-1}(X'X)^{-1})$, which becomes a multivariate normal as $\lambda \to \infty$ (the case of a normal error).

Now let us consider the inference of the parameters in a second order: λ, $\boldsymbol{\beta}$, σ. Here we use the second presentation of (10) and the variables of the error will be considered in the order $\mathbf{d}$, $\mathbf{t}$, s. In the case of Student's error (1), we have

$$f_\lambda(\mathbf{z})\,d\mathbf{z} = \frac{A_\lambda}{A_{\lambda+n}} \frac{|X'X|^{1/2} s^{n-1} \lambda^{-n/2}}{\left[1 + \lambda^{-1}(1 + \mathbf{t}'X'X\mathbf{t})s^2\right]^{(\lambda+n)/2}}\, ds\, d\mathbf{t}\, d\mathbf{a}$$

$$= \frac{A_\lambda A_n}{A_{\lambda+n}} \frac{(1 + \mathbf{t}'X'X\mathbf{t})^{n/2} s^{n-1} \lambda^{-n/2}}{\left[1 + \lambda^{-1}(1 + \mathbf{t}'X'X\mathbf{t})s^2\right]^{(\lambda+n)/2}}\, ds \tag{14}$$

$$\times \frac{A_{n-r}}{A_n} \frac{|X'X|^{1/2}}{\left[1 + \mathbf{t}'X'X\mathbf{t}\right]^{(n-r+r)/2}}\, d\mathbf{t} \tag{15}$$

$$\times \frac{1}{A_{n-r}}\, d\mathbf{a}, \tag{16}$$

and in the case of spherical error (3) we have

$$g(\mathbf{z}'\mathbf{z})\,d\mathbf{z} = A_n(1+\mathbf{t}'X'X\mathbf{t})^{n/2} g(\mathbf{z}'\mathbf{z}) s^{n-1} \lambda^{-n/2} ds \tag{14a}$$

$$\times \frac{A_{n-r}}{A_n} \frac{|X'X|^{1/2}}{(1+\mathbf{t}'X'X\mathbf{t})^{(n-r+r)/2}} d\mathbf{t} \tag{15a}$$

$$\times \frac{1}{A_{n-r}} da. \tag{16a}$$

The only observable of $\mathbf{z}$ is $\mathbf{d}(\mathbf{z})=\mathbf{d}(\mathbf{y})$ which has the uniform distribution (16), (16a) on the unit sphere in $\mathcal{L}^{\perp}(X)$. The analysis is the same as the first ordering.

For the tests and confidence regions of $\boldsymbol{\beta}$ given λ we use the relation

$$\frac{\mathbf{b}(\mathbf{y})-\boldsymbol{\beta}}{s(\mathbf{y})} = \mathbf{t},$$

where the variable $\mathbf{t}$ has distribution (12) which is exactly the same as (12a). Note that this distribution is a multivariate Student $(n-r;\ \mathbf{0},(X'X)^{-1})$, not depending on the shape parameter λ. It coincides with the result from normal theory and it agrees with Zellner's analysis for $\boldsymbol{\beta}$ in [8].

For the tests and confidence intervals of σ given $\lambda, \boldsymbol{\beta}$ we use the relation

$$\frac{s(\mathbf{y})}{\sigma} = s(\mathbf{z}),$$

where $s = s(\mathbf{z})$ has distribution (14) in the case of Student's error and (14a) in the case of spherical error. Note that the distribution (14) for s is equivalent to an $F(n,\lambda)$ distribution for $s^2(1+\mathbf{t}'X'X\mathbf{t})/n$.

This second order of examining may be the more natural way. It corresponds in the ordinary analysis of variance to the sequential pooling of error variance. For any amenable ordering of the parameters, the shape λ is first considered, using the likelihood function; and in the error distribution here we have the peculiarity that there is no observable information concerning λ alone.

Other than the two orders of parameters, one may examine σ using the first and examine $\boldsymbol{\beta}$ using the second. This does not control the overall confidence level as with the sequential procedures of either of the two orders (see [5] for general discussion), but it does allow the separate examination of the location and the scale parameters. Computer simulations to date give some preference for this alternative procedure.

3. Multivariate regression with linear scaling

The methods of presentation and analysis given for the univariate regression model can now be mechanized routinely to give the methods and distributions theory for multivariate regression. For this we distinguish two cases corresponding to different degree of identifiability of the error distribution involved. In this section we examine the case where the error distribution is identifiable only up to a positive linear transformation. Then in section 4 we consider the case in which the error distribution is further identified up to a positive triangular transformation; this corresponds to a special case in which there is a known ordering to the response variables and the distributional information is relative to this ordering.

Now consider the model

$$\begin{aligned} &Y = \mathfrak{B}X + \Gamma Z \\ &f_\lambda(Z)\,dZ \end{aligned} \tag{17}$$

where Y is a $p \times n$ matrix of observable responses, $\mathfrak{B}$ a $p \times r$ matrix of regression parameters, X a $r \times n$ design matrix of regressors with full rank $r < n$, Γ a $p \times p$ matrix of scaling parameters with $|\Gamma| > 0$, Z a $p \times n$ matrix of unobservable error variables with the density $f_\lambda(Z)$. We assume that the density is a canonical matrix student:

$$f_\lambda(Z) = \frac{A_\lambda^{(p)}}{A_{\lambda+n}^{(p)}} |I_p + Z'Z|^{-(\lambda+n)/2} \tag{18}$$

where we denote $A_k^{(p)} = A_k A_{k-1} \cdots A_{k-p+1}$ for the decreasing product of surface areas of the unit spheres in $\mathbf{R}^k, \mathbf{R}^{k-1}, \ldots, \mathbf{R}^{k-p+1}$ (please see (1)). Or more generally, we assume that the density has the spherical form:

$$f_\lambda(Z) = g(ZZ'), \tag{19}$$

where $g(\cdot)$ is *any* non-negative function over the $p \times p$ positive definite matrices such that $f_\lambda(Z)$ is a density. This is the general form for a distribution invariant under rotations through the sample: Z and ZO have identical distribution for any $n\times n$ orthogonal matrix O.

Following Section 1, we want to present the model (17) in a more transparent coordinates system in which the observable portion and the unobservable portion of the random error Z are readily recognized. For this purpose, it is convenient to view a matrix Z in $\mathbf{R}^{pn}$ as a sequence $\mathbf{z}_1,\mathbf{z}_2,\ldots,\mathbf{z}_p$ of p vectors in $\mathbf{R}^n$. Now for any Z, project an arbitrary but fixed sequence of p linearly independent vectors (say the first p of the n unit vectors in $\mathbf{R}^n$) onto the $(r+p)$-dimensional subspace $\mathcal{L}^+(\mathbf{x}_1,\ldots,\mathbf{x}_r;\mathbf{z}_1,\ldots,\mathbf{z}_p)$, or simply $\mathcal{L}^+(X;Z)$, spanned by the row vectors of X and Z together with that order as the positive orientation. The p projections are then orthonormalized in that sequence to obtain p orthonormal vectors $\mathbf{d}_1,\mathbf{d}_2,\ldots,\mathbf{d}_p$, which are orthogonal to $\mathbf{x}_1,\ldots,\mathbf{x}_r$. This procedure gives a basis $\mathbf{x}_1,\ldots,\mathbf{x}_r,\mathbf{d}_1,\ldots,\mathbf{d}_p$ for the subspace $\mathcal{L}^+(X;Z)$ except for a set of measure zero for which the projections are linearly dependent. Note that

$$D(Z)=\begin{bmatrix}\mathbf{d}_1\\ \vdots\\ \mathbf{d}_p\end{bmatrix}$$

depends on $\mathcal{L}^+(X;Z)$ but not otherwise on Z. Thus $D(Z)=D(Y)$ if and only if $\mathcal{L}^+(X;Z)=\mathcal{L}^+(X;Y)$.

We denote the regression coefficients of $\mathbf{z}_j$ on $\mathbf{x}_1,\ldots,\mathbf{x}_r$ and $\mathbf{d}_1,\ldots,\mathbf{d}_p$ by $\mathbf{b}_j=(b_{j1},\ldots,b_{jr})$ and $\mathbf{c}_j=(c_{j1},\ldots,c_{jp})$, for $j=1,\ldots,p$. Writing

$$B(Z)=\begin{bmatrix}\mathbf{b}_1\\ \vdots\\ \mathbf{b}_p\end{bmatrix},\qquad C(Z)=\begin{bmatrix}\mathbf{c}_1\\ \vdots\\ \mathbf{c}_p\end{bmatrix},$$

we have the relation between the old and new coordinates:

$$Z=B(Z)X+C(Z)D(Z)=BX+CD. \tag{20}$$

As Z varies in $\mathbf{R}^{pn}, D(\cdot)$ traces out smoothly the set of all p-dimensional subspaces of $\mathbf{R}^{n-r}$, which is a copy of the Grassman manifold $\mathfrak{D}_{p,n-r}$. Let dD denote the volume element of $\mathfrak{D}_{p,n-r}$ orthogonal to the subspaces

$\mathfrak{L}^+(X;Z)$, the relation among the volume element is

$$dZ = |XX'|^{p/2}|C|^{n-p-r}\,dB\,dC\,dD. \tag{21}$$

In terms of the new coordinates, the model (17) becomes

$$\begin{aligned}
&D(Y) = D(Z)\\
&B(Y) = \mathfrak{B} + \Gamma B(Z)\\
&C(Y) = \Gamma C(Z)\\
&f_\lambda(BX + CD)|XX'|^{p/2}|C|^{n-p-r}\,dB\,dC\,dD,
\end{aligned} \tag{22}$$

where we have abbreviated $D(Z)$, $B(Z)$, $C(z)$ by D, B, C in the probability element. The variable $D(Z)$ is directly observable, and, given data $\mathfrak{Y}^0$, the appropriate probabilistic description for the unobservable $B(Z)$, $C(Z)$ is the conditional distribution of B, C given $D = D(Y^0)$. This leads to the quantities

$$\begin{cases}\Gamma^{-1}(\mathfrak{B}(\mathfrak{Y}) - B) = B\\ \Gamma^{-1}C(\mathfrak{Y}) = C\end{cases} \quad \text{or} \quad \begin{cases}C^{-1}(Y)(B(Y) - \mathfrak{B}) = H\\ \Gamma^{-1}C(Y) = C\end{cases} \tag{23}$$

to be examined conditionally given $D = D(Y^0)$. Note that in the second presentation of (23), we have made a further change of variable: $H = C^{-1}B$.

We first examine the parameters in the order: λ, Γ, $\mathfrak{B}$. The corresponding order for the quantities is then: D, C, B. According to this order, the error distribution (18) is factored

$$f_\lambda(Z)\,dZ = \frac{A_\lambda^{(p)}}{A_{\lambda+n}^{(p)}}\frac{|X'X|^{p/2}|C|^{n-r-p}}{|I + CC' + BXX'B'|^{(\lambda+n)/2}}\,dB\,dC\,dD$$

$$= \frac{A_{\lambda+n-r}^{(p)}}{A_{\lambda+n}^{(p)}}\frac{|I + CC'|^{(\lambda+(n-r))/2}|XX'|^{p/2}}{|I + CC' + BXX'B'|^{(\lambda+(n-r)+r)/2}}\,dB \tag{24}$$

$$\times\frac{A_{\lambda:p}^{(p)}A_{n-r:p}^{(p)}}{A_{\lambda+n-r:p}^{(p)}}\frac{|C|^{n-r-p}}{|I + CC'|^{(\lambda+(n-r))/2}}\,dC \tag{25}$$

$$\times\frac{1}{A_{n-r:p}^{(p)}}\,dD, \tag{26}$$

where we have written

$$A^{(p)}_{m:k} = A^{(p)}_m / A^{(p)}_k.$$

For the spherical error (19), we have

$$g(ZZ')\mathrm{d}Z = h^{-1}(CC')g(CC' + BXX'B')|XX'|^{p/2}\mathrm{d}B \tag{24a}$$

$$\times A^{(p)}_{n-r:p} h(CC')|C|^{n-r-p}\mathrm{d}C \tag{25a}$$

$$\times \frac{1}{A^{(p)}_{n-r:p}}\mathrm{d}D, \tag{26a}$$

where

$$h(CC') = \int g(CC' + BXX'B')|X'X|^{p/2}\mathrm{d}B$$

is the norming constant of (24a)

In both cases, the marginal distribution for D is the uniform distribution with respect to the volume measure of the Grassman manifold, not depending on the shape parameter λ. The likelihood function of λ, like the case in section 2, is not informative.

For the parameter Γ given λ we use the relation

$$\Gamma^{-1}C(Y) = C(Z)$$

where the random variable $C = C(Z)$ has the distribution (25) in the case of Student's error (18), and (25a) in the case of spherical error (19). The distribution (25) is a positive root F distribution $F_p^{\frac{1}{2}}(n-r,\lambda)$ for the $p\times p$ square matrices C with positive determinant, in the sense that the positive definite matrix $F = CC'$ has a matrix-F distribution:

$$\frac{A^{(p)}_{n-r}A^{(p)}_{\lambda}}{A^{(p)}_{n-r+\lambda}} \cdot \frac{|F|^{(n-r)/2}}{|1+F|^{(n-r+\lambda)/2}} \cdot \frac{\mathrm{d}F}{2^p|F|^{(p+1)/2}} \tag{27}$$

For the population covariance matrix $\Sigma = \Gamma\Gamma'$, we note that $C'(Y)\Sigma^{-1}C(Y) = C'C$ and that $C'C$ has the same distribution as CC', which is just the matrix-F in (27).

For the parameter $\mathcal{B}$ given λ, Γ, we use

$$\Gamma^{-1}(B(Y) - \mathcal{B}) = B(Z),$$

where the random variable $B = B(Z)$ has the distribution (24) in the case of Student's error, and (24a) in the case of spherical error. Note that the distribution of (24) is a matrix-t distribution, $T_{p\times r}(n-r+\lambda; O, I+CC', (XX')^{-1})$.

Now we examine the parameters in a more natural order for the regression purpose: Γ, $\mathfrak{B}$, λ. The quantities in the second presentation of (23) are thus examined in the order: D, H, C. For the Student's error (18), we have

$$f_\lambda(Z)\,\mathrm{d}Z = \frac{A_\lambda^{(p)}}{A_{\lambda+n}^{(p)}} \frac{|XX'|^{p/2}|C|^{n-p}}{|I+C(I+HXX'H')C'|^{(\lambda+n)/2}}\,\mathrm{d}C\,\mathrm{d}H\,\mathrm{d}D \tag{28}$$

$$= \frac{A_{\lambda:p}^{(p)}A_{n:p}^{(p)}}{A_{\lambda+n:p}^{(p)}} \frac{|I+HXX'H'|^{n/2}|C|^{n-p}}{|I+C(I+HXX'H')C'|^{(\lambda+n)/2}}\,\mathrm{d}C \tag{29}$$

$$\times \frac{A_{n-r}^{(p)}}{A_n^{(p)}} \frac{|XX'|^{p/2}}{|I+HXX'H'|^{(n-r+r)/2}}\,\mathrm{d}H \tag{30}$$

$$\times \frac{1}{A_{n-r:p}^{(p)}}\,\mathrm{d}D, \tag{31}$$

where (28) is obtained using a change of variable $(B,C)\to(H,C)$ with $B = CH$,

$$Z = C(HX+D), \mathrm{d}Z = |XX'|^{p/2}|C|^{n-p}\,\mathrm{d}C\,\mathrm{d}H\,\mathrm{d}D. \tag{32}$$

And for the spherical error (19) we have

$$g(ZZ')\,\mathrm{d}Z$$

$$= A_{n:p}^{(p)}|I+HXX'H'|^{n/2}g_\lambda(C(I+HXX'H')C')|C|^{n-p}\,\mathrm{d}C \tag{29a}$$

$$\times \frac{A_{n-r}^{(p)}}{A_n^{(p)}} \frac{|XX'|^{p/2}}{|I+HXX'H'|^{(n-r+r)/2}}\,\mathrm{d}H \tag{30a}$$

$$\times \frac{1}{A_{n-r:p}^{(p)}}\,\mathrm{d}D \tag{31a}$$

In both error cases, the marginal distribution for the observable $D(Z) =$

$D(Y)$ is, as in the first order considered above, a uniform distribution on the Grassman manifold $\mathcal{D}_{p,n-r}$, not depending on the shape parameter λ. Hence the likelihood function of λ is a constant function, giving no discriminatory information about λ.

For the parameter $\mathcal{B}$ given λ, we use

$$C^{-1}(Y)(B(Y)-\mathcal{B})=H(Z),$$

where $H=(H(Z)$ has a matrix-t distribution, $T_{p\times r}(n-r;O,I,(XX')^{-1})$, given by (30) and (30a). Note that this distribution does not depend on λ and that the inference about $\mathcal{B}$ is identical with the case of normal error.

For the parameter of Γ given λ, $\mathcal{B}$, we use

$$\Gamma^{-1}C(Y)=C(Z),$$

where $C=C(Z)$ has the distribution (29) and (29a) respectively for the Student's and spherical error. Note that (29) is a positive root F distribution, $F_p^{\frac{1}{2}}(n,\lambda)$, as defined earlier.

In the second order of examining the parameters, the conditional tests and confidence regions for Γ correspond to the pooling of error variance. One may prefer the 'mixed' procedure of using the first order for the inference of Γ while using the second for the inference of $\mathcal{B}$; but then the overall confidence level cannot be controlled. With this approach we have a matrix-t distribution for the inference of $\mathcal{B}$ as in the normal case while we have a matrix-F distribution for the inference of $\Sigma^{-1}=(\Gamma\Gamma')^{-1}$ instead of a Wishart distribution as in the normal case.

4. Multivariate regression with triangular scaling

In this section we consider the multivariate regression model for the rather special case in which the response variables have a known sequential pattern and the error term can be identified up to a positive triangular transformation. The structural model has the form

$$Y=\mathcal{B}X+\mathcal{T}Z$$

$$f_\lambda(Z)\,dZ \tag{32}$$

where Y, $\mathcal{B}$, X and Z are matrices as defined in section 3, and $\mathcal{T}$ is a $p\times p$ positive lower triangular matrix of scaling parameters. The error distribution $f_\lambda(Z)$ is assumed to be (18) or (19) in section 3.

For the space $\mathbf{R}^{pn}$ of all $p\times n$ matrices Z, we change coordinates in the following way. Let $B(Z)$ be the $p\times r$ matrix of regression coefficients of the p row vectors of Z on X,

$$B(Z)=ZX'(XX')^{-1}. \tag{33}$$

The matrix $Z-B(Z)X$ then consists of the p residual vectors. We orthonormalize these p residual vectors and keep that orientation, getting

$$Z-B(Z)X=T(Z)D(Z), \tag{34}$$

where $T(Z)$ is a $p\times p$ positive lower triangular matrix and $D(Z)$ is a $p\times n$ semi-orthogonal matrix with row vectors orthogonal to the row vectors of X

$$D(Z)D'(Z)=I_p, \qquad XD'(Z)=O \tag{35}$$

so that

$$Z=B(Z)X+T(Z)D(Z). \tag{36}$$

As Z varies in $\mathbf{R}^{pn}$, $D(\cdot)$ traces out smoothly the set of all those p-frames (i.e. ordered sequences of orthonormal vectors) in $\mathbf{R}^n$ that are orthogonal to $\mathcal{L}(X)$. This set is a copy of the Stiefel submanifold in $\mathbf{R}^{p(n-r)}$ with dimension $p(n-r)-p(p+1)/2$. Let dD denotes its volume measure. The volume elements are then related

$$dZ=|XX'|^{p/2}|T|^{n-r}|T|_{\Delta}^{-1}\,dB\,dT\,dD, \tag{37}$$

where $|T|_{\Delta}=t_{11}t_{22}^2\cdots t_{pp}^p$ denotes the product of the diagonal elements of T in increasing power. Note that we have abbreviated $B(Z)$, $T(Z)$ and $D(Z)$ by B, T, D respectively. We do this whenever it is convenient.

The model (32) can now be presented as

$$\begin{aligned}&D(Y)=D(Z)\\&B(Y)=\mathfrak{B}+\mathfrak{T}B(Z)\\&T(Y)=\mathfrak{T}T(Z)\\&f_{\lambda}(BX+TD)|XX'|^{p/2}|T|^{n-r}|T|_{\Delta}^{-1}\,dB\,dT\,dD\end{aligned} \tag{38}$$

It is clear in this presentation that $D(Z)$ is the only observable portion and that the quantities

$$\begin{cases} B(Y)=\mathfrak{B}+\mathfrak{T}B \\ T(Y)=\mathfrak{T}T \end{cases} \quad \text{or} \quad \begin{cases} T^{-1}(Y)(B(Y)-\mathfrak{B})T^{-1}B=H \\ \mathfrak{T}^{-1}T(Y) \end{cases} \tag{39}$$

would be examined together with the conditional distribution of B, T given $D=D(Y)$.

We first examine the parameters in the order: λ, $\mathfrak{T}$, $\mathfrak{B}$. This leads to the sequence D, T, B for the error variable. The error distribution (18) becomes

$$f_\lambda(Z)\,\mathrm{d}Z=\frac{A^{(p)}_{\lambda+n-r}}{A^{(p)}_{\lambda+n}}\frac{|I+TT'|^{(\lambda+n-r)/2}|XX'|^{p/2}}{|I+TT'+BXX'B'|^{(\lambda+n-r+r)/2}}\,\mathrm{d}B \tag{40}$$

$$\times\frac{A^{(p)}_{n-r}A^{(p)}_{\lambda}}{A^{(p)}_{n-r+\lambda}}\frac{|T|^{n-r}|T|^{-1}_{\Delta}}{|I+TT'|^{(n-r+\lambda)/2}}\,\mathrm{d}T \tag{41}$$

$$\times\frac{1}{A^{(p)}_{n-r}}\,\mathrm{d}D; \tag{42}$$

and the distribution (19) becomes

$$g(ZZ')\,\mathrm{d}Z=h^{-1}(TT')g(TT'+BXX'B')|XX'|^{p/2}\,\mathrm{d}B \tag{40a}$$

$$\times A^{(p)}_{n-r}h(TT')|T|^{n-r}|T|^{-1}_{\Delta}\,\mathrm{d}T \tag{41a}$$

$$\times\frac{1}{A^{(p)}_{n-r}}\,\mathrm{d}D. \tag{42a}$$

The marginal distribution of the observable D is uniform with respect to the volume measure, in a Stiefel manifold providing no discriminatory information of the shape parameter λ.

For the parameter $\mathfrak{T}$ given λ, we use

$$\mathfrak{T}^{-1}T(Y)=T,$$

where $T=T(Z)$ is subject to the distribution (41) or (41a). The distribution (41) is a *triangular root-F* distribution $\Delta F_p^{\frac{1}{2}}(n-r,\lambda)$ in the sense that the positive definite matrix $F=TT'$ has the matrix-F distribution (27). For the covariance matrix $\Sigma=\mathfrak{T}\mathfrak{T}'$, we note that $T'(Y)\Sigma^{-1}T(Y)=T'T=V$ and that V has what we call here 'disguised matrix-F' distribution:

$$\frac{A_{n-r}^{(p)}A_{\lambda}^{(p)}}{A_{n-r+\lambda}^{(p)}}\frac{|V|^{(n-r)/2}}{|I+V|^{(n-r+\lambda)/2}}\frac{|T|_{\nabla}}{|T|_{\Delta}}\frac{\mathrm{d}V}{2^{p}|V|^{(p+1)/2}}, \tag{43}$$

where $|T|_{\nabla}=t_{11}^{p}t_{22}^{p-1}\cdots t_{pp}$ is the product of the diagonal elements of T in decreasing power. Note that the only difference between (27) and (43) is the 'disguising factor' $|T|_{\nabla}/|T|_{\Delta}$. So the inference of Σ^{-1} is based on a disguised matrix-F distribution.

For the parameter $\mathfrak{B}$ given λ, $\mathfrak{T}$, we use

$$\mathfrak{T}^{-1}(B(Y)-\mathfrak{B})=B(Z),$$

where $B=B(Z)$ is subject to the distribution (40) or (40a). The distribution (40) is a matrix-t distribution, $T_{p\times r}(n-r+\lambda;O,I+TT',(XX')^{-1})$.

At this point we introduce some notation developed in [1], for uses in the sequal. For a $p\times p$ matrix V let $V^{(k)}$ and $V_{(k)}$ be the upper-left and low-right $k\times k$ matrix and define respectively the decreasing determinant and increasing determinant of V as

$$|V|_{\nabla}=\prod_{k=1}^{p}|V^{(k)}|,|V|_{\Delta}=\prod_{k=1}^{p}|V_{(k)}|. \tag{44}$$

These definitions generalized the decreasing and increasing determinants of triangular matrices. If V is a positive definite matrix such that $V=LL'$, $V=T'T$, where both L and T are positive lower triangular, then

$$|V|_{\nabla}=|L|_{\nabla}^{2},\qquad |V|_{\Delta}=|T|_{\Delta}^{2} \tag{45}$$

$$|V|^{(p+1)/2}=|L|_{\Delta}|L|_{\nabla}=|T|_{\Delta}|T|_{\nabla},$$

so that

$$|L|_{\nabla}/|L|_{\Delta}=|V|_{\nabla}|V|^{-(p+1)/2}$$

$$|T|_{\nabla}/|T|_{\Delta}=|V|_{\nabla}^{-1}|V|^{(p+1)/2}. \tag{46}$$

For instance, the disguised matrix-F distribution (43) can be expressed completely in terms of V:

$$\frac{A_{n-r}^{(p)}A_{\lambda}^{(p)}}{A_{n-r+\lambda}^{(p)}}\cdot\frac{|V|^{(n-r)/2}}{|I+V|^{(n-r+\lambda)/2}}\cdot\frac{\mathrm{d}V}{2^{p}|V|_{\nabla}} \tag{47}$$

Consider now the parameters in the order: λ, $\mathfrak{B}$, $\mathfrak{T}$. The sequence of error variables is then: D, H, T, using the second presentation of (39). The error distribution (18) becomes

$$f_\lambda(Z)\,\mathrm{d}Z = \frac{A^{(p)}_{\lambda}}{A^{(p)}_{\lambda+n}} \cdot \frac{|XX'|^{p/2}|T|^n|T|_\Delta^{-1}}{|I+T(I+HXX'H')T'|^{(\lambda+n)/2}}\,\mathrm{d}T\,\mathrm{d}H\,\mathrm{d}D$$

$$= \frac{A^{(p)}_{\lambda}A^{(p)}}{A^{(p)}_{\lambda+n}} \frac{|I+HXX'H'|^{(n-p-1)/2}|1+HXX'H'|_\nabla|T|^n|T|_\Delta^{-1}}{|I+T(I+HXX'H')T'|^{(\lambda+n)/2}}\,\mathrm{d}T \tag{48}$$

$$\times \frac{A^{(p)}_{n-r}}{A^{(p)}_{n}} \frac{|XX'|^{p/2}}{|I+HXX'H'|^{n/2}} \frac{|I+HXX'H'|^{(p+1)/2}}{|I+HXX'H'|_\nabla}\,\mathrm{d}H \tag{49}$$

$$\times \frac{1}{A^{(p)}_{n-r}}\,\mathrm{d}D, \tag{50}$$

where the conditional distribution (48) for T is obtained by transforming the standard triangular root-F distribution (41). The distribution (49) is a 'disguised matrix-t' with 'disguising' factor

$$q = \frac{|I+HXX'H'|^{(p+1)/2}}{|I+HXX'H'|_\nabla}. \tag{51}$$

A disguised matrix-t distribution is related to a disguised Wishart distribution in the same way that a matrix-t distribution is related to a Wishart distribution; see [9], [4] and [1].

Similarly, the error distribution (19) becomes

$$g(ZZ')\,\mathrm{d}Z = A^{(p)}_n|I+HXX'H'|^{n/2}g(T(I + HXX'H')T')|T|^n|T|_\Delta^{-1}q^{-1}\,\mathrm{d}T \tag{48a}$$

$$\times \frac{A^{(p)}_{n-r}}{A^{(p)}_{n}} \frac{|XX'|^{p/2}}{|I+HXX'H'|^{n/2}} \cdot q\,\mathrm{d}H \tag{49a}$$

$$\times \frac{1}{A^{(p)}_{n-r}}\,\mathrm{d}D \tag{50a}$$

The marginal distribution for the observable D is uniform with respect to the volume measure in the Stiefel manifold as in the first order of parameters.

For the parameter $\mathfrak{B}$ given λ, we have the disguised matrix-t distribution (49), which is the same as (49a), together with the relation

$$T^{-1}(Y)(B(Y)-\mathfrak{B})=H.$$

This coincides with the corresponding result for a normal error.

For the parameter $\mathfrak{T}$ given λ, $\mathfrak{B}$, we have the distribution (48) or (48a), together with the relation

$$\mathfrak{T}^{-1}T(Y)=T.$$

As for the model in section 3, one may prefer the mixed procedure of using the first order for inference of $\mathfrak{T}$ and the second order for the inference of $\mathfrak{B}$. With this approach we will have a disguised matrix-t distribution for the inference of $\mathfrak{B}$ as in the normal case, while we will have a disguised matrix-F for $\Sigma^{-1}=(\mathfrak{T}\mathfrak{T}')^{-1}$ instead of a disguised Wishart distribution as in the normal case.

References

[1] Bishop, Fraser D. A. S., and Ng, K. W. (1979). Some decompositions of spherical distributions, Statistische Hefte, forthcoming.

[2] Fraser, D. A. S. and MacKay, J. (1975). Parameter factorization and inference based on significance, likelihood and objective prior. *Ann. Statist.* **3**, 559–572.

[3] Fraser, D. A. S. (1976). *Probability and Statistics*. North Scitnate, MA, Duxbury.

[4] Fraser, D. A. S. and Ng, K. W. (1977). Inference for the multivariate regression model. In *Multivariate Analysis—IV* (P. R. Krishnaiah, Ed.) North-Holland, Amsterdam.

[5] Fraser, D. A. S. (1979). *Inference and Linear Models*. McGraw Hill International, Dusseldorf.

[6] Johnson, N. L. and Kotz, S. (1972). *Distributions in Statistics: Continuous Multivariate Distributions*. Wiley, New York.

[7] Prentice, M. J. (1978), On invariant tests of uniformity for directions and orientations. *Ann. Statist.* **6**, 169–176.

[8] Zellner, A. (1976). Bayesian and non-Bayesian analysis of the regression model with multivariate student-t error terms. *J. Amer. Statist. Assoc.* **71**, 400–408.

[9] Tan, W. Y. and Guttman, I. (1971). A disguised Wishart variable and a related theorem, *J. Roy. Statist. Soc. B.* **33**, 147–152.

P. R. Krishnaiah, ed., *Multivariate Analysis–V*

SAMPLE REUSE SELECTION AND ALLOCATION CRITERIA

Seymour GEISSER*
University of Minnesota, Twin Cities, MN, U.S.A.

1. Introduction

We present several sample reuse criteria for selecting among alternative densities, or probability functions, those most appropriate for classifying or allocating new observations. In standard model selection problems, when densities are completely specified, the choice between models can appropriately be made to rest on a comparison of the alternative probability densities of the sample as determined by the models. Assume that there are two distinct populations π_1 and π_2 from which sets of values have been observed $X_1 = x_1$ and $X_2 = x_2$ respectively with designation for the joint set of random variables $X = (X_1, X_2)$ and $X_i = \{X_{i1}, \ldots, X_{iN_i}\}$, $i = 1,2$. Then we need to determine the most appropriate density $f_\omega(x|\pi_i)$, indexed by the double designator ω, $\omega \in \Omega$, which jointly specifies a pair of densities for π_1 and π_2, where Ω represents the totality of such pairs of potential predicting densities under consideration. This is accomplished by obtaining that ω which maximizes

$$p(\omega) f_\omega(x_1, x_2|\pi_1, \pi_2) \tag{1.1}$$

w.r.t. ω for $p(\omega)$ the prior probability that ω is the correct density pair.

Of course, if we know the true ω^*, then the allocation or diagnosis of a new observation $Z = z$ should depend on the posterior odds ratio

$$R(\omega^*, z) = \frac{q_1 f_{\omega^*}(z|\pi_1)}{q_2 f_{\omega^*}(z|\pi_2)} = \frac{\Pr[\pi_1|z, \omega^*]}{\Pr[\pi_2|z, \omega^*]} \tag{1.2}$$

where $q_i = \Pr[\pi_i]$, the prior probability that $z \in \pi_i$.

More generally, we can incorporate available probabilistic information

*This work was supported in part by a NIH-GMS research grant.

on ω. Assuming q_i is independent of ω and noting that

$$p(\omega|z)=f(z|\omega)p(\omega)/f(\omega),$$

then we can calculate

$$\Pr[\pi_i|z]=E_{\omega|z}[\Pr(\pi_1|z,\omega)]\propto q_i E_\omega[f_\omega(z|\pi_i)]$$

and

$$R(z)=\Pr[\pi_i|z]/\Pr[\pi_i|z],$$

where the first expectation is over the conditional probability of ω given z and the second is merely over the marginal probability function $p(\omega)$. In practice ω would usually range over a small finite number of discrete possibilities so that

$$\Pr[\pi_i|z]\propto q_i \sum_{k=1}^{K} f_{\omega_k}(z|\pi_i)p(\omega_k) \tag{1.3}$$

would be a simple mixture of densities. Although, assuming we know $p(\omega)$, there is no intrinsic difficulty in applying (1.3) directly for classification purposes, our goal is often a compromise in that we would both like to select a model and then use that model for classification. When this is the case we shall not average over ω, as it were, but make the choice of ω in keeping with our ultimate goal of classification.

To further complicate the issue it is a fact that in most instances a designator ω is a specification of the form of the density with values of the parameters unknown. Such cases then would require that prior distributions for the parameters be introduced and predictive (marginal) densities be calculated for X for each ω. Maximization of

$$p(\omega)f_\omega(x_1,x_2|\pi_1,\pi_2) \tag{1.4}$$

w.r.t. ω, where, assuming the X_{ij}'s are independently distributed,

$$f_\omega(x_1,x_2|\pi_1,\pi_2)=\int \prod_{i=1}^{2}\prod_{j=1}^{N_i} f_\omega(x_{ij}|\theta_i,\pi_i)g_\omega(\theta)\,\mathrm{d}\theta \tag{1.5}$$

and $g_\omega(\theta)$ is the prior density of the set of parameters specified by ω, π_1 and π_2 and $\theta=\theta_1\cup\theta_2$ represents the set of distinct parameters, leads to a full Bayesian solution for the selection problem. But this solution usually requires proper prior distributions with hyperparameters specified.

As before, if a choice of model $\omega=\omega^*$ is made and one uses this for allocation, the relevant posterior odds used for assigning a new observation z, are

$$R(\omega^*,z,x)=\frac{q_1 f_{\omega^*}(z|x,\pi_1)}{q_2 f_{\omega^*}(z|x,\pi_2)}, \tag{1.6}$$

where

$$f_\omega(z|x,\pi_i)=\int f_\omega(z|\theta_i,\pi_i)\,\mathrm{d}G_\omega(\theta|x) \tag{1.7}$$

and

$$\mathrm{d}G_\omega(\theta|x)\propto g_\omega(\theta)\prod_i\prod_j f_\omega(x_{ij}|\theta_i,\pi_i)\,\mathrm{d}\theta. \tag{1.8}$$

If the complete Bayesian solution is to be used for classification in the presence of a prior probability function for ω, then one needs to calculate

$$\Pr[\pi_i|z,x,q_i]\propto q_i E_\omega f_\omega(z|x,\pi_i)f_\omega(x_1,x_2|\pi_1,\pi_2) \tag{1.9}$$

as the basis for assigning z, where the expectation is over the prior distribution of ω. One can justify (1.9) in the following way: Since

$$\Pr[\pi_i|z,x,q_i,\omega]=\frac{q_i f_\omega(z|x,\pi_i)}{f_\omega(z|x)} \tag{1.10}$$

where

$$f_\omega(z|x)=q_1 f_\omega(z|x,\pi_1)+q_2 f_\omega(z|x,\pi_2),$$

then clearly

$$\Pr[\pi_i|z,x,q_i]\propto q_i \mathrm{E}_{\omega|z,x}\left(\frac{f_\omega(z|x,\pi_i)}{f_\omega(z|x)}\right) \tag{1.11}$$

i.e. the expectation on the right hand side is w.r.t. the distribution of ω

conditional on $Z=z$ and $X=x$. Now

$$p(\omega|z,x)=\frac{f_\omega(z|x)f_\omega(x)p(\omega)}{f(z|x)f(x)} \tag{1.12}$$

$$\left.\begin{aligned}
f_\omega(x)&=\int\prod_i\prod_j f_\omega(x_{ij}|\theta_i,\pi_i)\,\mathrm{d}G_\omega(\theta)=f_\omega(x_1,x_2|\pi_1,\pi_2)\\
p(\omega|x)&=\frac{f_\omega(x)p(\omega)}{f(x)}=\frac{f_\omega(x_1,x_2|\pi_1,\pi_2)p(\omega)}{f(x)}\\
f(x)&=\int f_\omega(x_1,x_2|\pi_1,\pi_2)\,\mathrm{d}P(\omega).
\end{aligned}\right\} \tag{1.13}$$

After evaluating (1.12), using the results of (1.13), we evaluate (1.11) and obtain (1.9).

Full Bayesian solutions require a body of prior knowledge that is often unavailable. Also, even when presumed available, the analysis may be highly sensitive to some of the assumptions which, in fact, may have been grossly violated.

For the aforementioned problem, data analytic solutions based on sample reuse techniques, in the spirit of Geisser and Eddy (1979), will be presented. We consider then the case of two distinct populations that have been sampled with respect to some set of p variables but there is some doubt as to the distributions which generated the samples.

Hence, to the usual classification problem there is added the uncertainty of distributional assumptions regarding the two populations and a goal of the model selection procedure is to optimize classification in some sense—i.e. select the single model which will do the best job of classifying future observations.

2. Criteria for model selection

Previously, Geisser and Eddy (1979) recommended for model selection geared to prediction, that a series of predicting densities $f_\omega(z|x,\pi_i)$ be established from which to compute $R(\omega^*,z)$ of (1.6). One then maximizes the reused product of conditional predicting densities of this form to obtain ω^*. In this case, this would be equivalent to maximizing w.r.t. ω, $p(\omega)L(\omega)$ where

$$L_\omega = \prod_{i=1}^{2} \prod_{j=1}^{N_i} f_\omega(x_{ij} | x_{(ij)}, \pi_i) \tag{2.1}$$

and $X_{(ij)} = x_{(ij)}$ the set of observations $X = x$ with $X_{ij} = x_{ij}$ deleted but of the same form as $f_\omega(z|x, \pi_i)$. This is certainly a useful procedure for rather tight specifications when they are met. A variant of this procedure which tends more to emphasize the ultimate goal, classification of a new observation, is maximization of $p(\omega)O(\omega)$ where

$$O(\omega) = \frac{L(\omega)}{\bar{L}(\omega)} \left(\frac{q_1}{q_2} \right)^{N_1 - N_2} \tag{2.2}$$

and

$$\bar{L}_\omega = \prod_{i=1}^{2} \prod_{j=1}^{N_i} f_\omega(x_{ij} | x_{(ij)}, \pi_{3-i}). \tag{2.3}$$

This reflects more directly the posterior odds ratio for a given ω which will be used to classify a new observation. Note that $(q_1/q_2)^{N_1 - N_2}$ is independent of ω and therefore does not effect comparisons for various ω.

Both (2.1) and (2.2) depend on products—perhaps too much so to be usefully robust to the presence of outliers, contaminants or inadvertantly misclassified observations in the initial or training samples. This has the effect that a single observation in a low density region has enormous influence on $L(\omega)$. Hence if a particular ω^* were true, the insinuation of a single wildly discrepant observation could so diminish $L(\omega^*)$ that other ω could wrongly come to the fore. The effect of such an observation is mitigated by the use of $O(\omega)$ if it is in a low density region for both specifications of ω^*, since the odds ratio for the discrepant observation would minimally influence $O(\omega^*)$. Hence the effect of a few discrepant observations of this type would be largely diluted. On the other hand observations that were actually misclassified in a high density region of one of the pair and in a low density region of the other would have enormous effect on $O(\omega)$ in the wrong direction—even more so than on L_ω. Hence both $L(\omega)$ and $O(\omega)$ are highly volatile criteria in that they tend to be somewhat oversensitive to aberrancies.

Because of this we present a third sample reuse criterion which is far less sensitive to the type of eccentricities previously discussed—in that an aberrant observation would have far less influence on the resolution of the appropriate ω.

Let

$$R_{ij}(\omega)=\frac{q_1 f_\omega(x_{ij}|x_{(ij)},\pi_1)}{q_2 f_\omega(x_{ij}|x_{(ij)},\pi_2)} \tag{2.4}$$

for $j=1,\ldots,N_i$; $i=1,2$. Thus x_{ij} would be assigned to π_1 or π_2 as $R_{ij}(\omega)\geqslant 1$ or <1. Hence for each ω, $R_{ij}(\omega)$ will correctly assign $n_i(\omega)$ of the x_{ij}'s to π_i. Optimization then requires that we choose that ω which maximizes $Q(\omega)=q_1N_1^{-1}n_1(\omega)+q_2N_2^{-1}n_2(\omega)$, or $Q(\omega)$ may be multiplied by $p(\omega)$, if a sensible $p(\omega)$ exists for the alternative ω, before maximizing. If q_i is unknown but estimable by $N_i/(N_1+N_2)$, then for constant $p(\omega)$, we are maximizing $n_1(\omega)+n_2(\omega)$, the total number correctly classified by designator ω. Note also that if there is an ω such that $n_i(\omega^*)\geqslant n_i(\omega)$, $i=1,2$, then irrespective of q_i and N_i this ω^* is optimal.

Although with probability 1, unique solutions for ω's specified by continuous densities can be guaranteed for (2.1) and (2.2), this is obviously not the case for (2.4). Hence if the maximum is achieved for several ω's using (2.4) these ties may be broken by use of (2.1) or (2.2) to discriminate amongst these ω's. Further even when unique solutions exist in regard to any of the selection criterion they may not be distinguishable for classification purposes. For example, assume $N_1=N_2=M$ and either $q_1=q_2$ or $\hat{q}_1=\hat{q}_2$ and the possible models are

(1) a pair of normal densities with differing unknown means but with the same but unknown variance that is estimated by insertion of the usual estimators in the normal densities,

$$f_{\omega_\varphi}(z|\pi_i)=\frac{1}{\sqrt{2(\pi s^2)}}\,e^{-1/(2s^2)(z-\bar{x}_i)^2} \qquad i=1,2 \tag{2.5}$$

where

$$s^2=(2M-2)^{-1}\sum_{i=1}^{2}\sum_{j=1}^{M}\left(x_{ij}-\bar{x}_i\right)^2, \quad \text{and} \quad \bar{x}_i=M^{-1}\sum_{j=1}^{M}x_j;$$

or (2) a pair of t densities (which can also be considered Bayesian estimates of the underlying normal pair, Geisser (1971), Aitchison (1975)),

$$f_{\omega_t}(z|\pi_i)=\left[\frac{M}{\pi(2M-2)(M+1)}\right]^{\frac{1}{2}}$$

$$\frac{\Gamma\left(\frac{1}{2}(2M-1)\right)}{s\Gamma\left(\frac{1}{2}(2M-2)\right)}\left[1+\frac{M(z-\bar{x}_i)^2}{(M+1)(2M-2)s^2}\right]^{-\frac{1}{2}(2M-1)} \tag{2.6}$$

It is then easy to demonstrate that for every z, $R(z,\omega_\varphi)$ results in the same allocation as $R(z,\omega_t)$ since

$$R(z,\omega_\varphi)=\frac{f_{\omega_\varphi}(z|\pi_1)}{f_{\omega_\varphi}(z|\pi_2)}>1 \quad \text{or} \quad \leqslant 1$$

implies that

$$R(z,\omega_t)=\frac{f_{\omega_t}(z|\pi_1)}{f_{\omega_t}(z|\pi_2)}>1 \quad \text{or} \quad \leqslant 1 \quad \text{respectively.}$$

It is to be noted, however, that the value of the odds ratio itself varies for the two alternative forms so that a different cutoff point would result in differing allocations. This would occur if either for known q_1 and q_2, $q_1 \neq q_2$ or if unknown and estimated $\hat{q}_1 \neq \hat{q}_2$.

3. An application to multivariate normal populations

Assume that under ω_1, $\mathbf{x}_1=\{\mathbf{x}_{11},\ldots,\mathbf{x}_{1N_1}\}$ and $\mathbf{x}_2=\{\mathbf{x}_{21},\ldots,\mathbf{x}_{2N}\}$ are respectively the observed values of independently distributed p-dimensional random variables which, respectively under π_1, arose from a $N(\boldsymbol{\mu}_1,\Sigma)$, and under π_2, arose from a $N(\boldsymbol{\mu}_2,\Sigma)$. Under ω_2 similarly the set of observations $\mathbf{x}_1$ and $\mathbf{x}_2$ are the observed sets of values which respectively arose as independent realizations of a $N(\boldsymbol{\mu}_1,\Sigma_1)$ under π_1 and a $N(\boldsymbol{\mu}_2,\Sigma_2)$ under π_2.

For the classification of a future vector observation $\mathbf{Z}=\mathbf{z}$, where $\Pr(\pi_1)$

$=q_1$, and $q_2=\Pr(\pi_2)$, $q_1+q_2=1$ and convenient prior density; under ω_1,

$$g_{\omega_1}(\boldsymbol{\mu}_1,\boldsymbol{\mu}_2,\Sigma^{-1})\propto|\Sigma|^{(p+1)/2},$$

Geisser (1964) obtained

$$\Pr(\pi_i|\mathbf{z},\omega_1)\propto q_i f_{\omega_1}(\mathbf{z}|\bar{\mathbf{x}}_i,S,N_i,N,\pi_i)$$

where

$$f_{\omega_1}(\mathbf{z}|\bar{\mathbf{x}}_i,S,N_i,N,\pi_i)=\left[\frac{N_i}{\pi(N_i-1)}\right]^{p/2}\frac{\Gamma\left(\frac{N-1}{2}\right)}{\Gamma\left(\frac{N-p-1}{2}\right)|(N-2)S|^{\frac{1}{2}}}$$

$$\left[1+\frac{N_i(\mathbf{z}-\bar{\mathbf{x}}_i)'S^{-1}(\mathbf{z}-\mathbf{x}_i)}{(N_i+1)(N-2)}\right]^{-(N-1)/2} \tag{3.1}$$

and

$$\bar{\mathbf{x}}_i=N_i^{-1}\sum_{j=1}^{N_i}\mathbf{x}_{ij},\ (N_i-1)S_i=\sum_{j=1}^{N_i}(\mathbf{x}_{ij}-\bar{\mathbf{x}}_i)(\mathbf{x}_{ij}-\bar{\mathbf{x}}_i)',$$

$$(N-2)S=(N_1-1)S_1+(N_2-1)S_2,\qquad N=N_1+N_2.$$

This predictive density (3.1), suggested by Geisser (1964, 1971) as a Bayesian estimate of the sampling density, was actually shown by Murray (1977) to be optimal in the frequency sense among all estimators of the sample density that are invariant under translations and non-singular linear transformations of the sample space using as a goodness of fit criterion the information measure of Kullback and Liebler (1951).

Under ω_2 we assume a similar prior density

$$g_{\omega_2}(\boldsymbol{\mu}_1,\boldsymbol{\mu}_2,\Sigma_1,\Sigma_2)\propto|\Sigma_1|^{(p+1)/2}|\Sigma_2|^{(p+1)/2}$$

and obtain

$$\Pr(\pi_i|\mathbf{z},\omega_2)\propto q_i f_{\omega_2}(\mathbf{z}|\bar{\mathbf{x}}_i,S_i,N_i,\pi_i)$$

where

$$f_{\omega_2}(\mathbf{z}|\bar{\mathbf{x}}_i, S_i, N_i, \pi_i) = \left(\frac{N_i}{\pi(N_i+1)}\right)^{p/2} \frac{\Gamma(N_i/2)}{\Gamma\left(\frac{N_i-p}{2}\right)|(N_i-1)S_i|^{\frac{1}{2}}}$$

$$\left[1+\frac{N_i(\mathbf{z}-\bar{\mathbf{x}}_i)'S_i^{-1}(\mathbf{z}-\bar{\mathbf{x}}_i)}{(N_i+1)(N_i-1)}\right]^{-N_i/2}. \tag{3.2}$$

To establish which of the two models, ω_1 or ω_2, is more appropriate for classification, we can apply one or more of the following methods.

Method 1. Let

$$P(\omega_k) = p(\omega_k)L(\omega_k) \qquad k=1,2$$

where

$$L(\omega_1) = \prod_{i=1}^{2} \prod_{j=1}^{N_i} f_{\omega_1}(\mathbf{x}_{ij}|\bar{\mathbf{x}}_{i(j)}, S_{(ij)}, N_i-1, N-1, \pi_i)$$

$$L(\omega_2) = \prod_{i=1}^{2} \prod_{j=1}^{N_i} f_{\omega_2}(\mathbf{x}_{ij}|\bar{\mathbf{x}}_{i(j)}, S_{i(j)}, N_i-1, \pi_i) \tag{3.3}$$

where $f_{\omega_1}(\cdot)$ and $f_{\omega_2}(\cdot)$ are defined as in (3.1) and (3.2) respectively and

$$\bar{\mathbf{x}}_{i(j)} = (N_i-1)^{-1}[N\bar{\mathbf{x}}_i - \mathbf{x}_{ij}]$$

$$(N-3)S_{(ij)} = (N_i-2)S_{i(j)} + (N_{3-i}-1)S_{3-i} \qquad i=1,2$$

$$(N_i-2)S_{i(j)} = \sum_{\substack{t=1\\t\neq j}}^{N_i} (\mathbf{x}_{it}-\bar{\mathbf{x}}_{i(j)})(\mathbf{x}_{it}-\bar{\mathbf{x}}_{i(j)})'.$$

Select that ω_k which maximizes $P(\omega_k)$.

Method 2. Define

$$\bar{L}(\omega_1)=\prod_{i=1}^{2}\prod_{j=1}^{N_i} f_{\omega_1}(\mathbf{x}_{ij}|\bar{\mathbf{x}}_{3-i},S_{(ij)},N_{3-i},N-1,\pi_i)$$

$$\bar{L}(\omega_2)=\prod_{i=1}^{2}\prod_{j=1}^{N_i} f_{\omega_2}(\mathbf{x}_{ij}|\bar{\mathbf{x}}_{3-i},S_{3-i},N_{3-i},\pi_i) \tag{3.4}$$

and

$$O(\omega_k)=\frac{L(\omega_k)}{\bar{L}(\omega_k)}\cdot\left(\frac{q_1}{q_2}\right)^{N_1-N_2} \qquad k=1,2 \tag{3.5}$$

select that ω_k which maximizes $p(\omega_k)O(\omega_k)$.

Method 3. Define

$$R_{1j}(\omega_1)=\frac{q_1 f_{\omega_1}(\mathbf{x}_{1j}|\bar{\mathbf{x}}_{1(j)},S_{(1j)},N_1-1,N-1,\pi_1)}{q_2 f_{\omega_1}(\mathbf{x}_{1j}|\bar{\mathbf{x}}_2,S_{(1j)},N_2,N-1,\pi_2)}$$

$$R_{2j}(\omega_1)=\frac{q_1 f_{\omega_1}(\mathbf{x}_{2j}|\bar{\mathbf{x}}_1,S_{(2j)},N_1,N-1,\pi_1)}{q_2 f_{\omega_1}(\mathbf{x}_{2j}|\bar{\mathbf{x}}_{2(j)},S_{(2j)},N_2-1,N-1,\pi_2)} \tag{3.6}$$

$$R_{1j}(\omega_2)=\frac{q_1 f_{\omega_2}(\mathbf{x}_{1j}|\bar{\mathbf{x}}_{1(j)},S_{1(j)},N_1-1,\pi_1)}{q_2 f_{\omega_2}(\mathbf{x}_{ij}|\bar{\mathbf{x}}_2,S_2,N_2,\pi_2)}$$

$$R_{2j}(\omega_2)=\frac{q_1 f_{\omega_2}(\mathbf{x}_{2j}|\bar{\mathbf{x}}_1,S_1,N_1,\pi_1)}{q_2 f_{\omega_2}(\mathbf{x}_{2j}|\bar{\mathbf{x}}_{2(j)},S_{2(j)},N_2-1,\pi_2)}. \tag{3.7}$$

Now calculate

$$Q(\omega_k)=q_1 n_1(\omega_k)N_1^{-1}+q_2 n_2(\omega_k)N_2^{-1}$$

where, in general,

$$n_i(\omega_k)=\sum_{j=1}^{N_i}\delta_{ij}(\omega_k) \tag{3.8}$$

and

$$\delta_{ij}(\omega_k)=\begin{cases}1 & \text{if } \left[R_{ij}(\omega_k)\right]^{3-2i}>1\\ 0 & \text{otherwise}\end{cases} \tag{3.9}$$

and incidentally

$$O(\omega_k)=\prod_{i=1}^{2}\prod_{j=1}^{N_i}\left[R_{ij}(\omega_k)\right]^{3-2i}. \tag{3.10}$$

Further choose that ω_k which maximizes $p(\omega_k)Q(\omega_k)$, or if $\hat{q}_i=N_i/N$ and one is a priori indifferent to a choice between ω_1 and ω_2, select that ω_k which maximizes

$$n(\omega_k)=n_1(\omega_k)+n_2(\omega_k) \tag{3.11}$$

the total number of $\mathbf{x}_{ij}$'s correctly classified using $R_{ij}(\omega_k)$.

4. Summary

The three sample reuse methods presented here approximately simulate different comparative measures. The first approximates the posterior probability assuming the source, π_i, of the omitted observation is known, which heavily emphasizes selecting the best model. The second attempts to approximate the posterior odds ratio that each observation is correctly classified and uses as a measure the product of these odds ratios.

The last initially treats the omitted observation as if its source were unknown and proceeds to simulate the classification scheme itself by assigning each omitted observation to π_1 or π_2 and then determines the number correctly classified for each ω_k. This permits each observation to contribute more equally to the selection measure.

Actually for any particular problem, given the kind of computations to be made, it would appear that all three methods can be simultaneously calculated and their results compared before a final conclusion is reached as to the choice of ω most suitable for allocating new observations.

References

[1] Aitchison, J. (1975). Goodness of prediction fit. *Biometrika*. **62**, 547–54.

[2] Geisser, S. (1964). Posterior odds for multivariate normal classifications. *J. Roy. Statist. Soc. B*. **25**, 368–376.

[3] Geisser, S. (1971). The inferential use of predictive distributions. *Foundations of Statistical Inference* (V. Godambe and D. Sprott, Eds.). Holt, Rinehart and Winston, 456–469.

[4] Geisser, S. and Eddy, W. F. (1979). A predictive approach to model selection, *J. Amer. Statist. Assoc*. **74**(365), pp. 153–160.

[5] Kullback, S. and Leibler, R. A. (1951). On information and sufficiency. *Ann. Math. Statist*. **22**, 79–86.

[6] Murray, G. D. (1977). A note on the estimation of probability density functions. *Biometrika*. **64**(1), 150–2.

P. R. Krishnaiah, ed., *Multivariate Analysis–V*

INTERVAL ESTIMATES FOR A BIVARIATE PRINCIPAL AXIS

A. T. JAMES and W. VENABLES,
The University of Adelaide, Australia.

1. Introduction

Suppose $\overset{2\times 2}{S} \sim W_2(n,\Sigma)$. We consider making inferences on the principal axes determined by Σ.

To distinguish two problems we define ψ as the angle between the *first* eigenvectors of S and Σ, with values in $-\pi/2<\psi\leqslant\pi/2$, and θ as the angle between the two axis systems defined by the eigenvectors of S and Σ, with values in $-\pi/4<\theta\leqslant\pi/4$.

Note that ψ and θ are either identical or differ by $\pi/2$.

Of course for large samples, and well separated eigenvalues of Σ, the two angles are almost certainly the same, as in Fig. 1(a), and the two problems may in practice be regarded as the same.

Take an axis system along the sample principal axes. In this system we have S diagonal;

$$S=\begin{pmatrix} l_1 & 0 \\ 0 & l_2 \end{pmatrix}, \quad l_1 \geqslant l_2 > 0.$$

An hypothetical value θ_0 for θ may be tested by rotating the axis system through an angle θ_0, thus transforming

$$S \to H_0' S H_0$$

where

$$H_0=\begin{pmatrix} \cos\theta_0 & -\sin\theta_0 \\ \sin\theta_0 & \cos\theta_0 \end{pmatrix},$$

and testing for independence of the two new variates. If r^2 is the sample correlation coefficient in the rotated Wishart matrix, a short calculation

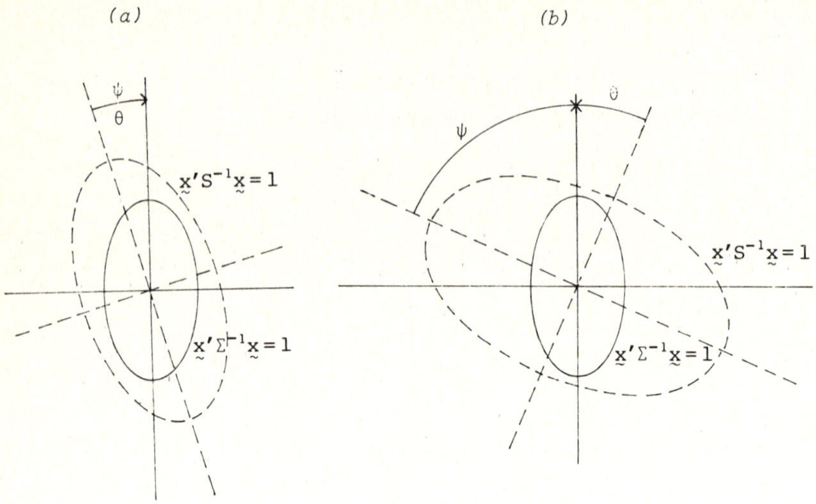

Fig. 1.

shows

$$\frac{r^2}{1-r^2} = \frac{(l_1-l_2)^2}{4l_1l_2}\sin^2 2\theta_0$$

and hence the hypothetical value of θ_0 may be tested using the fact that

$$F(\theta_0) = (n-1)\frac{(l_1-l_2)^2}{4l_1l_2}\sin^2 2\theta_0 \tag{1}$$

is an $F(1,n-1)$ variate if the hypothesis is true. This procedure has been suggested by Mallows (1961), Kshirsagar (1961), Vaughton (1970) and James (1975).

Conversely, if $\varphi(1,n-1,1-\alpha)$ is the $(1-\alpha)$-point of the $F(1,n-1)$ distribution the interval

$$\theta : (n-1)\frac{(l_1-l_2)^2}{4l_1l_2}\sin^2 2\theta \leqslant \varphi(1,n-1,1-\alpha) \tag{2}$$

is a valid $(1-\alpha)$-confidence interval for θ.

Although the test, (1), appears unexceptionable for individual hypotheses $\theta=\theta_0$ specified in advance, we contend that the confidence intervals (2), although valid, are not really appropriate, at least intuitively, as an inferential statement.

Our objection appears most clearly when the intervals are very wide. If the bivariate distribution is in fact *spherical*, *all* axis systems are equally valid as principal axis systems, and we would argue that unless this *sphericity hypothesis* can be rejected at an appropriate level of significance, an inferentially appropriate interval estimate for θ should not be less than the full range.

A very natural test for the sphericity hypothesis uses the *ellipticity statistic*

$$F^* = \tfrac{1}{2}(n-1)\frac{(l_1-l_2)^2}{4l_1l_2}$$

which has a null distribution, $F(2,n-1)$.

If we had a sample for which

$$\varphi(1,n-1,1-\alpha)<(n-1)\frac{(l_1-l_2)^2}{4l_1l_2}<2\varphi(2,n-1,1-\alpha)$$

we would have the odd situation in which the sphericity hypothesis could be *retained* at level α, and also the $(1-\alpha)$-confidence interval for the principal axis system *excludes* some possible axis systems. We argue that it is intuitively reasonable for an interval estimate for θ (or ψ) not to be less than the full range if the sphericity hypothesis can be retained at the appropriate level of significance.

It is also possible to give an intuitive argument to suggest that interval estimates for θ should closely agree with (1) if the ellipticity statistic is very large, however we shall not discuss this here.

The difficulties in the inference are similar to those which arise in the Creasy Fieller (1954) problem, see James, Wilkinson and Venables (1975). As this problem is a little simpler, the difficulties can be illustrated by a preliminary discussion of it.

One seeks a 95% confidence interval for the line through the origin and a parameter point $\boldsymbol{\mu}=[\mu_1\,\mu_2]'$ with an estimate $\boldsymbol{x}=[x_1\,x_2]'$ distributed as $N(\boldsymbol{\mu},I_2)$. The Fieller solution produces a 95% confidence interval bounded by lines

$$l_1x_1+l_2x_2=0, \qquad l_1^2+l_2^2=1 \tag{1'}$$

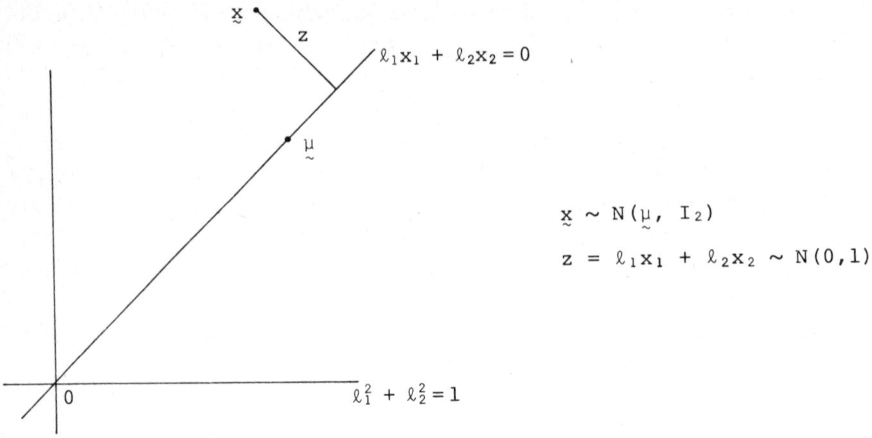

Fig. 2. Creasy–Fieller problem.

such that the perpendicular distance from $\boldsymbol{x}$

$$z = l_1 x_1 + l_2 x_2 = \chi \sin\theta, \qquad \chi^2 = x_1^2 + x_2^2$$

is 1.96, because if $\boldsymbol{\mu}$ lies on the line (1)′, z is a standard normal variate.

If $\chi^2 < (1.96)^2 = 3.84$, there is no line through the origin at distance 1.96 and the whole parameter space is taken as the 95% confidence interval. A 95% confidence interval which does not include the entire parameter space is called a "proper" confidence interval.

Three objections to the Fieller confidence intervals can be raised.

(1) They produce a proper 95% confidence interval for values of χ^2 between 3.84 and 5.99 for which the null hypothesis $\boldsymbol{\mu} = 0$ is not rejected at the 5% level.

(2) It would reasonably be required of a 95% confidence interval that all points of the complementary set in parameter space can be rejected at the 5% level by the statistic which gives rise to the confidence interval. For the line at right angles to the sample vector, i.e. the line most easily rejected, the statistic becomes

$$z^2 = x_1^2 + x_2^2 = \chi^2$$

The confidence interval procedure is treating z^2 as a χ^2 variable on 1 d.f. whereas, in this case, it clearly has 2 d.f.. Scheffé (1970) has noted this anomaly and its relation to the multiple comparisons problem.

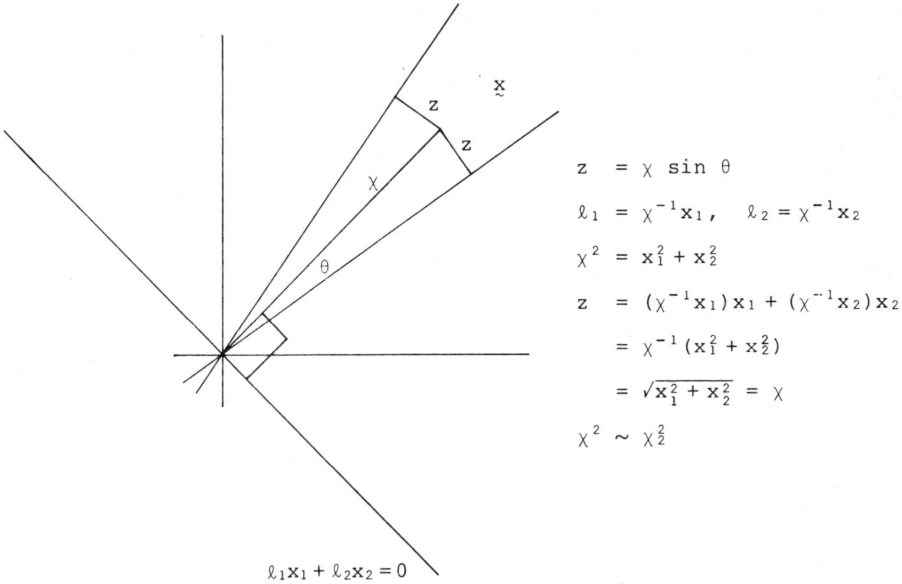

Fig. 3. Rejection of perpendicular to sample vector.

(3) The third difficulty is the most subtle but the most important. The improper 95% Fieller confidence intervals are of no practical use. We have no objection that they occur; the data may lack the information to allow any substantive statement with 95% confidence. However, when a person with a useful 95% proper confidence interval realizes that some of the 95% covering frequency is supplied by improper confidence intervals, he will query whether he really has 95% confidence in his own interval. He sees that conditional on his having a proper and sufficiently short interval to be practically useful and meaningful, the real confidence frequency may be substantially less than the unconditional 95% interval. The length depends upon the ancillary statistic χ^2 and he will want to condition on it.

If one has a proper ancillary statistic which is not required for the estimation of a nuisance parameter, one should take the distribution conditional on it as the basis of a confidence interval for the parameter in question. (For a definition of "proper" ancillary, see Wilkinson (1977).) Such conditional confidence intervals are now generally accepted.

Suppose now the person wants a 95% confidence interval conditional on χ^2. When he asks for an interval which has 95% covering frequency for all values of the parameter vector $\boldsymbol{\mu}$ he finds he can no longer have any sensible interval. For very small values of the parameter ω^2 for which the data is practically uninformative as to direction, only intervals covering practically 95% of parameter space will have the covering property and therefore, these must be his intervals.

In this case, one cannot have both the conditionality on the ancillary, and the confidence covering frequency for every conceivable value of the parameter vector. One must choose between these two desiderata.

In other words, in both the Creasy–Fieller and principal components problems there is a statistic whose distribution does not depend upon the parameter to be estimated, and consequently interval estimates should be conditional upon it. But the statistic is also needed for the estimation of a nuisance parameter. For a classical confidence interval, the requirement that it be conditional on the ancillary conflicts with the requirement that the covering property hold for all conceivable values of the parameters, in particular, for all values of the nuisance parameters; viz. one is asking for a confidence interval conditional upon the ancillary statistic, which has the covering property for values of the nuisance parameter in wide disagreement with the ancillary statistic. It is usually impossible to meet the two requirements simultaneously and a choice must be made between them.

A solution that overcomes these objections is proposed. Our approach owes much to the fiducial ideas of R. A. Fisher and to recent extensions of these ideas by G. N. Wilkinson (1977). Despite the admittedly controversial nature of some of these ideas, we offer this solution in the hope that its intuitively attractive properties at least will mean it does attract some interest in practice.

2. The proposed solution

We argue initially that since the problem is invariant under the same change of scale to both variables, i.e. under $S \to \mu^2 S$, $-\infty < \mu < \infty$, the appropriate starting point for inferences on ψ (or θ) is the marginal distribution of $(l_1/l_2, \psi)$.

It is slightly more convenient to put

$$W = \frac{l_1 - l_2}{l_1 + l_2} \quad \text{and} \quad \omega = \frac{\lambda_1 - \lambda_2}{\lambda_1 + \lambda_2},$$

(where $\lambda_1 \geqslant \lambda_2$ are the eigenvalues of Σ); the joint distribution of W and ψ

has density (James (1966)) $f(W^2,\psi)=\pi^{-1}(1-W\omega\cos 2\psi)^{-n}\cdot\frac{1}{2}(n-1)(1-\omega^2)^{n/2}(1-W^2)^{(n-3)/2}$.

Using the integral

$$ {}_2F_1\left(\tfrac{1}{2}n,\tfrac{1}{2}(n+1);1;\omega^2W^2\right)=\pi^{-1}\int_{-\pi/2}^{\pi/2}(1-\omega W\cos(2\psi))^{-n}\,d\psi $$

the conditional distribution of $\psi|W^2$ has density

$$ f(\psi|W^2)=\frac{\pi^{-1}(1-\omega W\cos 2\psi)^{-n}}{{}_2F_1\left(\frac{1}{2}n,\frac{1}{2}(n+1);1;\omega^2W^2\right)} \tag{3} $$

and the marginal density of W^2 is

$$ f(W^2)=(1-\omega^2)^{n/2}{}_2F_1\left(\tfrac{1}{2}n,\tfrac{1}{2}(n+1);1;\omega^2W^2\right) \times\tfrac{1}{2}(n-1)(1-W^2)^{(n-1)/2-1} \tag{4} $$

[from which it may be easily deduced, as claimed in Section 1, that if $\omega=0$, $\frac{1}{2}(n-1)W^2/(1-W^2)\sim F(2,n-1)$].

If ω^2 were *known*, then W^2 would be an ancillary statistic, and inferences on ψ using the conditional density (3) would be appropriate. This is now a widely accepted procedure. (See, for example, Cox and Hinkley (1975)).

Since ω^2 is not known, we cannot use (3) directly as it depends on ω^2, the parameter of secondary interest. However we still condition on W^2, and use (3) as a conditional inferential distribution for $\psi|\omega^2$. We now use (4) to supply a marginal inferential distribution for ω^2, and average (3) with respect to this distribution to obtain an inferential distribution for ψ.

Consider now the marginal inferential distribution for ω^2. The distribution function of W^2 may be written in mixture form as

$$ F(W^2)=(1-\omega^2)^{n/2}\sum_{k=0}^{\infty}\frac{\left(\frac{1}{2}n\right)_k}{k!}(\omega^2)^k I_{W^2}\left(k+1,\tfrac{1}{2}(n-1)\right). $$

We can now substitute for the incomplete beta-function with the formula

$$ I_{W^2}\left(k+1;\tfrac{1}{2}(n-1)\right)=1-(1-W^2)^{(n-1)/2}\sum_{j=0}^{k}\frac{\left[\frac{1}{2}(m-1)\right]_j}{j!}(W^2)^j, $$

and re-arranging the orders of summation gives

$$F(W^2)=1-(1-W^2)^{(n-1)/2}\sum_{j=0}^{\infty}\frac{\left[\frac{1}{2}(n-1)\right]_j}{j!}$$

$$\times(W^2)^j\left\{(1-\omega^2)^{n/2}\sum_{k=j}^{\infty}\frac{\left(\frac{1}{2}n\right)_k}{k!}(\omega^2)^k\right\}$$

$$=1-(1-W^2)^{(n-1)/2}\sum_{j=0}^{\infty}\frac{\left[\frac{1}{2}(n-1)\right]_j}{j!}(W^2)^j I_{\omega^2}\left(j;\tfrac{1}{2}n\right)$$

(where $I_{\omega^2}(0,\frac{1}{2}n)\overset{\text{def}}{=}1$). Taking $1-F(W^2)$ as a uniformly distributed pivotal quantity, the marginal inferential distribution for ω^2 is given by the distribution function

$$G(\omega^2)=1-F(W^2)=(1-W^2)^{(n-1)/2}\sum_{j=0}^{\infty}\frac{\left[\frac{1}{2}(n-1)\right]_j}{j!}I_{\omega^2}\left(j,\tfrac{1}{2}n\right).$$

The discrete mass of probability $p=(1-W^2)^{(n-1)/2}$ is left unassigned (or assigned to the parameter point $\omega^2=0$), and omitted from the averaging process. Accordingly the averaged inferential distribution for ψ is given by

$$f(\psi)=\int_{0+}^{1}f(\psi|\omega^2)\,\mathrm{d}G(\omega^2) \tag{5}$$

and is, of course, a defective distribution since

$$\int_{-\pi/2}^{\pi/2}f(\psi)\,\mathrm{d}\psi=1-(1-W^2)^{n/2-1}=1-p.$$

[The concept of unassigned probability, primarily due to G. N. Wilkinson, is discussed elsewhere.]

Our $(1-\alpha)$-interval estimate for ψ is then $\underline{\psi}_{\alpha/2}<\psi<\overline{\psi}_{\alpha/2}$ where $(\underline{\psi}_{\alpha/2},\overline{\psi}_{\alpha/2})$ are the upper- and lower-$\frac{1}{2}\alpha$-points of (5). By symmetry $\underline{\psi}_{\alpha/2}=-\overline{\psi}_{\alpha/2}$. A short table is given in the next section.

Because the distribution is defective our interval will be the full range, $-\pi/4<\psi\leqslant\pi/4$ unless

$$(1-W^2)^{(n-1)/2}<\alpha$$

that is, unless

$$\tfrac{1}{2}(n-1)\frac{(l_1-l_2)^2}{4l_1l_2}>\varphi(2,n-1,1-\alpha) \tag{6}$$

in other words the ellipticity statistic must be significant at level α.

An unavoidable weakness of the Mallows-Kshirsagar procedure is that it can only provide inferences on the angle θ rather than on the more precisely defined angle, φ. In this paper we confine the discussion to φ, but for comparative purposes at least it would be of interest to consider θ as well.

Our inferential density for θ is

$$g(\theta)=f(\theta)+f(\theta+\pi/2), \quad -\pi/4<\theta\leqslant\pi/4 \tag{7}$$

which again is defective, and provides intervals not less than the full range unless condition (6) is met.

3. Computation and tables

Calculating the percentage points of (5) is an awkward numerical problem. Our extempore approach is as follows. From (3) we can write

$$\mathrm{d}G(\omega^2)=(1-W^2)^{(n-1)/2}\left\{\Delta(0)+\sum_{j=1}^{\infty}\frac{\left[\frac{1}{2}(n-1)\right]_j}{j!}(W^2)^j\frac{\left(\frac{1}{2}n\right)_j}{(j-1)!}\right.$$
$$\left.\times(\omega^2)^{j-1}(1-\omega^2)^{\frac{1}{2}n-1}\right\}\mathrm{d}(\omega^2)$$

(where $\Delta(0)$ is Dirac's symbol)

$$=(1-W^2)^{(n-1)/2}\Delta(0)\,\mathrm{d}(\omega^2)+$$
$$+\tfrac{1}{4}n(n-1)\,W^2(1-W^2)^{(n-1)/2}(1-\omega^2)^{n/2-1}$$
$$\times{}_2F_1\left(\tfrac{1}{2}(n+1),\tfrac{1}{2}(n+2);2;\omega^2W^2\right)\mathrm{d}(\omega^2).$$

Removing the discrete mass, and integrating gives

$$f(\psi)=\tfrac{1}{4}n(n-1)W^2(1-W^2)^{(n-1)/2}\pi^{-1} \times \int_0^1 (1-W\omega\cos 2\psi)^{-n}(1-\omega^2)^{\frac{1}{2}n-1} \times \frac{{}_2F_1\left(\frac{1}{2}(n+1),\frac{1}{2}(n+2);2;W^2\omega^2\right)}{{}_2F_1\left(\frac{1}{2}n,\frac{1}{2}(n+1);1;W^2\omega^2\right)}\,d(\omega^2) \tag{6}$$

With this expression, we now solve the equation

$$\int_0^{\bar{\psi}_{\alpha/2}} f(\psi)\,d\psi=\tfrac{1}{2}(1-\alpha)$$

by a process of double numerical integration. The ratio of ${}_2F_1$-hypergeometric functions in the integrand of (6) has a useful Gauss-type terminating continued fraction expansion, namely,

$$\frac{{}_2F_1\left(\frac{1}{2}(n+1),\frac{1}{2}(n+2);2;\omega^2W^2\right)}{{}_2F_1\left(\frac{1}{2}(n+1),\frac{1}{2}n;1;\omega^2W^2\right)} = \cfrac{2}{2+\cfrac{(n+1)(n-2)\omega^2W^2}{4+\cfrac{(n+2)(n-3)\omega^2W^2}{6+\cfrac{(n+3)(n-4)\omega^2W^2}{8+\cdots}}}}$$

and just the first fifteen or so cycles seems to provide an accurate rational approximation.

The following table gives the outer extremes, $\bar{\psi}_{0.025}$, of the 0.95-interval estimate for ψ found by numerical integration of (5), and as a comparison the corresponding extreme of the confidence interval, (2), for θ. The horizontal argument is taken as $-\log_{10}$ (unassigned probability) = $(n-1)\log_{10}(\frac{1}{2}(l_1+l_2)/\sqrt{l_1l_2}\,)$. This seems to produce a reasonable comparison with different degrees-of-freedom (the vertical argument). As the variation between rows is quite small, interpolation should be no problem. The asymptotic value, $n=\infty$, is discussed in the next section.

As expected, our interval estimate for ψ is always wider than the confidence interval for θ, but the differences become very small as the ellipticity increases relative to its null distribution.

[The interval estimate for θ based on (7) is also wider than the confidence interval, and is always between the two tabulated values (see Table 1).]

Table 1
Outer extremes of the 0.95-interval estimate for ψ (in degrees)
(Confidence interval for θ in parentheses)

degrees of freedom n	$-\log_{10}(p)$ 1.5	2.0	2.5	3.0	3.5	4.0	5.0	6.0
10	34.57	19.78	14.96	12.12	10.12	8.59	6.37	4.81
	(22.29)	(17.19)	(13.96)	(11.64)	(9.86)	(8.44)	(6.31)	(4.79)
15	35.45	21.09	16.45	13.75	11.85	10.40	8.26	6.72
	(22.93)	(18.23)	(15.25)	(13.11)	(11.47)	(10.15)	(8.14)	(6.66)
20	35.86	21.72	17.18	14.55	12.72	11.33	9.27	7.78
	(23.24)	(18.72)	(15.87)	(13.84)	(12.27)	(11.02)	(9.11)	(7.68)
25	36.10	22.10	17.61	15.04	13.25	11.89	9.88	8.44
	(23.42)	(19.01)	(16.24)	(14.26)	(12.75)	(11.54)	(9.70)	(8.32)
30	36.28	22.34	17.90	15.35	13.59	12.26	10.30	8.89
	(23.54)	(19.20)	(16.48)	(14.54)	(13.07)	(11.89)	(10.09)	(8.76)
35	36.35	22.51	18.10	15.58	13.84	12.52	10.60	9.21
	(23.62)	(19.33)	(16.65)	(14.74)	(13.29)	(12.13)	(10.37)	(9.07)
40	36.43	22.64	18.25	15.75	14.03	12.72	10.82	9.46
	(23.69)	(19.43)	(16.78)	(14.89)	(13.46)	(12.32)	(10.58)	(9.30)
50	36.54	22.81	18.46	15.99	14.29	13.00	11.13	9.80
	(23.77)	(19.57)	(16.95)	(15.10)	(13.69)	(12.57)	(10.88)	(9.63)
60	36.61	22.93	18.60	16.14	14.46	13.19	11.34	10.03
	(23.83)	(19.66)	(17.07)	(15.24)	(13.85)	(12.74)	(11.08)	(9.85)
75	36.68	23.05	18.74	16.30	14.63	13.37	11.55	10.26
	(23.89)	(19.76)	(17.19)	(15.37)	(14.00)	(12.91)	(11.27)	(10.07)
100	36.76	23.17	18.88	16.45	14.80	13.56	11.76	10.49
	(23.94)	(19.85)	(17.30)	(15.51)	(14.16)	(13.08)	(11.47)	(10.29)
200	36.83	23.34	19.08	16.69	15.05	13.83	12.08	10.83
	(24.03)	(19.98)	(17.47)	(15.71)	(14.38)	(13.34)	(11.76)	(10.62)
∞	36.96	23.51	19.29	16.92	15.31	14.11	12.39	11.19
	(24.11)	(20.11)	(17.64)	(15.91)	(14.61)	(13.59)	(12.05)	(10.95)

4. A limiting case

If the degrees of freedom, n, increases for any fixed ellipticity the sampling (and inferential) distributions of ψ degenerate to a point distribution at $\psi=0$.

Suppose now we put $W^2=X/n$, $\omega^2=\delta/n$ and consider the limiting case in which $n\to\infty, X, \delta$ fixed. The density, (6), can be easily seen to approach

$$f_\infty(\psi)=2\cdot\frac{\frac{1}{2}Xe^{-X/2}}{2\pi}\int_0^\infty e^{-\delta/2+\sqrt{\delta X}\cos 2\psi}\frac{{}_0F_1(2;\frac{1}{4}\delta X)}{{}_0F_1(1;\frac{1}{4}\delta X)}\,d\delta,$$

$$-\pi/2<\psi<\pi/2. \tag{8}$$

In James, Wilkinson and Venables (1974) the authors considered the following version of the Creasy–Fieller problem. Suppose $\boldsymbol{x}\sim N(\boldsymbol{\mu}, I_2)$, and let ψ^* be the angle between the directed rays through $\mathbf{0}$ and $\boldsymbol{x}, \boldsymbol{\mu}$ respectively. Find an interval estimate for ψ^*. Using a parallel approach to that set out above we are led to an inferential density

$$f^*(\psi^*)=\frac{\frac{1}{2}Xe^{-\frac{1}{2}X}}{2\pi}\int_0^\infty e^{-\frac{1}{2}\delta+\sqrt{\delta X}\cos\psi^*}\frac{{}_0F_1(2;\frac{1}{4}\delta X)}{{}_0F_1(1;\frac{1}{4}\delta X)}\,d\delta,$$

$$-\pi<\psi^*<\pi \tag{9}$$

where $X=\boldsymbol{x}'\boldsymbol{x}$ and $\delta=\boldsymbol{\mu}'\boldsymbol{\mu}$.

By comparing (8) and (9) we see that the limiting density of 2ψ and of ψ^* are the same, and this gives the last line the table in the previous section. The independent calculation that produced this last line acts as a simple check on the accuracy.

Corresponding to this limiting result, the pivotal statistic, (1), which may be written

$$F(\theta_0)=\frac{(n-1)W^2}{1-W^2}\sin 2\theta_0\left(\approx X\sin^2 2\theta_0\right) \tag{10}$$

clearly has an asymptotic $\chi^2(1)$ distribution. If θ^* is the smallest angle between *undirected* rays through $\mathbf{0}$ and $\boldsymbol{x}, \boldsymbol{\mu}$ respectively, then the Fieller solution uses the pivotal statistic

$$F^*(\theta^*) = X \sin^2 \theta^*$$

which has a $\chi^2(1)$ distribution.

References

[1] Cox, D. R. and Hinkley, D. V. (1975). *Theoretical Statistics*. Chapman and Hall, London.

[2] Creasy, M. A. (1954). Limits for the ratio of means. *J. Roy. Statist. Soc.*, (B) **16**, 186–192.

[3] Fieller, E. C. (1954). Some problems in interval estimation. *J. Roy. Statist. Soc.*, (B) **16**, 175–186.

[4] James, A. T. (1966). Inference on latent roots by calculation of hypergeometric functions of matrix argument. *Multivariate Analysis I* (P. R. Krishnaiah, Ed.). Academic Press, New York.

[5] James, A. T. (1975). Test for a prescribed subspace of principal components. *Multivariate Analysis IV* (P. R. Krishnaiah, Ed.). Academic Press, New York.

[6] James, A. T., Wilkinson, G. N. and Venables, W. N. (1974). Interval estimates for a ratio of means. *Sankhyā* (A) **36**, 177–183.

[7] Kshirsagar, A. M. (1961). The goodness of fit of a single (non-isotropic) hypothetical principal component. *Biometrika* **48**, 397–406.

[8] Mallows, C. L. (1961). Latent vectors of random symmetric matrices. *Biometrika* **48**, 133–145.

[9] Vaughton, M. J. (1970). The statistical analysis of examination results. M.Sc. Thesis, The University of Adelaide.

[10] Wilkinson, G. N. (1977). On resolving the controversy in statistical inference (with discussion). *J. Roy. Statist. Soc.*, (B) **39**, 119–171.

P. R. Krishnaiah, ed., *Multivariate Analysis–V*

UNBIASEDNESS OF MULTIVARIATE TESTS: RECENT RESULTS*

Michael D. PERLMAN
Department of Statistics, University of Chicago, Chicago, IL 60637, U.S.A.

Results concerning the unbiasedness and/or power function monotonicity of invariant tests for various hypotheses concerning the parameters of one or more real or complex multivariate normal distributions are surveyed. Problems treated include MANOVA, independence of two or more sets of variables, equality of two or more covariance matrices, equality of several multivariate normal distributions, and testing for real or complex structure. In some problems the entire class of admissible invariant tests is treated, while in others only the likelihood ratio test (LRT) is considered.

1. Introduction

This paper presents a survey of results concerning the unbiasedness of invariant tests for many of the classical testing problems in multivariate analysis, as well as several less familiar problems. Emphasis is placed on the recent results of Perlman (1980a), Perlman and Olkin (1980), and Andersson and Perlman (1980).

The testing problems considered concern both mean vectors and covariance matrices, and have been divided into two types. Section 3 deals with problems which, after reduction to a maximal invariant statistic, involve testing a simple hypothesis $\theta=0$ against a one-sided alternative $\theta \geqslant 0$. Here, the vector $\theta \equiv (\theta_1, \ldots, \theta_m)$ represents a maximal invariant parameter and $\theta \geqslant 0$ indicates that $\theta_i \geqslant 0$, $1 \leqslant i \leqslant m$. For this type of problem, unbiasedness (respectively, monotonicity of the power function) can be shown for the entire class of invariant tests with monotone acceptance regions (respectively, with acceptance regions having convex sections). On the other hand, Section 4 treats testing problems which either involve a 'many-sided' alternative $\theta \neq \theta_0$, or else for which no such explicit representations of the maximal invariant statistic and parameter are even available. For this type

*Support for this research was provided in part by National Science Foundation Grant No. MCS76-81435 and by U.S. Department of Energy Contract No. EY-76-S-02-2751.*000.

of problem only the likelihood ratio test (LRT) is considered; indeed, no other unbiased invariant tests are known.

In Section 2 we discuss three main mathematical tools which have been used to prove the unbiasedness and monotonicity of invariant multivariate tests. These are

(i) the theorem of Anderson (1955) based on convexity;

(ii) the method of Pitman (1939), which essentially combines the Neyman-Pearson Lemma with an invariance argument; and

(iii) a method in Perlman and Olkin (1980) which applies the FKG inequality to multivariate distributions with monotone likelihood ratio. A fourth tool, the theory of majorization and Schur functions, is mentioned briefly in Section 4 but is applied much more extensively in the forthcoming paper of Perlman (1980b), which treats a number of testing problems solely concerning variances and covariance matrices.

2. Three lemmas

The first lemma is a familiar consequence of the theorem of Anderson (1955) concerning symmetric unimodal multivariate probability densities.

Lemma 2.1. *Suppose that the p-dimensional random vector U is distributed according to the multivariate normal distribution $N_p(t\xi,\Sigma)$, where $\xi: p\times 1$ is fixed and $t\geqslant 0$ is a scalar. Then for any convex set C symmetric about the origin in $\mathbf{R}^p$, $P[U\in C]$ is nonincreasing in t.*

The next lemma is implicit in the work of Pitman (1939) and is also used by Sugiura and Nagao (1968) and Perlman (1980a).

Lemma 2.2. *Let f be a probability density on a measure space $(\mathfrak{X},\mathfrak{B},\mu)$, where μ is a σ-finite measure. Let g be a one-to-one bimeasurable transformation of $\mathfrak{X}$ onto itself, and suppose that μ is invariant under g, i.e., $\mu(gB)=\mu(B)$ for every $B\in\mathfrak{B}$. For any $k>0$ let $A=\{f\geqslant k\}\in\mathfrak{B}$. Then*

$$\int_A f\,d\mu\geqslant\int_{gA} f\,d\mu.$$

Proof. Since $f \geqslant k$ on A and $f<k$ on A^c, the complement of A, we have

$$\left\{\int_A - \int_{gA}\right\} f\,\mathrm{d}\mu = \left\{\int_{A\cap(gA)^c} - \int_{(gA)\cap A^c}\right\} f\,\mathrm{d}\mu$$

$$\geqslant k\left\{\int_{A\cap(gA)^c} - \int_{(gA)\cap A^c}\right\} \mathrm{d}\mu$$

$$= k\left\{\int_A - \int_{gA}\right\} \mathrm{d}\mu$$

$$=0.$$

The last equality follows from the invariance of μ, while the next-to-last equality requires the finiteness of $\int_{A\cap(gA)^c}\mathrm{d}\mu$, which follows since

$$\int_{A\cap(gA)^c}\mathrm{d}\mu \leqslant \int_A \mathrm{d}\mu \leqslant \frac{1}{k}\int_A f\,\mathrm{d}\mu < \infty.$$

The third lemma, which generalizes a result of Lehmann (1955) for independent random variables, is given by Perlman and Olkin (1980). We restrict attention to random vectors $x \equiv (x_1,\ldots,x_q)$ assuming values in the positive orthant

$$\mathbf{R}^q_+ = \{y \equiv (y_1,\ldots,y_q)\,|\,y_i > 0, 1 \leqslant i \leqslant q\}. \tag{2.1}$$

A probability density function ϕ (with respect to Lebesgue measure, or, more generally, any product measure) on $\mathbf{R}^q_+$ is said *to satisfy the* FKG *condition* if

$$\phi(y \vee z)\phi(y \wedge z) \geqslant \phi(y)\phi(z) \tag{2.2}$$

for every $y \equiv (y_1,\ldots,y_q)$ and $z \equiv (z_1,\ldots,z_q)$ in $\mathbf{R}^q_+$, where

$$y \vee z = (y_1 \vee z_1,\ldots,y_q \vee z_q)$$

$$y \wedge z = (y_1 \wedge z_1,\ldots,y_q \wedge z_q),$$

and $y_i \vee z_i = \max(y_i, z_i)$, $y_i \wedge z_i = \min(y_i, z_i)$. The FKG *inequality* states that if the density ϕ of the random vector $\mathbf{x}$ satisfies the FKG condition, then the components $\mathbf{x}_1,\dots,\mathbf{x}_q$ are (positively) *associated*, in the sense that

$$E[g(\mathbf{x})h(\mathbf{x})] \geqslant E[g(\mathbf{x})]E[h(\mathbf{x})] \tag{2.3}$$

for every pair of integrable functions g,h which are nondecreasing in each argument. (The reader is referred to the papers of Kemperman (1977) and Perlman and Olkin (1980) for proofs, additional references, and stronger results including conditions for strict inequality in (2.3).) Finally, a subset $A \subseteq \mathbf{R}^q_+$ is said to be *monotone* if $y \in A$, $y \geqslant z$ (i.e., $y_i \geqslant z_i$, $1 \leqslant i \leqslant q) \Rightarrow z \in A$.

Lemma 2.3. *Let $x \equiv (x_1,\dots,x_q)$ be a random vector assuming values in $\mathbf{R}^q_+$, whose probability density ϕ_θ (with respect to Lebesgue measure, or any product measure) is of the form*

$$\phi_\theta(x) = F_\theta(x)\phi_0(x) \tag{2.4}$$

for $\mathbf{x} \in \mathbf{R}^q_+$ and $\theta \geqslant 0$. Suppose that $F_\theta > 0$ on $\mathbf{R}^q_+$, that $F_0 \equiv 1$, that $F_\theta(x)$ is nondecreasing in each argument x_i, and that the null density ϕ_0 satisfies the FKG *condition* (2.2). *Then for every $\theta \geqslant 0$,*

$$\mathrm{E}_\theta g(x) \geqslant \mathrm{E}_0 g(x) \tag{2.5}$$

for each integrable function g which is nondecreasing in each argument. In particular, every monotone acceptance region $A \subset \mathbf{R}^q_+$ is unbiased for testing $\theta = 0$ vs. $\theta \geqslant 0$ based on x.

Proof. From (2.4), the FKG inequality, and the fact that $\mathbf{E}_0[F_\theta(x)] = 1$, it follows immediately that

$$\mathbf{E}_\theta g(x) = \mathbf{E}_0[g(x)F_\theta(x)] \geqslant \mathbf{E}_0[g(x)].$$

Remark 2.4. In the applications in Section 3, the null density ϕ_0 is always of the special form

$$\phi_0(x) = \prod_{i=1}^{q} \alpha_i(x_i) \prod_{i<j} [\beta(x_i) - \beta(x_j)]^{\gamma}_{+}, \tag{2.6}$$

where β is a nondecreasing function on $[0, \infty)$, $\gamma \geqslant 0$, $a_+ = a \vee 0$, and the α_i are arbitrary nonnegative functions on $[0, \infty)$. It is easy to see (cf. Perlman and Olkin (1980)) that ϕ_0 given by (2.6) does satisfy the FKG condition.

3. Testing problems with one-sided alternatives

In this section eight testing problems involving the real and/or complex multivariate normal and Wishart distributions are listed in canonical form, and then two classes of invariant tests possessing monotonicity and/or unbiasedness properties are described for these problems. Each of the eight testing problems, after reduction by invariance, involves a simple null hypothesis and a one-sided alternative. The mathematical tools used are Lemmas 2.1 and 2.3.

3.1. *The* MANOVA *problem with known covariance matrix: real case*

Here, $X{:}p \times r$ is a normally distributed real random matrix with unknown mean $EX \equiv \mu$. The columns of X are independent with known real positive definite covariance matrix Σ_0, which we take to be the $p \times p$ identity matrix I_p without loss of generality. The problem of testing $\mu = 0$ vs. $\mu \neq 0$ remains invariant under orthogonal linear transformations $X \to \Psi X \Gamma'$, where $\Psi{:}p \times p$ and $\Gamma{:}r \times r$ are orthogonal matrices. A maximal invariant statistic under this group of transformations is the vector $l \equiv (l_1, \ldots, l_t)$, where $t = \min(p, r)$ and $l_1 > \cdots > l_t > 0$ are the nonzero characteristic roots of the noncentral Wishart matrix XX' (or, more generally, of $XX'\Sigma_0^{-1}$ when $\Sigma_0 \neq I_p$). An acceptance region $\mathcal{Q}$ in the X-space $\mathbf{R}^{pr}$ is invariant if $X \in \mathcal{Q} \Rightarrow \Psi X \Gamma' \in \mathcal{Q}$ for all Ψ, Γ. To each invariant acceptance region $\mathcal{Q}$ there corresponds a region $A \subset \mathbf{R}^t_+$ such that $X \in \mathcal{Q}$ iff $l \in A$, and the power function of the corresponding test depends on μ only through the vector of noncentrality parameters $\lambda \equiv (\lambda_1, \ldots, \lambda_t)$, where $\lambda_1 \geqslant \cdots \geqslant \lambda_t \geqslant 0$ are the nontrivial characteristic roots of $\mu\mu'$. After reduction to the maximal invariant statistic l and parameter λ, the testing problem can be restated as that of testing the simple hypothesis $\lambda = 0$ vs. the one-sided alternative $\lambda \geqslant 0$.

3.2. *The* MANOVA *problem with unknown covariance matrix: real case*

In canonical form (see Anderson (1958), Chapter 8), $Y{:}p \times r$ and $Z{:}p \times n$ are independent, normally distributed real random matrices whose columns are mutually independent with common covariance matrix Σ, assumed positive definite (and real) but otherwise unknown. Assume $EZ=0$, while $EY\equiv\mu$ is unknown. The problem of testing $\mu=0$ vs. $\mu\neq 0$ remains invariant under $(Y,Z)\to(BY\Gamma_1',BZ\Gamma_2')$, where $B:p\times p$ is nonsingular and $\Gamma_1:r\times r$ and $\Gamma_2:n\times n$ are orthogonal. When $n+r>p$, a nontrivial maximal invariant statistic is $d\equiv(d_{s+1},\ldots,d_t)$, where $s=\max(p-n,0)$ and $1=d_1=\cdots=d_s>d_{s+1}>\cdots>d_t>0$ are the nonzero characteristic roots of the noncentral multivariate beta matrix $YY'(YY'+ZZ')^{-1}$. An invariant acceptance region $\mathcal{Q}$ in the (Y,Z)-space $\mathbf{R}^{p(r+n)}$ corresponds to a region A in the d-space $\mathbf{R}^{t-s}$, and the power function of the corresponding test depends on (μ,Σ) only through $\delta\equiv(\delta_1,\ldots,\delta_t)$, where $\delta_1\geqslant\cdots\geqslant\delta_t\geqslant 0$ are the nontrivial characteristic roots of $\mu\mu'\Sigma^{-1}$. In terms of d and δ, the problem is to test $\delta=0$ vs. $\delta\geqslant 0$.

3.3. *Testing independence of two sets of variates (canonical correlations): real case*

Suppose that $X:(p_1+p_2)\times n$ is a normally distributed real random matrix with mean 0, whose columns are independent with common covariance matrix Σ, positive definite but unknown. Then $S\equiv XX'$ has the (central) Wishart distribution $W_{p_1+p_2}(n,\Sigma)$. Partition S as (S_{ij}), $1\leqslant i,j\leqslant 2$, with S_{ij}: $p_i\times p_j$, and partition Σ accordingly. The problem of testing $\Sigma_{12}=0$ vs. $\Sigma_{12}\neq 0$ remains invariant under $S_{ij}\to B_iS_{ij}B_j$, $1\leqslant i,j\leqslant 2$, where $B_i:p_i\times p_i$ is nonsingular. When $n>\max(p_1,p_2)$, a nontrivial maximal invariant statistic is the vector $r\equiv(r_{s+1},\ldots,r_t)$, where $s=\max(p_1+p_2-n,0)$, $t=\min(p_1,p_2)$, and $1=r_1=\cdots=r_s>r_{s+1}>\cdots>r_t>0$ are the nonzero sample canonical correlations; i.e., the r_i^2 are the nonzero characteristic roots of $S_{12}S_{22}^{-1}S_{21}S_{11}^{-1}$. The vector $\rho\equiv(\rho_1,\ldots,\rho_t)$ is a maximal invariant parameter, where $1\geqslant\rho_1\geqslant\cdots\geqslant\rho_t\geqslant 0$ are the population canonical correlations; i.e., the ρ_i^2 are the nontrivial characteristic roots of $\Sigma_{12}\Sigma_{22}^{-1}\Sigma_{21}\Sigma_{11}^{-1}$. In terms of r and ρ, the problem is to test $\rho=0$ vs. $\rho\geqslant 0$.

3.4. *The* MANOVA *problem with known covariance matrix: complex case*

This problem is identical to problem 3.1 except for the following changes. The columns of the random matrix X, now having complex entries, are distributed according to the complex multivariate normal distribution; the covariance matrix is again taken to be the identity matrix without loss of generality. The matrix of means μ now has complex entries, the matrices Ψ and Γ are now complex unitary matrices (rather than real orthogonal), and the problem remains invariant under $X\to\Psi X\Gamma^*$, where

Γ^* is the conjugate transpose of Γ. The maximal invariant statistic l and parameter λ still are vectors with nonnegative components, defined in terms of the characteristic roots of XX^* and $\mu\mu^*$ respectively; XX^* has a complex noncentral Wishart distribution. The testing problem remains that of testing $\lambda=0$ vs. $\lambda \geqslant 0$.

3.5. The MANOVA *problem with unknown covariance matrix: complex case*

This problem is identical to problem 3.2 except that the columns of Y and Z now have complex multivariate normal distributions with common covariance matrix Σ, now a complex Hermitian positive definite matrix but otherwise unspecified. The matrices μ, B, Γ_1, Γ_2 in problem 3.2 now have complex entries, with B nonsingular and Γ_1, Γ_2 unitary. The maximal invariant statistic d and parameter δ are defined in terms of the characteristic roots of $YY^*(YY^*+ZZ^*)^{-1}$ and $\mu\mu^*\Sigma^{-1}$, respectively, and the problem is to test $\delta=0$ vs. $\delta \geqslant 0$. This problem has been discussed by Pillai and Li (1970).

3.6. Testing idependence of two sets of variates (canonical correlations): complex case

This problem is identical to problem 3.3 except that X is now a normally distributed complex random matrix, $S \equiv XX^*$ has the complex Wishart distribution $CW_{p_1+p_2}(n,\Sigma)$ (cf. Goodman (1963)), and Σ is now Hermitian positive definite. The vectors r and ρ are defined as in problem 3.3, and the problem is to test $\rho=0$ vs. $\rho \geqslant 0$ (cf. Pillai and Li (1970)).

3.7. Testing for reality of the covariance matrix of a complex multivariate normal distribution

Let $X: p \times n (n \geqslant p)$ be a complex normally distributed random matrix with mean 0, whose columns are independent with common covariance matrix $\Sigma: p \times p$, Hermitian positive definite but unknown. Then $S \equiv XX^*$ has the complex Wishart distribution $CW_p(n,\Sigma)$. Express S and Σ in terms of their real and imaginary parts: $S = S_1 + iS_2$ and $\Sigma = \Sigma_1 + i\Sigma_2$, where S_1, Σ_1 are real symmetric positive definite and S_2, Σ_2 are real skew-symmetric ($S_2' = -S_2$). Khatri (1965) considered the problem of testing that Σ is real, i.e., that $\Sigma_2 = 0$, vs. $\Sigma_2 \neq 0$ (see also Andersson and Perlman (1980)). This problem remains invariant under *real* linear transformations, i.e., under $S \to BSB'$ where $B: p \times p$ is a real nonsingular matrix. A maximal invariant statistic is the vector $\mathbf{a} = (\mathbf{a}_1, \ldots, \mathbf{a}_t)$, where $t = [p/2]$ and $1 > a_1$

$>\cdots>a_t>0$ are the ordered nonzero characteristic roots (each with multiplicity two) of $S_1^{-1}S_2S_1^{-1}S_2'$. A maximal invariant parameter is the vector $\alpha=(\alpha_1,\ldots,\alpha_t)$, where $1\geqslant\alpha_1\geqslant\cdots\geqslant\alpha_t\geqslant 0$ are the ordered nontrivial characteristic roots (each with multiplicity two) of $\Sigma_1^{-1}\Sigma_2\Sigma_1^{-1}\Sigma_2'$. In terms of a and $\boldsymbol{\alpha}$, the problem is to test $\alpha=0$ vs. $\alpha\geqslant 0$.

3.8. *Testing for complex structure in a real multivariate normal distribution*

Here, X: $2p\times n(n\geqslant 2p)$ is a real normally distributed random matrix with mean 0, whose columns are independent with common covariance matrix Σ: $2p\times 2p$, real positive definite but unknown. Then $S\equiv XX'$: $2p\times 2p$ is distributed according to the real Wishart distribution $W_{2p}(n,\Sigma)$. Partition S and Σ as (S_{ij}) and (Σ_{ij}), $1\leqslant i,j\leqslant 2$, with each $S_{ij}{:}p\times p$ and each $\Sigma_{ij}{:}p\times p$. Andersson (1978) and Andersson and Perlman (1980) consider the problem of testing that Σ has complex structure, i.e., that $\Sigma_{11}=\Sigma_{22}$ and $\Sigma_{12}'=-\Sigma_{12}$, vs. the alternative that Σ is an arbitrary $2p\times 2p$ real symmetric positive definite matrix. This problem remains invariant under linear transformations of complex form, i.e., under $S\to BSB'$ where now B: $2p\times 2p$ is a real nonsingular matrix of the form

$$B=\begin{pmatrix} B_1 & B_2\\ -B_2 & B_1\end{pmatrix}$$

with $B_1{:}p\times p$ and $B_2{:}p\times p$. A maximal invariant statistic (cf. Andersson (1978)) is the vector $b\equiv(b_1,\ldots,b_p)$, where $b_1>\cdots>b_p>0$ are the ordered positive roots of the determinantal equation

$$\left|\begin{pmatrix} S_{12}'+S_{12} & S_{22}-S_{11}\\ S_{22}-S_{11} & -S_{12}'-S_{12}\end{pmatrix}-b\begin{pmatrix} S_{11}+S_{22} & S_{12}'-S_{12}\\ S_{12}-S_{12}' & S_{11}+S_{22}\end{pmatrix}\right|=0.$$

The maximal invariant parameter $\beta\equiv(\beta_1,\ldots,\beta_p)$ is defined similarly with the S_{ij} replaced by the Σ_{ij}. The problem can be restated as that of testing $\beta=0$ vs. $\beta\geqslant 0$.

3.9. *Monotonicity and unbiasedness results for problems 3.1–3.8*

The first general results make use of Lemma 2.1. We begin with problem 3.1. Conditioning on all column vectors of X except one and applying Lemma 2.1 yields the following result: if the invariant acceptance region $\mathcal{Q}\subset\mathbf{R}^{pr}$ has convex sections in each column vector of X with the remaining column vectors fixed, then the power function of the corresponding test is nondecreasing in each $\boldsymbol{\lambda}_i$. (Note that the invariance of $\mathcal{Q}$

implies that each section is symmetric about the origin in $\mathbf{R}^p$, so Lemma 2.1 is applicable.) Similarly, Das Gupta, Anderson and Mudholkar (1964) obtained the following result for problem 3.2: if the invariant acceptance region $\mathcal{A} \subset \mathbf{R}^{p(r+n)}$ has convex sections in each column vector of Y with the remaining column of Y fixed and with Z fixed, then the power function is nondecreasing in each δ_i. Next, since problem 3.3 can be reduced to problem 3.2 by conditioning on one set of variates (i.e., on S_{11} or on S_{22}) the preceding result immediately carries over to problem 3.3 (cf. Anderson and Das Gupta (1964) or Perlman (1974)). (We digress to mention that the above monotonicity result for the MANOVA problem 3.2 also easily carries over by a conditioning argument to the generalized MANOVA (growth curves) model; cf. Fujikoshi (1973).) Furthermore, each of the preceding results has its direct analog in the corresponding problem involving the *complex* multivariate normal and Wishart distributions (in particular, cf. Pillai and Li (1970) for problems 3.5 and 3.6). Finally, Lemma 2.1 also may be applied in problems 3.7 and 3.8 to obtain similar results (cf. Andersson and Perlman (1980) for details). (We remark that for each of the problems 3.1–3.8, the likelihood ratio test (LRT) belongs to the class of tests to which the results in this paragraph apply.)

Whereas the above results imply the unbiasedness of large classes of invariant tests for problems 3.1–3.8, these classes do not include all invariant tests of interest, i.e., all admissible invariant tests. In problem 3.2, for example, the following four invariant tests are commonly considered:

(i) Roy's test, with acceptance region $\{d_1 \leqslant c_\alpha\}$;

(ii) the Lawley–Hotelling test, with acceptance region $\{\sum_1^t d_i(1-d_i)^{-1} \leqslant c_\alpha\}$;

(iii) the LRT, with acceptance region $\{\prod_1^t(1-d_i) \geqslant c_\alpha\}$; and

(iv) the Bartlett–Nanda–Pillai test, with acceptance region $\{\sum_{s+1}^t d_i \leqslant c_\alpha\}$. (Note that tests (i), (ii), (iii) are defined only when $n \geqslant p$, i.e., when $s=1$.) Whereas the result of Das Gupta, Anderson and Mudholkar (1964) applies to tests (i), (ii) and (iii), it has been shown by Perlman (1974) that this result applies to test (iv) only if $c_\alpha \leqslant \max\{1, p-n\}$, which corresponds to $\alpha \geqslant \alpha^*$ for some $\alpha^* \equiv \alpha^*(n,p,r) > 0$. Since test (iv) is known to be admissible (Schwartz (1967b)), proper Bayes if $n \geqslant p$ (Kiefer and Schwartz (1965)), and locally best invariant (Schwartz (1967a)), it would be disturbing were its unbiasedness not resolved. Similar considerations apply to the analog of test (iv) in each of problems 3.3 and 3.5–3.6 (as well as in the generalized MANOVA model mentioned above).

In the recent papers of Perlman and Olkin (1980) and Andersson and Perlman (1980), however, Lemma 2.3 has been applied in problems 3.1–3.8

to demonstrate *the unbiasedness of all invariant tests with acceptance regions monotone in the maximal invariant statistics l, d, r, a, b*. Furthermore, it is pointed out that in each problem, *a necessary condition for admissibility of an invariant test is that its acceptance region be monotone in the maximal invariant statistic*. Therefore, it follows that *every admissible invariant test for problems 3.1–3.8 is unbiased*. (The unbiasedness, but not the admissibility result, carries over to the generalized MANOVA model as well.) In particular, test (iv) for problem 3.2 is unbiased, as are the corresponding tests in problems 3.3 and 3.5–3.6.

To illustrate how Lemma 2.3 is applied, consider problem 3.1 as treated by Perlman and Olkin (1980). Without loss of generality, assume that $p \leqslant r$, so that $t=p$. Then (cf. James (1961), (1964)) the joint density of $l \equiv (l_1,\ldots,l_t)$ is given by

$$\phi_\lambda(l)=\exp\left(-\frac{1}{2}\sum_{i=1}^{p}\lambda_i\right){}_0F_1\left(\frac{1}{2}r;\frac{1}{4}\Lambda,L\right)\phi_0(l), \tag{3.1}$$

where $\Lambda=\mathrm{diag}(\lambda_1,\ldots,\lambda_p)$, $L=\mathrm{diag}(l_1,\ldots,l_p)$, ${}_0F_1$ denotes the (real) hypergeometric function of two matrix arguments, and where

$$\phi_0(l)=k(p,r)\prod_{i=1}^{p}\left[l_i^{(r-p-1)/2}\mathrm{e}^{-l_i/2}\right]\prod_{i<j}(l_i-l_j)_+ \tag{3.2}$$

with $c_+\equiv\max\{c,0\}$. Since $\phi_\lambda(l)$ and $\phi_0(l)$ are precisely of the forms (2.4) and (2.6), respectively, Lemma 2.3 is applicable provided it can be shown that ${}_0F_1(\frac{1}{2}r;\frac{1}{4}\Lambda,L)$ is nondecreasing in each l_i. This may be seen in two different ways: first, from the zonal polynomial expansion (James (1964))

$${}_0F_1\left(\frac{1}{2}r;\frac{1}{4}\Lambda,L\right)=\sum_{k=0}\sum_{\kappa}\frac{C_\kappa\left(\frac{1}{4}\Lambda\right)C_\kappa(L)}{(r/2)_\kappa C_\kappa(I_p)k!} \tag{3.3}$$

and the fact (James (1968, Section 8)) that the (real) zonal polynomial $C\kappa(L)$ has nonnegative coefficients; alternatively, from the integral representation (James (1961), eqn. (8), and Perlman and Olkin (1980), equation (3.7))

$$\begin{aligned}{}_0F_1\left(\frac{1}{2}r;\frac{1}{4}\Lambda,L\right)&=\int_{O(p)}\int_{O(r)}\exp(\mathrm{tr}\,D_\lambda'\Psi D_l\Gamma')\,\mathrm{d}\Psi\,\mathrm{d}\Gamma\\&=\int_{O(p)}\int_{O(r)}\prod_{j=1}^{p}\left[\sum_{k=0}^{\infty}\frac{l_j^k}{(2k)!}\left(\sum_{i=1}^{t}\lambda_i^{1/2}\Psi_{ij}\Gamma_{ij}\right)^{2k}\right]\mathrm{d}\Psi\,\mathrm{d}\Gamma,\end{aligned} \tag{3.4}$$

where $\Psi \equiv (\Psi_{ij}) \in O(p)$ = the group of $p \times p$ orthogonal matrices, $d\Psi$ is the Haar probability measure on $O(p)$, $\Gamma \in O(r)$, and $D_l : p \times r$ is defined by

$$(D_l)_{ij} = \begin{cases} l_i^{1/2} & \text{if } i = j \\ 0 & \text{if } i \neq j. \end{cases}$$

Thus, Lemma 2.3 implies that the characteristic roots $l = (l_1, \ldots, l_p)$ of the noncentral Wishart matrix XX' are stochastically larger than those of the corresponding central Wishart matrix. In particular, any test with monotone acceptance region $A \subset \mathbf{R}_+^p$ in l-space is unbiased for testing $\lambda = 0$ vs. $\lambda \geqslant 0$.

Exactly parallel arguments can be applied in the remaining problems 3.2–3.8. The densities of l, d, and r in problems 3.1–3.3 and their complex versions 3.4–3.6 appear in James (1964), while the densities of a and b in problems 3.7 and 3.8 are given by Andersson and Perlman (1980). In each case the density is of the form (2.4) and (2.6), and in each case it can be shown that the likelihood ratio $F_\theta(x)$ is nondecreasing, so Lemma 2.3 can be applied. Alternatively, once the unbiasedness result has been established for problem 3.1, it can be extended to problem 3.2 and then 3.3 by conditional arguments—cf. Perlman and Olkin (1980); similarly, the result for problem 3.4 carries over to problems 3.5 and 3.6.

Although the results obtained by means of Lemma 2.3 apply to larger classes of tests than do the results obtained via Lemma 2.1, the conclusion is weaker, namely, unbiasedness rather than monotonicity. We conjecture that the stronger conclusion (monotonicity of the power function) also holds for the larger classes of tests (those with monotone acceptance regions). Perlman and Olkin (1980) point out that this will follow from Lemma 2.3 if in each problem the likelihood ratio $F_\theta(x)$ can be shown to have monotone likelihood ratio in each pair (x_i, θ_j) and in each pair (x_i, x_j).

By a different argument, Perlman and Olkin (1980) have obtained a partial monotonicity result for problems 3.1–3.6. They show that for rank one alternatives, i.e., alternatives of the form $(\lambda_1, 0, \ldots, 0)$, $(\delta_1, 0, \ldots, 0)$, or $(\rho_1, 0, \ldots, 0)$, the power of any test with monotone acceptance region is nondecreasing in the parameter λ_1, δ_1, or ρ_1. Their argument is based on representing the noncentral Wishart matrix XX' of problem 3.1 in terms of rectangular coordinates, i.e., $XX' = TT'$ where $T \equiv (T_{ij}) : p \times p$ is a lower triangular matrix with nonnegative diagonal elements. Since T may be defined so that $T_{11}^2 \sim \chi^2(\lambda_1)$, a noncentral χ^2-variate, while all other T_{ij} are independent and do not involve λ_1 (Farrell (1976), Section 11.4), and since it can be shown that each characteristic root of TT' is an increasing function of T_{11}, the partial monotonicity result for problem 3.1 follows.

The result extends to problems 3.2 and 3.3 by conditioning, and analogous arguments hold for the complex problems 3.4–3.6.

Lately, let us mention three other multivariate testing problems that involve a simple null hypothesis and a one-sided alternative after reduction by invariance. Let $S \sim W_p(n,\Sigma)$, $S_1 \sim W_p(n_1,\Sigma_1)$, and $S_2 \sim W_p(n_2,\Sigma_2)$ be independent Wishart matrices. Consider the problems of testing

(i) $\Sigma = \Sigma_0$ (fixed) vs. $\Sigma \geqslant \Sigma_0$ based on S,

(ii) $\Sigma = \Sigma_0$ vs. $\Sigma \leqslant \Sigma_0$ based on S, and

(iii) $\Sigma_1 = \Sigma_2$ vs. $\Sigma_1 \geqslant \Sigma_2$ based on S_1 and S_2. (Here, $\Sigma_1 \geqslant \Sigma_2$ indicates that $\Sigma_1 - \Sigma_2$ is positive semidefinite.) In problem (iii), for example, a maximal invariant statistic (respectively, parameter) is the vector $e \equiv (e_1,\ldots,e_p)$ (resp., $\varepsilon \equiv (\varepsilon_1,\ldots,\varepsilon_p)$) of characteristic roots of $S_1 S_2^{-1}$ (resp., $\Sigma_1 \Sigma_2^{-1}$), and the problem is to test $\varepsilon = (1,\ldots,1)$ vs. $\varepsilon \geqslant (1,\ldots,1)$. Anderson and Das Gupta (1964b) show that if an invariant test has a monotone acceptance region in e, then its power function increases monotonically in each ε_1; they also obtain similar results for problems (i) and (ii). Pillai and Li (1970) treat the complex versions of these problems. The essential difference between these problems and our problems 3.1–3.8 is that the former basically are (multivariate) scale-parameter problems, and a relatively simple inclusion argument yields the monotonicity results. By contrast, problems 3.1–3.8 essentially are location-parameter problems, or can be put in that form by conditioning, and require more complicated arguments to deduce unbiasedness and/or monotonicity.

4. Testing problems with many-sided alternatives

In this section four additional testing problems involving the real multivariate normal and Wishart distributions are discussed. The complex versions of these problems are not explicitly mentioned, but the results and methods carry over from the real case. In each of the four problems only the LRT (or modified LRT) is considered. The mathematical tools used are Lemmas 2.1 and 2.2, plus the well-known conditional properties of the multivariate normal and Wishart distributions.

4.1. *Testing independence of three or more sets of variates*

This problem is similar to problem 3.3, except that now $X:(p_1 + \cdots + p_k) \times n$ is partitioned into $k \geqslant 3$ sets of variates, with $n \geqslant p_1 + \cdots p_k$. Partition the (central) Wishart matrix $S \equiv XX'$ as (S_{ij}), $1 \leqslant i, j \leqslant k$, with

$S_{ij}:p_i \times p_j$, and partition Σ accordingly. The problem of testing

$$H_0: \Sigma_{ij}=0 (1 \leqslant i \neq j \leqslant k) \text{ vs. } H_1: \text{not } H_0$$

remains invariant under $S_{ij} \to B_i S_{ij} B_j$, $1 \leqslant i, j \leqslant k$, where $B_i: p_i \times p_i$ is nonsingular, but no explicit representation of a maximal invariant statistic or parameter is available. The LRT (cf. Anderson (1958), Chapter 9) rejects H_0 for small values of

$$|S| / \prod_1^k |S_{ii}|.$$

Express this statistic as

$$\frac{|S_{11\cdot(2)}|}{|S_{11}|} \frac{|S_{(22)}|}{\prod_2^k |S_{ii}|} = \frac{|S_{11\cdot(2)}|}{|S_{11\cdot(2)} + S_{1(2)} S_{(22)}^{-1} S'_{1(2)}|} \frac{|S_{(22)}|}{\prod_2^k |S_{ii}|},$$

where

$$S_{1(2)} = (S_{12}, \ldots, S_{1k}),$$

$$S_{(22)} = \begin{bmatrix} S_{22} \cdots S_{2k} \\ \vdots \quad \vdots \\ S_{k2} \cdots S_{kk} \end{bmatrix},$$

$$S_{11\cdot(2)} = S_{11} - S_{1(2)} S_{(22)}^{-1} S'_{1(2)}.$$

Conditioning on $S_{(22)}$, it is well-known (e.g., Anderson and Das Gupta (1964a), Perlman (1974)) that the conditional distribution of

$$\frac{|S_{11\cdot(2)}|}{|S_{11\cdot(2)} + S_{1(2)} S_{(22)}^{-1} S'_{1(2)}|}$$

is identical to the distribution of the LRT statistic for the MANOVA problem 3.2; the hypothesis $\mu=0$ in that problem corresponds to the hypothesis $\Sigma_{1(2)}=0$ here. Since it was shown in Section 3 that the LRT is unbiased in problem 3.2, and since the distribution of $S_{(22)}$ does not involve $\Sigma_{1(2)}$, it follows that the power function of the LRT for the present problem decreases if $\Sigma_{1(2)}$ is set equal to 0 and Σ_{11} replaced by $\Sigma_{11\cdot(2)}$. By repeating this argument after permuting the k groups of variates, it follows that the LRT is unbiased. This result was first established by Narain (1950).

4.2. Testing equality of two covariance matrices

Let $\{X_j^{(i)} \mid 1 \leqslant j \leqslant N_i\}$ be a sample of size N_i from the (real) p-variate normal distribution $N_p(\mu^{(i)}, \Sigma^{(i)})$, $1 \leqslant i \leqslant 2$. Each $\mu^{(i)}$ and $\Sigma^{(i)}$ is unknown, the latter positive definite. The problem considered here is that of testing

$$H_0 : \Sigma^{(1)} = \Sigma^{(2)} \quad \text{vs.} \quad H_1 : \Sigma^{(1)} \neq \Sigma^{(2)}.$$

The sample mean and sample covariance matrix from the ith sample are given by

$$\begin{aligned} \bar{X}^{(i)} &= \frac{1}{N_i} \sum_{j=1}^{N_i} X_j^{(i)} \\ S^{(i)} &= \sum_{j=1}^{N_i} \left(X_j^{(i)} - \bar{X}^{(i)}\right)\left(X_j^{(i)} - \bar{X}^{(i)}\right)' \sim W_p\left(n_i, \Sigma^{(i)}\right), \end{aligned} \tag{4.1}$$

where $n_i = N_i - 1$; assume that each $n_i \geqslant p$. The (unmodified) LRT (cf. Anderson (1958), Chapter 10), which rejects H_0 for large values of $|S|^N / \prod_1^2 |S^{(i)}|^{N_i}$ where $S = \sum_1^2 S^{(i)}$ and $N = \sum_1^2 N^{(i)}$, is well-known to be biased when the N_i are unequal—cf. Brown (1939), Sugiura and Nagao (1969), Das Gupta (1969). The modified LRT proposed by Bartlett rejects H_0 for large values of

$$|S|^n / \prod_1^2 |S^{(i)}|^{n_i},$$

where $n = \sum_1^2 n_i$.

Pitman (1939) established the unbiasedness of the modified LRT in the univariate case $p = 1$ (for two or more populations), while Suguira and Nagao (1968) adapted Pitman's method to treat the multivariate two-population problem considered here. The latter argument is now presented. The power function of the modified LRT, defined as

$$\pi(\Sigma^{(1)}, \Sigma^{(2)}) = P_{\Sigma^{(1)}, \Sigma^{(2)}}\left[|S|^n / \prod_1^2 |S^{(i)}|^{n_i} > c_\alpha \right], \tag{4.2}$$

satisfies the invariance condition

$$\pi(\Sigma^{(1)},\Sigma^{(2)})=\pi(B\Sigma^{(1)}B',B\Sigma^{(2)}B') \tag{4.3}$$

for all nonsingular $p\times p$ matrices B. Therefore, to demonstrate the unbiasedness of the modified LRT it suffices to show that

$$\pi(\Sigma^{(1)},I_p)\geqslant\pi(I_p,I_p) \tag{4.4}$$

for every $\Sigma^{(1)}$, where I_p is the $p\times p$ identity matrix. If we define

$$f(S^{(1)},S^{(2)})=\left[\prod_1^2|S^{(i)}|^{n_i}/|S|^n\right]^{1/2}, \tag{4.5}$$

then

$$1-\pi(\Sigma^{(1)},I_p)=c^*\cdot\int_{\{f(S^{(1)},S^{(2)})<c\}}|\Sigma^{(1)}|^{-n/2}\prod_1^2|S^{(i)}|^{(n_i-p-1)/2}$$
$$\times\exp\left[-\frac{1}{2}\operatorname{tr}(\Sigma^{(1)})^{-1}S^{(1)}+S^{(2)}\right]\mathrm{d}S^{(1)}\,\mathrm{d}S^{(2)}, \tag{4.6}$$

where $c=c_\alpha^{-1/2}$ and c^* is a generic positive constant. In (4.6) make the transformation $(S^{(1)},S^{(2)})\to(U_1,U_2)\equiv(S^{(1)},(S^{(1)})^{-1/2}S_2(S^{(1)})^{-1/2})$ and integrate out U_1. The Jacobian is $|S^{(1)}|^{(p+1)/2}$, so (4.6) becomes

$$1-\pi(\Sigma^{(1)},I_p)$$
$$=c^*\cdot\int_{\{f(I_p,U_2)\geqslant c\}}|\Sigma^{(1)}|^{-n_1/2}|U_2|^{n_2/2}|(\Sigma^{(1)})^{-1}+U_2|^{-(n_1+n_2)/2}\,\mathrm{d}\mu(U_2) \tag{4.7}$$

where $\mathrm{d}\mu(U_2)$ denotes the measure $|U_2|^{-(p+1)/2}\mathrm{d}U_2$ on the set of all $p\times p$ positive definite matrices. Now let $G=(\Sigma^{(1)})^{-1/2}$ and make the transformation $V=G^{-1}U_2(G^{-1})'$ in (4.7) to obtain

$$1-\pi(\Sigma^{(1)},I_p)=c^*\cdot\int_{\{f(I_p,GVG')\geqslant c\}}f(I_p,V)\,\mathrm{d}\mu(V). \tag{4.8}$$

Since the measure μ is invariant under $V \to GVG'$, we can now apply Lemma 2.2 with $A = \{V | f(I_p, V) \geqslant c\}$ to obtain (4.4).

4.3. Testing equality of three or more covariance matrices

This problem is identical to problem 4.2 except that now there are samples from $k(\geqslant 3)$ p-variate normal distributions $N_p(\mu^{(i)}, \Sigma^{(i)})$, $1 \leqslant i \leqslant k$, and it is wished to test

$$H_0: \Sigma^{(1)} = \cdots = \Sigma^{(k)} \quad vs. \quad H_1: \text{not}\, H_0.$$

The modified LRT statistic is now

$$|S|^n \Big/ \prod_1^k |S^{(i)}|^{n_i},$$

where $S = \Sigma_1^k S^{(i)}$, $n = \Sigma_1^k n_i$. Denote the power function of this test by

$$\pi(\Sigma^{(1)}, \ldots, \Sigma^{(k)}) = P_{\Sigma^{(1)}, \ldots, \Sigma^{(k)}} \left[|S|^n \Big/ \prod_1^k |S^{(i)}|^{n_i} \geqslant c_\alpha \right]. \tag{4.9}$$

If we try to prove the unbiasedness of the modified LRT by the same argument used in problem 4.2, we immediately encounter difficulty in the step from (4.6) to (4.7), since in general the equation

$$\operatorname{tr}(\Sigma^{(i)})^{-1}(S^{(1)})^{1/2} U_i (S^{(1)})^{1/2} = \operatorname{tr}(\Sigma^{(i)})^{-1} U_i S^{(1)} \tag{4.10}$$

is not valid. Notice, however, that in the special case where each $\Sigma^{(i)}$ is of the form $\Sigma^{(i)} = \sigma_i I_p$ for some scalar $\sigma_i > 0$, the equation (4.10) is valid and the argument used for problem 4.2 can be carried through, again invoking Lemma 2.2. Therefore, we have that

$$\pi(\sigma_1 I_p, \ldots, \sigma_k I_p) \geqslant \pi(I_p, \ldots, I_p). \tag{4.11}$$

Perlman (1980a) has proved the complete unbiasedness of the modified LRT by means of (4.11) and two additional steps. The following notation is needed. For each population covariance matrix $\Sigma^{(i)} \equiv (\Sigma_{\alpha\beta}^{(i)})$, $1 \leqslant \alpha, \beta \leqslant p$, let $\Sigma_{\alpha\alpha \cdot \beta}^{(i)}$ denote the conditional variance of the αth variate given the βth,...,pth variates, where $1 \leqslant \alpha < \beta \leqslant p$. Define the diagonal matrix $D_i \equiv$

$D(\Sigma^{(i)})$ and the scalar $\sigma_i \equiv \sigma(\Sigma^{(i)})$ by

$$D_i = \operatorname{diag}\left(\Sigma^{(i)}_{11\cdot 2}, \Sigma^{(i)}_{22\cdot 3}, \ldots, \Sigma^{(i)}_{p-1,p-1\cdot p}, \Sigma^{(i)}_{pp}\right)$$

$$\sigma_i = |\Sigma^{(i)}|^{1/p}, \tag{4.12}$$

and notice that

$$|\sigma_i I_p| = |D_i| = |\Sigma^{(i)}|.$$

It is shown in Perlman (1980) that

$$\pi(\Sigma^{(1)}, \ldots, \Sigma^{(k)}) \geqslant \pi(D_1, \ldots, D_k) \tag{4.13}$$

and

$$\pi(D_1, \ldots, D_k) \geqslant \pi(\sigma_1 I_p, \ldots, \sigma_k I_p). \tag{4.14}$$

Taken together with (4.11), these two inequalities imply the complete unbiasedness of the modified LRT. Inequality (4.13) is obtained by successively conditioning on each subset of $p-1$ variates simultaneously in all k populations and applying Lemma 2.1, while (4.14) follows by successively applying two-dimensional rotations to all pairs of coordinates simultaneously in all k populations and then repeating the preceding argument. The details, which are numerous and at one point rely on the theory of majorization, appear in Perlman (1980a).

4.4. *Testing equality of two or more multivariate normal populations*

As in problems 4.2 and 4.3, here we have samples of size N_i from $N_p(\mu^{(i)}, \Sigma^{(i)})$, $1 \leqslant i \leqslant k$, with $k \geqslant 2$. The problem is to test

$$H_0\colon \mu^{(1)} = \cdots = \mu^{(k)}, \quad \Sigma^{(1)} = \cdots = \Sigma^{(k)} \quad vs. \quad H_1\colon \text{not } H_0.$$

The (unmodified) LRT (cf. Anderson (1958), Chapter 10) rejects H_0 for large values of

$$|S+T|^N \Big/ \prod_1^k |S^{(i)}|^{N_i},$$

where $S=\Sigma_1^k S^{(i)}$, $N=\Sigma_1^k N_i$,

$$T=\sum_{i=1}^{k} N_i(\bar{X}^{(i)}-\bar{X})(\bar{X}^{(i)}-\bar{X})',$$

$$\bar{X}=\frac{1}{N}\sum_{i=1}^{k} N_i\bar{X}^{(i)}.$$

Anderson (1958) suggested the modified LRT statistic $|S+T|^n/\prod_1^k|S^{(i)}|^{n_i}$. However, Perlman (1980a) shows that it is the unmodified LRT that is unbiased for this problem. Three inequalities parallel to (4.11), (4.13), and (4.14) are established by means of arguments similar to those used in problem 4.3, with a few additional details. Denoting the power function by

$$\pi(\mu^{(1)},\ldots,\mu^{(k)};\Sigma^{(1)},\ldots,\Sigma^{(k)})$$
$$\equiv P_{\mu^{(1)},\ldots,\mu^{(k)};\Sigma^{(1)},\ldots,\Sigma^{(k)}}\left[|S+T|^N/\prod_1^k|S^{(i)}|^{N_i}\geqslant c_\alpha\right],$$

these inequalities are

$$\begin{aligned}\pi(\mu^{(1)},\ldots,\mu^{(k)};\Sigma^{(1)},\ldots,\Sigma^{(k)})&\geqslant\pi(0,\ldots,0;D_1,\ldots,D_k)\\&\geqslant\pi(0,\ldots,0;\sigma_1 I_p,\ldots,\sigma_k I_p)\\&\geqslant\pi(0,\ldots,0;I_p,\ldots,I_p),\end{aligned}$$

where D_i and σ_i are defined in (4.12).

References

[1] Anderson, T. W. (1955). The integral of a symmetric unimodal function over a symmetric convex set and some probability inequalities. *Proc. Amer. Math. Soc.* **6**, 170–176.

[2] Anderson, T. W. (1958). *An Introduction to Multivariate Statistical Analysis*. John Wiley and Sons, New York.

[3] Anderson, T. W. and Das Gupta, S. (1964a). Monotonicity of the power functions of some tests of independence between two sets of variates. *Ann. Math. Statist.* **35**, 206–208.

[4] Anderson, T. W. and Das Gupta, S. (1964b). A monotonicity property of the power functions of some tests of the equality of two covariance matrices. *Ann. Math. Statist.* **35**, 1059–1063.

[5] Andersson, S. A. (1978). Canonical correlations with respect to a complex structure. Technical Report No. 132, National Science Foundation Grant MPS 75-09450; also 33, Office of Naval Research Contract N00014-75-C-0442, Dept. of Statistics, Stanford University.

[6] Andersson, S. A. and Perlman, M. D. (1980). Two testing problems relating the real and complex multivariate normal distributions. In preparation.

[7] Brown, G. W. (1939). On the power of the L_1 test for equality of several variances. *Ann. Math. Statist.* **10**, 119–128.

[8] Das Gupta, S. (1969). Properties of power functions of some tests concerning dispersion matrices of multivariate normal distributions. *Ann. Math. Statist.* **40**, 697–701.

[9] Das Gupta, S., Anderson, T. W., and Mudholkar, G. (1964). Monotonicity of the power functions of some tests of the multivariate linear hypothesis. *Ann. Math. Statist.* **35**, 200–205.

[10] Farrell, R. (1976). *Techniques of Multivariate Calculation*. Lecture Notes in Mathematics **520**, Springer-Verlag, Berlin.

[11] Fujikoshi, Y. (1973). Monotonicity of the power functions of some tests in general MANOVA models. *Ann. Statist.* **1**, 388–391.

[12] Goodman, N. R. (1963). Statistical analysis based on a certain multivariate complex Gaussian distribution (an introduction). *Ann. Math. Statist.* **34**, 152–176.

[13] James, A. T. (1961). The distribution of noncentral means with known covariance. *Ann. Math. Statist.* **32**, 874–882.

[14] James, A. T. (1964). Distributions of matrix variates and latent roots derived from normal samples. *Ann. Math. Statist.* **35**, 475–501.

[15] James, A. T. (1968). Calculation of zonal polynomial coefficients by use of the Laplace-Beltrami operator. *Ann. Math. Statist.* **39**, 1711–1718.

[16] Kemperman, J. H. B. (1977). On the FKG-inequality for measures on a partially ordered space. *Indag. Math.* **39**, 313–331.

[17] Khatri, C. G. (1965). A test for reality of a covariance matrix in a certain complex Gaussian distribution. *Ann. Math. Statist.* **36**, 115–119.

[18] Kiefer, J. and Schwartz, R. (1965). Admissible Bayes character of T^2-, R^2-, and other fully invariant tests for classical multivariate normal problems. *Ann. Math. Statist.* **36**, 747–770.

[19] Lehmann, E. L. (1955). Ordered families of distributions. *Ann. Math. Statist.* **26**, 399–419.

[20] Narain, R. D. (1950). On the completely unbiased character of tests of independence in multivariate normal systems. *Ann. Math. Statist.* **21**, 293–298.

[21] Perlman, M. D. (1974). On the monotonicity of the power functions of tests based on traces of multivariate beta matrices. *J. Multivariate Anal.* **4**, 22–30.

[22] Perlman, M. D. (1980a). Unbiasedness of the likelihood ratio tests for equality of several covariance matrices and equality of several multivariate normal populations. To appear in *Ann. Statist.*

[23] Perlman, M. D. (1980b). Tests for variances and covariance matrices: unbiasedness and monotonicity of power. To appear in *Ann. Statist.*

[24] Perlman, M. D. and Olkin, I. (1980). Unbiasedness of invariant tests for MANOVA and other multivariate testing problems. Submitted for publication.

[25] Pillai, K. C. S. and Li, H. C. (1970). Monotonicity of the power functions of some tests of hypotheses concerning multivariate complex normal distributions. *Ann. Inst. Statist. Math.* **22**, 307–318.

[26] Pitman, E. J. G. (1939). Tests of hypotheses concerning location and scale parameters. *Biometrika* **31**, 200–215.

[27] Schwartz, R. (1967a). Locally minimax tests. *Ann. Math. Statist*. **38**, 340–359.

[28] Schwartz, R. (1967b). Admissible tests in multivariate analysis of variance. *Ann. Math. Statist*. **38**, 698–710.

[29] Sugiura, N. and Nagao, N. (1968). Unbiasedness of some test criteria for the equality of one or two covariance matrices. *Ann. Math. Statist*. **39**, 1686–1692.

[30] Sugiura, N. and Nagao, N. (1969). On Bartlett's test and Lehmann's test for homogeneity of variance. *Ann. Math. Statist*. **40**, 2018–2032.

PART VI

SIMULTANEOUS TEST PROCEDURES

P. R. Krishnaiah, ed., *Multivariate Analysis–V*

A STUDY ON FINITE INTERSECTION TESTS FOR MULTIPLE COMPARISONS OF MEANS*

C. M. COX, P. R. KRISHNAIAH, J. C. LEE, J. REISING and F. J. SCHUURMANN
University of Pittsburgh, Air Force Flight Dynamics Laboratory and Miami University

1. Introduction

In a number of situations, it is of interest to test simultaneously the hypotheses on various contrasts of means (mean vectors) of univariate (multivariate) normal populations. In the univariate case, the above hypotheses can be tested by using Scheffé's test, Krishnaiah's finite intersection tests and other procedures. For multiple comparison of mean vectors of multivariate normal populations, one may use (i) S. N. Roy's largest root test, (ii) $T^2_{\max}$ test, (iii) J. Roy's step-down procedure and (iv) Krishnaiah's finite intersection tests. The object of this paper is to study the merits of the finite intersection tests and the computation of the critical values associated with these procedures. For a review of the literature of the above simultaneous test procedures, the reader is referred to Krishnaiah (1979).

In Section 2 of this paper, we discuss the evaluation of the critical values associated with the finite intersection tests for multiple comparisons of means of univariate normal populations. These tests are compared with alternative procedures like Scheffé's test and Tukey's test in Section 3. Finite intersection tests for multiple comparisons of mean vectors of multivariate normal populations are discussed in Section 4. Also, we compare the above tests with some alternative procedures for multiple comparisons of mean vectors. In Section 5, we describe a simulated experiment conducted on the performance of pilots when they use different keyboards in the cockpit of an aircraft. The usefulness of the finite intersection tests is illustrated with the data obtained from the above experiment.

*Part of this work was sponsored by the Air Force Flight Dynamics Laboratory, Air Force Systems Command, under grant AFOSR 77-3239. The authors wish to thank Mr. C. Fang for some helpful discussions.

2. Finite intersection tests for multiple comparisons of means

We will first define a multivariate chi-square distribution and a multivariate F distribution since they are needed in the sequel. Let $\mathbf{x}'=(x_1,\ldots,x_q)$ be distributed as the multivariate normal with mean vector $\boldsymbol{\mu}'=(\mu_1,\ldots,\mu_q)$ and covariance matrix $\Sigma=\sigma^2\Omega$ where $\Omega=(\rho_{ij})$ is the correlation matrix. Then, it is known (see Krishnamoorthy and Parthasarathy (1951) and Krishnaiah and Rao (1961)) that the joint distribution of $x_1^2,\ldots,x_q^2$ is a central or non-central multivariate chi-square distribution with one degree of freedom and with Σ as the covariance matrix of the 'accompanying' multivariate normal. Next let s^2/σ^2 be distributed independent of $\mathbf{x}$ as chi-square with n degrees of freedom. Also, let $F_i=nx_i^2/s^2$ for $i=1,2,\ldots,q$. Then, the joint distribution of $F_1,\ldots,F_q$ is known (see Krishnaiah (1965a, 1965b)) to be a central or non-central multivariate F distribution with $(1,n)$ degrees of freedom and with Ω as the correlation matrix of the 'accompanying' multivariate normal. We will now discuss the finite intersection tests of Krishnaiah for multiple comparisons of means.

For $i=1,2,\ldots,k$, let $x_{i1},\ldots,x_{in_i}$ be a random sample from a normal population with mean μ_i and variance σ^2. Also, let $s^2=\Sigma_{i=1}^k\Sigma_{j=1}^{n_i}(x_{ij}-\bar{x}_{i.})^2$, $n_i\bar{x}_{i.}=\Sigma_{j=1}^{n_i}x_{ij}$, $\lambda_i=c_{i1}\mu_1+\cdots+c_{ik}\mu_k$, and $\hat{\lambda}_i=c_{i1}\bar{x}_{1.}+\cdots+c_{ik}\bar{x}_{k.}$. In addition, let $F_i=(n-k)\hat{\lambda}_i^2/s^2d_i$ where $n=n_1+\cdots+n_k$ and $d_i=\Sigma_{j=1}^{n_k}c_{ij}^2/n_j$. Also, let $H_i:\lambda_i=0$, $H=\bigcap_{i=1}^q H_i$ and $A_i:\lambda_i\neq 0$. According to the finite intersection tests, we accept or reject H_i against A_i according as

$$F_i \lesseqgtr F_\alpha,$$

where

$$P\left[F_i\leqslant F_\alpha;i=1,\ldots,q\,\middle|\,\bigcap_{i=1}^{q} H_i\right]=(1-\alpha). \tag{2.1}$$

Here we note that the lengths of the confidence intervals associated with the finite intersection test decrease as q decreases. When H is true, the joint distribution of $F_1,\ldots,F_q$ is the central q-variate F distribution with $(1,\nu)$ degrees of freedom and with Ω as the correlation matrix of the 'accompanying' multivariate normal where $\nu=n-k$. When $\rho_{ij}=\rho(i\neq j)$ values of F_α for given values of α can be obtained from the tables of Krishnaiah and Armitage (1965c, 1970) when $\nu\leqslant 35$. When ν is large, the distribution function of the multivariate F distribution can be approximated with the distribution function of the multivariate chi-square distribution. Next, let $t_i = x_i\sqrt{\nu}/s$ for $i=1,\ldots,q$. Then the joint distribution of $t_1,\ldots,t_q$ is a central or noncentral multivariate t distribution (see Dunnett and Sobel (1954)) with ν degrees of freedom according as $\boldsymbol{\mu}=\mathbf{0}$ or $\boldsymbol{\mu}\neq\mathbf{0}$. We now discuss the accuracy of certain bounds on the probability integral of the multivariate F distribution.

Using Poincare's formula, we know that

$$1-\sum_{i=1}^{q} P[F_i \geqslant c] \leqslant P[F_i \leqslant c; i=1,\ldots,q]$$

$$\leqslant 1-\sum_{i=1}^{q} P[F_i \geqslant c] + \sum_{i<j} P[F_i \geqslant c, F_j \geqslant c]. \tag{2.2}$$

Also, Sidak (1967) showed that

$$P[F_i \leqslant c; i=1,2,\ldots,q] \geqslant P[F_{i0} \leqslant c; i=1,\ldots,q], \tag{2.3}$$

where $F_{i0}=(w_i\nu)/s^2$, $w_i=y_i^2$ and $\mathbf{y}'=(y_1,\ldots,y_q)$ is distributed independent of s^2 as a multivariate normal with mean vector $\mathbf{0}$ and covariance matrix $\sigma^2 I_p$. Kimball (1951) showed that

$$P[F_{i0} \leqslant c; i=1,\ldots,q] \geqslant \prod_{i=1}^{q} P[F_{i0} \leqslant c]. \tag{2.4}$$

Let c_1, c_2, c_3, c_4 and c_5 be defined by the following equations:

$$1-\sum_{i=1}^{q} P[F_i \geqslant c_1] = (1-\alpha), \tag{2.5}$$

$$\prod_{i=1}^{q} P(F_i \leqslant c_2) = (1-\alpha), \tag{2.6}$$

$$P[F_{i0} \leqslant c_3; i=1,2,\ldots,q] = (1-\alpha), \tag{2.7}$$

$$1-\sum_{i=1}^{q} P[F_i \geqslant c_4] + \sum_{i<j=1}^{q} P[F_i \geqslant c_4, F_j \geqslant c_4] = (1-\alpha), \tag{2.8}$$

$$P[F_i \leqslant c_5; i=1,\ldots,q] = (1-\alpha). \tag{2.9}$$

Then c_1, c_2 and c_3 give upper bounds on c_5 whereas c_4 gives a lower bound on c_5.

We now discuss the relationship of the distribution of the studentized range with the multivariate F distribution. Let $u_1,\ldots,u_k$ be distributed independently as normal with mean 0 and variance σ^2. Also, let s^2/σ^2 be a chi-square variable with ν degrees of freedom distributed independently of $u_1,\ldots,u_k$. Then the statistic $R=(\max u_i-\min u_i)\sqrt{(\nu/s^2)}$ is known to be studentized range. Also

$$P[R^2\leqslant 2c]=P[F_i\leqslant c; i=1,\ldots,q] \tag{2.10}$$

when $q=k(k-1)/2$ and Ω has special structure; in this case the correlations ρ_{ij} $(i\neq j)$ take the values of 0.5, -0.5 or 0. Harter (1960) constructed exact tables for percentage points of the studentized range distribution. Tables 1–3 give a comparison of some of these values with bounds obtained by using the inequalities (2.2)–(2.4).

Now, let $R_0=\max(U,R)$ where $U=(\nu/s^2)^{1/2}\max(|u_1|,\ldots,|u_k|)$. Then, the statistic R_0 is known to be studentized augmented range. We know that

$$P[R_0^2\leqslant c]=P\left[F_i\leqslant c, i=1,\ldots,k;\ F_{ij}\leqslant \frac{c}{2}, i<j=1,\ldots,k\right] \tag{2.11}$$

where $F_i=\nu u_i^2/s^2$ and $F_{ij}=\nu(u_i-u_j)^2/2s^2$. The joint distribution of the q F-statistics $(F_1,\ldots,F_k,F_{12},F_{13},\ldots,F_{k-1,k})$ is a multivariate F distribution with $(1,\nu)$ degrees of freedom and with $\Omega=(\rho_{ij})$ as the correlation matrix of

Table 1
Comparison of bounds with exact percentage points of the studentized range square distribution
$k=3$

ν	α	c_1	c_2	c_3	c_4	c_5 (Exact)
5	0.05	12.49	12.38	11.55	10.51	10.59
10	0.05	8.24	8.18	8.00	7.51	7.52
15	0.05	7.26	7.21	7.12	6.75	6.75
20	0.05	6.83	6.79	6.73	6.40	6.40
30	0.05	6.43	6.40	6.36	6.08	6.08
40	0.05	6.25	6.21	6.19	5.92	5.92
5	0.01	27.54	27.42	26.02	24.23	24.33
10	0.01	14.65	14.61	14.45	13.89	13.89
15	0.01	12.14	12.11	12.05	11.69	11.69
20	0.01	11.09	11.09	11.04	10.76	10.76
30	0.01	10.17	10.16	10.14	9.92	9.92
40	0.01	9.75	9.73	9.73	9.54	9.54

Table 2
Comparison of bounds with exact percentage points of the studentized range square distribution
$k=4$

ν	α	c_1	c_2	c_3	c_4	c_5 (Exact)
5	0.05	17.80	17.62	15.43	10.80	13.61
10	0.05	10.74	10.65	10.24	8.84	9.36
15	0.05	9.22	9.16	8.96	8.04	8.31
20	0.05	8.57	8.52	8.40	7.65	7.83
30	0.05	7.98	7.93	7.87	7.27	7.39
40	0.05	7.71	7.66	7.62	7.08	7.19
5	0.01	37.68	37.50	33.75	26.57	30.45
10	0.01	18.14	18.13	17.68	16.23	16.64
15	0.01	14.61	14.61	14.44	13.63	13.79
20	0.01	13.18	13.16	13.08	12.50	12.59
30	0.01	11.93	11.91	11.89	11.47	11.52
40	0.01	11.37	11.37	11.34	10.99	11.03

Table 3
Comparison of bounds with exact percentage points of the studentized range square distribution
$k=5$

ν	α	c_1	c_2	c_3	c_4	c_5 (Exact)
5	0.05	22.79	22.56	18.58	—	16.09
10	0.05	12.83	12.73	12.02	8.63	10.83
15	0.05	10.80	10.72	10.42	8.64	9.54
20	0.05	9.94	9.88	9.70	8.39	8.95
30	0.05	9.18	9.12	9.03	8.06	8.41
40	0.05	8.83	8.78	8.71	7.89	8.16
5	0.01	47.18	47.27	40.00	—	35.46
10	0.01	21.04	20.98	20.28	17.42	18.83
15	0.01	16.59	16.56	16.32	14.95	15.43
20	0.01	14.82	14.80	14.68	13.75	14.01
30	0.01	13.29	13.28	13.23	12.60	12.74
40	0.01	12.61	12.58	12.57	12.06	12.16

the 'accompanying' multivariate normal and $q=k(k+1)/2$. If we partition Ω as

$$\Omega=\begin{pmatrix}\Omega_{11} & \Omega_{12}\\ \Omega_{21} & \Omega_{22}\end{pmatrix},$$

where Ω_{11} is of order $k\times k$, then $\Omega_{11}=I_k$, and the off-diagonal elements of

Table 4
Comparison of bounds with exact percentage points of the studentized augmented range square distribution

q	ν	α	c_2	c_3	c_4	c_5
3	16	0.05	10.12	10.05	9.47	9.49
3	20	0.05	9.71	9.65	9.14	9.14
6	20	0.05	13.95	13.81	12.62	12.89
6	30	0.05	13.07	12.99	12.03	12.22
3	30	0.01	15.55	15.55	15.30	15.30
3	20	0.01	16.80	16.80	16.41	16.44
6	20	0.01	22.38	22.27	21.45	21.57
6	16	0.01	24.02	23.91	22.73	22.96

Table 5
Comparison of bounds with exact percentage points of the multivariate F distribution in equicorrelated case
$q=3$

ν	α	ρ	c_1	c_2	c_3	c_4	c_5 (Exact)
5	0.05	0.1	12.49	12.38	11.55	11.42	11.53
10	0.05	0.5	8.24	8.18	8.00	7.51	7.61
20	0.05	0.7	6.83	6.79	6.73	5.94	6.13
35	0.05	0.9	6.32	6.29	6.26	4.51	5.16
5	0.01	0.9	27.54	27.42	26.02	17.27	20.63
10	0.01	0.7	14.65	14.61	14.45	13.10	13.40
20	0.01	0.5	11.09	11.08	11.04	10.76	10.79
30	0.01	0.1	10.17	10.16	10.14	10.13	10.13

Table 6
Comparison of bounds with exact percentage points of the multivariate F distribution in equicorrelated case
$q=5$

ν	α	ρ	c_1	c_2	c_3	c_4	c_5 (Exact)
5	0.05	0.1	16.26	16.09	14.36	13.64	14.30
10	0.05	0.5	10.04	9.96	9.63	8.34	8.94
20	0.05	0.7	8.10	8.04	7.95	5.62	6.95
30	0.05	0.9	7.56	7.52	7.46	—	5.69
5	0.01	0.9	34.73	34.69	31.56	—	22.59
10	0.01	0.7	17.17	17.19	16.80	13.03	15.00
20	0.01	0.5	12.62	12.58	12.53	11.94	12.10
30	0.01	0.1	11.46	11.44	11.42	11.41	11.41

Ω_{22} are 0.5, -0.5 or 0. The elements of Ω_{12} are $1/\sqrt{2}$, $-1/\sqrt{2}$ or 0. Table 4 gives a comparison of the bounds with the exact values of the percentage points of the distribution of R_0^2. The exact percentage points are obtained from the tables of Stoline (1978).

Krishnaiah and Armitage (1965c, 1970) compared the exact critical values for a few values of the parameters with bounds obtained by using Equation (2.8) for the equicorrelated case.

Tables 5 and 6 give a comparison of the exact percentage points c_5 of the multivariate F distribution with $(1,\nu)$ degrees of freedom when $\rho_{ij}=\rho$ for $i\neq j$ with the bounds c_1, c_2, c_3 and c_4. The exact percentage points are obtained from the tables of Krishnaiah and Armitage (1965c).

Now, consider the joint distribution of $x_1^2,\ldots,x_q^2$ when $\sigma^2=1$. Also, let the constants a_1, a_2, a_4 and a_5 be chosen such that

$$1-\sum_{i=1}^{q} P\left[x_i^2 \geqslant a_1\right]=(1-\alpha) \tag{2.12}$$

$$\prod_{i=1}^{q} P\left[x_i^2 \leqslant a_2\right]=(1-\alpha) \tag{2.13}$$

$$1-\sum_{i=1}^{q} P\left[x_i^2 \geqslant a_4\right]+\sum_{i<j} P\left[x_i^2 \geqslant a_4, x_j^2 \geqslant a_4\right]=(1-\alpha) \tag{2.14}$$

$$P\left[x_i^2 \leqslant a_5; i=1,\ldots,q\right]=(1-\alpha). \tag{2.15}$$

Next, let $R^*=(\max u_i-\min u_i)$, $\sigma^2=1$ and $q=k(k-1)/2$. Then

$$P\left[R^{*2} \leqslant 2c\right]=P\left[x_i^2 \leqslant c; i=1,\ldots,q\right] \tag{2.16}$$

where the correlations $\rho_{ij}(i\neq j)$ between x_i and x_j are equal to 0.5, -0.5 or 0. Table 7 gives a comparison of the bounds a_1, a_2 and a_4 with exact percentage points a_5 of R^{*2}. The exact percentage points are obtained from the tables of Harter (1960).

Next, let $R_0^*=\max(U^*,R^*)$ where $U^*=\max(|u_1|,\ldots,|u_k|)$ and $\sigma^2=1$. The statistic R_0^* is known to be the augmented range. There $q=k(k+1)/2$ and

$$P\left[R_0^{*2} \leqslant c\right]=P\left[\chi_i^2 \leqslant c, i=1,2,\ldots,k; \chi_{ij}^2 \leqslant \frac{c}{2}, i<j=1,\ldots,k\right] \tag{2.17}$$

Table 7
Comparison of bounds with exact percentage points of the range square distribution

q	α	$2a_1$	$2a_2$	$2a_4$	$2a_5$ (Exact)
3	0.05	11.46	11.40	10.99	10.99
6	0.05	13.92	13.84	13.09	13.20
10	0.05	15.76	15.68	14.64	14.88
3	0.01	17.23	17.18	16.98	16.98
6	0.01	19.77	19.76	19.36	19.38
10	0.01	21.66	21.64	21.14	21.19

Table 8
Bounds on the percentage points of the augmented range square distribution

q	3	6	10	15	3	6	10	15
α	0.05	0.05	0.05	0.05	0.01	0.01	0.01	0.01
a_2	8.25	11.54	13.90	15.70	13.44	17.27	19.77	21.68
a_4	7.95	10.94	13.01	14.55	13.36	16.95	19.38	21.09

Table 9
Comparison of bounds with exact percentage points of the multivariate chi-square distribution with one degree of freedom in equicorrelated case

q	ρ	α	a_1	a_2	a_4	a_5 (Exact)
3	0.1	0.05	5.73	5.70	5.70	5.70
5	0.5	0.05	6.64	6.60	6.20	6.31
6	0.6	0.05	6.96	6.92	5.98	6.41
8	0.1	0.01	10.42	10.40	10.39	10.40
6	0.5	0.01	9.89	9.88	9.63	9.68
4	0.8	0.01	9.14	9.14	8.07	8.47

where $\chi_i^2 = u_i^2$ and $\chi_{ij}^2 = (u_i - u_j)^2/2$. Table 8 gives a comparison of the bounds a_2 and a_4 where a_2 and a_4 were defined by equations (2.13) and (2.14) respectively. Table 9 gives a comparison of the bounds a_1, a_2, a_4 with exact percentage points of the multivariate chi-square distribution with one degree of freedom in the equicorrelated case.

Now, let the constants b_2, b_3, b_4 and b_5 be defined as follows:

$$\prod_{i=1}^{q} P[t_i \leqslant b_2] = (1-\alpha), \tag{2.18}$$

$$P[t_{i0} \leqslant b_3; i=1,2,\ldots,q]=(1-\alpha), \tag{2.19}$$

$$1-\sum_{i=1}^{q} P[t_i \geqslant b_4]+\sum_{i<j} P[t_i \geqslant b_4, t_j \geqslant b_4]=(1-\alpha), \tag{2.20}$$

$$P[t_i \leqslant b_5; i=1,\ldots,q]=(1-\alpha), \tag{2.21}$$

where the joint distribution of $t_1,\ldots,t_q$ is the central multivariate t distribution as defined earlier. Also, $t_{10},\ldots,t_{q0}$ are distributed jointly as multivariate t distribution with ν degrees of freedom and with $\rho_{ij}=0(i \neq j)$. Tables 10 and 11 give a comparison of the exact values b_5 with the bounds b_2, b_3 and b_4. The exact values are obtained from the tables of Krishnaiah and Armitage (1966).

Table 10

Comparison of bounds with exact percentage points of the multivariate t distribution in equicorrelated case

$q=3$

ν	α	ρ	b_2	b_3	b_4	b_5 (Exact)
5	0.05	0.2	2.90	2.84	2.78	2.79
10	0.05	0.5	2.46	2.44	2.31	2.34
20	0.05	0.7	2.28	2.27	2.07	2.12
35	0.05	0.9	2.21	2.20	1.73	1.93
5	0.01	0.9	4.45	4.39	3.44	3.82
10	0.01	0.7	3.41	3.40	3.18	3.23
20	0.01	0.5	3.03	3.02	2.97	2.97
35	0.01	0.2	2.88	2.88	2.87	2.87

Table 11

Comparison of bounds with exact percentage points of the multivariate t distribution in equicorrelated case

$q=5$

ν	α	ρ	b_2	b_3	b_4	b_5 (Exact)
5	0.05	0.2	3.34	3.23	3.08	3.15
10	0.05	0.5	2.75	2.72	2.42	2.56
20	0.05	0.6	2.52	2.51	2.16	2.34
35	0.05	0.9	2.43	2.42	—	2.03
5	0.01	0.9	5.02	4.90	—	4.00
10	0.01	0.7	3.71	3.69	3.13	3.44
20	0.01	0.5	3.25	3.24	3.13	3.16
35	0.01	0.2	3.08	3.08	3.07	3.07

Now, let b_2^*, b_4^* and b_5^* be defined as follows:

$$\prod_{i=1}^{q} P[x_i \leqslant b_2^*] = (1-\alpha), \tag{2.22}$$

$$1 - \sum_{i=1}^{q} P[x_i \geqslant b_4^*] + \sum_{i<j} P[x_i \geqslant b_4^*, x_j \geqslant b_4^*] = (1-\alpha) \tag{2.23}$$

$$P[x_i \leqslant b_5^*; i=1,\ldots,q] = (1-\alpha). \tag{2.24}$$

Table 12 gives a comparison of the exact percentage points b_5^* with bounds b_2^* and b_4^* in the equicorrelated case when $\sigma^2 = 1$.

In Table 13, we compare the exact percentage points of the multivariate F distribution with $(1,\nu)$ degrees of freedom with the corresponding percentage points of the multivariate chi-square distribution with 1 degree of freedom when $\rho_{ij} = 0.5$ for $i \neq j$. The values for the percentage points of

Table 12

Comparison of bounds with exact percentage points of the multivariate normal in equicorrelated case

q	α	ρ	b_2^*	b_4^*	b_5^* (Exact)
3	0.05	0.7	2.12	1.96	2.00
5	0.05	0.5	2.32	2.19	2.23
7	0.05	0.3	2.44	2.39	2.40
9	0.05	0.1	2.53	2.52	2.52
3	0.01	0.2	2.71	2.71	2.71
5	0.01	0.4	2.88	2.85	2.85
7	0.01	0.6	2.98	2.84	2.90

Table 13

Comparison of multivariate F distribution with multivariate chi-square distribution

(α,ν) \ q	3	4	6	9
(0.05, 40)	5.96	6.45	7.18	7.90
(0.05, 60)	5.81	6.30	6.97	7.67
(0.01, 40)	9.55	10.18	11.02	11.83
(0.01, 60)	9.18	9.73	10.57	11.36
(0.05, ∞)	5.52	5.96	6.59	7.22
(0.01, ∞)	8.50	8.99	9.68	10.36

Table 14

Comparison of multivariate t distribution with multivariate normal distribution

ν \ q	3	4	6	9
10	2.34	2.47	2.64	2.81
20	2.19	2.30	2.46	2.61
30	2.15	2.25	2.40	2.54
35	2.13	2.24	2.38	2.52
∞	2.06	2.16	2.29	2.42

the multivariate F distribution are taken from the tables of Dunnett (1964) for $\nu=40,60$ whereas the percentage points (entries corresponding to $\nu=\infty$) of the multivariate chi-square are obtained from the tables of Krishnaiah and Armitage (1965b).

In Table 14, we compare the percentage points of the multivariate t distribution with $(1,\nu)$ degrees of freedom with the corresponding values obtained from normal distribution when $\rho_{ij}=0.5(i\neq j)$ and $\alpha=0.05$.

The percentage points of multivariate t distribution for $\nu=10,20,30,35$ are taken from the tables of Krishnaiah and Armitage (1966) whereas the percentage points (entries for $\nu=\infty$) of the multivariate normal are taken from the tables of Gupta (1963).

3. Comparison of finite intersection tests with alternative procedures in univariate case

If we use finite intersection tests, the confidence intervals on the parametric functions λ_g are given by

$$\hat{\lambda}_g-\surd\left(F_\alpha d_g s^2/\nu\right)\leqslant\lambda_g\leqslant\hat{\lambda}_g+\surd\left(F_\alpha d_g s^2/\nu\right) \tag{3.1}$$

for $g=1,2,\ldots,q$ in the notation of Section 2. If we use Scheffé's test, the corresponding confidence intervals on λ_g are given by

$$\hat{\lambda}_g-\surd\left((k-1)F_{\alpha 1}d_g s^2/\nu\right)\leqslant\lambda_g\leqslant\hat{\lambda}_g+\surd\left((k-1)F_{\alpha 1}d_g s^2/\nu\right),$$

where

$$P\left[F_0\leqslant(k-1)F_{\alpha 1}|H\right]=(1-\alpha), \tag{3.2}$$

and $F_0 = \nu\Sigma_{i=1}^{k} n_i(\bar{x}_{i.} - \bar{x}_{..})^2/s^2$. Now, let r be the ratio of the length of the confidence interval on λ_g associated with the finite intersection test and the length of the corresponding confidence interval associated with Scheffé's test. Then $r = \{F_\alpha/(k-1)F_{\alpha 1}\}^{\frac{1}{2}}$. Tables 15–18 give the values of r^2 (relative efficiency of Scheffé's test) for different values of the parameters when the correlations between $\hat{\lambda}_g$'s are equal. The values $F_{\alpha 1}$ are obtained from the tables of F distribution with $(1,\nu)$ degrees of freedom whereas the values of F_α are obtained from the tables of Krishnaiah and Armitage (1970) for multivariate F distribution with $(1,\nu)$ degrees of freedom.

Table 15
Relative efficiency of Scheffé's test
when compared with finite intersection test
$\alpha = 0.05, \nu = 10$

k	q \ ρ	0.1	0.2	0.3	0.5	0.7	0.9
4	3	0.718	0.713	0.707	0.684	0.644	0.571
6	3	0.480	0.477	0.473	0.457	0.431	0.381
6	4	0.533	0.529	0.523	0.502	0.467	0.402
6	5	0.577	0.571	0.563	0.537	0.494	0.417
8	3	0.364	0.361	0.358	0.346	0.326	0.289
8	5	0.437	0.433	0.427	0.407	0.374	0.316
8	7	0.488	0.483	0.474	0.448	0.407	0.333
10	3	0.294	0.292	0.290	0.280	0.264	0.234
10	5	0.353	0.350	0.345	0.329	0.302	0.256
10	7	0.394	0.390	0.383	0.362	0.329	0.270
10	9	0.426	0.421	0.413	0.387	0.347	0.280

Table 16
Relative efficiency of Scheffé's test
when compared with finite intersection test
$\alpha = 0.01, \nu = 15$

k	q \ ρ	0.1	0.2	0.4	0.5	0.8	0.9
5	3	0.616	0.614	0.606	0.600	0.561	0.533
5	4	0.665	0.663	0.652	0.643	0.592	0.555
6	3	0.528	0.527	0.520	0.515	0.482	0.457
6	4	0.570	0.568	0.559	0.552	0.507	0.476
6	5	0.604	0.601	0.590	0.581	0.527	0.490
7	4	0.502	0.500	0.492	0.485	0.446	0.419
7	5	0.531	0.529	0.519	0.511	0.464	0.431
7	6	0.556	0.553	0.542	0.532	0.478	0.441
9	4	0.406	0.405	0.398	0.393	0.362	0.339
9	6	0.450	0.448	0.439	0.431	0.387	0.358
9	7	0.468	0.465	0.454	0.445	0.397	0.364
9	8	0.483	0.480	0.468	0.458	0.405	0.370

Table 17
Relative efficiency of Scheffé's test when compared with finite intersection test
$\alpha=0.05, \nu=20$

k	q	ρ 0.1	0.2	0.3	0.5	0.7	0.9
6	3	0.496	0.494	0.489	0.476	0.404	—
6	4	0.545	0.542	0.537	0.519	0.486	—
6	5	0.585	0.581	0.574	0.552	0.513	0.441
8	3	0.382	0.381	0.377	0.367	0.349	0.312
8	5	0.451	0.448	0.443	0.426	0.396	0.340
8	7	0.499	0.494	0.487	0.464	0.427	0.357
10	3	0.312	0.311	0.308	0.300	0.285	0.255
10	5	0.369	0.366	0.362	0.348	0.323	0.278
10	7	0.407	0.404	0.398	0.378	0.349	0.292
10	9	0.437	0.432	0.425	0.404	0.367	0.302

Table 18
Relative efficiency of Scheffe´'s test when compared with finite intersection test
$\alpha=0.01, \nu=30$

k	q	ρ 0.1	0.2	0.4	0.5	0.8	0.9
5	3	0.630	0.629	0.624	0.619	0.585	0.558
5	4	0.675	0.673	0.665	0.659	0.613	0.579
6	3	0.548	0.547	0.542	0.538	0.508	0.485
6	4	0.586	0.585	0.578	0.572	0.533	0.503
6	5	0.617	0.615	0.606	0.599	0.552	0.517
7	4	0.521	0.520	0.514	0.509	0.474	0.447
7	5	0.548	0.546	0.539	0.533	0.491	0.460
7	6	0.570	0.567	0.560	0.552	0.504	0.470
9	4	0.428	0.427	0.422	0.418	0.389	0.367
9	6	0.468	0.466	0.460	0.453	0.414	0.386
9	7	0.484	0.482	0.474	0.467	0.424	0.392
9	8	0.498	0.496	0.487	0.479	0.432	0.398

If we are interested in all possible pairwise comparisons of means, Krishnaiah's finite intersection test and Tukey's test are equivalent when the sample sizes are equal. But, if we are interested in testing a subset of all possible pairs when the sample sizes are equal, Krishnaiah's finite intersection test yields shorter confidence intervals than Tukey's test. Here we note that Krishnaiah's finite intersection test can be applied for pairwise comparisons of means even when the sample sizes are unequal. We will now make some comparison of the two test procedures.

Let us first consider the case when we are interested in testing the hypotheses $H_1,\ldots,H_{k-1}$ simultaneously against two-sided alternatives, where $H_i: \mu_i = \mu_{i+1}$ and the sizes of the samples are equal to m. In this case, the length (l_1) of the confidence interval on $\mu_i - \mu_{i+1}$ is $2\sqrt{(2F_\alpha s^2/\nu m)}$ if we use Krishnaiah's finite intersection test. Here, F_α is the upper $100\alpha\%$ point of the multivariate F distribution with $(1,\nu)$ degrees of freedom and with $\Omega = (\rho_{ij})$ as the correlation matrix of the 'accompanying' multivariate normal where ρ_{ij}'s $(i \neq j)$ are equal to $-\frac{1}{2}$ or 0 according as $|i-j|$ is equal to one or greater than one. The length (l_2) of the corresponding confidence interval associated with Tukey's test is given by $2\sqrt{2F_{\alpha 2}s^2/\nu m)}$ where

$$P\left[\frac{R^2}{2} \leqslant F_{\alpha 2} \middle| H\right] = (1-\alpha) \tag{3.3}$$

and R is the studentized range defined earlier. Then $r_1^2 = l_1^2/l_2^2 = F_\alpha/F_{\alpha 2}$. The values of $F_{\alpha 2}$ can be obtained from the tables of Harter (1960) but exact values of F_α are not available in the literature. Now, let $F_{\alpha \mathrm{L}}$ and $F_{\alpha \mathrm{U}}$ respectively denote the lower and upper bounds on F_α when we use equations (2.8) and (2.7) respectively. Also, let $r_{1\mathrm{L}}^2 = F_{\alpha \mathrm{L}}/F_{\alpha 2}$ and $r_{1\mathrm{U}}^2 = F_{\alpha \mathrm{U}}/F_{\alpha 2}$. Table 19 gives the values of $r_{1\mathrm{L}}^2$ and $r_{1\mathrm{U}}^2$ for different values of k and ν.

We now compare the finite intersection test with Tukey's test when it is of interest to compare various means with the mean of the control

Table 19

Relative efficiency of Tukey's test when compared with Krishnaiah's test for the equality of adjacent means

		$\alpha = 0.05$		$\alpha = 0.01$	
k	ν	$r_{1\mathrm{L}}^2$	$r_{1\mathrm{U}}^2$	$r_{1\mathrm{L}}^2$	$r_{1\mathrm{U}}^2$
3	6	0.871	0.903	0.886	0.910
6	12	0.769	0.799	0.796	0.817
4	12	0.824	0.856	0.851	0.871
5	15	0.794	0.825	0.826	0.844
6	18	0.775	0.803	0.809	0.824
4	16	0.829	0.858	0.858	0.875
5	20	0.799	0.827	0.833	0.847
3	15	0.882	0.907	0.905	0.919
4	20	0.832	0.859	0.863	0.877
6	30	0.783	0.806	0.819	0.831
5	30	0.805	0.830	0.907	0.921
3	18	0.883	0.907	0.840	0.852
3	30	0.886	0.908	0.912	0.923

population and the sample sizes are equal. In this special case, Krishnaiah's finite intersection test is equivalent to Dunnett's test (Dunnett (1955)). Let r_2 denote the ratio of the length of the confidence interval on $\mu_i - \mu_k (i=1,\ldots,k-1)$ associated with Dunnett's test and the length of the corresponding confidence interval associated with Tukey's test. Then r_2 is equal to $\sqrt{(F_\alpha / F_{\alpha 2})}$ where F_α is upper $100\alpha\%$ point of $(k-1)$-variate F distribution with $(1,\nu)$ degrees of freedom and with $\Omega=(\rho_{ij})$ as the correlation matrix of the 'accompanying' multivariate normal and $\rho_{ij}=0.5$ for $i \neq j$. Also, $F_{\alpha 2}$ is the upper $100\alpha\%$ value of $R^2/2$ where R is the studentized range defined in Section 2. Table 20 gives the values of r_2^2 for different values of k and ν.

Now, let $H_{ij}: \mu_i = \mu_j$ and $H_{i0}: \mu_i = 0$. Krishnaiah's finite intersection tests for testing the hypotheses $H_{ij}(i<j=1,2,\ldots,k)$ and $H_{i0}(i=1,\ldots,k)$ simultaneously is equivalent to Tukey's test based upon the studentized augmented range (Tukey (1953)) when the sample sizes are equal. But, if one is interested in testing a subset of the hypotheses $H_{ij}(i<j=1,\ldots,k)$ and $H_i(i=1,2,\ldots,k)$ simultaneously, the finite intersection tests of Krishnaiah yield shorter confidence intervals than Tukey's test based upon the

Table 20
Comparison of Tukey's test with Dunnett's test

k	ν	$\alpha=0.05$, r_2^2	$\alpha=0.01$, r_2^2
3	6	0.871	0.885
6	12	0.746	0.780
7	14	0.727	0.765
8	16	0.713	0.753
9	18	0.701	0.744
10	20	0.692	0.736
4	12	0.817	0.846
5	15	0.780	0.815
6	18	0.755	0.794
8	24	0.722	0.767
10	30	0.700	0.750
4	16	0.821	0.852
5	20	0.785	0.823
6	24	0.759	0.802
3	15	0.882	0.905
4	20	0.823	0.857
6	30	0.763	0.807
3	18	0.883	0.908
5	30	0.790	0.831
3	30	0.887	0.912

studentized augmented range. Here we note that the finite intersection tests of Krishnaiah can be applied even when the sample sizes are unequal whereas Tukey's test is restricted to the case when the sample sizes are equal.

We now compare finite intersection tests with Scheffé's test when σ^2 is known. In this case, the confidence intervals on the parametric functions λ_g are

$$\hat{\lambda}_g - \sqrt{(c_\alpha d_g \sigma^2)} \leqslant \lambda_g \leqslant \hat{\lambda}_g + \sqrt{(c_\alpha d_g \sigma^2)} \tag{3.4}$$

for $g=1,\ldots,q$ if we use finite intersection test. The corresponding confidence intervals associated with Scheffé's test are

$$\hat{\lambda}_g - \sqrt{((k-1)c_{\alpha 1} d_g \sigma^2)} \leqslant \lambda_g \leqslant \hat{\lambda}_g + \sqrt{((k-1)c_{\alpha 1} d_g \sigma^2)} \tag{3.5}$$

for $g=1,\ldots,q$. The critical value c_α is given by

$$P\left[\frac{\hat{\lambda}_g^2}{d_g \sigma^2} \leqslant c_\alpha;\ g=1,\ldots,q \,\middle|\, \bigcap_{g=1}^{q} H_g\right] = (1-\alpha) \tag{3.6}$$

whereas $(k-1)c_{\alpha 1}$ is the upper $100\alpha\%$ point of the univariate chi-square distribution with $k-1$ degrees of freedom. Tables 21 and 22 give the values of $c_\alpha/(k-1)c_{\alpha 1}$ when correlations between $\hat{\lambda}_g$'s are equal to ρ.

Table 21
Relative efficiency of Scheffé's test in comparison with finite intersection test when σ^2 is known
$\alpha = 0.01$

k	q	ρ 0.1	0.2	0.3	0.5	0.7
4	3	0.759	0.758	0.756	0.749	0.731
6	3	0.571	0.570	0.569	0.563	0.550
6	4	0.605	0.604	0.603	0.596	0.578
6	5	0.632	0.631	0.629	0.621	0.600
8	3	0.466	0.465	0.464	0.460	0.449
8	5	0.516	0.515	0.514	0.507	0.490
8	7	0.550	0.549	0.547	0.538	0.516
10	3	0.397	0.397	0.396	0.392	0.383
10	5	0.440	0.439	0.438	0.432	0.418
10	7	0.469	0.468	0.466	0.459	0.440

Table 22
Relative efficiency of Scheffé's test in comparison with finite intersection test when σ^2 is known
$\alpha=0.05$

k	q	ρ 0.1	0.2	0.3	0.5	0.7
4	3	0.729	0.726	0.722	0.706	0.676
6	3	0.514	0.513	0.510	0.500	0.477
6	4	0.560	0.557	0.553	0.540	0.510
6	5	0.595	0.592	0.588	0.570	0.536
8	3	0.405	0.403	0.401	0.392	0.375
8	5	0.468	0.466	0.462	0.448	0.422
8	7	0.511	0.508	0.503	0.485	0.452
10	3	0.337	0.335	0.333	0.326	0.312
10	5	0.389	0.388	0.384	0.373	0.351
10	7	0.425	0.422	0.418	0.404	0.376

Next, we compare Krishnaiah's finite intersection test with extended Tukey's test when the sample sizes of the k groups are equal to m. The confidence intervals on $\lambda_g(g=1,\ldots,q)$ associated with the extended Tukey's test (see Scheffé (1959), p. 74) are given by

$$\hat{\lambda}_g-\frac{sR_\alpha}{\sqrt{mv}}\left(\frac{1}{2}\sum_{i=1}^{k}|c_{gi}|\right)\leqslant\lambda_g\leqslant\hat{\lambda}_g+\frac{sR_\alpha}{\sqrt{mv}}\left(\frac{1}{2}\sum_{i=1}^{k}|c_{gi}|\right) \tag{3.7}$$

where R_α is the upper $100\alpha\%$ point of the distribution of the studentized range. Let r_3 be the ratio of the length of the confidence interval on λ_g associated with finite intersection test to the length of the corresponding confidence interval associated with the extended Tukey's test. Then

$$r_3^2=\frac{4F_\alpha\sum_{i=1}^{k}c_{gi}^2}{R_\alpha^2\left(\sum_{i=1}^{k}|c_{gi}|\right)^2} \tag{3.8}$$

where F_α was defined by equation (2.1).

Now, let r_{3L}^2 be the value obtained by replacing F_α with c_4 in r_3^2 where c_4 was defined by equation (2.8). Similarly, let r_{3U}^2 be the value obtained by replacing F_α with c_3 in r_3^2 where c_3 was defined by equation (2.7). Table 23 gives the values of the bounds r_{3L}^2 and r_{3U}^2 when $q=(k-1)$ and λ_g's are contrasts of the type $\mu_i-\mu_{i+1}(i=1,\ldots,k-1)$. In Table 24, the values of r_{3L}^2

Table 23
Relative efficiency of extended Tukey's test when compared with Krishnaiah's test for comparison of adjacent means

k	ν	α	r_{3U}^2	r_{3L}^2
4	24	0.05	0.860	0.834
5	30	0.05	0.829	0.805
6	30	0.05	0.807	0.783
4	24	0.01	0.879	0.863
5	30	0.01	0.852	0.838
6	30	0.01	0.831	0.819

Table 24
Relative efficiency of extended Tukey's test when compared with Krishnaiah's test for comparison of adjacent mean differences

k	ν	α	r_{3U}^2	r_{3L}^2
4	24	0.05	0.559	0.531
5	30	0.05	0.568	0.534
6	30	0.05	0.565	0.530
4	24	0.01	0.593	0.577
5	30	0.01	0.598	0.578
6	30	0.01	0.593	0.572

and r_{3U}^2 are given when λ_g's are contrasts of the type $\mu_i - 2\mu_{i+1} + \mu_{i+2}$ and $q = k-2$.

We now compare Krishnaiah's finite intersection tests with an extension of Dunnett's test when the sample sizes of different groups are equal to m. The confidence intervals on λ_g associated with the extended Dunnett's test (see Shaffer (1977)) are given by

$$\hat{\lambda}_g - sd_\alpha(2/m\nu)^{1/2} \sum_{i=1}^{k-1} |c_{gi}| \leq \lambda_g \leq \hat{\lambda}_g + sd_\alpha(2/m\nu)^{1/2} \sum_{i=1}^{k-1} |c_{gi}| \quad (3.9)$$

where d_α^2 is the upper $100\alpha\%$ point of the multivariate chi-distribution with $(1,\nu)$ degrees of freedom and with $\Omega = (\rho_{ij})$ as the correlation matrix of the accompanying multivariate normal where $\rho_{ij} = 0.5$ for $i \neq j$. Now, let r_4 denote the ratio of the length of the confidence interval on λ_g associated with the finite intersection test to the length of the corresponding confidence interval associated with the extended Dunnett's test. Then

$$r_4^2 = \frac{F_\alpha \sum_{i=1}^{k} c_{gi}^2}{2d_\alpha^2\left(\sum_{i=1}^{k-1} |c_{gi}|\right)^2}. \tag{3.10}$$

Let r_{4U}^2 and r_{4L}^2 denote the value of r_4^2 when F_α is replaced with c_3 and c_4 respectively where c_3 and c_4 were defined by (2.7) and (2.8). Table 25 gives the values of r_{4U}^2 and r_{4L}^2 when $q=(k-1)$ and λ_g's are contrasts of the type $\mu_i-\mu_{i+1}(i=1,\ldots,k-1)$. The columns corresponding to 'without μ_k' are the values of r_{4U}^2 or r_{4L}^2 for the contrasts $\mu_1-\mu_2,\mu_2-\mu_3,\ldots,\mu_{k-2}-\mu_{k-1}$ whereas the columns corresponding to 'with μ_k' are the values of $1/r_{4U}^2$ or $1/r_{4L}^2$ associated with the contrast of the type $\mu_{k-1}-\mu_k$. In Table 26, the values of r_{4U}^2 and r_{4L}^2 are given when λ_g's are contrasts of the type $\mu_i-2\mu_{i+1}+\mu_{i+2}(i=1,2,\ldots,k-2)$.

Table 25
Relative efficiency of extended Dunnett's test when compared with Krishnaiah's test for comparison of adjacent means

k	ν	α	r_{4U}^2 (without μ_k)	$1/r_{4U}^2$ (with μ_k)	r_{4L}^2 (without μ_k)	$1/r_{4L}^2$ (with μ_k)
4	24	0.05	0.260	0.961	0.252	0.992
5	30	0.05	0.262	0.954	0.255	0.982
6	30	0.05	0.264	0.946	0.256	0.974
4	24	0.01	0.256	0.978	0.251	0.996
5	30	0.01	0.257	0.975	0.252	0.992
6	30	0.01	0.257	0.972	0.254	0.986

Table 26
Relative efficiency of extended Dunnett's test when compared with Krishnaiah's test for comparison of adjacent mean differences

k	ν	α	r_{4U}^2 (without μ_k)	r_{4U}^2 (with μ_k)	r_{4L}^2 (without μ_k)	r_{4L}^2 (with μ_k)
4	24	0.05	0.193	0.342	0.183	0.325
5	30	0.05	0.195	0.346	0.183	0.326
6	30	0.05	0.197	0.349	0.184	0.327
4	24	0.01	0.190	0.338	0.185	0.329
5	30	0.01	0.191	0.340	0.185	0.329
6	30	0.01	0.192	0.342	0.185	0.330

From Tables 23 and 24, we observe that the finite intersection test is better than the extended Tukey's test for the cases considered. Table 25 indicates that the finite intersection test is better than the extended Dunnett's test for testing the hypotheses $\mu_i-\mu_{i+1}=0(i=1,\ldots,k-2)$ whereas extended Dunnett's test is slightly better than the finite intersection test when $\mu_{k-1}-\mu_k=0$ is tested, if the experimenter is interested in comparing the means of adjacent populations. If the experimenter is interested in testing the hypotheses of the form $\mu_i-2\mu_{i+1}+\mu_{i+2}(i=1,2,\ldots,k-1)$, Table 26 indicates that the finite intersection test is better than the extended Dunnett's test.

4. Simultaneous test procedures for multiple comparisons of mean vectors

For $i=1,2,\ldots,k$, let $\mathbf{x}_{i1},\ldots,\mathbf{x}_{in_i}$ be drawn from a p-variate normal population with mean vector $\boldsymbol{\mu}_i$ and covariance matrix Σ. Also, let $H_i:\boldsymbol{\lambda}_i=\mathbf{0}$, $A_i:\boldsymbol{\lambda}_i\neq\mathbf{0}$, $(i=1,\ldots,q)$ where $\boldsymbol{\lambda}_i=c_{i1}\boldsymbol{\mu}_1+\cdots+c_{ik}\boldsymbol{\mu}_k$ and $c_{i1},\ldots,c_{ik}$ are known. In addition, let $\chi_i^2=\hat{\boldsymbol{\lambda}}_i'\Sigma^{-1}\hat{\boldsymbol{\lambda}}_i/d_i$ where Σ is known, $\hat{\boldsymbol{\lambda}}_i=c_{i1}\bar{\mathbf{x}}_{1.}+\cdots+c_{ik}\bar{\mathbf{x}}_{k.}$, $n_i\bar{\mathbf{x}}_{i.}=\Sigma_{t=1}^{n_i}\mathbf{x}_{it}$ and $d_i=\Sigma_{t=1}^{k}c_{it}^2/n_t$. Then, we accept or reject H_i against A_i according as

$$\chi_i^2 \lessgtr b_\alpha$$

where

$$P\left[\chi^2_{\max}\leqslant b_\alpha \mid H\right]=(1-\alpha) \tag{4.1}$$

and $\chi^2_{\max}=\max(\chi_1^2,\ldots,\chi_q^2)$. When H is true, the joint distribution of $\chi_1^2,\ldots,\chi_q^2$ is a multivariate chi-square with p degrees of freedom and with $D=(d_{ij})$ as the covariance matrix of the 'accompanying' multivariate normal where $d_{ii}=1$ and $d_{ij}=\Sigma_{t=1}^{k}c_{it}c_{jt}/n_t$. The simultaneous confidence intervals associated with the above test are

$$\mathbf{a}'\hat{\boldsymbol{\lambda}}_g-\left\{b_\alpha\mathbf{a}'\Sigma\mathbf{a}d_g\right\}^{1/2}\leqslant\mathbf{a}'\boldsymbol{\lambda}_g\leqslant\mathbf{a}'\hat{\boldsymbol{\lambda}}_g+\left\{b_\alpha\mathbf{a}'\Sigma\mathbf{a}d_g\right\}^{1/2} \tag{4.2}$$

for all non-null $\mathbf{a}$. If we use Roy's largest root test when Σ is known, we accept or reject H_i according as

$$\chi_i^2\lessgtr(k-1)c_\alpha,$$

where

$$P\left[c_L\left(S_H\Sigma^{-1}\right)\leqslant(k-1)c_\alpha|H\right]=(1-\alpha), \tag{4.3}$$

$$S_H=\sum_{i=1}^{k} n_i(\bar{\mathbf{x}}_{i.}-\bar{\mathbf{x}}..)(\bar{\mathbf{x}}_{i.}-\bar{\mathbf{x}}..)',$$

$$n_i\bar{\mathbf{x}}_{i.}=\sum_{j=1}^{n_i}\mathbf{x}_{ij},\ n\bar{\mathbf{x}}..=\sum_i\sum_j\mathbf{x}_{ij}\quad\text{and}\quad n=n_1+\cdots+n_k.$$

When H is true, $S_0S_H\Sigma^{-1}$ is distributed as the central Wishart matrix with $(k-1)$ degrees of freedom and $\mathbf{E}(S_0/k-1)=I$. Tables for the values of c_α can be obtained from the tables of Hanumara and Thompson (1968) and Pillai and Chang (1970). The confidence intervals on $\mathbf{a}'\boldsymbol{\lambda}_g$ associated with Roy's largest root test are

$$\mathbf{a}'\hat{\boldsymbol{\lambda}}_g-\left\{d_gc_\alpha(k-1)\mathbf{a}'\Sigma\mathbf{a}\right\}^{1/2}\leqslant\mathbf{a}'\boldsymbol{\lambda}_g\leqslant\mathbf{a}'\hat{\boldsymbol{\lambda}}_g+\left\{d_gc_\alpha(k-1)\mathbf{a}'\Sigma\mathbf{a}\right\}^{1/2} \tag{4.4}$$

for all non-null $\mathbf{a}$. Let r_5 denote the ratio of the length of the confidence interval on $\mathbf{a}'\boldsymbol{\lambda}_g$ associated with $\chi^2_{\max}$ test to the length of the corresponding confidence interval associated with the largest root test. It is known that $r_2\leqslant 1$ where $r_5^2=b_\alpha/(k-1)c_\alpha$. Since exact values of b_α are not available, upper bound $b_{\alpha1}$ and lower bound $b_{\alpha2}$ on b_α are computed where $b_{\alpha1}$ and $b_{\alpha2}$ are given by the following equations:

$$1-\sum_{i=1}^{q}P\left[\chi_i^2\geqslant b_{\alpha1}|H\right]=(1-\alpha)$$

$$1-\sum_{i=1}^{q}P\left[\chi_i^2\geqslant b_{\alpha2}|H\right]+\sum_{i<j}P\left[\chi_i^2\geqslant b_{\alpha2},\chi_j^2\geqslant b_{\alpha2}|H\right]=(1-\alpha).$$

Now, let $r_{5U}^2=b_{\alpha1}/(k-1)c_\alpha$ and $r_{5L}^2=b_{\alpha2}/(k-1)c_\alpha$. Table 27 gives the values r_{5L}^2 and r_{5U}^2 when $H_i:\boldsymbol{\mu}_i=\boldsymbol{\mu}_{i+1}(i=1,\ldots,k-1)$ whereas Table 28 gives the corresponding values when $H_i:\boldsymbol{\mu}_i-\boldsymbol{\mu}_k=0(i=1,\ldots,k-1)$; in both of these cases, the sample sizes are assumed to be equal.

Table 27
Relative efficiency of Roy's largest root test when compared with χ^2_{max} test for comparing adjacent mean vectors

	$p=2$		$p=3$		$p=4$		$p=5$	
(k,α)	r^2_{5L}	r^2_{5U}	r^2_{5L}	r^2_{5U}	r^2_{5L}	r^2_{5U}	r^2_{5L}	r^2_{5U}
(3,0.05)	0.843	0.859						
(4,0.05)	0.747	0.762	0.768	0.781				
(5,0.05)	0.677	0.691	0.701	0.713	0.718	0.728		
(6,0.05)	0.624	0.636	0.649	0.659	0.667	0.676	0.682	0.690
(3,0.01)	0.866	0.871						
(4,0.01)	0.778	0.783	0.793	0.798				
(5,0.01)	0.712	0.716	0.730	0.734	0.744	0.748		
(6,0.01)	0.659	0.664	0.680	0.683	0.695	0.698	0.708	0.710

Table 28
Relative efficiency of Roy's largest root test when compared with χ^2_{max} test for comparisons with control population

	$p=2$		$p=3$		$p=4$		$p=5$	
(k,α)	r^2_{5L}	r^2_{5U}	r^2_{5L}	r^2_{5U}	r^2_{5L}	r^2_{5U}	r^2_{5L}	r^2_{5U}
(3,0.05)	0.843	0.859						
(4,0.05)	0.740	0.762	0.762	0.781				
(5,0.05)	0.665	0.691	0.691	0.713	0.710	0.728		
(6,0.05)	0.607	0.636	0.635	0.659	0.656	0.676	0.671	0.691
(3,0.01)	0.866	0.871						
(4,0.01)	0.775	0.783	0.791	0.798				
(5,0.01)	0.707	0.716	0.727	0.734	0.741	0.748		
(6,0.01)	0.653	0.664	0.675	0.683	0.691	0.698	0.704	0.710

We now compare T^2_{max} test with Roy's largest root test when Σ is unknown. According to the largest root test, we accept or reject H_g according as

$$T_g^2 \lessgtr (k-1)c_{\alpha 1},$$

where

$$T_g^2 = \hat{\lambda}_g' S^{-1} \hat{\lambda}_g (n-k)/d_g,$$

$$S = \sum_{i=1}^{k} \sum_{j=1}^{n_i} (\mathbf{x}_{ij} - \bar{\mathbf{x}}_{i.})(\mathbf{x}_{ij} - \bar{\mathbf{x}}_{i.})'. \tag{4.5}$$

and

$$P\left[(n-k)c_L(S_HS^{-1}) \leqslant (k-1)c_{\alpha 1}|H\right]=(1-\alpha). \tag{4.6}$$

The confidence intervals on $\mathbf{a}'\boldsymbol{\lambda}_g$ associated with the largest root test are

$$\begin{aligned} \mathbf{a}'\hat{\boldsymbol{\lambda}}_g - \left\{d_g c_{\alpha 1}(k-1)\mathbf{a}'S\mathbf{a}/(n-k)\right\}^{1/2} &\leqslant \mathbf{a}'\boldsymbol{\lambda}_g \\ &\leqslant \mathbf{a}'\hat{\boldsymbol{\lambda}}_g + \left\{d_g c_{\alpha 1}(k-1)\mathbf{a}'S\mathbf{a}/(n-k)\right\}^{1/2}. \end{aligned} \tag{4.7}$$

According to the $T^2_{\max}$ test, we accept or reject H_g according as

$$T_g^2 \lesseqgtr T_\alpha^2,$$

where

$$P\left[T^2_{\max} \leqslant T^2_\alpha | H\right]=(1-\alpha). \tag{4.8}$$

The confidence intervals on $\mathbf{a}'\boldsymbol{\lambda}_g$ associated with $T^2_{\max}$ test are given by

$$\begin{aligned} \mathbf{a}'\hat{\boldsymbol{\lambda}}_g - \left\{d_g T^2_\alpha \mathbf{a}'S\mathbf{a}/(n-k)\right\}^{1/2} &\leqslant \mathbf{a}'\boldsymbol{\lambda}_g \\ \mathbf{a}'\hat{\boldsymbol{\lambda}}_g + \left\{d_g T^2_\alpha \mathbf{a}'S\mathbf{a}/(n-k)\right\}^{1/2}&. \end{aligned} \tag{4.9}$$

Let r_6 be the ratio of the length of the confidence interval on $\mathbf{a}'\boldsymbol{\lambda}_g$ associated with the $T^2_{\max}$ test with the length of the corresponding confidence interval associated with the largest root test. Then $r_6=\{T^2_\alpha/(k-1)c_{\alpha 1}\}^{1/2}$. It is known (see Krishnaiah (1969)) that $r_6 \leqslant 1$. Exact values of T^2_α are not available and so we used the approximate values computed by Siotani (1960) in calculating r_6^2. Table 29 gives the approximate values of r_6^2 for $p=2$ when $\hat{\boldsymbol{\lambda}}_1,\ldots,\hat{\boldsymbol{\lambda}}_q$ are distributed independently.

We now compare finite intersection tests with alternative tests. Let $H_g=\bigcap_{j=1}^p H_{gj}$ where $H_{gj}:\lambda_{gj}=0$,

$$\lambda_{gj}=c_{g1}\eta_{1j}+\cdots+c_{gk}\eta_{kj}, \qquad \eta_{i,j+1}=\mu_{i,j+1}-\boldsymbol{\beta}_j'\boldsymbol{\theta}_{ij}$$

and $\boldsymbol{\theta}'_{ij}=(\mu_{i1},\ldots,\mu_{ij})$. Also, let $\mathbf{x}'_{it}=(x_{it1},\ldots,x_{itp})$, and $\mathbf{y}'_{i,j+1}=(x_{i1,j+1},\ldots,x_{in_ij+1})$. Then, the conditional distributions of the elements of

$\mathbf{y}_{i,j+1}$ given $\mathbf{y}_{i1},\ldots,\mathbf{y}_{ij}$ are distributed independently with variance σ_{j+1}^2 and means given by

$$\mathbf{E}_c(x_{it,j+1}) = \eta_{i,j+1} + (x_{it1},\ldots,x_{itj})\boldsymbol{\beta}_j, \tag{4.10}$$

where

$$\boldsymbol{\beta}_j = \Sigma_j^{-1}\begin{bmatrix}\sigma_{1,j+1}\\ \vdots \\ \sigma_{j,j+1}\end{bmatrix}, \qquad \sigma_{j+1}^2 = \frac{|\Sigma_{j+1}|}{|\Sigma_j|}$$

and Σ_j is the top $j\times j$ left hand corner of Σ. Also, the elements of $\mathbf{y}_{i1}$ are distributed independently with variance $\sigma_1^2\Sigma_1$ and means given by

$$\mathbf{E}(x_{it1}) = \eta_{i1} \tag{4.11}$$

Table 29
Relative efficiency of Roy's largest root test for $p=2$ when compared with $T^2_{\max}$ test

$n-k$ \ (k,α)	(3,0.5)	(4,0.05)	(6,0.05)	(8,0.05)	(3,0.01)	(4,0.01)	(6,0.01)	(8,0.01)
20	0.85	0.74	0.61	0.53	0.85	0.76	0.63	0.54
24	0.85	0.74	0.61	0.53	0.85	0.75	0.63	0.54
30	0.85	0.75	0.62	0.53	0.86	0.76	0.63	0.55
34	0.85	0.75	0.62	0.53	0.86	0.77	0.64	0.55
40	0.84	0.75	0.62	0.53	0.86	0.77	0.6	0.56
44	0.85	0.75	0.62	0.53	0.86	0.77	0.64	0.56
50	0.85	0.75	0.63	0.54	0.87	0.78	0.64	0.57

The model (4.10)–(4.11) can be rewritten as

$$\mathbf{E}_c(\mathbf{y}_{j+1}) = A_{j+1}\boldsymbol{\theta}_{j+1} \tag{4.12}$$

where $y'_{j+1} = (x_{11,j+1},\ldots,x_{1n_1,j+1},\ldots,x_{k,j+1},\ldots,x_{kn_k,j+1})$. Also, $\boldsymbol{\theta}'_{j+1} = (\eta_{1,j+1},\ldots,\eta_{k,j+1},\boldsymbol{\beta}'_j)$ for $j=1,2,\ldots,p-1$ and $\boldsymbol{\theta}'_1 = (\eta_{11},\ldots,\eta_{k1})$. In (4.12), the elements of A_1 are zero or one whereas the elements of $A_2,\ldots,A_p$ are zero, one or x_{itj}'s. The least square estimates of $\boldsymbol{\theta}'_1$ and $\boldsymbol{\theta}'_{j+1}$ $(j=1,\ldots,p-1)$ are given by $\hat{\boldsymbol{\theta}}'_1 = (\hat{\eta}_{11},\ldots,\hat{\eta}_{k1})$ and $\hat{\boldsymbol{\theta}}'_{j+1} = (\hat{\eta}_{1,j+1},\ldots,\hat{\eta}_{k,j+1},\hat{\boldsymbol{\beta}}'_j)$ where $\hat{\boldsymbol{\theta}}_1 = (A'_1A_1)^{-1}A_1\mathbf{y}_1, \hat{\boldsymbol{\theta}}_{j+1} = (A'_{j+1}A_{j+1})^{-1}A_{j+1}\mathbf{y}_{j+1}$. For given j, the variance covariance matrix of $\hat{\boldsymbol{\theta}}_j (j=1,\ldots,p)$ is given by $\sigma_j^2(A'_jA_j)^{-1}$.

When Σ is known, the two-sided $100(1-\alpha)\%$ confidence interval on λ_{gj} associated with the finite intersection test is given by

$$\hat{\lambda}_{gj}-\surd\left(d_{gj}\,\sigma_j^2 d_{\alpha j}\right)\leqslant\lambda_{gj}\leqslant\hat{\lambda}_{gj}+\surd\left(d_{gj}\,d_{\alpha j}\sigma_j^2\right) \tag{4.13}$$

when the hypotheses $\lambda_{gj}=0$ are tested simultaneously. Here, the critical values $d_{\alpha j}$ are given by

$$P\left[\chi_{gj}^2\leqslant d_{\alpha j};g=1,\ldots,q,j=1,\ldots,p|H\right]=(1-\alpha) \tag{4.14}$$

$\chi_{gj}^2=\hat{\lambda}_{gj}^2/d_{gj}\sigma_j^2, \hat{\lambda}_{gj}=c_{g1}\hat{\eta}_{1j}+\cdots+c_{gk}\hat{\eta}_{kj}$, and $d_{gj}\sigma_j^2$ is the variance of $\hat{\lambda}_{gj}$. But the confidence intervals on $\mathbf{a}'\boldsymbol{\lambda}_g$ associated with the largest root test for known Σ, are given by (4.2). In equation (4.2), let $\mathbf{a}'=(-\boldsymbol{\beta}_{j-1}',1,0,\ldots,0)$. Then, we obtain the following confidence intervals:

$$\hat{\lambda}_{gj}-\left\{d_g b_\alpha \sigma_j^2\right\}^{1/2}\leqslant\lambda_{gj}\leqslant\hat{\lambda}_{gj}+\left\{d_g b_\alpha\sigma_j^2\right\}^{1/2}. \tag{4.15}$$

Let r_7 be the ratio of the length of the confidence interval on λ_{gj} associated with the finite intersection test to the length of the corresponding confidence interval associated with the $\chi_{\max}^2$ test. Then $r_7^2=d_{gj}d_{\alpha 1}/d_g b_\alpha$ when $d_{\alpha j}=d_{\alpha 1}$.

Now, let $S=(s_{tu})$ where

$$s_{tu}=\sum_{i=1}^{k}\sum_{j=1}^{n_i}\left(x_{ijt}-\bar{x}_{i.t}\right)\left(x_{iju}-\bar{x}_{i.u}\right)$$

and $n_i\bar{x}_{i.t}=\Sigma_{j=1}^{n_i}x_{ijt}$. Also, let $s_1^2=s_{11}$, and

$$\mathbf{b}_j=S_j^{-1}\begin{bmatrix}s_{1,j+1}\\ \vdots\\ s_{j,j+1}\end{bmatrix},\; s_{j+1}^2=\frac{|S_{j+1}|}{|S_j|}$$

for $j=1,\ldots,p-1$, where S_j denotes the top $j\times j$ left hand corner of S. In addition, let $a=n-k+1$ and

$$F_{gj}=\frac{\hat{\lambda}_{gj}^2(a-j)}{s_j^2 d_{gj}}$$

where $d_{gj}\sigma_j^2$ is the variance of $\hat{\lambda}_{gj}$. Then, we accept or reject H_{gj} according as

$$F_{gj}\lesseqgtr F_{\alpha j},$$

where

$$\prod_{j=1}^{p} P\left[F_{gj} \leqslant F_{\alpha j}; g=1,\ldots,q|H\right]=(1-\alpha).$$

When H is true, the joint distribution of $F_{1j},\ldots,F_{qj}$ is a multivariate F distribution with $(1,a-j)$ degrees of freedom and with Ω_j as the correlation matrix of the 'accompanying' multivariate normal. Here we note that $\Omega_j = E_j^{-1/2} C D_j C' E_j^{-1/2}$ where $C=(c_{ij}): q\times k$, D_j is the top $k\times k$ left-hand corner of $\sigma_j^2 A_j' A_j)^{-1}$ and $\mathbf{E}_j$ is the diagonal matrix whose diagonal elements are the diagonal elements of CD_jC'. When $F_{\alpha j}=F_{\alpha 1}$, the ratio of the expected squared length of the confidence interval associated with the finite intersection test to the expected squared length of the corresponding confidence interval associated with the $T^2_{\max}$ test on λ_{gj} is given by $r_8^2 = d_{gj} F_{\alpha 1}/ d_g T_\alpha^2$.

It is known that $T^2_{\max}$ test yields shorter confidence intervals than the test based upon $c_L(S_H S^{-1})$ when Σ is unknown. Similarly, $\chi^2_{\max}$ test yields shorter confidence intervals than the test based upon $c_L(S_H \Sigma^{-1})$ when Σ is known. It is also known that the finite intersection tests yield shorter confidence intervals than the step-down procedure of J. Roy (1958). Mudholkar and Subbaiah (1979) made some comparisons of the finite intersection tests with the step-down procedure and largest root test.

5. Illustrative example

The Air Force Flight Dynamics Laboratory conducted an experiment to determine the optimal way of implementing the multifunction keyboard (MFK) in the cockpit of an aircraft. In this experiment, data was collected on the performance of some pilots in a cockpit simulator on four different MFK configurations. Two of these configurations were located on the left side of the front instrument panel, while the remaining two were located on the right side console and/or the right side of the front panel. The configurations utilized two types of hardware: projection switches (legend on the switch face) and CRT (legend presented on a display surface adjacent to the switch). Fig. 1 gives the MFK configurations whereas Fig. 2 gives the cockpit simulator used in the evaluation. In order to examine the ease of calling up data on MFK, two types of logic were used. They were branching logic (programmed in parallel with the subsystem operation) and tailored logic (programmed according to what functions are most likely to be used in the current phase of system operation). The pilot's performance in terms of flying the simulator and operating the MFK's was measured. The following flight parameters were recorded two times per

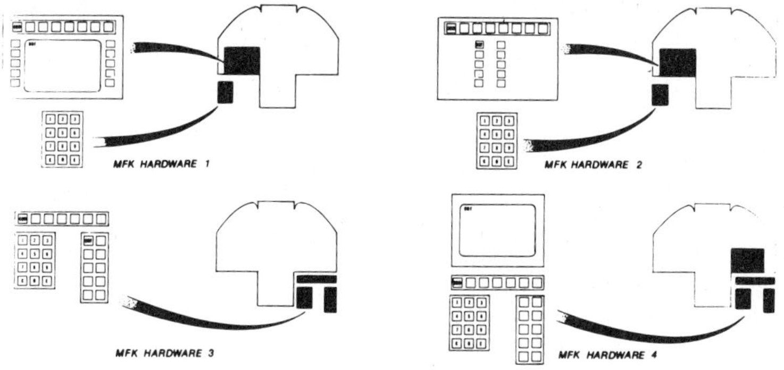

Fig. 1

Fig. 2

second on magnetic tape: groundspeed (knots), bank steering error (arbitrary units), and pitchsteering error. Keyboard task performance was evaluated by measuring (a) keyboard operation time to correctly complete an assigned task and (b) number of switch hits.

We analyze the data in the above experiment on the variables pitch steering error (variable 1), bank steering error (variable 2), and keyboard operation time (variable 3). The data was collected on 16 pilots. The same pilots were asked to fly the aircraft with different keyboards. So, there may be some correlation between the observations on different keyboards. However, for the purpose of illustration, we assume that the above correlation is zero.

In the notation defined in Section 4, let $k=4$ and $p=3$. Also, let x_{itj} denote the score for tth pilot on jth variable and the ith keyboard. In addition μ_{ij} denotes the mean score on the jth variable and ith keyboard. We analyze a subset of the data collected from the experiment described above. In the following analysis, we considered the data on 11 pilots on keyboard 1, 10 pilots on keyboard 2, 8 pilots on keyboard 3 and 10 pilots on keyboard 4.

$$\bar{\mathbf{x}}_{1.}=\begin{bmatrix}-2.1929\\ 7.0186\\ 11.5348\end{bmatrix} \qquad \bar{\mathbf{x}}_{2.}=\begin{bmatrix}0.6845\\ 3.3742\\ 10.1106\end{bmatrix}$$

$$\bar{\mathbf{x}}_{3.}=\begin{bmatrix}1.9544\\ 4.0973\\ 9.1930\end{bmatrix} \qquad \bar{\mathbf{x}}_{4.}=\begin{bmatrix}3.1894\\ -0.6383\\ 17.8321\end{bmatrix}$$

$$S=\begin{bmatrix}1355.5169 & 683.0879 & 48.9294\\ 683.0879 & 2646.2725 & 285.5629\\ 48.9294 & 285.5629 & 268.5165\end{bmatrix}$$

Now, let $H_{1j}:\eta_{1j}-\eta_{2j}=0$, $H_{2j}:\eta_{1j}-\eta_{3j}=0$, $H_{3j}:\eta_{1j}-\eta_{4j}=0$, $H_{4j}:\eta_{2j}-\eta_{3j}=0$, $H_{5j}:\eta_{2j}-\eta_{4j}=0$, $H_{6j}:\eta_{3j}-\eta_{4j}=0$, where η_{ij} was defined in the preceding section. Then the values of the test statistics F_{ij} are given in Table 30.

Table 30
Values of the test statistics F_{ij}

i \ j	1	2	3
1	1.11984	1.94556	0.60371
2	2.05699	1.62249	2.14934
3	3.91809	7.48069	28.45853
4	0.18506	0.00045	0.58215
5	0.81002	2.00819	43.47428
6	0.17502	1.87503	49.37588

For given j, the joint distribution of $F_{1j},\ldots,F_{6j}$ is a six variate F distribution with $(36-j)$ degrees of freedom. The critical values $F_{\alpha 1}$, $F_{\alpha 2}$, $F_{\alpha 3}$ are chosen such that

$$P\left[F_{ij}\leqslant F_{\alpha j};\ i=1,2,\ldots,6\,|\,H\right]=(1-\alpha)^{1/3},$$

for $j=1,2,3$, where $\alpha=0.05$. Using the inequality (2.3), we observe that the upper bounds $F^*_{\alpha 1}$, $F^*_{\alpha 2}$ and $F^*_{\alpha 3}$ on $F_{\alpha 1}$, $F_{\alpha 2}$ and $F_{\alpha 3}$ are 10.27, 10.31 and 10.36 respectively. Using the right side of the inequality (2.2), we observe that the lower bounds $F^{**}_{\alpha 1}$, $F^{**}_{\alpha 2}$ and $F^{**}_{\alpha 3}$ on $F_{\alpha 1}$, $F_{\alpha 2}$ and $F_{\alpha 3}$ are 9.80, 9.86 and 9.90 respectively. In Table 30, F_{i1} and F_{i2} are all less than $F^*_{\alpha 1}$ and $F^*_{\alpha 2}$ respectively for $i=1,2,\ldots,6$. Also, F_{13}, F_{23}, F_{43} are all less than $F^*_{\alpha 3}$ whereas F_{33}, F_{53} and F_{63} are all greater than $F^{**}_{\alpha 3}$. So, we reject the hypotheses H_{33}, H_{53} and H_{63} and accept the remaining hypotheses. We observe that $\mu_{ij}-\mu_{i'j}=0$ for $j=1$, 2 and $i\neq i'=1,2,3,4$. Also, $\mu_{13}=\mu_{23}$, $\mu_{13}=\mu_{33}$, $\mu_{13}\neq\mu_{43}$, $\mu_{23}=\mu_{33}$, $\mu_{23}\neq\mu_{43}$ and $\mu_{33}\neq\mu_{43}$.

To illustrate the finite intersection test for pairwise comparison of means in the univariate case for unequal sample sizes, we analyzed the data on the third variable for $n_1=11$, $n_2=10$, $n_3=8$, $n_4=10$; this data is a part of the data analyzed above. Now, let $H^*_{ij}: \mu_{i3}-\mu_{j3}=0$ (for $i\neq j=1,2,3,4$) and

$$F^*_{ij}=\frac{(\bar{x}_{i.3}-\bar{x}_{j.3})^2 35}{s^2\left(\dfrac{1}{n_i}+\dfrac{1}{n_j}\right)},$$

where

$$s^2=\sum_{i=1}^{3}\sum_{t=1}^{n_i}(x_{it3}-\bar{x}_{i.3})^2.$$

Then $F^*_{12}=1.38485$, $F^*_{13}=3.31063$, $F^*_{14}=27.07592$, $F^*_{23}=0.48775$, $F^*_{24}=38.85723$ and $F^*_{34}=43.23647$. The joint distribution of the test statistics is six variate F distribution with (1, 35) degrees of freedom and with $\Omega=(\rho_{ij})$ as the correlation matrix of the 'accompanying' multivariate normal where $\rho_{12}=0.448$, $\rho_{13}=0.476$, $\rho_{14}=0.482$, $\rho_{15}=0.512$, $\rho_{16}=0.0$, $\rho_{23}=0.448$, $\rho_{24}=0.567$, $\rho_{25}=0.0$, $\rho_{26}=0.567$, $\rho_{34}=0.0$, $\rho_{35}=0.512$, $\rho_{36}=0.482$, $\rho_{45}=0.471$, $\rho_{46}=0.556$, and $\rho_{56}=0.471$. Using the right side of the inequality (2.2), we obtain 7.14 as a lower bound on the critical value at 5% level, whereas we obtain 7.72 as an upper bound on the critical at 5% level using the inequality (2.3). Since the values of F^*_{12}, F^*_{13} and F^*_{23} are less than 7.14, we accept the hypotheses H^*_{12}, H^*_{13} and H^*_{23}. The hypotheses H^*_{14}, H^*_{24} and H^*_{34} are rejected since F^*_{14}, F^*_{24} and F^*_{34} are greater than 7.72.

We will now consider the problem of testing the hypotheses $H^*_{ij}(i\neq j=1,2,3,4)$ and $H^*_i(i=1,2,3,4)$ simultaneously by using finite intersection tests where $H^*_i: \mu_{i3}=0$. Now, let $F^*_i=\bar{x}^2_{i.3}n_i(n-4)/s^2$. Then $F^*_1=190.76827$,

$F_2^* = 133.24412$, $F_3^* = 88.12534$, $F_4^* = 414.47756$ and the values of F_{ij}^* are given in the preceding paragraph. The joint distribution of the statistics $F_{ij}^*(i \neq j = 1,2,3,4)$ and $F_a^*(a = 1,2,3,4)$ is 10-variate F distribution with (1, 35) degrees of freedom. Using the inequality (2.3), we obtain 8.85 as the upper bound on the critical value at 5% level, whereas we obtain 7.37 as the lower bound at 5% level by using the right side of the inequality (2.2). So, the hypotheses H_{14}^*, H_{24}^*, H_{34}^*, H_1^*, H_2^*, H_3^* and H_4^* are rejected and the hypotheses H_{12}^*, H_{13}^* and H_{23}^* are accepted.

The computations in this section were done by using a computer program developed by Cox and Fang (1979) for the finite intersection tests for multiple comparisons of mean vectors.

References

Armitage, J. V. and Krishnaiah, P. R. (1964). *Tables for the Studentized Largest Chi-Square Distribution and their Applications*. ARL 64-188. Wright–Patterson Air Force Base, Ohio.

Chang, T. C. (1974). Upper percentage points of the extreme roots of the MANOVA matrix. *Ann. Inst. Statist. Math.* (*Suppl.*) **8**, pp. 59–66.

Cox, C. M. and Fang, C. (1979). Computer program for Krishnaiah's finite intersection tests for multiple comparisons of mean vectors. Unpublished.

Dunn, O. J. (1961). Multiple comparisons among means. *J. Amer. Statist. Assoc.* **56**, pp. 52–64.

Dunnett, C. W. (1955). A multiple comparisons procedure for comparing several treatments with a control. *J. Amer. Statist. Assoc.* **50**, pp. 1096–1121.

Dunnett, C. W. (1964). New tables for multiple comparisons with a control. *Biometrics* **20**, pp. 482–491.

Dunnett, C. W. and Sobel, M. (1954). A bivariate generalization of Student's *t*-distribution, with tables for certain special cases. *Biometrika* **41**, pp. 153–169.

Dunnett, C. W. and Sobel, M. (1955). Approximations to the probability integral and certain percentage points of a multivariate analogue of Student's *t*-distribution. *Biometrika* **42**, pp. 258–260.

Gupta, S. S. (1963). Probability integrals of multivariate normal and multivariate *t*. *Ann. Math. Statist.* **34**, pp. 792–828.

Gupta, S. S. and Sobel, M. (1957). On a statistic which arises in selection and ranking problems. *Ann. Math. Statist.* **28**, pp. 957–967.

Hahn, G. J. and Hendrickson, R. W. (1971). A table of percentage points of the distribution of the largest absolute value of *k* Student *t* variates and its applications. *Biometrika* **58**, pp. 323–332.

Hanumara, R. C. and Thompson, W. A., Jr. (1968). Percentage points of the extreme roots of a Wishart matrix. *Biometrika* **55**, pp. 505–512.

Harter, H. L. (1960). Tables of range and studentized range. *Ann. Math. Statist.* **31**, pp. 1122–1147.

Harter, H. L. (1961). Use of tables of percentage points of range and studentized range. *Technometrics* **3**, pp. 407–411.

Kimball, A. W. (1951). On dependent tests of significance in the analysis of variance. *Ann. Math. Statist.* **22**, pp. 600–602.

Krishnaiah, P. R. (1963). *Simultaneous Tests and the Efficiency of Generalized Incomplete Block Designs.* ARL 63-174. Wright–Patterson Air Force Base, Ohio.

Krishnaiah, P. R. (1965a). On the simultaneous ANOVA and MANOVA tests. *Ann. Inst. Statist. Math.* **17**, pp. 35–53.

Krishnaiah, P. R. (1965b). Multiple comparison tests in multi-response experiments. *Sankhyā Ser. A.* **27**, pp. 65–72.

Krishnaiah, P. R. (1969). Simultaneous test procedures under general MANOVA models. In *Multivariate Analysis–II* (P. R. Krishnaiah, ed.), pp. 121–143. Academic Press, New York.

Krishnaiah, P. R. (1978). Some developments on real multivariate distributions. In *Developments in Statistics* (P. R. Krishnaiah, ed.), Vol. 1, pp. 135–169. Academic Press, New York.

Krishnaiah, P. R. (1979). Some developments on simultaneous test procedures. In *Developments in Statistics*, Vol. 2, pp. 157–201. Academic Press, New York.

Krishnaiah, P. R. and Armitage, J. V. (1965a). *Percentage Points of the Multivariate t Distribution.* ARL 65–199. Wright–Patterson Air Force Base, Ohio.

Krishnaiah, P. R. and Armitage, J. V. (1965b). Tables for the distribution of the maximum of correlated chi-square variates with one degree of freedom. *Trabajos Estadist.* **16**, pp. 91–96.

Krishnaiah, P. R. and Armitage, J. V. (1965c). *Probability Integrals of the Multivariate F Distribution, with Tables and Applications.* ARL 65–236. Wright–Patterson Air Force Base, Ohio.

Krishnaiah, P. R. and Armitage, J. V. (1966). Tables for multivariate t distribution. *Sankhya Ser. B.* **28**, pp. 31–56.

Krishnaiah, P. R. and Armitage, J. V. (1970). On a multivariate F distribution. In *Essays in Probability and Statistics* (R. C. Bose et al., eds.), pp. 439–468. Univ. of North Carolina Press, Chapel Hill.

Krishnaiah, P. R., Armitage, J. V. and Breiter, M. C. (1969a). *Tables for the Probability Integrals of the Bivariate t Distribution.* ARL 69-060. Wright–Patterson Air Force Base, Ohio.

Krishnaiah, P. R., Armitage, J. V. and Breiter, M. C. (1969b). *Tables for the Bivariate $|t|$ Distribution.* ARL 69-0210. Wright–Patterson Air Force Base, Ohio.

Krishnaiah, P. R. and Rao, M. M. (1961). Remarks on a multivariate gamma distribution. *Amer. Math. Monthly* **68**, pp. 342–346.

Krishnamoorthy, A. S. and Parthasarathy, M. (1951). A multivariate gamma distribution. *Ann. Math. Statist.* **22**, pp. 549–557; (1960) Correction **31**, p. 229.

Lee, J. C. (1979). On the computations of multivariate distributions done at the Aerospace Research Laboratories. *International Statistical Review*, **47**, pp. 57–65.

Mudholkar, G. S. and Subbaiah, P. (1975). A note on MANOVA multiple comparisons based upon step-down procedures. *Sankhyā Ser. B.* **37**, pp. 300–307.

Mudholkar, G. S. and Subbaiah, P. (1979). MANOVA multiple comparisons associated with finite intersection tests. To appear in *Multivariate Analysis–V* (P. R. Krishnaiah, ed.). North-Holland, Amsterdam.

Pillai, K. C. S. and Chang, T. C. (1970). An approximation to the cdf of the largest root of a covariance matrix. *Ann. Inst. Statist. Math.* (*Suppl.*) **6**, pp. 115–124.

Roy, J. (1958). Step-down procedure in multivariate analysis. *Ann. Math. Statist.*, **29**, pp. 1177–1187.

Roy, S. N. (1957). *Some Aspects of Multivariate Analysis*. Wiley, New York.

Scheffé, H. (1959). *The Analysis of Variance*. Wiley, New York.

Schuurmann, F. J., Krishnaiah, P. R. and Chattopadhyay, A. K. (1975). Tables for a multivariate F distribution. *Sankhya Ser. B* **37**, pp. 308–331.

Shaffer, J. P. (1977). Multiple comparisons emphasizing selected contrasts: an extension and generalization of Dunnett's procedure. *Biometrics* **33**, pp. 293–303.

Sidak, Z. (1967). Rectangular confidence regions for the means of multivariate normal distributions. *J. Amer. Statist. Assoc.* **62**, pp. 626–633.

Sidak, Z. (1968). On multivariate normal probabilities of rectangles: Their dependence on correlations. *Ann. Math. Statist.* **39**, pp. 1425–1434.

Sidak, Z. (1971). On probabilities of rectangles in multivariate student distributions: Their dependence on correlations. *Ann. Math. Statist.* **42**, pp. 169–175.

Siotani, M. (1959a). On the range in multivariate case. *Proc. Inst. Statist. Math.* **6**, pp. 155–156.

Siotani, M. (1959b). The extreme value of the generalized distances of the individual points in the multivariate normal sample. *Ann. Inst. Statist. Math.* **10**, pp. 183–207.

Siotani, M. (1960). Notes on multivariate confidence bounds. *Ann. Inst. Statist. Math.* **11**, pp. 167–182.

Siotani, M. (1961). The extreme value of generalized distances and its applications. *Bull. Internat. Statist. Inst.* **38**, pp. 591–599.

Stoline, M. R. (1978). Tables of the studentized augmented range and applications to problems of multiple comparison. *J. Amer. Statist. Assoc.* **37**, pp. 656–660.

Tong, Y. L. (1970). Some probability inequalities of multivariate normal and multivariate *t*. *J. Amer. Statist. Assoc.* **65**, 1243–1247.

Tukey, J. W. (1953). *The Problem of Multiple Comparisons*. Unpublished.

P. R. Krishnaiah, ed., *Multivariate Analysis–V*
© North-Holland Publishing Company (1980) 467–482.

MANOVA MULTIPLE COMPARISONS ASSOCIATED WITH FINITE INTERSECTION TESTS*

Govind S. MUDHOLKAR and Perla SUBBAIAH**
University of Rochester, Rochester, NY, U.S.A and Oakland University, Rochester, MI, U.S.A.

A finite intersection procedure consists of simultaneous tests based upon the variatewise finite decomposition of the MANOVA hypothesis (as in the step-down tests), which is also contrastwise finite. This procedure proposed by P. R. Krishnaiah [7] is stepwise in nature and involves, univariate contrastwise finite tests at each stage. In this paper the associated method for the multiple comparisons of the means is developed and illustrated.

1. Introduction

The problem of comparing the means of several populations is of common occurrence in statistical practice. It is generally formulated in the framework of analysis of variance (ANOVA) in case of univariate response, and of MANOVA when the response is multidimensional. The univariate problem is usually treated using variance ratio test or finite intersection methods based upon statistics, such as Tukey's studentized range. Best known methods for MANOVA include the invariant tests based upon statistics, such as Roy's largest root, Hotelling–Lawley's trace, trace of the beta-matrix and Wilks' likelihood ratio, and the class of step-down tests introduced by J. Roy [21] and Krishnaiah [6, 7]. The invariant tests treat the response variables symmetrically, thus ignoring their possibly unequal importance. Also in earlier times their distributions and associated critical constants were less readily available. In step-down tests the variables are ordered according to their scientific significance or relevance, and the analysis consists of a sequence of univariate tests. At any stage in the procedure, the data on a response are analyzed using the preceding variables as covariates, and the necessary null distributions are essentially

*Research sponsored in part by the Air Force Office of Scientific Research, Air Force Systems Command, USAF under Grant No. AFOSR-77-3360. The United States Government is authorized to reproduce and distribute reprints for Governmental purposes notwithstanding a any copyright notation hereon. Work started while visiting KUL, Belgium.
**Research supported in part by Oakland University Research Fellowship.

univariate. The invariant tests are better known and better understood theoretically (e.g. Kiefer [5], Perlman [17]), than the step-down methods (e.g. Das Gupta [2], Subbaiah and Mudholkar [23]).

Practically, a test of significance is only a starting point in the comparison of k means and the multiple comparisons following the test are more informative. The status of multiple comparisons in univariate and multivariate analyses is traced and reviewed by Krishnaiah [9] and Miller [11]. It is well known that in general the multiple comparisons related to a test of significance are readily obtained if the test is expressible as a union-intersection test (Roy, Gnanadasikan, and Srivastava [22]). Various sets of confidence bounds associated with invariant MANOVA tests are derived essentially in this manner by Mudholkar [13], Mudholkar, Davidson, and Subbaiah [14, 15], and Wijsman [24]. J. Roy [21] and Krishnaiah [7] both deal with the problem of multiple comparisons using their step-down MANOVA methods by deriving simultaneous confidence bounds on some measures of departures from the null hypotheses which involve not only the means, which are the parameters of main interest, but also the elements of unknown dispersion matrices as the nuisance parameters. Hence Rao [19, p. 356] observes that "the general drawback of the step-down procedures is that they do not enable us to study the configuration of the mean values of the different populations, which in practical problems is more important than merely establishing differences in the mean values." Mudholkar and Subbaiah [16] attempt to rectify this shortcoming and render J. Roy's step-down method suitable for post-hoc analysis by deriving the simultaneous confidence bounds on all linear functions of the means. In this paper we present and illustrate corresponding results for the finite intersection tests by Krishnaiah.

In Section 2 the MANOVA model and the step-down methods are summarized. The simultaneous confidence bounds associated with the finite intersection tests are derived in Section 3. In Section 4 the computational aspects of the multiple comparisons are discussed in the context of the one-way classification. In the final section the methods are illustrated and compared using Fisher's Iris data.

2. Preliminaries

In literature the general linear hypothesis is discussed either in the usual MANOVA model or a growth curve model which can be reduced to it (Potthoff and Roy [18], Krishnaiah [8]). In this section we present the

MANOVA model and summarize the step-down procedure due to J. Roy [21] and the finite intersection tests of Krishnaiah [6, 7].

2.1. *The MANOVA Model*

Let $\mathbf{Y}(n\times p)$ denote a matrix whose rows are independently distributed according to a p-variate normal distribution with the same positive definite dispersion matrix $\mathbf{\Sigma}$ and means given by

$$E(\mathbf{Y})=\mathbf{A}\mathbf{\Theta} \tag{2.1}$$

where $\mathbf{A}(n\times m)$ is a known design matrix of rank $r\leqslant n-p$, and $\mathbf{\Theta}(m\times p)$ is a matrix of unknown parameters. The MANOVA problem is that of testing the null hypothesis $H_0:\mathbf{B\Theta}=\mathbf{0}$ against the alternative $H_1:\mathbf{B\Theta}\neq\mathbf{0}$, where $\mathbf{B}(t\times m)$ is a matrix of full rank. This restriction on the rank of $\mathbf{B}$ will not be assumed in the following sections. The invariant tests of H_0 (Lehmann [10]) depend on the eigenvalues of $\mathbf{S}_h\mathbf{S}_e^{-1}$ where $\mathbf{S}_h$ and $\mathbf{S}_e$, the SSP matrices 'due to the hypothesis,' and 'due to error,' respectively, are given by

$$\mathbf{S}_h=\mathbf{Y}'\mathbf{A}(\mathbf{A}'\mathbf{A})^{-}\mathbf{B}'\left[\mathbf{B}(\mathbf{A}'\mathbf{A})^{-}\mathbf{B}'\right]^{-1}\mathbf{B}(\mathbf{A}'\mathbf{A})^{-}\mathbf{A}'\mathbf{Y},$$
$$\mathbf{S}_e=\mathbf{S}=\mathbf{Y}'\left[\mathbf{I}-\mathbf{A}(\mathbf{A}'\mathbf{A})^{-}\mathbf{A}'\right]\mathbf{Y}, \tag{2.2}$$

where $(\mathbf{A}'\mathbf{A})^{-}$ is a generalized inverse of $(\mathbf{A}'\mathbf{A})$. Roy's largest root $\lambda_{\max}(\mathbf{S}_h\mathbf{S}_e^{-1})$, Hotelling–Lawley's trace $\operatorname{tr}(\mathbf{S}_h\mathbf{S}_e^{-1})$, Wilks' likelihood ratio $|\mathbf{S}_e|/|\mathbf{S}_h+\mathbf{S}_e|$, and $\operatorname{tr}\mathbf{S}_h(\mathbf{S}_h+\mathbf{S}_e)^{-1}$ trace of the betamatrix are among the well known invariant test procedures.

2.2. *The step-down procedure*

Suppose that the p variables are arranged in a decreasing order with respect to their importance or relevance. Then H_0 has a response-wise finite decomposition which depends upon this order and the step-down procedure is the union-intersection consequence of this decomposition. Specifically the procedure is as follows:

Let us denote $\mathbf{Y}=[\mathbf{y}_1,\ldots,\mathbf{y}_p]$, $\mathbf{\Theta}=[\boldsymbol{\theta}_1,\ldots,\boldsymbol{\theta}_p]$, and $\mathbf{Y}_i=[\mathbf{y}_1,\ldots,\mathbf{y}_i]$, $\mathbf{\Theta}_i=[\boldsymbol{\theta}_1,\ldots,\boldsymbol{\theta}_i]$, for $i=1,\ldots,p$. Then the conditional distributions of the elements of $\mathbf{y}_{i+1}$ given $\mathbf{Y}_i$ are univariate normal with common variance $\sigma_{i+1}^2=|\mathbf{\Sigma}_{i+1}|/|\mathbf{\Sigma}_i|$ and means,

$$E(\mathbf{y}_{i+1}|\mathbf{Y}_i)=\mathbf{A}\boldsymbol{\eta}_{i+1}+\mathbf{Y}_i\boldsymbol{\beta}_i, \tag{2.3}$$

$i=0,1,\ldots,p-1$, where $\beta_0=0$, $\boldsymbol{\beta}_i=\boldsymbol{\Sigma}_i^{-1}(\sigma_{1,i+1},\ldots,\sigma_{i,i+1})'$, $\boldsymbol{\eta}_{i+1}=\boldsymbol{\theta}_{i+1}-\boldsymbol{\Theta}_i\boldsymbol{\beta}_i$, and $\boldsymbol{\Sigma}_i$ is the first principal minor of order i containing the first i rows and i columns of $\boldsymbol{\Sigma}$. It may be noted that the MANOVA hypothesis is true if and only if $\mathbf{BN}=\mathbf{0}$, where $\mathbf{N}=[\boldsymbol{\eta}_1,\boldsymbol{\eta}_2,\ldots,\boldsymbol{\eta}_p]$. That is,

$$H_0 \equiv \bigcap_{i=1}^{p} \{H_{0i}: \mathbf{B}\boldsymbol{\eta}_i=\mathbf{0}\}.$$

The step-down procedure consists of testing the hypotheses $H_{01},\ldots,H_{0p}$ sequentially by univariate analysis of covariance at each stage regarding the responses at the previous stages as the concomitant variables. If $\hat{\boldsymbol{\eta}}_i$ and $s_i^2/(n-r-i+1)$ are the Gauss–Markoff estimates of $\boldsymbol{\eta}_i$ and σ_i^2 respectively, and $\mathbf{C}_i\sigma_i^2$ is the variance covariance matrix of $\mathbf{B}\hat{\boldsymbol{\eta}}_i$, then the **F**-statistic for testing H_{0i} is

$$F_i=\frac{(\mathbf{B}\hat{\boldsymbol{\eta}}_i)'\mathbf{C}_i^{-1}(\mathbf{B}\hat{\boldsymbol{\eta}}_i)/t}{s_i^2/(n-r-i+1)}. \tag{2.4}$$

Under H_0, the p step-down statistics are independently distributed (J. Roy [21]), and therefore the hypothesis H_0 can be tested at significance level α by distributing the type I error such that $(1-\alpha)=\prod_{i=1}^{p}(1-\alpha_i)$, α_i being the level corresponding to H_{0i}. If f_i is the upper 100 α_i percentage point of the F distribution with d.f. $(t, n-r-i+1)$, then H_0 is accepted if and only if $F_i \leqslant f_i, i=1,2,\ldots,p$.

2.3. *The finite intersection tests*

The finite intersection tests of H_0 are based on the same response-wise finite decomposition of H_0 as above. Moreover the decomposition, unlike the decomposition in J. Roy's step-down procedure, is finite with respect to contrasts as well. That is,

$$\{H_0: \mathbf{B}\boldsymbol{\Theta}=\mathbf{0}\} \equiv \bigcap_{i=1}^{p} \{H_{0i}: \mathbf{B}\boldsymbol{\eta}_i=\mathbf{0}\}$$

$$\equiv \bigcap_{i=1}^{p} \bigcap_{j=1}^{t} \{H_{0ij}: \mathbf{b}_j'\boldsymbol{\eta}_i=0\} \tag{2.5}$$

where $\mathbf{b}_1',\ldots,\mathbf{b}_t'$ are the t rows of $\mathbf{B}$. The hypothesis H_{0ij} is tested using the statistic

$$F_{ij}=\frac{(\mathbf{b}_j'\hat{\boldsymbol{\eta}}_i)^2(n-r-i+1)}{c_{ij}s_i^2} \tag{2.6}$$

for $i=1,\ldots,p;\ j=1,\ldots,t$, where $\mathbf{b}_j'\hat{\boldsymbol{\eta}}_i$ is the least squares estimate of $\mathbf{b}_j'\boldsymbol{\eta}_i$ and $s_i^2/(n-r-i+1)$ is unbiased estimate of σ_i^2. Krishnaiah's finite intersection test procedure consists in accepting H_0 if and only if,

$$F_{ij}\leqslant f_i \qquad j=1,\ldots,t \qquad i=1,\ldots,p. \tag{2.7}$$

The critical constants $f_1,\ldots,f_p$ are obtained such that

$$\begin{aligned}
&\Pr\big(F_{ij}\leqslant f_i,\quad j=1,\ldots,t;\ i=1,\ldots,p\,|\,H_0\big)\\
&\quad=\prod_{i=1}^{p}\Pr\big(F_{ij}\leqslant f_i, j=1,\ldots,t\,|\,H_0\big)\\
&\quad=\prod_{i=1}^{p}(1-\alpha_i)=(1-\alpha).
\end{aligned} \tag{2.8}$$

under H_0, the t statistics $(F_{i1},\ldots,F_{it})$ occurring at the ith stage, $i=1,\ldots,p$, are distributed according to t-variate F-distribution with $(1,n-r-i+1)$ d.f., and the statistics at different stages are independent. Various approximations to the critical constants occurring at various stages are available and are known to be of reasonable accuracy (Armitage and Krishnaiah [1]).

Both J. Roy [21] and Krishnaiah [7] have constructed simultaneous confidence intervals on linear functions and norms of the parameters which involve elements of $\boldsymbol{\Sigma}$ as the nuisance parameters. In the following section we derive confidence bounds on similar functions of the original parameters $\boldsymbol{\theta}_i, i=1,\ldots,p$.

3. The multiple comparisons

The simple relationship between the step-down statistics occurring in J. Roy's procedure and the MANOVA likelihood ratio is used in Mudholkar and Subbaiah [16] for constructing the simultaneous confidence intervals for all linear functions of the original means. Because no such relationship in case of the finite intersection tests is available, an alternative approach is used in the following derivation of the multiple comparisons associated with Krishnaiah's procedure. This argument is however applicable for obtaining the multiple comparisons related to the step-down method also. The following lemmas are needed in the sequel.

Lemma 3.1. *If*

$$\mathbf{P}=\left[\begin{array}{c:c}\mathbf{P}_{11} & \mathbf{P}_{12}\\ \hdashline \mathbf{P}_{12}' & \mathbf{P}_{22}\end{array}\right]$$

is a symmetric nonsingular matrix, and $\mathbf{P}_{11}, \mathbf{P}_{22}$ *are square, then*

$$\mathbf{P}^{-1}=\mathbf{Q}=\left[\begin{array}{c:c}\mathbf{Q}_{11} & \mathbf{Q}_{12}\\ \hdashline \mathbf{Q}_{12}' & \mathbf{Q}_{22}\end{array}\right],$$

where

$$\mathbf{Q}_{11}=\mathbf{P}_{11}^{-1}+\mathbf{P}_{11}^{-1}\mathbf{P}_{12}(\mathbf{P}_{22}-\mathbf{P}_{12}'\mathbf{P}_{11}^{-1}\mathbf{P}_{12})^{-1}\mathbf{P}_{12}'\mathbf{P}_{11}^{-1}$$

$$\mathbf{Q}_{22}=(\mathbf{P}_{22}-\mathbf{P}_{12}'\mathbf{P}_{11}^{-1}\mathbf{P}_{12})^{-1}$$

$$\mathbf{Q}_{12}=-\mathbf{P}_{11}^{-1}\mathbf{P}_{12}(\mathbf{P}_{22}-\mathbf{P}_{12}'\mathbf{P}_{11}^{-1}\mathbf{P}_{12})^{-1}.$$

This result is well known (see e.g. Rao [20]). In singular case, the inverses are read as the generalized inverses.

Rewriting (2.3) in more familiar general linear model form

$$E(\mathbf{y}_{i+1}|\mathbf{Y}_i)=(\mathbf{A} \,\vdots\, \mathbf{Y}_i)\left(\begin{array}{c}\boldsymbol{\eta}_{i+1}\\ \hdashline \boldsymbol{\beta}_i\end{array}\right), \tag{3.1}$$

the least square estimates of $\boldsymbol{\eta}_{i+1}, \boldsymbol{\beta}_i$ may readily be seen as

$$\left[\begin{array}{c}\hat{\boldsymbol{\eta}}_{i+1}\\ \hdashline \hat{\boldsymbol{\beta}}_i\end{array}\right]=\left[\left(\begin{array}{c}\mathbf{A}'\\ \hdashline \mathbf{Y}_i'\end{array}\right)(\mathbf{A} \,\vdots\, \mathbf{Y}_i)\right]^{-1}\left(\begin{array}{c}\mathbf{A}'\\ \hdashline \mathbf{Y}_i'\end{array}\right)\mathbf{y}_{i+1}=\left[\begin{array}{c:c}\mathbf{A}'\mathbf{A} & \mathbf{A}'\mathbf{Y}_i\\ \hdashline \mathbf{Y}_i'\mathbf{A} & \mathbf{Y}_i'\mathbf{Y}_i\end{array}\right]^{-1}\left(\begin{array}{c}\mathbf{A}'\mathbf{y}_{i+1}\\ \hdashline Y_i'\mathbf{y}_{i+1}\end{array}\right). \tag{3.2}$$

As a consequence of Lemma 3.1, these least square estimates may alternatively be expressed as in the following:

Lemma 3.2. *If* $\hat{\mathbf{\Theta}}=(\hat{\theta}_1,\ldots,\hat{\theta}_p)=(\mathbf{A}'\mathbf{A})^-\mathbf{A}'\mathbf{Y}$, *and* $\mathbf{S}=\mathbf{Y}'[\mathbf{I}-\mathbf{A}(\mathbf{A}'\mathbf{A})^-\mathbf{A}']\mathbf{Y}$, *then*

$$\hat{\boldsymbol{\eta}}_{i+1}=\hat{\boldsymbol{\theta}}_{i+1}-\hat{\mathbf{\Theta}}_i\hat{\boldsymbol{\beta}}_i,$$

and

$$\hat{\boldsymbol{\beta}}_i=\mathbf{S}_i^{-1}(s_{i+1,1},\ldots,s_{i+1,i})'. \tag{3.3}$$

Lemma 3.3. *The variance covariance matrix of* $\mathbf{B}\hat{\boldsymbol{\eta}}_{i+1}$ *is*

$$\mathrm{Var}(\mathbf{B}\hat{\boldsymbol{\eta}}_{i+1})=\mathbf{B}\left[(\mathbf{A}'\mathbf{A})^-+\hat{\mathbf{\Theta}}_i\mathbf{S}_i^{-1}\hat{\mathbf{\Theta}}_i'\right]\mathbf{B}'\sigma_{i+1}^2 \tag{3.4}$$

Proof. The result follows immediately from

$$\mathrm{Var}(\hat{\boldsymbol{\eta}}_{i+1})=\left[(\mathbf{A}'\mathbf{A})^-+\hat{\mathbf{\Theta}}_i\mathbf{S}_i^{-1}\hat{\mathbf{\Theta}}_i'\right]\sigma_{i+1}^2$$

which is a consequence of the expression for $\mathbf{Q}_{11}$ used in Lemma 3.2.

Corollary 3.4. *As a particular case of* (3.4) *we have*

$$\mathrm{Var}(\mathbf{b}_j'\hat{\boldsymbol{\eta}}_{i+1})=\mathbf{b}_j'\left[(\mathbf{A}'\mathbf{A})^-+\hat{\mathbf{\Theta}}_i\mathbf{S}_i^{-1}\hat{\mathbf{\Theta}}_i'\right]\mathbf{b}_j\sigma_{i+1}^2 \tag{3.5}$$

and F_{ij} *in the Equation* (2.6) *can be expressed as*

$$F_{ij}=\frac{(\mathbf{b}_j'\hat{\boldsymbol{\eta}}_i)^2(n-r-i+1)}{\mathbf{b}_j'\left[(\mathbf{A}'\mathbf{A})^-+\hat{\mathbf{\Theta}}_{i-1}S_{i-1}^{-1}\hat{\mathbf{\Theta}}_{i-1}'\right]\mathbf{b}_j s_i^2}, \tag{3.6}$$

where $s_i^2=|S_i|/|S_{i-1}|$.

Lemma 3.5. *If* $\mathbf{D}_i=\mathrm{diag}(s_1^2,s_2^2,\ldots,s_i^2)$, $s_j^2=|\mathbf{S}_j|/|\mathbf{S}_{j-1}|$, *and* $\hat{\mathbf{N}}_i=[\hat{\boldsymbol{\eta}}_1,\hat{\boldsymbol{\eta}}_2,\ldots,\hat{\boldsymbol{\eta}}_i]$, *then*

$$\hat{\mathbf{\Theta}}_i\mathbf{S}_i^{-1}\hat{\mathbf{\Theta}}_i'=\hat{\mathbf{N}}_i\mathbf{D}_i^{-1}\hat{\mathbf{N}}_i'. \tag{3.7}$$

Proof. Let

$$L_i = \begin{bmatrix} 1 & -\hat{\beta}_{11} & \cdots & -\hat{\beta}_{i-1,1} \\ 0 & 1 & & -\hat{\beta}_{i-1,2} \\ \vdots & \vdots & & \vdots \\ 0 & 0 & & 1 \end{bmatrix}$$

be the upper triangular matrix with the kth column containing $-\boldsymbol{\beta}_{k-1}$ and 1 in the kth position and zeroes in the remaining $i-k$ elements. It may be verified that $\mathbf{L}_i'\mathbf{S}_i\mathbf{L}_i = \mathbf{D}_i$. The lemma then follows easily from the fact that $\hat{\mathbf{N}}_i = \hat{\boldsymbol{\Theta}}_i\mathbf{L}_i$, and $\mathbf{S}_i^{-1} = \mathbf{L}_i\mathbf{D}_i^{-1}\mathbf{L}_i'$.

As a consequence of Lemma 3.5 it can be shown that

$$F_{ij} = \frac{(\mathbf{b}_j'\hat{\boldsymbol{\eta}}_i)^2(n-r-i+1)}{\left[\mathbf{b}_j'(A'A)^-\mathbf{b}_j + \sum_{k=1}^{i-1}\frac{(\mathbf{b}_j'\hat{\boldsymbol{\eta}}_k)^2}{s_k^2}\right]s_i^2}. \tag{3.8}$$

Lemma 3.6. *If* $F_{ij} \leqslant f_i, j=1,\ldots,t;\ i=1,2,\ldots,p$, *then*

$$(\mathbf{b}_j'\hat{\boldsymbol{\eta}}_i)^2 \leqslant \left[\mathbf{b}'(A'A)^-\mathbf{b}\right]\cdot c_i^* s_i^2, \tag{3.9}$$

where $c_i = f_i/(n-r-i+1), i=1,\ldots,p$ *and* $c_1^* = c_1, c_i^* = c_i(1 + c_1^* + \cdots + c_{i-1}^*), i=2,\ldots,p$.

Proof. $F_{ij} \leqslant f_i$ implies that

$$(\mathbf{b}_j'\hat{\boldsymbol{\eta}}_i)^2 \leqslant \left[\mathbf{b}_j'(\mathbf{A}'\mathbf{A})^-\mathbf{b}_j + \sum_{k=1}^{i-1}\frac{(\mathbf{b}_j'\hat{\boldsymbol{\eta}}_k)^2}{s_k^2}\right]\cdot s_i^2 c_i, \tag{3.10}$$

because $c_i = f_i/(n-r-i+1)$. For $i=1$, we have

$$(\mathbf{b}_j'\hat{\boldsymbol{\eta}}_1)^2 \leqslant \mathbf{b}_j'(\mathbf{A}'\mathbf{A})^-\mathbf{b}_j\cdot s_1^2 c_1^* (\text{since } c_1^* = c_1), \tag{3.11}$$

and for $i=2$,

$$(\mathbf{b}_j'\hat{\boldsymbol{\eta}}_2)^2 \leqslant \left[\mathbf{b}_j'(\mathbf{A}'\mathbf{A})^-\mathbf{b}_j + \frac{(\mathbf{b}_j'\hat{\boldsymbol{\eta}}_1)^2}{s_1^2}\right] s_2^2 c_2$$

$$\leqslant \left[\mathbf{b}_j'(\mathbf{A}'\mathbf{A})^-\mathbf{b}_j\right](1+c_1^*)c_2 s_2^2$$

$$= \left[\mathbf{b}_j'(\mathbf{A}'\mathbf{A})^-\mathbf{b}_j\right] c_2^* s_2^2. \tag{3.12}$$

In a similar manner, more generally, we have

$$(\mathbf{b}_j'\hat{\boldsymbol{\eta}}_i)^2 \leqslant \left[\mathbf{b}_j'(\mathbf{A}'\mathbf{A})^-\mathbf{b}_j\right] c_i(1+c_1^*+\cdots+c_{i-1}^*)s_i^2$$

$$= \left[b_j'(\mathbf{A}'\mathbf{A})^-\mathbf{b}_j\right] c_i^* s_i^2.$$

Theorem 3.7. *If $F_{ij} \leqslant f_i, j=1,\ldots,t;\ i=1,\ldots,p$, then*

$$|\mathbf{b}_j'\hat{\boldsymbol{\Theta}}\mathbf{d}| \leqslant \left(\sum_{i=1}^{p} |h_i| \sqrt{c_i^*}\right)\left(\mathbf{b}_j'(\mathbf{A}'\mathbf{A})^-\mathbf{b}_j\right)^{\frac{1}{2}}, \tag{3.13}$$

where $\mathbf{h}=(h_1,\ldots,h_p)'=\mathbf{U}\mathbf{d}$, and $\mathbf{U}$ is the upper triangular matrix such that $\mathbf{S}=\mathbf{U}'\mathbf{U}$.

Proof. Because $\mathbf{U}$ is the upper triangular matrix such that $\mathbf{S}=\mathbf{U}'\mathbf{U}$ and $\mathbf{L}_p'\mathbf{S}\mathbf{L}_p=\mathbf{D}=\mathrm{diag}(s_1^2,\ldots,s_p^2)$, where $\mathbf{L}_p$ is as defined in Lemma 3.5, it follows that $\mathbf{U}=\mathbf{D}^{\frac{1}{2}}L_p^{-1}$. Hence,

$$\mathbf{b}_j'\hat{\boldsymbol{\Theta}}\mathbf{d}=\mathbf{b}_j'\hat{\mathbf{N}}L_p^{-1}\mathbf{d}$$

$$=\mathbf{b}_j'\hat{\mathbf{N}}\mathbf{D}^{-\frac{1}{2}}\mathbf{U}\,\mathbf{d}$$

$$=\mathbf{b}_j'\hat{\mathbf{N}}\mathbf{D}^{-\frac{1}{2}}\mathbf{h},$$

where $\mathbf{h}=\mathbf{U}\,\mathbf{d}=(h_1,\ldots,h_p)'$. Consequently we have

$$|\mathbf{b}_j'\hat{\boldsymbol{\Theta}}\,\mathbf{d}| = \left|\sum_{i=1}^{p} (\mathbf{b}_j'\hat{\boldsymbol{\eta}}_i)\cdot \frac{h_i}{s_i}\right|$$

$$\leqslant \left(\sum_{i=1}^{p} |h_i| \sqrt{c_i^*}\right)\left(\mathbf{b}_j'(A'A)^-\mathbf{b}_j\right)^{\frac{1}{2}}.$$

Corollary 3.8. *As a particular case of the above theorem we have*

$$|\mathbf{b}_j'\hat{\boldsymbol{\theta}}_i| \leqslant \left(\sum_{k=1}^{i} |u_{ki}| \sqrt{c_k^*} \right) (\mathbf{b}_j'(\mathbf{A}'\mathbf{A})^- \mathbf{b}_j)^{\frac{1}{2}}, \tag{3.14}$$

and replacing $\hat{\boldsymbol{\theta}}_i$ *by* $\hat{\boldsymbol{\theta}}_i - \boldsymbol{\theta}_i$ *in* (3.14), *the simultaneous confidence intervals with confidence coefficient* $(1-\alpha)$ *are*

$$\mathbf{b}_j'\boldsymbol{\theta}_i \varepsilon \mathbf{b}_j\hat{\boldsymbol{\theta}}_i \pm \left(\sum_{k=1}^{i} |u_{ki}| \sqrt{c_k^*} \right) (\mathbf{b}_j'(\mathbf{A}'\mathbf{A})^- \mathbf{b}_j)^{\frac{1}{2}}. \tag{3.15}$$

Remark 3.9. Equation (3.15) leads to the following confidence interval for arbitrary contrast $\mathbf{a}'\mathbf{B}\boldsymbol{\theta}_i = \Sigma_{j=1}^{t} a_j \mathbf{b}_j'\boldsymbol{\theta}_i$,

$$\mathbf{a}'\mathbf{B}\boldsymbol{\theta}_i \varepsilon \mathbf{a}'\mathbf{B}\hat{\boldsymbol{\theta}}_i \pm \sum_{j=1}^{t} |a_j| (\mathbf{b}_j'(\mathbf{A}'\mathbf{A})^- \mathbf{b}_j)^{\frac{1}{2}} \left(\sum_{k=1}^{i} |u_{ki}| \sqrt{c_k^*} \right). \tag{3.16}$$

Remark 3.10. Comparing equation (3.15) above with the result of Corollary 2.4. in Mudholkar and Subbaiah [16] it may be seen that the bounds on the contrasts $\mathbf{b}_j'\boldsymbol{\theta}_i$ associated with finite intersection tests differ from the corresponding confidence intervals obtained from step-down procedure only in the critical constants. Furthermore the step-down procedure confidence intervals on arbitrary linear functions have the same form as (3.13).

Remark 3.11. Pursuing Remark 3.10 in conjunction with Theorem 3.7 of Krishnaiah [8] it can be concluded that the confidence intervals for the basic contrasts $\mathbf{b}_j'\boldsymbol{\theta}_i$ obtained using the finite intersection tests are shorter than those obtained using the step-down procedure. Evidently this is not true in the case of arbitrary contrasts.

4. Computational aspects

In this section some computational considerations for the test statistics and the multiple comparisons are discussed. The details of the computation are given for both the finite intersection tests and the step-down method with the general model and are more explicitly described for the simplest yet practically most common one-way MANOVA model.

As is evident from (2.4), (3.8), (3.15) the test statistics in the finite intersection as well as the step-down procedure, and the associated confidence intervals depend upon the estimates $\hat{\boldsymbol{\Theta}}$ and $\mathbf{S}$ of $\boldsymbol{\Theta}$ and a multiple of $\boldsymbol{\Sigma}$ respectively. The computations may be conducted as follows:

(i) Construct $\mathbf{B}\hat{\boldsymbol{\Theta}}$, the $t\times p$ matrix consisting the estimates of the pt contrasts $\mathbf{b}_j'\hat{\boldsymbol{\theta}}_i, j=1,\ldots,t;\ i=1,\ldots,p$.

(ii) Augment the error SSP matrix $\mathbf{S}$ by $\hat{\boldsymbol{\Theta}}'\mathbf{B}'$ to obtain the $p\times(p+t)$ matrix $(\mathbf{S} \,\vdots\, \hat{\boldsymbol{\Theta}}'\mathbf{B}')$.

(iii) Sweep the rows of this matrix by square root method to obtain the Cholesky decomposition of $\mathbf{S}$, i.e., to reduce $\mathbf{S}$ to the upper triangular $\mathbf{U}$ of Theorem 3.7 with $\mathbf{S}=\mathbf{U}'\mathbf{U}$. Specifically,

$$
\begin{aligned}
&u_{11}=\sqrt{s_{11}},\ u_{1j}=s_{1j}/u_{11} && j=2,\ldots,p\\
&u_{ii}=\left(s_{ii}-\sum_{k=1}^{i-1}u_{ki}^2\right) && i=2,\ldots,p\\
&u_{ij}=\left(s_{ij}-\sum_{k=1}^{i-1}u_{ki}u_{kj}\right)\Big/u_{ii} && j=i+1,\ldots,p\quad i=2,\ldots,p.
\end{aligned}
\tag{4.1}
$$

The sweep reduces $\hat{\boldsymbol{\Theta}}'\mathbf{B}'$ to $\mathbf{U}'^{-1}\hat{\boldsymbol{\Theta}}'\mathbf{B}'=Z$, say. Schematically,

$$
(\mathbf{S} \,\vdots\, \hat{\boldsymbol{\Theta}}'\mathbf{B}')\rightarrow(U \,\vdots\, Z).
\tag{4.2}
$$

(iv) The test statistics F_{ij} in (2.6), (3.6), or (3.8) are then given by

$$
F_{ij}=\frac{Z_{ij}^2(n-r-i+1)}{\left(\mathbf{b}_j'(\mathbf{A}'\mathbf{A})^-\mathbf{b}_j+\sum_{k=1}^{i-1}Z_{kj}^2\right)}
\tag{4.3}
$$

(v) The confidence intervals (3.15) can be constructed using the p sums $\sum_{k=1}^{i}|u_{ki}|\sqrt{c_k^*}, i=1,\ldots,p$, the elements $\mathbf{b}_j'\hat{\boldsymbol{\theta}}_i$ of $\mathbf{B}\hat{\boldsymbol{\Theta}}$, and rows of $\mathbf{B}$. Here u_{ij} are elements of $\mathbf{U}$ obtained in (iii) and c_k^* are as in Lemma 3.6.

(vi) The step-down statistics F_i in (2.4) can be obtained directly by sweeping $\mathbf{S}_h+\mathbf{S}_e=\mathbf{S}_h+\mathbf{S}$ to obtain upper triangular $\mathbf{V}=(v_{ij})$ in the same manner as the reduction $\mathbf{S}\rightarrow\mathbf{U}$. The step-down statistics are then given by

$$
F_i=\frac{(n-r-i+1)}{t}\left(\frac{u_{ii}^2}{v_{ii}^2}-1\right),
\tag{4.4}
$$

where t is the rank of $\mathbf{B}$. The confidence intervals of Remark 3.10 discussed more explicitly in Mudholkar and Subbaiah [16] are obtained as in (3.15), except that the critical constants are obtained from the F distribution instead of multivariate F. Specifically $c_i = f_i/(n-r-i+1)$ as defined in Lemma 3.6, where f_i satisfies $\Pr(F_{ij} \leqslant f_i, j=1,\ldots,t|H_0)=1-\alpha_i$, should be replaced by $c_i = tf_i/(n-r-i+1)$, f_i being the upper 100 α_i percentage point of the F distribution with $(t, n-r-i+1)$ d.f.

Now consider the one-way MANOVA model; i.e., suppose that we have a random sample of n_i P-dimensional observations $\mathbf{Y}_{ij}, j=1,\ldots,n_i$ from the ith group, $i=1,\ldots,k$, such that

$$\mathbf{E}(\mathbf{Y}_{ij}) = \boldsymbol{\mu}_i \quad \text{and} \quad \operatorname{Var}(\mathbf{Y}_{ij}) = \boldsymbol{\Sigma}. \tag{4.5}$$

In this case the hypothesis H_0: $\boldsymbol{\mu}_1 = \boldsymbol{\mu}_2 = \cdots = \boldsymbol{\mu}_k$, the null hypothesis of interest has a number of meaningful and well known decompositions which yield different versions of $\mathbf{b}$'s in (2.5). For example, $H_0 = \cap_{i \neq j}\{H_{0ij}: \boldsymbol{\mu}_i = \boldsymbol{\mu}_j\}$ corresponds to the well known pairwise comparisons in the univariate case due to Tukey, and $H_0 = \cap_{i=1}^{k-1}\{H_{0ik}: \boldsymbol{\mu}_i = \boldsymbol{\mu}_k\}$ is multivariate analogue of Dunnett's comparisons of means with mean of a control population.

The computations described above may then be carried out (in the obvious notation) with $\hat{\boldsymbol{\Theta}}' = (\bar{\mathbf{y}}_1, \ldots, \bar{\mathbf{y}}_k)$, $\mathbf{b}$'s of the form $(0, 1, 0, -1, 0, \ldots)$, $n = \sum_i n_i$, $\mathbf{S} = \mathbf{S}_e = \sum_i \sum_j (\mathbf{y}_{ij} - \bar{\mathbf{y}}_{i.})(\mathbf{y}_{ij} - \bar{\mathbf{y}}_{i.})'$, $\mathbf{S}_h + \mathbf{S}_e = \sum_i \sum_j (\mathbf{y}_{ij} - \bar{\mathbf{y}}_{..})(\mathbf{y}_{ij} - \bar{\mathbf{y}}_{..})'$, etc. In this case $(\mathbf{A}'\mathbf{A})^{-1} = \operatorname{diag}(n_1^{-1}, n_2^{-1}, \ldots, n_k^{-1})$.

5. An illustrative example

The tests and the simultaneous confidence intervals discussed in the earlier sections are now illustrated using the well known Iris data considered by Fisher [3] when introducing the linear discriminant function. These data or the relevant sufficient statistics are widely available including in Kendall and Stuart [4], and Morrison [72].

The data consist of multivariate measurements on random samples of size 50 from each of the three iris species viz., Setosa, Versicolor, and Virginica. The variables measured on each flower are its sepal width (X_1), sepal length (X_2), petal width (X_3) and petal length (X_4). The problem is of comparing the population mean vectors for the species. The sample means and the sample error SSP matrix on which the rest of the analysis depends are as follows:

The matrix $\hat{\mathbf{\Theta}}$, of the means, is given by Table 1.

Table 1
Sample Means (in cms)

Variety	Sepal width X_1	Sepal length X_2	Petal width X_3	Petal length X_4
Setosa	3.428	5.006	0.246	1.462
Versicolor	2.770	5.936	1.326	4.260
Virginica	2.974	6.588	2.026	5.552

The error SSP matrix $\mathbf{S}$ is

$$\begin{array}{c} \\ X_1 \\ X_2 \\ X_3 \\ X_4 \end{array} \begin{array}{c} \begin{array}{cccc} X_1 & X_2 & X_3 & X_4 \end{array} \\ \begin{bmatrix} 16.9620 & 13.6300 & 4.8084 & 8.1208 \\ 13.6300 & 38.9562 & 5.6450 & 24.6246 \\ 4.8084 & 5.6450 & 6.1566 & 6.2718 \\ 8.1208 & 24.6246 & 6.2718 & 27.2226 \end{bmatrix} \end{array}.$$

The multivariate analysis of variance and multiple comparisons using Roy's largest root statistic for these data are given in Morrison [72] with $\alpha=0.01$. For the current illustration we take $\alpha=0.05$. The value 32.188 of the largest root of $\mathbf{S}_h\mathbf{S}_e^{-1}$ is significant at $\alpha=0.01$, and a fortiori so at $\alpha=0.05$.

For the purpose of illustrating the finite intersection tests and the multiple comparisons we focus upon the pairwise comparisons of the three species, that is take

$$\mathbf{b}_1'=(1,-1,0), \mathbf{b}_2'=(1,0,-1), \mathbf{b}_3'=(0,1,-1). \tag{5.1}$$

The estimates of the contrasts $\mathbf{B}\hat{\mathbf{\Theta}}$ are given in Table 2. The sweep operation of Cholesky decomposition (4.1) on the augmented matrix $(\mathbf{S} \mid \hat{\mathbf{\Theta}}'\mathbf{B}')$ gives $(\mathbf{U} \mid \mathbf{U}'^{-1}\hat{\mathbf{\Theta}}'\mathbf{B}')=(\mathbf{U} \mid \mathbf{Z})=$

$$\left[\begin{array}{cccc|ccc} 4.1185 & 3.3095 & 1.1675 & 1.9718 & 0.1598 & 0.1102 & -0.0495 \\ 0.0 & 5.2918 & 0.3366 & 3.4202 & -0.2757 & -0.3679 & -0.0922 \\ 0.0 & 0.0 & 2.1634 & 1.3028 & -0.5426 & -0.8250 & -0.2825 \\ 0.0 & 0.0 & 0.0 & 3.1527 & -0.4642 & -0.6262 & -0.1620 \end{array}\right].$$

The finite intersection test statistics F_{ij}, $j=1,2,3$, and $i=1,2,3,4$, computed with the matrix $(\mathbf{U} \mid \mathbf{Z})$ using (4.3), together with the step-down statistics $F_i, i=1,2,3,4$ computed as described at (4.4) are given in Table 2.

Table 2
The finite-intersection and the step-down test statistics

j	Step-down	Finite-intersection		
	F_i	F_{i1}	F_{i2}	F_{i3} [a]
1	44.0425	93.8063	44.6572	9.0166
2	20.2893	169.3110	378.8961	29.2547
3	15.2469	301.6160	526.4150	227.0562
4	48.1828	71.1764	65.0385	28.9141

[a] F_{ij} corresponds to the contrast $\boldsymbol{b}_j'\boldsymbol{\theta}_i$ of (5.1)

The well known largest root multiple comparisons (e.g., Morrison [12, p. 199]) with $\alpha=0.05$ are given in column 5 of Table 2. The comparisons associated with the step-down test are obtained according to Mudholkar and Subbaiah [23], or as in (3.15), with $\alpha_i=0.01274$, i.e., $\alpha=0.05, c_i=f_i t/(n-r-i+1)=f_i 2/(148-i), f_i$ being the upper α_i percentage point of the F-distribution with d.f. $(2,148-i)$, $i=1,2,3,4$. The centers and the half widths of these confidence intervals are given in Table 3.

The selection of the critical constants for the finite-intersection tests and therefore also for constructing the related simultaneous confidence intervals is a topic of current research (e.g., Armitage and Krishnaiah [1]). There are several practically adequate approximations and tables available for the purpose. In the present illustration we use the simplest Bonferroni-type approximation

$$(1-\alpha_i)=\Pr(F_{ij}\leqslant f_i, j=1,2,3)$$

$$\approx \sum_{j=1}^{3}\Pr(F_{ij}\leqslant f_i) \tag{5.2}$$

giving the upper $\alpha_i/3$ percentage point of the F-distribution with d.f. 1 and $(148-i)$, as the approximation for $f_i, i=1,2,3,4$. To achieve the overall (nominal) error rate 0.05, we take $\alpha_i=0.01274$. The resulting conservative confidence intervals are given in Table 3.

In general, for a given i, the width of a confidence interval for a contrast $\mathbf{b}_j'\boldsymbol{\Theta}_i$ would vary with respect to j. But in the present case, because of the equal number 50 of the observations in the three groups, the width of the intervals for the pairwise differences are the same for a given variable.

Table 3
Estimates and half-widths of confidence intervals for pairwise comparisons $\alpha = 0.05$

i	Estimates of contrasts			Half-widths		
	$b_1'\hat{\mathbf{\Theta}}_i$	$b_2'\hat{\mathbf{\Theta}}_i$	$b_3'\hat{\mathbf{\Theta}}_i$	Largest root	Step-down	Finite intersection
1	0.658	0.454	-0.204 [b]	0.3853	0.2037	0.1974
2	-0.930	-1.582	-0.652	0.2543	0.4342	0.4202
3	-1.080	-1.780	-0.700	0.3221	0.1893	0.1830
4	-2.798	-4.090	-1.292	0.1532	0.5137	0.4961

[b]Refer to Observation 5.1 concerning $b_3'\mathbf{\Theta}_1$.

Observation 5.1. It may be noted from Table 2 that all except one of the 12 comparisons are significant with respect to the largest root, the step-down and the finite intersection tests as the confidence intervals exclude zero. The difference $b_3'\mathbf{\Theta}_1$ between the mean sepal widths for Versicolor and Virginica is insignificant when examined using the largest root, almost significant with respect to the step-down method. But this difference is clearly significant with respect to the conservative finite intersection tests.

References

[1] Armitage, J. V. and Krishnaiah, P. R. (1978). Approximate percentage points for applications of finite intersection tests for multiple comparisons of means. Technical report.

[2] Das Gupta, S. (1970). Step-down multiple decision rules. In: *Essays in Probability and Statistics* (R. C. Bose et al., eds.) University of North Carolina Press, Chapel Hill, NC.

[3] Fisher, R. A. (1939). The use of multiple measurements in taxonic problems. *Ann. Eugen.* **7**, 179–188.

[4] Kendall, M. G., and Stuart A., (1968). *The Advanced Theory of Statistics*, Vol. 3, 2nd ed., Hafner Publishing Company, New York.

[5] Kiefer, J. C. (1966). Multivariate optimality results. In: *Multivariate Analysis* (P. R. Krishnaiah, ed.) Academic Press, New York.

[6] Krishnaiah, P. R. (1965a). On the simultaneous ANOVA and MANOVA tests. *Ann. Inst. Statist. Math.*, **17**, 35–53.

[7] Krishnaiah, P. R. (1965b). Multiple comparison tests in multiresponse experiments. *Sankhya, Ser. A.*, **27**, 65–72.

[8] Krishnaiah, P. R. (1969). Simultaneous test procedures under general MANOVA models. In *Multivariate Analysis—II* (P. R. Krishnaiah, ed.). Academic Press, New York.

[9] Krishnaiah, P. R. (1979). Some developments on simultaneous test procedures. In: *Developments in Statistics*, Vol. 2 (P. R. Krishnaiah, ed.). Academic Press, New York.

[10] Lehmann, E. L. (1959). *Testing Statistical Hypotheses*. John Wiley, New York.

[11] Miller, R. G. (1977). Developments in multiple comparisons 1966–1976. *J. Amer. Statist. Assoc.* **72**, 779–788.

[12] Morrison, D. F. (1976). *Multivariate Statistical Methods*, 2nd ed., McGraw-Hill, New York.

[13] Mudholkar, G. S. (1966). On confidence bounds associated with multivariate analysis of variance and non-independence between two sets of variates. *Ann. Math. Statist.* **37**, 1736–1746.

[14] Mudholkar, G. S., Davidson, M. L. and Subbaiah, P. (1974a). Extended linear hypotheses and simultaneous tests in multivariate analysis of variance. *Biometrika* **61**, 467–477.

[15] Mudholkar, G. S., Davidson, M. L. and Subbaiah, P. (1974b). A note on the union-intersection character of some MANOVA procedures, *J. Multivariate Anal.* **4**, 486–493.

[16] Mudholkar, G. S. and Subbaiah, P. (1975). A note on MANOVA multiple comparisons based upon step-down procedures. *Sankhya, Ser. B*, **37**, 300–307.

[17] Perlman, M. D. (1979). Multivariate tests: Unbiasedness and monotonicity of power. In *Multivariate Analysis—V* (P. R. Krishnaiah, ed.) pp. 399–411. [This volume]

[18] Potthoff, R. F. and Roy, S. N. (1964). A generalized multivariate analysis of variance model useful especially for growth curve problems. *Biometrika*, **51**, 313–326.

[19] Rao, C. R. (1964). The use and interpretation of principal component analysis in applied research. *Sankhya, Ser. B*, **26**, 329–358.

[20] Rao, C. R. (1973). *Linear Statistical Inference and Its Applications*, 2nd ed., John Wiley, New York.

[21] Roy, J. (1958). Step-down procedure in multivariate analysis. *Ann. Math. Statist.* **29**, 1177–1187.

[22] Roy, S. N., Gnanadesikan, R. and Srivastava, J. N. (1971). *Analysis and Design of Certain Quantitative Multiresponse Experiments*. Pergamon Press, Oxford.

[23] Subbaiah, P. and G. S. Mudholkar (1978). A comparison of two tests for the significance of a mean vector, *J. Amer. Statist. Assoc.* **73**.

[24] Wijsman, R. A. (1979). Constructing all smallest simultaneous confidence sets in a given class, with applications to MANOVA. To appear in *Ann. Statist.*

P. R. Krishnaiah, ed., *Multivariate Analysis–V*

SMALLEST SIMULTANEOUS CONFIDENCE SETS WITH APPLICATIONS IN MULTIVARIATE ANALYSIS*

Robert A. WIJSMAN
University of Illinois at Urbana-Champaign, U.S.A.

A method is given for generating smallest simultaneous confidence sets $\{A_i\}$ for a family of parametric functions $\{\psi_i(\gamma)\}$ of a parameter of interest $\gamma = \gamma(\theta) \in \Gamma$ ($\theta \in \Theta =$ parameter space) starting from a confidence set F_0 for γ, together with the confidence set $F_1 \supset F_0$ with respect to which the family $\{A_i\}$ is exact. When there is an invariance group that acts on Γ and is transitive over Θ, then an invariant pivotal quantity can be constructed which can be used to generate equivariant starting confidence sets F_0 for γ. The method is illustrated with several applications to problems in multivariate analysis, including MANOVA (both under full and under triangular group), regression matrix, and covariance matrix.

1. Introduction

This paper deals with some rather general considerations concerning the construction of simultaneous confidence sets for parametric functions. All applications that are presented are in the field of normal multivariate analysis, but the general theory is not limited to it.

Let X be a random variable, taking values in a space $\mathcal{X}$ and having distribution P_θ, $\theta \in \Theta$. Suppose one desires inference about a function γ of θ, taking values in a space Γ. Thus, $\gamma(\theta)$ is the parameter of interest, and the rest of θ may be regarded as a nuisance parameter. Let $F(X)$ be a confidence set for γ such that $P_\theta\{\gamma(\theta) \in F(X)\} \geqslant 1-\alpha$ for every $\theta \in \Theta$. Suppose there is given a family $\{\psi_i : i \in I\}$ of parametric functions $\psi_i : \Gamma \to \Psi$, where I is an arbitrary index set (the range space Ψ could depend on i but this will not be considered). Suppose one wants confidence sets $A_i(X)$ for the $\psi_i(\gamma)$ such that the following two events are equal no matter what the value of $\theta \in \Theta$ is:

$$[\gamma(\theta) \in F(X)] = [\psi_i(\gamma(\theta)) \in A_i(X) \forall i \in I]. \tag{1.1}$$

This can for instance be used if the hypothesis $\gamma = \gamma_0 \in \Gamma$ is being tested at the level α by rejecting the hypothesis whenever $\gamma_0 \notin F(X)$; then the blame

*Research supported by the National Science Foundation under grant MCS 75-07978.

for the falsity of the hypothesis can be put on those ψ_i for which $\psi_i(\gamma_0) \notin A_i(X)$. An extensive treatment of simultaneous testing and estimation procedures is given in [2], see also [12].

For brevity we shall often omit the argument X and write F, A_i. A family $\{A_i\}$ of simultaneous confidence sets for the ψ_i will be said to be *exact* with respect to F (abbreviated: e.w.r.t. F) if (1.1) holds (in [2] this property is called *coherent and consonant*). A confidence set F for γ will be called exact if there exists a family $\{A_i\}$ such that (1.1) holds. The family $\{A_i\}$ will be called *smaller* than the family $\{B_i\}$ if both are e.w.r.t. the same F, and $A_i(x) \subset B_i(x) \forall i \in I$, $x \in \mathcal{X}$.

In [12] it was shown how to construct a smallest family $\{A_i\}$ starting from a given family $\{B_i\}$. In the present paper a slightly different point of view will be taken. Given a confidence set F for γ, there is no guarantee it is exact. Thus, if γ is the matrix M of means in a MANOVA problem (Section 4.1) and the ψ_i are the real valued functions $a'Mb$, and if F is derived from an invariant test, then F is not exact unless derived from Roy's maximum root test [12]. In this paper it will be shown how one can start from an arbitrary confidence set F_0 for γ and construct the smallest exact confidence set F_1 that contains F_0, together with the family $\{A_i\}$ e.w.r.t. F_1. In addition, it turns out that $\{A_i\}$ is the smallest family e.w.r.t. F_1.

The considerations, sofar, are quite general and not restricted to multivariate normal distributions. For useful results, however, it is usually necessary to impose certain restrictions on the type of simultaneous confidence sets that one allows. This is discussed in some detail in [12]. A very convenient method of effecting a reasonable restriction exists if there is a group G of transformations that leaves the problem invariant. This method works especially well if G acts not only on $\mathcal{X}$ and Θ, but on Γ as well, and if G is transitive over Θ. In that case one can derive a pivotal quantity for γ, i.e., a function of both X and γ whose distribution does not depend on θ, and use it to derive a confidence set for γ. This method can be expected to find its applications mostly in multivariate normal problems, which usually exhibit a considerable amount of symmetry. Indeed, all applications presented in Section 4 are of that nature.

The method of Section 2 is generally applicable, but the method of Section 3 is likely to fail in problems where the family of parametric functions is finite, such as in [4] and [5]. Some results on smallest simultaneous confidence sets in MANOVA have been obtained by Mudholkar [7], by Gabriel [1] and by Jensen and Mayer [3]. However, in all these cases additional restrictions besides invariance were imposed that are not imposed in the present paper. The emphasis here is, first, on the construction of exact confidence sets for γ and the corresponding simultaneous confidence set for the $\psi_i(\gamma)$; and, second, to apply this within the class of equivariant estimators. No attempt is being made at exhaustiveness when it comes to the applications in Section 4, and the latter should be regarded mostly as a set of illustrations.

2. Construction of exact confidence sets: the operation T.

In this section the spaces $\mathfrak{X}$, Θ, Γ, Ψ, the function γ, the index set I, and the family of functions $\{\psi_i : i \in I\}$ will be considered given and fixed. With a confidence set F for γ will be meant, interchangeably, a family $\{F(x) : x \in \mathfrak{X}\}$ of subsets of Γ or a random subset $F(X)$ of Γ. In this section no requirements will be placed on the probability that $F(X)$ cover the true value of γ. Let $\mathfrak{F}$ be the family of all such confidence sets for γ. Similarly, let $\mathcal{C}$ be the family of all simultaneous confidence sets A for $\{\psi_i(\gamma) : i \in I\}$, where $A = \{A_i(x) : x \in \mathfrak{X}, i \in I\}$, is a family of subsets of Ψ. If $F_1, F_2 \in \mathfrak{F}$, then $F_1 \subset F_2$ shall mean $F_1(x) \subset F_2(x) \forall x \in \mathfrak{X}$, and if $A_1, A_2 \in \mathcal{C}$, then $A_1 \subset A_2$ shall stand for $A_{1i}(x) \subset A_{2i}(x) \forall x \in \mathfrak{X}$, $i \in I$.

For any $F \in \mathfrak{F}$ there is a natural corresponding $A \in \mathcal{C}$ defined by

$$A_i(x) = \psi_i F(x), \qquad x \in \mathfrak{X}, \quad i \in I. \tag{2.1}$$

Conversely, for any $A \in \mathcal{C}$ there is a natural corresponding $F \in \mathfrak{F}$ defined by

$$F(x) = \bigcap \{\psi_i^{-1} A_i(x) : i \in I\}, \qquad x \in \mathfrak{X}. \tag{2.2}$$

It is convenient to have symbols for the operations of passing from F to A and back. This leads to

Definition 2.1. If $F \in \mathfrak{F}$, then TF will mean the member $A \in \mathcal{C}$ defined by (2.1). If $A \in \mathcal{C}$, then $T^{-1}A$ will mean the member $F \in \mathfrak{F}$ defined by (2.2).

Note that neither $T^{-1}T$ nor TT^{-1} need be the identity. Obviously, $F_1 \subset F_2$ implies $TF_1 \subset TF_2$, and analogously for T^{-1}. The following relations hold between T and T^{-1}:

Lemma 2.2.

(i) $T^{-1}TF \supset F \quad \forall F \in \mathfrak{F};$

(ii) $TT^{-1}A \subset A \quad \forall A \in \mathcal{Q};$

(iii) $TT^{-1}T = T;$

(iv) $T^{-1}TT^{-1} = T^{-1}.$

Proof. Suppress the argument x. Then (i) follows from $\psi_i^{-1}\psi_i A_i \supset A_i \forall i \in I$. (ii) Put $F = T^{-1}A$ and fix $i \in I$. Then by (2.2), $F \subset \psi_i^{-1}A_i$, so $\psi_i F \subset \psi_i\psi_i^{-1}A_i = A_i$. (iii) Take any $F \in \mathfrak{F}$, then (i) implies $T(T^{-1}TF) \supset TF$, and (ii) implies $TT^{-1}(TF) \subset TF$. The proof of (iv) is entirely analogous.

Theorem 2.3.

(i) *If $F_0 \in \mathfrak{F}$ is a given confidence set for γ, then $F_1 = T^{-1}TF_0$ is the smallest exact confidence set for γ such that $F_0 \subset F_1$; $TF_0 = TF_1$ is the family of smallest simultaneous confidence sets for $\{\psi_i(\gamma)\}$ that are e.w.r.t. F_1.*

(ii) *If $B \in \mathcal{Q}$ and $F_1 \in \mathfrak{F}$, then B is e.w.r.t. F_1 if and only if $F_1 = T^{-1}B$; if F_1 is exact, then $A = TF_1$ is the family of smallest simultaneous confidence sets for $\{\psi_i(\gamma)\}$ that are e.w.r.t F_1.*

Proof. Part (ii) will be proved first. Suppress the argument x. Then $\{\gamma : \psi_i(\gamma) \in B_i \forall i \in I\} = \cap\{\psi_i^{-1}B_i : i \in I\} = T^{-1}B$ so that B is e.w.r.t. F_1 if and only if $F_1 = T^{-1}B$. Now let $F_1 \in \mathfrak{F}$ be exact, then by the part of (ii) just proved there exists $B \in \mathcal{Q}$ such that $F_1 = T^{-1}B$. Hence, with $A = TF_1$, we have the string of equalities $T^{-1}A = T^{-1}TF_1 = T^{-1}TT^{-1}B = T^{-1}B$ (using Lemma 2.2 (iv)) $= F_1$ which shows that A is e.w.r.t. F_1 by the first part of (ii). Furthermore, B is arbitrary subject to $T^{-1}B = F_1$, and $A = TT^{-1}B \subset B$ by Lemma 2.2 (ii) which shows that A is the smallest family e.w.r.t. F_1.

In order to prove part (i), first observe that $F_1 \supset F_0$ by Lemma 2.2 (i), and F_1 is exact by part (ii) since it is of the form $T^{-1}A, A \in \mathcal{Q}$. Now let $F_2 \in \mathfrak{F}$ be exact and $F_2 \supset F_0$. By part (ii), $F_2 = T^{-1}B$ for some $B \in \mathcal{Q}$. Then $TF_0 \subset TF_2 = TT^{-1}B \subset B$ by Lemma 2.2 (ii). Hence $F_1 = T^{-1}TF_0 \subset T^{-1}B = F_2$, showing that F_1 is the smallest exact confidence set containing F_0. Finally, $TF_0 = TF_1$ is a consequence of Lemma 2.2 (iii), and since F_1 is exact, the last part of (ii) shows that TF_1 constitutes the smallest confidence sets e.w.r.t. F_1.

Theorem 2.3 (ii) shows that $F \in \mathfrak{F}$ is exact if and only if $F = T^{-1}B$ for some $B \in \mathcal{Q}$. Using Lemma 2.2 (iv), F is of the form $T^{-1}B$ if and only if

$$T^{-1}TF = F \tag{2.3}$$

Property (2.3) of F may be termed *self-reproducing* relative to T (under the transformation $\mathfrak{F}\to\mathfrak{F}$ given by $F\to T^{-1}TF$). This will be abbreviated SR henceforth. Thus, $F\in\mathfrak{F}$ is exact if and only if F is SR.

In some of the applications each index i is really a double index, say (i,j), $i\in I$, $j\in J$, and each ψ_{ij} is a composition $\psi_{ij}(\gamma)=\lambda_j(\kappa_i(\gamma))$, where $\kappa_i:\Gamma\to K$ and $\lambda_j:K\to\Psi$. For instance, in the MANOVA problem in Section 4 the function $M\to a'Mb$ is the composition of $M\to a'M$ with $a'M\to a'Mb$. Besides the transformations T and T^{-1}, defined in terms of $\{\psi_{ij}\}$, we now also can define T_1 and T_1^{-1} in terms of $\{\kappa_i\}$, and T_2, T_2^{-1} in terms of $\{\lambda_j\}$. Since the exact confidence sets for γ are those that are SR, it is of considerable interest to find out under which conditions a confidence set for γ is SR relative to T if it is given to be SR relative to T_1. If these conditions are met, then most of the problem involving the ψ_{ij} is solved already by the problem involving only the κ_i.

Lemma 2.4. *Suppose that $F\in\mathfrak{F}$ SR relative to T_1 implies that $\kappa_i F$ is* SR *relative to T_2 $\forall i\in I$. Then $F\in\mathfrak{F}$ is* SR *relative to T if and only if it is* SR *relative to T_1.*

Proof. Let $F\in\mathfrak{F}$, then $T^{-1}TF=\bigcap_{i\in I}\bigcap_{j\in J}\kappa_i^{-1}\lambda_j^{-1}\lambda_j\kappa_iF=\bigcap_{i\in I}\kappa_i^{-1}\bigcap_{j\in J}\lambda_j^{-1}\lambda_j\kappa_iF=\bigcap_{i\in I}\kappa_i^{-1}T_2^{-1}T_2\kappa_iF\supset\bigcap_{i\in I}\kappa_i^{-1}\kappa_iF=T_1^{-1}T_1F\supset F$. If F is SR relative to T, then the extreme members in the string of equalities and inclusions are equal and it follows that $T_1^{-1}T_1F=F$. If F is SR relative to T_1, then the last inclusion is an equality, and by hypothesis so is the other inclusion, resulting in equality of the extreme members.

In some applications we may be concerned only with a particular F. The proof of Lemma 2.4 shows

Corollary 2.5. *If $F\in\mathfrak{F}$ is* SR *relative to T_1 and κ_iF is* SR *relative to $T_2\forall i\in I$, then F is* SR *relative to T.*

3. Invariance and pivotal quantities

If q is a function on $\mathfrak{X}\times\Gamma$ it is called a *pivotal quantity* for γ if the distribution of $q(X,\gamma(\theta))$ does not depend on $\theta\in\Theta$. A pivotal quantity is a very convenient device for constructing confidence sets for γ whose probability of covering the true value of $\gamma(\theta)$ is pre-assigned and free of θ.

Invariance considerations, if applicable, can often be used to great advantage to obtain pivotal quantities, as the next lemma shows. A group G is called a *group of invariance transformations*, or, simply, an *invariance group*, if it acts on $\mathfrak{X}$ and on Θ, if it transforms measurable subsets of $\mathfrak{X}$ into measurable sets, and if $P_{g\theta}(gA) = P_\theta A$ for every $g \in G$, $\theta \in \Theta$, and measurable set A [6]. The last condition can be stated, equivalently, $E_{g\theta}f(X) = E_\theta f(gX)$ for every real valued bounded measurable f. The two crucial conditions put on the invariance group G in the next lemma are, first, that G should act transitively on Θ, and, second, that an action of G on Γ can be defined that commutes with the function γ (i.e., γ is *equivariant*). As is well-known, the latter can be done if and only if $\gamma(\theta_1) = \gamma(\theta_2)$ implies $\gamma(g\theta_1) = \gamma(g\theta_2) \forall g \in G$, $\theta_1, \theta_2 \in \Theta$. If this condition is fulfilled, then for arbitrary $\gamma_0 \in \Gamma$ one can define $g\gamma_0 = \gamma_0(g\theta_0)$ by choosing any $\theta_0 \in \Theta$ such that $\gamma_0 = \gamma(\theta_0)$. We shall then simply say that G *acts on* Γ. The following lemma is essentially an extension of Theorem 3, Chapter 6, in [6].

Lemma 3.1. *Let G be an invariance group that acts on Γ and is transitive over Θ. Then any invariant function q on $\mathfrak{X} \times \Gamma$ is a pivotal quantity.*

Proof. Let f be a real valued bounded, measurable function on the range space of q. Let $\theta \in \Theta$ be arbitrarily chosen and fixed. For any $g \in G$ we have $E_{g\theta}(f(q(X, \gamma(g\theta))) = E_\theta f(q(gX, \gamma(g\theta)))$ since G is an invariance group. Then use $\gamma(g\theta) = g\gamma(\theta)$ since G acts on Γ, $q(gX, g\gamma(\theta)) = q(X, \gamma(\theta))$ since q is invariant, and finally the fact that G is transitive over Θ, to conclude that $E_\theta f(q(X, \gamma(\theta)))$ does not depend on $\theta \in \Theta$.

Remark. Possibly all known explicit pivotal quantities are obtainable through the use of Lemma 3.1. As an example consider the problem of estimating μ in a $N(\mu, \sigma^2)$ distribution from a sample with sample mean $\bar{x}$ and standard deviation s. The group of translations and multiplications satisfies Lemma 3.1 and leads to a maximal invariant $(\bar{x} - \mu)s^{-1}$, which has essentially a central t-distribution.

If g is a pivotal quantity for γ, then a $(1-\alpha)$-confidence set for γ can be constructed by choosing a measurable set Q_0 in the range space of q such that $P_\theta\{q(X, \gamma(\theta)) \in Q_0\} = 1 - \alpha$. Then let

$$F(X) = \{\gamma \in \Gamma : q(X, \gamma) \in Q_0\} \tag{3.1}$$

and it follows that

$$P_\theta\{\gamma(\theta) \in F(X)\} = 1 - \alpha, \qquad \theta \in \Theta, \tag{3.2}$$

i.e., F is a confidence set for γ with a constant probability of covering the true value of $\gamma(\theta)$. Note that if q is invariant, then F of (3.1) has the property

$$F(gX)=gF(X) \qquad \forall g\in G, \tag{3.3}$$

i.e., F is *equivariant*. Conversely, suppose that F satisfies (3.3), then F can be expressed in the form (3.1) with q invariant by taking $q(X,\gamma)=1$ if $\gamma\in F(X)$, $q=0$ otherwise, and $Q_0=\{1\}$. By Lemma 3.1, q is a pivotal quantity. Thus, $F\in\mathfrak{F}$ is expressible in terms of an invariant pivotal quantity if and only if F is equivariant.

Now suppose q is an invariant pivotal quantity and $F_0(X)$ is defined by the right hand side of (3.1) so that F_0 is equivariant. Starting from this F_0, Theorem 2.3 (i) can be used to construct simultaneous confidence sets TF_0 for $\{\psi_i : i\in I\}$ and the corresponding $F_1=T^{-1}TF_0$ with respect to which the family TF_0 is exact. There is a priori no guarantee that F_1 can be expressed in terms of a pivotal quantity. Yet, it is of great interest to be able to do so, since then also the left hand side of (3.2), with $F=F_1$, would have a constant value, say $1-\alpha_1$, with $\alpha_1\leqslant\alpha$, independent of θ. What we may hope for is that F_1 is equivariant whenever F_0 is, but whether this is true seems to depend on the nature of $\{\psi_i\}$. Below we state a sufficient condition on $\{\psi_i\}$, satisfied in all applications in Section 4, under which the desired result holds.

Lemma 3.2. *Let the conditions of Lemma* 3.1 *be met and let* $\{\psi_i : i\in I\}$ *have the property that an action of G on I can be defined such that for every* $i\in I$, $g\in G$, γ_1, $\gamma_2\in\Gamma$,

$$\psi_i(\gamma_1)=\psi_i(\gamma_2)\Rightarrow\psi_{gi}(g\gamma_1)=\psi_{gi}(g\gamma_2). \tag{3.4}$$

Then F_0 *equivariant implies* $F_1=T^{-1}TF_0$ *equivariant.*

Proof. From (3.4) it is easily established that for any $\gamma_1\in\Gamma$, $\psi_{gi}^{-1}\psi_{gi}(g\gamma_1)=g\psi_i^{-1}\psi_i(\gamma_1)$, and the same is therefore true if γ_1 is replaced by a subset of Γ. We have therefore $F_1(gX)=\bigcap_i\psi_i^{-1}\psi_iF_0(gX)=\bigcap_i\psi_{gi}^{-1}\psi_{gi}F_0(gX)=\bigcap_i\psi_{gi}^{-1}\psi_{gi}gF_0(X)=\bigcap_i g\psi_i^{-1}\psi_iF_0(X)=gF_1(X)$.

Condition (3.4) makes it possible to define, for each $g\in G$, a function $\Psi_i\to\Psi_{gi}$ where $\Psi_i=\psi_i\Gamma$: define $g\psi_i(\gamma)$ to be $\psi_{gi}(g\gamma)$. Then the confidence sets $A_i(X)=\psi_iF_0(X)$ for the $\psi_i(\gamma)$, with F_0 equivariant, are seen to transform according to

$$A_{gi}(gX) = gA_i(X). \tag{3.5}$$

This will be the case in all applications in Section 4.

4. Applications to some multivariate problems

The following notation will be used. Euclidean n-space is denoted $\mathbf{R}^n$, and $\mathbf{R}^n_{0+}$ is the 'ordered positive cone' of $\mathbf{R}^n$, consisting of those $x \in \mathbf{R}^n$ for which $x_1 > \cdots > x_n > 0$. $\mathbf{E}^n = \{x \in \mathbf{R}^n : \|x\| = 1\}$, and members of $\mathbf{E}^n$ will usually be written $\mathbf{e}$. In particular, $\mathbf{e}_i$ is the vector with 1 in the ith place, 0 elsewhere. $M(m,n)$ is the set of all $m \times n$ real matrices, and $M^r(m,n)$ is the set of all such matrices of rank $\leqslant r$. The set of all $n \times n$ nonnegative definite matrices is denoted NND(n), and the $n \times n$ identity matrix by I_n or by I. GL(n) is the group of all $n \times n$ real nonsingular matrices, O(n) the group of all $n \times n$ orthogonal matrices, and LT(n) the group of all $n \times n$ lower triangular matrices with positive diagonal elements. The vector of ordered characteristic roots of an $n \times n$ matrix B will be denoted $\lambda(B)$, with $\lambda_1(B) \geqslant \cdots \geqslant \lambda_n(B)$. Symmetric gauge functions [12] will be denoted φ, its polar by φ°, and the φ-norm of a matrix by $\| \ \|_\varphi$. The set of all φ that on $\mathbf{R}^n_{0+}$ depend only on the first r arguments is denoted Φ_r.

In all applications that follow Lemma 3.2 is applicable. However, this lemma will not explicitly be used to obtain the results, and the action of G on I will be indicated only in the first application in Subsection 4.1.

4.1. *MANOVA under full group*

This problem was treated in some detail in [12] from a slightly different point of view. Let $X : q \times p$, $Y : q_2 \times p$, and $Z : q_3 \times p$ be random matrices whose rows are independent p-variate normal with common nonsingular covariance matrix Σ, and mean matrices $\mathbf{E}X = M$, $\mathbf{E}Y = M_Y$, and $\mathbf{E}Z = 0$. Thus, one can take $\theta = (M, M_Y, \Sigma)$. Let the parameter of interest $\gamma(\theta)$ be M. The following transformations together constitute an invariance group G, which acts on Γ and is transitive over Θ: G_1: $X \to X + A$, $M \to M + A$, $A \in M(q,p)$; G_2: $Y \to Y + B$, $M_Y \to M_Y + B$, $B \in M(q_2,p)$; G_3: $X \to U_1 X$, $M \to U_1 M$, $U_1 \in \mathrm{O}(q)$; G_4: $Z \to U_2 Z$, $U_2 \in \mathrm{O}(q_3)$; G_5: $[X', Y', Z', M', M_Y'] \to C[X', Y', Z', M', M_Y']$, $\Sigma \to C \Sigma C'$, $C \in \mathrm{GL}(p)$. Actions that are trivial have been omitted from the list. Lemma 3.1 applies, and a maximal invariant on

$\mathfrak{X} \times \Gamma$ is $\lambda(WW')$, where

$$W = (M - X)S^{-\frac{1}{2}} \tag{4.1.1}$$

and $S = Z'Z$. Thus, $\lambda(WW')$ is an invariant pivotal quantity which, with probability one, takes its values in $\Lambda = \mathbf{R}^s_{o+} \cup \{0\}$ where $s = \min(p,q)$.

Any confidence set F for M corresponds, according to (4.1.1), to a set $F^* = (F - X)S^{-1/2}$ for W. Similar correspondences exist for various parametric functions. Thus, to a confidence set A_a for $a'M$ corresponds the set $A_a^* = (A_a - a'X)S^{-1/2}$ for $a'W$; to a set for Mb corresponds one for Wh, where $h = S^{1/2}b$; and to a set for $\operatorname{tr} NM$ corresponds one for $\operatorname{tr} S^{1/2}NW$. It is clear that for any of these families of functions, $F_1 = T^{-1}TF_0$ if and only if $F_1^* = T^{-1}TF_0^*$. All questions concerning exact confidence sets can therefore be investigated by considering only the sets F^*, etc., and it will be more convenient to do so. It will also be convenient to extend the terminology 'confidence set' and 'equivariance' to the sets F^*, etc. For notational ease the asterisk will henceforth be omitted.

Any equivariant F_0 is of the form

$$F_0 = \{W : \lambda(WW') \in Q_0\} \tag{4.1.2}$$

in which Q_0 is some measurable subset of Λ. Let $\lambda_0 = \sup\{\lambda_1 : \lambda \in Q_0\}$, i.e., the supremum of the maximum characteristic root of WW' as W runs through F_0. We may and shall assume that the supremum is a maximum, if necessary enlarging Q_0 by a nullset.

Consider first simultaneous confidence sets for all $a'W$, $a \in R^q$. Thus, the index set I is identified with $\mathbf{R}^q$. The action of G on I will be defined as follows: under G_3, $a \to U_1 a$; the actions of the other subgroups are trivial. Then (3.4) can be verified to hold so that starting from F_0 of (4.1.2), $F_1 = T^{-1}TF_0$ is necessarily equivariant by Lemma 3.2. The result of the operations T and $T^{-1}T$ depend on whether $p < q$ or $p \geqslant q$.

Case 1. $p < q$. Working forward with the operation T, starting from F_0, one finds

$$\begin{aligned} A_a &= \{a'W : W \in F_0\} \\ &= \{a'W : \|a'W\|^2 \leqslant \|a\|^2 \lambda_0\} \quad a \in R^q. \end{aligned} \tag{4.1.3}$$

Then going back with T^{-1}:

$$F_1=\{W:\lambda_1(WW')\leqslant\lambda_0\}. \tag{4.1.4}$$

Thus, the only exact equivariant confidence sets for W are of the form (4.1.4), i.e., derived from Roy's maximum root test. The simultaneous confidence sets for all $a'M$ corresponding to (4.1.3) are given in [12].

Case 2. $p\geqslant q$. The situation is more complicated here than in Case 1. Define

$$J=\bigcup\{[\lambda_q,\lambda_1]:\lambda\in Q_0\}, \tag{4.1.5}$$

then $J-\{0\}$ is the union of nondegenerate intervals (since $\lambda_q<\lambda_1$ for every $\lambda\in\Lambda-\{0\}$) so that J can be written as a countable union of disjoint intervals, say $J=J_1\cup J_2\cup\cdots$. Further, define

$$Q_1=\{\lambda\in\Lambda:[\lambda_q,\lambda_1]\subset J\}, \tag{4.1.6}$$

i.e., $\lambda\in Q_1$ if and only if λ_q and λ_1 belong to the same J_i, for some i. Then one finds

$$A_a=\{a'W:\|a'W\|^2\in\|a\|^2J\} \tag{4.1.7}$$

and

$$F_1=\{W:\lambda(WW')\in Q_1\}. \tag{4.1.8}$$

The results in Case 1 are also of this form, with J of the special form $[0,\lambda_0]$. In Case 2 the equivariant confidence sets for W have to be of the form (4.1.8), with Q_1 defined in (4.1.6). There is complete freedom in the choice of J, but inspection of (4.1.7) reveals that the confidence sets for the $a'M$ have undesirable properties unless $J=[0,\lambda_0]$ is chosen in which case (4.1.7) and (4.1.8) reduce to (4.1.3) and (4.1.4), respectively. Simultaneous confidence sets for the functions Wh, $h\in R^p$, can be handled analogously.

Next, consider the functions $a'Wh$, $a\in R^q$, $h\in R^p$. If $p<q$, identify $a'W$ with $\kappa_i(\gamma)$ of Lemma 2.4 and $a'Wh$ with λ_j $(\kappa_i(\gamma))$. It can easily be established that the condition of Lemma 2.4 holds. Starting with F_0 of (4.1.2), since $T_1^{-1}T_1F_0=F_1$ of (4.1.4), the conclusion of Lemma 2.4 implies that $T^{-1}TF_0$ is that same F_1. Hence, the only exact confidence sets for W with respect to all functions $a'Wh$ are those given by (4.1.4) and the

corresponding simultaneous confidence sets for the $a'Wh$ are the intervals

$$A_{a,h}=\{a'Wh:|a'Wh|\leqslant \|a\|\,\|h\|\lambda_0^{1/2}\}. \tag{4.1.9}$$

Substitution of (4.1.1) and $h=S^{1/2}b$ into (4.1.9) translates into the simultaneous confidence intervals for $a'Mb$ of [11]. The same conclusion is reached if $p>q$ by identifying the functions Wh with the $\kappa_i(\gamma)$ of Lemma 2.4. The case $p=q$ has to be treated directly, rather than by use of Lemma 2.4, but leads to the same conclusion.

Lastly, consider the functions $\operatorname{tr} NW$, $N\in M^r(p,q)$, with $1\leqslant r\leqslant s=\min(p,q)$. Starting from F_0 of (4.1.2) one finds that there is a symmetric gauge function $\varphi\in\Phi_r$, depending on Q_0, such that

$$F_1=\{W:\|W\|_\varphi\leqslant 1\} \tag{4.1.10}$$

and

$$A_N=\{\operatorname{tr} NW:|\operatorname{tr} NW|\leqslant \|N\|_{\varphi^0}\}. \tag{4.1.11}$$

For details see the proof of Theorem 4.2 in [12]. Thus, each exact F is determined by a $\varphi\in\Phi_r$, and conversely, each $\varphi\in\Phi_r$ determines an exact F by (4.1.10). For $r=s$ the simultaneous confidence intervals for all $\operatorname{tr} NM$, corresponding to (4.1.11) were derived by Mudholkar (1966). For $r=1$ the intervals (4.1.9) of Roy and Bose re-emerge.

4.2 *Regression matrix*

Let Y be $(p+q)$-variate $N(\mu,\Sigma)$ (Σ nonsingular) and let $\bar{Y}$ and S be the sample mean and covariance matrix in a sample of size $>p+q$. Let $Y^{(1)}$ be the first p and $Y^{(2)}$ the last q variates and partition Σ and S according to $Y^{(1)}$ and $Y^{(2)}$. Then $\theta=(\mu,\Sigma)$, or, equivalently, $\theta=(\mu,\Sigma_{11\cdot 2},\Sigma_{22},\boldsymbol{\beta})$, where $\Sigma_{11\cdot 2}=\Sigma_{11}-\Sigma_{12}\Sigma_{22}^{-1}\Sigma_{21}$ and $\boldsymbol{\beta}=\Sigma_{12}\Sigma_{22}^{-1}$ is the regression matrix of $Y^{(1)}$ on $Y^{(2)}$. Let $B=S_{12}S_{22}^{-1}$ be its sample counterpart. Take $\boldsymbol{\beta}$ as the parameter γ of interest. An invariance group G that acts on Γ and is transitive over Θ is generated by the following transformations of Y: G_1: $Y\to Y+a$, $a\in R^{p+q}$; G_2: $Y^{(1)}\to Y^{(1)}+CY^{(2)}$, $C\in M(p,q)$; G_3: $Y^{(1)}\to CY^{(1)}$, $C\in \mathrm{GL}(p)$; G_4: $Y^{(2)}\to CY^{(2)}$, $C\in\mathrm{GL}(q)$. Then the action of G on Γ is as follows: G_1 acts trivially; G_2: $\boldsymbol{\beta}\to\boldsymbol{\beta}+C$; G_3: $\boldsymbol{\beta}\to C\boldsymbol{\beta}$; G_4: $\boldsymbol{\beta}\to\boldsymbol{\beta}C^{-1}$. A maximal invariant on $\mathfrak{X}\times\Gamma$ is $\boldsymbol{\lambda}(WW')$, in which

$$W=S_{11\cdot 2}^{-1/2}(\boldsymbol{\beta}-B)S_{22}^{1/2} \tag{4.2.1}$$

Thus, $\lambda(WW')$ is an invariant pivotal quantity. Parametric functions of β, such as $a'\beta, \beta b, a'\beta b$, and $\operatorname{tr} N\beta$, translate into analogous functions of W. Since the pivotal quantity here is exactly the same function of W as it is in Subsection 4.1, all results from that subsection carry over (note, however, that p and q have been interchanged). In particular, simultaneous confidence intervals for all $\operatorname{tr} N\beta$ were derived by Mudholkar (1966).

4.3. *Covariance matrix.*

Let Y be p-variate $N(\mu, \Sigma)$ (Σ nonsingular) and let $\bar{Y}, S$ be the sample mean and covariance matrix in a sample of size $>p$ from this distribution. Let Σ be the parameter γ of interest. The transformations $Y \to Y+a$, $a \in \mathbf{R}^p$, and $Y \to CY$, $C \in \mathrm{GL}(p)$, generate an invariance group G that acts on Γ and is transitive over Θ. A maximal invariant on $\mathcal{X} \times \Gamma$ is $\lambda(\Sigma S^{-1})$. Let Q_0 be as in Subsection 4.1 and take

$$F_0 = \{\Sigma : \lambda(\Sigma S^{-1}) \in Q_0\}. \tag{4.3.1}$$

Suppose one is interested in simultaneous confidence sets for all $a'\Sigma a$, $a \in \mathbf{R}^p$. With the same J as in Subsection 4.1 and Q_1 defined in (4.1.6) one computes from (4.3.1)

$$F_1 = \{\Sigma : \lambda(\Sigma S^{-1}) \in Q_1\} \tag{4.3.2}$$

and the confidence sets for the $a'\Sigma a$ are

$$A_a = a' SaJ. \tag{4.3.3}$$

All equivariant simultaneous confidence sets for $a'\Sigma a$ are necessarily of the form (4.3.3) and they are e.w.r.t. the equivariant confidence set F_1 of (4.3.2). The set J may be any countable union of disjoint nondegenerate intervals, but the sets (4.3.3) presumably will be unacceptable to most statisticians unless J is chosen as an interval that contains the number 1.

4.4. *MANOVA under triangular group (stepdown procedure)*

In the problem of Subsection 4.1 suppose that in the subgroup G_5 the matrices C are restricted to $\mathrm{LT}(p)$, for instance because then the Hunt–Stein theorem [6] is applicable, or because one wants the confidence sets to be related to a step-down test [10]. Note that this smaller G is still transitive over Θ. Simultaneous confidence intervals for all $a'Mb$ were given by Mudholkar and Subbaiah [8]. Using the general methods of

Sections 2 and 3 it will be shown that one arrives at these intervals in a natural way. At the same time, it is found that these intervals are not the only equivariant simultaneous confidence sets. We shall also look at simultaneous confidence sets for all Mb. It is to be expected that there is much more leeway in the choice of equivariant confidence sets for M as a result of the choice of G, which is smaller here than in Subsection 4.1.

Employing the same notation as in Subsection 4.1, let $S=LL'$, $L\in LT(p)$, be the Cholesky decomposition of S and put

$$Z=(M-X)L'^{-1}, \tag{4.4.1}$$

then a maximal invariant on $\mathcal{X}\times\Gamma$ is $Z'Z$. This is therefore an invariant pivotal quantity. Study of confidence sets for M and functions of it will be done in terms of Z and functions of it, in the same way as this was done in Subsections 4.1 and 4.2.

Any equivariant F_0 is of the form

$$F_0=\{Z:Z'Z\in Q_0\} \tag{4.4.2}$$

in which Q_0 is an arbitrary measurable subset of NND(p). A natural choice for F_0 is the set derived from a stepdown procedure (see [8]). Write $Z=[z_1,\dots,z_p]$, $Z_i=[z_1,\dots,z_i]$, $i=1,\dots,p$, and $Z_0=0$. Then a stepdown F_0 has the form

$$F_0=\left\{Z:z_i'(I_q+Z_{i-1}Z_{i-1}')^{-1}z_i\leqslant c_i, i=1,\dots,p\right\} \tag{4.4.3}$$

in which the c_i are arbitrarily chosen positive numbers in terms of which the level of significance α of the corresponding step-down test is easily calculated.

First consider the functions Zh, $h\in R^p$. These correspond to the parametric functions Mb, with $h=L'b$. Let T be the corresponding transformation of Section 2. It will be shown in Lemma 4.4.3 that with F_0 of (4.4.3), $F_1=T^{-1}TF_0$ is given by

$$F_1=\left\{Z:\|z_i\|^2\leqslant c_i^*, i=1,\dots,p\right\}, \tag{4.4.4}$$

in which

$$c_1^*=c_1,\quad c_i^*=c_i\left(1+\sum_1^{i-1}c_j^*\right),\qquad i=2,\dots,p \tag{4.4.5}$$

(the notation is taken from [8]). For convenience put $d_i = (c_i^*)^{1/2}$, then the sets $A_h = \{Zh : Z \in F_1\}$ are easily seen to be

$$A_h = \left\{ Zh : \|Zh\| \leqslant \sum |h_i| d_i \right\}. \tag{4.4.6}$$

Via (4.4.1) this translates into simultaneous confidence sets for all Mb:

$$\left\{ Mb : \|Mb - Xb\| \leqslant \sum |(L'b)_i| d_i \right\}, b \in R^p. \tag{4.4.7}$$

Next consider the functions $a'Zh$. Associate $\kappa_i(\gamma)$ of Corollary 2.5 with Zh and take F in that corollary to be our present F_1 given by (4.4.4). It is easily established that every $\kappa_i F$, i.e., every A_h of (4.4.6), is SR relative to T_2. The conclusion of Corollary 2.5 then implies that F_1 is also exact with respect to the family $a'Zh$. Using (4.4.6), the corresponding confidence sets for the $a'Zh$ are the intervals

$$A_{a,h} = \left\{ a'Zh : |a'Zh| \leqslant \|a\| \sum |h_i| d_i \right\} \tag{4.4.8}$$

which translate into the confidence intervals (2.9) for $a'Mb$ in [8].

In the next three lemmas the notion of SR will be understood relative to T defined by the family of functions Zh, $h \in R^p$.

Lemma 4.4.1. *Let $Q \subset NND(p)$ and $F = \{Z \in M(q,p) : Z'Z \in Q\}$. Then F is* SR *if and only if for every $Z \in M(q,p)$:*

$$h'Z'Zh \in h'Qh \, \forall h \in R^p \Rightarrow Z \in F. \tag{4.4.9}$$

Proof. For $h \in R^p$ let $A_h = \{Zh : Z \in F\} = \{v \in R^q : \|v\|^2 \in h'Qh\}$. Then $T^{-1}TF = \{Z : Zh \in A_h \forall h \in R^p\} = \{Z : h'Z'Zh \in h'Qh \, \forall h \in R^p\}$ so that $T^{-1}TF = F$ if and only if (4.4.9) holds.

Lemma 4.4.2. *For $i = 1, \ldots, p$ let J_i be an arbitrary subset of* $[0, \infty)$. *If*

$$F = \left\{ Z = [z_1, \ldots, z_p] : \|z_i\| \in J_i, i = 1, \ldots, p \right\}, \tag{4.4.10}$$

then F is SR.

Proof. Let $Q = \{B \subset NND(p) : b_{ii}^{\frac{1}{2}} \in J_i,\ i = 1, \ldots, p\}$, then $F = \{Z : Z'Z \in Q\}$. Now suppose $Z = [z_1, \ldots, z_p]$ satisfies the left hand side of the implication (4.4.9). In particular, taking $h = \mathbf{e}_i$ one finds $\|z_i\| \in J_i$, $i = 1, \ldots, p$, so that $Z \in F$. Since therefore (4.4.9) is satisfied, F is SR by the conclusion of Lemma 4.4.1.

Lemma 4.4.3. *Let* F_0 *and* F_1 *be given by* (4.4.3) *and* (4.4.4), *respectively. Then* $T^{-1}TF_0 = F_1$.

Proof. The set F_1 of (4.4.4) is of the form (4.4.10), with $J_i = [0, d_i]$, so that by Lemma 4.4.2 F_1 is SR. Since $F_1 \supset F_0$ (see [8, Lemma 2.1]) and since $T^{-1}TF_0$ is the smallest SR set containing F_0 (Theorem 2.3(i)), it follows that $T^{-1}TF_0 \subset F_1$. It remains to be shown that $T^{-1}TF_0 \supset F_1$ and for this it suffices to exhibit a subset of F_0, say F_2, such that $T^{-1}TF_2 \supset F_1$. Take $F_2 = \cup_{j=1}^{p} F_{2j}$, in which F_{2j} consists of all $Z = [z_1, \ldots, z_p]$ with $z_i = \pm d_i e$, $i = 1, \ldots, j-1$, $z_j = \eta d_j \mathbf{e}$, $z_{j+1} = \cdots = z_p = 0$, $\mathbf{e} \in \mathbf{E}^q$, $|\eta| \leqslant 1$. Then $F_2 \subset F_0$, and $A_h = \{Zh : Z \in F_2\} = \cup_{j=1}^{p} \{Ah : Z \in F_{2j}\}$ is computed to be the same as A_h defined in (4.4.6). From this it follows that $T^{-1}TF_2 = F_1$.

Let α_0 be the level of significance that goes with F_0 of (4.4.3). Since $F_1 = T^{-1}TF_0$ is also equivariant, it, too, has a constant level, say α_1, independent of θ. Since F_0 is properly contained in F_1, it follows that $\alpha_1 < \alpha_0$. In principle, the constants c_i^* in (4.4.4) could be determined to produce a pre-chosen α_1. Admittedly, this is much harder than choosing the c_i in (4.4.3) to produce a given α_0.

It follows from Lemma 4.4.2 that F_1 of (4.4.4) is not the only type of SR set relative to the transformation T defined by the functions Zh, $h \in R^p$. Hence, the simultaneous confidence sets for the Mb given by (4.4.7) are not the only equivariant ones. It is of course possible that additional desirable conditions on the confidence sets for the Mb may limit them to (4.4.7). A characterization of SR sets F is provided by Lemma 4.4.1, but a convenient description of all such F is lacking. That there are SR sets F other than those given by (4.4.10) is exemplified by $F = \{Z = [\mathbf{e}, \ldots, \mathbf{e}] : \mathbf{e} \in \mathbf{E}^q\}$, which can be checked to satisfy (4.4.9).

Consider now the functions $a'Mb$, or, equivalently, the functions $a'Zh$, and the corresponding operation T. It could be hoped that now the only SR sets F are of the form (4.4.4), but a counter example shows that to be false. Hence the intervals (4.4.8) are not the only equivariant confidence sets for the $a'Zh$. In general the sets $A_{a,h}$ will be intervals symmetric with respect to 0, but their endpoints may depend in a more complicated way on h than those of (4.4.8). In addition, the confidence set $T^{-1}TF_0$ with respect to which the $A_{a,h}$ are exact may be intractable.

5. Limitations and prospects

The method of Section 2 is generally applicable, but presumably it is most useful in situations where the structure of Section 3 prevails so that an invariant pivotal quantity can be constructed with the help of Lemma 3.1. Here is an example where that method fails. Referring to Section 4.2, if γ stands for the vector of canonical correlations, then G does not act on Γ. Hence we are unable to find an invariant pivotal quantity for the canonical correlations. On the other hand, the method can be expected to work well in various other problems in multivariate analysis such as in comparing two covariance matrices, in simultaneous confidence sets for all AMB in MANOVA where A and B are matrices [1], and in generalized MANOVA [9].

References

[1] Gabriel, K. R. (1968). Simultaneous test procedures in multivariate analysis of variance. *Biometrika* **55**, 489–504.

[2] Gabriel, K. R. (1969). Simultaneous test procedures—some theory of multiple comparisons. *Ann. Math. Statist.* **40**, 224–250.

[3] Jensen, D. R. and Mayer, L. S. (1977). Some variational results and their applications in multiple inference. *Ann. Statist.* **5**, 922–931.

[4] Khatri, C. G. (1967). On certain inequalities for normal distributions and their applications to simultaneous confidence bounds. *Ann. Math. Statist.* **38**, 1853–1867.

[5] Krishnaiah, P. R. (1969). Simultaneous test procedures under general MANOVA models. *Multivariate Analysis II* (P. R. Krishnaiah, Ed.). Academic Press, New York.

[6] Lehmann, E. L. (1959). *Testing Statistical Hypotheses*. Wiley, New York.

[7] Mudholkar, Govind S. (1966). On confidence bounds associated with multivariate analysis of variance and non-independence between two sets of variates. *Ann. Math. Statist.* **37**, 1736–1746.

[8] Mudholkar, Govind S. and Subbaiah, Perla (1975). A note on MANOVA multiple comparisons based upon step-down procedure. Sankhyā B **37**, 300–307.

[9] Potthoff, R. F. and Roy, S. N. (1964). A generalized multivariate analysis of variance model useful especially for growth curve problems. *Biometrika* **51**, 313–326.

[10] Roy, J. (1958). Step-down procedure in multivariate analysis. *Ann. Math. Statist.* **29**, 1177–1187.

[11] Roy, S. N. and Bose, R. C. (1953). Simultaneous confidence interval estimation. *Ann. Math. Statist.* **24**, 513–536.

[12] Wijsman, R. A. (1979). Constructing all smallest simultaneous confidence sets in a given class, with applications to MANOVA. To appear in *Ann. Statist.*

PART VII

APPLICATIONS

P. R. Krishnaiah, ed., *Multivariate Analysis–V*

MULTIDIMENSIONAL SCALING WITH RESTRICTIONS ON THE CONFIGURATION

Jan DE LEEUW and Willem HEISER
Department of Data Theory, University of Leiden, Netherlands

A convergent algorithm model for multidimensional scaling with restrictions is described. The algorithm model is applied to accelerating MDS iterations, to fitting models similar to the ones used in the analysis of covariance structures, and to fitting individual differences models.

1. Introduction

It is difficult to define the precise methodological status of multidimensional scaling (MDS) techniques, and it is equally difficult to evaluate their usefulness. There are many applications of MDS, with results that are apparently regarded as useful by the people who have used the technique. Most of them, of course, have been in the social and behavioural sciences. Bibliographies covering most of these applications have been compiled by Bick et al [3] and by Nishisato [35]. Recently MDS has also been applied, apparently with success, in geography (Colledge and Rushton [11]), cartography (Gilbert [22]), genetics (Lalouel [32]), archeology (Kendall [27]), and biochemistry (Crippen [13, 14]).

If we study these applications carefully, it turns out that there are at least four different types. The first possibility is that a Euclidean spatial structure is known to exist, but is (partially) undetermined. We want to recover the structure 'optimally' by using MDS on the inexact or incomplete distance measurements. This type of application does not seem to occur in the behavioural sciences, but an example is Crippen's method for the conformation of molecules (Crippen [13, 14]). The second type of application arises if we know that a particular spatial structure exists (which is not necessarily Euclidean), and we want to compare an 'optimal' Euclidean map, based on dissimilarity data, with this known structure. In the behavioural sciences applications of type II occur in multidimensional psychophysics, more particularly in acoustics (an early example is Levelt et al. [33]). Most of the applications of MDS in geography and cartography are of this type. An impressive example is the 'maps from marriages' paper by Kendall (28). The third type of application starts with a theory about dissimilarity data, which implies that they are related in a well-defined way to Euclidean distances. We want to test the theory and perhaps find the 'hidden structure'. Applications of this type occur in mathematical psychology. According to Shepard [41, p. 374–375] this is the type of application he had in mind when he started developing MDS as we know it now. The developments in measurement theory (Beals et al. [1]) and in statistical methods for MDS (Ramsay [37, 38]) are mainly relevant for applications of this type. The problems of archeological seriation (Kendall [27], Wilkinson [45]), and of genetic mapping of linkage groups (Lalouel [32]) are also of type III.

For the fourth type of application there is no known a priori structure, and there is no theory which even tells us that a spatial representation is appropriate. We merely want to find an 'optimal' spatial representation of our dissimilarity data, and we apply MDS because 'a picture is worth a thousand numbers'. More sophisticated arguments for this approach are based on invariance (we get the same picture from a wide variety of dissimilarity data), stability (the spatial representation has greater statistical reliability than the individual dissimilarity estimates), or communicability (cf. Shepard [41, p. 375–376]). Approximately 90% of the applications in the behavioural sciences have been of this fourth type. Measurement theory and traditional statistical theory are not very relevant here, because they require the formulation of a definite algebraic or stochastic model. MDS, as used in type IV, is a sophisticated technique for making plots. It is obvious that the four types of applications differ because they require different levels of external information (or theory). Type I requires more theory than type II, type II more than type III, and type III more than type IV, which seems to require no theory at all. In all cases, however, the external information is used only *after* the MDS analysis has been completed. It is used in the interpretation of the results, to relate them to the already existing body of theory. It is not incorporated directly into the analysis, because most current MDS programs do not have the capability to incorporate restrictions.

If we analyze the type IV applications in detail, however, it turns out that even there some kind of external information often is available, which is again used in the interpretation phase. The theory is, of course, much less specific than the theory used in type I applications. As a consequence it is often difficult to relate MDS results to external information, and the interpretations can easily become far-fetched. In type I or type II applications the existing theory is so specific that interpretation of MDS results is unnecessary, they either make sense or they do not make sense. In type IV applications there is so much freedom that it is almost always possible to make some sort of sense out of MDS results. It follows from this discussion that it would be desirable to have MDS algorithms, which can incorporate all kinds of external information in the form of restrictions on the configuration. Such algorithms could be used in type I or type II situations, but there they would probably give the same results as unrestricted MDS methods. They will be especially useful in type III and type IV situations, in fact they tend to make type III applications out of type IV applications. We need 'confirmatory' MDS if we want to use our prior information efficiently.

2. Related work

For the very same reasons we discussed in the previous section a number of special methods have been proposed in the recent MDS literature which impose various kinds of restrictions on the solutions. In order to compare them with the method we are going to propose we need some terminology and notation. We follow Kruskal [31] as closely as possible. The data of a classical MDS problem are collected in a matrix $\Delta=\{\delta_{ij}\}$ of *dissimilarity measurements* between n objects. We want to find a *configuration matrix* X, of order $n\times p$, in such a way that $d_{ij}(X)$ is approximately equal to δ_{ij}, for all $i,j=1,\ldots,n$. Here $d_{ij}(X)$ is the Euclidean distance between rows x_i and x_j, interpreted as *points* in p-dimensional space. Thus

$$d_{ij}^2(X)=(x_i-x_j)'(x_i-x_j).$$

'Approximately equal to' is usually defined in terms of some real-valued *loss function*, which measures departure from perfect fit. If this is the case, then the MDS problem can be formulated in a more specific way: we want to find X in such a way that the loss function is minimized.

A more general problem occurs if we have m dissimilarity matrices Δ_k. We then want to find configuration matrices X_k, in such a way that $d_{ij}(X_k)$ is approximately equal to δ_{ijk} for all $i,j=1,\ldots,n$ and for all $k=1,\ldots,m$. If the X_k are otherwise unrestricted the problem is equivalent to m separate classical scaling problems, and consequently not of independent interest. It is therefore not surprising that the problem of imposing restrictions on MDS configurations started in the area of *individual differences scaling*.

The key paper in this area is Carroll and Chang [7]. They discuss restrictions of the form $X_k = YW_k$, with $Y \sqcup n \times p$, and $W_k \sqcup p \times p$ and diagonal. Moreover they present an elegant algorithm for fitting this model. The same model and algorithm were proposed, independently and in a different context, by Harshman [25]. We shall refer to this model as the INDSCAL model, using the name of the Carroll–Chang computer program. In the same paper Carroll and Chang also discuss (briefly) the more general model $X_k = YC_k$, with Y as before, and with C_k again $p \times p$, but not necessarily diagonal. This is the IDIOSCAL model, named after another Carroll-Chang computer program. In [26] Harshman proposes the PARAFAC-2 model, which is $X_k = YW_kZ'$, with Y as before, with W_k diagonal, and with Z another $p \times p$ matrix. We shall simply call this PARAFAC because in our context PARAFAC-1 is the same thing as INDSCAL. A general discussion of the algebraic properties of these models, their interrelationships, and of the corresponding algorithms, is given in Carroll and Wish [8], Kroonenberg and DeLeeuw [29], and De Leeuw and Pruzansky [20].

Existing MDS algorithms can be divided into two classes. In the first place there are the gradient algorithms. They include MDSCAL, KYST, SSA, and MINISSA. For MDSCAL and KYST we refer to Kruskal [30, 31], for SSA and MINISSA to Lingoes and Roskam [34]. And in the second place there are the alternating least squares algorithms. These methods divide the parameters of the problem into subsets, in such a way that minimizing the least squares loss function over each subset, with the other subsets fixed, is a comparatively simple problem. Kruskal [31, p. 309] calls such subproblems, and the parameter subsets which define them, 'nice'. The alternating least squares algorithms then cycles through its nice subproblems until convergence. Examples are INDSCAL and ALSCAL of Takane, Young, and De Leeuw [43].

The advantages of gradient methods in the MDS context have been reviewed quite thoroughly by Kruskal [31]. We mention two problems with the gradient method. The first one is that $d_{ij}(X)$ is not differentiable everywhere, if $x_i = x_j$ the partials do not exist. The second problem is the choice of step-size procedure in gradient methods. Kruskal [31, p. 315–319] has developed a step-size procedure based on some heuristic ideas, combined with a great deal of numerical experimentation. The method (also used in MINISSA) seems to work quite well in practice, but does not guarantee monotone convergence of loss function values. Because the step-size is a complicated function of all previous gradients and function values, it seems extremely difficult to give a precise analysis of convergence behaviour of the procedure.

The disadvantages of ALS (i.e. alternating least squares) are of a completely different nature. ALS procedures always start by transforming the model (Kruskal [31, p. 309] would say 'by neglecting errors'), either to squared distances (as in ALSCAL) or to inner products (as in INDSCAL). The resulting loss functions are differentiable everywhere, and by definition ALS gives monotone convergence. The main problem with ALS is that transforming the model may be undesirable. In the first place the transformations only make sense in the case of Euclidean distances, and do not generalize to other scaling models. In the second place the transformations may affect the errors in undesirable ways. If large dissimilarities have the largest measurement or sampling errors, for example, then it is statistically unwise to apply unweighted least squares to the squared dissimilarities. In fact Ramsay [37, 38] suggests applying unweighted least squares to the logarithms of the dissimilarities, because this makes more sense from the error theoretical point of view. Of course ALS cannot be applied if we use logarithms. If the dissimilarity measurements are independent, then the double-centering transformation that converts to scalar products in the error-free case introduces dependencies. Again this makes the use of unweighted least squares problematical.

There is a third type of algorithm, which does not use transformations, uses weighted least squares to incorporate possible error-theories, leads to monotone convergence without step-size choices, and has no problems with differentiability. The method was originally derived in the unweighted case by Guttman [24]. He assumed differentiability, however, and interpreted the method as a gradient method with constant step-size. He did not show that the method worked, but empirical studies of Lingoes and Roskam [34] indicated that it seemed to converge in practice. De Leeuw [17] studied the method in the weighted least squares case, generalized it to nonmetric scaling, discussed non-Euclidean models, and gave the first formal convergence proof. An alternative derivation will be presented in the next section of this paper.

The forms of restricted scaling that have been proposed in the literature strongly depend on the type of algorithm that is used. Some constraints fit nicely into a gradient framework, others can easily be combined with ALS. In gradient algorithms, for example, it is very simple to fix some of the coordinates of the configuration at constant values. In fact this possibility is already built into MINISSA. Bentler and Weeks [2] generalized this considerably. The coordinates x_{is} can be fixed, free, or proportional. By proportional we mean that the restrictions are of the form $x_{is} = u_{is}x_{jt}$, with the u_{is} known, and with x_{jt} a free coordinate. Bloxom [4] fits the general individual differences model $X_k = Y_k C_k$. The parameters in Y_k and C_k can be either fixed or free, and some free parameters can be restricted to be equal to others. This clearly has INDSCAL and IDIOSCAL as special cases. It seems to us that Bloxom's work cannot be generalized very much any more. More general restrictions can only be fitted into a gradient framework by using the much more complicated feasible directions method of nonlinear programming.

Alternating least squares algorithms usually solve subproblems which are linear regression problems. This is not true in all cases, the ALSCAL subproblems for example amount to solving cubic equations, but generally nice sets of variables enter linearly into the model equations. This implies that ALS is especially valuable in fitting multilinear models, and that the natural constraints that can be used with ALS are linear constraints. The combination of multilinear models (principal component analysis, three-mode component analysis, transformed Euclidean scaling) with linear constraints has recently been studied by Carroll, Green, and Carmone [9] and by Carroll and Pruzansky [10]. They show that incorporating linear restrictions is possible by transforming the input matrix, and by applying the CANDECOMP algorithm of Carroll and Chang [7] to this transformed matrix. In the MDS context the constraints imposed by Carroll et al are of the type $X_k = YCW_k$, with Y known, C and W_k unknown, and W_k diagonal for each k. More general constraints are possible, but these seem to be the most natural ones.

In the next section we show that our new algorithm combines the most convenient features of gradient and ALS methods. We also show that it can incorporate a very general class of constraints in a natural way.

3. Algorithm model

We want to minimize

$$\sigma(X) = \sum_{i=1}^{n} \sum_{j=1}^{n} w_{ij}(\delta_{ij} - d_{ij}(X))^2,$$

over all X in Ω, a given subset of R^{np}, the space of all $n \times p$ matrices. The

w_{ij} are given non-negative weights, if $w_{ij}=0$ we treat δ_{ij} as missing. Without loss of generality we can assume that the dissimilarities are normalized by

$$\sum_{i=1}^{n} \sum_{j=1}^{n} w_{ij}\delta_{ij}^2 = 1.$$

Then

$$\sigma(X) = 1 - 2\rho(X) + \eta^2(X),$$

where

$$\rho(X) = \sum_{i=1}^{n} \sum_{j=1}^{n} w_{ij}\delta_{ij}d_{ij}(X), \quad \text{and} \quad \eta^2(X) = \sum_{i=1}^{n} \sum_{j=1}^{n} w_{ij}d_{ij}^2(X).$$

The function η^2 is quadratic in X. If we define the matrix $V=\{v_{ij}\}$ by

$$v_{ij} = -(w_{ij}+w_{ji}) \quad \text{for } i \neq j, \qquad v_{ii} = \sum_{j\neq i}^{n} (w_{ij}+w_{ji}),$$

then

$$\eta^2(X) = \operatorname{tr} X'VX.$$

If $w_{ij}=1$ for all $i \neq j$, then $V=2(nI-ee')$. If X is centered, i.e. if the centroid of the n points x_i is the origin, then in this case $\eta^2(X)=2n \operatorname{tr} X'X$. In the general weighted case we assume that V is irreducible [17, p. 138], which implies that, if V^- is the Moore–Penrose inverse of V, then $X=VV^-X=V^-VX$ for all centered X. The function ρ causes the trouble with differentiability in gradient methods. Consequently we use another regularity property. Because the Euclidean distance is a convex and positively homogeneous function, the same thing is true for ρ. Observe that this result is true for much more general definitions of d_{ij}, in fact it remains true for $d_{ij}(X)=\mu(x_i-x_j)$, with μ any gauge. By using this fact we can extend at least some of our results to general Minkovski geometries, in which distance is not necessarily symmetric. For convex functions the notion of a gradient is replaced by that of a subgradient. The subgradient $\partial\rho(X)$ of ρ at X is a non-empty, convex, compact set, defined by $Y \in \partial\rho(X)$ if $\rho(Z) \geq \rho(X) + \operatorname{tr} Y'(Z-X)$ for all Z in R^{np}. If ρ is differentiable at X, with gradient $\nabla\rho(X)$, then the subgradient is the singleton $\{\nabla\rho(X)\}$. For proofs of these results we refer to Rockafellar [39, part V].

Theorem 1. *For all X, Y in R^{np} and for all $Z \in V^{-}\partial\rho(Y)$ we have*

$$\sigma(X) \leqslant 1 - \eta^2(Z) + \eta^2(X - Z).$$

For all X in R^{np} and for all $Z \in V^{-}\partial\rho(X)$ we have

$$\sigma(X) = 1 - \eta^2(Z) + \eta^2(X - Z).$$

Proof. Suppose $U \in \partial\rho(X)$. By definition $\rho(Z) \geqslant \rho(X) + \operatorname{tr} U'(Z - X)$ for all Z in R^{np}, and consequently also for all Z of the form αX, with $\alpha \geqslant 0$. Using the homogeneity of ρ gives $(1-\alpha)\rho(X) \leqslant (1-\alpha)\operatorname{tr} U'X$ for all $\alpha \geqslant 0$, which implies that $\rho(X) = \operatorname{tr} U'X$. Substitute this again in the definition of the subgradient. This gives $\rho(Z) \geqslant \operatorname{tr} U'Z$. If we substitute the last equation and inequality in the formula $\sigma(X) = 1 - 2\rho(X) + \eta^2(X)$, and simplify, we obtain the results stated in the theorem.

In the same way we can prove that $Z \in \partial\rho(X)$ implies that Z is centered. By definition we must have $\rho(X + e\mu') \geqslant \rho(X) + \operatorname{tr} Z'e\mu'$ for all p-vectors μ, i.e. for all translations. Because the distances are translation invariant this simplifies to $e'Z\mu \leqslant 0$ for all μ, which implies $e'Z = 0$.

Theorem 2. *If $Z \in V^{-}\partial\rho(X)$, then $\eta^2(Z) \leqslant 1$.*

Proof. Because the dissimilarities are normalized applying Cauchy-Schwartz to the definition of ρ gives $\rho(X) \leqslant \eta(X)$ for all X in R^{np}. The second part of Theorem 1 gives $\rho(X) = \operatorname{tr} Z'VX \leqslant \eta(Z)\eta(X)$. The first part gives $\rho(Z) \geqslant \operatorname{tr} Z'VZ = \eta^2(Z)$. If we combine the three inequalities, we obtain the chain

$$\frac{\rho(X)}{\eta(X)} \leqslant \eta(Z) \leqslant \frac{\rho(Z)}{\eta(Z)} \leqslant 1.$$

This is more than enough to prove the theorem.

We also define the metric projection P_Ω, in the metric defined by V, as

$$P_\Omega(X) = \left\{ Y \in \Omega \,|\, \eta^2(Y - X) = \min_{Z \in \Omega} \eta^2(Z - X) \right\}.$$

We assume that Ω is defined in such a way that $P_\Omega(X)$ is non-empty for all X in R^{np}. In the terminology used in nonlinear approximation theory we assume that Ω is *proximinal.* It is sufficient for proximinality that Ω is either compact or closed and convex, but these conditions are by no means necessary. Using the metric projection we can now describe our basic algorithm model. We start with some X^0 in Ω, and in each iteration we find the update X^+ of our current solution by the rule

$$X^+ \in P_\Omega(V^- \partial\rho(X)).$$

There is one exception to this rule. If our current solution satisfies $X \in P_\Omega(V^- \partial\rho(X))$, then the algorithm stops. We call this an algorithm model because it is incomplete. We have not specified how to select an element from the subgradient, and we have not specified how we intend to solve the metric projection problem. It is obvious that the practical simplicity of our algorithm depends on the ease with which the metric projection problem can be solved. To use the terminology of Kruskal once again, some projection problems are nice while others are not so nice. We now prove a first convergence theorem for the case in which the algorithm generates an infinite sequence X^k, with corresponding loss function values $\sigma(X^k)$.

Theorem 3. *$\sigma(X^k)$ is a decreasing sequence, and consequently converges.*

Proof. Suppose $\bar{X} \in V^- \partial\rho(X)$. By part 1 of Theorem 1

$$\sigma(X^+) \leqslant 1 - \eta^2(\bar{X}) + \eta^2(X^+ - \bar{X}).$$

By assumption $X \notin P_\Omega(V^- \partial\rho(X))$, and thus, by the definition of the metric projection,

$$1 - \eta^2(\bar{X}) + \eta^2(X^+ - \bar{X}) < 1 - \eta^2(\bar{X}) + \eta^2(X - \bar{X}).$$

By part 2 of Theorem 1:

$$1 - \eta^2(\bar{X}) + \eta^2(X - \bar{X}) = \sigma(X).$$

If we combine the results we find $\sigma(X^+) < \sigma(X)$, and because the sequence is bounded below by zero it converges.

Theorem 3 is reassuring, but not entirely satisfactory because it tells us nothing about the behaviour of the sequence X^k. Before we study this in more detail we prove a simple necessary condition for a minimum of σ.

Theorem 4. *If X minimizes σ on Ω, then $X \in P_\Omega(V^- \partial\rho(X))$.*

Proof. Suppose $X \notin P_\Omega(V^- \partial\rho(X))$ and $Y \in P_\Omega(V^- \partial\rho(X))$. By exactly the same argument that proved Theorem 3 we show that $\sigma(Y) < \sigma(X)$.

From now on we call points satisfying the necessary condition of theorem 4 *desirable*. We already have the trivial result that if the algorithm stops, it stops at a desirable point. What happens if it does not stop? To study this we need some preliminary results.

Theorem 5. *All X^k are in the compact set $\{X \mid \eta(X) \leq 1 + \sqrt{\sigma(X^0)}\}$.*

Proof. From the proof of Theorem 3, using the same notation,

$$1 - \eta^2(\bar{X}) + \eta^2(X^+ - \bar{X}) < \sigma(X) \leq \sigma(X^0).$$

By using Theorem 2

$$\eta^2(X^+ - \bar{X}) \leq \sigma(X^0).$$

But, by Cauchy–Schwartz,

$$(\eta(X^+) - \eta(\bar{X}))^2 \leq \eta^2(X^+ - \bar{X}),$$

which implies, by using Theorem 2 again, that

$$\eta(X^+) \leq 1 + \sqrt{\sigma(X^0)}.$$

Theorem 6. *The point-to-set map $X \to P_\Omega(V^- \partial\rho(X))$ is closed.*

Proof. The subgradient map is closed (Rockafellar [39, p. 233]). The set of all $V^- \partial\rho(X)$, as X varies over R^{np}, is compact by Theorem 2. The map P_Ω is closed, and consequently the composition is closed (Zangwill [46]).

Theorem 7. *The sequence X^k has convergent subsequences. Each subsequential limit is a desirable point. All subsequential limits have the same value of σ.*

Proof. This follows from convergence Theorem A of Zangwill [46, p. 91].

The type of convergence described in Theorem 7 is still quite weak. We can prove a considerably stronger result by making the additional assumption that Ω is convex.

Theorem 8. *If Ω is convex, then $\eta^2(X^{k+1}-X^k)$ converges to zero.*

Proof. It follows from the proof of Theorem 3 that, if $\overline{X}\in V^-\partial\rho(X)$, $\eta^2(X-\overline{X})-\eta^2(X^+-\overline{X})=\eta^2(X^+-X)+2\,\mathrm{tr}(X^+-\overline{X})'V(X-X^+)$ converges to zero. If Ω is convex then, by convex approximation theory, the last term on the right is nonnegative. The first term on the right is positive, and thus both terms converge to zero.

The theorem does not say that the sequence X^k converges. It tells us, by a familiar theorem of Ostrowski, that either the sequence X^k converges or that the sequence has a continuum of accumulation points. Convergence follows if we make additional uniqueness or smoothness assumptions. From a practical point of view, however, the result of Theorem 8 is strong enough. If we define an ε-desirable point as a configuration X for which the distance between X and the set $P_\Omega(V^-\partial\rho(X))$ is less than ε, then the theorem tells us that the algorithm finds an ε-desirable point in a finite number of steps, no matter how small we choose the positive number ε. We also observe that the assumption that Ω is convex is not necessary for $\eta^2(X^{k+1}-X^k)\to 0$. It is often possible to prove this result from other uniqueness or smoothness assumptions.

There is a generalization of the algorithm which can be extremely important in those cases in which the metric projection problem is not nice. Suppose Q is a closed point-to-set map from $R^{np}\times\Omega$ into subsets of Ω. Suppose moreover that $Y\in Q(Z,X)$ implies that $\eta^2(Y-Z)\leqslant\eta^2(X-Z)$, with equality if and only if $X\in P_\Omega(Z)$. The new algorithm is $X^+\in Q(V^-\partial\rho(X),X)$. Convergence Theorems 3 and 7 remain true for this modified algorithm, which does not solve the metric projection problem completely, but merely takes a step in the right direction. In many examples the map Q can be defined by using alternating least squares.

There are some obvious relationships of our algorithm with alternating least squares theory. In each iteration we have to solve a least squares projection problem, if this subproblem is not easy to solve we can often use one or more ALS cycles to improve our current best solution. One interpretation of our algorithm is that the basic majorization result proved in theorem 1 makes it possible to apply the powerful machinery of ALS, without first having to transform to squared distances or inner products. There are also some relationships with gradient theory. Suppose σ is differentiable at X. Then $\nabla\sigma(X)=2VX-2\nabla\rho(X)$, and thus the algorithm can be written as a gradient projection method of the form $X^+\in P_\Omega(X-\frac{1}{2}V^-\nabla\sigma(X))$. If we want to relate our algorithm to the earlier work of Guttman [24] we first have to specify how we select an element from the subgradient of ρ. Define the matrix $B(X)$ by

$$b_{ij}(X)=-(w_{ij}\delta_{ij}+w_{ji}\delta_{ji})s_{ij}(X) \quad \text{if } i\neq j,$$

with $s_{ij}(X)=1/d_{ij}(X)$ if $d_{ij}(X)\neq 0$, and $s_{ij}(X)=0$ otherwise. Moreover

$$b_{ii}(X)=-\sum_{j\neq i}^{n} b_{ij}(X).$$

By simple algebra we find $\rho(X)=\operatorname{tr} X'B(X)X$, and from the Cauchy–Schwarz inequality $B(X)X\in\partial\rho(X)$. Thus one possible specification of our algorithm is $X^+\in P_\Omega(V^-B(X)X)$. If ρ is differentiable at X, then $B(X)X=\nabla\rho(X)$. Guttman studied the case in which Ω is R^{np}, and in which the off-diagonal weights are all unity. In this case we find $X^+=(1/2n)B(X)X$, which is basically Guttman's correction matrix algorithm. Observe that the mapping B is not continuous at X if some of the $d_{ij}(X)$ are zero. This implies that the limits of convergent subsequences generated by the algorithm do not necessarily satisfy $X\in P_\Omega(V^-B(X)X)$, although they must always satisfy $X\in P_\Omega(V^-\partial\rho(X))$.

4. Unrestricted scaling

It is convenient to introduce the *Guttman-transform* $\overline{X}$ of a configuration matrix X as $\overline{X}=V^-B(X)X$. If Ω is R^{np}, then the basic algorithm is $X^+=\overline{X}$. The chain in the proof of Theorem 3 becomes

$$\sigma(X^+)\leqslant 1-\eta^2(\overline{X})+\eta^2(X^+-\overline{X})=1-\eta^2(\overline{X})$$

$$\leqslant 1-\eta^2(\overline{X})+\eta^2(X-\overline{X})=\sigma(X).$$

This immediately implies that $\eta^2(X^+-X)$ converges to zero. If $\sigma(X^k)$ decreases to σ_∞, then $\eta^2(X^k)$ increases to $\eta^2_\infty=1-\sigma_\infty$. Moreover from the proof of theorem 2 we find that $\rho(X^k)$ increases to $\rho_\infty = \eta^2_\infty$, and that $\lambda(X^k) = \rho(X^k)/\eta(X^k)$ increases to $\lambda_\infty = \eta_\infty$.

As an example of a more general algorithm using the map $Q(\overline{X}, X)$ we have $X^+ = \overline{X} + \alpha(X - \overline{X})$. We find $\sigma(X^+) \leqslant \sigma(X) + (\alpha^2 - 1)\eta^2(X - \overline{X})$, and thus we have convergence if $-1 < \alpha < +1$. Moreover $\eta^2(X^+ - X) = (1-\alpha)2\eta^2(X - \overline{X})$ also converges to zero in this case. The reasons for choosing $\alpha \neq 0$ are as follows. Very slow linear convergence is typical for MDS algorithms, and precise solutions are impossible if we do not apply some sort of acceleration device. A more precise analysis of the convergence rate of our MDS algorithms will be given elsewhere. For the moment we merely observe that using $\alpha = \varepsilon_0 - 1$, with ε_0 a very small positive number, preserves global convergence and approximately halves the number of iterations required to obtain a given precision, at no extra cost. The explanation is, roughly, as follows. The basic algorithm $X^+ = \overline{X}$ converges linearly with rate $\kappa_0 = 1 - \varepsilon_1$, with ε_1 another small positive number. If we use any $\alpha \neq 0$ we obtain the rate $\kappa_\alpha = \kappa_0 + \alpha(1 - \kappa_0)$, and if $\alpha = \varepsilon_0 - 1$ then this is equal to $(1-\varepsilon_1)^2 + \varepsilon_1(\varepsilon_0 - \varepsilon_1)$, which is approximately $(1-\varepsilon_1)^2$. Thus the convergence rate is squared, the number of iterations is halved.

A further generalization of our basic algorithm can be used to derive versions of the optimal gradient, memory gradient, and conjugate gradient methods. In the unweighted case we have at a point where σ is differentiable that $X^+ = \overline{X} + \alpha(X - \overline{X})$ can also be written as $X^+ = X - \mu\nabla\sigma(X)$ with $\mu = (1-\alpha)/4n$. If $\alpha = 0$ then $\mu = 1/4n$, if $\alpha = -1$ then $\mu = 1/2n$. We can also interpret $\sigma(X - \mu\nabla\sigma(X))$ as a function of μ and minimize it over μ with our algorithm. To see how this is done consider the more general problem of minimizing $\sigma(X)$ over all X of the form $Y_0 + \sum \mu_s Y_s$, where the Y_s, including Y_0, are fixed matrices. This is a problem with linear restrictions. The solution by projection can be computed from $\hat{\mu} = C^{-1}b$, where C contains the elements $\operatorname{tr} Y_s' V Y_t$, and b the elements $\operatorname{tr} Y_s' V(\overline{X} - Y_0)$. We then use $\hat{\mu}$ to compute X^+, compute its Guttman-transform, and repeat the procedure until convergence. Observe that by using a relaxation parameter again, we can also set $\hat{\mu} = 2C^{-1}b$, and still obtain convergence.

In Table 1 we have collected the results of a number of experiments with these step-size techniques. There are five different data sets.

$K1$: Nine Dutch political parties, data collected by Van der Kamp, initial configuration Kruskal's L-shape (cf. [30]).

$K2$: Same data, initial configuration Young–Householder–Torgerson.

$F1$: Thirteen ethnic groups, from Funk et al. [21], initial L.

$F2$: Same data, YHT initial configuration.

D: Ten points, $\delta_{ij} = 1$ for all $i \neq j$, initial L.

Table 1
Speed of convergence of various step-size procedures

Example	No it	loss	rate
$K1, \alpha=0$	99	0.0444477	0.945592
$K1, \alpha=-1$	57	0.0444380	0.888642
$K1, \alpha=\text{opt}$	47	0.0444365	0.855814
$K2, \alpha=0$	62	0.0446179	0.912394
$K2, \alpha=-1$	35	0.0446135	0.852053
$K2, \alpha=\text{opt}$	32	0.0446115	0.831023
$F1, \alpha=0$	201	0.0655642	0.940073
$F1, \alpha=-1$	111	0.0655561	0.876544
$F1, \alpha=\text{opt}$	107	0.0655556	0.872367
$F2, \alpha=0$	67	0.0602583	0.713715
$F2, \alpha=-1$	36	0.0602560	0.507205
$F2, \alpha=\text{opt}$	35	0.0602564	0.527138
$D, \alpha=0$	123	0.0111065	0.927797
$D, \alpha=-1$	72	0.0111058	0.853318
$D, \alpha=\text{opt}$	59	0.0111057	0.812373

All analyses were in two dimensions, and unweighted. We iterated until $\sigma-\sigma^{+}<10^{-6}$. For each example there were three runs, one with $\alpha=0$, one with $\alpha=-1$, and one which uses inner iterations for the optimal step until $|\alpha^{+}-\alpha|<10^{-3}$. The table lists the number of major iterations, the minimum value of the loss function, and an empirical estimate of the linear convergence rate $(\sigma^{+}-\sigma^{++})/(\sigma-\sigma^{+})$. The examples show that indeed $\alpha=-1$ approximately halves the number of iterations and approximately squares the convergence rate of $\alpha=0$. Using the optimal step-size gives a slightly better convergence rate, and fewer major iterations. But because inner iterations are about as expensive as major iterations, computing the optimal α is not worthwhile (the argument is the same as in [31, p. 315–316]). Even an 'ideal' step-size routine which finds the optimal α in a single inner iteration would be as expensive as using $\alpha=0$ throughout. The step-size routine we use needs on the average between five and fifteen inner iterations, which makes the complete algorithm about five times as expensive as the one with $\alpha=0$. Thus $\alpha=-1$ is by far the most efficient choice, and we have adopted this in our FORTRAN computer program SMACOF1 for metric MDS. The examples also illustrate the completely different point that using a good start can reduce the number of iterations considerably, and that different starts can lead to different local optima.

5. Linear restrictions

We have already analyzed one particular example of using linear restrictions on X in the previous section. A more interesting example from a practical point of view is $X=YC$, with Y a given $n\times q$ matrix, and C an unknown $q\times p$ matrix. This can also be fitted with the gradient method of Bloxom [4], and in an inner product framework by the ALS method of Carroll et al [9]. Our algorithm simply gives

$$X^+=YC^+=Y(Y'VY)^{-1}Y'V\bar{X}=Y(Y'VY)^{-1}Y'B(X)YC,$$

provided that the inverse exists. It is clear that we can require without loss of generality that $Y'VY=I$, which simplifies matters even more. In applications Y can be an ANOVA-type design matrix, it can also be a matrix with real valued measurements on q 'independent' variables. The last case occurs, for example, in multidimensional psychophysics. Here Y contains physical characteristics of the stimuli, such as frequency characteristics of synthetically generated vowels.

In many special cases C can be restricted to be diagonal. In this case the update is, supposing $y_s'Vy_s=1$ for all $s=1,\ldots,q$,

$$c_s^+=y_s'V\bar{x}_s=c_sy_s'B(X)y_s.$$

Observe that if $c_s\geqslant 0$, then $c_s^+\geqslant 0$. And if $c_s=0$ then $c_s^+=0$. The model with C diagonal can also be fitted with the gradient method of Bentler and Weeks (2). A particularly interesting class of examples has both a diagonal C and a binary Y. Here the *squared* distance has a set theoretical interpretation. If H_i is the set $\{s|y_{is}=1\}$, and the measure of a subset of $\{1,2,\ldots,q\}$ is defined as $m(H)=\sum\{c_s^2|s\in H\}$, then

$$d_{ij}^2(X)=m(H_i\Delta H_j),$$

with Δ the symmetric difference. Thus in this case the squared Euclidean distance is a metric too (cf. the discussion of the ADCLUS model by Shepard [41, p. 414–417]). In some cases a graph theoretical interpretation is possible. If we are fitting a simplex, for example, we use the matrix Y defined by $y_{is}=1$ if $i\geqslant s$, and $y_{is}=0$ otherwise. Call this matrix Y_S. In this case $d_{ij}^2(X)$ is the path length distance in a simple chain, i.e. $d_{ij}^2(X)$ can be represented as a Euclidean distance in one dimension, while $d_{ij}(X)$ itself needs $n-1$ dimensions. This was already observed by Guttman [23]. Another possibility is fitting a circumplex, with width (k_1,k_2), where

$1 \leqslant k_1 \leqslant k_2 \leqslant n$. Then $y_{is}=1$ if $k_1 \leqslant |i-s| \leqslant k_2$, and $y_{is}=0$ otherwise. We call a matrix of this class Y_C. This makes $d_{ij}^2(X)$ the path length distance in a circular graph, consisting of a single cycle. In the same way we can construct for each tree a binary matrix Y such that $d_{ij}^2(YC)$ is the path length distance in the tree [6]. In the particular case $Y=I$ the tree is a star. In factor analytic terminology $Y=I$ can also be interpreted as no common, only unique variance [2, p. 140].

The theory generalizes without any further complications to mixed models in which parts are unrestricted and other parts are restricted. A particularly attractive class of models is $X=|X_1|YC_1|C_2|$ in which X_1 is $n \times q$ and unrestricted, Y is either Y_S or one of the Y_C, and C_1 and C_2 are both square and diagonal. The models can be coded by a three digit number $q_1q_2q_3$. Here q_1 is the number of unrestricted dimensions, if $q_1=0$ the X_1 part is missing from the model. If $q_2=0$ the YC_1 part is missing, if $q_2=1$ then $Y=Y_S$, if $q_2=2$ then Y is one of the Y_C. If $q_3=0$ the C_2 part is missing, if $q_3=1$ it is not missing. Thus $q01$ is the q-dimensional MDS model with uniquenesses, also discussed by Bentler and Weeks [2], model 010 is a simplex, 011 is a quasi-simplex, and so on.

In many types of applications unrestricted MDS finds semi-circular, parabolic or horse-shoe shaped configurations in two dimensions. By using the simplex and circumplex we can often find more parsimonious representations. This is illustrated by the results in Table 2. We applied the computerprogram SMACOF2, which fits these three-digit models, to a number of dissimilarity matrices.

Table 2
Minimum loss for various models and various examples

Model	*K*	*S*	*L*	*C*	*DP*	*DK*	*DV*
100	0.4167	0.3493	0.3270				
200	0.2214	0.1370	0.1772	0.1075	0.1068	0.1404	0.0784
010	0.2016	0.1029	0.1990	0.0681	0.0836	0.1755	0.0864
020				0.2051	0.2391	0.0985	0.2211
001	0.1588	0.2125		0.3181			
101	0.0640	0.0430					
201	0.0543			0.0388			
011	0.1001	0.0513	0.1933	0.0561	0.0666	0.1314	0.0816
021				0.1245	0.1841	0.0829	0.2090
110	0.1466	0.0536	0.1644				
210	0.0848	0.0221		0.0535			
111		0.0185					

K: Nine Dutch political parties, data collected by Van der Kamp in 1968, averaged similarity ratings, subjects 100 students.
S: Six Swedish political parties, data taken from Sjöberg [42].
L: 15 musical intervals, taken from Levelt et al [33].
C: Nine number ability tests, data from Coombs [12].
DP: 13 Dutch political parties, data from De Gruyter [15]. Data averaged over students who vote PvdA (social democrat).
DK: Same data, but averaged over students who vote KVP (christian democrat).
DV: Same data, but averaged over students who vote VVD (liberal party).

Table 2 lists the minimum loss function values for various models. In this table we use the square root of the loss, which is more convenient. For *K*, *S*, and *C* the simplex 010 is considerably better than two-dimensional MDS 200, despite the fact that 010 has only n-1 parameters while 200 has $2n-3$ parameters. For *K* and *S* the quasi-simplex 011 is again considerably better, and the scaling models with uniqueness 101 and 201 also perform well. For *L* we would also prefer 010 to 200. In the data *DP*, *DK*, and *DV* we expect that people from the middle will think that extremist left and extremist right are quite alike, while people from the moderate left and moderate right will think this to a lesser degree. This would imply that the *DK*-data are more circumplex-like, while the *DP* and *DV*-data must be more simplex-like. We do find something like this in Table 2.

6. Nonlinear restrictions

There are many models possible which impose nonlinear restrictions. We only discuss some examples, but do not work them out in detail. In nonlinear problems computing the metric projection may not be simple. Thus we may have to use alternating least squares inner iterations.

For the first example we discuss inner iterations are unnecessary. The restrictions are $x_s \in K_s$, with K_s a given convex cone. The subproblem is to solve p cone projection problems, one for each dimension. If K_s is the cone of all vectors with a given ordering of the elements, for example, then the subproblems are monotone regression problems. This problem was also studied by Noma and Johnson [36], who used a completely different algorithm. More generally the requirements $x_s \in K_s$ can be used to scale dimensions which are defined nominally, ordinally, or numerically. These extensions of MDS are closely related to recent extensions of principal component analysis such as PRINCIPALS [44] or HOMALS [16].

In the second example (inspired by Borg [5, p. 638]) we want to restrict the representation of a subset of the objects in such a way that they are on a straight line in two-space, or, more generally, on a q-dimensional linear manifold in p-space. The corresponding projection problem is finding the best rank q approximation to a submatrix, which is easy to solve by computing the singular value decomposition. More generally it may be interesting to require that some or all of the objects are on a nonlinear manifold of some sort. In Levelt et al. [33], for example, it would make sense to require that the tone intervals are on a quadratic manifold, in colour studies we can require that the objects are on a circle. This generally requires more complicated iterative techniques than the linear case.

The third example we discuss is $X = YC$, with Y binary and C diagonal as before, but now with both Y and C unknown. This is the ADCLUS model of Arabee and Shepard (discussed in [41]). We have to use ALS here. Computing the optimal C for fixed Y has been explained in the previous section, and computing the optimal binary Y for fixed C is also quite simple. After one or more inner ALS cycles we compute a new Guttman-transform, and start a new major iteration. The major problem with the algorithm is not the amount of computation, but the fact that there are so many desirable points. This is due to the discreteness of the restrictions, and cannot be helped.

7. Individual differences

The theory of Section 13 generalizes in an obvious way to individual differences scaling. For each iteration we have to minimize a function of the form

$$\sum_{k=1}^{m} \operatorname{tr}\left(Z_k - \overline{X}_k\right)' V_k \left(Z_k - \overline{X}_k\right)$$

over the Z_k, where $\overline{X}_k$ are the Guttman-transforms of the X_k from the previous iteration, i.e. $\overline{X}_k = V_k^- B_k(X_k)X_k$. In the general weighted case we need ALS techniques to fit the individual differences models discussed in Section 2. This is true in particular if we also want to impose restrictions of the form discussed by Bloxom [4]. If the weights are all equal the computations simplify considerably. We discuss the three most important models in some detail, and show how the subproblems can be solved.

For IDIOSCAL we define the $n \times mp$ supermatrix $\bar{X}$, containing the current $\bar{X}_k$, and the $p \times mp$ supermatrix C, containing the C_k. The problem can be rewritten as minimization of $\operatorname{tr}(\bar{X} - YC)'(\bar{X} - YC)$, which means that we can set Y equal to the normalized eigenvectors corresponding with the p largest eigenvalues of $\bar{X}\bar{X}'$, and set $C_k = Y'\bar{X}_k$. For INDSCAL we can proceed by fitting one dimension at a time. Define the $n \times m$ matrices $\bar{X}_s$, which contain column s of each of the $\bar{X}_k$, and define the vectors c_s, which contain diagonal element s of each of the C_k. Then we must minimize, for each s separately, $\operatorname{tr}(\bar{X}_s - y_s c_s')'(\bar{X}_s - y_s c_s')$. Thus we can set y_s equal to the normalized eigenvector corresponding with the largest eigenvalue of $\bar{X}_s \bar{X}_s'$, and we set $c_s' = y_s' \bar{X}_s$.

For PARAFAC we have to minimize, in each subproblem,

$$\sum_{k=1}^{m} tr\left(\bar{X}_k - YC_k Z'\right)'\left(\bar{X}_k - YC_k Z'\right).$$

But this is exactly the problem solved by the CANDECOMP algorithm of Carroll and Chang [7]. From the general theory we know that it suffices to perform just one CANDECOMP cycle before computing new Guttman-transforms.

8. Nonmetric scaling

All the techniques discussed in this paper are for metric MDS, in which the dissimilarities are either completely known or completely unknown (if δ_{ij} is unknown we set $w_{ij} = 0$). If the dissimilarities are partially known, for example up to a linear transformation or up to a monotone transformation, then we have to apply nonmetric MDS. The loss is now a function of the configuration *and* of the *disparities*, which are admissible transformations of the dissimilarities. De Leeuw [17] shows that in the unrestricted case the basic algorithm $X^+ = V^- B(X)X$ is still convergent, provided we replace the normalized dissimilarities by the normalized disparities in each iteration. The disparities are computed, in the classical case, by monotone regression. This result remains true in the case in which there are restrictions on the configuration. De Leeuw and Heiser [19] study more general partitioned loss function for nonmetric MDS, and show that the basic algorithm can still be used, but that we must not only change the dissimilarities in each iteration but possibly also the weights. Again this result remains true in the restricted case. Consequently for nonmetric MDS with restrictions only the computation of the Guttman-transform becomes more complicated, the metric projection problem remains exactly the same.

References

[1] Beals, R., Krantz, D. H., and Tversky, A. (1968). Foundations of multidimensional scaling. *Psychol. Rev.* **75**, 127–142.

[2] Bentler, P. M., and Weeks, D. G. (1978). Restricted multidimensional scaling models. *J. Math. Psychol.* **17**, 138–151.

[3] Bick, W., Bauer, H., Mueller, P. J., and Gieseke, O. (1977). Multidimensional scaling and clustering techniques (theory and applications in the social sciences). A Bibliography. Institut für angewandte Sozialforschung. Universität zu Köln.

[4] Bloxom, B. (1978). Constrained multidimensional scaling in *N* spaces. *Psychometrika* **43** 397–408.

[5] Borg, I. (1977). Geometric representations of individual differences, In *Geometric Representations of Relational Data*. Mathesis Press. Ann Arbor.

[6] Bunemann, P. (1971). The recovery of trees from measures of dissimilarity. In *Mathematics in the archeological and historical sciences*. University of Edinburgh Press, Edinburgh.

[7] Carroll, J. D., and Chang, J. J. (1970). Analysis of individual differences in multidimensional scaling via an *N*-way generalization of "Eckart-Young" decomposition. *Psychometrika* **35**, 283–319.

[8] Carroll, J. D., and Wish, M. (1974). Models and methods for three-way multidimensional scaling. In *Contemporary Developments in Mathematical Psychology*. Freeman, San Francisco.

[9] Carroll, J. D., Green, P. E., and Carmone, F. J. (1976). CANDELINC (*CAN*onical *DE*composition with *LIN*ear *C*onstraints): a new method for multidimensional analysis with constrained solutions. Paper presented at *International Congress of Psychology*, Paris, France.

[10] Carroll, J. D., and Pruzansky, S. (1977). MULTILINC: MULTIWAY CANDELINC (*CAN*onical *DE*composition with *LIN*ear *C*onstraints). Paper presented at *American Psychological Association* meeting, San Francisco.

[11] Colledge, R. G., and Rushton, G. (1972). Multidimensional scaling: review and geographical applications. *Geographic technical papers series*, no. 10. Association of American geographers. Washington.

[12] Coombs, C. H. (1941). A factorial study of number ability. *Psychometrika* **6**, 161–189.

[13] Crippen, G. M. (1977). A novel approach to calculation of conformation: distance geometry. *J. Computational Physics* 24, 96–107.

[14] Crippen, G. M. (1978). Rapid calculation of coordinates from distance measures. *J. Computational Physics* **26**, 449–452.

[15] De Gruyter, D. N. M. (1967). The cognitive structure of Dutch political parties in 1967. *Report EO1-67*. Psychological Institute. University of Leiden. The Netherlands.

[16] De Leeuw, J. (1976). HOMALS. Paper presented at *Psychometric Society meeting*, Murray Hill, N. J.

[17] De Leeuw, J. (1977). Applications of convex analysis to multidimensional scaling. In

Progress in Statistics. North Holland Publishing Company, Amsterdam.

[18] De Leeuw, J., Young, F. W., and Takane, Y. (1976). Additive structure in qualitative data: an alternating least squares method with optimal scaling features. Psychometrika **41**, 471–503.

[19] De Leeuw, J., and Heiser, W. (1977). Convergence of correction matrix algorithms for multidimensional scaling. In *Geometric Representations of Relational Data*. Mathesis Press, Ann Arbor.

[20] De Leeuw, J., and Pruzansky, S. (1978). A new computational method to fit the weighted Euclidean distance model. *Psychometrika* **43** 479–490.

[21] Funk, S., Horowitz, A., and Young, F. W. (1976). The perceived structure of American ethnic groups: the use of multidimensional scaling in stereotype research. *Sociometry* 39, 116–130.

[22] Gilbert, E. N. (1974). Distortion in maps. *SIAM Review* **16**, 47–62.

[23] Guttman, L. (1955). An additive metric from all the principal components of a perfect scale. *British J. Math. Statist. Psychol.* **8**, 17–24.

[24] Guttman, L. (1968). A general nonmetric technique for finding the smallest coordinate space for a configuration of points. *Psychometrika* **33**, 469–506.

[25] Harshman, R. A. (1970). Foundations of the PARAFAC procedure: models and conditions for an explanatory multi-modal factor analysis. *Unpublished thesis*, UCLA.

[26] Harshman, R. A. (1972). PARAFAC2: mathematical and technical notes. *Working papers in phonetics*, no. 22, UCLA.

[27] Kendall, D. G. (1971). Construction of maps from "odd" bits of information. *Nature* **231**, 158–159.

[28] Kendall, D. G. (1971). Maps from marriages: an application of nonmetric multidimensional scaling to parish register data. In *Mathematics in the archeological and historical sciences*. University of Edinburgh Press, Edinburgh.

[29] Kroonenberg, P., and De Leeuw, J. (1977). TUCKALS2: a principal component analysis of three mode data. *RN 01-77*. Department of Data Theory. University of Leiden. The Netherlands.

[30] Kruskal, J. B. (1964). Nonmetric multidimensional scaling. *Psychometrika* **29**, 1–27 and 115–129.

[31] Kruskal, J. B. (1977). Multidimensional scaling and other methods for discovering structure. In *Mathematical methods for digital computers*, vol III, New York, Wiley.

[32] Lalouel, J. M. (1977). Linkage mapping from pairwise recombination data. *Heredity* **38**, 61–77.

[33] Levelt, W. J. M., Van de Geer, J. P., and Plomp, R. (1966). Triadic comparisons of musical intervals. *British J. Math. Statist. Psychol.* **19**, 163–179.

[34] Lingoes, J. C., and Roskam, E. E. (1973). A mathematical and empirical analysis of two mutlidimensional scaling algorithms, *Psychometrika*, **38**. Monograph supplement.

[35] Nishisato, S. (1978). Multidimensional scaling: a historical sketch and bibliography. Ontario institute for studies in education. Toronto.

[36] Noma, E., and Johnson, J. (1977). Constrained nonmetric multidimensional scaling configurations. *Tech. Rep.* **60**. Human Performance Center, University of Michigan. Ann Arbor.

[37] Ramsay, J. O. (1977). Maximum likelihood estimation in multidimensional scaling. *Psychometrika* **42**, 241–266.

[38] Ramsay, J. O. (1978). Confidence regions for multidimensional scaling analysis. *Psychometrika* **43**, 145–160.

[39] Rockafellar, R. T. (1970). *Convex Analysis*. Princeton University Press, Princeton.

[40] Schönemann, P. H. (1972). An algebraic solution for a class-of subjective metrics models. *Psychometrika* **37**, 441–451.

[41] Shepard, R. N. (1974). Representation of structure in similarity data: problems and prospects. *Psychometrika* **39**, 373–421.

[42] Sjöberg, L. (1975). Choice frequency and similarity. *Psych. Rep.* **23**, University of Göteborg. Sweden.

[43] Takane, Y., Young, F. W., and De Leeuw, J. (1977). Nonmetric individual differences multidimensional scaling: an alternating least squares method with optimal scaling features. *Psychometrika* **42**, 7–67.

[44] Takane, Y., Young, F. W., and De Leeuw, J. (1978). The principal components of mixed measurement level multivariate data: an alternating least squares method with optimal scaling features. *Psychometrika* **43**, 279–282.

[45] Wilkinson, E. M. (1970). Techniques of data analysis: seriation theory. Unpublished dissertation, Cambridge University, G. B.

[46] Zangwill, W. I. (1969). *Nonlinear Programming: a Unified Approach*. Prentice Hall, Englewood Cliffs, NJ.

P. R. Krishnaiah, ed., *Multivariate Analysis–V*

MULTISTATE RELIABILITY MODELS: A SURVEY

Emad EL-NEWEIHI* and Frank PROSCHAN**
The Florida State University, Department of Statistics, Tallahassee, FL 32306, U.S.A.

For a long time, the vast majority of the models in reliability theory have concentrated on the case in which both components and systems assume only two possible states: functioning and failed. Unfortunately, this represents a gross oversimplification of the many real life situations in which both components and systems actually assume a variety of states ranging from perfect operation to complete failure.

More recently, papers have appeared which treat the more sophisticated and more realistic situations in which components and systems may assume many states. In the present paper, a survey is made of earlier work, but more especially of some quite recent work (some completed, some still in progress) by the relatively small number of researchers active in this important new area of reliability. It is becoming apparent that this research will generate results not only of value in reliability applications, but also of independent interest in multivariate statistical analysis.

1. Introduction

The vast majority of reliability analyses assume that components and system are in either of the two states: functioning or failed. In many situations one is capable of distinguishing between various 'levels of performance' for both the system and its components. For such cases, the existing dichotomous model is a gross oversimplification of the real situation, whereas models representing multistate systems and components are much more suitable.

Until recently, very little work has been done on this more general problem of multistate systems. Some earlier work treated only very specialized aspects of multistate systems, but no comprehensive treatment of these models was available. Among the earlier papers are [11–14, 16 and 17]. However, there has been recently a growing interest in this important new area of reliability theory. More sophisticated and comprehensive work

*University of Kentucky. Research supported by the Air Force Office of Scientific Research under AFOSR Grant 77-3322.

**Florida State University. Research supported by the Air Force Office of Scientific Research under AFOSR Grant 74-2581D.

on multistate models has been performed by Barlow [2], El-Neweihi, Proschan and Sethuraman [8] and Ross [15]. In this expository paper a survey is made of the various treatments of multistate models. We briefly mention the earlier work, but we concentrate on the more recent and more comprehensive treatments of multistate models performed by the relatively small number of researchers active in this important area of reliability.

We now summarize the contents of this paper. Our terminology and notation are similar to that of Barlow and Proschan [1] for the two state case. In Section 2 we present the notation and terminology used throughout the paper. In Section 3 we consider a system of n components. For each component and for the system itself, we can distinguish among different 'levels of performance' represented by a state space S. For component i, x_i denotes the corresponding state, $i=1,\ldots,n$; the vector $\mathbf{x}=(x_1,\ldots,x_n)$ denotes the vector of states of components $1,\ldots,n$. The state of the system is assumed to be a deterministic function ϕ of the states of the components from S^n, the nth Cartesian power of S, into S. Thus $\phi(\mathbf{x})$ is the state of the system corresponding to the component state vector $\mathbf{x}$. We then survey the various choices of state space S, and various definitions of the structure function ϕ presented in different treatments of multistate models. We investigate structural properties of the various models, occasionally comparing and contrasting them.

In Section 4 we investigate probabilistic aspects of multistate models. We survey the relationship (in a probabilistic sense) between the performance of the system and the performances of its components. For instance, system performance is, as expected, a monotone increasing function of component performances. When the exact value of system performance is difficult to compute, bounds are provided.

In Section 5, we survey dynamic aspects of multistate systems. In earlier sections, it is tacitly assumed that time is fixed. In Section 5, multistate systems are viewed as operating over time. At time 0, the system and each of its components are at the maximal 'level of performance.' As time passes, the performance levels of components (and consequently of the system) deteriorate to successively lower levels until finally level 0 (complete failure) is reached. Concepts of IFRA and NBU stochastic processes, analogous to the corresponding lifelength distributions in the binary case are defined and studied by various researchers. Some generalized IFRA and NBU closure theorems are presented.

Finally in Section 6, we show by means of two examples how theories of multistate systems may be applied to existing binary reliability models.

2. Notation and terminology

The vector $\mathbf{x} \equiv (x_1,\ldots,x_n)$ denotes the vector of states of components $1,\ldots,n$.

$$C = \{1,2,\ldots,n\} \text{ denotes the set of component indices.}$$

$$(j_i,\mathbf{x}) \equiv (x_1,\ldots,x_{i-1},j,x_{i+1},\ldots,x_n), \quad \text{where } j=0,1,\ldots,M.$$

$$(\cdot_i,\mathbf{x}) \equiv (x_1,\ldots,x_{i-1},\cdot,x_{i+1},\ldots,x_n).$$

$$\mathbf{j} \equiv (j,\ldots,j).$$

$\mathbf{y} < \mathbf{x}$ means that $y_i \leqslant x_i$ for $i=1,\ldots,n$ and $y_i < x_i$ for some i. $\boldsymbol{\alpha} = (\alpha_0,\ldots,\alpha_M)$ is a *probability vector* means that $0 \leqslant \alpha_j \leqslant 1$, $j=0,1,\ldots,M$, and $\sum_{j=0}^{M} \alpha_j = 1$.

$\boldsymbol{\alpha} \leqslant_{\text{st}} \boldsymbol{\alpha}'$, where both $\boldsymbol{\alpha}, \boldsymbol{\alpha}'$ are probability vectors, means that $\sum_{K=j}^{M} \alpha_k \leqslant \sum_{k=j}^{M} \alpha'_k, j=0,1,\ldots,M$.

$$x \vee y \text{ denotes } \max(x,y).$$

$$\mathbf{x} \vee \mathbf{y} \equiv (x_1 \vee y_1,\ldots,x_n \vee y_n).$$

$$x \wedge y \text{ denotes } \min(x,y).$$

$$\mathbf{x} \wedge \mathbf{y} \equiv (x_1 \wedge y_1,\ldots,x_n \wedge y_n).$$

'Increasing' is used in place of 'nondecreasing' and 'decreasing' is used in place of 'nonincreasing'. When we say $f(x_1,\ldots,x_n)$ is *increasing* we mean f is increasing in each argument.

Given a univariate distribution F, its complement $1-F$ is denoted by $\bar{F}$. Given a set S, S^n denotes its nth Cartesian power.

3. Deterministic properties of multistate coherent systems

Consider a system of n components. We assume that the performance of the system depends deterministically on the performances of the components. Thus given $\mathbf{x}$, the vector of component states, we may determine $\phi(\mathbf{x})$, the system state. The function ϕ is called the *structure function* of the system. In the binary case, it is assumed that both components and system are in either of two states: functioning or failed. The variables $x_i, i=1,\ldots,n$, as well as $\phi(\mathbf{x})$, assume their values in the state space $S=\{0,1\}$, where 0 denotes failure and 1 denotes functioning. The structure function ϕ is then a map from $\{0,1\}^n$ into $\{0,1\}$. The structure function ϕ satisfies certain conditions that represent intuitively reasonable properties of systems encountered in practice. The following two conditions are required for a binary system to be a *coherent structure* [1, Def. 2.1, p. 6]:

(i) The function $\phi(\mathbf{x})$ is increasing.

(ii) Each component is *relevant* to the system, i.e., for each i there exists a vector $(\cdot_i, \mathbf{x})$ such that $\phi(1_i, \mathbf{x}) > \phi(0_i, \mathbf{x})$. This means that the function ϕ is not constant in its ith argument, $i=1,\ldots,n$.

Condition (i) embodies the reasonable assumption that improving the performance of a component is not harmful to system performance. Condition (ii) eliminates from consideration components which have no effect on system performance. The theory of binary coherent structures has served as a unifying foundation for a mathematical and statistical theory of reliability for the dichotomous case.

The binary model, however, is an oversimplification in describing a situation in which either the system or its components (or both) are capable of assuming a whole range of levels of performance, varying from perfect functioning to complete failure. In these situations, models representing *multistate* systems and multistate components are much more useful in describing system performance in terms of component performances. Naturally, the first step in constructing such models is to provide useful definitions of state spaces, representing the sets of levels of performance, and of the structure function ϕ, that relates the performance of the system to the performances of its components. A theory of multistate structures can then serve as a unifying foundation for a mathematical and statistical theory of reliability in the multistate case. Among the earlier attempts of this type, we mention the following two examples:

Hirsch et al. [11], in a treatment of 'cannibalization', consider a system of n components; the state of component i is represented by the binary variable x_i assuming the values 1 or 0 according to whether component i is functioning or failed, $i=1,\ldots,n$. However the system itself can be in any of $M+1$ states representing various levels of performance. The set of possible performance levels is assumed to be totally ordered and is then represented, without loss of generality, by $S=\{0,1,\ldots,M\}$, where 0 denotes complete failure and M denotes perfect performance. The structure function ϕ is a map from $\{0,1\}^n$ into S. Notice that S can have at most 2^n

elements. The structure function ϕ is assumed to be monotone increasing, with $\phi(\mathbf{0})=0$, $\phi(\mathbf{1})=M$.

The authors, however, do not attempt a general treatment of multistate models. Their main concern rather is to investigate the mathematical model for cannibalization to determine how components may be exchanged to improve system performance.

The following multistate model is presented by Postelnicu [14]: Consider a system of n components in which the state space for the system and for each of its components is the unit interval $[0,1]$, representing a continuous range of performance from perfect performance (1) to complete failure (0). The structure function ϕ: $[0,1]^n \to [0,1]$ satisfies the following conditions:

(a) $\phi(\mathbf{1})=1$.

(b) $\phi(\mathbf{0})=0$.

(c) $\phi(\mathbf{x})$ is monotone increasing.

(d) $\phi(\mathbf{c}) \geqslant c$, $0 \leqslant c \leqslant 1$.

(e) $\phi(c_i, \mathbf{0}) \leqslant c$, $0 \leqslant c \leqslant 1$.

The author does not attempt a comprehensive treatment of such multistate structures, but rather investigates some very special applications.

More recent and more comprehensive research in multistate systems has been performed by Barlow [2], El-Neweihi, Proschan and Sethuraman [8] (hereafter referred to as EPS [8]) and Ross [15]. The definition given by Barlow [2] for the multistate structure is set-theoretical, based on the concept of min path sets and min cut sets of binary coherent structures. Consider a system of n components. Assume that the state space for each of the components as well as for the system is the set $S=\{0,1,\ldots,M\}$, where 0 denotes the failed state and M denotes the maximal or perfect state. Let $P_1,\ldots,P_r$ be non-empty subsets of C such that $\bigcup_{j=1}^{r} P_j = C$ and $P_i \not\supset P_j$, $i \neq j$. The structure function ϕ: $S^n \to S$ is defined by

$$\phi(\mathbf{x}) = \max_{1 \leqslant j \leqslant r} \min_{i \in P_j} x_i, \tag{3.1}$$

where $\mathbf{x} \in S^n$ is the vector representing the states of components $1,2,\ldots,n$. In the binary case the structure function given in (3.1) is the most general coherent structure [1, Ch. 1], and the sets $P_1,\ldots,P_r$ are its min path sets. Let ϕ' be the binary coherent structure associated with $P_1,\ldots,P_r$. The multistate coherent structure ϕ specified in (3.1) can then be expressed in terms of the corresponding binary coherent structure ϕ' as follows: For each $i=1,\ldots,n$, let

$$y_{ij}=\begin{cases}1 & \text{if } x_i \geqslant j\\ 0 & \text{o.w.}\end{cases}$$

and let $\mathbf{y}_j=(y_{1j},\ldots,y_{nj})$, $j=0,1,\ldots,M$. It is fairly easy to see that $\phi(\mathbf{x})\geqslant j$ iff $\phi'(\mathbf{y}_j)=1$, and

$$\phi(\mathbf{x})=\sum_{j=1}^{M}\phi'(\mathbf{y}_j). \tag{3.2}$$

Thus the multistate coherent structure given by Barlow [2], is very closely related to a corresponding binary coherent structure. Exploiting this relationship makes it easy to extend results from the binary case to the multistate case.

A more general approach has been taken by EPS [8] to define multistate coherent structures. The common state space for each of the components and for the system is again taken to be the set $S=\{0,\ldots,M\}$, representing the $M+1$ levels of performance ranging from complete failure (0) to perfect functioning (M). The structure function ϕ: $S^n\rightarrow S$ is assumed to satisfy three conditions.

Definition 3.1. A system of n components is said to be a *multistate coherent system* (MCS) if its structure function ϕ satisfies:

(i)′ ϕ is increasing;

(ii)′ for level j and component i, there exists a vector $(\cdot_i,\mathbf{x})$ such that $\phi(j_i,\mathbf{x})=j$ while $\phi(l_i,\mathbf{x})\neq j$ for $l\neq j$, $i=1,\ldots,n$, and $j=0,1,\ldots,M$.

(iii)′ $\phi(\mathbf{j})=j$ for $j=0,1,\ldots,M$.

The three axioms embodied in Definition 3.1 extend the standard notion of a binary coherent system to the new notion of a multistate coherent system. Note that conditions (i)′ and (ii)′ generalize conditions (i) and (ii) in the binary case. Condition (iii)′ is automatically satisfied in the binary case, but is not implied in the present multistate case by (i)′ and (ii)′. Also note that since the structure function in (3.1), defined by Barlow [2], satisfy conditions (i)′, (ii)′, and (iii)′ of Definition 3.1, they constitute a subclass of the MCS class. For instance, for a two component system, only two distinct systems satisfy (3.1); namely, the parallel and the series system, regardless of the cardinality of S. However, for $S=\{0,1,2\}$, there are more than 12 MCS's.

The definition given by Ross [16] for a multistate system is less structured than either the Barlow [2] specification or the EPS [8] specification. The state space S is taken to be $[0,\infty)$ and the structure function ϕ is any monotone increasing function from $[0,\infty)^n$ into $[0,\infty)$. Ross [15] has not attempted to investigate structural properties of his model; rather, he concentrates on the stochastic properties of his model when observed either at a fixed point in time, or when observed at different points in time (dynamic models). Results of this type will be surveyed in the next two sections.

In the remainder of this section we present various structural properties of the multistate structures given by Barlow [2] and EPS [8]. These properties extend well-known results in the binary case [1, Ch. 1] to the more general multistate case.

The following theorem gives simple bounds on MCS performance:

Theorem 3.1. *Let ϕ be the structure function of an* MCS *of n components. Then*

$$\min_{1<i\leqslant n} x_i \leqslant \phi(\mathbf{x}) \leqslant \max_{1<i\leqslant n} x_i. \tag{3.3}$$

Theorem 3.1 states that a parallel system yields the best performance of an MCS, and a series system yields the worst performance. Using this theorem, EPS [8] establish probabilistic bounds on system reliability.

As in the binary case, the following lemma in EPS [8] gives a decomposition identity useful in carrying out inductive proofs. It holds for any multistate structure, not just for the MCS.

Lemma 3.1. *The following identity holds for any n-component structure function ϕ:*

$$\phi(\mathbf{x}) = \sum_{j=0}^{M} \phi(j_i, \mathbf{x}) I_{[x_i=j]} \qquad \text{for } i=1,\ldots,n, \tag{3.4}$$

where

$$I_{[x_i=j]} = \begin{cases} 1 & \text{if } x_i = j \\ 0 & \text{if } x_i \neq j. \end{cases}$$

As in the binary case, EPS [8], define a dual structure for each multistate structure.

Definition 3.2. Let ϕ be the structure function of a multistate system. The *dual structure function* ϕ^D is given by:

$$\phi^D(\mathbf{x}) = M - \phi(M - x_1, M - x_2, \ldots, M - x_n). \tag{3.5}$$

It is easy to verify that the dual of an MCS is an MCS.

Example 3.1. The dual of a series (parallel) system is a parallel (series) system. More generally, the dual of a k-out-of-n system is an $(n-k+1)$-out-of-n system, where a k-out-of-n system is given by $\phi(\mathbf{x}) = x_{(n-k+1)}$.

Design engineers have used the well-known principle that redundancy at the component level is preferable to redundancy at the system level. This principle is presented by EPS [8] in mathematical form in (i) of the following theorem; (ii) is a dual result.

Theorem 3.2. Let ϕ be a structural function of an MCS. Then
(i) $\phi(\mathbf{x} \vee \mathbf{y}) \geqslant \phi(\mathbf{x}) \vee \phi(\mathbf{y})$.
(ii) $\phi(\mathbf{x} \wedge \mathbf{y}) \leqslant \phi(\mathbf{x}) \wedge \phi(\mathbf{y})$.
Equality holds in (i) for all $\mathbf{x}$ and $\mathbf{y}$ if and only if the structure is parallel. Equality holds in (ii) for all $\mathbf{x}$ and $\mathbf{y}$ if and only if the structure is series.

Parts (i) and (ii) of Theorem 3.2 are also proved by Barlow [2].

In binary coherent structures the concepts of minimal path vectors and minimal cut vectors play a crucial role. The analogue in MCS theory is the concept of critical connection vectors. This concept is defined by EPS [8] in the following:

Definition 3.3. A vector $\mathbf{x}$ is said to be a *connection vector to level j* if $\phi(\mathbf{x}) = j, j = 0, 1, \ldots, M$.

Definition 3.4. A vector $\mathbf{x}$ is said to be an *upper critical connection vector* to level j if $\phi(\mathbf{x}) = j$ and $\mathbf{y} < \mathbf{x}$ implies $\phi(\mathbf{y}) < j, j = 1, \ldots, M$.

A lower critical connection vector to level j can be defined in a dual manner, $j = 0, 1, \ldots, M-1$.

The existence of such critical connection vectors is guaranteed by the conditions of Definition 3.1.

Let $\mathbf{x}$ be an upper critical connection vector to level j. Define $C_j(\mathbf{x}) = \{i: x_i \geqslant j\}$. Obviously $C_j(\mathbf{x})$ is a non-empty subset of $C = \{1, \ldots, n\}$. For

$j=1,\ldots,M$, let $\mathcal{C}_j=\{C_j(\mathbf{x})$: $\mathbf{x}$ is an upper critical connection vector to level $j\}$. Then the following lemma by EPS [8], shows that $\mathcal{C}_j$ enjoys a property similar to that enjoyed by the minimal path sets and minimal cut sets in the binary case.

Lemma 3.2. *For* $j=1,\ldots,M$,

$$\bigcup \mathcal{C}_j=\{1,\ldots,n\}.$$

For $j=1,\ldots,M$, let $\mathbf{y}_1^j,\ldots,\mathbf{y}_{n_j}^j$ be the upper critical connection vectors to level j, where $\mathbf{y}_r^j=(y_{1r}^j,\ldots,y_{nr}^j)$, $1\leqslant r\leqslant n_j$. The following theorem by EPS [8], expresses the state of an MCS using its upper critical connection vectors.

Theorem 3.3. *Let ϕ be the structure function of an* MCS. *Let $\mathbf{y}_1^j,\ldots,\mathbf{y}_{n_j}^j$ be its upper critical connection vectors to level $j, j=1,\ldots,M$. Then $\phi(\mathbf{x})\geqslant j$ if and only if $\mathbf{x}\geqslant y_l^t$ for some $j\leqslant t\leqslant M$ and some $1\leqslant l\leqslant n_t$.*

The above theorem is utilized to establish bounds on the system performance distribution, as will be shown in the next section.

4. Probabilistic properties of multistate coherent systems

The deterministic relationships between the performance of a multistate system and that of its components are exploited by the various researchers in the field to investigate the probabilistic properties of multistate systems. In this section we survey important relationships between the stochastic performance of the system and the stochastic performances of its components. These results provide bounds on system performance which are particularly useful when exact system performance is difficult to evaluate.

Let X_i denote the random state of component $i, i=1,\ldots,n$. Let $X=(X_1,\ldots,X_n)$ be the random vector representing the states of components $1,\ldots,n$, where the X_i's are assumed to be stochastically mutually independent. Then $\phi(\mathbf{X})$ is the random variable representing the system state, where ϕ is the structure function of the system. Naturally, the random variables assume their values in the state space S according to certain probability laws. In the model described by Postelnicu [14], $X_1,\ldots,X_n$ as well as $\phi(\mathbf{X})$ are distributed in the unit interval $[0,1]$, with cumulative distribution functions $F_1,\ldots,F_n$ and F respectively. Postelnicu [14] discusses briefly bounds on F in terms of $F_1,\ldots,F_n$. In the models described by Barlow [2] and EPS [8], the random variables $X_1,\ldots,X_n$ and $\phi(X)$

assume their values in $S=\{0,\ldots,M\}$, with

$$\begin{aligned} &P[X_i=j]=P_{ij} && P[\phi(\mathbf{X})=j]=P_j \\ &P[X_i\leqslant j]=P_i(j) && P[\phi(\mathbf{X})\leqslant j]=P(j), \end{aligned} \tag{4.1}$$

for $j=0,1,\ldots,M$, and $i=1,2,\ldots,n$. P_i represents the performance distribution of component i, while P represents the performance distribution of the system. Clearly,

$$P_i(j)=\sum_{k=0}^{j} P_{ik}, \qquad P_i(M)=1,$$

for $i=1,\ldots,n$. Similar relationships hold for P. Let $h=\mathbf{E}\phi(\mathbf{X})$; we may express h as follows:

$$h\equiv h\boldsymbol{p}(P_1,\ldots,P_n),$$

since h is a function of the $P_1,\ldots,P_n$. Alternatively, we may express h as follows:

$$h=h\mathbf{p}(\boldsymbol{P}_1,\ldots,\boldsymbol{P}_n),$$

where $\boldsymbol{P}_i\equiv(P_{i0},P_{i1},\ldots,P_{iM})$ for $i=1,\ldots,n$. In either case, EPS [8] calls h the *performance function* of the system.

Using Lemma 3.1, EPS [8], expresses the performance function of a system of n components in terms of performance functions of systems of $n-1$ components. Such a decomposition identity is useful in carrying out a proof by induction and in deriving properties of h.

Lemma 4.1. *The following identity holds for* h:

$$h(\boldsymbol{P}_1,\ldots,\boldsymbol{P}_n)=\sum_{j=0}^{M} p_{ij}h(j_i,\boldsymbol{p}_1,\ldots,\boldsymbol{p}_n), \qquad i=1,\ldots,n, \tag{4.2}$$

where $h(j_i,\boldsymbol{p}_1,\ldots,\boldsymbol{p}_n)=E\phi(j_i,\boldsymbol{X})$.

The following theorem due to EPS [8] shows that h is strictly increasing in each p_{ij}, $j>0$. This property generalizes the corresponding well known property of h in the binary case.

Theorem 4.1. *Let $h(\mathbf{p}_1,\ldots,\mathbf{p}_n)$ be the performance function of an* MCS. *Let $0<p_{ij}<1$ for $i=1,\ldots,n, j=0,1,\ldots,M$. Then $h(\mathbf{p}_1,\ldots,\mathbf{p}_n)$ is strictly increasing in $p_{ij}, i=1,\ldots,n, j=1,\ldots,M$.*

Properties of h as a function of $P_1,\ldots,P_n$ are also investigated by Barlow [2] and EPS [8]. The following theorem due to EPS [8], shows that $h(P_1,\ldots,P_n)$ is monotone increasing with respect to stochastic ordering. A similar result is proved by Barlow [2] for his subclass of the MCS (see (3.1)) using a different proof. The same property is also proved by Ross [15] for his multistate model.

Theorem 4.2. *Let P_i, P_i' be two possible performance distributions for component $i, i=1,\ldots,n$. Assume $P_i(j)\geqslant P_i'(j)$ for $j=0,1,\ldots,M, i=1,\ldots,n$. Let $P(P')$ be the corresponding system performance distribution. Then*

(i) $\quad P(j)\geqslant P'(j)$ *for* $j=0,1,\ldots,M$,

(ii) $\quad h(P_1,\ldots,P_n)\leqslant h(P_1',\ldots,P_n')$. $\quad$ (4.3)

A useful decomposition identity is given by EPS [8] for $P[\phi(X)\geqslant l]$, namely

Theorem 4.3. *Let ϕ be a multistate structure function. Then*

$$P[\phi(X)\geqslant l]=\sum_{j=0}^{M} p_{ij}P[\phi(j_i,X)\geqslant l], \qquad l=1,\ldots,M. \tag{4.4}$$

Relation (4.4) expresses the survival probability of a structure of n components in terms of survival probabilities of structures of $n-1$ components.

Using Theorem 3.1, EPS [8] obtain the following useful bounds on P and h in terms of $P_1,\ldots,P_n$:

Let P be the performance distribution and p be the performance function of an MCS. Let P_i be the ith component performance distribution, $i=1,\ldots,n$. Then for $j=0,1,\ldots,M-1$:

$$\prod_{i=1}^{n} P_i(j)\leqslant P(j)\leqslant 1-\prod_{i=1}^{n}\bar{P}_i(j),$$

$$\sum_{j=1}^{M}\prod_{i=1}^{n}\bar{P}_i(j-1)\leqslant h\leqslant \sum_{j=1}^{M}\left[1-\prod_{i=1}^{n}P_i(j-1)\right],$$

where $\bar{P}_i(j)=1-P_i(j)$.

The concept of upper connection critical vectors introduced by EPS [8] is exploited to establish further bounds on P and h. Let $\mathbf{y}_1^j,\ldots,\mathbf{y}_{n_j}^j$ be the upper critical connection vectors to level $j, j=1,\ldots,M$ (see Definition 3.4). Let A_r^j denote the event $[\mathbf{X} \geqslant \mathbf{y}_r^j]$, $r=1,\ldots,n_j$. By Theorem 3.3,

$$P[\phi(\mathbf{X}) \geqslant j] = P\left[\bigcup_{t=j}^{M} \bigcup_{r=1}^{n_t} A_r^t\right].$$

Now using the well known inclusion–exclusion principle the authors establish upper and lower bounds on $P[\phi(\mathbf{X}) \geqslant j] = \bar{P}(j-1)$. Note that

$$P(A_r^j) = P[\mathbf{X} \geqslant \mathbf{y}_r^j] = \prod_{i=1}^{n} P[X_i \geqslant y_{ir}^j]$$

for $1 \leqslant r \leqslant n_j$ and $j=1,\ldots,M$.

An interesting generalization of the Moore–Shannon Theorem [1, Theorem 5.4] is obtained by Barlow [2] using the close relationship between his definition of a multistate coherent system and that of the binary coherent system. Recall that corresponding to every multistate structure function ϕ defined by Barlow [2], there is a binary coherent structure ϕ' closely related to ϕ (see (3.1) and (3.2)). Let h' be the binary reliability function associated with ϕ', i.e., $h' = E\phi'(\mathbf{Y})$, where $\mathbf{Y}=(Y_1,\ldots,Y_n)$ is a random vector whose components are Bernoulli random variables. In view of (3.2), it is easily verified that

$$P[\phi(\mathbf{X}) \geqslant j] = E\phi'(\mathbf{Y}_j) = h'(\mathbf{q}_j), \tag{4.5}$$

where $\mathbf{q}_j = (q_{1j},\ldots,q_{nj})$, and $q_{ij} = \sum_{k=j}^{M} p_{ik}, i=1,\ldots,n$.

Recall that Moore and Shannon show that binary coherent reliability functions are S-shaped in the sense that if all components function with probability $\mathbf{p}$, either $h(\mathbf{p}) \geqslant p$ or $h(\mathbf{p}) \leqslant \mathbf{p}$ for all $0 \leqslant \mathbf{p} \leqslant 1$, or there exists $0 < \mathbf{p}_0 < 1$ such that $h(\mathbf{p}) \leqslant \mathbf{p}$ for $0 \leqslant \mathbf{p} \leqslant \mathbf{p}_0$, while $h(\mathbf{p}) \geqslant \mathbf{p}$ for $1 \geqslant \mathbf{p} \geqslant \mathbf{p}_0$. Barlow [2] gives a natural generalization of this result to the multistate case with respect to stochastic ordering.

Theorem 4.4. *Let $\mathbf{p}_i = \boldsymbol{\alpha} = (\alpha_0,\ldots,\alpha_M)$ for $i=1,\ldots,n$. Assume $h'(\mathbf{p}_0) = \mathbf{p}_0$ $(0 < \mathbf{p}_0 < 1)$. Let $\boldsymbol{\alpha}^* = (1-p_0, 0,\ldots,0,p_0)$. Then*

$$\boldsymbol{\alpha} \leqslant_{\text{st}} \boldsymbol{\alpha}^* \text{ implies that } \mathbf{p} \leqslant_{\text{st}} \boldsymbol{\alpha}, \tag{a}$$

$$\boldsymbol{\alpha} \geqslant_{\text{st}} \boldsymbol{\alpha}^* \text{ implies that } \mathbf{p} \geqslant_{\text{st}} \boldsymbol{\alpha}, \tag{b}$$

where $\mathbf{P}=(P_0,\ldots,P_M), P_j=P[\phi(\boldsymbol{X})=j], j=0,\ldots,M$, *and* $\boldsymbol{\alpha}' \leqslant_{\mathrm{st}} \boldsymbol{\alpha}''$ *means that*

$$\sum_{k=j}^{M} \alpha_k' \leqslant \sum_{k=j}^{M} \alpha_k'', \qquad j=0,1,\ldots,M.$$

Note that (4.5) is central to the proof of the above theorem.

Finally, the model proposed by Ross [16], $X_i, i=1,\ldots,n$, and $\phi(\boldsymbol{X})$ are non-negative random variables with distribution functions $F_i, i=1,\ldots,n$, and F respectively. The function $r(\bar{F}_1,\ldots,\bar{F}_n)$ is defined by

$$r(\bar{F}_1,\ldots,\bar{F}_n)=\mathbf{E}\phi(\boldsymbol{X}),$$

where $\bar{F}_i=1-F_i, i=1,\ldots,n$.

Using an extension of Lemma 2.3, of Barlow and Proschan [1, 10.84], Ross [15] proves the following:

Theorem 4.5. *If* ϕ *is a binary function then*

$$r(\bar{F}_1^{\alpha},\ldots,\bar{F}_n^{\alpha}) \geqslant \left[r(\bar{F}_1,\ldots,\bar{F}_n)\right]^{\alpha} \tag{4.6}$$

for all $0 \leqslant \alpha \leqslant 1$.

As a consequence of the above theorem, Ross [15] proves:

Corollary 4.1. *Let* $X_1,\ldots,X_n$ *be independent* IFRA *random variables. Then*

$$\sum_{i=1}^{n} X_i \text{ is IFRA.} \tag{a}$$

$$P\left\{\prod_{i=1}^{n} X_i > a\alpha^n\right\} \geqslant \left(P\left\{\prod_{i=1}^{n} X_i > a\right\}\right)^{\alpha}, \qquad 0 \leqslant \alpha \leqslant 1. \tag{b}$$

Recall that a distribution function F with $F(0)=0$ is said to be an *increasing failure rate average* (IFRA) *distribution* if

$$\bar{F}(\alpha x) \geqslant \left[\bar{F}(x)\right]^{\alpha} \qquad \text{for all } 0 \leqslant \alpha \leqslant 1, \quad x \geqslant 0.$$

Observe that part (a) of Corollary 4.1 represents the well-known property of the closure of the IFRA distributions under the convolution operation.

Ross [15] also utilizes Theorem 4.5 in proving a generalized IFRA closure theorem which is presented in the next section.

5. Dynamic models for multistate coherent systems

In previous sections, we consider deterministic and probabilistic properties of multistate systems at a fixed point in time. In this section we survey some dynamic aspects of multistate structures. We now consider multistate systems as operating over time. At time 0 the system and each of its components are at their maximal level of performance. As time passes, the performance levels of components (and consequently of the system) deteriorate to lower levels until finally level 0 (complete failure) is reached.

In the binary case, the length of time during which a component (system) functions is called the *lifelength* of the component (system); each lifelength is a nonnegative random variable. The corresponding lifelength distribution has been classified according to various notions of aging. See, e.g., [1]. Two of the important classes of life distributions are the increasing failure rate average (IFRA) class and the new better than used (NBU) class. Closure of these classes under various basic reliability operations, such as convolution of distributions and formation of binary coherent systems, is demonstrated in [1]. The counterparts of these concepts in the multistate case have been investigated by Barlow [2], EPS [8], and Ross [15].

Let $\{X_i(t), t \geqslant 0\}$ denote the decreasing stochastic process representing the state of component i at time t, where t ranges over the non-negative real numbers for $i=1,\ldots,n$. The stochastic process $\{\phi(\boldsymbol{X}(t)), t \geqslant 0\}$ is also decreasing and represents the corresponding system state as time varies, where $\boldsymbol{X}(t)=(X_1(t),\ldots,X_n(t))$. The processes $\{X_i(t), t \geqslant 0\}, i=1,\ldots,n$, are assumed to be mutually independent.

In Barlow's model where the state space is $\{0,1,\ldots,M\}$, let us call $\{j, j+1,\ldots,M\}$ the 'good' states. Assume that $[P\{X_i(t) \geqslant j\}]^{1/t}$ is decreasing in $t \geqslant 0$ for fixed j. It is easily verified that $[P\{\phi(\boldsymbol{X}(t)) \geqslant j\}]^{1/t}$ is decreasing in $t \geqslant 0$ for fixed j. Thus the above result states that if the length of time spent by each component in the 'good' states is an IFRA random variable, then the corresponding length of time spent by the multistate system in the 'good' states is also an IFRA random variable. In the binary case this represents the so-called IFRA closure (under formation of binary coherent systems) theorem. Note that from (4.5) the proof of the IFRA closure theorem for Barlow's model is immediate.

The following definition is due to Ross [15].

Definition 5.1. The stochastic process $\{X(t), t \geqslant 0\}$ is said to be an *IFRA process* if $T_a = \inf\{t : X(t) \leqslant a\}$ is an IFRA random variable for every $a \geqslant 0$.

Having introduced this definition, Ross [15] then proves the following generalized IFRA closure theorem.

Theorem 5.1. *Let* $\{X_i(t), t \geqslant 0\}$, $i = 1, \ldots, n$, *be independent* IFRA *processes and* ϕ *a multistate structure function. Then* $\{\phi(X(t)), t \geqslant 0\}$ *is an* IFRA *process.*

The crucial tool in proving the above theorem is Theorem 4.5.

Ross [15] also defines an NBU process and proves a generalized NBU closure theorem. First let us recall the definition of an NBU random variable:

Definition 5.2. A nonnegative random variable Y with distribution function F is said to be *new better than used* (*NBU*) if $\bar{F}(s+t) \leqslant \bar{F}(s)\bar{F}(t)$ for all $s \geqslant 0, t \geqslant 0$, where $\bar{F} = 1 - F$.

Now, Ross [15] gives the following definition of an NBU process.

Definition 5.3. The decreasing stochastic process $\{X(t), t \geqslant 0\}$ is said to be *NBU* if with probability 1,

$$P\{T_a > s + t \mid X(u) > a, 0 \leqslant u \leqslant s\} \leqslant P\{T_a > t\}$$

for all $s, t, a \geqslant 0$, where T_a is as in Definition 5.1.

Using his definition, Ross proves:

Theorem 5.2. *If the component processes are independent* NBU *processes, then* $\{\phi(\boldsymbol{X}(t)), t \geqslant 0\}$ *is also* NBU.

Another definition of an NBU process is given by EPS [8], and then a simple characterization for this NBU process is derived. Using their characterization, they give a simple proof of a generalized NBU closure theorem. The EPS definition of an NBU process is as follows:

Definition 5.4. The stochastic process $\{X_i(t), t \geqslant 0\}$ is an *NBU stochastic process* if $T_{i,j} = \inf\{t : X_i(t) \leqslant j\}$ is an NBU random variable for $j = 0, \ldots, M - 1$.

Recall that the state space for the EPS [8] model is the set $\{0, \ldots, M\}$. The following lemma gives a simple characterization of an NBU process.

Lemma 5.1. *The stochastic process* $\{X_i(t), t \geqslant 0\}$ *is* NBU *if and only if for all* $s \geqslant 0, t \geqslant 0$,

$$X_i(s+t) \leqslant_{\text{st}} \min(X_i'(s), X_i'(t)),$$

where $X_i'(s)$ *and* $X_i'(t)$ *are two independent random variables having the same distributions as* $X_i(s), X_i(t)$ *respectively.*

Using their Lemma 5.1, EPS [8], prove the following generalized NBU closure theorem.

Theorem 5.3. *Let* ϕ *be the structure function of an* MCS *having n components and* $\{X_i(t), t \geqslant 0\}$ *be the* i*th component performance process,* $i = 1, \ldots, n$. *Let* $\{X_i(t), t \geqslant 0\}, i = 1, \ldots, n$, *be mutually independent* NBU *processes. Then* $\{\phi(\boldsymbol{X}(t)), t \geqslant 0\}$ *is an* NBU *stochastic process.*

Remark 5.1. The useful characterization of Lemma 5.1, adapted to the binary case, yields a simpler proof of the NBU closure theorem than the proof given in [1].

6. Applications of multistate reliability models

In this section we illustrate by means of two examples that the theory of multistate reliability models provides useful and new treatments of some existing binary reliability models. This shows that not only do the multistate reliability models provide more realistic analyses of many real life situations, but they also permit us to obtain a better understanding and a more efficient treatment of existing models in the two-state case. Example 1 appears in El-Neweihi [9].

Example 1. EPS [7] study the following model. A series–parallel system consists of $k+1$ subsystems $C_0, C_1, \ldots, C_k$, also called cut sets. Cut set C_i contains n_i components arranged in parallel, $i=0,1,\ldots,k$. No two cut sets have a component in common. Components fail one at a time, and after t components have failed, each of the remaining components is equally likely to fail, $t=0,1,\ldots$. The system fails upon failure of any of the cut sets; a cut set fails when all of its components fail. This model has many applications in the study of reliability, extinction of species, inventory depletion, urn sampling, among others.

In [7, Part I], EPS study the probability $P(m_0;n)$ that the system fails because a specified cut set, say C_0, fails first. Several alternative expressions and recurrence relations for this probability are obtained. Some of these formulae are useful in the computation of desired quantities, while others are used to demonstrate qualitative features like monotonicity, Schur-concavity, etc., and to derive asymptotic limits. Similar results are also obtained for the more general model in which an 'alarm' rings when a cut set size first reaches a, where a is a specified positive integer.

In [7, Part II], the authors study the probability distribution, frequency function, and failure rate of the lifelength of series-parallel systems, where system lifelength refers to the number of components that have failed at the time of the system failure.

We now show the relationship between the above model and the multistate models. Assume $n_i = M, i=0,\ldots,k$. Let the state of component i of an MCS be defined to be the number of functioning components in cut set C_i. Thus the above model may be viewed as a series MCS. Let $X_1^i, \ldots, X_M^i$ be the random variables representing the lifelengths of components in cut set $C_i, i=0,1,\ldots,k$. For multistate component i, the lengths of time spent in state $M, M-1, \ldots, 1$ are given by

$$X_{(1)}^i, X_{(2)}^i - X_{(1)}^i, \ldots, X_{(M)}^i - X_{(M-1)}^i, \qquad i=0,\ldots,k$$

where $X_{(1)}^i, X_{(2)}^i, \ldots, X_M^{(i)}$ are the M order statistics of $X_1^i, \ldots, X_M^i$.

Such an identification relating the multistate model and the binary model permits us to answer a host of questions concerning one of the two models using results obtained for the other. For instance one can find the probability that component 0, say, reaches an 'alarm' state j, say, first. Such information is helpful in planning maintenance and replacement policies.

Example 2. Consider a series system of n binary components. Assume we have $M-1$ spares for each of the n components. A failed component is instantaneously replaced by one of its spares. When the original component i is functioning (and thus none of the spares has been used), we consider that component i is in state M. Upon failure of the original component one of its spares is used to replace it, and so the component now enters state $M-1$, etc. Thus we can view the system with its spares as a multistate series system. Let $X_1^i,\dots,X_M^i$ be random variables representing the lifelengths of component i and of its spares, $i=1,\dots,n$. Assume that all the random variables are mutually independent. Obviously, the length of time spent by component i in a single 'state' or in a group of 'states' can be expressed in terms of a sum of an appropriate subset of $X_1^i,\dots,X_M^i, i=1,\dots,n$. Thus we may view a binary system with spares as a multistate system. Again, such an identification is mutually beneficial in the study of both models.

References

[1] Barlow, R. E. and Proschan, F. (1975). *Statistical Theory of Reliability and Life Testing.* Holt, Rinehart and Winston, New York.

[2] Barlow, R. E. (1978). Coherent systems with multi-state components. *Mathematics of Operation Research* (to appear).

[3] Birnbaum, Z. W., Esary, J. D. and Saunders, S. C. (1961). Multicomponent systems and structures and their reliabilities. *Technometrics* **3**, 55–77.

[4] Birnbaum, Z. W. and Esary, J. D. (1965). Some inequalities for reliability functions. *Proc. of the Fifth Berkeley Symposium on Math. Statistics and Probability*, pp. 271–283.

[5] Birnbaum, Z. W. (1960). On the probabilistic theory of complex structures. *Proc. of the Fourth Berkeley Symposium on Math. Statistics and Probability*, pp. 49–55.

[6] Birnbaum, Z. W., Esary, J. D., and Marshall, A. W. (1966). Stochastic characterization of wearout for components and systems. *Ann. Math. Statist.* **37**, 816–825.

[7] El-Neweihi, E., Proschan, F., and Sethuraman, J. (1978). A simple model with applications in structural reliability, extinction of species, inventory depletion, and urn sampling. *Advances in Applied Prob.* (to appear).

[8] El-Neweihi, E., Proschan, F., and Sethuraman, J. (1978). Multistate coherent systems. *J. Appl. Prob.* (to appear).

[9] El-Neweihi, E. (1978). On a simple model with applications in structural reliability, extinction of species, inventory depletion, and urn sampling (in preparation).

[10] Esary, J. D. and Proschan, F. (1963). Coherent structures of non-identical components. *Technometrics*, **5**, 191–209.

[11] Hirsch, W. M. et al. (1968). Cannibalization in multicomponent systems and the theory of reliability. *Naval Research Logistics Quarterly* **15**, 331–359.

[12] Hochberg, M. (1973). Generalized multicomponent systems under cannibalization. *Naval Research Logistics Quarterly* **20**, 585–605.

[13] Murchland, J. D. (1975). Fundamental concepts and relations for reliability analysis of multistate systems. *Reliability and Fault Tree Analysis* (R. E. Barlow, J. Fussell, and N. D. Singpurwalla, Eds.), pp. 581–1618.

[14] Postelnicu, V. (1970). Nondichotomic multicomponent structures. *Bulletin de la Societé des Sciences Mathematiques de la R. S. Roumaine, Tom. 14 (62), nv. 2* 209–217.

[15] Ross, S. (1977). Multi-valued state component reliability systems. Technical Report, Department of Industrial Engineering and Operations Research, University of California, Berkeley.

[16] Simon, R. M. (1970). Optimal cannibalization policies for multicomponent systems. *SIAM J. Appl. Math.* **19** 700–711.

[17] Simon, R. M. (1972). The reliability of multicomponent systems subject to cannibalization. *Naval Research Logistics Quarterly* **19** 1 – 14.

P. R. Krishnaiah, ed., *Multivariate Analysis–V*
© North-Holland Publishing Company (1980) 543–560.

DESIGNING SURVEYS FOR MEASURING CHANGE IN CATEGORICAL DATA STRUCTURES OVER TIME*

Stephen E. FIENBERG and Richard R. PICARD
Department of Applied Statistics, School of Statistics, University of Minnesota, St. Paul, MN, U.S.A.

In many surveys involving the study of characteristics of a population over time, data may not be collected on all individuals at all time points in the survey. A methodology for the analysis of several such categorical data structures with loglinear models is outlined, using the two-wave, two-variable panel study as an illustration. An important outgrowth of this methodology is in evaluation of the efficiency in allocating resources in panel studies. Simulations are performed for the two-wave, two-variable panel study to provide some insight into the issues involved.

1. Longitudinal structures and multivariate methods for sample survey data

Statistical methods for the design and analysis of sample survey data have traditionally been geared toward the estimation of aggregate quantities for single points in time, such as population rates or averages, and the changes in these quantities from one point in time to another. The development of optimal or efficient sampling designs for such quantities, balancing costs of various sorts and accuracy of estimation, has often led to surveys which involve the repeated measurement of the same individuals or households over time.

For example, ongoing surveys conducted by the U.S. Bureau of the Census, such as The Current Population Survey (CPS) and the National Crime Survey (NCS), are based on several panels of households that are interviewed multiple times. The panel structure is typically balanced so that for any survey period one panel is being interviewed for the last time, and is then replaced by a new panel for the next survey period. Thus in the NCS each household is interviewed 7 times, and then its panel or rotation group is dropped from the sample. The reasons for the rotation group

*This work was supported in part by grant NIE-G-76-0096 from the National Institute of Education, U.S. Department of Health, Education, and Welfare. The opinions expressed herein do not necessarily reflect the positions or policies of the National Institute of Education.

structure are primarily related to cost efficiency relative to sample recruitment and the collection of background information. Of secondary, although not negligible importance are potential gains in efficiency of estimation for the measurement of change as a result of the correlation structure between repeated survey items (see U.S. Bureau of the Census, 1978). These surveys make no other use of the longitudinal or panel structure of the data collected.

There are other surveys where the longitudinal structure is of major importance (see e.g. the National Longitudinal Survey of Labor Force Experience (Parnes, 1973), and the Wisconsin Youth Panel (Sewell and Hauser, 1975)), and data are collected and analyzed to measure changes over time. In these surveys, individuals with a set of common characteristics such as cohort membership are followed over time, and multivariate methods are typically used in their analysis.

In both types of surveys, detailed analysis has focussed on one-dimensional quantities, especially if they are categorical in nature. There were two reasons for this focus. First was the lack of broad-based multivariate methods for categorical data analysis, a lack that has been remedied during the past decade (e.g. see Bishop, Fienberg, and Holland (1975), Fienberg (1977), Haberman (1974), and Plackett (1974)). Second was the fact that most multivariate methods, including those for measurement data, are appropriate primarily for data arising out of simple random sampling, not for the complex sample structures involving multiple levels of stratification and clustering which occur in practice. Most attempts to get around this second difficulty have focussed on the use of design effects (e.g. see Kish and Frankel (1974)), which are essentially adjustment factors used to produce sample sizes equivalent to those based on simple random samples, or the use of individual sample weights to carry out weighted analyses (see e.g. Koch and Lemeshow (1972) and Koch, Freeman and Freeman (1975)). The justification for both these approaches for several classes of statistical models is tenuous at best. Design effects, while conceptually appealing, tend to vary from one problem to the next depending on the substantive context. The relevance of sample weights for analyses depends on the statistical models being considered, and their use typically requires greater justification than authors have given in the past (see Porter (1972) and DuMouchel and Duncan (1977)).

The use of superpopulation models has offered yet another way around the problems of the multivariate analysis of complex sample survey data, but those models often lead to new classes of statistical problems that have not been widely explored. For categorical data situations, the recent work of Brier (1979) on models for the analysis of cluster samples is a major development, but his methods are still not directly applicable to national surveys as complex as those carried out by most major survey organizations.

We thus divide the work to be done in the development of methods for measuring change in survey structures involving categorical data over time into two parts:

(i) the development of methods based on simple random sampling, and
(ii) the extension of these methods to the complex sample surveys.

This paper is concerned with the first of these problems. The actual utility of the approach we propose will inevitably depend on its appropriateness when extensions to complex sample surveys are considered.

2. Design of longitudinal surveys involving categorical variables

In this paper we consider a methodology for the analysis of categorical data structures over time based on loglinear models. In particular we consider situations where some sample individuals have complete longitudinal records, while others have less than complete records, due to absence from the sample. For concreteness we describe our approach in the relatively simple situation involving two dichotomous variables measured at two points in time, the so-called two-wave, two-variable panel study. In this situation we can have data on individuals for both time points, for only the first, or for only the second. A sample design in this context involves specifying which combination of the three forms of data is to be used and a mechanism for selecting the individuals for the three 'subsamples.'

The key to our approach is how we divide the data into complete and incomplete longitudinal observations. When the only difference between the two types of observations is that for the latter some data is missing by design, then the categorical data model for the parameters of the complete data structure reduces to what Haberman (1977) refers to as a special case of product models for frequency tables involving indirect observation. We outline Haberman's approach as applied to our specific problem in Section 4, using his coordinate-free notation so that the results can be generalized to other situations with relative ease. The reduction of our problem to the product model structure follows because the likelihood function associated with our model factors into two components, one involving the parameters of the complete data structure and the other involving the assignment or sample design parameters. Thus we can examine the two parts separately from a likelihood perspective.

For our problem we can estimate and draw inferences about parameters using all of the data (both complete and incomplete) because the sample assignment mechanism has either a known probabilistic structure, or an unknown one which we can model separately in our analysis. There is a very close analogy between the type of assignment structure assumed here, and that considered by Rubin (1978) as being 'ignorable' for the purposes of deriving inferences from a Bayesian perspective for causal effects.

When the sample assignment mechanism is based on some other type of stochastic mechanism which is a function of unmeasured or unmeasurable variables, our approach is inapplicable. For example in the NCS, households that are highly victimized tend to move (and thus leave the sample) at higher rates than those who are not victimized at all. These 'movers' produce incomplete longitudinal data compared with those who do not move, and the incompleteness is beyond the control of the survey designer. Great care is needed in modelling such data since the dropout mechanism is related in unknown ways not only to the key variable of interest (i.e. victimization) but also to other measured sociodemographic variables, as well as other unmeasured variables and unestimable parameters. The current mechanism used to replace household dropouts in the CPS and NCS is a form of matching, i.e., the location stays in the sample and the dropout household is replaced by a new one which moves into that location. The resulting data are not directly amenable to analysis by the methods described here.

In practice we need to distinguish between sample assignment mechanisms that are known, and those that are unknown but separately capable of being modelled. If the mechanism is known precisely, then we can assess the fit of our models to the data directly. If the mechanism requires modelling, then we need to have estimates of the parameters associated with the mechanism in order to assess the goodness-of-fit. We explore some of these features in the following sections.

3. The two-wave, two-variable panel study

Consider a study involving the repeated measurement on a sample of individuals of two dichotomous variables, e.g. victimization status (victim, nonvictim) and employment status (employed, unemployed). We refer to these variables as A_1 and B_1 at time 1 and A_2 and B_2 at time 2. We envision collecting data related to this study in two ways:

(i) a longitudinal sample involving data on A and B at both times 1 and 2;

(ii) supplementary cross-sectional samples, one for time 1 only and one for time 2 only.

Data from (i) form counts in the form of a 2^4 contingency table with entries $\mathbf{x}=\{\mathbf{x}_{ijkl}\}$ (where i and j correspond to the states of A and B at time 1, and k and l to the states of A and B at time 2). These counts correspond to cell probabilities $\boldsymbol{\pi}=\{\pi_{ijkl}\}$. The mechanism for generating the supplementary data of (ii) is as follows. Originally we have N individuals from which we can in principle get complete data. For an individual whose classification is (i,j,k,l) we chose not to collect the data at time 2 with probability $\lambda_{1(i,j)}$, or not to collect the data at time 1 with probability $\lambda_{2(k,l)}$. Thus with probability $(1-\lambda_{1(i,j)}-\lambda_{2(k,l)})$ we collect the complete longitudinal data. Note that if $\{\lambda_{1(i,j)}\}$ are to represent or include differential probabilities for individuals dropping out of the survey, then the model specifies that the dropout mechanism can depend only on the values for A_1 and B_1, and not on any other information.

The partially categorized data for time 1 form a 2×2 supplemental margin with counts $\{y_{ij}\}$, and that for time 2 a second 2×2 supplemental margin with counts $\{z_{kl}\}$. The result of this allocation mechanism is a 24-cell multinomial with cell probabilities and corresponding counts:

$$\begin{aligned}&(1-\lambda_{1(i,j)}-\lambda_{2(k,l)})\pi_{ijkl}\leftrightarrow x_{ijkl},\\ &\lambda_{1(i,j)}\pi_{ij++}\leftrightarrow y_{ij},\\ &\lambda_{2(k,l)}\pi_{++kl}\leftrightarrow z_{kl},\end{aligned}\tag{1}$$

where a + indicates the summation over the corresponding subscript.

The model we have just described is a special version of one first considered by Chen (1972), and adapted to two-dimensional tables in Chen and Fienberg (1974). The likelihood function for the 24-cell multinomial is proportional to

$$L\propto\prod_{ijkl}\left[(1-\lambda_{1(i,j)}-\lambda_{2(k,l)}\pi_{ijkl}\right]^{x_{ijkl}}\prod_{ij}\left[\lambda_{1(i,j)}\pi_{ij++}\right]^{y_{ij}}\prod_{kl}\left[\lambda_{2(k,l)}\pi_{++kl}\right]^{z_{kl}}\tag{2}$$

Finding maximum likelihood estimates of the $\boldsymbol{\pi}$ and $\boldsymbol{\lambda}$ is not difficult. The likelihood function factors, i.e. $L=f_1(\boldsymbol{\pi})f_2(\boldsymbol{\lambda})$, and thus the log likelihood is separable. This means we can find maximum likelihood estimates of $\boldsymbol{\pi}$ without regard to $\boldsymbol{\lambda}$ and vice versa.

We begin with the likelihood equations for $\boldsymbol{\pi}$. These reduce to

$$N\hat{\pi}_{ijkl}=x_{ijkl}+\hat{\pi}_{ijkl}\left(\frac{y_{ij}}{\hat{\pi}_{ij++}}+\frac{z_{kl}}{\hat{\pi}_{++kl}}\right)$$

$$=\mathbf{E}(m_{ijkl}|\mathbf{x},\mathbf{y},\mathbf{z},\hat{\boldsymbol{\pi}}) \tag{3}$$

where $m_{ijkl}=N\pi_{ijkl}$. If we impose the usual log-linear structure $\boldsymbol{\pi}$, i.e.

$$\log\pi_{ijkl}=u+u_{1(i)}+u_{2(j)}+u_{3(k)}+u_{4(l)}+u_{12(ij)}+\cdots+u_{1234(ijkl)} \tag{4}$$

where each u-term sums to zero over each subscript, we can find maximum likelihood estimates for $\boldsymbol{\pi}$ under various models for the u-terms in a manner similar to the standard complete-data situation. In the latter, the likelihood equations are found by setting the minimal sufficient statistics equal to their expected values. Here a similar situation results in that we set the corresponding margin totals of $\mathbf{E}(m_{ijkl}|\mathbf{x},\mathbf{y},\mathbf{z},\hat{\boldsymbol{\pi}})$ in (3) equal to their expected values. We illustrate by an example.

Suppose our model posits 1st-order interactions between A_1 and B_1, B_1 and A_2, B_1 and B_2, and A_2 and B_2, as well as a 2nd-order interaction involving B_1, A_2, and B_2. In the notation of Fienberg (1977) the minimal sufficient statistics in the complete data situation are described by $[A_1B_1]$ $[B_1A_2B_2]$, and the likelihood equations are:

$$x_{ij++}=\hat{m}_{ij++} \quad \forall i,j,$$

$$x_{+jkl}=\hat{m}_{+jkl} \quad \forall j,k,l. \tag{5}$$

In the partially categorized data situation, the corresponding likelihood equations are

$$\sum_{k,l}\mathbf{E}(m_{ijkl}|\mathbf{x},\mathbf{y},\mathbf{z},\hat{\boldsymbol{\pi}})=\hat{m}_{ij++} \quad \forall i,j,$$

$$\sum_{i}\mathbf{E}(m_{ijkl}|\mathbf{x},\mathbf{y},\mathbf{z},\hat{\boldsymbol{\pi}})=\hat{m}_{+jkl} \quad \forall j,k,l. \tag{6}$$

In general, the likelihood equations for $\boldsymbol{\pi}$ in the partially categorized data situation must be solved iteratively. In certain cases closed form estimates do exist but, unlike the standard case, there are no simple rules for detecting when this is so. An iterative procedure to solve for $\hat{\boldsymbol{\pi}}$ under various models for the u-terms suggested by Chen (1972) is quite similar to the iterative proportional fitting scheme used in the standard case. For the above example, the algorithm would proceed as follows: Let $\pi_{ijkl}^{(0)}=1/16$. The nth cycle of the iteration consists of the following steps:

$$\begin{aligned}
\pi_{ij++}^{(4n-3)} &= \frac{1}{N}\sum_{k,l}\mathbf{E}\left(m_{ijkl}\,|\,\mathbf{x},\mathbf{y},\mathbf{z},\boldsymbol{\pi}^{(4n-4)}\right), && \forall i,j\\
\pi_{ijkl}^{(4n-2)} &= \pi_{ijkl}^{(4n-4)}\frac{\pi_{ij++}^{(4n-3)}}{\pi_{ij++}^{(4n-4)}}, && \forall i,j,k,l\\
\pi_{+jkl}^{(4n-1)} &= \frac{1}{N}\sum_{i}\mathbf{E}\left(m_{ijkl}\,|\,\mathbf{x},\mathbf{y},\mathbf{z},\boldsymbol{\pi}^{(4n-2)}\right), && \forall j,k,l\\
\pi_{ijkl}^{(4n)} &= \pi_{ijkl}^{(4n-2)}\frac{\pi_{+jkl}^{(4n-1)}}{\pi_{+jkl}^{(4n-2)}}, && \forall i,j,k,l.
\end{aligned} \tag{7}$$

This algorithm can be viewed as a simple variant on the EM-algorithm of Dempster, Laird, and Rubin (1977), and their proof of convergence is applicable here.

Next we turn to the maximization of that component of the likelihood function involving the λ-parameters. The likelihood equations are:

$$\begin{aligned}
\hat{\lambda}_{1(i,j)} &= \frac{y_{ij}}{\displaystyle\sum_{k,l}\frac{x_{ijkl}}{1-\hat{\lambda}_{1(i,j)}-\hat{\lambda}_{2(k,l)}}},\\
\hat{\lambda}_{2(k,l)} &= \frac{z_{kl}}{\displaystyle\sum_{i,j}\frac{x_{ijkl}}{1-\hat{\lambda}_{1(i,j)}-\hat{\lambda}_{2(k,l)}}}.
\end{aligned} \tag{8}$$

As with the $\boldsymbol{\pi}$ parameters, these equations must also be solved iteratively. An algorithm to do this is given in Chen and Fienberg (1974). If the λ's have a 'nice' structure closed-form estimates can exist. In particular if we get supplemental data 'at random', i.e.

$$\lambda_{1(i,j)} \equiv \lambda_1 \qquad \forall i,j,$$

$$\lambda_{2(k,l)} \equiv \lambda_2 \qquad \forall k,l, \tag{9}$$

then $\hat{\lambda}_1 = y_{++}/N$ and $\hat{\lambda}_2 = z_{++}/N$.

Finally we can combine the MLE's of the λ's and $\boldsymbol{\pi}$ as in (1) to produce estimated expected values for the 24-cell multinomial. These are needed if we are to check the goodness-of-fit of the overall model in the usual manner.

One way to compare different sample allocation mechanisms is by the precision with which we estimate the various u-terms in our loglinear model for $\boldsymbol{\pi}$. To compute the estimated asymptotic covariance matrix of $\hat{u}$-terms for a given loglinear model, we can compute the information matrix $I(\mathbf{u})$, substitute maximum likelihood estimates $\hat{\mathbf{u}}$ for $\mathbf{u}$, and invert, getting $[I(\hat{\mathbf{u}})]^{-1}$, an estimate of the asymptotic covariance matrix of $\hat{\mathbf{u}}$. Details are described in the Appendix.

There do not appear to be many shortcuts to this procedure. This is unfortunate in that computing $I(\mathbf{u})$ directly, a fair amount of algebra is required–even for the simple 2^4 case. Basically the approach is to express the likelihood in terms of the u parameters directly, using the fact that

$$\pi_{ijkl} = \frac{e^{\pm u_{1(1)} \pm u_{2(1)} \pm \cdots}}{\sum_{i,j,k,l} e^{\pm u_{1(1)} + \cdots}}, \tag{10}$$

where the + or − is determined by (i,j,k,l). (For example, if $i=1$ and $j=2$, we have $+u_{1(1)} - u_{2(1)} - u_{12(11)} \cdots$). The matrix of second partials can be computed directly and expectations taken, and thus $I(\mathbf{u})$ can be derived. For larger problems this approach seems intractable. Haberman (1977) gives general results for asymptotic properties of $\hat{\boldsymbol{\pi}}$ under models which include those mentioned here, but they do not appear to lead to any computational simplifications.

A computer program to do the calculations outlined in the Appendix is available from the authors.

4. Estimation using product model formulation

We can recast some of the results of the previous section to utilize results on product models for frequency tables due to Haberman (1977). Even though the likelihood function in expression (2) is not exactly of product model form, when we look at it in the factored form

$$L = f_1(\boldsymbol{\pi}) f_2(\boldsymbol{\lambda}) \tag{11}$$

$f_1(\boldsymbol{\pi})$ does have the product model structure. This means that we can perform the more difficult parts of the likelihood maximization using Haberman's coordinate-free framework. In particular, if we assume the simplest model for the λ-terms ('missing at random'), then we can apply Haberman's results directly. In what follows, we describe the two-wave, two-variable panel study in this fashion. Generalizations to more complex problems (e.g. more than two time points or more than two categories per variable of interest) are straightforward.

For the 2^4 table situation, there exists an underlying (partially unobserved) frequency table $\mathbf{n}_{(48\times 1)}$ which consists of three independent multinomial vectors $\mathbf{n}_1$, $\mathbf{n}_2$, and $\mathbf{n}_{3(16\times 1)}$ with sample sizes N_1, N_2, and N_3, corresponding to the 'complete,' 'time 1 only,' and 'time 2 only' observations, respectively. There is a probability vector $\mathbf{p}(\boldsymbol{\pi})$ corresponding to $\mathbf{n}$, the ith entry of which is of product model form:

$$p_i(\boldsymbol{\pi}) = d_i \prod_{h \in H} \pi_h^{c(h,i)} \qquad i = 1, 2, \ldots, 48. \tag{12}$$

Here H is an index set with 16 elements (corresponding to all possible cells (i, j, k, l)) and the $c(h, i)$ are either 0 or 1. For fixed i, exactly one $c(h, i)$ is 1, indicating which π_{ijkl} corresponds to the ith (possibly unobserved) cell. For the simplest model for the $\boldsymbol{\lambda}$ terms, we may set $d_i \equiv 1$.

The entire vector $\mathbf{n}$ is not observed, but instead we see $\mathbf{n}^*_{24\times 1}$, a somewhat collapsed version of $\mathbf{n}$. The first 16 components of $\mathbf{n}^*$ are the same as the first 16 of $\mathbf{n}$—the 'complete observations.' The next 4 components of $\mathbf{n}^*$ correspond to the next 16 components of $\mathbf{n}$, collapsed over the time 2 responses. The final 4 components of $\mathbf{n}^*$ correspond to the last 16 components of $\mathbf{n}$, collapsed over the time 1 responses. Let $\mathbf{p}^*(\boldsymbol{\pi})_{24\times 1}$ be $(p_1(\boldsymbol{\pi}), \ldots, p_{48}(\boldsymbol{\pi}))$ collapsed in the same manner as $\mathbf{n}$ is to give $\mathbf{n}^*$. For

$$m_h(\boldsymbol{\pi} | \mathbf{n}^*) \qquad h = 1, \ldots, 16,$$

the conditional expected values of $y_h = \Sigma_{i=1}^{48} c(h,i) n_i$, the likelihood equations are

$$m_h(\boldsymbol{\pi}|\mathbf{n}^*) = \sum_{j=1}^{24} n_j^* \left\{ \sum_{i \in J_j} c(h,i) \frac{p_i(\boldsymbol{\pi})}{p_j^*(\boldsymbol{\pi})} \right\} \qquad h = 1, \ldots, 16 \tag{13}$$

where $J_j = \{\text{cells in } \mathbf{n} \,|\, \text{those cells are collapsed to give } n_j^*, \text{ the } j\text{th entry of } \mathbf{n}^*\}$. (14)

In the notation of Section 3, we can express the right hand side of expression (13) as

$$\mathbf{E}(m_{ijkl} | \mathbf{x}, \mathbf{y}, \mathbf{z}, \hat{\boldsymbol{\pi}}) = x_{ijkl} \cdot 1 + y_{ij} \frac{\pi_{ijkl}}{\pi_{ij++}} + z_{kl} \frac{\pi_{ijkl}}{\pi_{++kl}}$$

$$= x_{ijkl} + \pi_{ijkl} \left(\frac{y_{ij}}{\pi_{ij++}} + \frac{z_{kl}}{\pi_{++kl}} \right). \tag{15}$$

For $N_k / \Sigma_l N_l \to \tau_k$ as $N = \Sigma N_l \to \infty$, the asymptotic properties of $\hat{\boldsymbol{\pi}}$ for the simplest $\boldsymbol{\lambda}$-model depend on its asymptotic mean and covariance:

$$m(\boldsymbol{\pi}|\boldsymbol{\mu}^*)_{16 \times 1} = \{ m_h(\boldsymbol{\pi}|\boldsymbol{\mu}^*) \} = \{ \lim_{N \to \infty} m_h(\boldsymbol{\pi}|\mathbf{n}^*) \} \tag{16}$$

and

$$C(\boldsymbol{\pi}|\boldsymbol{\mu}^*)_{16 \times 16} = \lim_{N \to \infty} C(\boldsymbol{\pi}|\mathbf{n}^*), \tag{17}$$

where

$$\boldsymbol{\mu}^* = \{ \tau_k p_j^*(\boldsymbol{\pi}) \,|\, j \in A_k, k = 1,2,3 \}, \tag{18}$$

$$A_k = \{ j \,|\, J_j \text{ is a subset of the } k\text{th multinomial} \}, \tag{19}$$

and

$$[C(\boldsymbol{\pi}|n^*)]_{hh'} = \sum_{j=1}^{24} n_j^* \left(\left[\sum_{i \in J_j} c(h,i) c(h',i) \frac{p_i(\boldsymbol{\pi})}{p_j^*(\boldsymbol{\pi})} \right] \right.$$

$$\left. - \left[\sum_{i \in J_j} c(h,i) \frac{p_i(\boldsymbol{\pi})}{p_j^*(\boldsymbol{\pi})} \right] \left[\sum_{i \in J_j} c(h',i) \frac{p_i(\boldsymbol{\pi})}{p_j^*(\boldsymbol{\pi})} \right] \right). \tag{20}$$

To set up the asymptotic structure we require the following notation:

$$B(\boldsymbol{\pi}|\boldsymbol{\mu}^*)_{16\times 16}=\mathrm{diag}\{m_h(\boldsymbol{\pi}|\boldsymbol{\mu}^*)\},$$

$$\Pi^-(\boldsymbol{\pi})_{16\times 16}=\mathrm{diag}\{\pi_h^{-1}\},$$

$$\mathbf{E}(\boldsymbol{\pi}|\boldsymbol{\mu}^*)_{16\times 16}=\Pi(\pi)[B(\boldsymbol{\pi}|\boldsymbol{\mu}^*)-C(\boldsymbol{\pi}|\boldsymbol{\mu}^*)]\Pi(\pi),$$

Ξ is an affine subspace of the space $\mathbf{R}^H$ of functions from H to R in which $\boldsymbol{\pi}$ is assumed to lie,

$$\Omega=\{z-w|z,w\in\Xi\},$$

$$\Omega'(\boldsymbol{\pi})=\{x\in\Omega|\pi_h=0\Rightarrow x_h=0\},$$

$P(\boldsymbol{\pi}|\boldsymbol{\mu}^*)$ the projection on $\Omega'(\boldsymbol{\pi})$ with respect *to* $\mathbf{E}(\boldsymbol{\pi}|\boldsymbol{\mu}^*)$,

$$\Sigma(\boldsymbol{\pi}|\boldsymbol{\mu}^*)=P(\boldsymbol{\pi}|\boldsymbol{\mu}^*)(\mathbf{E}(\boldsymbol{\pi}|\boldsymbol{\mu}^*))^-(P(\boldsymbol{\pi}|\boldsymbol{\mu}^*))^A,$$

where $(\ )^-$ denotes a generalized inverse, and $(\ )^A$ denotes an adjoint. Then, subject to appropriate regularity conditions, Haberman (1977) shows that

$$\sqrt{(N)}(\hat{\boldsymbol{\pi}}-\boldsymbol{\pi})\xrightarrow{d}\mathfrak{N}(\mathbf{0},\Sigma(\boldsymbol{\pi}|\boldsymbol{\mu}^*)). \tag{21}$$

Haberman suggests solving the likelihood equations using a Newton-Raphson procedure, with a typical iteration of the form:

$$\boldsymbol{\pi}^{(\nu+1)}=\boldsymbol{\pi}^{(\nu)}+\Sigma(\boldsymbol{\pi}^{(\nu)}|\mathbf{n}^*)\Pi^-(\boldsymbol{\pi}^{(\nu)})m(\boldsymbol{\pi}^{(\nu)}|\mathbf{n}^*). \tag{22}$$

Although more difficult to implement computationally than the functional iterative algorithm described in Section 3, this method can result in a savings of computer time in that it converges quadratically. A bonus is that an estimate of the large sample covariance matrix is obtained. If sample sizes are small, however, convergence is not assured.

This approach can be easily adapted to the estimation of $\boldsymbol{\pi}$ for rotating panel structure such as that used in the surveys described in Section 1.

5. Examining alternative sample allocation schemes

A major use of the modelling results of the preceding sections is in the evaluation of alternative sample allocation schemes. For example, in the two-wave, two-variable panel study, we would like to be able to choose from among alternative allocations between longitudinal and cross-sectional data. In this section we illustrate by means of a series of examples how a choice between such alternatives might be made. A more detailed study of such a choice utilizing Monte Carlo methods is beyond the scope of this paper, and we will report on it at a later time.

For the two-wave, two-variable panel study we intend initially to compare allocation schemes which involve interviews with the same number of individuals at both time points. Thus N (the sum of the total number of individuals interviewed) will vary from one allocation scheme to the next. In Table 1 we list three allocations to be examined here. In each case 300 individuals are to be interviewed at each time point.

Table 1
Three allocation schemes used to examine two-wave, two-variable panel study

Allocation Scheme	Number of complete observations	Number of observations in each supplemental margin	N
1.	100	200	500
2.	150	150	450
3.	200	100	400

In Table 2 we list 6 comparisons among the three allocation schemes of Table 1, based on 6 different choices of parameter values for the model:

$$\log \pi_{ijkl} = u + u_{1(i)} + u_{2(j)} + u_{3(k)} + u_{4(l)} + u_{12(ij)} + u_{13(ik)}$$

$$+ u_{14(il)} + u_{23(jk)} + u_{24(jl)} + u_{34(kl)}.$$

For each comparison we generated a single random sample for each allocation scheme, and computed the estimated u-terms and their estimated asymptotic variance matrix, assuming the true model. The $\hat{u}$-terms and the estimates of their asymptotic variances are listed in Table 2.

Table 2
Six comparisons among the allocation schemes of Table 1

(a)	u-term	value of u-term	$\hat{u}$ 1	$\hat{u}$ 2	$\hat{u}$ 3	$\hat{var}(\hat{u})$ ($\times 100$) 1	$\hat{var}(\hat{u})$ ($\times 100$) 2	$\hat{var}(\hat{u})$ ($\times 100$) 3
	1	0.2	0.23	0.17	0.26	0.53	0.45	0.56
	2	0.2	0.12	0.01	0.03	0.54	0.47	0.53
	3	0.2	0.23	0.22	0.28	0.48	0.44	0.45
	4	0.2	0.22	0.21	0.29	0.50	0.43	0.49
	12	0.2	0.26	0.22	0.39	0.45	0.40	0.47
	13	0.1	0.11	0.06	0.13	1.27	0.86	0.71
	14	0.1	−0.04	0.13	−0.21	1.32	0.85	0.79
	23	0.1	0.09	0.16	0.04	1.28	0.83	0.69
	24	0.1	0.18	0.09	0.21	1.25	0.84	0.70
	34	0.2	0.13	0.22	0.17	0.44	0.44	0.45

(b)	u-term	value of u-term	$\hat{u}$ 1	$\hat{u}$ 2	$\hat{u}$ 3	$\hat{var}(\hat{u})$ ($\times 100$) 1	$\hat{var}(\hat{u})$ ($\times 100$) 2	$\hat{var}(\hat{u})$ ($\times 100$) 3
	1	0.2	0.23	0.13	0.20	0.61	0.63	0.52
	2	0.2	0.15	0.24	0.30	0.63	0.60	0.53
	3	0.2	0.15	0.30	0.11	0.52	0.61	0.58
	4	0.2	0.17	0.20	0.27	0.54	0.55	0.57
	12	0.4	0.57	0.29	0.29	0.56	0.55	0.49
	13	0.1	0.10	0.29	0.09	1.67	1.08	0.78
	14	0.1	0.04	0.05	0.09	1.72	1.07	0.81
	23	0.1	0.06	0.16	0.16	1.67	1.15	0.81
	24	0.1	0.17	0.17	0.14	1.66	1.04	0.85
	34	0.3	0.23	0.30	0.39	0.44	0.59	0.49

(c)	u-term	value of u-term	$\hat{u}$ 1	$\hat{u}$ 2	$\hat{u}$ 3	$\hat{var}(\hat{u})$ ($\times 100$) 1	$\hat{var}(\hat{u})$ ($\times 100$) 2	$\hat{var}(\hat{u})$ ($\times 100$) 3
	1	0.2	0.16	0.15	0.30	0.64	0.84	0.58
	2	0.2	0.28	0.18	0.14	0.62	0.87	0.59
	3	0.2	0.04	0.31	0.21	0.83	0.69	0.63
	4	0.2	0.23	0.27	0.21	0.75	0.71	0.63
	12	0.5	0.51	0.66	0.47	0.59	0.60	0.51
	13	0.1	0.19	0.03	0.08	2.15	1.82	1.06
	14	0.1	0.05	0.00	0.11	2.24	1.86	1.05
	23	0.1	0.14	0.03	0.09	2.25	1.89	1.00
	24	0.1	0.09	0.22	0.05	2.29	1.86	1.00
	34	0.6	0.60	0.63	0.55	0.62	0.64	0.53

Table 2 (Continued)

(d)	u-term	value of u-term	$\hat{u}$			$\hat{\text{var}}(\hat{u})$ ($\times$ 100)		
			1	2	3	1	2	3
	1	0.2	0.18	0.09	0.21	0.51	0.50	0.48
	2	0.2	0.09	0.19	0.16	0.61	0.50	0.48
	3	0.2	0.26	0.29	0.20	0.45	0.45	0.46
	4	0.2	0.16	0.22	0.28	0.57	0.45	0.47
	12	0.2	0.19	0.24	0.15	0.58	0.42	0.45
	13	0.1	0.19	0.16	0.16	1.27	0.84	0.63
	14	0.1	0.13	0.03	0.19	1.37	0.84	0.67
	23	0.1	−0.07	0.08	0.06	1.43	0.89	0.63
	24	0.2	0.37	0.16	0.19	1.29	0.84	0.65
	34	0.1	0.11	0.12	0.07	0.62	0.45	0.47

(e)	u-term	value of u-term	$\hat{u}$			vâ($\hat{u}$) ($\times$ 100)		
			1	2	3	1	2	3
	1	0.2	0.15	0.23	0.28	0.59	0.64	0.56
	2	0.2	0.33	0.06	0.15	0.62	0.67	0.53
	3	0.2	0.25	0.29	0.19	0.55	0.49	0.44
	4	0.2	0.10	0.28	0.11	0.68	0.56	0.53
	12	0.4	0.31	0.40	0.43	0.62	0.53	0.54
	13	0.1	0.16	0.12	0.03	1.38	1.06	0.80
	14	0.1	0.19	0.20	0.31	1.42	1.07	0.78
	23	0.1	0.03	0.08	0.06	1.58	1.03	0.72
	24	0.3	0.31	0.25	0.20	1.48	1.02	0.75
	34	0.1	0.03	0.10	0.04	0.60	0.56	0.47

(f)	u-term	value of u-term	$\hat{u}$			$\hat{\text{var}}(\hat{u})$ ($\times$ 100)		
			1	2	3	1	2	3
	1	0.2	0.14	0.12	0.08	0.97	0.71	0.73
	2	0.2	0.02	0.42	0.25	1.48	1.67	0.89
	3	0.2	0.28	0.30	0.23	0.53	0.56	0.49
	4	0.2	0.42	0.00	0.27	0.99	1.57	0.69
	12	0.5	0.64	0.57	0.61	1.26	1.70	0.84
	13	0.1	−0.14	0.23	0.03	2.13	1.20	0.94
	14	0.1	0.11	0.08	0.15	2.63	2.14	1.12
	23	0.1	0.23	−0.16	0.21	2.51	2.32	1.18
	24	0.6	0.60	0.92	0.52	2.68	2.34	1.18
	34	0.1	0.15	0.19	0.08	1.11	1.35	0.71

From the limited examination of these three allocation schemes in Table 2, two features are clear.

(i) The variance estimates of $\hat{u}$-terms measuring the cross-time links (e.g. u_{13}), are monotonic functions of the number of longitudinal observations.

(ii) The variance estimates of main effects (e.g. u_1), and of same-time interactions (e.g. u_{12}) seem to vary relatively little from one scheme to the next, and from one parameter set to the next.

A cursory look at the estimated correlation matrices (not listed here) exhibit the following qualities:

(i) remarkable similarities among the matrices in both magnitude of entries and sign patterns,

(ii) off diagonal elements are primarily in the $(-0.4, 0.2)$ range,

(iii) in each matrix, the largest (absolute value) off-diagonal entries are almost exclusively negative.

(iv) the same time interactions tend to have small correlation with each other and larger, negative correlation with the cross-time links,

(v) the correlations between $\hat{u}_{13}$ and $\hat{u}_{14}$, $\hat{u}_{13}$ and $\hat{u}_{23}$, and $\hat{u}_{14}$ and $\hat{u}_{24}$ are, on average, larger in absolute value than those between $\hat{u}_{13}$ and $\hat{u}_{24}$, and $\hat{u}_{14}$ and $\hat{u}_{23}$.

Several outstanding questions remain. For example,

(i) How much information is contained in the cross-sectional data regarding parameters involving cross-time links?

(ii) Are the asymptotic variances of the $\hat{u}$-terms for cross-time links approximately inversely proportional to the number of complete observations?

These and other questions will be explored in a Monte Carlo study.

Appendix. Calculation of the information matrix for π

From expression (2), we have

$$\log L = \sum_{i,j,k,l} x_{ijkl} \log \pi_{ijkl} + \sum_{i,j} y_{ij} \log \pi_{ij++}$$

$$+ \sum_{k,l} z_{kl} \log \pi_{++kl} + \left(\begin{array}{c}\text{terms not}\\ \text{involving } \pi\end{array}\right)$$

To compute the information matrix of $u' = (u_{1(1)}, u_{2(1)}, \ldots)$ we take partial derivatives with respect to each of the three terms above, or

$$\frac{\partial^2 \log L}{\partial \mathbf{u}\, \partial \mathbf{u}} = \frac{\partial^2}{\partial \mathbf{u}\, \partial \mathbf{u}} \sum x_{ijkl} \log \pi_{ijkl} + \frac{\partial^2}{\partial \mathbf{u}\, \partial \mathbf{u}} \sum y_{ij} \log \pi_{ij++} + \frac{\partial^2}{\partial \mathbf{u}\, \partial \mathbf{u}} \sum z_{kl} \log \pi_{++kl}.$$

These can be evaluated directly by the use of expression (10), yielding

$$\begin{aligned}
\frac{\partial^2 \log L}{\partial \mathbf{u}\, \partial \mathbf{u}} &= \frac{\partial^2}{\partial \mathbf{u}\, \partial \mathbf{u}} \sum_{i,j,k,l} x_{ijkl} \log\left(e^{\pm u_{1(1)} \pm \cdots}\right) \\
&\quad - \frac{\partial^2}{\partial \mathbf{u}\, \partial \mathbf{u}} \sum_{i,j,k,l} x_{ijkl} \log\left(\sum_{\substack{\text{all}\\ \text{cells}}} e^{\pm u_{1(1)} \pm \cdots}\right) \\
&\quad + \frac{\partial^2}{\partial \mathbf{u}\, \partial \mathbf{u}} \sum_{i,j} y_{ij} \log\Bigg(\underbrace{\sum_{\substack{k,l\\ \text{fixed } ij}} e^{\pm u_{1(1)} \pm \cdots}}_{(=\pi_{ij++})}\Bigg) \\
&\quad - \frac{\partial^2}{\partial \mathbf{u}\, \partial \mathbf{u}} \sum_{i,j} y_{ij} \log\left(\sum_{\substack{\text{all}\\ \text{cells}}} e^{\pm u_{1(1)} \pm \cdots}\right) \\
&\quad + \frac{\partial^2}{\partial \mathbf{u}\, \partial \mathbf{u}} \sum_{k,l} z_{kl} \log\left(\sum_{\substack{i,j\\ \text{fixed } k,l}} e^{\pm u_{1(1)} \pm \cdots}\right) \\
&\quad - \frac{\partial^2}{\partial \mathbf{u}\, \partial \mathbf{u}} \sum_{k,l} z_{kl} \log\left(\sum_{\substack{\text{all}\\ \text{cells}}} e^{\pm u_{1(1)} \pm \cdots}\right) \\
&= \frac{\partial^2}{\partial \mathbf{u}\, \partial \mathbf{u}} \Bigg\{ \left(\sum_{i,j} y_{ij} \log\left[\sum_{\substack{k,l\\ \text{fixed } i,j}} e^{\pm u_{1(1)} \pm \cdots}\right]\right) \\
&\quad + \left(\sum_{k,l} z_{kl} \log\left[\sum_{\substack{i,j\\ \text{fixed } k,l}} e^{\pm u_{1(1)} \pm \cdots}\right]\right) - N \log \sum_{\substack{\text{all}\\ \text{cells}}} e^{\pm u_{1(1)} \pm \cdots} \Bigg\}
\end{aligned} \tag{A1}$$

The information matrix is just the negative expectation of the above. We can see from (A1) how this compares to the information matrix if all N observations were completely categorized. The third term,

$$-N \log \sum_{\substack{\text{all} \\ \text{cells}}} e^{\pm u_{1(1)} \pm \cdots} \tag{A2}$$

corresponds to the information when all observations are completely categorized, while the first two terms represent the penalty, or loss in information, due to the partial categorizations that exist.

References

[1] Bishop, Y. M. M., Fienberg, S. E. and Holland, P. W. (1975). *Discrete Multivariate Analysis: Theory and Practice*. M.I.T. Press, Cambridge, MA.

[2] Brier, S. S. (1979). Categorical data models for complex sampling schemes. Unpublished Ph.D. dissertation, School of Statistics, University of Minnesota.

[3] Chen, T. (1972). Mixed-up frequencies and missing data in contingency tables. Unpublished Ph.D. dissertation, Department of Statistics, University of Chicago.

[4] Chen, T. and Fienberg, S. E. (1974). Two-dimensional contingency tables with both completely and partially cross-classified data. *Biometrics* **30**, 629–642.

[5] Dempster, A. P., Laird, N. W., and Rubin, D. B. (1977). Maximum likelihood from incomplete data via the EM algorithm. (with discussion). *J. Roy. Statist. Soc. B* **39**, 1–38.

[6] Dumouchel, Wm. H. and Duncan, G. J. (1977). Using sample survey weights to compare various linear regression models. Department of Statistics, University of Michigan, Technical Report No. 72.

[7] Fienberg, S. E. (1977). *The Analysis of Cross-classified Categorical Data*. M.I.T. Press, Cambridge, MA.

[8] Haberman, S. J. (1974). *The Analysis of Frequency Data*. University of Chicago Press, Chicago, IL.

[9] Haberman, S. J. (1977). Product models for frequency tables involving indirect observation. *Ann. Statist.* **5**, 1124–1147.

[10] Kish, L. and Frankel, M. R. (1974). Inference from complex samples (with discussion). *J. Roy. Statist. Soc. B* **36**, 1–37.

[11] Koch, G. G. and Lemeshow, S. (1972). An application of multivariate analysis to complex sample survey data. *J. Amer. Statist. Assoc.* **67**, 750–782.

[12] Koch, G. G., Freeman, D. H., Jr., and Freeman, J. L. (1975). Strategies in the multivariate analysis of data from complex surveys. *Int. Statist. Rev.* **43**, 59–78.

[13] Parnes, H. S. (1975). Sources and uses of panels of microdata—The National Longitudinal Surveys: new vistas for labor market research. *Amer. Econ. Rev.* **65**, 244–249.

[14] Plackett, R. L. (1974). *The Analysis of Categorical Data*. Hafner Press, New York.

[15] Porter, R. D. (1972). On the use of survey sample weights in the linear model. *Ann. Econ. Soc. Meas.* **1**, 141–158.
[16] Rubin, D. B. (1978). Bayesian inference for causal effects: The role of randomization. *Ann. Statist.* **6**, 34–58.
[17] Sewell, W. and Hauser, R. (1975). *Education, Occupation and Earnings*. Academic Press, New York.
[18] U.S. Bureau of the Census (1978). The Current Population Survey: Design and Methodology. Technical Paper 40.

P. R. Krishnaiah, ed., *Multivariate Analysis–V*

STOCHASTIC TREE LANGUAGES AND THEIR APPLICATIONS TO PICTURE PROCESSING*

K. S. FU
School of Electrical Engineering, Purdue University, West Lafayette, Indiana 47907

1. Introduction

Recently, formal (nonstochastic) languages and stochastic languages have been used in the description of pattern structures [1,2]. A pattern is described as a composition of its components, called subpatterns and pattern primitives (the simplest subpatterns). This approach draws an analogy between the structure of patterns in terms of components and relations among components and the syntax of a language in terms of grammar rules. Using a string of primitives and relations to describe a pattern, we can construct a set of grammar rules to characterize the string representations of the patterns in a class. Such a grammar is often called a "pattern grammar". For one-dimensional signal and line patterns, the one-dimensional string representation appears to be quite natural and efficient. However, when the patterns are high dimensional, such as two-dimensional pictures and three-dimensional scenes, an extension from the one-dimensional string language approach to higher dimensional will often result in a more efficient representation.

One natural extension of one-dimensional string languages to high dimensional is tree languages. A string could be regarded as a single-branch tree. The capability of having more than one branch often gives trees a more efficient pattern representation. Interesting applications of tree languages to picture recognition include the classification of bubble chamber events, the recognition of fingerprint patterns, and the interpretation of LANDSAT data [3–7]. In some practical applications, pattern distortion and measurement noise often exist. In order to describe noisy and distorted patterns, the use of stochastic languages has been proposed [1]. With probabilities associated with grammar rules, a stochastic grammar generates strings with a probability distribution. The probability distribution of the strings representing patterns can be used to model the

*This work was supported by the NSF grant ENG 78-16970.

noisy situations. In this paper, we present some results on stochastic tree languages and their application to picture processing, in particular to texture modeling.[1]

2. Stochastic tree grammars and languages

Definition 1. A stochastic tree grammar G_S is a 4 tuple $G_S=(V,r',P,S)$ over ranked alphabet $\langle V_T,r\rangle$ where $\langle V,r'\rangle$ is a finite ranked alphabet such that $V_T\subseteq V$ and $r'|V_T=r$. The elements of V_T and $V-V_T=V_N$ are called terminals and nonterminal symbols respectively. P is a finite set of stochastic production rules of the form $\phi\overset{p}{\rightarrow}\psi$ where ϕ and ψ are trees over $\langle V,r'\rangle$ and $0\leqslant p\leqslant 1$. $S\subseteq T_V$ is a finite set of start symbols, where T_V is the set of all trees over V.

Definition 2. $\alpha\underset{a}{\overset{p}{\rightarrow}}\beta$ is in G_S if and only if there is a production $\phi\overset{p}{\rightarrow}\psi$ in P such that $\alpha|a=\phi$ and $\beta=(a\overset{p}{\leftarrow}\psi)\alpha$; i.e., ϕ is a subtree of α at "a" and β is obtained by replacing the occurrence of ϕ at "a" by ψ. We write $\alpha\overset{p}{\rightarrow}\beta$ in G_S if and only if there exists $a\in D_\alpha$, the domain of α, such that $\alpha\underset{a}{\overset{p}{\rightarrow}}\beta$.

Definition 3. If there exists a sequence of trees $t_0,t_1,\ldots,t_m$ such that

$$\alpha=t_0,\qquad \beta=t_m,\qquad t_{i-1}\overset{p_i}{\rightarrow}t_i,\quad i=1,\ldots m$$

then we say that α generates β with probability $p=\Pi_{i=1}^{i=m}p_i$ and denote this derivation by $\alpha\overset{p}{\mapsto}\beta$ or $\alpha\overset{p}{\Rightarrow}\beta$.

The sequence of trees $t_0,\ldots,t_m$ is called a derivation of β from α. The probability associated with this derivation is equal to the product of the probabilities associated with the sequence of stochastic productions used in the derivation.

Definition 4. The language generated by stochastic tree grammar G_S is

$$L(G_s)=\left\{(t,p(t))|t\in T_{V_T},S\overset{p_j}{\Rightarrow}t,j=1,\ldots,k\text{ and }p(t)=\sum_{j=1}^{k}p_j\right\}$$

[1]An application of stochastic tree languages to the classification of bubble chamber pictures can be found in [8].

where T_{V_T} is the set of all trees over V_T, k is the number of all distinctly different derivation of t from S and p_j is the probability associated with the jth distinct derivation of t from S.

Definition 5. A stochastic tree grammar $G_s=(V,r',P,S)$ over $\langle V_T,r\rangle$ is simple if and only if all rules of p are of the form

$$X_0 \xrightarrow{p} \begin{array}{c} x \\ / \ \backslash \\ X_1,\ldots,X_{r(x)} \end{array}, \quad X_0 \xrightarrow{q} X_1, \quad \text{or} \quad \begin{array}{c} x \\ / \ \backslash \\ X_1,\ldots,X_{r(x)} \end{array} \xrightarrow{r} X_0$$

where $X_0,X_1,\ldots,X_{r(x)}$ are nonterminal symbols and $x\in V_T$ is a terminal symbol and $0<p,q,r\leqslant 1$. A rule with the form

$$X_0 \rightarrow \begin{array}{c} x \\ / \ \backslash \\ X_1,\ldots,X_{r(x)} \end{array}$$

can also be written as $X_0 \rightarrow x\ X_1\ldots X_{r(x)}$.

Lemma 1. *Given a stochastic tree grammar $G_s=(V,r,P,S)$ over $\langle V_T,r\rangle$, one can effectively construct a simple stochastic tree grammar $G_s'=(V',r',P',S')$ over V_T which is equivalent to G_s.*

Proof. To construct G_s', examine each rule $\phi_i \xrightarrow{p} \psi_i$ of P. Introduce new symbols U_a^i and V_b^i for each $a\in D_{\phi_i}$ and $b\in D_{\psi_i}$.

Let P' consist of

(1) rules that contract the tree ϕ_i a level at a time having the form $xU_{a\cdot 1}^1\cdots U_{a\cdot n}^i \rightarrow U_o^i$ where $\phi_i(a)=x\in V_{T_n}$ and V_{T_n} is the set of terminal symbols with rank n,

(2) the rule $U_0^i \xrightarrow{p} V_0^i$,

(3) rules of the form $V_a^i \xrightarrow{1} xV_{a\cdot 1},\ldots,V_{a\cdot n}^i$ that expand V_0^i to the tree ψ_i.

The construction of G_s' is clearly effective. We must now show $L(G_s')=L(G_s)$.

Note that $\phi_i \overset{p}{\vdash} \psi_i$ in G_s' for each rule $\phi_i \xrightarrow{p} \psi_i$ in P.

Suppose that $\alpha \underset{a}{\overset{p}{\rightarrow}} \beta$ in G_s. Then for some rule $\phi_i \xrightarrow{p} \psi_i$ in G_s, $\alpha|a=\phi_i$ and $\beta=(a\overset{p}{\leftarrow}\psi_i)\alpha$ i.e., ϕ_i is a subtree of α at a and β is obtained by replacing the occurence of ϕ_i at a by ψ_i with probability p.

By the above argument $\alpha|a=\phi_i \overset{p}{\vdash} \psi_i=\beta|a$ in G_s' hence

$$\alpha=\left(a\overset{1}{\leftarrow}\alpha|a\right)\alpha \overset{p}{\vdash} \left(a\overset{1}{\leftarrow}\beta|a\right)\alpha=\left(a\overset{p}{\leftarrow}\psi_1\right)\alpha=\beta \quad \text{in } G_s'$$

Thus $\alpha \overset{p}{\vdash} \beta$ in G_s implies $\alpha \overset{p}{\vdash} \beta$ in G_s' and $L(G_s)\subseteq(G_s')$.

For the converse, suppose

$$\alpha\in L(G_s'), \quad \text{i.e.,} \quad S\overset{p}{\vdash} \alpha \quad \textit{in } G_s' \textit{ and } \alpha\in T_{V_T}$$

A deduction of $S\overset{p}{\vdash} \alpha$ in G_s may be constructed as follows: Examine the deduction $S\overset{p}{\vdash} \alpha$ in G_s'. Each time a rule $U_0^i \overset{p}{\rightarrow} V_0^i$ is applied at b, apply the rule $\phi_i \overset{p}{\rightarrow} \psi_i$ at b. The result will be a deduction of $S\overset{p}{\vdash} \alpha$ in G_s, since if the rule $U_0^i \overset{p}{\rightarrow} V_0^i$ can be applied at b, all contracting rules in P^i (i.e. those involving U_a^i, $a\in D_i$) must have been applied previously at the corresponding addresses $b\cdot a$ and all expanding rules of P^i (i.e., those involving V_a^i, $a\in D_i$) must be applied later at $b\cdot a$, since all symbols U_0^i and V_0^i are elements of $V'-V_T$.

Note that the application of a single rule $\phi_i \overset{p}{\rightarrow} \psi_i$ in G_s simulates the application of all rules of P'. An example should make this very clear.

Example 1. Let

$$G_s=(V,r,P,S)$$

where

$$V=\{+,x,S\}, \qquad V_T=\{+,x\}, \qquad r(+)=2, \qquad r(x)=0$$

and

$$P{:}\,(1)\; S\overset{p}{\rightarrow} \begin{array}{c} + \\ \diagup\;\diagdown \\ S\quad\;\; x \end{array}, \qquad (2)\; S\overset{q}{\rightarrow} x, \quad p+q=1.$$

In this case P':

$$(1')\ S \xrightarrow{p} U_0^1, \qquad (5')\ V_2^1 \xrightarrow{1} x,$$

$$(2')\ U_0^1 \xrightarrow{1} V_0^1, \qquad (6')\ S \xrightarrow{q} U_0^2,$$

$$(3')\ V_0^1 \xrightarrow{1} +V_1^1V_2^1, \qquad (7')\ U_0^2 \xrightarrow{1} V_0^2,$$

$$(4')\ V_1^1 \xrightarrow{1} S, \qquad (8')\ V_0^2 \xrightarrow{1} x.$$

Note that productions $(1'),(2'),\ldots,(5')$ in P' are the result of production (1) in P and productions $(6'),(7'),(8')$ are due to production (2) in P.

A simple deduction in G_s' is as follows

$$S \xrightarrow{p} U_0^1 \xrightarrow{1} V_0^1 \xrightarrow{1} +V_1^1V_2^1 \xrightarrow{1} +SV_2^1 \xrightarrow{1} +Sx \xrightarrow{q} +U_0^2x \xrightarrow{1} +V_0^2x \xrightarrow{1} +xx,$$

$$S \overset{pq}{\vdash} +xx.$$

The corresponding deduction in G_s is

$$S \xrightarrow{p} +Sx \xrightarrow{q} +xx, \qquad S \overset{pq}{\vdash} +xx.$$

Note that if the tree ϕ_i on the left hand side of the production rule is a single symbol of alphabet V, we will have no contracting production rules in our grammar.

Definition 6. A stochastic tree grammar $G_s=(V,r,P,S)$ over $\langle V_T,r\rangle$ is expansive if and only if each rule in P is of the form

$$X_0 \xrightarrow{p} \begin{array}{c} x \\ / \quad \backslash \\ X_1 \ldots X_{r(x)} \end{array} \quad \text{or} \quad X_0 \xrightarrow{p} x \quad \text{where } x \in V_T$$

and $X_0, X_1, \ldots, X_{r(x)}$ are non-terminal symbols contained in $V-V_T$.

Example 2. Following is a stochastic expansive tree grammar.

$$G_s=(V,r,P,S) \quad \text{over } \langle V_T,r\rangle$$

where

$$V_N = V - V_T = \{S, A, B, C\}, \qquad V_T = \{a, b, \$\},$$

$$r(a) = r(b) = \{2, 0\}, \qquad r(\$) = 2.$$

$$P:\ (1)\ S \xrightarrow{1.0} \begin{array}{c} \$ \\ / \quad \backslash \\ A \qquad B \end{array}, \qquad (4)\ B \xrightarrow{q} \begin{array}{c} b, \\ | \\ C \end{array}$$

$$(2)\ A \xrightarrow{p} \begin{array}{c} a \\ / \quad \backslash \\ A \qquad B \end{array}, \qquad (5)\ B \xrightarrow{1-q} b,$$

$$(3)\ A \xrightarrow{1-p} a, \qquad (6)\ C \xrightarrow{1.0} a,$$

$$0 \leqslant p \leqslant 1, \ 0 \leqslant q \leqslant 1.$$

Definition 7. Define a mapping h: $T_{V_T} \to V_{T_0}^*$ as follows

(i) $h(t) = x$ if $t = x \in V_{T_0}$.

Obviously, $p(t) = p(x)$.

(ii) $$h\left(\begin{array}{c} x \\ / \quad \backslash \\ t_1 \cdots t_n \end{array}\right) = h(t_1) \cdots h(t_n) \quad \text{if } x \in V_{T_n}, n > 0.$$

Obviously,

$$p\left(\begin{array}{c} x \\ / \quad \backslash \\ t_1 \cdots t_n \end{array}\right) = p(x) p(t_1) \cdots p(t_n).$$

The function h forms a string in $V_{T_0}^*$ obtained from a tree t by writing the frontier of t. Note frontier is obtained by writing in order the images (labels) of all end points of tree "t".

Theorem 1. *If L_T is a stochastic tree language, then $h(L_T)$ is a stochastic context-free language with the same probability distribution on its strings as* *the trees of L_T. Conversely, if $L(G_s')$ is a stochastic context-free language, then there is a stochastic tree language L_T such that $L(G_s')=h(L_T)$ and both languages have the same probability distribution.*

Proof. By Lemma 1, if L_T is a stochastic tree language, there is a simple stochastic tree grammar $G_s=(V,r,P,S)$ such that $L_T=L(G_s)$. Let

$$P'=\left\{X_0\xrightarrow{p}X_1\cdots X_n \,\middle|\, x\in V_{T_n}, n>0, X_0\xrightarrow{p}\begin{array}{c}x\\ \diagup\ \diagdown\\ X_1\ldots X_n\end{array}\in P\right\}$$

$$\cup\left\{X_0\xrightarrow{p}x \,\middle|\, X\xrightarrow{p}x\in P, x\in V_{T_0}\right\}$$

Then if G_s' is the stochastic context-free grammar $(V-V_T, V_T, P', S)$

$$L(G_s')=h(L(G_s))=h(L_T)$$

For the converse, suppose $L(G_s')$ is generated by the stochastic context-free grammar $G_s'=(V'-V_T', V_T', P', S)$. It may be assumed that all rules of G_s' are of the form (Chomsky Normal Form)

$$X_0\xrightarrow{p}X_1X_2 \text{ or } X_0\xrightarrow{p}x$$

where X_0, X_1, $X_2\in(V'-V_T')$ and $x\in V_{T_0}'$

Let $V_T=V_T'\cup\{+\}$ and $V=V'\cup\{+\}$ where $+\notin V_N'$. Let $r(x)=0$, $x\in V_T'$ and $r(+)=2$. Let

$$P=\left\{X_0\xrightarrow{p}\begin{array}{c}+\\ \diagup\ \diagdown\\ X_1\qquad X_2\end{array}\,\middle|\, X_0\xrightarrow{p}X_1X_2\in P'\right\}\cup\left\{X_0\xrightarrow{p}x \,\middle|\, X_0\xrightarrow{p}X_1X_2\in P'\right\}$$

Let $G_s=(V,r,P,S)$, then if $L_T=L(G_s)$, $L(G_s')=h(L(G_s))=h(L_T)$. Hence, the proof.

Definition 8. By a consistent stochastic representation for a language $L(G_s)$ generated by a stochastic tree grammar G_s, it means that the following condition is satisfied.

$$\sum_{t\in L(G_s)} p(t)=1$$

where t is a tree generated by G_s; $L(G_s)$ is the set of trees generated by G_s; $p(t)$ is the probability of the generation of tree "t".

The set of consistency conditions for a stochastic tree grammar G_s is the set of conditions which the probability assignments associated with the set of stochastic tree productions in G_s must satisfy such that G_s is a consistent stochastic tree grammar. The consistency conditions of stochastic context-free grammars can be found in Fu [1]. Since non-terminals in an intermediate generating tree appear only at its frontiers, they can be considered to be causing further branching. Thus, if only the frontier of an intermediate tree is considered at levels of branching and, due to Theorem 1, the consistency conditions for stochastic tree grammars are exactly the same as that for stochastic context-free grammars and the tree generating mechanism can be modelled by a generalized branching process [8, 9].

Let

$$P=\Gamma_{A_1} \cup \Gamma_{A_2} \cup \cdots \cup \Gamma_{A_K}$$

be the partition of P into equivalent classes such that two productions are in the same class if and only if they have the same premise (i.e., same left-hand side non-terminal). For each Γ_{Aj} define the conditional probabilities $\{p(t|A_j)\}$ as the probability that the production rule $A_j \to t$, where t is a tree, will be applied to the non-terminal symbol A_j where

$$\sum_{\Gamma_{A_j}} p(t|A_j)=1$$

Let $r_{jl}(t)$ denote the number of times the variable A_l appears in the frontier of tree "t" of the production $A_j \to t$.

Definition 9. For each Γ_{A_j}, $j=1,\ldots,K$, define the K-argument generating function $g_j(S_1, S_2, \ldots, S_K)$ as

$$g_j(S_1, S_2, \ldots, S_K)=\sum_{\Gamma_{A_j}} p(t|A_j) S_1^{r_{j,1}(t)} \ldots S_K^{r_{j,K}(t)},.$$

Example 3. For the stochastic tree grammar G_s in Example 2.

$$g_1(S_1, S_2, S_3, S_4)=p\Big(\begin{array}{c}\$\\ \diagup\ \diagdown\\ A\quad B\end{array}\Big|S\Big) S_2 S_3=S_2 S_3,$$

$$g_2(S_1, S_2, S_3, S_4)=p\Big(\begin{array}{c}a\\ \diagup\ \diagdown\\ A\quad B\end{array}\Big|A\Big) S_2 S_3+p(a|A)=pS_2 S_3+(1-p),$$

$$g_3(S_1,S_2,S_3,S_4)=p\left(\begin{array}{c} b \\ | \\ C \end{array}\Big| B\right)S_4+p(b|B)=qS_4+(1-q),$$

$$g_4(S_1,S_2,S_3,S_4)=p(a|C)=1.0.$$

These generating functions can be used to define a generating function that describes all ith level trees.

Note that for statistical properties, two ith level trees are equivalent if they contain the same number of non-terminal symbols of each type in the frontiers.

Definition 10. The ith level generating function $F_i(S_1,S_2,\ldots,S_K)$ is defined recursively as

$$F_0(S_1,S_2,\ldots,S_K)=S_1$$

$$F_1(S_1,S_2,\ldots,S_K)=g_1(S_1,S_2,\ldots,S_K)$$

$$\vdots$$

$$F_i(S_1,S_2,\ldots,S_K)=F_{i-1}[\,g_1(S_1,S_2,\ldots,S_K),$$
$$g_2(S_1,S_2,\ldots,S_K),\ldots,g_K(S_1,S_2,\ldots,S_K)\,].$$

$F_i(S_1,S_2,\ldots,S_K)$ can be expressed as $F_i(S_1,S_2,\ldots,S_K)=G_i(S_1,S_2,\ldots,S_K)+C_i$ where $G_i(\cdot)$ does not contain any constant term. The constant term C_i, corresponds to the probability of all trees $t\in L(G_s)$ that can be derived in i or fewer levels.

Theorem 2. *A stochastic tree grammar G_s with unrestricted probabilistic representation R is consistent if and only if*

$$\lim_{i\to\infty} C_i=1.$$

Proof. If the above limit does not equal to 1, this means that there is a finite probability that the generation process enters a generating sequence that has a finite probability of never terminating. Thus, the probability measure defined upon $L(G_s)$ will always be less than 1 and R will not be consistent. On the other hand, if the limit is 1, this means that no such infinite generation sequence exists since the limit represents the probability measure of all trees that are generated by the application of a finite number of production rules. Consequently R is consistent.

Definition 11. The expected number of occurrences of non-terminal symbol A_j in the production set Γ_{A_i} is

$$e_{ij} = \left| \frac{\partial g_i(S_1, S_2, \ldots, S_K)}{\partial S_j} \right|_{S_1, S_2, \ldots, S_K = 1}.$$

Definition 12. The first moment matrix $\boldsymbol{E}$ is defined as

$$\boldsymbol{E} = [e_{ij}] \quad 1 \leqslant i, j \leqslant K.$$

Lemma 2. A stochastic tree language with probabilistic representation R is consistent if all the eigenvalues of $\boldsymbol{E}$ are smaller than 1. Otherwise, it is not consistent.

Example 4. In this example, consistency conditions for the stochastic tree grammar G_s in Example 2 (as verified in part (a)) are found, and thus consistency criterion verified.

(a) The set of trees generated by G_s is as follows.

Tree (t) — *Probability of generation* [p(t)]

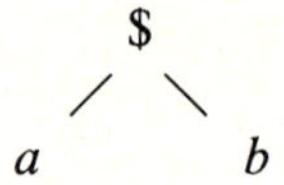

$(1-p)(1-q)$

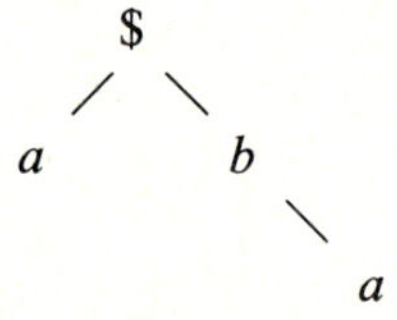

$(1-p)q$

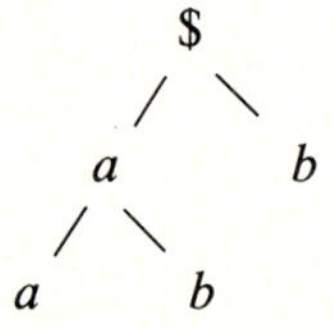

$p(1-p)(1-q)^2$

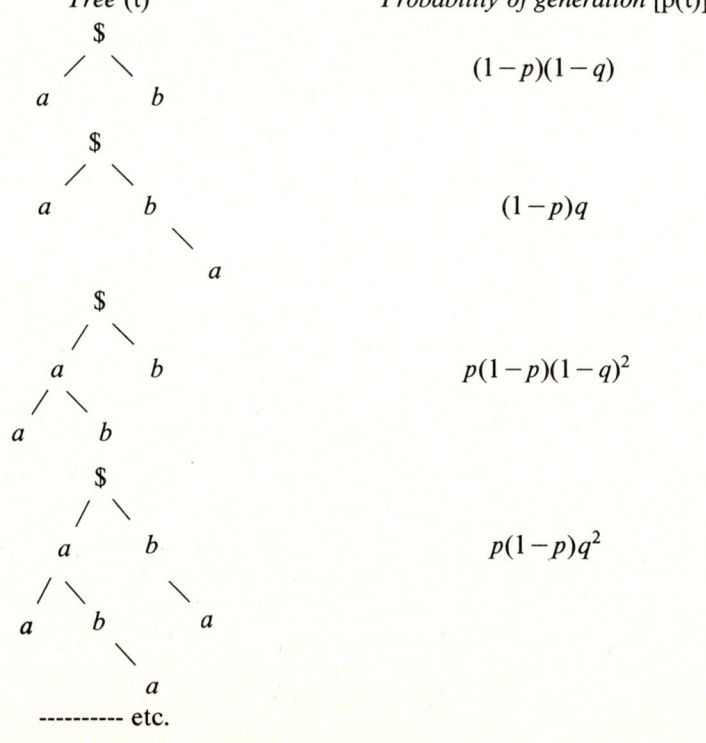

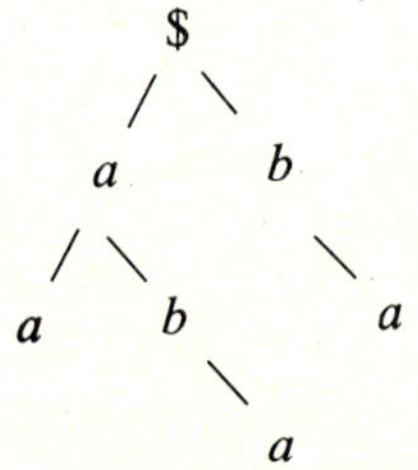

$p(1-p)q^2$

---------- etc.

In all the above trees, production (1) is always applied. If production (2) is applied $(n-1)$ times, there will be one 'A' and n 'B's in the frontier of such obtained tree. Production (3) is then applied when no more production (2) is needed. In the n 'B's in the frontier, any one, two, three or all n 'B's may have production (4) applied, and to the rest of 'B' production (5) is applied. Production (6) always follows production (4).

Thus we have

$$\begin{aligned}\sum_{t\in L(G_s)} p(t) = {} & (1-p)p^0\left[{}^1C_0(1-q)+{}^1C_1q\right] \\ & +(1-p)p^1\left[{}^2C_0(1-q)^2+{}^2C_1q(1-q)+{}^2C_2q^2\right] \\ & +(1-p)p^2\left[{}^3C_0(1-q)^3+{}^3C_1q(1-q)^2+{}^3C_2q^2(1-q)+{}^3C_3q^3\right] \\ & \cdots \\ & +(1-p)p^{n-1}\left[{}^nC_0(1-q)^n+{}^nC_1(1-q)^{n-1}q\right. \\ & \left.+\cdots+{}^nC_r(1-q)^{n-r}q^r+\cdots+{}^nC_nq^n\right]+\cdots .\end{aligned}$$

Note that power of p in the above terms shows the number of times production (2) has been applied before applying production (3). So

$$\begin{aligned}\sum_{t\in L(G_s)} p(t) = {} & (1-p)\left[\overline{1-q}+q\right]+(1-p)p\left[\left(\overline{1-q}+q\right)^2\right] \\ & +(1-p)p^2\left[\left(\overline{1-q}+q\right)^3\right]+\cdots \\ & +(1-p)p^{n-1}\left[\left(\overline{1-q}+q\right)^n\right]+\cdots\end{aligned}$$

or

$$\begin{aligned}\sum_{t\in L(G_s)} p(t) & = (1-p)+(1-p)p+\cdots+(1-p)p^{n-1}+\cdots \\ & = (1-p)\left[1+p^1+p^2+\cdots+p^{n-1}+\cdots\right] \\ & = (1-p)\cdot\frac{1}{1-p} \quad [\text{if } p<1] \\ & = 1.\end{aligned}$$

Hence, G_s is consistent for all values of p such that $0 \leqslant p \leqslant 1$. (b) Let us find the consistency condition for the grammar G_s using Lemma 2 and verify the consistency criterion. From Example 3, we obtain

$$E = \begin{bmatrix} 0 & 1 & 1 & 0 \\ 0 & p & p & 0 \\ 0 & 0 & 0 & q \\ 0 & 0 & 0 & 0 \end{bmatrix}.$$

The characteristic equation for $\boldsymbol{E}$ is

$$\phi(\tau) = (\tau - p)\tau^3.$$

Thus, the probability representation will be consistent as long as $0 \leqslant p < 1$. The value of q is constrained only for the normalization of production probabilities.

Hence, G_s is consistent.

3. Application of stochastic tree grammars to texture modelling

Research on texture modelling in picture processing has received increasing attention in recent years [10]. Most of the previous research has concentrated on the statistical approach [11, 12]. An alternative approach is the structural approach [13]. In the structural approach, a texture is considered to be defined by subpatterns which occur repeatedly according to a set of well-defined placement rules within the overall pattern. Furthermore, the subpatterns themselves are made of structural elements.

We have proposed a texture model based on the structural approach [14]. A texture pattern is divided into fixed-size windows. Repetition of subpatterns or a portion of a subpattern may appear in a window. A windowed pattern is treated as a subpattern and is represented by a tree. Each tree node corresponds to a single pixel or a small homogeneous area of the windowed patterns. A tree grammar is used to characterize windowed patterns of the same class. Two convenient tree structures and their corresponding windowed patterns are illustrated in Fig. 1. The advantage of the proposed model is its computational simplicity. The decomposition of a pattern into fixed-size windows and the use of a fixed tree structure for representation make the texture analysis procedures and its implementation very easy. However, the proposed model is very sensitive to local noise and structural distortion such as shift, rotation and

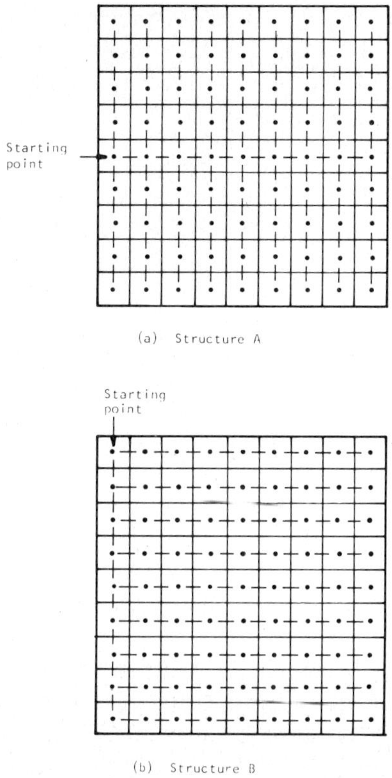

Fig. 1. Two tree structures for texture modeling. (a) Structure A, (b) Structure B.

fluctuation. In this section, we will describe the use of stochastic tree grammars and high level syntax rules to model local noise and structural distortions.

Figures 2a, and 2b are digitized pictures of the patterns D22, and D68 from Brodatz' book *Textures* [15]. For simplicity, we use only two primitives, black as primitive "1", and white as primitive "0". For pattern D22, the reptile skin, we may consider that it is the result of twisting the regular tessellation such as the pattern shown in Fig. 3. The regular tessellation pattern is composed of two basic subpatterns shown in Fig. 4. A distorted tessellation can result from shifting a series of basic subpatterns in one direction. Let us use the set of shifted subpatterns as the set of primitives. There will be 81 such windowed pattern primitives. Fig. 5 shows several of them. A tree grammar can be constructed for the generation of the 81

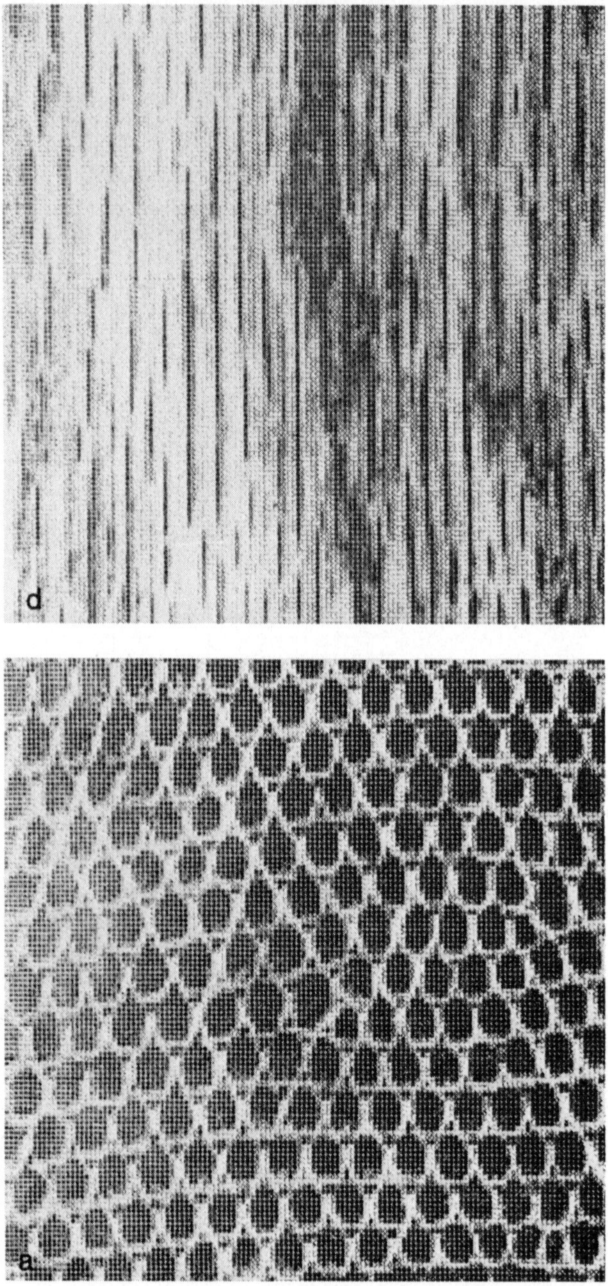

Fig. 2. Texture patterns. D22-Reptile skin, and D68-Woodgrain.

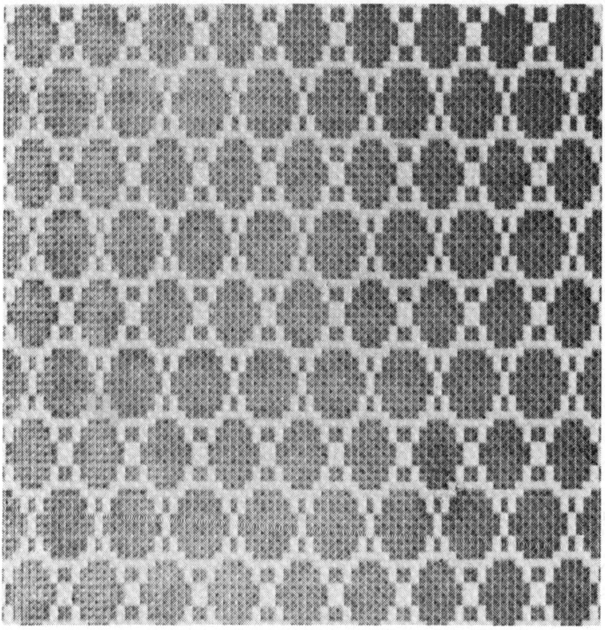

Fig. 3. The ideal texture of pattern D22.

(a) (b)

Fig. 4. Basic pattern of Fig. 3.

windowed patterns [14]. Local noise and distortion of the windowed patterns can be taken care of by constructing a stochastic tree grammar. The procedure of inferring a stochastic tree grammar from a set of texture patterns is described in [6]. A tree grammar for the placement of the 81 windowed patterns can then be constructed for the twisted texture pattern. A generated pattern D22 using a stochastic tree grammar is shown in Fig. 6.

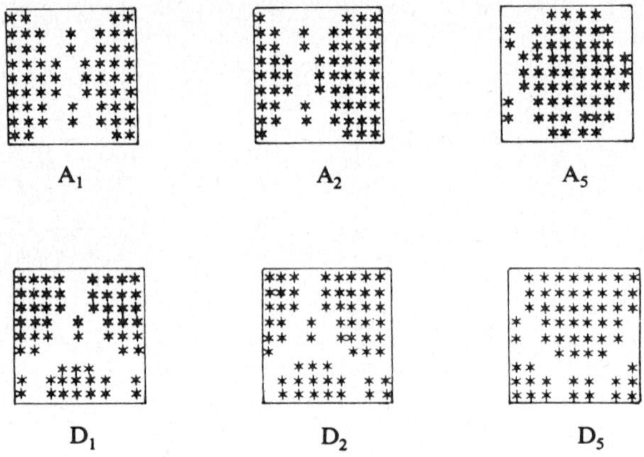

Fig. 5. Windowed pattern primitives.

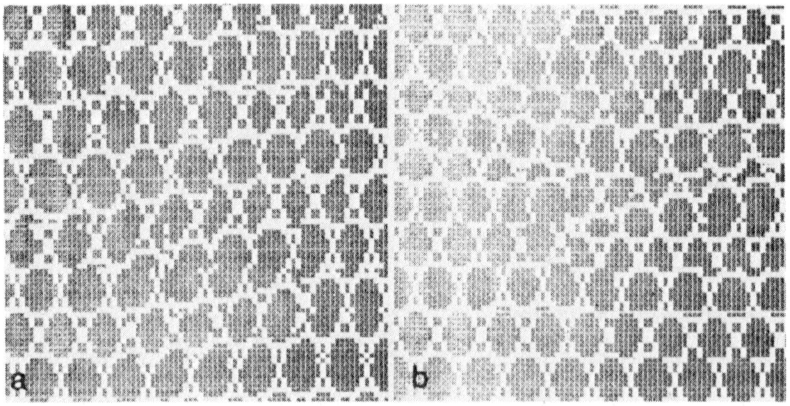

Fig. 6. Synthesis results for pattern D22.

The texture pattern D68, the wood grain pattern, consists of long verticle lines. It shows a higher degree of randomness than D22. No clear tessellation or subpattern exists in the pattern. Using verticle lines as subpatterns we can construct a stochastic tree grammar G_{68} to characterize the repetition of the subpatterns. The density of verticle lines depends on the probabilities associated with production rules. Fig. 7 shows two patterns generated from G_{68} using different sets of production probabilities.

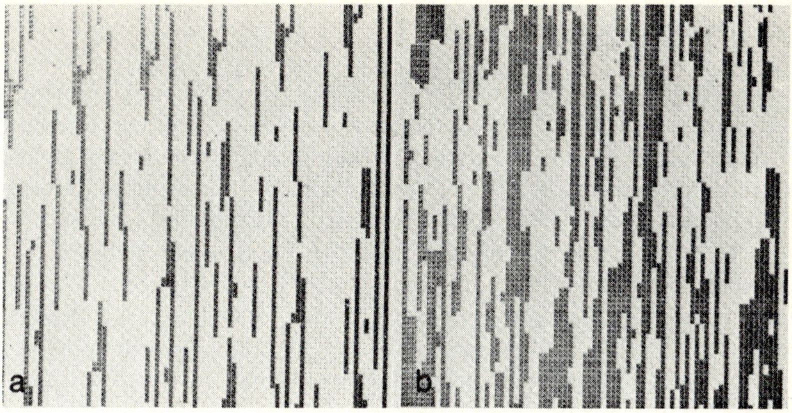

Fig. 7. Synthesis results for pattern D68.

$$G_{68} = (V, r, P, S),$$

where $V = \{S, A, B, 0, 1\}$, $V_T = \{0, 1\}$, $r(0) = r(1) = \{0, 1, 2, 3\}$

and P:

$$S \overset{0.5}{\rightarrow} \begin{array}{c} 0 \\ / \mid \backslash \\ A \quad S \quad A \end{array} \qquad S \overset{0.09}{\rightarrow} \begin{array}{c} 0 \\ / \mid \backslash \\ A \quad S \quad B \end{array} \qquad A \overset{0.05}{\rightarrow} \begin{array}{c} 0 \\ \mid \\ B \end{array}$$

$$S \overset{0.05}{\rightarrow} \begin{array}{c} 0 \\ / \quad \backslash \\ A \qquad A \end{array} \qquad S \overset{0.09}{\rightarrow} \begin{array}{c} 1 \\ / \mid \backslash \\ B \quad S \quad B \end{array} \qquad A \overset{0.05}{\rightarrow} 0$$

$$S \overset{0.09}{\rightarrow} \begin{array}{c} 0 \\ / \mid \backslash \\ B \quad S \quad A \end{array} \qquad S \overset{0.09}{\rightarrow} \begin{array}{c} 1 \\ / \mid \backslash \\ A \quad S \quad B \end{array} \qquad B \overset{0.85}{\rightarrow} \begin{array}{c} 1 \\ \mid \\ B \end{array}$$

$$S \overset{0.09}{\rightarrow} \begin{array}{c} 1 \\ / \mid \backslash \\ B \quad S \quad A \end{array} \qquad A \overset{0.90}{\rightarrow} \begin{array}{c} 0 \\ \mid \\ A \end{array} \qquad B \overset{0.10}{\rightarrow} \begin{array}{c} 1 \\ \mid \\ A \end{array}$$

$$B \overset{0.05}{\rightarrow} 1.$$

Concluding remarks

In this paper, we have introduced stochastic tree grammars and studied some of their properties. Tree grammars have been used in the description and modelling of fingerprint patterns, bubble chamber pictures, highway and rivers patterns in LANDSAT images, and texture patterns. In practical applications, noise and distortions often exist in the process under study. In order to describe and model the real world patterns more realistically, stochastic tree grammars have been suggested. We have briefly presented some recent results in texture modelling using stochastic tree grammars. For a given stochastic tree grammar describing a set of patterns, we can construct a stochastic tree automaton which will accept the set of patterns with their associated probabilities [8]. In the case of multiclass recognition problem, the maximum-likelihood or Bayes decision rule can be used to decide the class label of an input pattern representing by a tree [1].

In order to characterize the patterns of interest realistically, it would be nice to have the stochastic tree grammar actually inferred from the available pattern samples. Such an inference procedure requires the inference of both tree grammar and its porduction probabilities. Unfortunately, a general inference procedure for stochastic tree grammars is still a subject of research. Only some very special cases in practice have been discussed [16, 17].

References

[1] Fu, K. S. (1974). *Syntactic Methods in Pattern Recognition*. Academic Press, New York.

[2] Fu, K. S. (ed.) (1977). *Syntactic Pattern Recognition Applications*. Springer-Verlag, Berlin.

[3] Fu, K. S. and Bhargava, B. K. (1973). Tree systems for syntactic pattern recognition. *IEEE Trans. Computers* **C-22**.

[4] Moayer, B. and Fu, K. S. (1976). A tree system approach for fingerprint pattern recognition. *IEEE Trans. Computers* **C-25**, No. 3.

[5] Li, R. Y. and Fu, K. S. (1976). Tree system approach to LANDSAT data interpretation. *Proc. Symposium on Machine Processing of Remotely Sensed Data, June* 29–*July* 1, Lafayette, IN.

[6] Keng, J. and Fu, K. S. (1976). A syntax-directed method for land use classification of LANDSAT images. *Proc. Symposium on Current Mathematical Problems in Image Science. Nov.* 10–12, Monterey, CA.

[7] Fu, K. S. (1976). Tree languages and syntactic pattern recognition. In: *Pattern Recognition and Artificial Intelligence* (C. H. Chen, ed.). Academic Press, New York.

[8] Bhargava, B. K. and Fu, K. S. (1974). Stochastic tree system for syntactic pattern recognition. *Proc. 12th Annual Allerton Conference on Communication, Control and Computing, October*, Monticello, IL.

[9] Harris, T. E. (1963). *The Theory of Branching Processes*. Springer-Verlag.

[10] Zucker, S. W. (1976). Toward a model of texture. *Computer Graphics and Image Processing* **5**, 190–202.

[11] Haralick, R. M., Shammugam, K. and Dinstein. (1973). Texture features for image classification. *IEEE Trans. SMC*, **SMC-3**.

[12] Weszka, J. S., Dyer, C. R. and Rosenfield, A. (1976). A comparative study of texture measures for terrain classification. *IEEE Trans. SMC*, **SMC-6**.

[13] Lipkin, B. S. and Rosenfeld (eds.) (1970). *Picture Processing and Psychopictories*, pp. 289–381. Academic Press, New York.

[14] Lu, S. Y. and Fu, K. S. (1978). A syntactic approach to texture analysis. *Computer Graphics and Image Processing* **7**.

[15] Brodatz, P. (1966). *Textures*. Dover, New York.

[16] Lu, S. Y. and Fu, K. S. (1979). Stochastic tree grammar inference for texture synthesis and discrimination. *Computer Graphics and Image Processing* **8**.

[17] Brayer, J. M. and Fu, K. S. (1977). A note on the k-tail method of tree grammar influence. *IEEE Trans. on System, Man and Cybernetics*, **SMC-7**, No. 4.

P. R. Krishnaiah, ed., *Multivariate Analysis–V*
© North-Holland Publishing Company (1980) 581–591.

MULTIVARIATE GROUP JUDGMENTS BY QUALITATIVE CONTROLLED FEEDBACK

S. James PRESS
Department of Statistics, University of California, Riverside, CA, U.S.A.

1. Introduction

This paper continues the line of research reported on in Press (1978). That work proposed a new methodology for forming group judgments and for developing group decisions. The new procedure, called qualitative controlled feedback (QCF), was modeled for the univariate case involving one question of interest. An empirical application of QCF is described in Press et al. [3]. A Bayesian approach is given in Press [6]. In this paper we extend the results to the (more realistic) multivariate case where there are many related questions of simultaneous interest.

Suppose, for example, the Department of Energy would like to resolve some public policy issues that are of interest to the nation in various ways. They would like to determine how to allocate the resources in their research and development budget so that 'appropriate' funding is devoted to development of potential sources of energy, consistent with environmental considerations, political considerations, economic feasibility, engineering and scientific constraints, and perhaps other factors as well. These factors affect most people in some possibly indirect way, and no one person is likely to be knowledgeable in all related areas.

It is decided to adopt a QCF procedure to assist the policy makers in generating the factors that argue for one energy source over another. A sample of panelists is taken from some well defined population in which the Department of Energy is interested; each member of the panel is presented with a survey instrument that includes a battery of questions to which answers are desired. It may be of interest for example, to know how important the environmental affects of "strip mining" are, compared to the beneficial effects of tapping this energy source. Simultaneously, they may want to know the likelihood of a solar source of energy, such as the photovoltaic cell, being developed as a realistic source of energy by the year 2,000. It is of course likely that there would be many such questions of interest, and all of them could be included within the same instrument.

The survey instrument could be administered by mail, by telephone, by on-line computer, or whatever. In any case, the data collection protocol of QCF requires that each panelist respond to the questions independently of all other panelists, and without any panelists knowing the identity of any other panelists. Thus, the social pressures of face-to-face confrontation in a room, perhaps at the expense of logical reasoning, are avoided.

In applying a QCF procedure, each respondent is typically asked to answer a set of basic questions. In addition, the subject is asked to provide distinct reasons for each answer that will help justify the subject's answers. He is also likely to be asked to answer some subsidiary questions that will serve to provide demographic and attitudinal information about the degree of expertise of the subject, his likely institutional biases, etc.

An intermediary collects all the answers and forms a merged composite of the reasons provided by the panel for the answer to each question asked. This merging can be carried out with the aid of a computer editor. Reasons can be coded and classified into some basic minimum number (many reasons are probably just paraphrases of one another). The result is a composite of reasons corresponding to each question and answer pair.

The collection of composites of reasons is now presented to each panelist in a simple form (such as a checklist), and each panelist is asked to answer the same set of questions a second time; but this time, the panelist can see the reasoning used by all other panelists. The actual answers given by the other panelists is not made available to any of the subjects, nor do they receive summary descriptive statistics, such as sample group mean vectors for the battery of answers. Only the composites of reasons are fed back. For this reason, the second stage response of a panelist is likely to differ from his first stage response only because he feels he has ignored some arguments used by other panelists. Since the panelist is not even told the proportion of panelists who gave a particular reason, he does not know how much to weigh each reason in his own thinking, other than by adopting his own weighing system according to his own perceptions of value and importance.

The entire procedure is repeated for as many stages as is necessary for the process to stabilize, in the sense that respondents are not changing their responses very much from stage to stage.

Some background on earlier research involving group decision making and judgment formulation, and the effects of social interaction pressures, is summarized in Press [2]. In the next section we present a model that can be used for studying the relationships between responses to the questions, and the rationale the panel feels is most important to explain the answers.

It can also be used for predicting the next round's responses (in many situations, for economic or other reasons, it may be prohibitive to carry out the process for one more stage).

2. Multiple question model

2.1. *First Stage*

Let $z_{in}(j)$ denote the numerical response of subject i, on stage n, to question j; $i=1,2,\dots,N$; $j=1,2,\dots,q$. Let F_n denote the totality of information obtained on state n and fed back to each panelist at the beginning of stage $(n+1)$. Let $F^{(n)}$ denote the n-vector (F_j). Finally, let $X:N\times r$ denote a regressor matrix of explanatory variables observed for the N panelists (these are answers to subsidiary questions).

For the first stage model we adopt a simple regression with uncorrelated errors (subjects respond independently on the first stage). Accordingly, assume

$$z_1(j)|X = X\beta(j)+u_1(j),$$

$$\mathrm{E}(u_1)=0, \mathrm{var}[u_1(j)]=\sigma_1^2(j)I_N,$$

where

$$z_1(j)=[z_{11}(j),\dots,z_{N1}(j)]', \quad u(j)=[u_{11}(j),\dots,u_{N1}(J)]'.$$

$u_{il}(j)$ denotes an error term, and $\beta(j)$ denotes an $r\times l$ vector of unknown coefficients. For convenience, take

$$\underset{(N\times q)}{V} = \left[\underset{(N\times 1)}{u_1(1)},\dots,\underset{(N\times 1)}{u_1(q)}\right], \quad \underset{(q\times N)}{V'} = \left[\underset{(q\times 1)}{v_1},\dots,\underset{(q\times 1)}{v_N}\right],$$

and assume

$$\mathrm{E}(v_i v_j') = \begin{cases} \Phi, & i=j \\ 0, & i\neq j \end{cases}.$$

If

$$\underset{(N\times q)}{Z_1} \equiv \left[\underset{(N\times 1)}{z_1(1)},\dots,\underset{(N\times 1)}{z_1(q)}\right], \quad \underset{(r\times q)}{B} \equiv \left[\underset{(r\times 1)}{\beta(1)},\dots,\underset{(r\times 1)}{\beta(q)}\right],$$

the model may be written in the compact form

$$\underset{(N\times q)}{Z_1} = \underset{(N\times r)}{X}\ \underset{(r\times q)}{B} + \underset{(N\times q)}{V}, \tag{1}$$

where

$$\mathbf{E}(V)=0, \qquad \operatorname{cov}(v_i,v_j)=0, \quad i\neq j, \qquad \operatorname{var}(v_i)=\Phi.$$

The model of course represents a classical multivariate regression. The Gauss-Markov estimator of B is therefore

$$\hat{B}=(X'X)^{-1}X'Z_1. \tag{2}$$

Thus, the response matrix for a new sample of subjects from the same population, but with regressor matrix X^*, is predicted from eqns. (1) and (2) as $Z_1^*=X^*\hat{B}$.

$$Z_1^*=X^*\hat{B}.$$

2.2. Feedback stages ($n \geqslant 2$)

For later stages, beyond the first, the model must change. This is because of the composites of reasons fed back to each respondent. Since they all get the same feedback, their responses on the next stage are likely to be similarly correlated (homogeneous, or intraclass correlation structure). Moreover, their answers on stage two are likely to be related to their answers on stage one. Adopt the autoregressive model

$$\Delta z_{in}(j)\equiv\left[z_{in}(j)|F^{(n-1)}\right]-\left[z_{i,n-1}(j)|F^{(n-2)}\right]$$

$$=\sum_{\alpha=1}^{R_{n-1}(j)} c_i^{(\alpha)}(j)\left[1-\delta^{(\alpha)}b\text{-}\delta_{i,n-1}^{(\alpha)}(j)\right]p_n^{(\alpha)}(j)+u_{in}(j), \tag{3}$$

where $R_n(j)$ denotes the number of distinct reasons given by the panel (this is the number of reasons in the composite) for the answer to question j, on stage n; $\delta_{i,n}^{(\alpha)}(j)$ is unity or zero, depending upon whether or not respondent i records reason α for his answer to question j, on stage n; $c_i^{(\alpha)}(j)$ is an unknown constant of proportionality (to be estimated); and $p_n^{(\alpha)}(j)$ denotes the proportion of respondents who record reason α for question j, on stage n (this will be interpreted as the weight or importance the panel gives to this reason). Note that the panel members do not know $p_n^{(\alpha)}(j)$, but it can nevertheless be used in our model since it is known to the intermediary in the data collection process.

The model in equation (3) may be interpreted in the following way. $\Delta z_{in}(j)$ represents the change in response for subject i, on question j, from stage $(n-1)$ to stage (n). This change results from an incremental effect attributable to each reason (linear combination of effects). If the subject gave that reason on the last stage, there is of course no effect, while if he didn't give it, the effect is proportional to the importance of the reason (as measured by the proportion of panelists who gave the reason).

2.3. *Error Structure* $(n \geqslant 2)$

Define

$$\underset{(q\times 1)}{u_{in}} = [u_{in}(1), \ldots, u_{in}(q)]'$$

and assume

(1) $\mathbf{E}(u_{in}) = 0$

(2) $\mathrm{var}(u_{in}) = \underset{(q\times q)}{\Sigma_n}$

(3) $$\mathrm{cov}(u_{in}, u_{jm}) = \begin{cases} \underset{(q\times q)}{\Lambda_n}, & i \neq j, \quad n = m, \\ 0, & n \neq m. \end{cases}$$

For compactness, let

$$\underset{(Nq\times 1)}{u_n} \equiv [u'_{1n}, \ldots, u'_{Nn}]'.$$

Then, $\mathbf{E}(u_n) = 0$, and

$$\mathrm{var}(u_n) = \underset{(Nq\times Nq)}{\Omega_n} = \begin{bmatrix} \Sigma_n & & \Lambda_n \\ & \ddots & \\ \Lambda_n & & \Sigma_n \end{bmatrix}.$$

Ω_n is seen to be a matrix intraclass covariance matrix. Some of its properties are given, e.g., in Press [1, pp. 21, 48, 49]; see also [4]. The fact that Ω_n is assumed to be block diagonal means we are assuming multivariate homoscedasticity. The fact that all off-diagonal elements of the $q\times q$ blocks are assumed to be identical (Λ_n) reflects the assumption that in many situations it is reasonable to expect that the panel will be constituted with members who are sufficiently homogeneous in background so that a pattern of homogeneous correlation is reasonable.

2.4. *Transformations to canonical form*

Let

$$\underset{(q\times 1)}{\Delta z_{in}} = [\Delta z_{in}(1), \dots, \Delta z_{in}(q)]',$$

and assume

$$c_i^{(\alpha)}(j) = \underset{(1\times r)}{x_i'} \cdot \underset{(r\times 1)}{a_\alpha(j)}, \tag{4}$$

where x_i denotes the $(r\times 1)$ vector of explanatory variables for subject i, and $a_\alpha(j)$ denotes an $(r\times 1)$ vector of unknown weights. For compactness, let

$$\underset{[R_{n-1}(j)\times 1]}{c_i(j)} = [c_i^{(1)}(j), \dots, c_i^{(R_{n-1}(j))}(j)]',$$

and

$$\underset{[rR_{n-1}(j)\times 1]}{a^{(n-1)}(j)} = [a_1'(j), \dots, a_{R_{n-1}(j)}'(j)]'$$

so that

$$c_i(j) = (I \otimes x_i')a^{(n-1)}(j),$$

where $\otimes$ denotes the direct product. We next combine all the observable explanatory data into one matrix. Define

$$\underset{[rR_{n-1}(j)\times 1]}{w_{in}(j)} = (I \otimes X_i)t_{in}(j),$$

where

$$t_{in}^{(\alpha)}(j) \equiv [1 - \delta_{i,n-1}^{(\alpha)}(j)]p_n^{(\alpha)}(j),$$

$$\underset{[1\times R_{n-1}(j)]}{t_{in}'(j)} \equiv [t_{in}^{(1)}(j), \dots, t_{in}^{(R_{n-1}(j))}(j)]$$

and define

$$\underset{(q\times h_{n-1})}{W_{in}} = \begin{bmatrix} W'_{in}(1) & & 0 \\ & \ddots & \\ 0 & & W'_{in}(q) \end{bmatrix}$$

and

$$\underset{(h_{n-1}\times 1)}{a^{(n-1)}} = \left[a^{(n-1)'}(1),\dots,a^{(n-1)'}(q)\right]',$$

where

$$h_{n-1} \equiv r\sum_{j=1}^{q} R_{n-1}(j).$$

The model now becomes

$$\underset{(q\times 1)}{\Delta z_{in}} = \underset{(q\times h_{n-1})}{W_{in}} \underset{(h_{n-1}\times 1)}{a^{(n-1)}} + \underset{(q\times 1)}{u_{in}}. \tag{5}$$

Combining all subjects, (5) becomes

$$\underset{(Nq\times 1)}{\Delta z_{n}} = \underset{(Nq\times h_{n-1})}{W_{n}} \underset{(h_{n-1}\times 1)}{a^{(n-1)}} + \underset{(Nq\times 1)}{u_{n}}, \tag{6}$$

where

$$\Delta z_n = (\Delta z'_{1n},\dots,\Delta z'_{Nn})', \quad W_n = (W'_{1n},\dots,W'_{Nn}).$$

Iterating over the n stages gives

$$\underset{(Nq\times 1)}{z} \equiv z_n - z_1 = \underset{(Nq\times h)}{W} \cdot \underset{(h\times 1)}{a} + \underset{(Nq\times 1)}{u}, \tag{7}$$

where for $h = \sum_{j=1}^{n-1} h_j$,

$$\underset{(h\times 1)}{a} \equiv (a^{(1)'},\dots,a^{(n-1)'})',$$

and

$$\underset{(Nq\times h)}{W} = (W_2,\dots,W_n), \qquad \underset{(Nq\times 1)}{u} \equiv \sum_{i=1}^{n-1} u_i.$$

The transformed error vector in (7) satisfies

$$\mathbf{E}(u)=0, \qquad \operatorname{var}(u)=\Omega=\begin{bmatrix} \Sigma & & & \Lambda \\ & \ddots & & \\ & & \ddots & \\ \Lambda & & & \Sigma \end{bmatrix},$$

where

$$\Sigma \equiv \sum_{j=2}^{n-1} (\Sigma_j), \qquad \Lambda \equiv \sum_{j=2}^{n-1} (\Lambda_j).$$

2.5 *Estimation of model*

The model in (7) and (8) is now estimated in two stages. For the first stage we assume that Ω is known. Then, the generalized least squares estimator (Aitken, 1935) is the uniformly minimum variance linear unbiased estimator given by

$$\hat{a}=(W'\Omega^{-1}W)^{-1}W'\Omega^{-1}z. \tag{9}$$

If Ω is unknown, but can be consistently estimated, we can use (9) to generate an alternative estimator which would become increasingly efficient in larger samples. Accordingly, suppose $\tilde{\Omega}$ denotes a consistent estimator of Ω. Define

$$\tilde{a}=(W'\tilde{\Omega}^{-1}W)^{-1}W'\tilde{\Omega}^{-1}z. \tag{10}$$

Note that

$$\tilde{\Omega}\xrightarrow{P}\Omega, \qquad (\tilde{a}-\hat{a})\xrightarrow{P}0,$$

where the arrows denote convergence in probability. It only remains to find a consistent estimator of Ω.

First iterate Equation (5) over n to get

$$\underset{(q\times 1)}{z^{(i)}} = \underset{(q\times h)}{W^{(i)}}\ \underset{(h\times 1)}{a} + \underset{(q\times 1)}{u^{(i)}}, \tag{11}$$

where

$$\underset{(q\times 1)}{u^{(i)}} \equiv \sum_{k=1}^{n-1} u_{ik}, \quad W^{(i)} \equiv (W_{i2}, \ldots, W_{in}),$$

$$\mathbf{E}(u^{(i)}) = 0, \quad \operatorname{var}(u^{(i)}) = \Sigma, \quad \mathbf{E}(u^{(i)}u^{(j)\prime}) = \Lambda, \quad i \neq j.$$

The model in eqn. (11) may now be written in the alternative form

$$\underset{(N\times q)}{Z} = \underset{(N\times qh)}{\tilde{X}} \cdot \underset{(qh\times q)}{B} + \underset{(N\times q)}{U}, \tag{12}$$

where

$$\underset{(N\times q)}{Z} \equiv \Big(\underset{(N\times 1)}{\zeta_1}, \ldots, \underset{(N\times 1)}{\zeta_q} \Big), \qquad \underset{(N_q\times 1)}{z} \equiv \Big(\underset{(N\times 1)}{\zeta_1'}, \ldots, \underset{(N\times 1)}{\zeta_q'} \Big)',$$

$$\underset{(N\times qh)}{\tilde{X}} \equiv \Big(\underset{(N\times h)}{A_1}, \ldots, \underset{(N\times h)}{A_q} \Big), \qquad \underset{(N_q\times h)}{W} \equiv (A_1', \ldots, A_q')',$$

$$\underset{(qh\times q)}{B} = I_q \otimes \underset{(h\times 1)}{a} = \begin{pmatrix} a. & 0 \\ 0 & \cdot a \end{pmatrix},$$

$$\underset{(N\times q)}{\mathsf{U}} = \Big(\underset{(N\times 1)}{g_1}, \ldots, \underset{(N\times 1)}{g_q} \Big), \qquad \underset{(N_q\times 1)}{U} = (g_1', \ldots, g_1')'.$$

If we ignore the fact that B is not a full matrix but has a very simple structure, and the fact that the rows (and columns) of U are correlated, we can calculate the ordinary least squares estimator of B as

$$\tilde{B} = (\tilde{X}'\tilde{X})^{-1}\tilde{X}'Z. \tag{13}$$

This estimator is known to be consistent, even though it is clearly inefficient.

The residuals from this regression are

$$\underset{(N\times q)}{\tilde{U}} = Z - \tilde{X}\tilde{B} \equiv (\tilde{g}_1, \ldots, \tilde{g}_q).$$

So the corresponding residual estimator of U is given by

$$\underset{(Nq\times 1)}{\tilde{U}} = \Big(\underset{(N\times 1)}{\tilde{g}_1'}, \ldots, \underset{(N\times 1)}{\tilde{g}_q'} \Big) \equiv \Big(\underset{(q\times 1)}{\tilde{U}^{(1)\prime}}, \ldots, \underset{(q\times 1)}{\tilde{U}^{(N)\prime}} \Big)'.$$

Since consistent estimators of Σ, and Λ are readily seen to be

$$\tilde{\Sigma} = \frac{1}{N}\sum_{i=1}^{N} \tilde{U}^{(i)}\tilde{U}^{(i)\prime}, \tilde{\lambda} = \frac{1}{N(N-1)} \sum_{\substack{i=1 \\ i\neq j}}^{N} \sum_{j=1}^{N} \tilde{u}^{(i)}\tilde{u}^{(j)\prime},$$

a consistent estimator of Ω that can be used in eqn. (10) is

$$\tilde{\Omega} = \begin{pmatrix} \tilde{\Sigma} & \ddots & \tilde{\Lambda} \\ \tilde{\Lambda} & & \tilde{\Sigma} \end{pmatrix}, \tag{14}$$

where $(\tilde{\Sigma}, \tilde{\Lambda})$ are given immediately above. In fact, once a is estimated from equation (1), we can use the residuals from the regression in equation (7) to update the $\tilde{\Omega}$ estimator in equation (14). Repeating this iterative process of updating yields an increasingly efficient estimator of a.

The estimator of a found from equation (10) may be used to forecast all subjects' responses to all questions for a later stage, given their responses on earlier stages. Such forecasts would be useful when, for economic, political, or logistical reasons, it is not desirable or feasible to carry out the feedback process one more stage. For example, the third stage responses could be predicted from only the first and second stage responses.

3. A stopping rule

It is useful sometimes to have a measure of when to stop iterating the feedback process. Accordingly, we define the round-to-round variance

$$Q_n = \frac{1}{Nq}\sum_{i=1}^{N}\sum_{j=1}^{q}\left[\left(z_{in}(j)\,|\,F^{(n-1)}\right) - \left(z_{i,n-1}(j)\,|\,F^{(n-2)}\right)\right]^2$$

$$Q_n = \frac{1}{Nq}\sum_{i=1}^{N}\sum_{j=1}^{q}\left[\Delta z_{in}(j)\right]^2$$

$$Q_n = \frac{1}{Nq}(\Delta z_n)'(\Delta z_n); \qquad \Delta z_n : Nq \times 1.$$

Thus, if Q_n is 'close' to zero it is time to bring the process to a halt. The rate of decrease of Q_n over n can be studied in order to reach a realistic decision. Of course the panel may stop at a consensus point, or at a point where they have stopped changing their views on all issues, but their collections of views may be widely discrepant. Both types of results would have important implications in any type of public policy issue problem.

References

[1] Press, S. James (1972). *Applied Multivariate Analysis*. Holt, Rinehart and Winston, New York.

[2] Press, S. James (1978). Qualitative controlled feedback for forming group judgments and making decisions. *J. Amer. Statist. Assoc.*, **73** (363), 526–535.

[3] Press, S. James, M. W. Ali and Chung-fang, Elizabeth Yang (1979). An empirical study of a new method for forming group judgements: Qualitative controlled feed back. *Technological forecasting and social change*, in press. Also: Tech. Rept. No. 46, Dept. of Statistics, University of California, Riverside CA 92521, U.S.A., Jan. 1979.

[4] Press, S. James (1979). Matrix intraclass covariance matrices. Tech. Rept. No. 49, Dept. of Statistics, University of California, Riverside, CA 92951, U.S.A., Jan. 1979.

[5] Aitkon, A. C. (1935). On least squares and linear combinations of observations. *Proc. Roy. Soc. Edinburgh* **55**, 42–48.

[6] Press, S. James (1979). Bayesian interference in group judgement formulation and decision making using qualitative controlled feedback. *Trabajos Estadíst.*, to appear. Also: Tech. Rept. No. 47 Dept. of Statistics, University of California, Riverside, CA 92521, U.S.A., Feb. 1979.

P. R. Krishnaiah, ed., *Multivariate Analysis-V*

LEVEL SPACING DISTRIBUTIONS FOR ELECTRONS CONFINED BY IRREGULAR SURFACES

Keith F. RATCLIFF
Department of Physics, State University of New York at Albany, Albany, NY 12222, U.S.A.

Very small metallic particles (consisting of 10^4–10^5 atoms) have their thermodynamic properties strongly influenced by the shape of the surface of the particle. The surface confines the motion of the free conduction electrons and thereby determines their spectrum of energy eigenvalues. We investigate the hypothesis that irregularities in the surface of the particle will perturb these eigenvalues in such a manner as to yield a spacing distribution which is consistent with the predictions of random matrix theory. The problem of free electrons confined by an irregular surface is mapped into a problem of electrons moving in an effective potential but confined by a perfect sphere. A perturbation theory is then developed for the analysis of the transformed problem. The irregular surface, which defines the mapping, is described by a linear superposition of spherical harmonics. For an ensemble of such particles, the coefficients in the superposition are independently chosen random variables subject only to the constraint that the resulting surface be real. Analysis of the eigenvalue spacing distribution of the ensemble supports the predictions of random matrix theory for maximally rough surfaces. However, as the surfaces grow progressively smoother, by causing the low order spherical harmonics to become dominant, the spacing distribution becomes highly structured reflecting the emergence of a highly correlated electron energy spectrum.

1. Electron states in a small metallic particle

In a metal the least bound electrons ('valence' or 'conduction' electrons) are no longer associated with a single atom but instead become constituents of the solid as a whole. At very low temperature and in simplest approximation, these conduction electrons may be considered to move freely, confined only to remain within the volume of the solid by Coulomb forces. These conduction electrons are responsible for most of the physical properties of the metal.

If these non-interacting electrons are confined within a perfect sphere of radius R_0 then the individual states available to the electrons are eigenfunctions of the kinetic energy Hamiltonian,

$$T\psi = E\psi$$

where

$$T=-(\hbar^2/2m)\nabla^2. \tag{1.1}$$

The energy eigenvalues and eigenfunctions are simply given by

$$E_{nL}=Z_{nL}^2\hbar^2/2mR_0^2$$

$$\psi_{nLM}=\bar{N}j_L(Z_{nL}r/R_0)\,Y_{LM}(\theta,\phi) \tag{1.2}$$

where $\hbar$ is Planck's constant, m is the mass of the electron, $\bar{N}$ is a normalization constant, j_L is the Lth order spherical Bessel function whose nth zero we denote by Z_{nL}, and Y_{LM} is the spherical harmonic.

Due to the spherical symmetry of the bounding surface, these eigenfunctions of the kinetic energy are simultaneously eigenfunctions of the square of the total orbital angular momentum operator, $\hat{L}^2$, and of its z-component, $\hat{L}_z$,

$$\hat{L}^2\psi_{nLM}=L(L+1)\hbar^2\psi_{nLM},$$

$$\hat{L}_z\psi_{nLM}=M\hbar\psi_{nLM}, \tag{1.3}$$

where

$$\hat{L}^2=-\hbar^2\left\{\frac{1}{\sin\theta}\frac{\partial}{\partial\theta}\sin\theta\frac{\partial}{\partial\theta}+\frac{1}{\sin^2\theta}\frac{\partial^2}{\partial\phi^2}\right\},$$

$$\hat{L}_z=-i\hbar\frac{\partial}{\partial\theta}. \tag{1.4}$$

The spherical symmetry is further reflected in the independence of E_{nL} on the so-called 'magnetic quantum number', M. It then follows that each energy level E_{nL} is $2(2L+1)$-fold degenerate where the additional factor of 2 is due to the spin degeneracy of spin $\frac{1}{2}$ fermions. The low lying energy spectrum for electrons confined within a spherical particle is shown in Fig. 1. At very low temperatures these levels would be occupied by the conduction electrons in order of increasing energy to form the ground state of the many-electron metallic particle.

The scale of the energy eigenvalues is seen to depend inversely on the square of the radius of the particle. For macroscopic particles, taking into account finite temperatures and interactions which split and broaden the

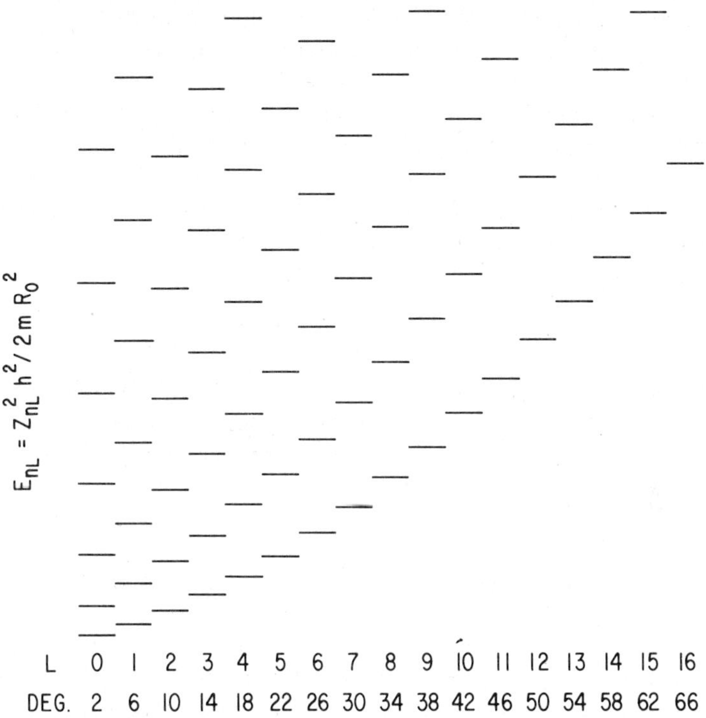

Fig. 1. Energy levels of an electron confined by a spherical surface. Each level of orbital angular momentum L has a degeneracy of $2(2L+1)$.

levels, the single electron states are so close together that the spectrum may be regarded as continuous. It is only recently that techniques have been developed that permit the production and analysis of particles so small (consisting of 10^4–10^5 atoms) that very low temperature measurements of thermodynamic properties are sensitive to the discrete nature of the electron spectrum [1, 2]. It is precisely these quantum size effects, which are not present in matter in bulk, which has excited interest in the small metallic particle problem over the years [1–12]. The statistical character of the discrete spectrum determines the partition function from which are derived thermodynamic properties such as the specific heat at constant volume and the magnetic susceptibility.

The shape of the surface of the particle will determine the energy eigenfunctions and eigenvalues of the conduction electrons. These surfaces cannot however be controlled in the production of the particles. The experiments of Meier and Weider [2] utilized indium particles with diameters of 20 to 100 Å. These particles appeared spherical to within the 5 Å resolution of the electron microscope with which they were examined. However the residual crystalline structure of the particles will certainly give rise to departures from sphericity and these irregularities will in turn destroy the degeneracy of the levels in Fig. 1. Theoretical attention has focused on the influence of the surface irregularities upon the statistical distribution of the energy levels and in particular upon the level spacing distribution. We order the energy eigenvalues

$$E_1 \leqslant E_2 \leqslant \cdots E_i \leqslant E_{i+1} \leqslant \cdots \tag{1.5}$$

denote the ith spacing by $S_i = E_{i+1} - E_i$, and define a normalized spacing by $X_i = S_i / D$ where D is the local average spacing and therefore D^{-1} is a measure of the local level density. The level spacing distribution is then characterized by a frequency function, $\mathbf{P}(X)$, such that $\mathbf{P}(X)\mathrm{d}X$ is the probability that $X_i = (E_{i+1} - E_i)/D$ will be found in the interval X to $X + \mathrm{d}X$. In the original work of Kubo [4], the energy eigenvalues themselves were assumed to be random which leads to a Poisson distribution for $P(X)$ and therefore a maximum for zero spacing. Later Gor'kov and Eliasberg [8] and Kubo [7] argued, without proof, that the irregularities of the surface could be expressed as an effective interaction seen by the electrons and that the matrix elements of this interaction could be considered as random variables. They thereby assumed applicability of the results of the well-established Random Matrix Theory [13–15] which has proven so successful in the statistical analysis of highly excited spectra of nuclei and atoms. The principal idea is that the presence of the off-diagonal elements of the Hamiltonian matrix causes a repulsion of the eigenvalues, as was first observed by Wigner [16], and thereby causes $\mathbf{P}(X)$ to vanish at the origin. Wigner surmised that the correct distribution should be

$$\mathbf{P}(X) = \frac{\pi}{2} X \exp\left(-\frac{\pi}{4} X^2\right) \tag{1.6}$$

and this surmise was not only found to be in excellent agreement with subsequently analyzed nuclear spectra, but was in excellent agreement with the subsequently derived [15] exact distribution which exists in the form of a power series in X.

To date, however, there has been no demonstration of just how irregularities in the geometrical shape of the surface of the particle do manifest themselves in the distribution of level spacings. The objectives of the present investigation [17, 18] is to first establish how the surface irregularities can be incorporated into a Hamiltonian matrix with an effective interaction. Then the surface may be characterized by a set of random variables which in turn determines the frequency function $\mathbf{P}(X)$ for the spacing distribution of the energy eigenvalues.

2. The equivalence of two Schröedinger equations

The surface of the irregular particle will be located in the direction (θ', ϕ') by a displacement vector

$$\mathbf{R}' = \mathbf{R}_0[1 + f(\theta', \phi')]. \tag{2.1}$$

In fact we introduce a mapping

$$\begin{aligned} r' &= r[1 + f(\theta, \phi)] \\ \theta' &= \theta \\ \phi' &= \phi \end{aligned} \tag{2.2}$$

which maps the entire space S confined with a perfect sphere of radius R_0 into the space S' confined by the irregular surface. Under the assumption that this mapping is one-to-one and invertible, the inverse of this mapping is then used to map the space S' onto the space S as is shown in Fig. 2. The Jacobian of this inverse transformation is denoted by J^{-1}. We may use this mapping to establish equivalent Schröedinger equations in the two spaces. The problem we would like to solve is the Schröedinger equation for the free particle in S'

$$H(x')\psi(x') = E\psi(x') \tag{2.3}$$

where H is the kinetic energy operator T of Equation (1.1) subject to the eigenfunction satisfying Dirichlet boundary conditions ($\psi = 0$) on the irregular surface. The Jacobian of the inverse transformation can be used, as shown in Fig. 2, to define a new Hamiltonian operator $\bar{H}(x)$ and eigenfunction $\bar{\psi}(x)$ in S and where $\bar{\psi}(x)$ satisfies the same Dirichlet boundary condition ($\bar{\psi} = 0$) on the spherical surface $r = R_0$. The new Schröedinger

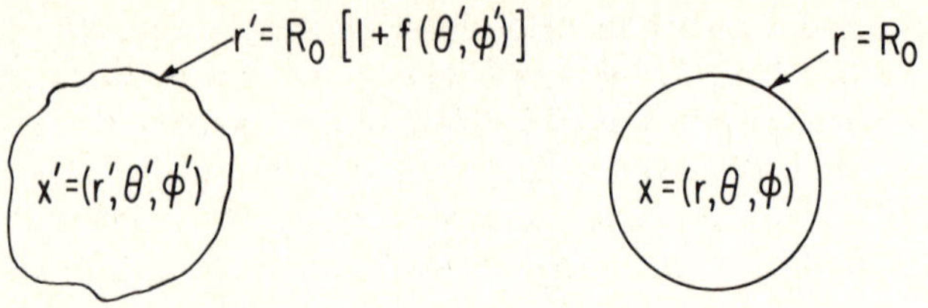

$$H(x')\psi(x') = E\psi(x') \qquad \bar{H}(x)\bar{\psi}(x) = E\bar{\psi}(x)$$

$$\bar{\psi}(x) = J^{-1/2}(x)\,\psi(x'(x))$$

$$\bar{H}(x) = J^{-1/2}(x)\,H(x'(x))\,J^{1/2}(x)$$

$$r = \frac{r'}{1+f(\theta',\phi')}$$

$$\theta = \theta'$$

$$\phi = \phi'$$

$$J^{-1}(x) = \left|\frac{\partial(r',\theta',\phi')}{\partial(r,\theta,\phi)}\right|$$

Fig. 2. The equivalence of the Schröedinger equations related by the mapping of the deformed region onto the spherical region.

equation in S

$$\bar{H}(x)\bar{\psi}(x) = E\bar{\psi}(x) \tag{2.4}$$

yields a spectrum identical to that of Equation (2.3).

3. The perturbation expansion

We arrive at a perturbation approach to the solution of Eq. (2.4) by introducing a parameter η to describe a continuous mapping

$$\mathbf{R}'(\eta) = \mathbf{R}_0[1+\eta f(\theta',\phi')] \tag{3.1}$$

of $\mathbf{R}_0$ onto the final $\mathbf{R}'$ as η varies from zero to unity. We then expand the

Hamiltonian and Jacobian in powers of η:

$$J^{\frac{1}{2}}(\eta)=1+\frac{1}{2}\eta\left(\frac{\partial J}{\partial\eta}\right)_0-\frac{1}{8}\eta^2\left(\frac{\partial J}{\partial\eta}\right)_0^2+\frac{1}{4}\eta^2\left(\frac{\partial^2 J}{\partial\eta^2}\right)_0+\cdots$$

$$J^{-\frac{1}{2}}(\eta)=1-\frac{1}{2}\eta\left(\frac{\partial J}{\partial\eta}\right)_0+\frac{3}{8}\eta^2\left(\frac{\partial J}{\partial\eta}\right)_0^2-\frac{1}{4}\eta^2\left(\frac{\partial^2 J}{\partial\eta^2}\right)_0+\cdots$$

$$H(\eta)=T+\eta\left(\frac{\partial H}{\partial\eta}\right)_0+\frac{1}{2}\eta^2\left(\frac{\partial^2 H}{\partial\eta^2}\right)_0+\cdots \tag{3.2}$$

and combine these to yield

$$\begin{aligned}\overline{H}(\eta)&=J^{-\frac{1}{2}}(\eta)H(\eta)J^{\frac{1}{2}}(\eta)\\&=T+\eta\left\{-\frac{1}{2}\left(\frac{\partial J}{\partial\eta}\right)_0T+\left(\frac{\partial H}{\partial\eta}\right)_0+\frac{1}{2}T\left(\frac{\partial J}{\partial\eta}\right)_0\right\}\\&\quad+\eta^2\left\{\frac{3}{8}\left(\frac{\partial J}{\partial\eta}\right)_0^2T-\frac{1}{4}\left(\frac{\partial^2 J}{\partial\eta^2}\right)_0T-\frac{1}{8}T\left(\frac{\partial J}{\partial\eta}\right)_0^2\right.\\&\qquad+\frac{1}{4}T\left(\frac{\partial^2 J}{\partial\eta^2}\right)_0+\frac{1}{2}\left(\frac{\partial^2 H}{\partial\eta^2}\right)_0-\frac{1}{2}\left(\frac{\partial J}{\partial\eta}\right)_0\left(\frac{\partial H}{\partial\eta}\right)_0\\&\qquad\left.+\frac{1}{2}\left(\frac{\partial H}{\partial\eta}\right)_0\left(\frac{\partial J}{\partial\eta}\right)_0-\frac{1}{4}\left(\frac{\partial J}{\partial\eta}\right)_0T\left(\frac{\partial J}{\partial\eta}\right)_0\right\}\\&\quad+\eta^3\cdots.\end{aligned} \tag{3.3}$$

For the mapping on Fig. 2, we then have

$$\begin{aligned}\left(\frac{\partial J}{\partial\eta}\right)_0&=-f\\\left(\frac{\partial H}{\partial\eta}\right)_0&=-2fT-\left[\tilde{L}^2,f\right]\tilde{R}\\\left(\frac{\partial^2 H}{\partial\eta^2}\right)_0&=6f^2T+5f\left[\tilde{L}^2,f\right]\tilde{R}+\left[\tilde{L}^2,f\right]f\tilde{R}+2\left\{\frac{1}{2}\left[\tilde{L}^2(f^2)\right]-f\left[\tilde{L}^2(f)\right]\right\}\tilde{R}^2\\\left(\frac{\partial^2 J}{\partial\eta^2}\right)_0&=2f^2\end{aligned} \tag{3.4}$$

where $\tilde{L}^2 = \hat{L}^2/2mr^2, \tilde{R} = r\partial/\partial r$, [,] is the usual commutator and $[\tilde{L}^2(\ )]$ indicates that $\tilde{L}^2$ only operates on what is contained in ( ) so that [ ] is a multiplicative factor. The final form of the Hamiltonian is then

$$\begin{aligned}\overline{H}(\eta) = T + \eta\Big\{ -\frac{3}{2}fT - \frac{1}{2}Tf - [\tilde{L}^2, f]\tilde{R}\Big\} \\ + \eta^2\Big\{\frac{15}{8}f^2T + \frac{3}{4}fTf + \frac{3}{8}Tf^2 \\ + (-2f^2\tilde{L}^2 + f\tilde{L}^2f + \tilde{L}^2f^2)\tilde{R} \\ + \Big(\frac{1}{2}[\tilde{L}^2(f^2)] - f[\tilde{L}^2(f)]\tilde{R}^2\Big\} \\ + \eta^3 \cdots . \end{aligned} \tag{3.5}$$

The zeroth order solution of $\overline{H}(x,\eta)\overline{\psi}(x,\eta) = E(\eta)\overline{\psi}(x,\eta)$ is therefore seen to be that of the free particle confined by a spherical surface whose highly degenerate spectrum was displayed in Fig. 1. The unusual feature of a perturbation expansion arrived at through such a mapping procedure is that the perturbed operator in the eigenvalue problem is itself a power series in η. If we write $\overline{H}(\eta) = T + V(\eta)$, then $V(\eta)$, as realized in Equation (3.5), may be regarded as the effective interaction experienced by electrons confined by a spherical surface which is the equivalent of the irregular surface experienced by otherwise free electrons. For reasons of space limitations, we shall report results in which we retain only the linear term in Equation (3.5). Our calculations have been carried out [17, 19] through second order and do not yield results qualitatively different from the first order results.

4. Matrix elements of the effective interaction

At very low temperatures (few degrees Kelvin) the electrons may be considered to occupy only the states of lowest energy consistent with the Pauli exclusion principle. In such a small particle, the conduction electrons with highest energy may then be considered to be distributed only over the subset of states which arise from the splitting of the highest lying of the occupied degenerate levels of the sphere. This level E_{nL} of Equation (1.2) will be characterized by a fixed value of the orbital angular momentum quantum number L. The degenerate states ψ_{nLM} of the zeroth order solution of Equation (3.5), i.e. Equation (1.2), are distinguished by the $2L+1$ possible values of the magnetic quantum number $M(M=-L,-L+1,\ldots,L-1,L)$. We shall ignore the 2-fold spin degeneracy in our analysis since that degree of freedom of the electron is unaffected by deformations in the geometry of the surface of the particle. Thus the single electron energy levels of interest to us are obtained in first order by the diagonalization of

$$V_1(\eta)=\eta\left\{-\frac{3}{2}fT-\frac{1}{2}Tf-\left[\tilde{L}^2,f\right]\tilde{R}\right\} \tag{4.1}$$

in the degenerate subspace spanned by the $2L+1$ degenerate eigenvectors ψ_{nLM} of T. Since all the states of the subspace are eigenstates of both T and $\hat{L}^2$ the matrix elements of $V_1(\eta)$ reduce simply to

$$\langle nLM|V_1(\eta)|nLM'\rangle=-2\eta E_{nL}\langle LM|f|LM'\rangle \tag{4.2}$$

where

$$\langle LM|f|LM'\rangle=\int d\Omega\, Y^*_{LM}f(\theta,\phi)\,Y_{LM'}, \tag{4.3}$$

the rows and columns of the matrix being labeled by the magnetic quantum numbers M and M'.

The function $f(\theta,\phi)$ completely characterizes the irregular surface of the particle through Equation (3.1). We must choose a general form for $f(\theta,\phi)$ capable of describing arbitrary deformations. We expand $f(\theta,\phi)$ as a linear combination of spherical harmonics

$$f(\theta,\phi)=\sum_{\lambda=1}^{\infty}\sum_{\mu=-\lambda}^{\lambda}C_{\lambda\mu}Y_{\lambda\mu}(\theta,\phi) \tag{4.4}$$

where the reality of $f(\theta,\phi)$ imposes the conditions

$$\begin{aligned}\operatorname{Re}C_{\lambda-\mu}&=(-1)^{\mu}\operatorname{Re}C_{\lambda\mu}\\ \operatorname{Im}C_{\lambda-\mu}&=(-1)^{\mu+1}\operatorname{Im}C_{\lambda\mu}.\end{aligned} \tag{4.5}$$

Subject to these conditions, the coefficients $\operatorname{Re}C_{\lambda\mu}(\mu=0,1,\ldots\lambda)$ and $\operatorname{Im}C_{\lambda\mu}(\mu=1,2,\ldots,\lambda)$ are then chosen as independent random variables for each λ. The orthonormal property of the spherical harmonics guarantees that by omission of the term with $\lambda=0$ in Equation (4.4), the value of

$\mathbf{R}'(\theta',\phi')$ in Equation (3.1) averaged over all θ' and ϕ' will be $\mathbf{R}_0$ so that the volume of space seen by the electron in the deformed and spherical particles will be essentially unchanged. The spherical harmonic expansion is further chosen so that in each order of perturbation theory the matrix element of $V(\eta)$ may be reduced using the well-known Racah algebra of the full rotation group in three dimensions. Therefore in first order we have

$$\langle nLM|V_1(\eta)|nLM'\rangle$$

$$= -\pi^{-\frac{1}{2}}\eta E_{nL}\sum_{\lambda\mu} C_{\lambda\mu}(2\lambda+1)^{\frac{1}{2}}\langle LM'\lambda\mu|LM\rangle\langle L0\lambda0|L0\rangle. \tag{4.6}$$

The multiplicative factors, $-\pi^{-\frac{1}{2}}\eta E_{nL}$, are common to every matrix element and as such will not influence the distribution of normalized spacings, $\mathbf{P}(X)$. The $\langle|\rangle$ are the usual Clebsch–Gordan coefficients, the real elements of the unitary transformation relating coupled and uncoupled representations of the rotation group. The presence of $\langle L0\lambda0|L0\rangle$ guarantees that in first order only even values of λ will contribute for reasons of parity. The presence of $\langle LM'\lambda\mu|LM\rangle$ fixes the value of $\mu=M-M'$ which, for each λ, will then be common to every element along a ray parallel to the diagonal. The Clebsch–Gordan coefficients likewise place an upper limit on the λ sum, $\lambda_{max}=2L$.

5. Results and analysis

We report results [17, 19] of an analysis of the splitting of an $L=13$ level by the deformation of the surface of the particle. The choice of $L=13$ was determined by an existing analysis in the literature [2] of the magnetic susceptibility based upon the assumption of perfectly spherical particles with only the $L=13$ level being magnetically active. Each particle's surface is then characterized by a choice of 195 values for $\operatorname{Re} C_{\lambda\mu}$ and 182 values for $\operatorname{Im} C_{\lambda\mu}$ for even values of $\lambda \leqslant 26$. These 377 variables are chosen randomly from a Gaussian distribution with zero centroid and unit width. The resulting $2(13)+1=27$ dimensional matrix is then diagonalized to yield 26 spacings S_i which are normalized, $X_i=S_i/D$, and combined with the results from an ensemble of such particles to produce the frequency function $\mathbf{P}(X)$ of the nearest neighbor spacing distribution. The required normalization was achieved in two distinct ways. In the first, which we refer to as the *matrix average* (MA), D is computed for the ith spacing of the jth member of the ensemble by averaging over a consecutive set of seven spacings of the jth matrix centered on the ith spacing. Thus

$$D_{\mathrm{MA}}^{(i)} = \sum_{k=i-3}^{k=i+3} S_k^{(j)}/7. \tag{5.1}$$

Alternately the local average spacing can be computed as an average over a fixed spacing of all members of the ensemble. In this way we define the *ensemble average* (EA),

$$D_{\mathrm{EA}}^{(i)} = \sum_{j=1}^{W} S_i^{(j)}/W \tag{5.2}$$

where W is the total number of particles making up the ensemble.

In Fig. 3 we display the results for an ensemble of 200 particles which give rise to 5200 normalized spacings. This histogram was prepared using the 'matrix average' for D but is essentially identical with that using the

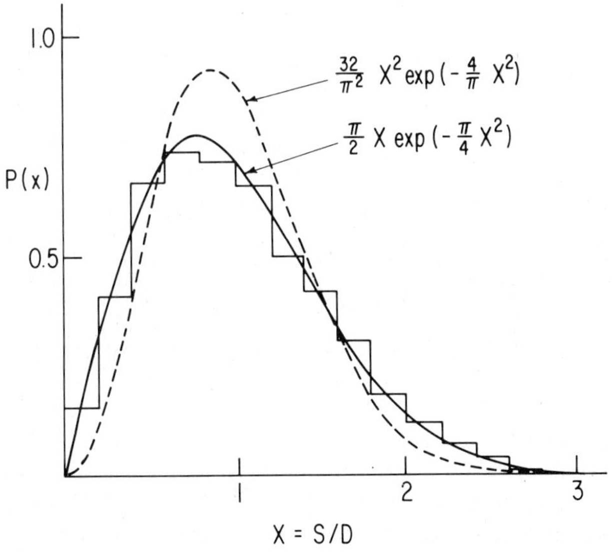

Fig. 3. Nearest neighbor spacing distribution for an ensemble of 200 irregular particles. Superimposed are the predictions of random matrix theory for real symmetric (solid) and complex (dashed) hermitian matrices.

'ensemble average'. The histogram is seen to be in excellent agreement with the Wigner surmise indicated by the solid line. The agreement with the Wigner surmise is perhaps somewhat surprizing in that the Wigner surmise is an excellent approximation to the true spacing distribution for the Hamiltonian of a system that is time reversal invariant and rotationally invariant [13] as is true for electromagnetic and nuclear interactions. Thus for atoms and nuclei the Hamiltonian matrix can be expressed as a real, symmetric matrix. However, if the system is not time reversal invariant then the Hamiltonian matrix is hermitian but its elements are complex. In this case the spacing distribution changes [13] and is well approximated by

$$\mathbf{P}(X)=\frac{32}{\pi^2}X^2\exp\left(-\frac{4}{\pi}X^2\right) \tag{5.3}$$

as is also shown in Fig. 3 by the dashed line. This prediction is in definite disagreement with the historgram despite the fact that the elements of our matrix are complex. We have no explanation of this fact but would observe that the number (377) of independent random variables which determine our matrices is much closer to the number, $\frac{1}{2}(27)(27-1)=351$, of real numbers that make up a real, symmetric 27×27 random matrix than the number (675) of real scalars that make up a complex, hermitian 27×27 random matrix.

The case just considered may be referred to as that of the 'maximally rough surface' because every spherical harmonic $Y_{\lambda\mu}$ was present in $f(\theta,\phi)$ with equal weight. The larger the value of λ, the more rapid is the oscillating behavior of $Y_{\lambda\mu}$. To consider the effect of the surface becoming progressively smoother we performed a series of calculations in which $V_1(\eta)$ was altered:

$$\langle nLM|V_1(\eta)|nLM'\rangle$$
$$=\pi^{-\frac{1}{2}}\eta E_{nL}\sum_{\lambda\mu}\lambda^{-N}C_{\lambda\mu}(2\lambda+1)^{\frac{1}{2}}\langle LM'\lambda\mu|LM\rangle\langle L0\lambda0|L0\rangle \tag{5.4}$$

by the incorporation of a weighting factor, λ^{-N}, which surpresses the rapidly varying high order spherical harmonics. This was found to have a strong effect upon the spacing distribution. In Table 4 we display the 26 spacings averaged over the ensemble of 50 particles for each value of N. These spacings have been renormalized for display so that unity represents the average of the 26 ensemble average spacings. We note that the ensemble average spacing varies in a systematic way upon location in the

Table 1.
The normalized ensemble averages for each spacing arising from the eigenvalues of Equation (5.4).

Spacing	$N=0$	$N=1$	$N=2$	$N=3$	$N=4$	$N=5$
1	1.606	1.097	0.118	0.00991	0.000945	0.0000870
2	1.291	2.157	2.883	2.757	2.759	2.724
3	1.199	0.956	0.122	0.0121	0.00198	0.000251
4	1.051	1.631	2.308	2.377	2.491	2.461
5	1.078	0.757	0.139	0.0289	0.00976	0.00776
6	0.961	0.999	1.876	2.009	2.212	2.176
7	0.950	0.883	0.187	0.0862	0.0444	0.0481
8	0.793	0.913	1.548	1.715	1.894	1.840
9	0.915	0.697	0.246	0.203	0.142	0.161
10	0.860	0.728	1.231	1.447	1.581	1.458
11	0.718	0.745	0.440	0.397	0.335	0.371
12	0.773	0.694	1.003	1.072	1.118	1.030
13	0.862	0.795	0.699	0.696	0.710	0.702
14	0.874	0.736	0.709	0.703	0.710	0.644
15	0.757	0.779	0.989	1.088	1.036	1.135
16	0.910	0.667	0.455	0.401	0.368	0.298
17	0.830	0.692	1.285	1.454	1.477	1.575
18	0.684	0.760	0.317	0.196	0.140	0.118
19	0.846	0.900	1.724	1.889	1.895	1.952
20	0.773	0.688	0.178	0.0791	0.0340	0.0462
21	1.035	1.004	1.887	2.058	2.245	1.843
22	1.041	0.902	0.141	0.0311	0.00564	0.00382
23	0.922	1.279	2.219	2.376	2.011	2.555
24	1.144	0.942	0.112	0.0161	0.00114	0.000342
25	1.234	2.536	3.113	2.907	2.855	2.829
26	1.892	1.067	0.115	0.0118	0.000854	0.000137

spectrum which is why a local average D must be used to normalize the spacings. But the most noticeable feature is that as N increases a systematic pattern develops of alternately broad and narrow spacings. The ratio of a broad spacing to that of its neighboring narrow spacing increases rapidly with N. Finally we note that this same ratio also increases rapidly as we move from the center of the spectrum in the direction of either extreme.

The development of this pattern as the particle surface becomes smoother by increasing N is immediately reflected in the nearest neighbor spacing distribution. In Fig. 4 we consider six ensembles each consisting of 50 particles and utilize the matrix average definition of D. Strong departures from the Wigner surmise are found beginning with $N=2$. This bimodal frequency function is easily understood on the basis of the

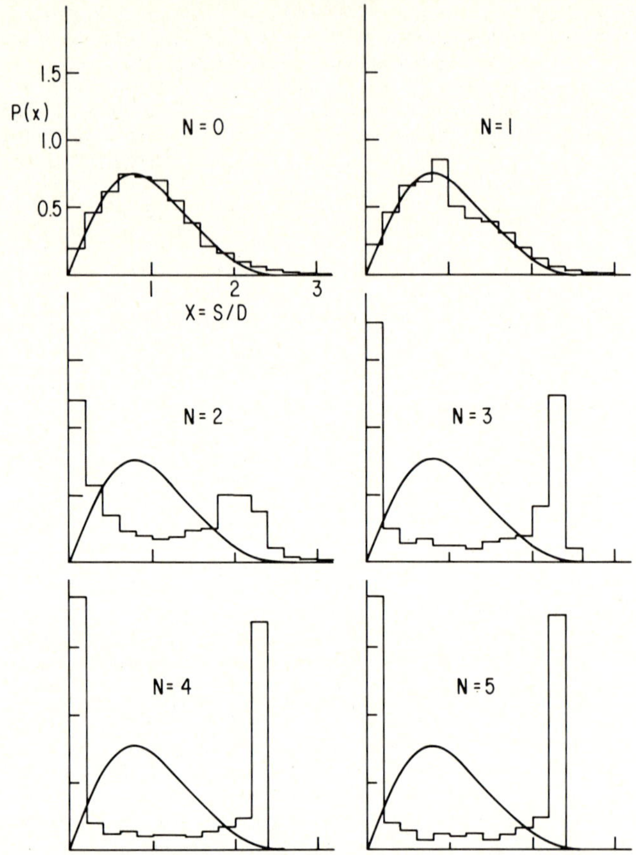

Fig. 4. Nearest neighbor spacing distributions for ensembles of 50 irregular particles. The ensembles differ by the weighting factor λ^{-N} in Equation (5.4) so that the surface is getting smoother as N increases. Matrix averaging is used in the definition of D.

definition of D_{MA} in Eq. (5.1). A narrow spacing will naturally contribute only at the origin. For a broad spacing, D_{MA} will consist of the average of 3 broad and 4 narrow spacings and therefore S/D_{MA} may be expected to lie in the neighborhood of $7/3$ as is seen. Therefore the location of the second peak is a function of the number of consecutive spacings used in the definition of D_{MA}.

In Fig. 5 we consider the same six ensembles of 50 particles as was used in Fig. 4 but this time we utilize the ensemble average definition of D. Since D_{EA} in Eq. (5.2) employs the same spacing of each matrix in the

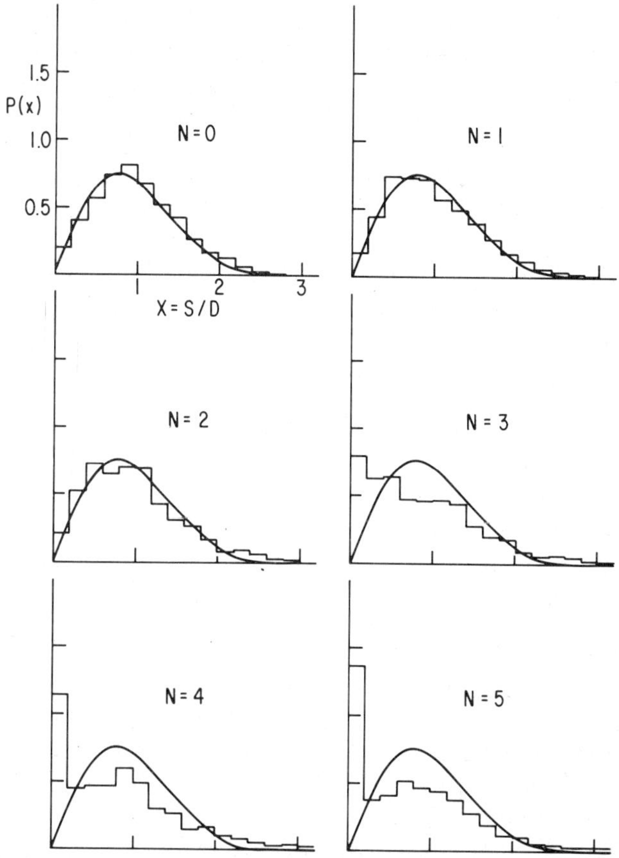

Fig. 5. The same case considered in Fig. 4 but with ensemble averaging used for D.

ensemble there is no a priori reason why the narrow spacings should make their contribution at the origin of $\mathbf{P}(X)$ while the broad spacings make their contribution at large values of X. Indeed Fig. 5 shows that departure from the Wigner surmise occurs more slowly with increasing N. However a strong peak at the origin does develop. Upon detailed investigation this peak is indeed found to have its contributions come from the small spacings but the explanation is more subtle than was true in Fig. 4. If we refer back to Table 1, we see that the ensemble average of all spacings is essentially symmetric about the geometrical center of the spectrum and further we had noted that the size of the narrow spacings decreases rapidly with distance from the "spectral center." The individual members of the ensemble share this second property however the location of this "spectral center" need not coincide with the geometrical center between spacings 13 and 14. Indeed the individual matrices may exhibit a spectral center displaced five or six spacings from the geometrical center. It then follows that D_{EA}, for a small spacing, can be dominated by the one small spacing which arises from that member of the ensemble whose spectral center lies closest to the small spacing in question. This will then cause contributions from the other members of the ensemble, whose narrow spacing at this point is an order of magnitude or more smaller than that of the dominant narrow spacing, to be made near the origin $X=0$ resulting in the peak in $\mathbf{P}(X)$ in Fig. 5.

When a physicist encounters a nearest neighbor spacing distribution which does not exhibit level repulsion by vanishing at the origin, the immediate suggestion of Random Matrix Theory [13, 14] is that one is looking at the merged spectra arising from different representations of exact or nearly exact symmetries of the Hamiltonian operator. This is not however the case here. Rosenzweig [14] has discussed how the value of $\mathbf{P}(X)$ at $X=0$ will depend upon the merger of the spectra arising from uncorrelated subspaces. In the limit as the number of such subspaces increases, $\mathbf{P}(X)$ at $X=0$ is bounded by unity in which case $\mathbf{P}(X)$ becomes the Poisson distribution. This bound on $\mathbf{P}(0)$ is strongly violated in both Figs. 4 and 5—we are looking at a highly correlated spectra in the limit as the particle becomes smoother. At this point we should mention that Figs. 4 and 5 are altered very little when we perform the second order perturbation calculation.

To see the origin of this correlation we recognize that in the limit as N becomes large, Eq. (4.6) becomes dominated by the $\lambda=2$ term and we have

$$\begin{aligned}&\langle nL=13M|V_1(\eta)|nL=13M'\rangle \\ &\qquad \simeq(\text{constant})\; C_{2M-M'}\langle 13M'2M-M'|13M\rangle.\end{aligned} \tag{5.5}$$

Therefore the 27×27 matrix collapses to a band of elements consisting of 5 rays centered on the diagonal. The number of important random variables reduces to 5, one for the real number C_{20} which determines the diagonal, and two each for the complex numbers C_{21} and C_{22} which determine the next two rays parallel to the diagonal. All other matrix elements essentially vanish. In Fig. 6 we show the variation of the Clebsch–Gordan coefficients and therefore of the matrix elements as we move along the diagonal and along the next two rays parallel to the diagonal. It is this smooth, systematic variation of the matrix elements which induces the strong

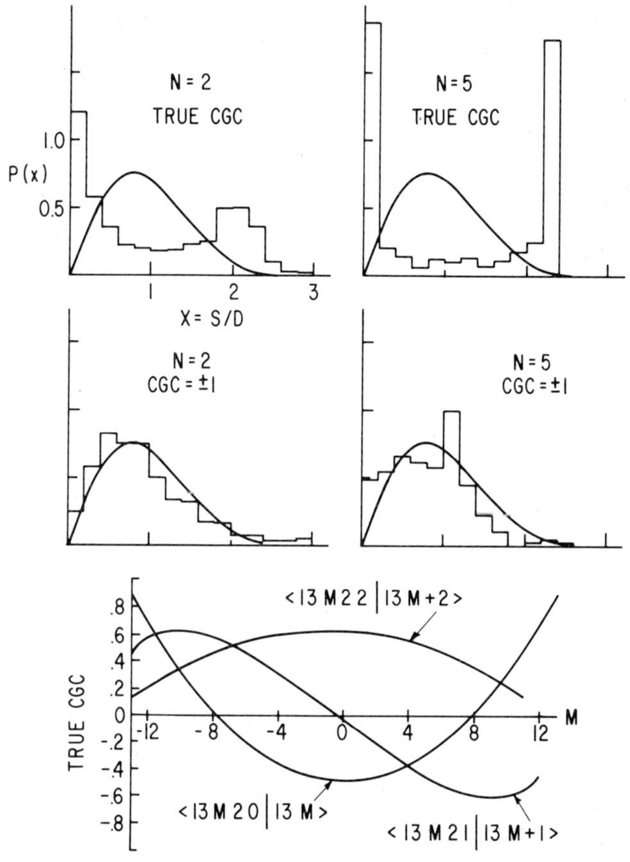

Fig. 6. Matrix averaged, nearest neighbor spacing distributions using true Clebsch–Gordan coefficients and using Clebsch–Gordan coefficients set equal to unity with random sign. At the bottom we show the variation of the Clebsch–Gordan coefficients with the magnetic quantum number.

correlation in the spectrum. This is reminiscent of a similar behavior seen by Wilkinson [20] in the case of tri-diagonal matrices. In fact for pure $\lambda=2$ (corresponding to $N\rightarrow\infty$), double precision calculations show that the ratio of large to neighboring small spacing can become as large as 10^9 when $L=13$. Corresponding behaviors are found for other values of L.

We stress the point that the correlation of the spectra is the result of the smooth, systematic variation of the Clebsch-Gordan coefficients and is not the product of having so many vanishingly small off-diagonal matrix elements. To make this clear we return to Fig. 6 where we record the nearest neighbor spacing distribution in the case of two ensembles of 50 particles ($N=2$ and $N=5$) in which the true Clebsch–Gordan coefficients have been replaced by unity with a randomly chosen sign. The correlation, which is reflected in $\mathbf{P}(X)$ when the true Clebsch–Gordan coefficients are used, is now seen to be completely destroyed.

Our interest at present centers on how such strong correlations will affect the partition function and thereby the thermodynamic properties of small metallic particles. I wish to express my appreciation to J. Tavel, M. Tavel, N. Rosenzweig, P. Kahn, D. Fox, L. Roth and J. Corbett for illuminating discussions of this problem.

References

[1] Novotny, V., Meincke, P. P. M. and Watson, J. H. P. (1972). Effect of size and surface on the specific heat of small lead particles. *Phys. Rev. Lett.*, **28** (14) 901–903.

[2] Meier, F. and Wyder, P. (1973). Magnetic moment of small indium particles in the quantum size-effect regime. *Phys. Rev. Lett.*, **30**, (5) 181–184.

[3] Frohlich, H. (1937). The specific heat of the electrons of small metallic particles at low temperatures. *Physica*, **4** (5) 406–412.

[4] Kubo, R. (1962). Electronic properties of metallic fine particles I. *J. Phys. Soc. Japan*, **17** (6) 975–986.

[5] Kubo, R. and Kawabata, A. (1966). Electronic properties of fine metallic particles II. Plasma Resonance Absorption. *J. Phys. Soc. Japan*, **21** (9) 1765–1771.

[6] Kawabata, A. (1970). Electronic properties of fine metallic particles III. E.S.R. Absorption Line Shape. *J. Phys. Soc. Japan*, **29**, (4) 902–911.

[7] Kubo, R. (1969). Electrons in small metallic particles in "Polarization matiere et rayonnement". *Livre de Jubile L'Honneur du Prof. A. Kastler*, pp. 325–339. Presses Universitaires de France, Paris.

[8] Gor'kov, L. P. and Éliashberg, G. M. (1965). Minute metallic particles in an electromagnetic field. *Soviet Physics JETP*, **21** (5) 940–947.

[9] Denton, R., Mühlschlegel, B. and Scalapino, D. J. (1971). Electronic heat capacity and susceptibility of small metal particles. *Phys. Rev. Lett.*, **26**, (12) 707–711.

[10] Mühlschlegel, B., Scalapino, D. J. and Denton, R. (1972). Thermodynamic properties of small superconducting particles. *Phys. Rev. B.*, **6** (5) 1767–1777.

[11] Denton, R., Mühlschlegel, B. and Scalapino, D. J. (1973). Thermodynamic properties of electrons in small metal particles. *Phys. Rev. B.*, **7** (8) 3589–3607.

[12] Barojas, J., Cota, E., Blaisten-Barojas, E., Flores, J. and Mello, P. A. (1977). Studies on the problem of small metallic particles I.-Spectrum fluctuations in a two-dimensional model and the associated specific heat. *Ann. Physics*, **107**, 95–109.

[13] Porter, C. E. (1965). *Statistical Theories of Spectra: Fluctuations*. Academic Press, New York. (This volume also collects reprints of all papers in Random Matrix Theory through 1963.)

[14] Rosenzweig, N. (1963). Statistical mechanics of equally likely quantum systems, In

Statistical Physics—1962 Brandeis Lectures, Vol. 3, pp. 93–158. W. A. Benjamin, New York.

[15] Mehta, M. L. (1967). *Random Matrices and the Statistical Theory of Energy Levels*. Academic Press, New York.

[16] Wigner, E. P. (1957). Statistical properties of real symmetric matrices with many dimensions. In *Proc. Canadian Math. Congr.* pp. 174–184. University of Toronto Press, Toronto (This paper is reprinted in [13].)

[17] Tavel, J. (1978). Ph.D. dissertation, State University of New York at Albany.

[18] Tavel, J., Ratcliff, K. F. and Rosenzweig, N. To be published.

[19] Tavel, J., Ratcliff, K. F. and Tavel, M. To be published.

[20] Wilkinson, J. H. (1965). *The Algebraic Eigenvalue Problem*. See especially pp. 308–312. Clarendon Press, Oxford.

P. R. Krishnaiah, ed., *Multivariate Analysis–V*

LATENT TRAIT THEORY AND ITS APPLICATIONS*

Fumiko SAMEJIMA
University of Tennessee, Knoxville, TN, U.S.A.

Latent trait theory has been developed in the area of psychology and sociology, from the necessity of estimating latent parameters from observable phenomena. It includes various mathematical models, which deal with both discrete and continuous responses, in both uni-dimensional and multi-dimensional latent spaces. As a mental test theory, the most popular family of models may be those with the uni-dimensional latent space and binary test items, in which a response to a test item is treated as a Bernoulli trial. For binary responses, the item characteristic function plays an important role, as does the operating characteristic of the item response category for graded responses, and the operating density characteristic of the item response for continuous responses. The theory is much more generalized and has broader areas of application when we turn to the graded response models, the continuous response models, and the multi-dimensional latent space.

The theory enables us to investigate both item parameters and individual subject parameters. One of its greatest contributions in application may be found in tailored testing, or computerized adaptive testing. It is a method of measuring an individual's ability efficiently, by selecting an optimal set of test items to an individual subject from a large item pool, mostly with the aid of an electronic computer. The accuracy of estimation of the individual parameter can be controlled individually by an effective use of information functions.

1. Introduction

Latent trait theory deals with the situation in which there are a certain number of *items*, the scores on which are the sole indicator of the individual's *latent trait*, which is not directly observable. If, for instance, we wish to measure a certain mathematical ability of tenth graders, we are likely to develop and administer a test consisting of a certain number of suitable mathematical problems. In this example, the latent trait is the mathematical ability that we are interested in, and the items are the mathematical problems that we have developed. To give another example, Roche, Wainer and Thissen have developed a scale for measuring the skeletal maturity of children of eighteen years of age or younger, using latent trait theory (Roche, Wainer and Thissen, 1975). In this research,

*Research is partly supported by the Office of Naval Research under Contract N00014-77-C-360, NR. 150-402.

they used the knee joint as the biological indicator of the skeletal maturity, i.e., the latent trait, and the items are various radiographs of the knee joint.

The statistical relationship between the set of item scores and the latent trait takes an essential role in the latent trait theory, therefore, and through this relationship the individual's position in the latent space is estimated. In so doing, the *item characteristic function*, the *operating characteristic of the graded item response category*, or the *operating density characteristic of the continuous item response*, plays an important role, depending upon the nature of the item and its scoring strategy.

Latent trait theory is basically a *population-free* theory, in the sense that the distribution of the item score is specified conditionally, for a specific point in the latent space. Fisher type information functions enable us to apply the theory interestingly and meaningfully, in combination with the maximum likelihood estimation of the individual's latent trait.

2. Rationale

Let θ be the *latent trait*, such that

$$\theta' = (\theta_1, \theta_2, \ldots, \theta_s, \ldots, \theta_k), \tag{2.1}$$

where

$$-\infty < \theta_s < \infty, \qquad s = 1, 2, \ldots, k. \tag{2.2}$$

When $k = 1$, the *latent space* is uni-dimensional, and multi-dimensional otherwise. Let g $(= 1, 2, \ldots, n)$ denote an *item*. We assume a continuous process behind the set of item scores for item g, and it is denoted by X_g. The response to item g can be scored continuously, or into a finite or countable number of categories, depending upon the nature of the item and the scoring strategy. We call the former situation the *continuous response level*, and the latter the *discrete response level*. The discrete response level is further classified into the *nominal response level*, on which these categories are not ordered, and the *graded response level*, on which they are (Samejima, 1972). The graded response level includes the *dichotomous response level* as a special case, in which there are only two score categories.

On the continuous response level, the item score z_g is a strictly increasing function of the continuous process X_g. Let $\mathbf{V}$ be a *response pattern*, or a vector of n item scores such that

$$\mathbf{V}'=(z_1,z_2,\ldots,z_g,\ldots,z_n). \tag{2.3}$$

Thus the specification of the conditional joint distribution of the response pattern $\mathbf{V}$, given a specific point in the latent space, is the crucial issue in the theory. On the graded response level, z_g is replaced by the graded response category x_g $(=0,1,2,\ldots,m_g)$.

A general model for the continuous response level defines the conditional distribution function of z_g, given θ, by $[1-P^*_{z_g}(\theta)]$, where

$$P^*_{z_g}(\theta)=\int_{-\infty}^{a_g'(\theta-b_{z_g})}\psi_g(u)\,\mathrm{d}u, \tag{2.4}$$

and a_g is the item parameter vector of order k, b_{z_g} is the item response parameter vector of order k, and ψ_g is any probability density function. The *operating density characteristic* of z_g is given by

$$H_{z_g}(\theta)=\psi_g\big[a_g'(\theta-b_{z_g})\big]\left[a_g'\frac{\mathrm{d}}{\mathrm{d}z_g}b_{z_g}\right] \tag{2.5}$$

(Samejima, 1973, 1974). On the graded response level, if we replace $P^*_{z_g}(\theta)$ by $P^*_{x_g}(\theta)$ and b_{z_g} by b_{x_g} in (2.4), we have, for the conditional distribution function of x_g, $[1-P^*_{(x_g+1)}(\theta)]$ in the *homogeneous case*, which is defined in contrast with the *heterogeneous case* (Samejima, 1972). The *operating characteristic* of x_g is derived such that

$$P_{x_g}(\theta)=\int_{a_g'(\theta-b_{x_g+1})}^{a_g'(\theta-b_{x_g})}\psi_g(u)\,\mathrm{d}u \tag{2.6}$$

(Samejima, 1969, 1972). On the dichotomous response level, (2.6) for $x_g=1$ is known as the *item characteristic function* (Lord and Novick, 1968).

It is assumed that the n item scores are conditionally independent, so that the conditional joint operating density characteristic $H_V(\theta)$ can be written as

$$H_V(\theta)=\prod_{z_g\in V}H_{z_g}(\theta). \tag{2.7}$$

This is also the *likelihood function* in estimating the individual's latent trait from his response pattern.

When $\psi_g(u)$ is specified by the standard normal density such that

$$\psi_g(u)=\frac{1}{\sqrt{2\pi}}\exp[-u^2/2], \tag{2.8}$$

the model is called the *normal ogive model*. It has been shown that on the continuous response level, there exists the *vector of sufficient statistics*, $t(V)$, in this model, such that

$$t(V)=\sum_{z_g\in V} a_g a_g' b_{z_g}, \tag{2.9}$$

and the *maximum likelihood estimator*, $\hat{\theta}$, is given by

$$\hat{\theta}=\left[\sum_{g=1}^{n} a_g a_g'\right]^{-1} t(V) \tag{2.10}$$

(Samejima, 1974). Moreover, it has been shown that, under a general condition, each element of $\hat{\theta}$ is, given θ, conditionally unbiased with the minimum mean square error subject to unbiasedness (Samejima, 1974), like the Bartlett's factor score estimator (Bartlett, 1937; Lawley and Maxwell, 1971) in linear factor analysis.

The *item response information matrix* $I_{z_g}(\theta)$, is defined as the (kxk) symmetric matrix whose element in the sth row and tth column is written as

$$-\frac{\partial^2}{\partial\theta_s\partial\theta_t}\log H_V(\theta), \tag{2.11}$$

on the continuous response level (Samejima, 1974). The *item information matrix*, $I_g(\theta)$, can be written as

$$I_g(\theta)=\mathbf{E}[I_{z_g}(\theta)], \tag{2.12}$$

and we have for the *response pattern information matrix* and the *test information matrix*

$$I_V(\theta)=\sum_{z_g\in V} I_{z_g}(\theta) \tag{2.13}$$

and

$$I(\theta)=E[I_V(\theta)]=\sum_{g=1}^{n} I_g(\theta). \tag{2.14}$$

In the normal ogive model, we can write

$$I_{z_g}(\theta)=I_g(\theta)=a_g a_g' \tag{2.15}$$

and

$$I_V(\theta)=I(\theta)=\sum_{g=1}^{n} a_g a_g'. \tag{2.16}$$

On the graded response level, these matrices are obtained by replacing $H_V(\theta)$ by $P_V(\theta)$ in (2.11), and z_g by x_g in (2.12) and (2.13). There are no simplified results for these functions in the normal ogive model, however, as there are on the continuous response level.

When the latent space is uni-dimensional, the above matrices are simply known as *information functions*. In the context of mental test theory, *weakly parallel tests* have been defined (Samejima, 1977d) as any two or more sets of test items whose test information functions are identical, regardless of the number of items included by each test, and so forth. This enables us to control the local accuracy of estimation of the latent trait, over the repetition of testing. It is also a convenient and useful characteristic that the maximum likelihood estimate distributes asymptotically normally with θ and $I(\theta)^{-1}$ as the mean and the variance, for any fixed value of θ (Samejima, 1975).

As a function of the latent trait θ, the *item information function*, and hence the *test information function*, always assumes a higher value for the continuous scoring than for the graded scoring. In fact, the item information function on the continuous response level has the meaning of the asymptote of the item information function on the graded response level.

3. Tailored testing

Tailored testing is one of the areas in which latent trait theory is most useful. In a typical tailored testing situation, we have a set of, say, several hundred test items, whose operating characteristics are known. From this *item pool*, we select one item at a time and present it to an individual examinee. After the examinee has responded to the item and his response has been scored, another item is selected for the subject, depending upon his previous item score. The third item is selected on the basis of his responses to the previous two items, and so on. The purpose of this type of sequential testing is to measure the examinee's ability, or latent trait, as accurately as we wish, without using too many items. Thus for each individual subject a test is 'tailored' by selecting an optimal set of items out of the item pool.

This process is facilitated when we use an electronic computer for item selection and presentation, with latent trait theory as its rationale. The scoring can be either continuous or graded. In practice, however, it is more common to use the graded scoring, and especially the dichotomous scoring. This makes the ability estimation procedure more complicated, for on the graded response level we do not have the vector of sufficient statistics like the one in the normal ogive model on the continuous response level, except for Birnbaum's sufficient statistic in the *logistic model* on the dichotomous response level (Birnbaum, 1968). When the latent space is uni-dimensional, however, the numerical process to obtain the maximum likelihood estimate is fairly simple. A sufficient condition that the likelihood function has a unique maximum for every response pattern has been investigated on the graded response level (Samejima, 1969, 1972). Using one of the models which satisfy this condition, like the normal ogive model or the logistic model, we let the computer produce the *basic function* such that

$$A_{x_g}(\theta)=\frac{\partial}{\partial\theta}\log P_{x_g}(\theta), \tag{3.1}$$

for the examinee's response x_g to item g. If the model satisfies the sufficient condition which was mentioned earlier, this basic function is strictly decreasing in θ. After each presentation of an item and the examinee's response, the basic function for this specific response is added to the sum total of the basic functions for his previous responses, and the value of θ which makes this sum equal to zero is searched by the computer. The resulting value of θ is the current maximum likelihood estimate of the examinee.

It has been shown by Monte Carlo data that the approximation by $N(\theta,I(\theta)^{-1})$ to the cumulative frequency distribution of one hundred maximum likelihood estimates using twenty four test items is fairly good when $I(\theta)=22.08$, or the standard error is 0.2128, and even when $I(\theta)$ is as small as 7.36, and the standard error is 0.3685 (Samejima, 1975, 1977a). It also has been shown that, with hypothetical one hundred examinees whose θ distributes $N(0,1)$, the best fitted line for their maximum likelihood estimates obtained on 35 hypothetical graded items turned out to be $0.962\theta+0.002$, and for those obtained on 20 hypothetical graded items it turned out to be $0.977\theta-0.003$, both of which are very close to θ (Samejima, 1977c). These results support the idea of using $N(0,I(\theta)^{-1})$ as the approximation to the error distribution, and $I(\theta)^{-\frac{1}{2}}$ as the standard error of estimation of the maximum likelihood estimate in the tailored testing situation, when the eventual amount of test information is reasonably large.

On the graded response level, the *item response information function* is given by

$$I_{x_g}(\theta)=-\frac{\partial^2}{\partial\theta^2}\log P_{x_g}(\theta)=-\frac{\partial}{\partial\theta}A_{x_g}(\theta) \tag{3.2}$$

in the uni-dimensional latent space, and for the *item information function* we have

$$I_g(\theta)=\mathbf{E}\left[I_{x_g}(\theta)\right]=\sum_{x_g=0}^{m_g}I_{x_g}(\theta)P_{x_g}(\theta). \tag{3.3}$$

Thus the current amount of test information is the sum of $I_g(\theta)$ over all the item g's presented to the individual so far.

In estimating the individual's θ, it is undoubtedly desirable to select items which have high values of item information at that value of θ, in order to perform an efficient testing without using so many test items. Since the *individual parameter* θ is not known, however, we need some trials and errors in selecting a desirable set of items. The actual procedure used is to select the item out of the item pool whose amount of item information is greatest at the current maximum likelihood estimate of the individual, which has been computed on his response pattern for the items presented earlier.

It is common among psychometricians and applied psychologists in measurement to use the closeness of successive maximum likelihood estimates as the criterion for terminating the presentation of new items. It is more theoretically sound, however, to use the current amount of test information as the criterion, in the sense that we can control the standard error of estimation in so doing. Since the individual parameter is not known, again this has to be approximated by the current amount of test information at the current value of the maximum likelihood estimate. In practice, this automatically makes a right selection of an item in most cases, if the item pool is well constructed (Samejima, 1977b).

Fig. 1 shows the shift of the current maximum likelihood estimate for five hypothetical subjects selected from the author's previous study of simulated tailored testing (Samejima, 1977b). The number accompanied with a solid horizontal line is the value of the subject's individual parameter, in each graph. The hypothetical item pool consists of nine sets of equivalent items, which are scored either 1 (correct) or 0 (incorrect). The model used is the normal ogive model on the dichotomous response level, whose item characteristic function $P_g(\theta)$ is defined by

$$P_g(\theta) = \frac{1}{\sqrt{(2\pi)}} \int_{-\infty}^{a_g(\theta - b_g)} \exp[-u^2/2]\,du \qquad (3.4)$$

The two item parameters, a_g and b_g, in (3.4) are presented in Table 1 for each of the nine sets of items. There are two sessions for each subject, and

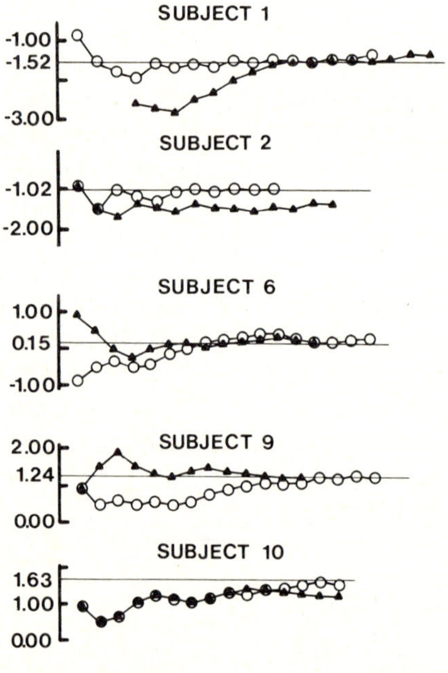

Fig. 1. Graphic presentation of the change of the maximum likelihood estimate after each item has been presented and responded to, for five hypothetical subjects in two sessions of tailored testing. The initial lack of estimates for Subject 1 in one session indicates that these estimates were negative infinity.

Table 1
Item parameters in the normal ogive model for nine binary item types

Item type	Discrimination parameter a_g	Difficulty parameter b_g
1	1.20	−2.00
2	1.60	−1.50
3	2.00	−1.00
4	1.40	−0.50
5	1.80	0.00
6	1.30	0.50
7	1.70	1.00
8	1.90	1.50
9	1.50	2.00

results are plotted by hollow circles and solid triangles respectively. The criterion for terminating the presentation of new items is the attainment to $I(\hat{\theta})=20$, which means that the value of the standard error of estimation is approximately 0.224. As we can see in Fig. 1, the stability of the current maximum likelihood estimate and the attainment to a sufficiently small standard error of estimation are two different things (e.g., Subject 9).

In the above example, it is assumed that we need the same accuracy of estimation for all the individuals. In some situations, however, this is not necessary. For instance, if we are in the position of selecting a small number of college seniors for some privileged graduate school fellowship, all we need is to classify all the examinees into two groups, the group of eligible people and that of non-eligible people. To make this classification efficiently, it is best to prearrange the *criterion curve*, which is not horizontal along the abscissa of θ but assumes high values of test information around the boarder line of the eligibility and non-eligibility with respect to θ, and gradually decreases as θ departs further from that critical value. The use of such a prearranged criterion curve for terminating the presentation of new items in tailored testing enables us to control and predict the error of classification as well (Samejima, 1977b).

4. Estimation of the operating characteristics of item response categories

There are two distinct standpoints in estimating the operating characteristics of item response categories. One is to assume some mathematical model, like the normal ogive model, and estimate the operating characteristics accordingly. On this standpoint, the estimation is reduced to that of the item parameters. The other is to approach the operating characteristics *without assuming any mathematical models*. While many works have been done on the first standpoint (e.g., Bock and Lieberman, 1970; Lord, 1968), there have been only a few made on the second standpoint. One of the reasons why the research on the second standpoint is more difficult is that we usually need a large number of subjects for our data, in addition to the theoretical difficulty. Lord used subjects on the order of a hundred thousand in his research on the second standpoint (Lord, 1970), and only institutes like Educational Testing Service can provide such data. In an ordinary university setting, it is fairly difficult to collect data for even one thousand subjects. The effective use of information functions will solve this problem to some extent, however.

Data used here are simulated data for 500 hypothetical subjects, whose response patterns on a set of 35 graded items, each having three item score categories, were calibrated by the Monte Carlo method, and also whose response patterns on a set of 10 binary items were calibrated in a similar manner. The model used for both sets of items is the normal ogive model. The test information of the first set of items is constant (21.63) for the range of θ in which all the individual parameters are located, which are actually from -2.475 to 2.475 with the interval width of 0.05 and with five individuals at each point of θ. It is assumed that the operating characteristics for each of the thirty five items in the first set are known, and those of the ten binary items in the second set are unknown, and we are to estimate the latter. The maximum likelihood estimate $\hat{\theta}$ of each of the 500 individuals is obtained using the method described in Section 3.

The method of moments (Elderton and Johnson, 1969), converted to apply for the set of observations instead of the frequency distribution to approximate the density function $g(\hat{\theta})$, was used to fit polynomials of degree 3, 4 and 5 respectively. Figure 2 presents the resulting polynomials, together with $g(\hat{\theta})$ obtained through

$$g(\hat{\theta}) = \int_{-\infty}^{\infty} \psi(\hat{\theta}|\theta) f(\theta)\, d\theta, \tag{4.1}$$

where $f(\theta)(=0.2)$ is the probability density function of θ, and $\psi(\hat{\theta}|\theta)$ is the conditional density function of $\hat{\theta}$, given θ, which is approximated by $n(\theta, 21.63^{-1})$ by virtue of the asymptotic property of the maximum likelihood estimate.

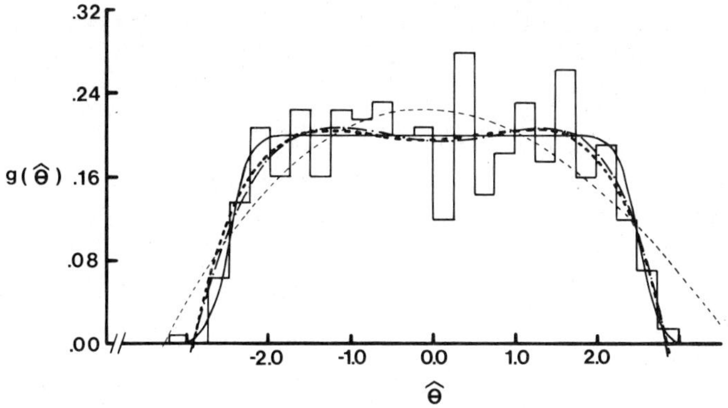

Fig. 2. The probability density function of $\hat{\theta}$ (solid line) and its approximations by the polynomials of degree 3 (thin, dashed line), of degree 4 (thick, dashed line) and of degree 5 (broken and dotted line), obtained by the method of moments. The frequency distribution of the five hundred maximum likelihood estimates is also drawn.

The conditional expectation of θ, given $\hat{\theta}$, and the second to fourth conditional moments about the mean are as follows.

$$\mathbf{E}(\theta|\hat{\theta}) = \hat{\theta} + \sigma^2 \frac{\mathrm{d}}{\mathrm{d}\hat{\theta}} \log g(\hat{\theta}). \tag{4.2}$$

$$\mu_2(\theta|\hat{\theta}) = \sigma^2\left[1 + \sigma^2 \frac{\mathrm{d}^2}{\mathrm{d}\hat{\theta}^2} \log g(\hat{\theta})\right]. \tag{4.3}$$

$$\mu_3(\theta|\hat{\theta}) = \sigma^6\left[\frac{\mathrm{d}^3}{\mathrm{d}\hat{\theta}^3} \log g(\hat{\theta})\right]. \tag{4.4}$$

$$\mu_4(\theta|\hat{\theta}) = \sigma^4\left[3 + 6\sigma^2\left\{\frac{\mathrm{d}^2}{\mathrm{d}\hat{\theta}^2} \log g(\hat{\theta})\right\}\right.$$
$$\left. + 3\sigma^4\left\{\frac{\mathrm{d}^2}{\mathrm{d}\hat{\theta}^2} \log g(\hat{\theta})\right\}^2 + \sigma^4\left\{\frac{\mathrm{d}^4}{\mathrm{d}\hat{\theta}^4} \log g(\hat{\theta})\right\}\right]. \tag{4.5}$$

In these formulae, σ^2 is the conditional variance of $\hat{\theta}$, given θ, which equals 21.63^{-1}. These conditional moments can be estimated by substituting one of the polynomials fitted to the set of 500 maximum likelihood estimates for $g(\hat{\theta})$. Using these estimated conditional moments, Pearson's criterion κ (Elderton and Johnson, 1969) such that

$$\kappa = \beta_1(\beta_2+3)^2[4(2\beta_2-3\beta_1-6)(4\beta_2-3\beta_1)]^{-1}, \tag{4.6}$$

where

$$\beta_1 = \mu_3(\theta|\hat{\theta})^2 \mu_2(\theta|\hat{\theta})^{-3} \tag{4.7}$$

and

$$\beta_2 = \mu_4(\theta|\hat{\theta}) \mu_2(\theta|\hat{\theta})^{-2}, \tag{4.8}$$

was computed for every individual, to decide which Pearson type distribution should be used for approximating the conditional distribution of θ, given $\hat{\theta}$. After this classification, the parameters of each distribution were computed, if necessary. For instance, when the distribution is of Pearson's Type I, it has a Beta density function such that

$$\hat{\phi}(\theta|\hat{\theta}) = [B(p_{\hat{\theta}}, q_{\hat{\theta}})]^{-1}(\theta - a_{\hat{\theta}})^{p_{\hat{\theta}}-1}(b_{\hat{\theta}} - \theta)^{q_{\hat{\theta}}-1}(b_{\hat{\theta}} - a_{\hat{\theta}})^{-(p_{\hat{\theta}}+q_{\hat{\theta}}-1)}, \tag{4.9}$$

and the four parameters, $a_{\hat{\theta}}$, $b_{\hat{\theta}}$, $p_{\hat{\theta}}$ and $q_{\hat{\theta}}$, are estimated from the four conditional moments (Johnson and Kotz, 1970; Samejima, 1978b). Then the estimated item characteristic function of each of the ten binary items is given by

$$\hat{P}_g(\theta) = \sum_{s \in G} \hat{\phi}(\theta|\hat{\theta}_s) \left[\sum_{s=1}^{N} \hat{\phi}(\theta|\hat{\theta}_s) \right]^{-1}, \tag{4.10}$$

where N is the number of examinees (=500) and G is the set of individuals who answered correctly to item g. The type of approach using (4.10) for the estimate of the operating characteristic is called the Conditional P.D.F. Method (Samejima, 1978a, 1978b), and this particular method using the Pearson System distributions is called the Pearson System Method (Samejima, 1978b). The results of this approach will be influenced by the sampling fluctuation of the binary item score for item g. For this reason, replacing $\hat{\phi}(\theta|\hat{\theta})$ in (4.10) by $\phi(\theta|\hat{\theta})$ such that

$$\phi(\theta|\hat{\theta}) = \psi(\hat{\theta}|\theta)f(\theta)[g(\hat{\theta})]^{-1}, \tag{4.11}$$

we call the resulting function of θ the *criterion operating characteristic*. This gives us a criterion for evaluating $\hat{P}_g(\theta)$ in the sense that it shows the maximal possible result obtainable by the Conditional P.D.F. Method. Figure 3 presents a typical result obtained by the Pearson System Method, together with the theoretical item characteristic function and the criterion item characteristic function.

To avoid the estimates of higher order conditional moments, in the Normal Approach Method (Samejima, 1978b), a normal density function is used for $\hat{\phi}(\theta|\hat{\theta})$ in (4.10), using only the first two conditional moments of θ, given $\hat{\theta}$. To ameliorate the lack of variation of the normal density curve, in the Two-Parameter Beta Method (Samejima, 1977e), we use a Beta

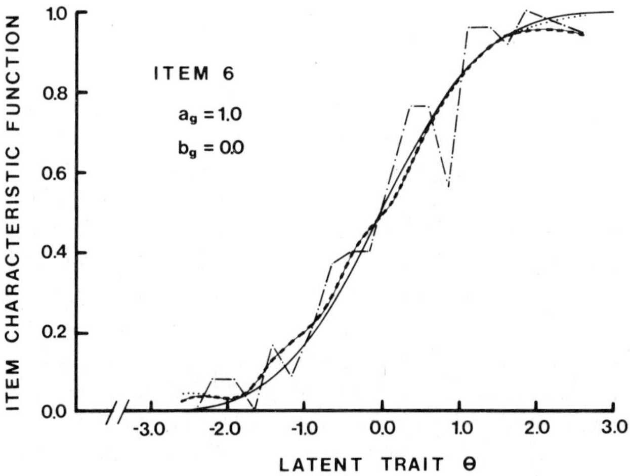

Fig. 3. Results of Pearson System Method, plotted with the true item characteristic function (thick, solid line) and the frequency ratio of the individuals who answered item 6 correctly to the total of 25 examinees in each subinterval of θ with the width of 0.25 (broken and dotted line). The estimated item characteristic function based on the polynomial of degree 3 for $\hat{g}(\hat{\theta})$ (thick, dashed line) and the one based on the polynomial of degree 4 for $\hat{g}(\hat{\theta})$ (dotted line) are practically identical with the criterion item characteristic function (thin, solid line) except for the range of θ outside the interval $(-2.0, 2.0)$. A typical example selected out of the results of the ten binary items.

density function for $\hat{\phi}(\theta|\hat{\theta})$ in (4.10), with a priori set two parameters, $a_{\hat{\theta}}$ and $b_{\hat{\theta}}$, and the two other parameters estimated from the first two conditional moments. Both methods have been tried on the same simulated data of 500 hypothetical subjects, and the results turned out to be just as good.

It is meaningful that these methods turned out to be successful using data of only 500 subjects. These methods depend upon the first set of items whose operating characteristics are known and which provide us with a uniform test information. If we use continuous response items, a uniform test information is automatically realized (Samejima, 1973). Otherwise, it is not very easy to construct a test having a uniform test information, especially in the paper-and-pencil testing situation. In computerized adaptive testing, however, it will be realized easily by setting an appropriate uniform test information function as the criterion curve for terminating the presentation of new items, as was described in Section 3. The present methods of estimating the operating characteristics of graded response categories are most useful, therefore, in tailored testing.

5. Discussion

Rationale in latent trait theory was introduced, and two applications of the theory in tailored testing and the estimation of the operating characteristics of graded response categories were introduced and discussed. They are still in the process of developing, and more useful devices for application, as well as further development of the theory, will come out in the near future. The next step in the estimation of the operating characteristics will be to find a way to overcome the limitation of the criterion operating characteristic. One possibility is to treat the set of maximum likelihood estimates separately for different item scores of item g, and approximate the bivariate density of θ and $\hat{\theta}$ for each score group, instead of approximating the conditional density function of θ, given $\hat{\theta}$. This idea is close to the Normal Approximation Method (Samejima, 1977c), which turned out to be fairly successful with the same simulated data.

References

[1] Bartlett, M. S. (1937). The statistical conception of mental factors. *British Journal of Psychology*, **28**, 97–104.

[2] Birnbaum, A. (1968). Some latent trait models and their use in inferring an examinee's ability. In F. M. Lord and M. R. Novick, *Statistical Theories of Mental Test Scores*. Addison-Wesley, Chapters 16–20.

[3] Bock, R. D. and Lieberman, M. (1970). Fitting a response model for *n* dichotomously scored items. *Psychometrika*, **35**, 179–97.

[4] Elderton, W. P. and Johnson, N. L. (1969). *Systems of Frequency Curves*. Cambridge University Press.

[5] Johnson, N. L. and Kotz, S. (1970). *Continuous Univariate Distributions*, Vol. 2. Houghton Mifflin.

[6] Lawley, D. N. and Maxwell, A. E. (1971). *Factor Analysis as a Statistical Method.* (2nd Edition) Butterworth.

[7] Lord, F. M. (1968). An analysis of the verbal scholastic aptitude test using Birnbaum's three-parameter logistic model. *Educational and Psychological Measurement*, **28**, 989–1020.

[8] Lord, F. M. (1970). Estimating item characteristic curves without knowledge of their mathematical form. *Psychometrika*, **35**, 43–50.

[9] Lord, F. M. and Novick, M. R. (1968). *Statistical Theories of Mental Test Scores*. Addison-Wesley.

[10] Roche, A. F., Wainer, H. and Thissen, D. (1975). *Skeletal Maturity. The Knee Joint as a Biological Indicator*. Plenum.

[11] Samejima, F. (1969). Estimation of latent ability using a response pattern of graded scores. *Psychometrika Monograph*, No. 17.

[12] Samejima, F. (1972). A general model for free-response data. *Psychometrika Monograph*, No. 18.

[13] Samejima, F. (1973). Homogeneous case of the continuous response model. *Psychometrika*, **38**, 203–219.

[14] Samejima, F. (1974). Normal ogive model on the continuous response level in the multidimensional latent space. *Psychometrika*, **39**, 111–121.

[15] Samejima, F. (1975). Graded response model of the latent trait theory and tailored testing. *Proceedings of the First Conference on Computerized Adaptive Testing*. Civil Service Commission and Office of Naval Research, pp. 5–17.

[16] Samejima, F. (1977a). Effects of individual optimization in setting the boundaries of dichotomous items on accuracy of estimation. *Applied Psychological Measurement*, **1**, 77–94.

[17] Samejima, F. (1977b). A use of the information function in tailored testing. *Applied Psychological Measurement*, **1**, 233–247.

[18] Samejima, F. (1977c). A method of estimating item characteristic functions using the maximum likelihood estimate of ability. *Psychometrika*, **42**, 163–191.

[19] Samejima, F. (1977d). Weakly parallel tests in latent trait theory with some criticisms of classical test theory. *Psychometrika*, **42**, 193–198.

[20] Samejima, F. (1977e). Estimation of the operating characteristics of item response categories I: Introduction to the Two-Parameter Beta Method. *Office of Naval Research*, Research Report, 77-1.

[21] Samejima, F. (1978a). Estimation of the operating characteristics of item response categories II: Further development of the Two-Parameter Beta Method. *Office of Naval Research*, Research Report, 78-1.

[22] Samejima, F. (1978b). Estimation of the operating characteristics of item response categories III: The normal approach method and the Pearson system method. *Office of Naval Research*, Research Report, 78-2.

P. R. Krishnaiah, ed., *Multivariate Analysis–V*

SINGULAR MOMENT MATRICES IN APPLIED ECONOMETRICS*

Henri THEIL and Kenneth LAITINEN
University of Chicago, Chicago, IL, USA

1. Introduction

It happens on numerous occasions that the inverse of a moment matrix is needed but that this inverse does not exist because the matrix is singular. For example, Deaton [3] estimated a linear expenditure system for 37 groups of goods. This requires the inverse of a residual moment matrix, but this matrix is singular because Deaton had only 17 years of data. A second example is three-stage least-squares (3SLS) estimation of a system of simultaneous equations. The third stage involves the inverse of the moment matrix of the 2SLS residuals, but this matrix is singular when the number of stochastic equations exceeds the number of observations, which may occur when the system is large. A third example is 2SLS estimation of an equation system when the number of predetermined variables exceeds the number of observations. The moment matrix of these variables is then singular and its inverse is needed for the computation of the 2SLS estimates. This problem arises even for medium-sized systems.

In all these examples the problem is formally a small-sample problem, but it is actually in most cases more a problem of a large system than of a small number of observations. Various solutions have been proposed, many of which are arbitrary to some degree. For example, the 3SLS problem can be handled if we are willing to assume that the disturbance covariance matrix of the stochastic equations is appropriately block-diagonal. Deaton also imposed a particular structure on the disturbance covariance matrix of the linear expenditure system. Numerous proposals have been made for the problem of how to handle the excess of the number of predetermined variables over the number of observations.[1] In

*Research supported in part by NSF Grant SOC 76-82718.

[1]Several proposals for handling this problem of 'undersized samples,' including those of Dhrymes [4], Brundy and Jorgenson [2], and Fisher [5], use information from equations other than that which is estimated, whereas other procedures such as that of Kloek and Mennes [9] are in the limited information tradition. The approach of this paper is in the same tradition (see Appendix B). We prefer this approach to a method formulated by one of us [15] because the former is equivalent to 2SLS for a sufficiently large sample. See also Hendry [7] and Sargan [13].

this paper we suggest that the ordinary sample moment matrix is not the only possible estimator of the population moment matrix. An alternative estimator is proposed which is nonsingular. To make the paper more readable, most of the derivations are given in Appendix A.

2. An alternative procedure

Although moment matrices refer to several random variables, it will be convenient to start with one such variable, X, which takes the values $x_1,\dots,x_n$. We write $f(\ )$ for the estimated density function of X, which is to be derived, and

$$H=-\int_{-\infty}^{\infty} f(x)\log f(x)\,\mathrm{d}x \tag{1}$$

for the associated entropy. Since we are interested in the second moment, it is natural to impose constraints on moments of lower order (see below). Subject to these constraints we select $f(\ )$ by maximizing H. Hence the second-moment estimator to be derived is a constrained maximum-entropy estimator.[2]

[2]The entropy has been widely used as a measure of uninformativeness. For example, when Bayesian statisticians translate prior judgments on parameters into a prior distribution, they frequently do so by selecting the distribution for which the entropy is maximized subject to certain constraints which represent their prior knowledge of the parameter values. The entropy criterion is by no means the only possible choice. We refer to Wegman [17] for a survey of methods of density estimation proposed before 1970. Boneva, Kendall and Stefanov [1] propose a technique using splines, of which several variations based on natural spline interpolation have been formulated (e.g., Kuhn [11, 12]). Good and Gaskins [6] propose a method based on maximizing the difference between the log-likelihood function and a 'roughness penalty.' All these techniques are based on smoothing out the spiky observed density which arises from the data. Each method involves some arbitrary choice: a function or set of functions, a 'smoothing' criterion (whose interpretation is not always obvious), or at least one parameter. For example, the simple histogram requires a specification of the cell width and a location parameter. Some of the techniques are designed to be used over a finite range, which necessitates the choice of end points. Also, the spline techniques (and some of those discussed by Wegman) may yield negative estimates of the density. Most of these techniques are generalizable to multivariate distributions, but there are few cases in which an explicit attempt has been made to do so (see e.g., Kuhn [12], Kuelbs [10] and Good and Gaskins [6]). Application to multivariate data appears to be rare. Very little attention has been paid to properties of the moments of the estimated density function (but several techniques are mean-preserving). These moments are important for our purpose, given that our problem is focused on second moments rather than on the density function.

We indicate by superscripts that the observed values are arranged in ascending order, $x^1 < x^2 < \cdots < x^n$, and define intermediate points between successive observations,

$$\xi_i = \xi(x^i, x^{i+1}) \qquad i = 1, \ldots, n-1 \tag{2}$$

where $\xi(\ )$ is a symmetric differentiable function of its two arguments whose value is not outside the range defined by its arguments. Each pair of successive ξ_i's defines an interval, $I_2 = (\xi_1, \xi_2), \ldots, I_{n-1} = (\xi_{n-2}, \xi_{n-1})$. There is also an interval I_1 to the left of ξ_1 and an interval I_n to the right of ξ_{n-1}. Hence there are n intervals, each containing one x^i and thus a fraction $1/n$ of the mass of the observed distribution. We impose on $f(\ )$ that it preserves these observed fractions,

$$\int_{I_i} f(x)\,\mathrm{d}x = \frac{1}{n} \qquad i = 1, \ldots, n \tag{3}$$

which is a mass-preserving constraint (a constraint on moments of order zero).

The other constraint is a mean-preserving constraint (a constraint on moments of order one), which amounts to the following. We define the x^i's associated with each interval as those which determine its end points. These end points are ξ_1 and ξ_{n-1} for the open-ended intervals I_1 and I_n, respectively, so that x^1 and x^2 are associated with I_1 [see (2)] and x^{n-1} and x^n are associated with I_n. In the same way x^{i-1}, x^i, and x^{i+1} are associated with I_i for $i = 2, \ldots, n-1$. The mean-preserving constraint states that $f(\ )$ must be constructed so that the mean of each interval is a homogeneous linear function of the associated x^i's and so that the overall mean is preserved. Thus, subject to this constraint and (3) we seek the $f(\ )$ which maximizes (1).

This problem is solved in Appendix A with the following results. First, the intermediate points (2) become midpoints, $\xi_i = \frac{1}{2}(x^i + x^{i+1})$, which is illustrated in the upper part of Fig. 1. Second, the density function $f(\ )$ is such that the adjusted distribution is uniform in each closed interval $(I_2, \ldots, I_{n-1})$ and exponential in the two open-ended intervals (I_1 and I_n). This is illustrated in the lower part of Figure 1. When we integrate the density picture of that figure, we obtain a cumulated distribution function which is continuous everywhere, piecewise linear around the observed values $x^2, \ldots, x^{n-1}$, and exponential around x^1 and x^n. Third, the means of

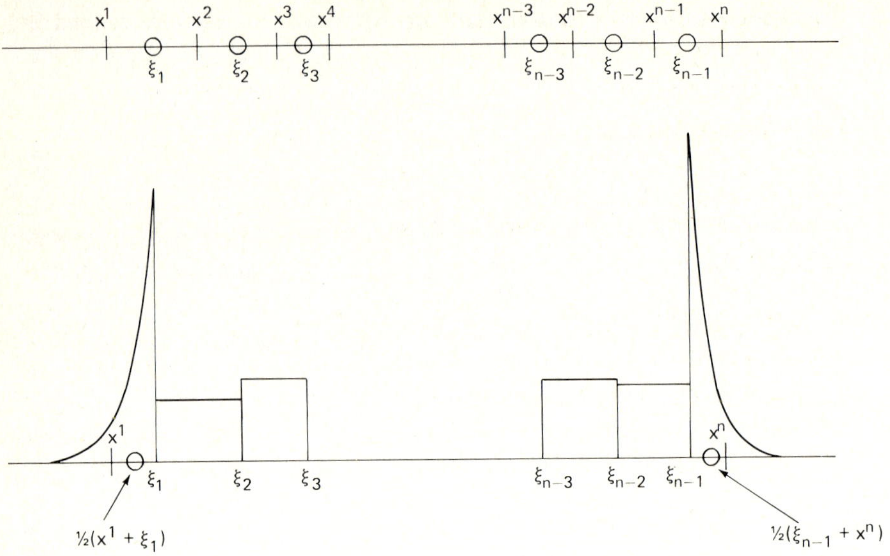

Fig. 1. The density function of the univariate maximum-entropy distribution

the individual intervals are

$$\begin{aligned} \mathrm{E}(X|X\in I_i) &= \tfrac{3}{4}x^1+\tfrac{1}{4}x^2 && \text{if } i=1\\ &= \tfrac{1}{4}x^{i-1}+\tfrac{1}{2}x^i+\tfrac{1}{4}x^{i+1} && \text{if } i=2,\dots,n-1\\ &= \tfrac{1}{4}x^{n-1}+\tfrac{3}{4}x^n && \text{if } i=n \end{aligned} \tag{4}$$

Fourth, the variance of the distribution with density $f(\ )$ equals

$$\operatorname{var}X = \text{sample variance} - \frac{1}{4n}\sum_{i=1}^{n-1}(x^{i+1}-x^i)^2 - \frac{1}{24n}\sum_{i=2}^{n-1}(x^{i+1}-x^{i-1})^2 \tag{5}$$

where the sample variance is the mean square of $x^1-\bar{x},\dots,x^n-\bar{x}$ with $\bar{x}$ = sample mean (which is also the mean of the maximum-entropy or ME distribution). Hence (5) implies that the ME variance is smaller, not larger, than the sample variance.[3]

3. A bivariate extension

Let (X, Y) be a random pair with observations $(x_1,y_1),\ldots,(x_n,y_n)$. Our objective is now to design a joint density function $f(x,y)$ so that the joint entropy,

$$H=-\int_{-\infty}^{\infty}\int_{-\infty}^{\infty} f(x,y)\log f(x,y)\,dx\,dy \tag{6}$$

is maximized subject to constraints which are extensions of the corresponding univariate constraints. As before, we indicate by superscripts the observed x's arranged in ascending order; we proceed similarly for the y's, $y^1<y^2<\cdots<y^n$. In addition to the intermediate points in (2), we construct similar points for the y^i's:

$$\eta_i=\eta(y^i,y^{i+1}) \qquad i=1,\ldots,n-1 \tag{7}$$

The situation is illustrated in Fig. 2 for seven observation pairs, (x^1, y^4), (x^2, y^1), (x^3, y^3), (x^4, y^6), (x^5, y^5), (x^6, y^2), and (x^7, y^7), which yield the $n^2=49$ cells shown in the figure. Out of these, $n=7$ cells (shaded) contain one observation each and $n(n-1)=42$ cells contain no observation. So we extend (3) by assigning probability $1/n$ to each shaded cell and zero probability to all others. Within each shaded cell the entropy is maximized when X and Y are stochastically independent. Four shaded cells in Fig. 2 are bounded on all sides; their ME distribution is the bivariate uniform distribution. Two shaded cells are open-ended on one side, with an ME distribution equal to the exponential (for the open-ended variable) multiplied by the uniform (for the other variable). The shaded cell upper right is open-ended on two sides; its ME distribution is the product of two exponentials.

[3]Note that this is in contrast to the ridge regression procedure [8], which solves the singularity problem of the moment matrix of independent variables of a regression by raising the diagonal elements of the matrix by positive amounts. On the other hand, when we use the transformed data points to be discussed in the next section, the ME variance exceeds the value obtained by putting $X\equiv Y$ in (10), which agrees with the ridge regression philosophy (see Appendix A).

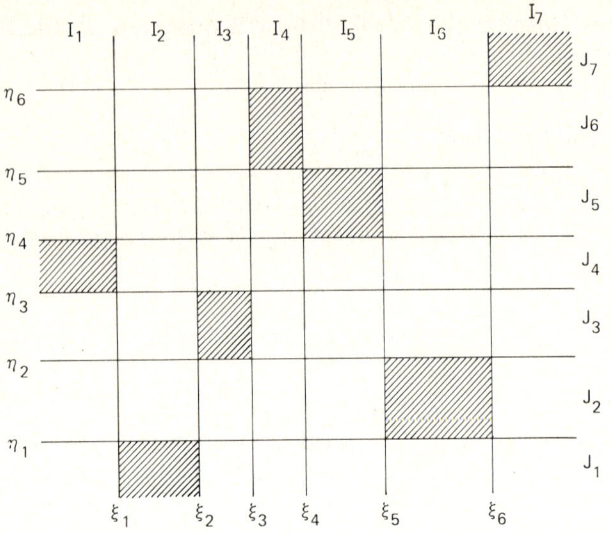

Fig. 2. A grid for a bivariate distribution

This bivariate specification yields marginal density functions of X and Y which are both piecewise uniform except for exponentials at the tails. The mean-preserving constraint applies to both marginal distributions, so that $\eta(\ )$ in (7) becomes the arithmetic mean of its two arguments. The intervals implied by $\eta_1,\ldots,\eta_{n-1}$ are indicated as $J_1,\ldots,J_n$ in Fig. 2.

The problem that remains is the ME covariance of X and Y,

$$\operatorname{cov}(X,Y)=\int_{-\infty}^{\infty}\int_{-\infty}^{\infty}(x-\bar{x})(y-\bar{y})f(x,y)\,\mathrm{d}x\,\mathrm{d}y \tag{8}$$

To solve this problem we must recognize that the n expectations in (4) can be viewed as transformations of the observed values $x_1,\ldots,x_n$, the latter being arranged in the order of successive sample values (in contrast to the x^i's, which are arranged in ascending order). Since each interval I_i contains one observed value x^i, which is x_k for some k, we can define

$$x_k^* = \mathbf{E}(X|X\in I_i) \qquad \text{if } x^i = x_k \tag{9}$$

and, similarly, $y_k^*=\mathbf{E}(Y|Y\in J_j)$ if $y^j=y_k$. Returning to (8), we note that since $f(x,y)$ vanishes identically in all non-shaded cells of Figure 2, we can write the double integral as the sum of n terms, one for each shaded cell.

Hence, if $(x_k, y_k) = (x^i, y^j)$, this term for the kth shaded cell takes the form

$$\int_{I_i}\int_{J_j} (x-\bar{x})(y-\bar{y}) f(x,y)\,\mathrm{d}x\,\mathrm{d}y$$

which equals $[\mathbf{E}(X|X\in I_i)-\bar{x}][\mathbf{E}(Y|Y\in J_j)-\bar{y}] = (x_k^* - \bar{x})(y_k^* - \bar{y})$ because X and Y are stochastically independent within each shaded cell. Therefore, given that each such cell is assigned probability $1/n$,

$$\operatorname{cov}(X,Y) = \frac{1}{n}\sum_{k=1}^{n} (x_k^* - \bar{x})(y_k^* - \bar{y}) \tag{10}$$

The above approach can be easily generalized for three or more variables, but since we are only interested in variances and covariances, there is no reason to pursue this here. The ME approach always yields a positive definite covariance matrix because the mass of each individual observation is spread out over a cell which has two (or more) dimensions.[4] See Appendix A for further details.

4. An example

Numerical applications are straightforward. ME variances can be obtained directly from (5). For ME covariances we compute transformed data points first, using (9) and (4) and similarly for y_k^*, and then apply (10).

Table 1 shows the results for log-changes (changes in natural logarithms), multiplied by 100, of price indexes of 14 Dutch commodity groups from 1954–55 until 1962–63.[5] Hence there are nine observations and the sample covariance matrix in the first 14 lines of Table 1 is singular. This is confirmed by its latent roots, of which six are zero and the others are

61.5 42.1 15.6 10.2 7.12 4.77 1.51 0.833

The ME covariance matrix is shown in the next 14 lines of the table. The diagonal elements are all a bit smaller than the corresponding elements in

[4]For example, for $n=2$ and two variables the sample covariance matrix is singular and the sample correlation coefficient (r) is ± 1. It can be easily shown that the ME correlation coefficient is $\frac{1}{2}$ when $r=1$ and $-\frac{1}{2}$ when $r=-1$, so that the 2×2 ME covariance matrix is far from ill-conditioned.

[5]The data used are the last nine rows of [16, Table 6.4]; the commodity groups are listed in [16, Table 6.1].

Table 1
Covariance matrices of 14 price log-changes

1	2	3	4	5	6	7	8	9	10	11	12	13	14	
					Sample covariance matrix									
5.49	2.68	3.03	2.08	0.18	4.37	1.56	1.32	0.63	4.58	−1.47	−2.06	1.70	1.15	1
	6.48	2.62	−3.53	−0.12	3.83	6.43	5.57	3.04	5.57	0.02	−5.65	2.34	1.93	2
		9.18	−3.78	2.77	−0.16	3.22	1.90	−0.26	3.40	0.71	−3.16	5.23	−0.29	3
			52.12	8.98	15.15	−0.89	−1.02	−6.31	−3.45	−1.90	5.47	3.84	−2.94	4
				12.36	1.47	−0.68	−0.51	−2.17	−3.17	−1.20	2.74	1.86	−3.01	5
					9.51	4.28	3.51	0.21	4.52	−1.01	−2.57	2.31	1.18	6
						10.80	9.00	2.96	5.32	−0.52	−6.52	6.08	3.15	7
							8.02	3.07	3.87	−1.20	−5.06	4.94	2.97	8
								2.80	2.54	−0.76	−3.02	0.02	1.58	9
									8.26	0.13	−6.66	1.64	1.99	10
										2.74	−0.59	−0.67	−0.66	11
											6.87	−2.23	−1.82	12
												6.71	1.30	13
													2.20	14
					Maximum-entropy covariance matrix									
5.06	2.59	2.59	1.61	0.41	3.83	1.25	1.41	0.87	4.60	−1.16	−1.31	1.58	1.16	1
	5.94	1.83	−3.54	−0.36	3.05	5.22	4.88	2.95	4.42	−0.10	−4.26	1.18	2.03	2
		8.49	−5.08	2.94	−0.54	1.53	0.97	0.05	3.33	1.11	−2.57	4.31	−0.27	3
			46.98	7.89	14.29	−2.28	−1.40	−4.71	−2.08	−3.00	3.73	3.49	−2.95	4
				11.32	1.64	−1.41	−0.94	−1.44	−1.63	−1.59	0.79	1.33	−2.84	5
					8.72	2.71	2.68	0.09	3.84	−1.12	−1.31	2.03	0.97	6
						8.61	7.09	3.03	2.96	−0.53	−4.21	3.90	2.96	7
							6.92	3.04	2.30	−1.07	−3.49	3.40	2.71	8
								2.49	1.58	−0.65	−2.28	0.12	1.43	9
									7.21	0.61	−4.00	1.07	1.64	10
										2.40	−0.77	−0.47	−0.33	11
											5.21	−1.16	−1.08	12
												6.07	0.98	13
													2.04	14
					Inverse of the maximum-entropy covariance matrix									
3.62	−0.22	−1.67	0.07	0.30	−1.37	0.66	0.76	−0.53	−1.16	2.50	−0.49	0.36	−1.57	1
	3.49	−0.51	0.26	−0.63	−1.47	−0.13	−0.83	−2.60	0.11	−1.50	0.17	0.84	−0.76	2
		2.64	0.29	−0.43	0.89	0.09	0.06	0.16	−0.34	−1.47	0.19	−2.15	2.07	3
			0.44	0.16	−0.88	0.02	−0.04	0.01	0.10	0.22	0.08	−0.45	1.24	4
				0.80	−0.43	−0.19	−0.18	0.99	0.19	1.17	−0.13	0.20	1.63	5
					3.14	−0.10	−0.23	1.25	−0.52	−0.94	−0.10	−0.08	−1.18	6
						1.25	−0.40	−0.09	−0.28	0.13	0.28	−0.45	−1.02	7
							2.63	−1.08	0.17	0.46	0.21	−0.84	−1.52	8
								5.92	0.37	2.36	0.81	0.48	1.43	9
									1.74	−0.38	0.80	0.52	−0.20	10
										4.06	0.10	0.53	0.59	11
											1.38	−0.04	−1.02	12
												2.67	−1.88	13
													10.19	14

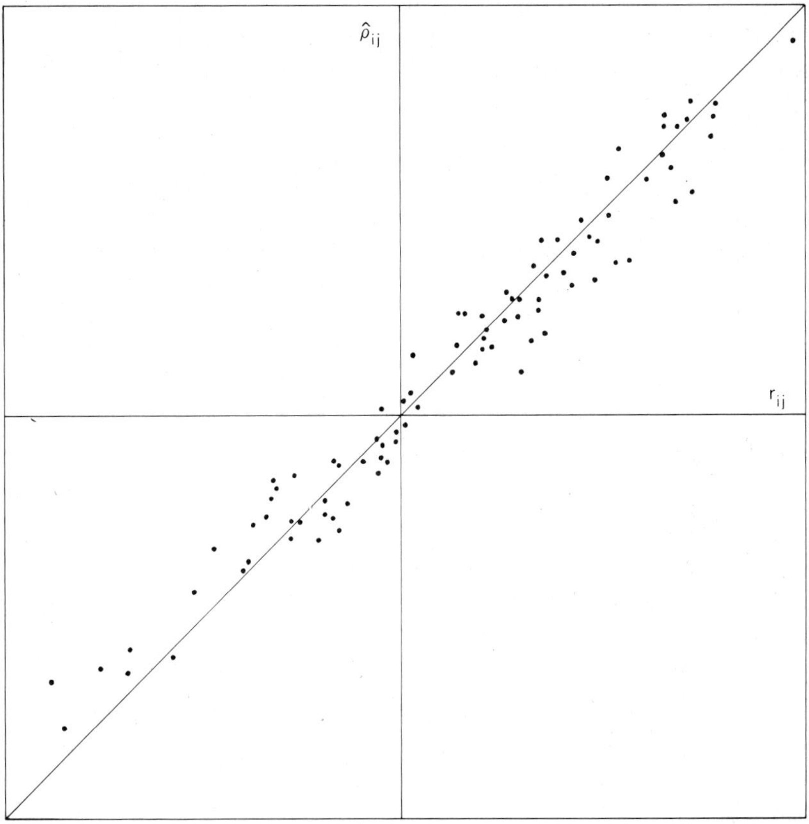

Fig. 3. Sample and maximum-entropy correlation coefficients of 14 price log-changes

the first 14 lines, which agrees with (5). The eight largest latent roots of the ME covariance matrix are

$$55.7 \quad 32.3 \quad 14.5 \quad 9.89 \quad 6.91 \quad 4.43 \quad 1.64 \quad 0.867$$

which should be compared with the nonzero latent roots of the sample covariance matrix shown above. The six smallest latent roots of the ME covariance matrix are 0.386, 0.304, 0.188, 0.131, 0.103, and 0.0765, so that the ratio of the largest to the smallest root is $55.7/0.0765 \approx 7 \times 10^2$, which is not particularly large. The inverse of the ME covariance matrix is shown in the lower part of the table.

Sample correlation coefficients (r_{ij}) can be computed from the upper part of the table and ME correlation coefficients ($\hat{\rho}_{ij}$) from the middle part. Fig. 3 displays all 91 pairs ($r_{ij}, \hat{\rho}_{ij}$). The points are close to the 45 degree line through the origin in spite of the fact that they are based on only nine observations. The figure suggests that for the points far from the origin $\hat{\rho}_{ij}$ tends to be a little smaller in absolute value than r_{ij}. This is not surprising, given that the ME approach pushes the observed distribution in the direction of stochastic independence.

We made experiments with larger numbers of variables and observations. For each of 100 variables, 50 observations were generated by means of random normal sample numbers. The latent roots of the ME covariance matrix (of order 100×100) indicate that this matrix is far from ill-conditioned. Since these results are bulky, they are not reproduced here. The application of the ME approach to 2SLS estimation is described in Appendix B.

Appendix

A. *Derivations*

A.1. *Derivation of the* ME *density function*

The mean-preserving constraint consists of a homogeneity requirement in each interval,

$$\begin{aligned} \mathbf{E}(X\,|\,X\in I_i) &= ax^1+(1-a)x^2 && \text{if } i=1\\ &= bx^{i-1}+(1-b-c)x^i+cx^{i+1} && \text{if } i=2,\ldots,n-1\\ &= dx^n+(1-d)x^{n-1} && \text{if } i=n \end{aligned} \tag{A.1}$$

together with preservation of the overall mean, which is equivalent [see (3)] to

$$n\mathbf{E}(X)=\sum_{i=1}^{n}\mathbf{E}(X\,|\,X\in I_i)=\sum_{i=1}^{n}x^i \tag{A.2}$$

It follows from (A.2) that the sum of the weights attached to x^i in (A.1) must be unity for each i. This is trivially satisfied for $2<i<n-1$. For $i=1$, 2, $n-1$, and n we obtain $a+b=1$; $1-a+(1-b-c)+b=1$; $c+(1-b-c)+1-d=1$; $c+d=1$. This implies $a=d=1-c$ and $b=c$, so that (A.1) is

simplified to

$$\begin{aligned} \mathbf{E}(X|X\in I_i) &= (1-c)x^1+cx^2 && \text{if } i=1 \\ &= cx^{i-1}+(1-2c)x^i+cx^{i+1} && \text{if } i=2,\dots,n-1 \\ &= (1-c)x^n+cx^{n-1} && \text{if } i=n \end{aligned} \tag{A.3}$$

Maximizing the entropy (1) subject to (3) and (A.3) can be performed for each of the intervals separately. The intervals I_i for $i=2,\dots,n-1$ are all bounded above and below, so that the maximum-entropy criterion yields a uniform distribution for each with mean $\frac{1}{2}(\xi_{i-1}+\xi_i)$. On combining this with (2) and (A.3), we obtain

$$\xi(x^{i-1},x^i)+\xi(x^i,x^{i+1})=2cx^{i-1}+2(1-2c)x^i+2cx^{i+1}$$

for $i=2,\dots,n-1$. Differentiation of both sides with respect to x^{i-1} gives $2c$ on the right, and with respect to x^i gives $2(1-2c)$. But it follows from the symmetry of $\xi(\ )$ in its two arguments that, on the left, the latter derivative must be twice as large as the former, $2(1-2c)=4c$, which yields $c=\frac{1}{4}$. Therefore, the expectation (A.3) takes the value shown in (4). For $i=2,\dots,n-1$ this can also be written as

$$\frac{1}{2}\left(\frac{x^{i-1}+x^i}{2}+\frac{x^i+x^{i+1}}{2}\right)$$

which shows that $\xi(\ )$ in (2) becomes the arithmetic mean of its arguments.

Given that $c=\frac{1}{4}$, the mean of the distribution in I_n is as shown in (4) for $i=n$. The distribution with this mean which maximizes the entropy is the negative exponential. For I_1 we obtain an analogous result with a tail in the direction of $-\infty$.

A.2. *Derivation of the* ME *variance*

Three results will be needed. First, for any real numbers a and b,

$$ab=\tfrac{1}{2}a^2+\tfrac{1}{2}b^2-\tfrac{1}{2}(a-b)^2 \tag{A.4}$$

Second, if X has a continuous distribution with density

$$\frac{1}{\beta-\alpha}\exp\left\{-\frac{x-\alpha}{\beta-\alpha}\right\} \qquad \beta>\alpha \tag{A.5}$$

for $x>\alpha$ and zero density for $x\leqslant\alpha$, then

$$\mathbf{E}X=\beta \quad \text{and} \quad \mathbf{E}(X^2)=\beta^2+(\alpha-\beta)^2 \tag{A.6}$$

Third, if X has a continuous distribution with density

$$\frac{1}{\alpha-\beta}\exp\left\{\frac{x-\alpha}{\alpha-\beta}\right\} \qquad \beta<\alpha \tag{A.7}$$

for $x<\alpha$ and zero density for $x\geqslant\alpha$, then the moments (A.6) also apply.

For $i=2,\dots,n-1$, the distribution of X in I_i (see Fig. 1) is uniform over the interval (ξ_{i-1},ξ_i), so that

$$\mathbf{E}(X^2|X\in I_i)=\tfrac{1}{3}(\xi_{i-1}^2+\xi_i^2+\xi_{i-1}\xi_i) \qquad i=2,\dots,n-1$$

We sum this over i,

$$\begin{aligned}\sum_{i=2}^{n-1}\mathbf{E}(X^2|X\in I_i)&=\frac{1}{3}\sum_{i=2}^{n-1}\xi_{i-1}^2+\frac{1}{3}\sum_{i=2}^{n-1}\xi_i^2+\frac{1}{3}\sum_{i=2}^{n-1}\xi_{i-1}\xi_i\\&=\frac{1}{2}\sum_{i=2}^{n-1}\xi_{i-1}^2+\frac{1}{2}\sum_{i=2}^{n-1}\xi_i^2-\frac{1}{6}\sum_{i=2}^{n-1}(\xi_i-\xi_{i-1})^2\end{aligned} \tag{A.8}$$

where the second step is based on (A.4). Also, since $\mathbf{E}(X|X\in I_i)$ is equal to $\frac{1}{2}(\xi_{i-1}+\xi_i)$ for $i=2,\dots,n-1$, we have

$$\begin{aligned}\sum_{i=2}^{n-1}[\mathbf{E}(X|X\in I_i)]^2&=\frac{1}{4}\sum_{i=2}^{n-1}\xi_{i-1}^2+\frac{1}{4}\sum_{i=2}^{n-1}\xi_i^2+\frac{1}{2}\sum_{i=2}^{n-1}\xi_{i-1}\xi_i\\&=\frac{1}{2}\sum_{i=2}^{n-1}\xi_{i-1}^2+\frac{1}{2}\sum_{i=2}^{n-1}\xi_i^2-\frac{1}{4}\sum_{i=2}^{n-1}(\xi_i-\xi_{i-1})^2\\&=\sum_{i=2}^{n-1}\mathbf{E}(X^2|X\in I_i)-\frac{1}{12}\sum_{i=2}^{n-1}(\xi_i-\xi_{i-1})^2\end{aligned} \tag{A.9}$$

where the second step is based on (A.4) and the third on (A.8). The definition of x_k^* in (A.9) implies

$$\sum_{k=1}^{n}x_k^{*2}=\sum_{i=1}^{n}[\mathbf{E}(X|X\in I_i)]^2 \tag{A.10}$$

It follows from (A.9) that the right-hand side of (A.10) is equal to the sum of

$$\sum_{i=2}^{n-1} \mathbf{E}(X^2|X \in I_i) - \frac{1}{12}\sum_{i=2}^{n-1}(\xi_i - \xi_{i-1})^2 \tag{A.11}$$

and

$$\left[\mathbf{E}(X|X \in I_1)\right]^2 + \left[\mathbf{E}(X|X \in I_n)\right]^2 \tag{A.12}$$

Since the square of $\mathbf{E}(X|X \in I_i)$ is smaller than $\mathbf{E}(X^2|X \in I_i)$ for both $i=1$ and $i=n$, this shows that the sum of (A.11) and (A.12), and hence also $\sum_k x_k^{*2}$ in (A.10), are smaller than $\sum_i \mathbf{E}(X^2|X \in I_i) = n\mathbf{E}(X^2)$, which proves the proposition in footnote 3.

To prove (5) it is sufficient to show that $\mathbf{E}(X^2|X \in I_i)$ summed over all i equals

$$\sum_{i=1}^{n} \mathbf{E}(X^2|X \in I_i)$$
$$= \sum_{i=1}^{n}(x^i)^2 - \frac{1}{4}\sum_{i=1}^{n-1}(x^{i+1} - x^i)^2 - \frac{1}{24}\sum_{i=2}^{n-1}(x^{i+1} - x^{i-1})^2 \tag{A.13}$$

We start with I_1, for which Fig. 1 shows that the distribution takes the form of (A.7) with $\alpha = \xi_1$ and $\beta = \frac{1}{2}(x^1 + \xi_1)$. So, using (A.6), we have

$$\mathbf{E}(X^2|X \in I_1) = \tfrac{1}{4}(x^1 + \xi_1)^2 + \tfrac{1}{4}(x^1 - \xi_1)^2 = \tfrac{1}{2}(x^1)^2 + \tfrac{1}{2}\xi_1^2$$

When we proceed similarly for I_n, we obtain

$$\mathbf{E}(X^2|X \in I_1) + \mathbf{E}(X^2|X \in I_n) = \tfrac{1}{2}\left[(x^1)^2 + (x^n)^2 + \xi_1^2 + \xi_{n-1}^2\right] \tag{A.14}$$

Next we use $\xi_i = \frac{1}{2}(x^i + x^{i+1})$ and $\xi_i - \xi_{i-1} = \frac{1}{2}(x^{i+1} - x^{i-1})$ for the relevant i's to write the right-hand side of (A.8) as

$$\frac{1}{4}\sum_{i=1}^{n-1}(x^i + x^{i+1})^2 - \frac{1}{2}(\xi_1^2 + \xi_{n-1}^2) - \frac{1}{24}\sum_{i=2}^{n-1}(x^{i+1} - x^{i-1})^2 \tag{A.15}$$

The first term of (A.15) can be written as

$$\frac{1}{4}\sum_{i=1}^{n-1}(x^i)^2+\frac{1}{4}\sum_{i=1}^{n-1}(x^{i+1})^2+\frac{1}{2}\sum_{i=1}^{n-1}x^ix^{i+1}$$

$$=\frac{1}{2}\sum_{i=1}^{n-1}(x^i)^2+\frac{1}{2}\sum_{i=1}^{n-1}(x^{i+1})^2-\frac{1}{4}\sum_{i=1}^{n-1}(x^{i+1}-x^i)^2$$

$$=\sum_{i=1}^{n}(x^i)^2-\frac{1}{2}(x^1)^2-\frac{1}{2}(x^n)^2-\frac{1}{4}\sum_{i=1}^{-1}(x^{i+1}-x^i)^2$$

where the first equal sign is based on (A.4). So when the expression in the last line is substituted for the first term of (A.15), we obtain

$$\sum_{i=2}^{n-1}\mathbf{E}(X^2|X\in I_i)=\sum_{i=1}^{n}(x^i)^2-\frac{1}{2}(x^1)^2-\frac{1}{2}(x^n)^2$$

$$-\frac{1}{4}\sum_{i=1}^{2}(x^{i+1}-x^i)^2-\frac{1}{2}(\xi_1^2+\xi_{n-1}^2)$$

$$-\frac{1}{24}\sum_{i=2}^{n-1}(x^{i+1}-x^{i-1})^2$$

The proof of (A.13) is completed by adding (A.14).

B. *A 2SLS application*

B.1. *Outline of the procedure*

We consider a complete linear system of L structural equations, the jth of which is written as

$$y_j=Z_j\delta_j+\varepsilon_j \tag{B.1}$$

where y_j is the n-element vector of observations of the left-hand variable, Z_j is the $n\times N_j$ matrix of observations on the right-hand variables, δ_j is the N_j-element parameter vector, and ε_j is the n-element vector of disturbances. These disturbances are assumed to be independently and identically distributed with zero mean and variance σ_{jj}.

Let there be K predetermined variables in the system, the observations on which are arranged in an $n\times K$ matrix X.[6] It is well-known that, under standard conditions and if (B.1) is identified within the system, δ_j is estimated consistently by the 2SLS estimator

$$d_j=\left[Z_j'X(X'X)^{-1}X'Z_j\right]^{-1}Z_j'X(X'X)^{-1}X'y_j \tag{B.2}$$

and that the estimated asymptotic covariance matrix of d_j is given by

$$s_{jj}\left[Z_j'X(X'X)^{-1}X'Z_j\right]^{-1} \quad \text{where } s_{jj}=\frac{1}{n}(y_j-Z_jd_j)'(y_j-z_jd_j) \tag{B.3}$$

If the number of predetermined variables is greater than the number of observations, $K>n$, then the matrix $X'X$ is singular so that the 2SLS estimator (B.2) does not exist. We write $\hat{Q}$ for the matrix of ME sample moments, which may then be regarded as a nonsingular estimator of $(1/n)X'X$. Substitution of $n\hat{Q}$ for $X'X$ in (B.2) and (B.3) then yields an ME 2SLS estimator

$$\hat{\delta}_j=\left(Z_j'X\hat{Q}^{-1}X'Z_j\right)^{-1}Z_j'X\hat{Q}^{-1}X'y_j \tag{B.4}$$

with an estimated asymptotic covariance matrix of the form

$$\hat{\sigma}_{jj}\left(n^{-1}Z_j'X\hat{Q}^{-1}X'Z_j\right)^{-1} \quad \text{where } \hat{\sigma}_{jj}=\frac{1}{n}(y_j-Z_j\hat{\delta}_j)'(y_j-Z_j\hat{\delta}_j) \tag{B.5}$$

B.2. Reestimation of Klein's Model I

We proceed to apply the maximum-entropy approach to Klein's Model I. This model consists of a consumption function, an investment function, and a demand-for-labor function:

$$C=\beta_0+\beta_1P+\beta_2P_{-1}+\beta_3(W+W')+\varepsilon \tag{B.6}$$

$$I=\beta_0'+\beta_1'P+\beta_2'P_{-1}+\beta_3'K_{-1}+\varepsilon' \tag{B.7}$$

$$W=\beta_0''+\beta_1''X+\beta_2''X_{-1}+\beta_3''(t-1931)+\varepsilon'' \tag{B.8}$$

[6]The matrix X occurs in $Y\Gamma+XB=E$, which is an $n\times L$ matrix equation with a square parameter matrix Γ and a rectangular parameter matrix B. Equation (B.1) is the jth column of this matrix equation with the understanding that variables with zero coefficients are deleted. For further details see [14, Sections 10.1 to 10.3].

Here C=consumption, P=profits, P_{-1}=profits lagged one year, W= private wage bill, W'=government wage bill, I=net investment, K_{-1}= capital stock at the beginning of the year, X=private output, and t= calendar time, while the β's are parameters and the ε's are disturbances. The predetermined variables of the model include three lagged variables (P_{-1}, K_{-1} and W_{-1}) as well as W', t, G (government nonwage expenditure) and T (business taxes). In addition, there is the constant-term variable which takes the unit value for each observation, so that there are eight predetermined variables as a whole.

The annual observations are from 1921 to 1941, so that $n=21$. Hence the sample is not undersized, but it is nevertheless appropriate to compare the conventional approach with that of maximum entropy. The matrix upper left in Table 2 is $(1/21)X'X$, the sample moment matrix of the predetermined variables, and the matrix upper right is the corresponding maximum-entropy moment matrix $\hat{Q}$. The first rows of these two matrices contain the sample means and are hence identical because of the mean-preserving constraint imposed on the maximum entropy approach. The submatrix of $(1/21)X'X$ which is obtained by striking the first row and column is equal to $\bar{x}\bar{x}'$ + the sample covariance matrix of the predetermined variables (other than the constant-term variable), where $\bar{x}$ is the vector of sample means of these variables. In the upper-right matrix the sample covariance matrix is replaced by the maximum-entropy covariance matrix. This yields smaller elements in the diagonal except for calendar time ($t-1931$). This exception results from the fact that time is not a continuously distributed exogenous variable; it takes one particular value with unit probability at any given moment, so that it must be treated as a dummy variable.[7] The inverses of the two moment matrices are shown at the top of Table 3; the elements of these inverses all differ pairwise, but the differences are mostly small. The seven matrices below those at the top in Tables 2 and 3 are discussed in the next subsection.

[7]The ME procedure is to retain the sample variance for discrete variables. Covariances are calculated using (10) without transforming the discrete variables. For continuous variables, rounding of the data may produce ties, i.e. $x^i = x^{i+1}$ for some i. If it is assumed that such a tie represents ignorance of which data point is larger, it can be shown that the appropriate way to deal with it is to force the corresponding transformed data points in (9) also to be tied. This problem arises mainly for 2SLS, since the other applications of the ME covariance matrix described in Section 1 are concerned with moment matrices of residuals. In accordance with the usual procedure, the lagged endogenous variables of Klein's model have been treated in the same way as exogenous variables.

Table 2

Moment matrices of the predetermined variables of Klein's Model I

1	W'	T	G	$t-1931$	P_{-1}	K_{-1}	X_{-1}	1	W'	T	G	$t-1931$	P_{-1}	K_{-1}	X_{-1}	
Sample moment matrix, $n=21$								Maximum-entropy moment matrix, $n=21$								
1	5.12	6.80	4.80	0.00	16.38	200.50	57.99	1	5.12	6.80	4.80	0.00	16.38	200.50	57.99	1
	29.85	37.58	27.30	11.33	83.15	1032.53	303.07		29.81	37.48	27.24	11.27	83.08	1032.24	302.64	W'
		50.24	36.04	8.38	111.83	1369.82	401.74			50.11	35.74	8.11	111.78	1369.49	400.92	T
			28.40	8.75	81.22	968.71	290.91				27.77	8.69	80.98	968.53	289.76	G
				36.67	−0.57	28.12	23.60					36.67	−0.81	27.33	22.25	$t-1931$
					283.63	3289.22	978.20						283.22	3289.92	977.86	P_{-1}
						40292.03	11665.94							40290.16	11664.24	K_{-1}
							3438.10								3436.38	X_{-1}
Sample moment matrix, $n=11$								Maximum-entropy moment matrix, $n=11$								
1	5.08	6.90	5.16	0.00	16.24	200.09	57.74	1	5.08	6.90	5.16	0.00	16.24	200.09	57.74	1
	29.50	37.75	30.32	11.95	82.37	1024.53	301.93		29.35	37.39	30.01	11.68	82.14	1024.27	299.96	W'
		51.68	40.73	7.69	112.90	1382.37	407.12			51.32	39.87	6.88	112.50	1381.33	404.00	T
			35.19	12.38	87.79	1042.13	316.74				33.67	11.97	86.39	1041.05	311.46	G
				40.00	0.27	32.56	30.04					40.00	−0.63	31.76	24.32	$t-1931$
					280.49	3252.37	968.06						279.24	3253.65	966.61	P_{-1}
						40133.08	11596.78							40126.95	11595.28	K_{-1}
							3417.39								3410.97	X_{-1}
Sample moment matrix, $n=7$								Maximum-entropy moment matrix, $n=7$								
1	5.40	7.17	5.51	1.00	17.17	201.64	59.91	1	5.40	7.17	5.51	1.00	17.17	201.64	59.91	1
	33.16	42.32	35.19	17.27	92.54	1095.24	333.53		32.82	41.73	34.54	16.67	93.25	1095.08	332.55	W'
		56.73	46.33	17.76	122.93	1451.84	440.09			55.90	44.30	16.71	122.27	1451.99	436.52	T
			42.49	21.50	99.00	1120.74	353.65				39.36	20.35	98.23	1119.98	348.43	G
				37.00	16.40	225.81	89.91					37.00	18.94	225.74	88.97	$t-1931$
					308.65	3460.41	1053.65						307.07	3461.67	1050.57	P_{-1}
						40736.98	12110.16							40726.03	12105.41	K_{-1}
							3666.91								3657.56	X_{-1}

Table 3
Inverses of moment matrices of predetermined variables

1	W′	T	G	t−1931	P_{-1}	K_{-1}	X_{-1}	1	W′	T	G	t−1931	P_{-1}	K_{-1}	X_{-1}	
Inverse of sample moment matrix, $n=11$								Inverse of maximum-entropy moment matrix, $n=21$								
4409	−328	13.68	−3.32	101.2	−31.9	−15.503	14.20	4717	−362	18.25	−4.12	110.9	−32.2	−16.255	14.09	1
	28	−1.68	0.35	−8.5	2.6	1.076	−1.13		32	−2.08	0.48	−9.5	3.0	1.182	−1.30	W′
		0.83	−0.36	0.4	−0.1	−0.054	0.06			0.83	−0.39	0.5	−0.2	−0.073	0.12	T
			0.67	−0.1	0.3	0.045	−0.23				0.75	−0.1	0.3	0.042	−0.22	G
				2.7	−0.6	−0.319	0.22					2.9	−0.7	−0.350	0.27	t−1931
					1.2	0.171	−0.62						1.2	0.163	−0.61	P_{-1}
						0.063	−0.09							0.063	−0.08	K_{-1}
							0.35								0.34	X_{-1}
Inverse of sample moment matrix, $n=21$								Inverse of maximum-entropy moment matrix, $n=11$								
17665	−1448	−62.85	117.67	392.7	−194.0	−62.638	90.11	3194	−217	−13.74	17.94	62.7	−23.0	−11.734	11.00	1
	131	1.34	−9.80	−36.0	13.4	4.682	−5.76		23	−0.78	−1.24	−6.4	0.8	0.573	−0.25	W′
		2.33	−0.95	−0.2	1.7	0.353	−0.94			1.25	−0.68	0.3	0.3	0.082	−0.16	T
			1.60	2.7	−1.0	−0.367	0.35				0.92	0.3	−0.1	−0.042	−0.04	G
				10.0	−3.2	−1.229	1.31					1.9	−0.0	−0.155	−0.03	t−1931
					4.0	0.876	−2.08						1.1	0.163	−0.59	P_{-1}
						0.248	−0.44							0..055	−0.09	K_{-1}
							1.14								0.33	X_{-1}
								Inverse of maximum-entropy moment matrix, $n=7$								
								2367	−130	−13.43	5.76	44.7	−12.9	−8.289	4.14	1
									12	0.54	−0.58	−3.7	0.7	0.350	−0.23	W′
										0.89	−0.42	−0.2	0.3	0.029	−0.09	T
											0.74	0.0	−0.1	0.008	−0.05	G
												1.3	−0.1	−0.122	0.04	t−1931
													0.6	0.061	−0.24	P_{-1}
														0.035	−0.03	K_{-1}
															0.14	X_{-1}

The results of estimation are shown in Table 4. The first column contains the 2SLS estimates obtained from (B.2) with asymptotic standard errors in parentheses. These standard errors are square roots of diagonal elements of the matrix (B.3). The second column of Table 4 contains the estimates (B.4) and their asymptotic standard errors obtained from (B.5). Evidently, the point estimates and asymptotic standard errors in these two columns are virtually identical.

B.3. Reducing the data base for Klein's Model I

A sample of $n=21$ observations is not undersized for Klein's Model I, but it is not difficult to obtain this situation. Imagine that we delete the second-last observation (1940), the fourth last (1938), and so on. This yields a sample of size $n=11$. Alternatively, we delete the second and third last observations (1939–40), the fifth and sixth last (1936–37), and so on, so that the sample size becomes $n=7$. This is undersized, given that the model contains eight predetermined variables.

The results are shown in the last three columns of Table 4. The point estimates of the two alternative procedures for $n=11$ are close to each other, although not as close as those for $n=21$; this difference reflects the fact that the two procedures are equivalent only for large n. For $n=7$ the approach based on sample moments yields no results, but the ME approach presents no problems. A comparison of the last column of Table 4 with the first two shows differences of an order of magnitude that could be expected, given that the last column represents an effort to estimate four parameters of each of three equations from only seven observations.[8]

[8]The behavior of the asymptotic standard errors is rather irregular in some cases, which is primarily due to the lack of stability of s_{jj} and $\hat{\sigma}_{jj}$ in (B.3) and (B.5). The variance estimates (in the order ε, ε', ε'') from 21 observations are 1.04, 1.38, .48 for the sample moment approach and 1.09, 1.39, .48 for the ME approach. For $n=11$ the two sets of estimates are 1.21, .34, .57 and 1.18, .41, .57, respectively. Hence the variance estimate of the investment function (B.7) is substantially smaller for $n=11$ than for $n=21$, which explains why the coefficients of this function have asymptotic standard errors that are smaller in the third and fourth columns of Table 4 than in the first two in spite of the fact that the former columns use fewer observations than the latter. For $n=7$ the ME estimates of the σ_{jj}'s are .62, 1.67, .30. A comparison with the corresponding figures 1.18, .41, .57 for $n=11$ explains the decline of the asymptotic standard errors in the last column of Table 4 (relative to the second-last column) of some of the coefficients of the consumption function (B.6) and the demand-for-labor function (B.8). The instability of the variance estimates for small n is not surprising. Under the standard linear model with normally distributed disturbances, the LS estimator of the variance σ^2 of the disturbances has a sampling variance equal to $2\sigma^4/(n-K)$, where K is the number of coefficients adjusted in the equation (see [14, p. 128]). This sampling variance implies a very small "t ratio" of the variance estimate when $n-K$ is small.

Table 4
Parameter estimates of Klein's Model I

	Sample moments		Maximum entropy		Sample moments		Maximum entropy		Maximum entropy	
	$n=21$				$n=11$				$n=7$	
β_0	16.55	(1.32)	16.75	(1.32)	17.89	(1.73)	18.24	(1.59)	17.57	(1.65)
β_1	0.02	(0.12)	0.00	(0.12)	−0.13	(0.26)	−0.12	(0.26)	−0.04	(0.15)
β_2	0.22	(0.11)	0.23	(0.11)	0.28	(0.20)	0.24	(0.21)	0.43	(0.20)
β_3	0.81	(0.04)	0.81	(0.04)	0.81	(0.06)	0.81	(0.05)	0.71	(0.04)
β_0'	20.28	(7.54)	20.97	(7.66)	12.62	(3.80)	14.51	(4.15)	12.00	(12.64)
β_1'	0.15	(0.17)	0.15	(0.18)	0.39	(0.11)	0.29	(0.13)	0.62	(0.32)
β_2'	0.62	(0.16)	0.62	(0.17)	0.45	(0.11)	0.53	(0.12)	0.23	(0.42)
β_3'	−0.16	(0.04)	−0.16	(0.04)	−0.12	(0.02)	−0.13	(0.02)	−0.12	(0.07)
β_0''	1.50	(1.15)	1.35	(1.14)	0.91	(1.70)	0.63	(1.63)	1.58	(1.59)
β_1''	0.44	(0.04)	0.44	(0.03)	0.46	(0.05)	0.45	(0.05)	0.48	(0.03)
β_2''	0.15	(0.04)	0.15	(0.04)	0.14	(0.06)	0.16	(0.06)	0.10	(0.05)
β_3''	0.13	(0.03)	0.12	(0.03)	0.15	(0.04)	0.14	(0.04)	0.10	(0.04)

Tables 2 and 3 are also revealing. The four moment matrices for $n=11$ and 7 in Table 2 are not very much different from either the sample moment matrix for $n=21$ or the ME moment matrix for $n=21$, but such minor differences do not imply that the inverses of these moment matrices are close to each other. Table 3 shows that the inverse of the ME moment matrix for $n=11$ is rather close to that for $n=21$, and also to the inverse of the sample moment matrix for $n=21$. The inverse of the ME moment matrix for $n=7$ shows larger differences; still, this matrix is well-behaved. However, the inverse of the sample moment matrix for $n=11$ is roughly four times larger than that for $n=21$, which indicates that the sample moment matrix becomes ill-conditioned even before the sample becomes undersized.

References

[1] Boneva, L. I., Kendall, D. and Stefanov, I. (1971). Spline transformations: three new diagnostic aids for the statistical data-analyst. *J. Roy. Statist. Soc. Ser. B* **33**, 1–37.

[2] Brundy, J. M. and Jorgenson, D. W. (1971). Efficient estimation of simultaneous equations by instrumental variables. *Review of Economics and Statistics* **53**, 207–244.

[3] Deaton, A. S. (1975). *Models and Projections of Demand in Post-War Britain*. Chapman and Hall, London.

[4] Dhrymes, P. J. (1971). A simplified structural estimator for large-scale econometric models. *Austral. J. Statist.*, **13**, 168–175.

[5] Fisher, F. M. (1965). Dynamic structure and estimation in economy-wide econometric models. Chapter 15 of *The Brookings Quarterly Econometric Model of the United States*, edited by J. S. Duesenberry, G. Fromm, L. R. Klein and E. Kuh. North-Holland Publishing Company, Amsterdam.

[6] Good, I. J. and Gaskins, R. A. (1971). Nonparametric roughness penalties for probability densities. *Biometrika*, **58**, 255–277.

[7] Hendry, D. F. (1976). The structure of simultaneous equations estimators. *Journal of Econometrics*, **4**, 51–88.

[8] Hoerl, A. E. and Kennard, R. W. (1970). Ridge regression: biased estimation for nonorthogonal problems. *Technometrics*, **12**, 55–67.

[9] Kloek, T. and Mennes, L. B. M. (1960). Simultaneous equations estimation based on principal components of predetermined variables. *Econometrica*, **28**, 45–61.

[10] Kuelbs, J. (1976). Estimation of the multi-dimensional probability density function. Mathematics Research Center Technical Summary Report No. 1646. University of Wisconsin, Madison.

[11] Kuhn, R. M. (1974). Reproducing kernels and natural spline density estimation. Statistics Department Technical Report No. 411. University of Wisconsin, Madison.

[12] Kuhn, R. M. (1974). Reproducing kernel Hilbert spaces applied to bivariate natural spline density estimation. Statistics Department Technical Report No. 435. University of Wisconsin, Madison.

[13] Sargan, J. D. (1975). Asymptotic theory and large models. *International Economic Review*, **16**, 75–91.

[14] Theil, H. (1971). *Principles of Econometrics*. Wiley, New York.

[15] Theil, H. (1973). A simple modification of the two-stage least-squares procedure for undersized samples. In *Structural Equation Models in the Social Science* (A. S. Goldberger and O. D. Duncan, Eds.) Seminar Press, New York.

[16] Theil, H. (1975–76). *Theory and Measurement of Consumer Demand*. Two volumes. Elsevier/North Holland, New York and North-Holland, Amsterdam.

[17] Wegman, E. J. (1972). Nonparametric probability density estimation: I. a summary of available methods. *Technometrics*, **14**, 533–546.

P. R. Krishnaiah, ed., *Multivariate Analysis–V*

TOWARD UNIVERSAL MACRO-DENOTATIVE MEANING SYSTEMS VIA A CROSS-CULTURAL MULTIVARIATE QUANTIFICATION PROCEDURE*

Oliver C. S. TZENG
Indiana University and Purdue University at Indianapolis, IN, U.S.A.

Charles E. OSGOOD and William H. MAY
University of Illinois, Urbana IL, U.S.A.

For compilation of the Atlas of Affective Meaning, Osgood et al. (1975) obtained subject ratings of 100 cross-culturally common concepts on 1250 indigenous semantic differential scales, 50 each from 25 different language/culture communities. In order to identify both the affective and denotative meaning systems of conceptions in general or broad life experiences, group mean ratings of concepts were analyzed by a pan-cultural factorization procedure, yielding strong evidence for three massive affective dimensions. The resultant factor matrix was then analyzed by a cross-cultural quantification procedure with seven clearly macro-denotative dimensions marked by five salient scales from each community. The generality and utility of these findings were further evaluated for future research in cross-cultural psychology and the general social and behavioral sciences.

1. Introduction

One of the most important tasks facing all researchers in the social and behavioral sciences is the study of psychosemantics involved in conceptions of all entities (objects) or signs (concepts) in human experiences. The acquisition of the meanings of an object or concept reflects an individual's learning experiences in the environment. Due to different processes involved in formulating psychological dispositions and conceptions of human beings, there are two types of psychological meanings that can be

*This research was part of the project "Studies in Comparative Psycholinguistics" supported by grants from the National Institute of Mental Health (NIMH 07705) and the National Science Foundation (NSF GS 2012X) to the University of Illinois. Principal Investigator is Professor Charles E. Osgood. In addition, this research was also supported in part by grants from Spencer Foundation (44-402-77) and Indiana University 1977 Summer Faculty Research Fellowship to the first author. The authors wish to express their gratitude to our foreign colleagues (cf. Table 1) who participated in data collection and preparation of this study.

differentiated: affect and denotation (Tzeng [7]). While *affect* reflects a person's emotional, autonomic reactions toward an object (e.g., distinguishing an object as being *good* or *bad*), *denotation* represents a person's descriptive attributes, or implicit theories, about the categorization of the object (e.g., characterizing an individual as being *tangible*, *familiar*, *predictable*, *comprehensible* and *usual* on the Predictability component in Japanese, cf. Tzeng [8].)

Furthermore, since the primitive affective congruence of human beings determines an individual's mental process in metaphorizing new concepts or objects with old experiences, affective meaning possesses semantic generalizations across a wide range of stimuli, responses and situations. On the other hand, the characteristics or features of denotative meanings are usually unique to specific contexts with different compositions of objects or conceptions. Therefore, denotative meaning systems can further be divided into general (macro) meaning, concerning highly diversified objects and concepts in life experiences, and specific (micro) meaning, concerning a particular homogeneous concept domain (e.g., personalities, cf. Tzeng [7]).

Over the past two decades, Osgood [3–6] has investigated in depth the affective meaning systems of human conceptions. Compelling evidence was found for three massive and universal psychosemantic constructs of Evaluation (GOODNESS), Potency (POWERFULNESS) and Activity (LIVELINESS). Also over the past few years, Tzeng [9] has formulated new research strategies and analytic methodologies for identifying *specific denotative meaning systems* in various restricted homogeneous context domains, e.g., personalities in social environment (Tzeng [9]), concepts referring to natural and artificial spaces in macroenvironments (Hogenraad and Tzeng [2]), the stimulus objects of all chairs in the near environment (Alexander, Alexander and Tzeng [1]; Tzeng and Landis [11]), and the concepts specifically referring to various aspects of self (Tzeng [10]).

However, no attempt has been made in the literature to investigate the macro-denotative meaning systems associated with the conceptions of heterogeneous objects or concepts in human experiences. The present study thus applied the theory of affect-denotation dichotomy and the related cross-cultural quantification procedures to the data collected by Osgood et al. [6] in order to identify, as well as characterize, the nature of universal systems or dimensions of macro-denotative meanings in a pancultural semantic space across 25 different language/culture communities.

2. Cross-cultural ratings of heterogeneous human conceptions

The meanings of an object or concept can be characterized by qualifiers (adjectives). Therefore, in order to identify the psychosemantic features in conceptions of macroenvironments of human beings with different nationalities and language specificities, Osgood et al. [6] adopted the following research procedures. First, a set of 100 diversified object names (nouns) were sampled mainly from lists developed by Glottochronologists, including very familiar notions of various environments, events, and conceptions encountered in daily life and social behaviors (from natural concrete concepts such as WATER and BIRD to abstractions such as HOPE and GUILT.) These concepts, being highly representative of human experiences and also being functionally equivalent in usage and translation between English and other languages, provided a common reference context for the subsequent elicitation and analysis of qualifiers from 25 indigenous communities (cf., Table 1).

Second, the above concepts were then used in a restricted adjective-noun association test to elicit a repertoire of 10,000 qualifier tokens from a group of 100 teenage boys within each community. These qualifiers were evaluated in each native language in terms of their overall frequencies of usage by all subjects, diversities of association with the 100 concepts, and congruence relationships with other qualifiers across all subjects and concepts. The first 50–80 most productive qualifiers were retained and used to construct 50 indigenous cultural semantic differential bi-polar scales (e.g., *good/bad* with seven steps in between.)

Third, another group of teenage male subjects in each community was sampled in order to rate the same 100 culture-common concepts against the 50 indigenous cultural SD scales with at least 20 replications for each scale by concept combination. The within-cultural mean ratings of 50 scales by 100 concepts from each of the 25 communities were then augmented to yield a data matrix of 1250 scales ($=50\times25$) by 100 concepts for subsequent analyses.

3. Pan-Cultural factorization: identification of affective markers

The above data from the 25 language/culture communities can be represented by matrix X of order $\mathbf{n}$ by $\mathbf{m}$, where $\mathbf{n}$ stands for 1250 scales and $\mathbf{m}$ for 100 heterogenous concepts. Entries in X, with values ranging

Table 1
General indices and two-letter key for 25 language/culture communities

Key	Location, Language	Site of Location	Language Family	Colleagues in the Field
FR	France, French	Paris, Strasbourg	Indo-European (Romance)	Abraham A. Moles; Francoise Enel
BF	Belgium, Flemish	Brussels	Indo-European (Germanic)	Herbert Rigaux; Robert Hogenraad
ND	Netherlands, Dutch	Amsterdam	Indo-European (Germanic)	Mathilda Jansen; A.J. Smolenaars
FF	Finland, Finish	Helsinki	Finno-Ugric	Metti Havio; Pertti Ounap
SW	Sweden, Swedish	Uppsala	Indo-European (Germanic)	Ulf Himmelstrand
AE	United States, American English	Illinois State (White Subjects)	Indo-European (Germanic)	Center for Comparative Psycho-linguistics
BE	United States, American English	Trenton, Chicago (Black Subjects)	Indo-European (Germanic)	Dan Landis; James E. Savage; Tulsi B. Saral
MS	Mexico, Spanish	Mexico City	Indo-European (Romance)	Rogelio Diaz-Guerrero
BP	Brazil, Portuguese	Sao Paulo	Indo-European (Romance)	Silvia T. Maurer-Lane
HM	Hungary, Magyar	Budapest	Finno-Ugric	Jenö Putnoky
YS	Yugoslavia, Serbo-Croatian	Belgrade	Indo-European (Slavic)	Tomislav Tomeković
IT	Italy, Italian	Padova	Indo-European (Romance)	Giovanni d'Arcais; Dora Capozza
GK	Greece, Greek	Athens	Indo-European (Greek)	Vasso Vassiliou
TK	Turkey, Turkish	Istanbul	Altaic	Beglan B. Tôgrol; Dogan Cuceloglu
LA	Lebanon, Arabic	Beirut	Afro-Asiatic (Semitic)	Lutfy N. Diab; Levon Melikian
IF	Iran, Farsi	Teheran	Indo-European	Tehran Research Unit (U. of III.)
AD	Afghanistan, Dari	Kabul	Indo-European (Iranic)	Noor Almad Shaker
AP	Afghanistan, Pashtu	Kabub, Kandahar	Indo-European (Iranic)	Noor Almad Shaker
DH	Delhi (India), Hindi	Delhi	Indo-European (Indic)	Krishna Rastogi; Ladli C. Singh
CB	Calcutta (India), Bengali	Calcutta	Indo-European (Indic)	Rhea Das; Alokananda Mitter
MK	Mysore (India), Kannada	Mysore City, Bangalore	Dravidian	A. Shanmugan; B. Kuppuswamy
MM	Malaysia, Malay	Kuala Lumpur	Malgyo-Polynesian	Jerry Boucher; Wong Fong Tong
TH	Thailand, Thai	Bangkok	Kadai	Jantorn Rufener; W. Wichiarajote
HC	Hong Kong, Cantonese	Hong Kong	Sino-Tibetan	Anita K. Li; Brian M. Young
JP	Japan, Japanese	Tokyo	Japanese	M. Asai; Y. Tanaka; Y. Iwamatsu

from +3 to −3, were group mean ratings of concepts on scales computed within each culture.

Let Z of order **r** by **m** be a new data matrix transformed from X. That is, entries within each row (scale) were group mean ratings of all concepts, but standardized with respect to the scale's mean and standard deviation across all 100 concepts. Furthermore, let Z be rescaled by the number of scales **m** as

$$Z^* = \frac{1}{\sqrt{m}} Z. \tag{1}$$

Then, Z^* can be defined as the product of the following three matrices:

$$Z^* = U' L^{\frac{1}{2}} W, \tag{2}$$

where U' is an $\mathbf{n} \times \mathbf{r}$ columnwise orthonormal matrix, L is an $\mathbf{r} \times \mathbf{r}$ diagonal matrix, and W is an $\mathbf{r} \times \mathbf{m}$ rowwise orthonormal matrix. In addition, **r**, the significant dimensions of Z^*, would be $\leqslant$ Min(**m**, **n**) and $UU' = WW' = I_r$.

Therefore, the cross-products among **n** scales in Z^* can be defined as matrix $R = Z^* Z^{*\prime}$. This matrix should be inter-item correlations among all 1250 scales computed across the 100 culture-common concepts. The relation of R with U and L in (2) would be

$$\begin{aligned} R = Z^* Z^{*\prime} &= U' L^{\frac{1}{2}} W W' L^{\frac{1}{2}} U \\ &= U' L U. \end{aligned} \tag{3}$$

Similarly, the cross-products of standard scores among **m** concepts can be computed across all 1250 scales as $P = Z^{*\prime} Z^*$. Then the relation of P with W and L in (2) can be expressed as

$$\begin{aligned} P = Z^{*\prime} Z^* &= W' L^{\frac{1}{2}} U U' L^{\frac{1}{2}} W \\ &= W' L W. \end{aligned} \tag{4}$$

Let matrix A of order **n** by **r** be the desired factor matrix of all 1250 scales on **r** cross-cultural common dimensions which can be derived from (2) as

$$A = U' L^{\frac{1}{2}}. \tag{5}$$

Such a matrix can in theory be obtained by solving for the eigenvalues and eigenvectors of the inter-scale correlation matrix R in (3). However, it is usually impractical to analyze such a large intercorrelation matrix of order 1250 by 1250. Alternatively, the cross-products matrix P of order 100 by 100 can be used to solve for its eigenvector matrix W of $\mathbf{r}$ by $\mathbf{m}$ in (4) which can subsequently be applied to (2) as

$$Z^* W' = U' L^{\frac{1}{2}} = A. \tag{6}$$

For simplicity, the resultant factor matrix A can further be rotated to A^* by a transformation matrix T of order $\mathbf{r}$ by $\mathbf{r}^*$ as

$$A^* = AT. \tag{7}$$

On the basis of the above factorization procedure, a pan-cultural semantic space was obtained that significantly accounted for the variances of the entire 1200 scale vectors of the 25 communities. Specifically, the first

Table 2

Salient scales of five indigenous communities on pan-cultural *E-P-A* factors[a]

Communities	Evaluation	Potency	Activity
AE	0.94 Nice	0.68 Big	0.61 Fast
(America/	0.92 Good	0.68 Powerful	0.55 Alive
English)	0.90 Sweet	0.57 Strong	0.44 Young
	0.89 Helpful	0.57 Deep	0.42 Noisy
BF	0.91 Good	0.57 Strong	0.69 Quick
(Belgium/	0.89 Magnificent	0.57 Big	0.65 Active
Flemish)	0.88 Agreeable	0.54 Heavy	0.42 Bloody
	0.88 Beautiful	0.50 Deep	0.40 Impetuous
TK	0.91 Beautiful	0.67 Big	0.50 Fast
(Turkey/	0.90 Good	0.58 Heavy	0.47 Living
Turkish)	0.90 Tasteful	0.53 Large	0.43 Flexible
	0.90 Pleasant	0.51 High	0.42 Young
DH	0.83 Glad	0.47 Strong	0.47 Gay
(Delhi/	0.83 Good	0.47 Brave	0.36 Thin
Hindi)	0.81 Ambrosial	0.46 Heavy	0.34 Soft
	0.80 Superion	0.44 Difficult	0.30 Loquacious
JP	0.93 Good	0.66 Heavy	0.48 Noisy
(Japan/	0.92 Pleasant	0.63 Big	0.45 Active
Japanese)	0.91 Comfortable	0.59 Difficult	0.44 Soft
	0.91 Happy	0.56 Brave	0.42 Fast

[a]In this table, all terms have been translated into English from four other languages. Values in the table are loadings.

three factors which dominated the entire structure were identified on the basis of their salient scales as being affective dimensions of Evaluation (E), Potency (P) and Activity (A). The 'positive' terms of the four highest-loading scales for E, P and A respectively, are displayed in Table 2 for five communities. A scanning of the scales in the first column is supportive of the universality of the Evaluative factor, and the loadings are all in the upper 80's and 90's. Such scales translating as *good, beautiful, pleasant* and *nice* appear repeatedly. The scale loadings for the Potency factor in column 2 are somewhat lower. However, common strengths and magnitude flavors are evident in scales like *strong, big, powerful*, and *brave*. Similarly, the loadings of the Activity scales in column 3 are relatively lower, but the liveliness flavor is clear and evident in such common scales as *fast, quick, alive* and *active*.

Within each community, the four scales with the highest loadings among the 50 indigenous scales on each pan-cultural factor were chosen as *affective markers* as in Table 2. These markers from the 25 communities were further considered as being functionally equivalent in representing the *same* universal E, P, and A factors in a cross-culturally common psychosemantic space. Osgood et al. [6] utilized these pan-cultural markers in the compilation of an *Atlas of Affective Meanings* in which the affective aspects of meanings of some 620 concepts were measured from 30 language/culture communities.

4. Exploration of universal macro-denotative components

In order to identify the characteristics of the remaining configuration of all 1250 scale vectors beyond the dominance of the E-P-A constructs in the pan-cultural semantic space, the method developed for identifying the *specific* denotative systems within an indigenous culture was applied in the present context. That is, given the factor matrix A (or A^*) of $\mathbf{n}\times\mathbf{r}$ ($=1250\times 10$) from the above factorization, the markers of all 25 communities can be augmented as A_1 of 300 by $\mathbf{r}$ (where 300 are 12 E-P-A markers from the 25 communities), whereas the remaining scales can be augmented as A_2 of order 950 by $\mathbf{r}$. Then the cross-products among $\mathbf{r}$ dimensions within the marker domain, Q, can be derived from A_1 by

$$Q=A_1'A_1. \tag{8}$$

Matrix Q can be decomposed to solve for its eigenvalues and eigenvectors as in matrices V and S respectively:

$$Q = VSV' = (V_1 : V_2) \begin{bmatrix} S_1 & 0 \\ 0 & S_2 \end{bmatrix} \begin{bmatrix} V_1' \\ V_2' \end{bmatrix} \tag{9}$$

where V_1 is **r** by 3, V_2 is **r** by 7, S_1 is 3 by 3, and S_2 is 7 by 7. Since Q is defined by the markers, the diagonal matrix S_2 should in theory have null entries. This implies that the matrix V can be used as the transformation cosines to rotate the entire 1250 scale vectors, within the 10 dimensions space, onto the three-dimensional affective subspace which is explicitly defined by the 300 functionally equivalent markers. That is, let

$$T = V \Leftrightarrow (V_1 : V_2) = (T_1 : T_2) \tag{10}$$

and

$$A^* = AT = \begin{bmatrix} A_1 \\ A_2 \end{bmatrix} (T_1 : T_2) = \begin{bmatrix} A_{11}^* & \vdots & A_{12}^* \\ A_{21}^* & \vdots & A_{22}^* \end{bmatrix} = (A_1^* \;\vdots\; A_2^*) \tag{11}$$

The above procedure will accomplish the separation of *E-P-A* affective subspace (A_1^*) from the remianing subspace (A_2^*) of 7 dimensions.

For simplicity, matrix A_{11}^*, containing the loadings of the markers in the affective *E-P-A* subspace, should further be rotated by some transformation matrix T_{11}^* that can be derived from various rotation procedures. That is,

$$A_{11}^{**} = A_{11}^* T_{11}^*. \tag{12}$$

Similarly, matrix A_{22}^*, the loadings of non-markers in the non-*E-P-A* subspace, can be rotated to simple structure by T_{22}^* as

$$A_{22}^{**} = A_{22}^* T_{22}^*. \tag{13}$$

Furthermore, in order to identify the characteristics of the 950 *non*-markers in the *rotated affective subspace* (A_{11}^{**}), the transformation cosines T_{11}^* derived from Equation (12) should be applied to A_{21}^* as follows:

$$A_{21}^{**} = A_{21}^* T_{11}^*. \tag{14}$$

Similarly, in order to identify the characteristics of the 300 *markers* in the *rotated non-affective subspace* (A^{**}_{22}), the transformation cosines T^{*}_{22} derived from Equation (13) should be applied to A^{*}_{12} as follows:

$$A^{**}_{12} = A^{*}_{12} T^{*}_{22}. \tag{15}$$

On the basis of the quantitative as well as psychological assessments of the results from the above four equations, the following conclusions can be made: (a) In the affective subspace of 300 markers by 3 dimensions (in A^{**}_{11}), the loadings of all markers are "pure" with extremely high loadings on the universal *E-P-A* dimensions. As a consequence, these 300 markers had close to zero or negligible loadings on the remaining seven dimensions (in A^{**}_{12}). (b) As expected, some of the non-markers also had high loadings on the *E-P-A* dimensions (in A^{**}_{21}). This would reinforce the notion of the affective dominance in human conceptions. (c) The remaining seven dimensions in A^{**}_{22} had apparently different characteristics from the *E-P-A* dimensions. For purposes of identifying their characteristics, five salient scales on each dimension were selected for all the 25 communities, as given in Table 3. Since the structures of these seven dimensions have been 'purified', *without* reference to the affective components, their meanings should represent pan-cultural affect free (macro-denotative) systems in conceptions of 100 heterogeneous concepts in the macroenvironments of human beings.

However, in order to derive accurate interpretations about these dimensions that would be consistent with the cultural backgrounds of all 25 communities, two colleagues from each site where data were collected were asked to name, independently, these seven dimensions and also to provide us with information regarding their naming processes (e.g., numbers of different cultures results consulted in naming each dimension, perception of each dimension on the continuum of the affect/denotation dichotomy, and the comfortableness level about each factor's name when applied to all 25 communities as a whole.) In addition, other pertinent information was also obtained from each colleague, e.g. the numbers of foreign countries and continents traveled, attitudes toward factor analysis and the semantic differential technique in cross-cultural research, and the experiences in naming factors in social and psychological data. At the time of preparing the present report, we have received responses from colleagues of more than 10 communities. It is encouraging that although some deviations did occur between *actual words* used for representing each dimension, the interpretations of the nature and characteristics of each dimension were highly congruent across communities. Furthermore, as given in Table 3,

Table 3

Salient loadings of indigenous scales in a pancultural macro-denotative semantic subspace

(**Note:** Decimal points in this table were omitted. Therefore, each loading should be divided by 100.)

Dimension 1 (Realism/Idealism)	Dimension 2 (Uniqueness/ Commonality)	Dimension 3 (Hot/Cold)	Dimension 4 (Age Developmental)	Dimension 5 (Immediacy/ Remoteness)	Dimension 6 (Artificiality/ Natural)	Dimension 7 (Instability/ Stability)
BP (BRAZIL, PORTUGUESE)						
48 concrete-abstract	49 humane-inhuman	51 warm-cold	42 near-far	46 near-far	52 artificial-natural	33 many-few
43 hard-soft	42 deep-shallow	45 colorful-plain	34 colorful-plain	32 easy-difficult	47 cold-warm	26 difficult-easy
31 enduring-fleeting	39 abstract-concrete	34 artificial-natural	32 long-short	30 strong-weak	31 weak-powerful	23 young-old
30 short-long	32 short-long	23 unfaithful-faithful	28 young-old	28 plain-colorful	29 green-red	22 pleasant-wicked
30 strong-weak	24 near-far		25 square-round	26 round-square	25 dispensable-indispensable	22 long-short
		23 brilliant-opaque				
AE (AMERICA, ENGLISH)						
40 hard-soft	46 few-many	58 burning-freezing	30 dark-light	45 known-unknown	44 square-round	42 wet-dry
38 tough-tender	42 serious-funny	52 hot-cold	28 square-round		37 freezing-burning	26 rich-poor
30 low-high	29 hot-cold	45 dry-wet	28 many-few	23 dull-shiny	35 cold-hot	25 full-empty
26 short-long	28 smart-dumb	32 sharp-dull	22 black-white	21 loud-soft	18 loud-soft	23 many-few
26 stale-fresh	26 square-round	25 momentary-everlasting	22 heavy-light	20 square-round	17 happy-sad	18 fresh-stale
				20 momentary-everlasting		

FR (FRANCE, FRENCH)

52 hard-soft	44 lowpitch-highpitch	47 hot-cold	34 mortal-immortal	29 allowed-prohibited	54 artificial-natural	29 blind-clear sighted
46 mortal-immortal	33 rare-abundant	47 red-green	30 brunette-blonde	23 found-lost	33 superficial-deep	23 difficult-easy
30 solid-fragile	27 modern-old	37 colorful-pale	27 black-white	23 indispensable-superficial	32 difficult-easy	21 fragile-solid
24 pale-colorful	23 deep-superficial	30 fat-thin	22 solid-fragile	21 useful-useless	31 rare-abundant	20 severe-indulgent
21 superficial-deep	20 pale-colorful	29 blonde-brunette	22 indulgent-severe	19 healthy-unhealthy	30 insignificant-important	20 low-high

BF (BELGIUM, FLEMISH)

51 tough-tender	42 little-much	37 red-green	48 young-old	37 cheap-expensive	45 blue-yellow	39 changeable-unchangeable
46 bound-free	39 rare-common	33 warm-cold	36 modern-old fashioned	33 realistic-romantic	38 green-red	33 low-high
42 unchangeable-changeable	34 modern-old fashioned	32 frivolous-serious	36 new-old	25 changeable-unchangeable	28 modern-old fashioned	32 changing-steady
41 hard-soft	33 difficult-easy	30 yellow-blue	27 much-little	24 sanguine-phlegmatic	26 new-old	29 cold-warm
40 realistic-romantic	33 expensive-cheap	27 changing-steady	24 dark-light	20 common-rare	25 unknown-known	24 unpredictable-predictable

ND (NETHERLANDS, DUTCH)

44 firm-loose	52 special-common	51 dry-wet	31 firm-loose	29 known-unknown	44 angular-round	47 wet-dry
42 deep-shallow	38 clever-stupid	44 yellow-blue	29 modern-old fashioned	22 trivial-impressive	38 unnatural-natural	48 changing-steady
38 constant-changeable	36 little-much	41 red-green	28 new-old	21 common-special	37 blue-yellow	42 changeable-constant
36 monocolored-multicolored	36 new-old	38 warm-cold	28 much-little	21 empty-full	36 loud-soft	37 unpredictable-predictable
36 tame-wild	31 difficult-easy	29 multicolored-monocolored	25 long-short	20 dull-shiny	33 modern-old fashioned	31 wild-tame

Table 3 (Continued)

Dimension 1	Dimension 2	Dimension 3	Dimension 4	Dimension 5	Dimension 6	Dimension 7
SW (SWEDEN, SWEDISH)						
51 hard-soft	35 sensitive-insensitive	43 red-blue	36 young-old	35 close-distant	45 angular-round	33 full-empty
50 stable-unstable	33 empty-full	40 warm-cold	31 dark-light	33 smoky-clear	40 blue-red	27 unreliable-reliable
42 shallow-deep	29 red-blue	36 red-green	29 stable-unstable	21 empty-full	36 green-red	26 cold-warm
37 reliable-unreliable	27 hollow-solid	31 light-dark	27 close-distant	19 dull-shiny	34 cold-warm	25 deep-shallow
36 tough-tender	26 red-green	26 shallow-deep	22 light-difficult	19 sour-sweet	31 sharp-blunt	24 tender-tough
FF (FINLAND, FINNISH)						
46 hard-soft	28 red-blue	42 red-blue	29 dark-light	42 near-distant	41 blue-red	27 unfaithful-faithful
40 scanty-abundant	25 wise-stupid	41 hot-cold	28 getting bigger-getting smaller	33 boring-interesting	39 angular-rounded	22 dangerous-safe
39 steady-capricious	20 short-long	31 lively-tired	28 near-distant	26 thin-thick	33 voluntary-compulsory	19 rich-poor
31 shallow-deep	19 deep-shallow	29 light-dark	26 black-white	23 brave-timid	33 cold-hot	19 getting bigger-getting smaller
28 compulsory-voluntary	19 difficult-easy	27 high-low	25 multicolored-unicolored	20 shallow-deep	32 multicolored-unicolored	19 deep-shallow
HH (HUNGARY, HUNGARIAN)						
53 hard-soft	32 red-green	40 red-green	30 near-far	34 near-far	52 blue-yellow	35 unsure-sure
50 stiff-loose	29 few-many	38 yellow-blue	22 yielding-rigorous	31 frequent-rare	41 angular-round	29 faithless-faithful
40 tightened-loosened	27 deep-shallow	36 bright-dim	21 faithful-faithless	29 loud-still	41 green-red	29 unreliable-reliable
38 shallow-deep	25 tiring-restful	30 exploding-suppressed	20 angular-round	23 tedious-interesting	35 talkative-wordless	22 blind-seeing
35 sure-unsure	24 clever-stupid	29 loud-still	20 many-few	22 tiring-restful	34 pied-unicolored	21 cold-warm

MS (MEXICO, SPANISH)

43 tiny-immense	40 scarce-abundant	58 dry-wet	22 long-short	26 broken-whole	43 artificial-natural	35 wet-dry
31 short-long	25 responsible irresponsible	46 hot-cold	22 dark-light	22 known-unknown	35 cold-hot	30 fresh-suffocating
28 superficial-profound	25 artificial-natural	28 superficial-profound	22 black-white	15 artificial-natural	30 unknown-known	29 tasty-distasteful
27 little-much	23 profound-superficial	27 red-blue	16 hot-cold	13 boring-amusing	28 false-true	24 cold-hot
23 tired-rested	23 short-long	26 artificial-natural	15 fresh-suffocating	12 dull-bright	26 blue-red	21 much-little

IT (ITALY, ITALIAN)

55 solid-liquid	37 deep-superficial	39 artificial-natural	36 stable-unstable	40 near-far	45 artificial-natural	27 rich-poor
45 precise-vague	32 scarce-abundant	38 colored-uncolored	36 near-far	25 broken-unbroken	19 hard-tender	24 fragile-sturdy
39 limited-unlimited	31 sparse-dense	35 superficial-deep	32 long-short	24 uncolored-colored	18 difficult-easy	29 dense-sparse
38 stable-unstable	31 uncolored-colored	32 solid-liquid short-long	30 solid-liquid	23 dense-sparse	18 sparse-dense	22 unstable-stable
36 hard-tender	30 fragile-sturdy		29 black-white	19 masculine-feminine	18 uncolored-colored	20 impetuous-calm

YS (YUGOSLAVIA, SERBO-CROATIAN)

44 opaque-transparent	46 political-apolitical	35 warm-cold	30 black-white	24 ramified-constricted	34 transparent-opaque	36 ramified-constricted
30 limited-unlimited	40 humane-inhumane	28 shallow-deep	29 elongated-round	22 guileless-cunning	27 thin-fat	27 fresh-stale
30 same-different	30 unlimited-limited	25 round-elongated	22 dark-bright	22 empty-full	25 different-same	25 cold-warm
26 calm-agitated	28 round-elongated	22 general-specific	21 ramified-constricted	19 agitated-calm	23 clever-stupid	24 agitated-calm
25 elongated-round	26 clever-stupid	22 white-black	17 heavy, difficult-light, easy	18 rough-gentle	21 political-apolitical	24 full-empty

Table 3 (Continued)

Dimension 1	Dimension 2	Dimension 3	Dimension 4	Dimension 5	Dimension 6	Dimension 7
GK (GREECE, GREEK)						
42 stable-unstable	49 scarce-abundant	49 hot-cold	29 hospitable-inhospitable	28 brunette-blond	47 cold-hot	37 full-empty
35 short-long	39 sincere-insincere	28 brilliant-dark	28 brunette-blond	27 indifferent-curious	31 brunette-blond	27 many-few
33 clean-dirty	37 superior-inferior	25 blond-brunette	27 false-true	23 simple minded-cunning	24 unripe-(immature) ripe (mature)	26 cold-hot
33 energetic-lazy	33 honest-dishonest	23 joyful-sad	22 temporary-eternal	22 temporary-eternal	23 open-closed	25 abundant-scarce
26 worthy-unworthy	31 right-wrong	21 shallow-deep	22 tall-short	19 shallow-deep	20 loud-soft	21 unfaithful-faithful
TK (TURKEY, TURKISH)						
34 calm-rough	37 expensive-cheap	55 dry-wet	26 expensive-cheap	29 cheap-expensive	46 straight-round	40 new-old
27 difficult-easy	33 effective-ineffective	35 colorful-colorless	24 dull-shiny	24 easy-difficult	24 double-single	37 wet-dry
27 yellow-red	31 terrific-ordinary	31 short-long, tall	19 finite-infinite	23 ordinary-terrific	20 cold-hot	33 many-few
26 full-empty	30 infinite-finite	29 hot-cold	19 closed-open	22 shallow-deep	19 laughable-terrifying	30 difficult-easy
25 careful-careless	29 wise-stupid	28 emotional-unemotional	14 long, tall-short	21 red-yellow	17 open-closed	30 dangerous-safe
			14 hot-cold			
LA (LEBANON, ARABIC)						
47 sturdy-not sturdy	46 learned-ignorant	22 young-old	31 near-far	37 matte-bright, brilliant	22 rare-plentiful	34 rich-poor
33 male-female	46 little-much	19 bright, brilliant-matte	28 much-little	34 indolent-industrious	21 imaginary-real	30 full-empty
31 real-imaginary	36 rare-plentiful	16 empty-full	20 young-old	30 passive-active	17 learned-ignorant	28 young-old
27 orderly-scattered	34 hidden-apparent	12 dead-alive	17 orderly-scattered	27 tumultuous-calm	16 civilized-barbaric	14 bright, brilliant-matte
27 permanent-temporary	29 particular-general	12 rich-poor	17 simple-difficult	24 near-far	12 male-female	14 much-little

IF (IRAN, FARSI)

22 whole-torn	35 burning-frozen	38 burning-frozen	36 long-round	31 tiny-large	41 frozen-burning	50 fresh-old
17 tall, long-short	33 influential-uninfluential	36 colorful-colorless	27 black-white	27 colorless-colorful	26 unreal-real	39 stout-slim
16 harmless-sly	28 unique-commonplace	23 round-long	22 dark-light	27 long-round	22 dirty-clean	36 much, many-little, few
16 calm-uncalm	27 colorless-colorful	21 badly colored-well colored	18 national-governmental	26 terminable-interminable	20 uncertain-certain	35 influential-uninfluential
15 rough-smooth, tender	26 round-long	18 fresh-old	18 narrow-broad	24 well colored-badly colored	20 torn-whole	34 broad-narrow

AD (AFGHANISTAN, DARI)

49 firm-loose	53 humane-inhumane	49 burning-frozen	23 fat-thin	41 slow-fast	27 cold-warm	39 fresh-fetid
33 powerful-powerless	48 learned-ignorant	43 colorful-colorless	17 firm-loose	37 timely-untimely	23 frozen-burning	37 very much-little
32 strong-weak	44 religious-irreligious	36 warm-cold	15 religious-irreligious	32 tiny-huge	20 thin-fat	34 many-few
31 whole-broker	48 moslem-neathen	24 new-old	14 dull-bright	28 sufficient-excessive	20 brotherly-unbrotherly	32 green-yellow
28 fast-slow	40 brotherly-unbrotherly	23 magnificent-paltry	13 empty-overflowing	27 dull-bright	19 powerless-powerful	32 overflowing-empty

AP (AFGHANISTAN, PASHTU)

51 solid-liquid	37 short-long	58 dry-wet	38 under-upon	32 thin-thick	46 artificial-natural	43 full-empty
46 unbroken-broken	32 scattered-thick	57 hot-cold	34 black-white	31 black-white	41 cold-hot	40 wet-dry
36 strong-weak	27 acquainted-stranger	35 solid-liquid	29 solid-liquid	29 light-heavy	19 weak-strong	35 cold-hot
34 complete-incomplete	27 less-much, more	33 artificial-natural	28 young-old	26 broken-unbroken	18 lean-fat	29 younger-elder
33 domestic, internal-foreign	27 expensive-cheap	30 short-long	22 expensive-cheap	26 incomplete-complete	17 black-white	21 much, more-less

Table 3 (Continued)

Dimension 1	Dimension 2	Dimension 3	Dimension 4	Dimension 5	Dimension 6	Dimension 7
DH (DELHI, HINDI)						
34 strong-weak (of its kind)	34 closed-open	56 hot-cold	30 pointed-rounded	51 twofolded-unfolded	34 cold-hot	25 much, many-little, few
25 scattered-dense	30 expensive-inexpensive	25 rounded-pointed	26 stale-fresh	38 with somebody-alone	31 pointed-rounded	24 fresh-stale
22 extraordinary-ordinary	28 heartfelt-pretentious	23 unskilled-skilled	25 insufficient-sufficient	35 dense-scattered	19 tall-short	24 fast-slow
22 constructive-destructive	25 true-false	22 pretentious-heartfelt	24 closed-open	32 strong-weak (of its kind)	17 twofolded-unfolded	22 new-old
19 skilled-unskilled	24 not current-current	22 with somebody-alone	24 black-white	28 sweet-bitter	17 with somebody-alone	21 disingenuous-ingenuous
CB (CALCUTTA, BENGALI)						
40 shallow-deep	50 few-many	60 hot-cold	29 many-one	35 ordinary-extraordinary	36 many-one	34 many-one
39 difficult-easy	50 scarce-numerous	43 red-blue	28 heavy-light	31 pale, faded-colored	32 many-few	33 much-little
37 hard-soft	43 new-old	40 colored-pale, faded	24 shallow-deep	23 torn-intact	29 cold-hot	28 numerous-scarce
36 shut-open	43 one-many	34 viscous-nonviscous	23 tall-short	19 light-heavy	29 uncertain-certain	25 extraordinary-ordinary
35 intact-torn	38 little-much	32 high-low	23 black-white	19 thin-thick	25 mild-intense	24 low-big
MK (MYSORE, KANNADA)						
34 common-uncommon	31 insufficient-plenty	30 temporary-permanent	31 colorful-plain	32 harty-(skimpy?)	22 best-mean	32 encroaching-protecting'
28 strong-weak	27 secret-public	28 harty-(skimpy?)	19 plenty-little	28 strong-weak	18 public-secret	27 harty-(skimpy?)
24 useful-useless	26 useless-useful	24 pretentious-humble	19 piercing-not piercing	21 noisy-quiet	18 attractive-repulsive	27 detachment-attachment
21 harsh-melodius	26 few-many	24 hard-soft	17 unselfish-selfish	20 constructive-destructive	18 black-red	25 pitiless-piteous
20 rough-soft	23 slim-fatty	21 rough-soft	13 expected-unexpected	18 dim-bright	16 wide-narrow	24 long-short

TH (THAILAND, THAI)

31 short-long	49 human-animal	45 dry-wet	23 long-short	45 near-far	35 much-little	46 wet-dry
29 difficult-easy	34 costly-inexpensive	31 hot-cold	23 costly-inexpensive	26 inexpensive-costly	34 new-old	30 difficult-easy
28 costly-inexpensive	30 depleted-plentiful	24 new-old	19 whole-broken	25 mine, ours-his, hers, theirs	32 cold-hot	27 new-old
22 short-tall	22 mine, ours-his, hers, theirs	24 bright-dark	16 new-old	24 easy-difficult	27 father-mother	24 much-little
19 mine, ours-his, hers, theirs	20 little-much	17 stay-leave	15 opaque-transparent	22 much-little	24 transparent-opague	21 beginning-ending

MM (MALAYSIA, MALAY)

36 hard-soft (gentle)	62 human-animal	53 hot-cold	34 young-old	45 near-far (distant)	38 imitation-original	39 much (many)-slight (few)
27 strong-weak	43 wide apart-dense	52 fiery-watery	29 oval-round	34 light-heavy	34 new-old	33 excessive-lacking
20 skillful (clever)-unskillful	36 new-old	34 excessive-lacking-	28 wood-stone	23 new-old	27 thin-thick	29 cold-hot
20 dense-wide apart	35 cooked (ripe)-uncooked (unripe)	30 external internal (deep)	28 dark-bright (clear)	22 human-animal	24 human-animal	29 internal (deep)-external
20 fast-slow	34 delicate (fine)-rough	26 imitation-original	26 new-old	20 cooked (ripe)-uncooked (unripe)	22 cold-hot	25 watery-fiery

HC (HONG KONG, CANTONESE)

52 solid-hollow	52 human-bestial	47 scorching hot-cool	37 old-new	27 human-bestial	38 square-round	39 "pushy"-yielding
49 limited-infinite	37 deep-superficial	29 "pushy"-yielding	29 powerless-omnipotent	26 strong willed-weak willed	31 cool-scorching hot	32 extreme-balanced (the golden mean)
33 difficult-easy	36 wonderful-ordinary	28 dead-alive	26 dirty-clean	22 powerful-powerless	31 open-secretive	29 difficult-easy
27 ordinary-wonderful	35 deep-shallow	25 transitory-eternal	25 square-round	21 new-old	18 human-bestial	25 new-old
26 strong willed-weak willed	31 extreme-balanced (the golden mean)	23 square-round	25 ordinary-wonderful	21 succeeding-failing	16 refined-common	25 secretive-open

Table 3 (Continued)

Dimension 1	Dimension 2	Dimension 3	Dimension 4	Dimension 5	Dimension 6	Dimension 7
JP (JAPAN, JAPANESE)						
37 solid-fragile	50 few-many	45 red-blue	34 near-far	50 near-far	38 blue-red	34 new-old
34 strong-weak	43 rare-common	43 cheerful-lonely	29 black-white	35 cheerful-lonely	35 thin-thick	28 cold-hot
34 narrow-wide	41 slow-fast	36 hot-cold	21 many-few	28 hot-cold	22 cold-hot	19 deep-shallow
32 tight-loose	38 late-early	31 shallow-deep	17 excessive-plain	26 many-few	18 skillful-unskillful	18 tense-relaxed
28 short-long	32 hot-cold	26 new-old	17 old-new	23 new-old	13 wide-narrow	18 delicious-not delicious
BE (AMERICAN BLACK ENGLISH)						
48 tied down-free	47 peaceful-ferocious	54 hot-cold	43 in-out	27 mellow-yellow	39 angular-rounded	40 ferocious-peaceful
47 tight-loose	42 black-white	40 peaceful-ferocious	36 sharp-ragged	21 dark-red	28 cool-warm	18 dry-funny
37 passive-active	32 mellow-yellow	34 tight-loose	32 together-dizzy	21 uptight-badly	29 cold-hot	18 silly-cool
32 chump-powerful	32 dark-red	27 uptight-cool	32 mellow-yellow	21 passive-active	23 hip-dry	16 hot-cold
28 uptight-cool	29 honest-slick	24 red-dark	31 hot-cold	20 low-high	19 slick-honest	16 phat (fat)-slim

markers within each dimension (column) appeared highly consistent across the entire 25 communities. Therefore, on the basis of the suggestions from our foreign colleagues and the apparent characteristics and occuring frequencies of the markers across different communities, the tentative names of these universal macro-denotative dimensions were derived as given at the top of the table.

Dimension 1 is dominated, in the order of the frequencies appearing in various communities, by the leading scales referred to in one way or another as hardness vs softness (e.g., *hard* and *tough* vs *soft* and *tender*), broad vs narrow sizes (*shallow* and *short* vs *deep* and *long*), stableness vs unstableness (*unchangeable* and *constant* vs *changeable*), forcefulness (*strong* vs *weak*), static condition (*calm* vs *uncalm* and *agitated*), wholesomeness level (*whole* and *intact* vs *broken* and *torn*) and difficulty level (*difficulty* vs *easy*). In addition, other salient scales also suggest that this dimension carries the "real vs unreal" (or enduring vs fragility) flavor. Therefore, it may be called a "Realism/Idealism" factor following Osgood et al. [6].

The second factor is characterized by the following six leading clusters of scales, in the order of *few* vs *many*, *modern* vs *tradition*, *human* vs *inhuman*, *clever* vs *stupid*, *sparse* vs *dense*, and *expensive* vs *cheap*. It clearly reflects the semantic properties of unique vs common objects in macroenvironments. It is referred to as a Uniqueness/Commonality dimension. Factor 3 seems to be readily identifiable as a Hot/Cold dimension in the natural environments, with salient scales like *hot*, *warm*, *dry*, *burning*, *colored*, *brilliant*, *red* and *yellow* on one end, and *cold*, *wet*, *freezing*, *opaque*, *unicolored* and *blue* on the other end.

Dimension 4, characterized by the leading scales in the order of *young*, *black*, *dark*, *many*, *close* and *tall* versus their opposites *old*, *white*, *light*, *few*, *distant* and *short*, respectively, seems to be a time sequence factor in differentiations of macroenvironmental objects on the continuum of birth vs death, mature vs immature, and attainable vs unattainable conditions. However, for the sake of simplicity, it will be identified as an Age (or Developmental) factor, even though more than human age is involved.

The most dominating scale on Factor 5 is *near-far* (also its synonyms, e.g., *close-distant*). Other salient scales, with the left poles associated with *near*, include *dull*, *plain*, *strong*, *cheap*, *changeable*, *common*, *broken* and *boring* versus their opposites *shiny*, *colorful*, *weak*, *expensive*, *unchangeable*, *rare*, *whole* and *interesting*. The remaining salient scales also reinforce the notion of ordinary immediacy from their left poles. Therefore, this factor will be called as a Immediacy/Remoteness factor. Dimension 6, defined by such higher-loading scales as *artificial*, *frozen*, *cold*, *green*, *blue* and *angular* versus *natural*, *burning*, *warm*, *red* and *round*, seems to characterize environments as being either man-made (superficial) or natural (causal). It will be called as an Artificiality/Natural dimension.

The scales loaded most heavily on dimension 7, with higher frequencies across different communities, include the left-hand terms in the order of *many* (including *much*, *abundant*), *changeable* (*unpredictable*, *unstable*), *wet*, *rich*, *fresh*, *new* and *cold*, and the right-hand terms of *few* (*less*, *scarce*), *unchangeable* (*predictable*, *stable*), *dry*, *poor*, *stale*, *old* and *warm*. This dimension seems to suggest a clear inverse relationship between the qualitative condition of objects (i.e., stable vs unstable) and the quantitative level of attainment—i.e., *rich*, *fresh* and *many* BUT *unstable*, and *poor*, *stale* and *few* BUT *stable*. Thercfore, it will be called as a Instability/Stability dimension.

5. Conclusion

The above seven dimensions seems to characterize the affect free natures of diversified objects in macroenvironments and thus should represent macro-denotative meaning systems. Although the factor loadings of their salient scales in Table 3 are in general much less than the loadings of the *E-P-A* markers in the affective subspace, the dominating scale-vectors identified from each indigenous community appeared consistently across most of the 25 communities. It is reasonable to conclude that these seven macro-denotative meaning systems are pancultural in nature, regardless of the linguistic, cultural or even geographic differences of our communities.

Finally, since these meaning systems represent the global 'shared semantic characters' of various objects or entities in macroenvironments, they would definitely function as the common frames of psychosemantic references, in addition to the pancultural *E-P-A* dimensions, for both cross-cultural and cross-group communications. Under such circumstances, future research in cross-cultural psychology and the general social and behavioral sciences can utilize such a denotative semantic framework as objective indicators for investigating cross-cultural similarities as well as differences in conceptions of various macroenvironmental objects. In addition, such indicators can further be linked to thesubject cultures that have been measured by the pancultural *E-P-A* constructs.

References

[1] Alexander, H. H., Alexander, M. A. and Tzeng, O. C. S. (1978). Designing semantic differential scales for a universe of the near environment-Chair. *Home Economics Research Journal* **6**(4), 293–304.

[2] Hogenraad, R. and Tzeng, O. C. S. (1976). Affective and denotative meaning systems in natural and built space conceptions. Indiana University—Purdue University at Indianapolis, Department of Psychology.

[3] Osgood, C. E. (1962). Studies on the generality of affective meaning systems. *American Psychologist* **17**, 10–28.

[4] Osgood, C. E. (1964). Semantic differential technique in the comparative study of culture. *American Anthropologist* **66**(3), 171–200.

[5] Osgood, C. E. (1971). Exploration in semantic space: a personal diary. *Journal of Social Issues*. **27**(4), 5–64.

[6] Osgood, C. E., May, W. H. & Miron, M. S. (1975). *Cross-Cultural Universals of Affective Meaning*. Illinois, Urbana: University of Illinois Press.

[7] Tzeng, O. C. S. (1975). Differentiation of affective and denotative meaning systems and their influence in personality ratings. *Journal of Personality and Social Psychology* **32**(6), 978–988.

[8] Tzeng, O. C. S. (1977). Differentiation of affective and denotative semantic subspaces. In L. L. Adler (Ed.) *Issues in cross-cultural research. Annals of the New York Academy of Sciences* (L. L. Adler, Ed.) **285**, 476–504.

[9] Tzeng, O. C. S. (1977). A quantitative method for separation of semantic subspaces. *Applied Psychological Measurement* **2**, 171–184.

[10] Tzeng, O. C. S. (1977). Individual differences in self-conception: a multivariate approach. *Perceptual and Motor Skills* **45**, 1119–1124.

[11] Tzeng, O. C. S. & Landis, D., (1977). Measurement of meanings—illustration of a research methodology and procedures. Paper presented at the annual meeting of the American Psychological Association in San Francisco.

LIST OF CONTRIBUTED PAPERS

K. Alam and J. S. Hawkes (Clemson University): Minimax Property of Stein's Estimator with Respect to Non-Quadratic Loss.

M. M. Ali, N. N. Mikhail and M. S. Haq (University of Western Ontario, Canada): A Class of Bivariate Distributions Including the Bivariate Logistic.

J. A. Anderson (University of Newcastle Upon Tyne, United Kingdom): Multivariate Logistic Compounds.

R. F. Anderson and W. E. Winkler (University of Pittsburgh): On Optimal Estimation in a Poisson Cost Problem.

R. Atkin (Carnegie-Mellon University): An Analysis of Career Salary Progression.

B. Baldessari (University of Rome, Italy): On Some Measure of Concordance.

A. P. Basu (University of Missouri-Columbia) and J. K. Ghosh (Indian Statistical Institute, India): Identifying of the Multinomial and Other Distributions Under Competing Risk Model.

H. W. Block and T. W. Savits (University of Pittsburgh): Multivariate IFRA Distributions.

W. F. Caselton and T. Husain (University of British Columbia, Canada): Multivariate Continuous Distributions and Hydrologic Networks.

H. J. Chen (University of Georgia): Interval Estimation for the Largest Component Mean for a Multivariate Normal Distribution.

M. J. Cleary (Wright State University): Application of Regression to a Machine Shop Problem.

F. Critchley (University of Glasgow, United Kingdom): Multidimensional Scaling—a Short Critique and a New Method.

H. L. Crutcher (University of North Carolina at Asheville) and L. M. Falls (University of Alabama): Testing for Multivariate Normality.

J. Delaney and A. S. Paulson (Rensselaer Polytechnic Institute): Robust Estimation of Mean Vector and Covariance Matrix of the Multivariate Normal Distribution.

P. D. Doreian and N. P. Hummon (University of Pittsburgh): On Providing Estimates for Differential Equation Models of Social Phenomena: An Essay on the Mathematical Study of Change.

W. Gersch (University of Hawaii): Discrimination in Stationary Gaussian and Non-Gaussian Time Series by Kullback-Leibler Number Rules.

R. D. Gupta and D. Richards (University of West Indies, Jamaica): Calculation of Zonal Polynomials of 3×3 Positive Definite Symmetric Matrices.

H. L. HARTER (Air Force Flight Dynamics Laboratory): Some Early Applications of Order Statistics to Multivariate Analysis.

S. T. HUANG (University of Cincinnati): On the Representation of Nonlinear Systems with Gaussian Inputs.

A. KAPTEYN, T. WANSBEEK and J. BUYZE (Leiden University, The Netherlands): Errors in Variables: Consistent Adjusted Least Squares (CALS) Estimation.

D. KAZAKOS (State University of New York at Buffalo) and T. PAPANTONI-KAZAKOS (Bell Laboratories and Rice University): Asymptotic Discrimination Between Stationary Gaussian Processes.

R. KULPERGER (Carleton University, Canada): Inference for a Simple Branching Diffusion Process.

J. C. LEE (Wright State University and Air Force Flight Dynamics Laboratory): A Review of Computations of Multivariate Distributions done at the Aerospace Research Laboratories.

L. C. MARSH and S. S. MODAK (Notre Dame University): Testing General Linear Multiple Mixtures of Equality and Inequality Restrictions with Applications to Regression Analysis.

M. MAZUMDAR (Westinghouse Research and Development Center): An Optimum Component Testing Procedure for a Series-Parallel System.

G. V. L. NARASIMHAM (United States Department of Commerce): Multiple Time Series Models for Forecasting Economic Time Series.

A. RUKHIN (Purdue University): Universal Estimation in the Multivariate Analysis.

S. SCLOVE (University of Illinois at Chicago Circle): Testing Independence of Variates in an Infinitely Divisible Random Vector, with Applications to Stochastic Processes.

T. SEIDENFELD (University of Pittsburgh): A Reconstruction of Fisher's Fiducial Argument.

G. A. SHEA (University of Texas at Austin): Linear Discrimination and Canonical Factor Analysis.

R. S. SINGH (University of Guelph, Canada): Speed of Convergence in Nonparametric Estimation of Mixed Partial Derivatives of a Multivariate Density.

W. SMOCK (Wayne State University): Multivariate Model for Child Care.

F. E. SYMONS and P. L. DARIUS (Katholieke Universiteit, Leuven, Belgium): A Comparison of $\det(W)$, $\operatorname{tr}(W)$ and $\operatorname{tr}(W^{-1}B)$ as Criterion-Functions in Cluster Analysis.

J. TAMA (Bell Telephone Laboratories) and D. F. MORRISON (University of Pennsylvania): An Error Rate Criterion for Step-Wise Linear Discriminant Analysis.

S. B. TAN (University of Pittsburgh): Maximum Likelihood Estimation for an Autoregressive Process with Missing Observations.

D. S. TRACY (University of Windsor, Canada): Estimation of Multivariate Moments of Finite Populations.

C. T. TU and C. P. HAN (Iowa State University): Discriminant Analysis Based on Binary and Continuous Variables.

D. TYLER (University of Pittsburgh): Asymptotic Inference for Eigenvectors.

G. W. WALSTER (University of Wisconsin) and M. J. TRETTER (Pennsylvania State University): Exact Noncentral Distribution of Wilks' Λ and Wilks–Lawley U Criteria as Mixtures of Incomplete Beta Functions: For Three Tests.

T. WANSBEEK and A. KAPTEYN (Leiden University, The Netherlands) A General Analysis of Covariance Structures of Longitudinal Data.

S. J. WOLFE (University of Delaware): On the Unimodality of Multivariate Infinitely Divisible Distribution Functions.

W. A. WOYCZYNSKI (Cleveland State University): Central Limit Theorems vs. Geometric Structures in Banach Spaces.

H. YANAI (Chiba University, Japan): Invariance Property of Canonical Correlation Analysis and its Applications.

H. YASSAEE (Arya-Mehr University of Technology, Iran): On Analysis of Multidimensional Contingency Tables: Multiple Linear Logic Model.

S. ZACKS (Case Western University): Simultaneous Estimation of Interaction Parameters in $2 \times k$ Binomial Experiments.

TITLES OF PRESENTATIONS IN CLINICAL SESSIONS

N. Cobean (Army Natick Research and Development Command): The Analysis of Response Data in Army Mess Hall Preference.

H. L. Crutcher (University of North Carolina at Asheville): Some Unsolved Meteorological Multivariate Problems.

S. Dubey (Food and Drug Administration): Some Multivariate Statistical Problems in Research on Drugs.

G. T. Duncan (Carnegie-Mellon University): Elicitation for Dirichlet–Multinomial Sampling.

I. I. Gringorten (Air Force Geophysics Laboratory): Conditional Probabilities of Ceilings and Visibilities at a Point, along a Line, and in an Area.

J. E. Michalek (Air Force School of Aerospace Medicine): The West Point Follow-Up Study.

M. G. Yochmowitz (USAF School of Aerospace Medicine): Analysis of Primate Flight Simulator Data.

AUTHOR INDEX